D1104877

Methods in Enzymology

Volume 246
BIOCHEMICAL SPECTROSCOPY

METHODS IN ENZYMOLOGY

EDITORS-IN-CHIEF

John N. Abelson   Melvin I. Simon

DIVISION OF BIOLOGY
CALIFORNIA INSTITUTE OF TECHNOLOGY
PASADENA, CALIFORNIA

FOUNDING EDITORS

Sidney P. Colowick and Nathan O. Kaplan

Methods in Enzymology

Volume 246

Biochemical Spectroscopy

EDITED BY

Kenneth Sauer

DEPARTMENT OF CHEMISTRY
AND STRUCTURAL BIOLOGY DIVISION
LAWRENCE BERKELEY LABORATORY
UNIVERSITY OF CALIFORNIA, BERKELEY
BERKELEY, CALIFORNIA

ACADEMIC PRESS

San Diego   New York   Boston   London   Sydney   Tokyo   Toronto

This book is printed on acid-free paper. ∞

Copyright © 1995 by ACADEMIC PRESS, INC.

All Rights Reserved.
No part of this publication may be reproduced or transmitted in any form or by any means, electronic or mechanical, including photocopy, recording, or any information storage and retrieval system, without permission in writing from the publisher.

Academic Press, Inc.
A Division of Harcourt Brace & Company
525 B Street, Suite 1900, San Diego, California 92101-4495

United Kingdom Edition published by
Academic Press Limited
24-28 Oval Road, London NW1 7DX

International Standard Serial Number: 0076-6879

International Standard Book Number: 0-12-182147-1

PRINTED IN THE UNITED STATES OF AMERICA
95 96 97 98 99 00 EB 9 8 7 6 5 4 3 2 1

Table of Contents

C. Linear Dichroism and Fluorescence

Section II. Vibrational Spectroscopy

Section III. Magnetic Resonance Spectroscopy, X-Ray Spectroscopy

Section IV. Special Topics

Contributors to Volume 246

Article numbers are in parentheses following the names of contributors.
Affiliations listed are current.

ROBERT H. AUSTIN (7), *Department of Physics, Princeton University, Princeton, New Jersey 08544*

GARY W. BRUDVIG (22), *Department of Chemistry, Yale University, New Haven, Connecticut 06511*

PETER S. BRZOVIĆ (8), *Department of Biochemistry, University of Washington, Seattle, Washington 98195*

THERESE M. COTTON (28), *Department of Chemistry, Iowa State University, Ames, Iowa 50011*

ROMAN S. CZERNUSZEWICZ (18), *Department of Chemistry, University of Houston, Houston, Texas 77204*

VICTORIA J. DEROSE (23), *Department of Chemistry, Northwestern University, Evanston, Illinois 60208*

SIMON DELAGRAVE (29), *Department of Biology, Brandeis University, Waltham, Massachusetts 02234*

SHAOJUN DONG (28), *Laboratory of Electroanalytical Chemistry, Changchun Institute of Applied Chemistry, Chinese Academy of Sciences, Changchun 130022, China*

MICHAEL F. DUNN (8), *Department of Biochemistry, University of California, Riverside, Riverside, California 92521*

SHYAMSUNDER ERRAMILLI (7), *Department of Physics, Princeton University, Princeton, New Jersey 08544*

WAYNE R. FIORI (24), *Department of Chemistry, Sinsheimer Laboratories, University of California, Santa Cruz, Santa Cruz, California 95064*

JOSEF FRIEDRICH (10), *Physikalisches Institut, Universität Bayreuth, D-95440 Bayreuth, Germany*

DANIEL R. GAMELIN (5), *Department of Chemistry, Stanford University, Stanford, California 94305*

ELLEN GOLDMAN (29), *Curagen Corporation, Branford, Connecticut 06405*

DONALD M. GRAY (3), *Department of Molecular and Cell Biology, University of Texas at Dallas, Richardson, Texas 75083*

BRIAN M. HOFFMANN (23), *Department of Chemistry, Northwestern University, Evanston, Illinois 60208*

ALFRED R. HOLZWARTH (14), *Max-Planck-Institut für Radiation Chemistry, D-45470 Mülheim/Ruhr, Germany*

SU-HWI HUNG (3), *Department of Molecular and Cell Biology, University of Texas at Dallas, Richardson, Texas 75083*

DAVID M. JAMESON (12), *Department of Biochemistry and Biophysics, University of Hawaii, John A. Burns School of Medicine, Honolulu, Hawaii 96822*

KENNETH H. JOHNSON (3), *Baylor College of Medicine, Center for Biotechnology, The Woodlands, Texas 77381*

JAMES R. KINCAID (19), *Department of Chemistry, Marquette University, Milwaukee, Wisconsin 53233*

MARTIN L. KIRK (5), *Department of Chemistry, Stanford University, Stanford, California 94305*

MELVIN P. KLEIN (21), *Structural Biology Division, Lawrence Berkeley Laboratory, University of California, Berkeley, Berkeley, California 94720*

SUE LEURGANS (27), *Department of Preventive Medicine, Rush Medical College, Chicago, Illinois 60612*

AUGUST H. MAKI (25), *Department of Chemistry, University of California, Davis, Davis, California 95616*

RICHARD A. MATHIES (16), *Department of Chemistry, University of California, Berkeley, Berkeley, California 94720*

SIOBHAN M. MIICK (24), *Department of Chemistry, Sinsheimer Laboratories, University of California, Santa Cruz, Santa Cruz, California 95064*

GLENN L. MILLHAUSER (24), *Department of Chemistry, Sinsheimer Laboratories, University of California, Santa Cruz, Santa Cruz, California 95064*

JIANJUN NIU (28), *Laboratory of Electroanalytical Chemistry, Changchun Institute of Applied Chemistry, Chinese Academy of Sciences, Changchun 130022, China*

WARNER L. PETICOLAS (17), *Department of Chemistry, University of Oregon, Eugene, Oregon 97403*

SABINE PULVER (5), *Department of Chemistry, Stanford University, Stanford, California 94305*

ROBERT T. ROSS (27), *Department of Biochemistry, Ohio State University, Columbus, Ohio 43210*

KENNETH SAUER (1), *Department of Chemistry, and Structural Biology Division, Lawrence Berkeley Laboratory, University of California, Berkeley, Berkeley, California 94720*

WILLIAM H. SAWYER (12), *Russell Grimwade School of Biochemistry, University of Melbourne, Parkville, Victoria 3052, Australia*

HUGO SCHEER (30), *Botanisches Institut der Universität München, D-80683 München, Germany*

PAUL R. SELVIN (13), *Structural Biology Division, Calvin Laboratory, Lawrence Berkeley Laboratory, and Department of Chemistry, University of California, Berkeley, Berkeley, California 94720*

F. SIEBERT (20), *Institute of Biophysics and Radiation Biology, University of Freiburg, D-7800 Freiburg, Germany*

EDWARD I. SOLOMON (5), *Department of Chemistry, Stanford University, Stanford, California 94305*

THOMAS G. SPIRO (18), *Department of Chemistry, Princeton University, Princeton, New Jersey 08544*

WALTER S. STRUVE (11), *Ames Laboratory–United States Department of Energy, and Department of Chemistry, Iowa State University, Ames, Iowa 50011*

JOHN C. SUTHERLAND (6), *Department of Biology, Brookhaven National Laboratory, Upton, New York 11973*

IGNACIO TINOCO, JR. (2), *Department of Chemistry and Chemical Biodynamics Laboratory, University of California, Berkeley, Berkeley, California 94720*

HERBERT VAN AMERONGEN (9, 11), *Department of Physics and Astronomy, Free University of Amsterdam, 1081 HV Amsterdam, The Netherlands*

RIENK VAN GRONDELLE (9), *Department of Physics and Astronomy, Free University of Amsterdam, 1081 HV Amsterdam, The Netherlands*

ALAN WAGGONER (15), *Department of Biological Sciences, and Center for Light Microscope Imaging and Biotechnology, Carnegie Mellon University, Pittsburgh, Pennsylvania 15213*

ROBERT W. WOODY (4), *Department of Biochemistry and Molecular Biology, Colorado State University, Fort Collins, Colorado 80523*

VITTAL K. YACHANDRA (26), *Structural Biology Division, Lawrence Berkeley Laboratory, University of California, Berkeley, Berkeley, California 94720*

MARY M. YANG (29), *KAIROS Inc., San Jose, California 95136*

DOUGLAS C. YOUVAN (29), *Palo Alto Institute of Molecular Medicine, Mountain View, California 94043*

Preface

The use of spectroscopic methods to examine biomolecules has a long and rich history. Such methods have the virtue of being largely noninvasive and capable of probing living materials as well as subcellular preparations and isolated biomolecules. The information gained is interpretable in terms of structural parameters and intramolecular interactions. Using time-resolved approaches, dynamics can be explored readily over a time range from less than a picosecond to seconds and longer. This permits ready exploration of intermolecular interactions and intramolecular motion relevant to biological processes.

Advances in technology and methodology in spectroscopy have moved the field forward at a breathtaking pace in recent years. We have come a long way from the era when cytochrome oxidation state changes were monitored visually using a hand spectroscope or when absorption spectrometry was done using photographic detection. In this volume the reader can learn about instrumentation that uses diode array detectors to monitor absorption or emission spectral properties with great precision at hundreds of wavelengths simultaneously or Fourier transform methods that provide a significant increase in the efficiency of collecting and analyzing spectroscopic information distributed over a wide wavelength band. Mode-locked lasers and associated pulse-compression and continuum-generation techniques allow pulse-probe measurements of fast (to 10 fsec = 10^{-14} sec) absorption changes or fluorescence relaxation in the picosecond regime. Single-photon counting methods have greatly improved the signal-to-noise of optical detection systems for steady-state spectroscopy and especially for time-resolved fluorescence measurements. Pulsed lasers have advanced the application of time-resolved Raman spectroscopy and the ability to discriminate between Raman and fluorescence signals. In combination with the use of ultralow temperatures, intense monochromatic laser sources can be used in hole-burning experiments to probe chromophore local environments and the modes by which the chromophores interact with their surroundings. Computation intensive methods, such as Fourier transform infrared, electron spin echo (ESE), pulsed electron nuclear double resonance (ENDOR), and digital imaging optical spectroscopy, have provided entirely new approaches to data collection and processing. Major developments in radiation sources, such as synchrotrons, have opened entirely new areas of investigation, including X-ray absorption spectroscopy (XAS) and extended X-ray absorption fine structure (EXAFS). The reader will find descriptions and examples of each of these new methodologies in the chapters that follow.

The audience for this volume includes the current generation of graduate students and professional scientists involved in biological or biochemical studies seeking an introduction to modern spectroscopic methods and instrumentation. To help the interested reader develop a deeper background and understanding of these methodologies and approaches, the authors of the individual chapters have cited general references and published reviews of the individual topics.

The chapters fall into major sections covering optical spectroscopy, vibrational spectroscopy, electron paramagnetic resonance, and X-ray spectroscopy. Areas such as nuclear magnetic resonance which have been described extensively in recent volumes of this series have not been included. A general overview of the contents of this volume and examples of problems or situations to which the different approaches have been applied are described in the first chapter.

I am indebted to many of my colleagues and students who have offered suggestions, have read and provided valuable comments on many of the contributed chapters, and have helped with the assembly of this volume. In particular, I wish to thank several Berkeley colleagues, including Judith Klinman who was instrumental in initiating this project. Ignacio Tinoco, Richard Mathies, and Melvin P. Klein have not only written comprehensive overviews of the broader spectroscopic fields, but have also reviewed the contributed chapters included in each section. Very helpful insights for several chapters were provided by Joy Andrews, Martin Debreczeny, and Mary Talbot, based on their perspective as graduate students in my research group. It is my hope that this volume will be informative and beneficial to them and to their contemporaries and successors in research laboratories throughout the scientific community.

KENNETH SAUER

METHODS IN ENZYMOLOGY

VOLUME XXXVI. Hormone Action (Part A: Steroid Hormones)
Edited by BERT W. O'MALLEY AND JOEL G. HARDMAN

VOLUME XXXVII. Hormone Action (Part B: Peptide Hormones)
Edited by BERT W. O'MALLEY AND JOEL G. HARDMAN

VOLUME XXXVIII. Hormone Action (Part C: Cyclic Nucleotides)
Edited by JOEL G. HARDMAN AND BERT W. O'MALLEY

VOLUME XXXIX. Hormone Action (Part D: Isolated Cells, Tissues, and Organ Systems)
Edited by JOEL G. HARDMAN AND BERT W. O'MALLEY

VOLUME XL. Hormone Action (Part E: Nuclear Structure and Function)
Edited by BERT W. O'MALLEY AND JOEL G. HARDMAN

VOLUME XLI. Carbohydrate Metabolism (Part B)
Edited by W. A. WOOD

VOLUME XLII. Carbohydrate Metabolism (Part C)
Edited by W. A. WOOD

VOLUME XLIII. Antibiotics
Edited by JOHN H. HASH

VOLUME XLIV. Immobilized Enzymes
Edited by KLAUS MOSBACH

VOLUME XLV. Proteolytic Enzymes (Part B)
Edited by LASZLO LORAND

VOLUME XLVI. Affinity Labeling
Edited by WILLIAM B. JAKOBY AND MEIR WILCHEK

VOLUME XLVII. Enzyme Structure (Part E)
Edited by C. H. W. HIRS AND SERGE N. TIMASHEFF

VOLUME XLVIII. Enzyme Structure (Part F)
Edited by C. H. W. HIRS AND SERGE N. TIMASHEFF

VOLUME XLIX. Enzyme Structure (Part G)
Edited by C. H. W. HIRS AND SERGE N. TIMASHEFF

VOLUME L. Complex Carbohydrates (Part C)
Edited by VICTOR GINSBURG

VOLUME LI. Purine and Pyrimidine Nucleotide Metabolism
Edited by PATRICIA A. HOFFEE AND MARY ELLEN JONES

VOLUME LII. Biomembranes (Part C: Biological Oxidations)
Edited by SIDNEY FLEISCHER AND LESTER PACKER

VOLUME LIII. Biomembranes (Part D: Biological Oxidations)
Edited by SIDNEY FLEISCHER AND LESTER PACKER

VOLUME LIV. Biomembranes (Part E: Biological Oxidations)
Edited by SIDNEY FLEISCHER AND LESTER PACKER

[1] Why Spectroscopy? Which Spectroscopy?

By KENNETH SAUER

Biochemists and scientists in related fields have a "need to know" about information covering an enormous range of disciplines. They are interested in the involvement and fates of very small molecules like O_2, CO_2, H_2O, and N_2, as well as those of the largest proteins and nucleic acids. Scientists also want to know about the interactions of the small molecules with the large ones: not only to view the needle in the haystack, but to determine whether "it is threaded and ready to use." These studies may involve "simple" reactions like the interchange between dissolved CO_2 gas and bicarbonate ion in aqueous solution, as well as the role of enzymes that can speed up such processes by orders of magnitude.

The context in which biochemical processes occur is essential to defining mechanisms and functions. The component proteins and nucleic acids exhibit dramatic changes in properties when they are incorporated into ribosomes or chromosomes or functional membranes. Even then, it is not sufficient to determine a static or average structure of the components in such complex assemblies, because the function typically depends on essential structural fluctuations or driven conformational changes. In this way time enters as a significant variable. Short-term events (picosecond time scale and shorter) can involve vibrational motion to dissipate heat energy or to initiate conformational events that trigger vision processes or start the electron transfer reactions of photosynthesis. Diffusion of small substrate molecules is a few orders of magnitude slower. The motion of major portions of proteins may require only microseconds, whereas the complex decision-making required for enzymes to function selectively may require times in the range of seconds.

Such an enormous range of conditions and interests places enormous demands on the panoply of experimental approaches needed to address the related issues. In a recent single issue of the American Chemical Society journal *Biochemistry* there were 25 articles, of which all but 4 explicitly involved some form of spectroscopy among the methods that were cited. In many cases two or three spectroscopic methods were used in the same study, and none of these entailed simply using absorbance at a particular wavelength to monitor the concentration of a particular component. Even the other 4 articles, which included one study involving X-ray crystallography and another that followed the course of a radioisotope, contained procedures that implied the application of spectroscopic methods in the preparation and purification of materials.

Numerous, wide-ranging spectroscopic techniques will be presented in this volume, with the exception of nuclear magnetic resonance (NMR),[1] which was the subject of Volumes 176, 177, and 239 of *Methods in Enzymology,* and mass spectrometry,[2] which was the subject of Volume 193. Examples of techniques from each of three major areas, ultraviolet/visible spectroscopy, vibrational spectroscopy, and electron or electron/nuclear magnetic resonance, are presented in this volume. Also included are special topics like rapid-scan diode-array spectroscopy, terbium labeling of chromopeptides, and deconvolution of complex spectra that are covered in chapters in Section IV of this volume.

Why Spectroscopy?

The range of areas of scientific interest represented in the *Biochemistry* issue mentioned above was extremely broad. A sampling includes the following: the interaction with DNA of a cobalt probe that resulted in shifts in the electron paramagnetic resonance (EPR) spectrum as well as changes in the circular dichroism, a study of protein oligomerization, premelting flexibility of a synthetic oligonucleotide, the formation of crosslinks between DNA and protein using platinum and ruthenium compounds, enzyme–substrate interactions, the structure and dynamics of channel-forming membrane proteins, the dynamics of unfolding and refolding of a redoxactive protein, and the use of a fluorescent amino acid reporter to send signals from the channel of a transporter protein. Spectroscopic methods can be used to examine the behavior of cell organelles or macroscopic assemblies that are not yet amenable to interpretation at the molecular level.

One of the virtues of spectroscopic approaches is that they are commonly nondestructive and noninvasive, although some procedures involve the attachment of labels chemically at specific sites. The former approaches are particularly valuable for *in vivo* studies. In laboratory experiments, the amounts of materials required range from the femtomole (10^{-15} mol) level for some particularly sensitive fluorescence approaches to milligrams or more for typical EPR, infrared, or X-ray spectroscopy. Ongoing requirements in instrumentation and techniques are continually whittling away at the limits to the amount of material required.

---

[1] N. J. Oppenheimer and T. L. James (eds.), this series, Vols. 176, 177, and 239: Nuclear Magnetic Resonance.
[2] J. A. McCloskey (ed.), this series, Vol. 193: Mass Spectrometry.

Which Spectroscopy?

One of the most important questions facing an experimentalist at the outset of a research project is the decision about which of the available (or potentially available) spectroscopic methods is most suitable to a particular problem or investigation. The decision is complex, in that it involves consideration of the nature of the biological or chemical materials being investigated, the particular goals of the research, the availability of suitable spectroscopic tools, and the ability to interpret the results in a meaningful way. The latter requirement implies the need for at least a vestige of a theoretical framework in which to formulate the interpretation and, in many cases, access to an "expert" who is knowledgeable about the relation between the experimentally observed quantities and their origin at the molecular or electronic level. One important source of such expertise is a suitable text or reference book that describes the underlying principles in terms that are both informative and understandable. Although the relation between underlying principles and spectroscopic measurements properly involves subjects such as quantum mechanics, statistical mechanics, and molecular dynamics, one does not necessarily need to be an authority in these fields to be sufficiently conversant to make intelligent use of the spectroscopic observables.

Each of the individual chapters in this volume presents references to authoritative monographs or compendia of relevant information on the main topic covered. There are, in addition, introductory books that describe a broad range of relevant methods; these books are particularly helpful in making the initial decision as to which spectroscopic methods to choose for a particular problem.[3,4] Sections that involve spectroscopy are included in a variety of texts in biochemistry,[5] physical chemistry for biological scientists,[6-10] or biophysics. At a somewhat more advanced

[3] R. E. Hester and R. B. Stirling, "Spectroscopy of Biological Molecules." CRC Press, Boca Raton, Florida, 1991.

[4] I. D. Campbell and R. A. Dwek, "Biological Spectroscopy." Benjamin/Cummings, Menlo Park, California, 1984.

[5] L. Stryer, "Biochemistry," 3rd Ed. Freeman, San Francisco, California, 1988.

[6] I. Tinoco, Jr., K. Sauer, and J. C. Wang, "Physical Chemistry—Principles and Applications in Biological Sciences," 3rd Ed. Prentice-Hall, Englewood Cliffs, New Jersey, 1995.

[7] D. Eisenberg and D. Crothers, "Physical Chemistry, with Applications to the Life Sciences." Benjamin/Cummings, Menlo Park, California, 1979.

[8] A. G. Marshall, "Biophysical Chemistry—Principles, Techniques and Applications." Wiley, New York, 1978.

[9] D. Freifelder, "Physical Biochemistry—Applications to Biochemistry and Molecular Biology," 2nd Ed. Freeman, San Francisco, California, 1982.

[10] K. E. Van Holde, "Physical Biochemistry," 2nd Ed. Prentice-Hall, Englewood Cliffs, New Jersey, 1985.

level, the monograph by Cantor and Schimmel[11] is a particularly valuable resource for this purpose. Greater detail in the theoretical underpinnings of many of the types of spectroscopy used by biochemists is covered in an excellent book by Struve.[17] Nevertheless, it is inevitable that some of the topics covered in the present volume are not to be found or are only briefly mentioned in the broader surveys. This occurs either because the fields, or the applications to biological problems, are relatively new [X-ray spectroscopy, hole burning, Fourier transform infrared spectroscopy (FTIR)] or because until recently the applications to biology were construed to be highly specialized [electron nuclear double resonance spectroscopy (ENDOR), spin labels, digital imaging spectroscopy]. Several of these methods have sufficient potential that they should be considered to be a part of the reserve toolkit of experimental biochemists and biologists.

Spectrometric methods rely on the availability of reference data or relevant properties of well-examined biochemical materials. Such data are abundant and widespread in the literature, but compilations of such spectrometric data are not nearly so available as are those in the areas of chemistry or physics. Even the few compilations that exist[13,14] are not particularly recent or up-to-date. This is a shortcoming that the leaders of the discipline should endeavor to remedy.

To aid the reader in the process of choosing suitable tools from the toolkit, the following paragraphs focus on a variety of objectives that are typical of biochemical and biological investigations and how each of the objectives may be productively illuminated by the methods described in detail in the subsequent chapters.

Structure of Biomolecules

The distinction between monomer, duplex, and triplex helices of DNA or RNA can be established by a comparison of absorption and circular dichroism (CD) spectroscopy. A clear difference in the CD is seen that distinguishes triplexes of RNA from those of DNA.[15,16] The CD spectra

[11] C. R. Cantor and P. R. Schimmel, "Biophysical Chemistry—Part II: Techniques for the Study of Biological Structure and Function." Freeman, San Francisco, California, 1980.
[12] W. S. Struve, "Fundamentals of Molecular Spectroscopy." Wiley (Interscience), New York, 1989.
[13] G. D. Fasman (ed.), "Handbook of Biochemistry and Molecular Biology, Vol. 1: Physical and Chemical Data," 3rd Ed. CRC Press, Cleveland, Ohio, 1976.
[14] G. D. Fasman (ed.), "Practical Handbook of Biochemistry and Molecular Biology." CRC Press, Boca Raton, Florida, 1989.
[15] D. M. Gray, S.-H. Hung, and K. H. Johnson, this volume [3].
[16] R. W. Woody, this volume [4].

of proteins and polypeptides are also useful in determining the content of α helix, β sheet, and β turns.[16] The coordination geometry and spin state of transition metal centers in biomolecules can be characterized using CD and magnetic circular dichroism (MCD) measurements.[17] The conformation of macromolecules in solution is sensitively measured by excitation transfer between fluorescence donor–acceptor molecular pairs that are covalently or noncovalently attached.[18,19] Detailed information about the backbone conformation of proteins or nucleic acid helices can be read from Raman[20] or infrared[21] spectra. Resonance Raman spectra are useful in documenting the subtle distinction between the α- and β-subunit hemes in hemoglobin.[22] Changes in the oxidation state of iron- or copper-containing proteins allow ENDOR spectroscopy to elucidate the coordination environment of the redox-active cofactors.[23] Rotational diffusion constants that are sensitive to molecular extension can be determined using spin-labeled macromolecules. The difference between a peptide in the α helix or 3_{10} helix configuration is readily seen in the EPR spectra of doubly spin-labeled molecules.[24]

Changes in Structure or Conformation

The transformation of polyproline from a right-handed to a left-handed 3-fold helix can be followed using CD.[16] MCD spectra are sensitive to the conformation of the heme–imidazole ligand interaction in cytochromes.[25] Hole-burning measurements provide sensitive tools for detecting differences in structure produced on thermal cycling or as a consequence of increasing pressure on horseradish peroxidase solutions.[26] Attachment of a fluorescence label[27] provides a useful indicator of the loss of activity of human immunodeficiency virus (HIV) protease,[18] and lifetime measurements can be used to determine the binding of ligands, of drugs to proteins, or the association of proteins with membranes.[19] Similarly, the decreasing content of free sulfhydryl groups that accompanies aging is detectable by Raman spectroscopy using deuterium exchange methods.[20] Raman spectra

[17] E. I. Solomon, M. L. Kirk, D. R. Gamelin, and S. Pulver, this volume [5].
[18] P. R. Selvin, this volume [13].
[19] A. R. Holzwarth, this volume [14].
[20] W. L. Peticolas, this volume [17].
[21] F. Siebert, this volume [20].
[22] J. R. Kincaid, this volume [19].
[23] V. J. DeRose and B. M. Hoffman, this volume [23].
[24] G. L. Millhauser, W. R. Fiori, and S. M. Miick, this volume [24].
[25] J. C. Sutherland, this volume [6].
[26] J. Friedrich, this volume [10].
[27] A. Waggoner, this volume [15].

also can be used to follow the course of DNA or RNA melting.[20] The involvement of aspartate in the photocycling of bacteriorhodopsin can be followed using difference infrared studies of isotopically substituted molecules.[21] Electron spin-labeled proteins exhibit striking changes in EPR spectra on undergoing structural transformations.[24]

Composition of Biocomplexes

The stoichiometry of RNA in duplex or DNA in triplex complexes is discernible from CD spectra, as are hybrid complexes between RNA and DNA.[15] The amino acid content of proteins and polypeptides is reflected in the UV absorption and CD spectra, as is the presence of cofactors such as heme or flavin. The CD spectrum is also sensitive to interactions between proteins and nucleic acids.[16] In general, cofactors that involve transition metals with incomplete d electron shells exhibit characteristic features in absorption, CD, and MCD spectra.[17,25] The presence of tryptophan gives rise to a distinctive band in the MCD spectrum of any protein.[25] The binding of substrate analogs like ε-AMP to carbamoyl-phosphate synthase can be detected by the influence on the anisotropy of the fluorescence. Protein–DNA interactions have been monitored using the anisotropy of fluorescence of Trp of the Ada protein and the *lac* promoter with cyclic AMP in the presence of receptor protein.[28] Added fluorescent dyes can be used to tag DNA molecules, even during electrophoresis.[18] FTIR spectra reflect the interaction between certain drugs and DNA, and they characterize caged Ca^{2+} and caged ATP as well.[21] Stacking interactions in protein–DNA complexes have been investigated using optically detected magnetic resonance (ODMR).[29] Photosynthetic complexes that contain antenna complexes and reaction centers exhibit complex fluorescence relaxation that reflects the spectroscopic heterogeneity of the complexes.[19] The identity and integrity of photosynthetic reaction center complexes are readily determined using chromatography with a diode-array detector, as is the degradation of isolated photosynthetic pigments.[30]

Orientation of Component Structures

The orientation of structural components in proteins can be characterized using polarized absorption or emission spectroscopy. Where single crystals are not available or convenient, anisotropic macromolecules may be oriented using gel compression, electric or magnetic fields, or stretched

[28] D. M. Jameson and W. H. Sawyer, this volume [12].
[29] A. H. Maki, this volume [25].
[30] H. Scheer, this volume [30].

films.[31] The ordering of component proteins in oriented membranes and the degree of ordering of component lipids can be studied using polarized FTIR.[21] In the case of macromolecules labeled with paramagnetic probes, the anisotropy of the electron spin tensor can be used to provide information about absolute and relative orientations.[24]

Local Probes of Macromolecules

For enzymes like liver alcohol dehydrogenase (LADH), where the transition metal cofactor Zn is colorless, replacing it with Co allows the spectroscopic detection of intermediates.[32] Conversely, replacing the colored but strongly quenching Fe in myoglobin by Mg allows fluorescence to be used as a spectroscopic probe.[33] MCD is particularly useful in distinguishing among the configurations of imidazole ligands to the heme groups of different cytochromes.[25] Adsorbed fluorescent dyes are useful for labeling particular DNA oligonucleotides.[18] Alternatively, resonance Raman spectroscopy helps to elucidate the interaction of O_2 and other ligands to myoglobin and hemoglobin, and it distinguishes between the hemes of the α and β subunits of hemoglobin.[22,34] The technique is particularly useful in probing the coordination environment of metals like Fe in FeS proteins (sulfite reductase, rubrerythrin, rubredoxin, and adrenodoxin), cytochrome P450, and oxyhemerythrin and Cu in azurin, plastocyanin, and hemocyanin.[34] In many cases the transition metal centers are paramagnetic, as in blue Cu proteins, azurin, Fe-containing methane monooxygenase, and nitrile hydratase, which permits sensitive investigation of ligand interactions using ENDOR spectroscopy.[23] EPR spectra of paramagnetic intermediates, like the S_2 state of the Mn-containing photosynthetic water oxidation complex or the heme of oxidized cytochrome c, provide information about coupling to paramagnetic nuclei in the immediate vicinity.[35] Spin labels directed at specific macromolecular sites serve as reporter groups that are sensitive to changes in the structure of the binding site.[24] Detailed and precise structural parameters, like bond lengths and coordination environments, of metal centers in FeS proteins, the Fe-Mo cofactor of nitrogenase, and the Mn-containing photosynthetic water oxidation complex are accessible from X-ray spectroscopy measurements of biomolecules,[36] even when crystal structures are not yet available.

[31] H. van Amerongen and W. S. Struve, this volume [11].
[32] P. S. Brzović and M. F. Dunn, this volume [8].
[33] R. H. Austin and E. Shyamsunder, this volume [7].
[34] T. G. Spiro and R. S. Czernuszewicz, this volume [18].
[35] G. W. Brudvig, this volume [22].
[36] V. K. Yachandra, this volume [26].

Surface Accessibility

Antibodies labeled with fluorescent chromophores greatly increase the sensitivity of detecting the interaction with antigens or with antigenic sites, especially when effects on the fluorescence lifetimes are monitored.[19] Electron spin labels attached to different amino acids in membrane proteins like bacteriorhodopsin give spectra that distinguish between locations in surface loops or transmembrane helices.[24] The use of heavy atom probes in the medium together with phosphorescence ODMR distinguishes tryptophan residues with regard to proximity to the accessible surface of proteins.[29] Similarly, the binding of a fluorescent dye to suspensions of chloroplast membranes is readily monitored using quenching by terbium ions added to the medium.[37]

Spatial Resolution

Spectroscopic signatures combined with high spatial resolution permit the localization of specific molecules in structurally complex biological materials. This technique has been developed using Raman spectroscopy to determine the configuration of oligonucleotides in microcrystals that are too small to study using X-ray crystallography.[20] Imaging of colonies of the photosynthetic bacterium *Rhodobacter capsulatus* in terms of fluorescence emission spectra is an efficient method for rapidly scanning for mutants that exhibit interesting differences in function and/or pigmentation.[38] Even time-resolved fluorescence can be applied to two-dimensional imaging.[19]

Function

The redox properties of transition metal complexes involving nonheme iron, like that in pyrocatechase, correlate with observable spectroscopic splittings.[17] The migration of electronic excitation and the subsequent transfer of electrons in photosynthetic reaction centers are readily monitored using time-resolved optical absorption spectrometry following flash excitation.[39] When these processes are studied at low temperature using hole-burning techniques, the mechanism of excitation delocalization can be investigated for photosynthetic reaction centers, bacteriochlorophyll antenna proteins, and phycobiliproteins.[26] Also using low temperature measurements, dephasing times for the excited state of the reaction center can be related to relaxation processes in the protein matrix. Similarly, early events in the photodissociation of carbonmonoxymyoglobin are re-

[37] R. T. Ross and S. Leurgans [27].
[38] D. C. Youvan, E. Goldman, S. Delagrave, and M. M. Yang, this volume [29].
[39] H. van Amerongen and R. van Grondelle, this volume [9].

solvable.[33] The latter process can be investigated also by transient resonance Raman spectroscopy.[22] FTIR can be used to study intermediates involving structural changes during the phototransformation of bacteriorhodopsin.[21] Time-resolved fluorescence measurements can also provide useful information about the structure and function of photoreceptors.[19] Where paramagnetic intermediates are produced, as in the electron transfer reactions involved in photosynthesis, EPR is useful for characterizing species like chlorophyll radical cations and tyrosine radicals.[35] Tyrosyl radicals have also been investigated using ENDOR spectroscopy,[23] as have ribonucleotide reductase and the interaction of the substrate citrate with aconitase. For processes not susceptible to photon stimulation, rapid mixing techniques allow one to follow the appearance and disappearance of reaction intermediates like those seen in the reduction of aldehydes by the enzyme LADH. Using cobalt-substituted LADH, intermediates in the oxidation of alcohols can be resolved, as can those involved in the action of tryptophan synthase.[32] Similarly, the reaction of O_2 with hemoglobin, myoglobin, or cytochrome oxidase can be followed using transient resonance Raman measurements.[22] The involvement of Mn oxidation during the evolution of O_2 from photosynthetic membranes has been confirmed by changes in the manganese X-ray absorption edge energies associated with individual steps in the process.[36]

Dynamics: Motion and Relaxation

The dissipation of excess energy and interaction with the matrix following photon absorption by molecular chromophores is reflected in the temperature dependence of the shapes of absorption and fluorescence bands.[33] Quantitative measurement of rotational diffusion processes is accessible using time-resolved fluorescence relaxation[19] and depolarization.[18,31] The binding of melittin to membrane-associated Ca^{2+}-ATPase results in a decrease in mobility that can be monitored using time-resolved fluorescence measurements.[19] Studies of the relaxation dynamics of tryptophan fluorescence can be used to derive information about the conformational flexibility of proteins.[19] The motional flexibility of proteins is alternatively reflected in the dynamics of hydrogen/deuterium (H/D) exchange that is detectable by Raman spectroscopy.[20] Alternatively, electron spin labels can be used to give information about rotational diffusion, as well as about the local flexibility at the labeling site.[24]

Spectrum Analysis and Relation to Theoretical Foundations

Crystal field theory and ligand field theory are indispensible tools for interpreting the spectra of transition metal complexes in terms of coordination interactions and geometry.[17] MCD is useful for identifying electronic transitions that are unresolved in absorption spectra, including those associated with degenerate ground or excited electronic states.[25] Investigations of spectroscopic line width in terms of homogeneous or inhomogeneous broadening give insight into the distribution of local chromophore environments and the rates at which they change.[26,33] Analysis of the relation between polarized optical properties and orientation provides useful structural information. Furthermore, applying time-resolved measurements allows one to explore coherently excited electronic states and the associated dephasing times.[31] The theoretical relations developed by Förster to describe inductive resonance energy transfer are useful to explore distances of separation and relative orientations of donor–acceptor pairs within or attached to macromolecules.[18] In every type of spectroscopy described in this volume, one may encounter situations where complex, overlapping spectra require resolution to facilitate interpretation. Although there is no single fail-safe method for doing this, deconvolution techniques are essential items for the spectroscopist.[37]

Section I

Ultraviolet/Visible Spectroscopy
Article 2

A. Absorption and Circular Dichroism
Articles 3 through 7

B. Transient Absorption and Kinetics
Articles 8 through 10

C. Linear Dichroism and Fluorescence
Articles 11 through 15

[2] Optical Spectroscopy: General Principles and Overview

By Ignacio Tinoco, Jr.

Introduction

In this volume Section I focuses on the use of the absorbance and fluorescence of light to learn about structures and dynamics of biological macromolecules. Native states of proteins and nucleic acids can be distinguished from denatured states by measurement of the absorbance of unpolarized light. The ultraviolet absorbance of a double-stranded, or triple-stranded, polynucleotide can be 20–30% less than that of the single strands. Helical polypeptides absorb significantly less in the amide absorbance region than coil polypeptides. This is the well-known hypochromicity caused by induced dipole–induced dipole interactions among the stacked bases of the nucleic acids, or among the helical array of the peptide bonds. Interactions among transition dipoles, which give rise to the hypochromicity, will be a recurring theme throughout this section.

The absorbance of linearly polarized light reveals the orientation of chromophores and the interactions among them. Molecules can be partially oriented by flow, stretched films, compressed gels, or electric or magnetic films. Linear dichroism measures the orientation of the molecules and the arrangement of chromophores relative to the orientation axis. Alternatively, an unoriented sample can be prepared in an anisotropic state by excitation with a linearly polarized light pulse. The transient linear dichroism produced can reveal transfer of excitation among transition moments. This is especially useful for macromolecular structures with several similar chromophores, such as photosynthetic pigments.

If right and left circularly polarized light is used, and the differential absorbance—the circular dichroism—is measured, the circular dichroism spectrum provides a very sensitive measure of conformation. Secondary structure elements, such as helices, β sheets, and turns in proteins, or double strands and loops in nucleic acids, can be quantitatively assessed. In nucleic acids, not only can one-, two-, and three-stranded structures be distinguished, but different conformations for each can be identified. The right-handed B-form typical of duplex DNA and the right-handed A-form typical of duplex RNA are easily distinguished. The left-handed Z-forms of both RNA and DNA have very different circular dichroism spectra from the right-handed forms. In general, circular dichroism is a very convenient way of following conformational changes in nucleic acids as a function of base sequence and environmental conditions such as temperature, pH, and solvent. For proteins, circular dichroism in the amide absorption region from 230 to 180 nm is used to determine percent α helix, β sheet, β turns, and other. In addition to the amides, other chromophores in proteins, such as phenylalanine, tyrosine, tryptophan, and cystine, serve as specific probes for studying the folding of particular domains in a protein. Non-amino acid ligands which are part of the active site of a protein, or are cofactors in an enzymatic reaction, can also introduce useful chromophores into a protein.

Only chiral structures—those different from their mirror images—differentially absorb circularly polarized light. However, in the presence of an external magnetic field all matter becomes chiral. Right and left circularly polarized light incident parallel to a magnetic field is differentially absorbed by all materials. This is magnetic circular dichroism; it can be especially helpful for studying transitions which involve magnetic moments in the ground or excited states.

The kinetics and mechanisms of reactions can be studied by absorption and circular dichroism. Disappearance of reactants and formation of products provides rates and rate constants, but spectra of transient intermediates give valuable clues about mechanisms. Combination of rapid scanning spectroscopy with stopped-flow kinetic methods can characterize steps in enzyme-catalyzed reactions which occur on the millisecond time scale. Faster reactions (femtoseconds to nanoseconds) which occur, for example, in the first steps of visual excitation can be studied with pump–probe methods. A short pulse of light starts the reaction; a probe pulse detects and characterizes the time dependence of intermediates. Hole-burning spectroscopy in which a sharp laser pulse is used to excite a specific transition and thus remove a narrow region of an absorption spectrum can also be used to study kinetics of fast chemical and physical processes. As a complement to using fast detection methods for studying kinetics, the rates of reactions can be slowed by lowering the temperature. Thus, low temperature spectroscopy is a useful addition to the methods for characterizing biological reactions.

Fluorescence adds a very powerful set of spectroscopic parameters to relate to structure and dynamics. The excitation spectrum (the absorption spectrum which leads to fluorescence at a chosen wavelength) and the fluorescence emission spectrum can identify and characterize fluorophores in complex mixtures. Time-resolved fluorescence spectra provide greater discrimination by introducing time as another variable.

Fluorescence energy transfer measures the efficiency of transfer of excitation from a donor group to an acceptor group. The efficiency of transfer decreases with the inverse sixth power of the distance between the groups. This clearly provides a sensitive and selective method of measuring distances in macromolecules. Often only qualitative information is very useful. For example, the kinetics of cleavage or ligation of a bond in a protein or nucleic acid can be monitored by the presence or absence of energy transfer.

Use of polarized light to excite fluorescence, and measurement of the state of polarization of the emitted light introduce another set of measurable parameters that can characterize structures and dynamics of molecules. The anisotropy of the polarization of fluoresence after excitation by linearly polarized light provides the rotational diffusion coefficient, or rotational correlation time, of the fluorophore. When there is fluorescence energy transfer, analysis of the anisotropy of both donor and acceptor can reveal the relative orientation, and the relative motion. Measurement of fluorescence after excitation by circularly polarized light provides the fluorescence-detected circular dichroism. This measurement characterizes the chiral environment of the ground state of the fluorophore. If the circular polarization of the fluorescence is measured, the circularly polarized luminescence is obtained. This measurement characterizes the chirality of the excited state.

Methods that use ultraviolet and visible light to learn about structures, dynamics, and therefore functions of biological molecules are presented in this section. The general theories are given, experimental methods are described, and specific applications are given to provide a good understanding of the wide range of methods available.

Absorption and Circular Dichroism

The application of absorption and circular dichroism spectra to the study of nucleic acids is discussed by Gray et al. [3]. Examples of the advantages of circular dichroism spectra over absorption spectra in analysis of the stoichiometry and structure of DNA, RNA, and hybrid duplexes and triplexes are given.

Woody [4] defines circular dichroism and circular intensity differential scattering and describes the instrumentation to measure them. He relates the experimental properties to the rotational strength, namely, the product of the electric dipole transition moment and the magnetic dipole transition moment. The main applications are to secondary and tertiary structures of proteins, protein folding, and protein–ligand interactions. The special problems of proteins in membranes are assessed. Applications not considered by Gray et al. in [3] to nucleic acids, such as binding of small molecule ligands and protein–nucleic acid complexes, are also discussed.

Solomon *et al.* [5] consider the application of the spectroscopic methods described in this book to metalloproteins. The emphasis is on understanding function by determining metal ion geometry and electronic structure, particularly in d^n electronic configurations. The fundamentals of crystal field theory (electrostatic interactions between the d orbitals and the ligands) and ligand field theory (including all types of interactions) are reviewed first. This provides a theoretical basis for understanding the geometry of a complex in the ground state, and for interpreting the $d \rightarrow d$ transitions. The mechanism of the nonheme iron enzyme metapyrocatechase, which catalyzes the oxidation of catechol, is discussed. Circular dichroism, magnetic circular dichroism, electron paramagnetic resonance, and Mössbauer studies are used to establish the structure of the metal site and to understand its interaction with various ligands.

The theory and practice of magnetic circular dichroism is presented by Sutherland [6]. Natural circular dichroism, which requires a chiral structure, is most useful for studying the overall conformation of a protein or nucleic acid. Magnetic circular dichroism is most useful in studying the individual peptides, amino acid chromophores, or nucleic acid bases. Instrumentation and applications are described for each region of the spectrum from the infrared to the X-ray (10 μm to 10 nm). Metalloproteins in general and those containing porphyrins in particular are the most usually studied. Interpretations of the magnetic circular dichroism spectra in terms of their shapes compared to the absorption spectra are described.

A method which can be used to give complementary information about conformations and kinetics is low temperature spectroscopy described by Austin and Shyamsunder [7]. Experimental methods for measuring absorption and fluorescence, theoretical principles involved in their interpretation, and recombination kinetics of carbon monoxide and other ligands with the heme iron in myoglobin are discussed.

Transient Absorption and Kinetics

Measurements of absorption as a function of time can provide a wide range of useful information. Brzović and Dunn [8] describe instrumentation for measuring the time dependence of absorption spectra after rapid mixing of reactants. Several rapid-scanning stopped-flow instruments are commercially available; reactions that take place in a millisecond or longer can be studied. Enzyme-catalyzed reactions with natural chromophores, such as NADH, are discussed, and the substitution of a colored metal center [Co(II)] for a colorless one [Zn(II)] are also described. Detailed mechanistic conclusions for horse liver alcohol dehydrogenase (LADH) are given.

Van Amerongen and van Grondelle [9] describe pump–probe methods for studying processes that occur in the range of femtoseconds or slower. After an intense short pulse of light (the pump) is incident on the sample, the changes in absorption can be followed by probe pulses. These changes are due to loss of ground state species, production of excited state species, production of new molecules, or the influence of the excitation on neighboring chromophores. Instrumentation for slow (nano- to millisecond) and fast (femto- to nanosecond) time regimes are described. Study of the efficient transfer of excitation from antenna pigments to the photosynthetic reaction center, where charge separation occurs, represents a useful application of transient absorption measurements.

Hole burning, reviewed by Friedrich [10], is the decrease in absorbance in a narrow wavelength region of a spectrum caused by an intense laser pulse. As in the pump–probe experiments, the change in absorbance can be caused by the depletion of a ground state species. In a broad spectrum caused by chromophores in slightly different environments, that is, an inhomogeneously broadened spectrum, one of the chromophores can be specifically excited, causing a sharp dip in the broad spectrum. Friedrich discusses the theoretical and experimental aspects of the method and concentrates on two applications: photosynthesis and protein physics. The first conclusive evidence for the presence of coherent, exciton-like states in the antenna pigments of photosynthetic species came from hole-burning experiments. Coherent and incoherent energy transfer are both very important in photosynthesis. The many conformational states that a protein can exit in at equilibrium means that the free energy minimum is actually a broad, rough trough with many local minima. Hole burning is an excellent way of detecting these slightly different conformations, and of measuring rates of interconversion.

Linear Dichroism and Fluorescence

Van Amerongen and Struve [11] give a thorough theoretical analysis of spectroscopy using linearly polarized light. They consider unoriented samples which are prepared in an anisotropic excited state by absorption of linearly polarized light. The state can be investigated by a polarized probe pulse or by the polarization of emitted fluorescent light. Samples that are partially oriented by compression or expansion of gels, by stretching of films, or by external electric or magnetic fields can also be studied by polarized absorption and emission. The linear dichroism and fluorescence anisotropy methods reveal the orientations of ground state and excited state transition moments, and the interactions among them. Coherent and incoherent states are possible; their properties are very different. With gel- and film-oriented samples the orientation of transition dipoles relative to molecular shapes are obtained. The permanent dipole vector, the electric polarizability tensor, and the magnetic susceptibility tensor are key for orientation by electric and magnetic fields. Clearly, a vast amount of useful data can be obtained about macromolecules and macromolecular assemblies.

Applications of fluorescence anisotropy to study macromolecular interactions is described by Jameson and Sawyer [12]. The measured anisotropy defined as

$$r = \frac{I_{\parallel} - I_{\perp}}{I_{\parallel} + 2I_{\perp}}$$

where $I_{\parallel}$ is the intensity of fluorescent light parallel to incident polarization and $I_{\perp}$ is the intensity of fluorescent light perpendicular to incident polarization, is independent of concentration. This greatly simplifies the analysis of mixtures of species because the measured anisotropy is a weighted sum of the contribution from each component. Binding isotherms for use in Scatchard plots, for example, are directly obtained. Studies of interactions of proteins with small ligands, peptides, other proteins, and nucleic acids have been done. Nucleic acids can be studied if extrinsic fluorophores are added. Hydrodynamic properties in the form of diffusion coefficients of the molecules and their complexes can be obtained from the magnitudes of the anisotropies.

Selvin [13] reviews fluorescence resonance energy transfer and its use as a molecular ruler. He provides a thorough discussion of the theory with emphasis on understanding the concepts. The effect of each molecular variable on the apparent donor–acceptor distance is assessed; this leads to the conclusion that the method is best for relative distances. Criteria for choosing dyes for the fluorescence energy measurement are given. The advantages and disadvantages of the various methods which have been used to measure transfer efficiency are discussed. The methods include decreases in fluorescence, lifetime, or photobleaching of the donor and increases in fluorescence of the acceptor. A wide range of applications is described including the structure of four-way junctions in DNA, the kinetics of protease cleavage of a vital human immunodeficiency virus (HIV) protein, and the measurement of translational diffusion rates.

Holzwarth [14] describes techniques for measuring and analyzing time-resolved fluorescence; he discusses applications to dynamics of proteins, nucleic acids, and membranes. Time-resolved two-dimensional microscopic images show promise of revealing detailed information about the location and motions of fluorophores in cells.

The article by Waggoner [15] is pertinent for everyone who uses fluorescence in research. It describes criteria for choosing a fluorophore for labeling a biomolecule, and gives detailed protocols for labeling proteins and nucleic acids.

[3] Absorption and Circular Dichroism Spectroscopy of Nucleic Acid Duplexes and Triplexes

By DONALD M. GRAY, SU-HWI HUNG, and KENNETH H. JOHNSON

Introduction

Nucleic acid duplexes and triplexes are of interest because nucleic acid oligomers can be used as antisense and antigene inhibitors of gene expression.[1-3] Ultraviolet absorption and circular dichroism (CD) measurements are extremely useful in the study of duplexes and triplexes. This chapter focuses on the use of these techniques to (a) monitor mixtures of complementary oligomer sequences and determine the stoichiometry of the strands in paired complexes and (b) characterize the resulting structures. The stoichiometry of the paired strands may be obtained from mixing curves, in which the optical property at a given wavelength is plotted as a function of the mole fraction of each strand,[4] and from isodichroic and isoabsorptive points. Absorption measurements have classically been used for mixing curves. The combination of absorption and CD spectra provides a more confident determination of complex formation and strand stoichiometry than is provided by absorption spectra alone. In addition, CD spectra are particularly valuable in determining (a) whether the individual strands themselves are self-complexed, (b) whether the individual strands alter their conformations during duplex or triplex formation, and (c) how the conformations of hybrid complexes, with mixed DNA and RNA strands, are related to those having all DNA or all RNA strands. Examples will be given for mixtures of oligomers that are 24 nucleotides long, with alternating A-G or C-T(U) sequences.

Methods

The availability of nucleic acid solutions that are uniform and free of precipitated material is crucial to the success of making mixtures having

[1] J. S. Cohen (ed.), "Oligodeoxynucleotides: Antisense Inhibitors of Gene Expression." CRC Press, Boca Raton, Florida, 1989.

[2] E. Wickstrom (ed.), "Prospects for Antisense Nucleic Acid Therapy of Cancer and AIDS." Wiley, New York, 1991.

[3] J. A. H. Murray (ed.), "Antisense RNA and DNA," Modern Cell Biology Series, Vol. 11. Wiley, New York, 1992.

[4] C. R. Cantor and P. R. Schimmel, "Biophysical Chemistry, Part III: The Behavior of Biological Macromolecules," p. 1135. Freeman, San Francisco, California, 1980.

METHODS IN ENZYMOLOGY, VOL. 246

well-defined, paired structures. For absorption and CD measurements, the solvent has to be a buffer, such as sodium phosphate, that is transparent in the ultraviolet. Molar concentrations of individual strands have to be determined, and mixtures of known molar ratios must be carefully made. Then, any competing intrastrand complexes have to be melted to allow the mixed strands to pair, and the mixtures have to be equilibrated before making the spectral measurements.

Extinction Coefficients

A key requirement is to know the molar concentrations of the individual oligonucleotide strands that are to be mixed. Concentrations can be readily obtained from the absorbance at 260 nm, A_{260}, of the oligomer solutions, if the extinction coefficients, ε, of the oligomer sequences are known:

$$C\ [M] = A_{260}/(\varepsilon[M^{-1}\,cm^{-1}]l[cm])$$

where l is the path length of the sample cuvette. The concentration is best determined in terms of moles of nucleotide subunits, rather than in terms of moles of oligomer sequence. This allows one to compare concentrations of bases in solutions of oligomers of different lengths, and it allows absorption and CD spectra of nucleotides, oligomers, and polymers to be on the same scale. Oligomer extinction coefficients, per mole of nucleotide, can be calculated from tabulated values of monomer and dimer extinction coefficients with the reasonable assumption that absorption is a nearest-neighbor property.[5] Table I gives an updated set of extinction coefficients for use in nearest-neighbor calculations of oligomer extinction coefficients.

An example is the nearest-neighbor calculation of the extinction coefficient for the DNA tetramer ApTpCpG, using data from Table I:

$$\varepsilon(ApTpCpG) = (1/4)[2\varepsilon(ApT) + 2\varepsilon(TpC) + 2\varepsilon(CpG) - \varepsilon(Tp) - \varepsilon(Cp)]$$
$$= 10,405\ M^{-1}\,cm^{-1}$$

Extinction coefficients of the dimers having the same nearest neighbors as in the oligomer are added (each multiplied by 2 because there are 2 mol of monomer per 1 mol of interacting bases in the dimer). Then, the extinction coefficients of the monomers, excluding the end monomers, are subtracted, because these each have been included twice, once with each neighboring dimer. Finally, the sum is divided by the number of monomers in the oligomer (i.e., 4 for a tetramer) to obtain the extinction coefficient per mole of nucleotide.

[5] C. R. Cantor and I. Tinoco, Jr., *J. Mol. Biol.* **13,** 65 (1965).

TABLE I
MOLAR EXTINCTION COEFFICIENTS OF NUCLEOTIDES AND
DINUCLEOSIDE PHOSPHATES[a,b,c]

Phosphates	ε (260), M^{-1} cm^{-1}	
	RNA	DNA
Monomer		
Ap	15,340	15,340
Cp	7,600	7,600
Gp	12,160	12,160
Up (dT)	10,210	8,700
Dimer		
ApA	13,650	13,650
ApC	10,670	10,670
ApG	12,790	12,790
ApU (ApT)	12,140	11,420
CpA	10,670	10,670
CpC	7,520	7,520
CpG	9,390	9,390
CpU (CpT)	8,370	7,660
GpA	12,920	12,920
GpC	9,190	9,190
GpG	11,430	11,430
GpU (GpT)	10,960	10,220
UpA (TpA)	12,520	11,780
UpC (TpC)	8,900	8,150
UpG (TpG)	10,400	9,700
UpU (TpT)	10,110	8,610

[a] At 260 nm. The values are needed to calculate the molar extinction coefficients of single-stranded DNA and RNA. All units are per molar concentration of *monomer*.

[b] Values were derived from the extinction coefficients of the 3'-monophosphate ribonucleotides (M. Alexis, Ph.D. Thesis, Univ. of London, 1978) at the wavelength maxima, absorption spectra of the 3'-monophosphates to convert peak values to those at 260 nm, and hypochromicity values for the RNA dimers (M. M. Warshaw, Ph.D. Thesis, Univ. of California, Berkeley, 1966). Monomer Ap, Cp, and Gp extinction coefficients and dimer hyperchromicity values were assumed to be the same for DNA and RNA; the extinction coefficient for dTMP was taken to be the same as that for dT [C. R. Cantor, M. M. Warshaw, and H. Shapiro, *Biopolymers* **9**, 1059 (1970)]. Values are for 0.1 M ionic strength, pH 7.0, 25°.

[c] Monomer and dimer extinction coefficients are estimated to be accurate to ± 100 M^{-1} cm^{-1} and $\pm 4\%$, respectively.

Such calculated extinction coefficients will be approximately correct only for an oligomer that is single-stranded and in which the conformations of neighboring bases are similar to those in the constituent dimers. Calculated extinction coefficients for sequences that may form intrastrand base pairs, such as sequences that are rich in G or C residues or repetitive sequences like $d(AG)_n$, will be incorrect. Melting profiles on individual oligomers will help determine whether there is intrastrand pairing. Incorrect extinction coefficients will also lead to apparent oligomer stoichiometries that are not integral $1:1$ or $1:2$ stoichiometries. In these cases, the extinction coefficient may have to be determined by hydrolysis or phosphate analysis. If mixtures are to be made at acid pH to induce triplex formation, the strand concentrations must first be determined by absorption measurements at neutral pH using the calculated extinction coefficients. Then, the pH can be adjusted, keeping track of the dilution and change in concentration. Oligomers containing 50% or more cytosines, such as $d(CT)_n$, readily form protonated $C^+ \cdot C$ base pairs at reduced pH. Nevertheless, triplexes can sometimes be formed with mixtures of such oligomers, as illustrated below.

Example of Mixing Procedure

The procedure that we use to obtain samples typically starts with the dissolution of synthesized oligomer samples in distilled, deionized water to obtain samples of 10^{-3} to 10^{-2} M in nucleotide concentration. The oligomer stock solutions are diluted by a factor of about 10^{-2}, with thorough mixing, in sodium phosphate buffer (50 mM Na^+, pH 7) to give working solutions having nucleotide concentrations in the range of 5×10^{-5} to 10^{-4} M. Absorption spectra are obtained at 20°, and these are used, together with oligomer extinction coefficients estimated as described above, to determine the concentrations of the oligomers. The oligomer solutions are adjusted as needed to make them equimolar in nucleotide subunit concentrations. If mixtures are to be made under different solution conditions (e.g., at pH 5.6 to form triplexes), the pH of the solutions of the two individual oligomer strands is accordingly adjusted by adding a small volume of 6 M HCl, with negligible dilution, before the mixtures are made.

The solutions are mixed in weighed proportions to give mixtures of known nucleotide molar ratios. Typically, seven mixtures are made with molar ratios of the strands of approximately $80:20, 67:33, 60:40, 50:50,$ $40:60, 33:67,$ and $20:80$. Together with samples of the individual strands, this gives nine samples, all of the same, known concentration, to measure for each set of experiments. Powder-free gloves should be worn to assure that the samples are nuclease-free. To overcome any intrastrand pairing, the samples, including the individual strands, are heated to 80° or 100° in a water bath for 1.5 or 2.0 min, the lower temperature and shorter time being used for samples containing RNA.

Samples are allowed to stand 24 to 48 hr at room temperature, or until no further change occurs in the spectra. The integrity of the samples is indicated by the cooperativity of thermal melting profiles of the duplex and triplex samples (after absorption and CD spectra are measured), by the presence of single bands by gel electrophoresis, and by the reproducibility of results from multiple mixing experiments.

Instrumentation

Absorption and CD spectra should be acquired with instruments that are capable of achieving spectral bandwidths of 1 nm or less over an ultraviolet wavelength range of 190 to 350 nm. Sample temperatures must be controlled. Spectra shown in this chapter have been obtained at 20° ± 5° with a Cary Model 118 spectrophotometer and a Jasco Model J500A spectropolarimeter. Typically, the spectrophotometer is operated with a scan speed of 0.5 nm/sec and a period of 1 sec, in the auto slit mode with the gain set so that the spectral bandwidth is always less than 1 nm. The spectropolarimeter is set to a time constant of 1 sec, a scan speed for single scans of 10 nm/min, and a spectral bandwith of 1 nm.

Calibration of CD instruments has to be maintained for comparison of CD spectra taken at different times or in different laboratories. The CD instrument should be calibrated to give a value of the ellipticity, Θ, that is within 2% of the value of 0.336° at 290.5 nm for a 0.1% (w/v), or 4.305×10^{-3} M, solution of (+)-10-camphorsulfonic acid (CSA) at 20°, as recommended by Chen and Yang.[6] Our procedure is to make an approximate 0.1% (w/v) solution of CSA (Aldrich Chemical Co., Milwaukee, WI; molecular weight 232.3) by dissolving a weighed amount (0.1 g) in 100 ml of distilled, deionized water. A more exact concentration is then determined by measuring the absorption spectrum of the solution in a 5-cm path length cell and using an extinction coefficient of 34.5 M^{-1} cm^{-1} at 285 nm.[7] For example, if the A_{285} of the CSA solution in a 5-cm cell equals 0.690, the CD instrument is set to read, for the CSA solution in a 1-cm cell,

$$\Theta(290.5) = 0.336 \times A_{285}/(5 \times 34.5 \times 4.305 \times 10^{-3})$$
$$= 312 \text{ [degrees ellipticity]}$$

[6] G. C. Chen and J. T. Yang, *Anal. Lett.* **10**, 1195 (1977).
[7] T. M. Lowry, "Optical Rotatory Power," p. 407. Dover, New York, 1964.

The instrumental variation of ellipticity with wavelength is also checked with the same CSA solution in a 0.1-cm cell. The ratio of ellipticity values at 192.5 (a negative band) and 290.5 nm (a positive band) is -2.00 ± 0.05.

The absorption and CD spectra of sample solutions are all stored in digitized form at 1-nm intervals. After subtracting the baselines, absorption and CD spectra are further smoothed by a sliding 13-point quadratic-cubic function.[8] Absorption spectra are converted to molar ε values, based on the concentrations of the single strands. CD data are acquired as ellipticity values, Θ. The CD spectra are converted to molar $\varepsilon_L - \varepsilon_R$ values (in units of $M^{-1} cm^{-1}$) by the following formula:

$$\varepsilon_L - \varepsilon_R = \Theta/[32.98lC]$$

where l is the path length in cm and C is the molar concentration of monomer subunits. C is the same for all of the mixtures made in a given series and is equal to the concentration of nucleotides in the solutions of the separate oligomer strands.

Results

Two examples of the absorption and CD spectra obtained from mixing experiments of oligomers are discussed. The first example is of mixtures of $r(AG)_{12}$ and $r(CU)_{12}$ at pH 7 to form an RNA duplex. The second example involves mixtures of $d(AG)_{12}$ and $d(CT)_{12}$ at pH 5.6 to form a DNA triplex.

Duplex of $r(AG)_{12} \cdot r(CU)_{12}$

Figure 1 shows seven of the CD and absorption spectra of mixtures of two oligomers, $r(AG)_{12}$ and $r(CU)_{12}$, each 24 nucleotides long. The solid curves are spectra of the strand containing pyrimidines, $r(CU)_{12}$, and the dashed curves are spectra of the strand containing purines, $r(AG)_{12}$. If mixtures of the two strands had consisted simply of the two noninteracting strands, the spectra of the mixtures, which have symbols, would all have been within the envelope formed by the solid and dashed curves. In addition, the spectra of the mixtures would have been spaced proportionally between the spectra of the individual strands, and any isodichroic or isoabsorptive points (wavelengths of equal CD or absorption magnitude) for the individual strands would have been isodichroic or isoabsorptive points for all of the mixtures.

Inspection of the spectra in Fig. 1 shows that the $r(AG)_{12}$ and $r(CU)_{12}$

[8] A. Savitzky and M. J. E. Golay, *Anal. Chem.* **36,** 1627 (1964).

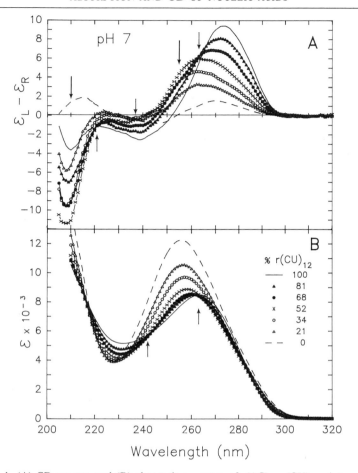

Fig. 1. (A) CD spectra and (B) absorption spectra of $r(AG)_{12} : r(CU)_{12}$ mixtures in the molar proportions of $0:100$ (—), $19:81$ (▲), $32:68$ (●), $48:52$ (×), $66:34$ (○), $79:21$ (△), and $100:0$ (– – –). Solution conditions (50 mM sodium phosphate buffer, pH 7) were such that an $r(AG)_{12} \cdot r(CU)_{12}$ duplex formed. Long arrows in (A) at 255 and 210 nm mark the maximum deviation in the CD spectrum of the duplex from the averaged spectra of the individual strands. Short arrows in (A) and (B) indicate isodichroic points in the CD spectra and isoabsorptive points in the absorption spectra.

strands did interact with a $1:1$ stoichiometry to form a duplex under the experimental conditions (50 mM sodium phosphate buffer, pH 7, 20°). CD spectra of the mixtures obviously lay outside the envelope of the spectra of the single strands at 255 and 210 nm. The deviation was maximal for the mixture containing close to the same proportions of the two strands, 48 mol % $r(AG)_{12}$ and 52 mol % $r(CO)_{12}$. The difference CD spectra in

Fig. 2 show that the spectrum of an equimolar mixture of r(AG)$_{12}$ and r(CU)$_{12}$ deviated most from the average of the spectra of the individual strands at these wavelengths. CD values at such wavelengths were therefore especially useful for plotting mixing curves. The mixing curve in Fig. 3A was plotted using the CD data at 255 nm from the spectra in Fig. 1A. The data were fitted by two straight lines having a breakpoint at a 50:50 molar ratio of the two strands, providing evidence that a duplex with A · U and G · C base pairs was formed. Intermediate mixtures probably had a simple proportion of the duplex and one of the single strands, although this was not guaranteed by the absence of additional break points in a single mixing curve. A mixing curve from the absorption spectra, using data at the peak absorbance value of 256 nm for r(AG)$_{12}$, is plotted in Fig. 3B. Again, there was a clear break point close to a 50:50 ratio of the two strands. The absorbance data of the mixtures did not exhibit an extremum at this wavelength because the absorption spectra of the mix-

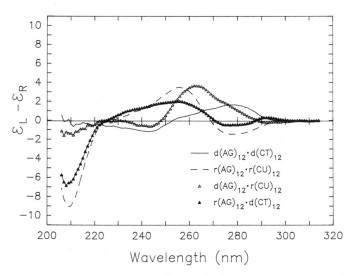

FIG. 2. Difference CD spectra of four oligomer duplexes, a DNA–DNA duplex (—), an RNA–RNA duplex (---), and the two analogous DNA–RNA hybrid duplexes (△, ▲). The averaged spectra of the individual single strands were subtracted from the spectra of 50:50 mixtures, under the conditions used to obtain the data for Fig. 1, except that the spectrum of single-stranded d(AG)$_{12}$ was taken at 90°. The difference spectra from two to six experiments have been averaged for each spectrum shown. Wavelengths of maximal difference, such as 255 and 210 nm for the r(AG)$_{12}$ · r(CU)$_{12}$ duplex, were good choices of wavelengths at which to plot mixing curves. Sensitive mixing curves could also be plotted with CD data from an isodichroic wavelength at which there was a substantial CD difference, such as at 263 nm for r(AG)$_{12}$ · r(CU)$_{12}$ (see Fig. 1A).

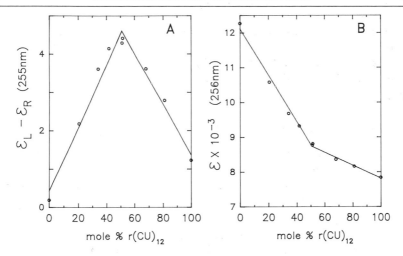

FIG. 3. (A) CD and (B) absorption mixing curves at 255 and 256 nm, respectively, for mixtures of r(AG)$_{12}$ and r(CU)$_{12}$. Data are plotted as a function of the mole percent of r(CU)$_{12}$. The mole percent of r(AG)$_{12}$ is 100 minus the percentage of r(CU)$_{12}$ in each mixture. Data were taken from Fig. 1, plus a mixture containing about 40 mol % of r(CU)$_{12}$ and an additional mixture with approximately 50 mol % of each oligomer. The break points at about 50 mol % of r(CU)$_{12}$ showed that a duplex was formed.

tures all lay between those of the individual strands in this region of the spectrum (Fig. 1B).

Although the mixing curves at single wavelengths indicated the formation of a duplex structure in an equimolar mixture of the strands, they did not exclude the possibility of complexes being formed with other stoichiometries, such as with a 1:2 molar ratio of r(AG)$_{12}$:r(CU)$_{12}$, in the other mixtures. Spectral features other than deviations from the average of the single strands were used to check for the formation of additional stoichiometric complexes. For example, as seen in Fig. 1A there were isodichroic points at wavelengths of 262–265 nm and near 221 nm at which only samples having 50% or more of the r(CU)$_{12}$ strand had the same CD value. There was an isodichroic point at about 237 nm at which samples having 50% or more of the r(AG)$_{12}$ strand had the same CD value. Iso-absorptive points in the absorption spectra in Fig. 1B were at about 242 and 263 nm, points at which the duplex and samples having an excess of the r(CU)$_{12}$ strand had the same absorption. In other words, the spectra all fell into two families, those with 50% or more of the purine-containing strand and those with 50% or more of the pyrimidine-containing strand. There was no evidence for the formation of complexes other than the duplex, to within the error of the measurements.

A method such as the singular value decomposition method[9] can be used to further confirm the number of independent components, but the application of such methods should not replace a direct inspection of the spectra for features that are consistent with a simple stoichiometry of base pairing.

Triplex of $d(C^+T)_{12} \cdot d(AG)_{12} \cdot d(CT)_{12}$

The possibility that triplex structures can form in solutions of $d(AG)_n$ plus $d(CT)_n$ sequences was inferred by Wells et al.[10] The formation of triplexes by the addition of a polypyrimidine strand to polypurine · polypyrimidine sequences has great potential as a means of controlling gene expression.[1-3] The third strand in classic triplex structures has C and T residues that are Hoogsteen-paired with G and A residues, the third strand being antiparallel with the pyrimidine strand that has Watson–Crick base pairing. The cytosines of the third strand are protonated in the Hoogsteen type of pairing (see Ref. 1), and triplex formation with C-containing sequences is enhanced at low pH. However, a polymeric triplex of $d(C^+T)_n \cdot d(AG)_n \cdot d(CT)_n$ can be stable at neutral pH. Antao et al.[11] showed that rearrangements of $d(AG)_{12} \cdot d(CT)_{12}$, including triplex formation, can be deciphered by CD measurements.

Seven CD and absorption spectra from a mixing experiment with $d(AG)_{12}$ and $d(CT)_{12}$, at 50 mM sodium phosphate buffer, pH 5.6, 20°, are shown in Fig. 4. It is obvious that the strands interacted under these conditions, as the spectra of the mixtures deviated from the averaged spectra of the individual strands. For example, at 210 nm, the CD spectra of three of the mixtures lay far outside the envelope formed by the spectra of the individual strands (the solid and dashed curves, Fig. 4), with the maximum deviation being for the mixture with 68% pyrimidines. A mixing curve from the CD data at 210 nm is shown in Fig. 5.

The CD and absorption spectra both fell into two families of spectra, as shown by two sets of isodichroic and isoabsorptive points. Spectra of samples containing 68% or more of $d(CT)_{12}$ had isodichroic points at 269 and 231 nm and isoabsorptive points at 258 and 238 nm. CD spectra of samples with 68% or less of $d(CT)_{12}$ showed an isodichroic point at 240 nm, and intersections of the spectra clustered at 273–275 nm. Absorption spectra of these samples had isoabsorptive points close to 268 and 230 nm. Slight variations of 1–2 nm in the wavelengths of equal CD or

---

[9] W. C. Johnson, Jr., this series, Vol. 210, p. 426.
[10] R. D. Wells, J. E. Larson, R. C. Grant, B. E. Shortle, and C. R. Cantor, J. Mol. Biol. **54,** 465 (1970).
[11] V. P. Antao, D. M. Gray, and R. L. Ratliff, Nucleic Acids Res. **16,** 719 (1988).

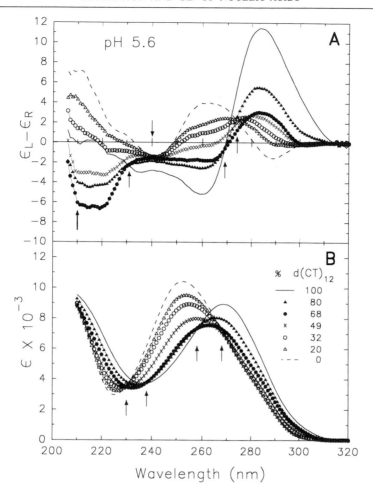

FIG. 4. (A) CD spectra and (B) absorption spectra of $d(AG)_{12}:d(CT)_{12}$ mixtures in the molar proportions of $0:100$ (—), $20:80$ (▲), $32:68$ (●), $51:49$ (×), $68:32$ (○), $80:20$ (△), and $100:0$ (---). Solution conditions (50 mM sodium phosphate buffer, pH 5.6) were such that a $d(C^+T)_{12} \cdot d(AG)_{12} \cdot d(CT)_{12}$ triplex formed. The long arrow in (A) at 210 nm marks the maximum deviation in the CD spectrum of the triplex from the weighted average of the spectra of the individual strands. Short arrows in (A) and (B) indicate isodichroic points in the CD spectra and isoabsorptive points in the absorption spectra.

equal absorption were within the reproducibility of repeat experiments under these conditions. Therefore, it appeared that a triplex of $d(C^+T)_{12} \cdot d(AG)_{12} \cdot d(CT)_{12}$ formed in mixtures of the two oligomers, without significant formation of a $d(AG)_{12} \cdot d(CT)_{12}$ duplex.

The experiment in Fig. 4 illustrates an additional important fact. Neither of the individual strands was single-stranded under the acidic condi-

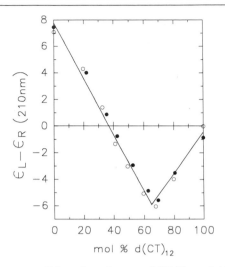

FIG. 5. CD mixing curve at 210 nm for mixtures of d(AG)$_{12}$ and d(CT)$_{12}$ under the same conditions used to obtain data for Fig. 4A. Data from Fig. 4A (○) were included plus data (●) from a duplicate experiment. The break point near 67 mol % of d(CT)$_{12}$ showed that a triplex was formed that contained two strands of d(CT)$_{12}$ and one strand of d(AG)$_{12}$.

tions. The large CD bands at 284 and 260 nm for the d(CT)$_{12}$ oligomer were indicative of C$^+$ · C base pairs and the formation of a self-complexed structure.[12] The red-shifted absorption curve of d(CT)$_{12}$ also indicated cytosine protonation. [See Gray *et al.*[12] and compare the spectrum of r(CU)$_{12}$ at pH 7 in Fig. 1B and the spectrum of d(CT)$_{12}$ at pH 5.6 in Fig. 4B.] The positive CD band at 260 nm and the negative CD band at 240 nm in the spectrum of d(AG)$_{12}$ were opposite in sign to the bands in alternating d(AG)$_n$ sequences that are single-stranded[11] and showed that this oligomer was also self-complexed. [The latter complex even existed at neutral pH. The extinction coefficient of d(AG)$_{12}$ at neutral pH, $\varepsilon(260)$, was estimated to be 9300 M^{-1} cm^{-1}, from the absorbance of a sample needed to form a duplex with d(CT)$_{12}$ at neutral pH. Note that the molar absorption spectra of r(AG)$_{12}$ and d(AG)$_{12}$ have different magnitudes in Figs. 1B and 4B.]

The self-complexes of the separate strands did not prevent the formation of a triplex structure under the conditions and with the procedure used to obtain the spectra in Fig. 4. However, in other circumstances the

[12] D. M. Gray, R. L. Ratliff, V. P. Antao, and C. W. Gray, *in* "Structure and Expression, Volume 2: DNA and Its Drug Complexes" (M. H. Sarma and R. H. Sarma, eds.), p. 147. Adenine Press, Schenectady, New York, 1988.

existence of self-complexes of the individual strands may hinder or prevent the formation of stoichiometric, paired complexes.

Characterization of Hybrids and Triplexes by Circular Dichroism Spectra

CD spectra contain more features than do absorption spectra, which makes CD spectra especially useful in distinguishing and characterizing the conformations of different oligomer complexes. Figure 6 shows the

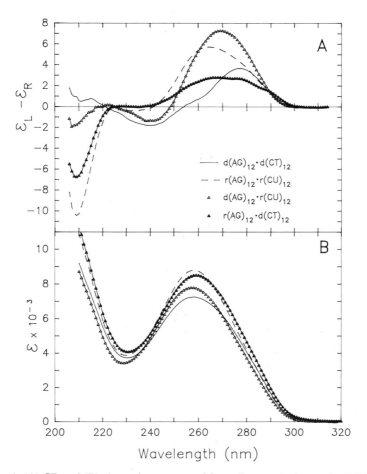

FIG. 6. (A) CD and (B) absorption spectra of four oligomer duplexes, the DNA–DNA duplex $d(AG)_{12} \cdot d(CT)_{12}$ (—), the RNA–RNA duplex $r(AG)_{12} \cdot r(CU)_{12}$ (---), and the two analogous hybrid duplexes $d(AG)_{12} \cdot r(CU)_{12}$ ($\triangle$) and $r(AG)_{12} \cdot d(CT)_{12}$ ($\blacktriangle$).

CD and absorption spectra of four duplexes with different mixtures of complementary DNA and RNA strands, each duplex containing alternating A · T and G · C base pairs (except that T in the DNA strands is replaced by U in the analogous RNA strands). The RNA duplex $r(AG)_{12} \cdot r(CU)_{12}$ had a negative band at 210 nm and a larger positive band above 250 nm than did the DNA duplex $d(AG)_{12} \cdot d(CT)_{12}$. These CD features are characteristics that generally distinguish the A-RNA conformation from the B-DNA conformation.[13,14]

The DNA–RNA hybrids had intermediate CD features and were not fully in the A-conformation. Even hybrids that adopt the A-conformation in crystals may be only partially in the A-form in solution.[15] The CD spectrum of the $r(AG)_{12} \cdot d(CT)_{12}$ hybrid had a strong 210 nm negative band but did not have a large positive band at long wavelengths. The $d(AG)_{12} \cdot r(CU)_{12}$ hybrid, on the other hand, was similar to the RNA duplex in having a spectrum with a large positive band above 250 nm. The difference CD spectra of Fig. 2 show that the RNA duplex and the hybrid duplex containing the $r(AG)_{12}$ strand, $r(AG)_{12} \cdot d(CT)_{12}$, both underwent similar CD changes on base pairing. The long-wavelength differences between the CD spectra of this hybrid and the RNA duplex (Fig. 6A) reside in the structural and optical differences between the $d(CT)_{12}$ and $r(CU)_{12}$ strands. Among the CD changes that occurred on pairing in the two duplexes, the 210 nm band was a dominant feature that was correlated with the presence of the $r(AG)_{12}$ strands within the duplexes. Simple base pairing of an alternating AG sequence to a CU or CT sequence was not sufficient to give a 210 nm CD band in the DNA duplex or the hybrid duplex containing a $d(AG)_{12}$ strand.

The CD spectra of three triplexes are given in Fig. 7A. Two interesting conclusions were drawn from the spectra. First, the CD spectrum of the DNA triplex was much different from those of the RNA triplex and the one hybrid triplex that could be formed under the conditions (see legend to Fig. 4). This was consistent with molecular mechanics calculations that showed that DNA helices with alternating $C^+ \cdot G \cdot C$ and $T \cdot A \cdot T$ base triples can have predominately C2'-*endo* sugars, except for cytidines in the third strand.[16] Second, the CD spectra of the other two triplexes were similar and A-like, indicating that the conformation of the hybrid triplex was close to that of the RNA triplex. The hybrid triplex had a $d(AG)_{12}$

[13] D. M. Gray, F. D. Hamilton, and M. R. Vaughan, *Biopolymers* **17,** 85 (1978).
[14] D. M. Gray, J.-J. Liu, R. L. Ratliff, and F. S. Allen, *Biopolymers* **20,** 1337 (1981).
[15] M. Egli, N. Usman, and A. Rich, *Biochemistry* **32,** 3221 (1993).
[16] M. Ouali, R. Letellier, F. Adnet, J. Liquier, J.-S. Sun, R. Lavery, and E. Taillandier, *Biochemistry* **32,** 2098 (1993).

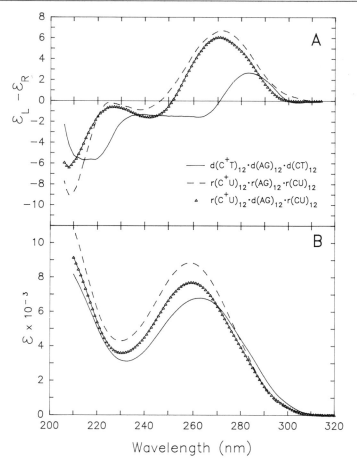

FIG. 7. (A) CD and (B) absorption spectra of three oligomer triplexes, the DNA triplex $d(C^+T)_{12} \cdot d(AG)_{12} \cdot d(CT)_{12}$ (—), the RNA triplex $r(C^+U)_{12} \cdot r(AG)_{12} \cdot r(CU)_{12}$ (---), and the hybrid triplex $r(C^+U)_{12} \cdot d(AG)_{12} \cdot r(CU)_{12}$ (△).

strand and a negative CD band at 210 nm. We inferred that the purine-containing DNA strand of the triplex hybrid was induced to assume a conformation more similar to that of the purine-containing RNA strands of the duplex hybrids, whose spectra were discussed above.

Absorption spectra differed for the DNA triplex and the other triplexes by a slight red shift of the DNA spectrum relative to the others (Fig. 7B).

Summary

Absorption and CD measurements of complementary oligomers and mixtures are described. The concentrations of oligomers may be estimated from absorption measurements and nearest-neighbor calculations of molar extinction coefficients. Interactions between complementary strands in mixtures can lead to obvious differences between measured CD spectra and the average of the spectra of the individual strands. CD spectra also allow an assessment of whether the individual strands are in self-complexes, which could compete with duplex or triplex formation. Isodichroic and isoabsorptive points provide important indicators of the stoichiometry of the strands in base-paired complexes. CD spectra provide an important means of characterizing differences in the conformations of DNA, RNA, and hybrid duplexes or triplexes having analogous sequences.

Acknowledgments

The work with oligomers was performed by S.-H. H. in partial fulfillment of the requirements for the Ph.D. degree in the Program in Molecular and Cell Biology at the University of Texas at Dallas. We thank Marilyn R. Vaughan for technical and drafting expertise. This work was supported by National Institutes of Health Research Grant GM19060 and by Grant AT-503 from the Robert A. Welch Foundation. This material is also based in part upon work supported by Grant 9741-036 from the Texas Advanced Technology Program.

[4] Circular Dichroism

By Robert W. Woody

Introduction

Circular dichroism (CD) is a spectroscopic method which depends on the fact that certain molecules interact differently with right and left circularly polarized light. Circularly polarized light is chiral, that is, occurs in two nonsuperimposable forms which are mirror images of one another, and is described further below. To discriminate between the two chiral forms of light, a molecule must be chiral, which includes the vast majority of biological molecules.

A method which can discern the subtle differences between nonsuperimposable mirror image molecules (enantiomers) must be highly sensitive to the three-dimensional features of molecules, that is, to conformation.

This expectation has been amply verified and forms the basis for the many applications of CD in biochemistry. For example, the CD spectrum of a polypeptide chain differs greatly depending on whether the polypeptide is an α helix, β sheet, or random coil. DNA molecules in the A-, B-, or Z-conformation have readily distinguishable CD spectra. Many ligands which bind to proteins and nucleic acids are achiral and thus exhibit no CD. However, when protoheme IX binds to globin, NADH to a dehydrogenase, or ethidium bromide to DNA, the absorption bands associated with the ligands exhibit strong CD. Binding of ligands or protein–protein and protein–DNA interactions can also alter the circular dichroism spectrum of the protein and/or nucleic acid. These changes in CD can be used to determine equilibrium constants, and they can also provide evidence for conformational changes. Thus, CD can provide information about the secondary structure of proteins and nucleic acids and about the binding of ligands to these types of macromolecules.

Several excellent reviews are available describing the general applications of CD to biochemistry.[1-4] Reviews of specific aspects are cited in the appropriate sections of this chapter.

Phenomenon of Circular Dichroism

Plane (linearly) polarized light can be shown to consist of two circularly polarized components of equal intensity. In each circularly polarized component, the electric vector rotates about the direction of propagation, undergoing a full rotation in a distance equal to the wavelength of the light, or a time equal to the period (reciprocal of the frequency) of the light. A snapshot of a circularly polarized light wave in which the electric vector could be visualized would show that the tip of the electric vector follows a helix, which is right-handed for right circularly polarized light (rcpl) and left-handed for lcpl. An alternative way of defining the sense of circularly polarized light is that for rcpl an observer looking toward the light source will see the electric vector rotating in a clockwise sense, whereas for lcpl it will appear to rotate counterclockwise. (An excellent

[1] D. W. Sears and S. Beychok, in "Physical Principles and Techniques of Protein Chemistry" (S. J. Leach, ed.), Part C, p. 445. Academic Press, New York, 1973.

[2] C. R. Cantor and P. R. Schimmel, "Biophysical Chemistry, Part II: Techniques for the Study of Biological Structure and Function," p. 409. Freeman, San Francisco, California, 1980.

[3] P. Bayley, in "An Introduction to Spectroscopy for Biochemists" (S. B. Brown, ed.), p. 148. Academic Press, London, 1980.

[4] W. C. Johnson, Jr., in "Methods of Biochemical Analysis" (D. Glick, ed.), Vol. 31, p. 61. Wiley, New York, 1985.

treatment of the properties and spectroscopic applications of circularly polarized light is given by Kliger et al.[5])

The absorption of unpolarized light by a sample is described by the absorbance, A, defined as

$$A = \log(I_0/I) \tag{1}$$

where I_0 is the intensity of the incident light and I is the intensity after the light has traveled a distance l through the medium. According to the Beer–Lambert law, if the absorbing species has a molar concentration c and the sample thickness is l, the absorbance is given by

$$A = \varepsilon c l \tag{2}$$

where ε is defined as the molar decadic absorption coefficient or, more commonly, the molar extinction coefficient.

Molar extinction coefficients can be defined in the same way for rcpl and lcpl; they are denoted by ε_R and ε_L, respectively. The molar circular dichroism, $\Delta\varepsilon$,

$$\Delta\varepsilon = \varepsilon_L - \varepsilon_R = (A_L - A_R)/cl \tag{3}$$

is defined as the difference between the extinction coefficients for the two types of circularly polarized light. The molar extinction coefficient for unpolarized light is the average of ε_L and ε_R:

$$\varepsilon = (\varepsilon_L + \varepsilon_R)/2 \tag{4}$$

Generally, $\Delta\varepsilon$ is small compared with ε_L, ε_R, and ε. The ratio

$$g = \Delta\varepsilon/\varepsilon = 2(\varepsilon_L - \varepsilon_R)/(\varepsilon_L + \varepsilon_R) \tag{5}$$

was defined by Kuhn[6] and is called the anisotropy factor or dissymmetry factor. It is a useful measure of how difficult it is to measure the CD in the region of a given absorption band, because $\Delta\varepsilon/\varepsilon$ is proportional to the signal/noise ratio. $\Delta\varepsilon$ determines the signal and ε determines the noise. The anisotropy factor is rarely larger than 10^{-2} and more commonly is around 10^{-4}.

All commercially available CD instruments measure $\Delta A = A_L - A_R$ which is readily converted to $\Delta\varepsilon$ by Eq. (3). However, most instruments are calibrated in terms of molar ellipticity, an alternative measure of CD which is mainly of historical interest. The ellipticity is an angular measure which is related to ΔA as follows:

[5] D. S. Kliger, J. W. Lewis, and C. E. Randall, "Polarized Light in Optics and Spectroscopy," p. 13. Academic Press, San Diego, 1990.
[6] W. Kuhn, Trans. Faraday Soc. 46, 293 (1930).

$$\theta = 32.98 \ \Delta A \tag{6}$$

where θ is in degrees. To eliminate the effects of path length and concentration, the molar ellipticity is defined as

$$[\theta] = \frac{100\theta}{lc} = 3298 \ \Delta\varepsilon \tag{7}$$

The units for molar ellipticity are degrees cm^2/dmol. Logically, it would be better to use $\Delta\varepsilon$ consistently in reporting CD. However, $[\theta]$ is firmly entrenched in the literature, especially in the area of biochemistry. Fortunately, conversion between the two measures of CD is simple.

This chapter concentrates on CD studies in the visible and ultraviolet, regions in which CD results from electronic excitations. There has been outstanding progress in extending CD measurements into the infrared, thus providing information about vibrational excitations in the form of vibrational circular dichroism (VCD). VCD instrumentation is currently available only in a few laboratories which have constructed custom-made VCD spectrometers. For a detailed discussion of VCD, the reader is referred to several reviews.[7-9]

Other Chiroptical Spectroscopy Methods

Circular dichroism is closely related to the more familiar phenomenon of optical rotation, namely, the rotation of the plane of polarization of plane-polarized light as it passes through a medium containing chiral molecules. Optical rotation results from a difference in refractive indices for lcpl and rcpl, n_L and n_R, respectively. The dependence of optical rotation on wavelength is called optical rotatory dispersion (ORD). Two excellent reviews[10,11] should be consulted for a description of ORD and references to the older literature.

CD and ORD are the two most familiar and widely used examples of chiroptical spectroscopy. However, there are other forms of chiroptical spectroscopy which can provide useful information about biological molecules. Excited states of chiral molecules emit rcpl and lcpl with different probabilities, thus giving rise to circularly polarized emission (CPE) or

[7] T. B. Freedman and L. A. Nafie, *Topics Stereochem.* **17**, 113 (1987).

[8] T. A. Keiderling, *in* "Practical Fourier Transform Infrared Spectroscopy" (J. R. Ferraro and K. Krishnan, eds.), p. 203. Academic Press, San Diego, 1990.

[9] T. A. Keiderling, *in* "Circular Dichroism: Interpretation and Applications" (K. Nakanishi, N. Berova, and R. W. Woody, eds.), p. 497. VCH, New York, 1994.

[10] K. Imahori and N. A. Nicola, *in* "Physical Principles and Techniques of Protein Chemistry" (S. J. Leach, ed.), Part C, p. 357. Academic Press, New York, 1973.

[11] J. T. Yang and T. Samejima, *Prog. Nucleic Acid Res. Mol. Biol.* **9**, 223 (1969).

circularly polarized luminescence.[12,13] Whereas CD and ORD give information about the ground state conformation of molecules, CPE provides a probe of excited state conformation.

Chiral molecules also scatter rcpl and lcpl to differing extents, both elastically in Rayleigh scattering and inelastically in Raman scattering. Chiral molecules which have dimensions comparable to the wavelength of light can give detectable differential Rayleigh scattering of rcpl and lcpl, which depends on the scattering angle. This phenomenon, called CIDS (circular intensity differential scattering), is described by the following parameter:

$$\text{CIDS} = (I_L - I_R)/(I_L + I_R) \tag{8}$$

where I_L and I_R are, respectively, the intensity of lcpl and rcpl scattered at a particular angle. CIDS was first recognized as a potential source of artifacts in CD studies of membrane fragments.[14] Tinoco, Maestre, Bustamante, and coworkers have studied the phenomenon extensively, both experimentally and theoretically, and developed methods for obtaining useful information from CIDS measurements on large DNA and nucleoprotein particles.[15,16] That work led to the development of a CD/CIDS microscope which has been used to study the condensation of chromatin in a single cell at various stages in the cell cycle.[16,17]

The differential Raman scattering of rcpl and lcpl provides an alternative way of studying vibrational optical activity and is called Raman optical activity (ROA). An expression identical to Eq. (8) is used to quantitate ROA. ROA was first observed in 1973 by Barron et al.,[18] on a pure organic liquid. Requirements for high concentration imposed by the combination of weak Raman scattering and small differential scattering have until relatively recently precluded applications to proteins and nucleic acids. Improvements in instrumentation, especially the introduction of highly sensitive charge-coupled device (CDC) detectors and a different scattering geometry (backscattering, i.e., scattering at 180°), have made ROA mea-

[12] J. P. Riehl and F. S. Richardson, *Chem. Rev.* **89**, 1 (1986).

[13] H. P. J. M. Dekkers, *in* "Circular Dichroism: Interpretation and Applications" (K. Nakanishi, N. Berova, and R. W. Woody, eds.), p. 121. VCH, New York, 1994.

[14] T. H. Ji and D. W. Urry, *Biochem. Biophys. Res. Commun.* **34**, 404 (1969).

[15] I. Tinoco, Jr., C. Bustamante, and M. F. Maestre, *Annu. Rev. Biophys. Bioeng.* **9**, 107 (1980).

[16] I. Tinoco, Jr., W. Mickols, M. F. Maestre, and C. Bustamante, *Annu. Rev. Biophys. Biophys. Chem.* **16**, 319 (1987).

[17] M. F. Maestre, G. C. Salzman, R. A. Tobey, and C. Bustamante, *Biochemistry* **24**, 5152 (1985).

[18] L. D. Barron, M. P. Bogaard, and A. D. Buckingham, *J. Am. Chem. Soc.* **95**, 603 (1973).

surements on biopolymers feasible.[19] These and other developments in ROA have been reviewed.[20,21]

Theoretical Aspects

The theoretical basis of CD is discussed in several reviews.[22-25] A quantity called the rotational strength, which is a property of a given electronic or vibrational transition, generally provides the connection between theory and experiment in CD. The theoretical definition of the rotational strength is

$$R_{oa} = \text{Im}(\boldsymbol{\mu}_{oa} \cdot \mathbf{m}_{ao}) \tag{9}$$

where $\boldsymbol{\mu}_{oa}$ and $\mathbf{m}_{ao}$ are the electric and magnetic dipole transition moments, respectively, for the transition $o \rightarrow a$, and $\text{Im}(z)$ is the imaginary part of the complex variable $z = x + iy$, that is, it is equal to y.

Equation (9) has a simple physical interpretation. The electric dipole transition moment for an electronic transition is defined as

$$\boldsymbol{\mu}_{oa} = \int \Psi_o^* \boldsymbol{\mu} \Psi_a \, d\tau = e \int \Psi_o^* (\Sigma_i \mathbf{r}_i) \Psi_a \, d\tau \tag{10}$$

where Ψ_o and Ψ_a are the wave functions for the ground and excited states, respectively; $\boldsymbol{\mu}$ is the electric dipole moment operator; $\mathbf{r}_i$ is the position of electron i; the $*$ on Ψ_o^* indicates that the complex conjugate of Ψ_o is to be taken; and the integration is to be taken over all electron coordinates. The electric dipole transition moment can be pictured as describing the linear oscillation of charge induced by light with a frequency resonant with the transition $o \rightarrow a$. The magnetic dipole transition moment of an electronic transition is defined as

$$\mathbf{m}_{ao} = \int \Psi_a^* \mathbf{m} \Psi_o \, d\tau = \frac{e}{2mc} \int \Psi_a^* (\Sigma_i \mathbf{r}_i \times \mathbf{p}_i) \Psi_o \, d\tau \tag{11}$$

Here, $\mathbf{m}$ is the magnetic moment operator, and $\mathbf{p}_i$ is the linear momentum of electron i. (If Ψ_o and Ψ_a are real wave functions, as is generally the case, the magnetic dipole transition moment is an imaginary quantity,

[19] L. Hecht, L. D. Barron, A. R. Gargaro, Z. Q. Wen, and W. Hug, *J. Raman Spectrosc.* **23**, 401 (1992).

[20] L. D. Barron and L. Hecht, *Adv. Spectrosc.* (*Chichester, U.K.*) **21**, 235 (1993).

[21] L. D. Barron and L. Hecht, *in* "Circular Dichroism: Interpretation and Applications" (K. Nakanishi, N. Berova, and R. W. Woody, eds.), p. 179. VCH, New York, 1994.

[22] A. Moscowitz, *Adv. Chem. Phys.* **4**, 67 (1962).

[23] I. Tinoco, Jr., *Adv. Chem. Phys.* **4**, 113 (1962).

[24] J. A. Schellman, *Chem. Rev.* **75**, 323 (1975).

[25] A. E. Hansen and T. D. Bouman, *Adv. Chem. Phys.* **44**, 545 (1980).

while the electric dipole transition moment is real. Thus, the product of $\boldsymbol{\mu}_{oa}$ and $\mathbf{m}_{ao}$ is purely imaginary. We must take the imaginary part to obtain R_{oa}, which must be real because it is an observable quantity.) Magnetic dipoles always arise from a circulation of charge, so the magnetic dipole transition moment can be visualized as the circular motion of charge induced by light with a frequency resonant with the $o \rightarrow a$ transition. According to Eq. (9), the rotational strength will be nonzero if the transition $o \rightarrow a$ is associated with both a linear motion of charge in a particular direction and a circular motion of charge about that direction. Such a combined linear and circular motion corresponds to a helical motion of charge. The sense of this helix determines whether rcpl or lcpl will be absorbed more strongly.

The units of rotational strength are erg cm^3 (cgs) or J m^3 (SI). Because electric dipole transition moments are of the order of 1 D (10^{-18} esu cm) and magnetic dipole transition moments are of the order of 1 Bohr magneton (0.9273×10^{-20} erg/gauss), a convenient unit for rotational strength is the Debye–Bohr magneton (1 DBM = 0.9273×10^{-38} erg cm^3 = 0.9273×10^{-51} J m^3).

Experimentally, the rotational strength is determined from the area of the CD band associated with a given transition $o \rightarrow a$, where o is the ground state and a is the excited state:

$$R_{oa} = \frac{6909 hc}{32\pi^3 N_o} \int_o^\infty \frac{\Delta\varepsilon_{oa}\, d\lambda}{\lambda} \tag{12}$$

Here $\Delta\varepsilon_{oa}$ is the contribution of the transition $o \rightarrow a$ to the CD spectrum, obtained as the total CD for a well-resolved transition or from a curve resolution procedure in a region of overlapping bands.

Experimental Aspects

Instrumentation

All commercially available CD instrumentation uses the modulation technique introduced by Grosjean and Legrand.[26-28] Light from a monochromator is linearly polarized, then passes through a modulating device, today a photoelastic modulator (PEM),[29,30] which converts the plane-polar-

[26] M. Grosjean and M. Legrand, *C. R. Acad. Sci.* (*Paris*) **251**, 2150 (1960).
[27] L. Velluz, M. Legrand, and M. Grosjean, *in* "Optical Circular Dichroism, Principles, Measurements and Applications." Verlag Chemie, Weinheim, 1965.
[28] A. Drake, *J. Phys. E: Sci. Instrum.* **19**, 170 (1986).
[29] M. Billardon and J. Badoz, *C. R. Acad. Sci.* (*Paris*) **262**, 1672 (1966).
[30] J. C. Kemp, *J. Opt. Soc. Am.* **59**, 950 (1969).

ized light to circularly polarized light, alternating between lcpl and rcpl at the modulation frequency. The light incident on the sample is modulated between left and right circular polarization, so if the sample exhibits CD, the light intensity detected by the photomultiplier will also be modulated at the same frequency. The current in the photomultiplier circuit will therefore consist of a small alternating current and a larger direct current. The ac and dc components are separated and the ac component is amplified. The ratio of the ac and dc currents is directly proportional to the absorbance difference for rcpl and lcpl. The dc component is maintained at a constant level by a servo system which adjusts the high voltage applied to the photomultiplier. Thus, ΔA is directly proportional to the ac component and hence to the gain of the amplifer.

A CD instrument must be calibrated regularly using a standard sample. The most commonly used standard for visible and UV measurements is (+)-10-camphorsulfonic acid (CSA), for which careful measurements have given a $\Delta\varepsilon_{290.5}$ value of 2.36 M^{-1} cm^{-1} and a $\Delta\varepsilon_{192.5}$ of -4.9 M^{-1} cm^{-1}. One complicating factor is that CSA is somewhat hygroscopic. Therefore, it is best to check the concentration of a CSA solution made by weight using ORD if a spectropolarimeter is available, or by absorbance at 290 nm. These and other aspects of instrument calibration have been discussed by Chen and Yang.[31]

There are four manufacturers of CD instrumentation. Aviv and Associates (Lakewood, NJ) has been modifying Cary 61 and 62 CD instruments, replacing the Pockels cell with a PEM, modernizing the electronics, and computerizing the instrument. Aviv has also been building complete new instruments, but still retaining the original Cary monochromator design. JASCO (Easton, MD; Tokyo) manufactures two very similar instruments, the J710 and J720, which were introduced in 1990. Jobin-Yvon (Longjumeau, France) makes the JY Mark VI. OLIS (On-Line Instrument Systems, Bogart, GA) also modifies Cary 61 and 62 CD instruments. Readers considering the purchase of CD instruments are urged to compare these available instruments by submitting representative samples to be run on each of them. Better yet, the prospective purchaser should visit as many as possible of the vendors or laboratories with well-maintained instruments from the various sources.

CD instruments using other basic designs have been constructed. VCD instrumentation using both dispersive and Fourier transform designs are described in several reviews.[8,32] Fluorescence-detected CD (FDCD)[33]

[31] G. C. Chen and J. T. Yang, *Anal. Lett.* **10**, 1195 (1977).
[32] T. A. Keiderling, *Appl. Spectrosc. Rev.* **17**, 189 (1981).
[33] D. H. Turner, this series, Vol. 49, p. 199.

measures CD by detecting the modulated intensity of light emitted by one or more fluorophores. This method has the advantage of being selective for fluorescent chromophores, although energy transfer can complicate the interpretation and photoselection can lead to artifacts.[34] CD can also be measured by photoacoustic detection,[35] which is advantageous for strongly scattering and/or absorbing samples.

The requirements for material and sample concentrations for CD measurements are generally similar to those required to measure an absorption spectrum. In general, the total sample absorbance (solute plus solvent and cell windows) should be kept below 1.0, as the signal/noise (S/N) ratio deteriorates sharply and artifacts arising from stray light can occur with higher absorbances. For regions where solvent absorbance is negligible, the sample absorbance can be adjusted by varying the solute concentration, and cells with convenient path lengths of 1 cm and 1 mm can be used. For measurements in the far-UV, solvent absorbance is no longer negligible, even with aqueous solutions. Cells with fixed path lengths of 0.1 or 0.05 mm are needed for measurements at wavelengths below 190 nm. Such measurements require higher solute concentrations, but of course the volume of the solution required is small.

Buffers, salts, and other additives must be selected with care for far-UV measurements. Dilute (10 mM) phosphate buffer is the optimum buffer from the standpoint of transparency, but other widely used buffers such as Tris can be tolerated at similar low concentrations. If higher ionic strengths are needed, fluoride, perchlorate, and sulfate salts should be used. (Of course, fluorides should not be used at acidic pH.) For further details on sample handling and techniques for enhancing UV penetration, the review of Johnson[36] is recommended. The wavelength limits imposed by various organic solvents[36] and buffer components[37] have been tabulated.

Sample cells for measuring CD should be checked for strain, and those which give significant blank values and a curving baseline should be discarded. Even with low-strain cells, it is essential to run a baseline spectrum and subtract it from the spectrum of the sample. The baseline should be run under the same conditions (solvent, gain, time constant, spectral band width, scanning speed) as the sample.

[34] I. Tinoco, Jr., B. Ehrenberg, and I. Z. Steinberg, *J. Chem. Phys.* **66,** 916 (1977).

[35] R. A. Palmer, J. C. Roark, and J. C. Robinson, *in* "Stereochemistry of Optically Active Transition Metal Compounds" (B. E. Douglas and Y. Saito, eds.), ACS Symp. Ser. Vol. 219, p. 375. American Chemical Society, Washington, D.C., 1980.

[36] W. C. Johnson, Jr., *Proteins: Struct. Funct. Genet.* **7,** 205 (1990).

[37] F. X. Schmid, *in* "Protein Structure: A Practical Approach" (T. E. Creighton, ed.), p. 258. IRL Press, Oxford, 1989.

Biochemical samples frequently pose challenges because of relatively poor S/N ratios. This may be due either to a low $\Delta\varepsilon/\varepsilon$ value or to operating near the wavelength limits of the instrument. Low S/N ratios can be overcome by carrying out multiple scans. Assuming negligible instrumental drift and that no time-dependent processes occur in the sample or instrument, accumulation of N scans will lead to an N-fold increase in the signal but, because of its random character, only an $N^{1/2}$-fold increase in the noise, thus improving the S/N ratio by a factor of $N^{1/2}$. Modern CD instruments are interfaced to microcomputers, and software is available from the manufacturers for processing multiple scans. Further, the trend is toward faster scanning rates. For example, with the JASCO J-720, it is possible to scan from 250 to 180 nm in less than 1 min. Thus, in a 10-min period, ten such scans can be accumulated, leading to a 3-fold improvement in S/N ratio.

Proteins and Peptides

Chromophores

Proteins and peptides have a number of groups which have characteristic absorption bands in specific areas of the visible and ultraviolet region. Such groups are called chromophores. The most numerous and characteristic chromophore in proteins is the peptide group linking amino acid residues. Amides have two well-characterized low-energy electronic transitions: the $n\pi^*$ transition, weak in absorption ($\varepsilon_{max} \sim 100$) but often strong in CD, at about 220 nm, and the $\pi\pi^*$ transition near 190 nm (200 nm in tertiary amides formed by the α-imino group of Pro), which is strong in both absorption ($\varepsilon_{max} \sim 7000$) and in CD. The two transitions dominate the CD and absorption spectra of proteins in the far-ultraviolet. The CD spectrum resulting from the peptide groups is characteristic for various types of secondary structure, as discussed below, and this forms the basis for analyzing protein secondary structure by CD.

Most peptides and nearly all proteins have one or more aromatic residues (Phe, Tyr, Trp). The aromatic groups have characteristic $\pi\pi^*$ absorption bands in the near-UV between 250 and 300 nm, and in the far-UV. In CD, the far-UV bands of the aromatic side chains are generally small in comparison with those of the peptide groups, but in some cases the aromatic groups make detectable contributions to the far-UV CD.[38] In the near-UV, CD bands are generally dominated by aromatic side-chain contributions, with contributions from disulfides and prosthetic groups,

[38] R. W. Woody, *Biopolymers* **17**, 1451 (1978).

if present, as described below. The near-UV, therefore, is sensitive to the environment of aromatic side chains and therefore to tertiary stuctural features in proteins.

The disulfide group present in many extracellular proteins and some peptides has two well-characterized $n\sigma^*$ electronic transitions in the near-UV.[39] The wavelengths of the corresponding absorption bands depend on the dihedral angle of the disulfide. When χ_{SS} is approximately 90°, as is generally the case in proteins, the two $n\sigma^*$ transitions are degenerate, giving rise to a single broad absorption band near 260 nm. As the dihedral angle opens or closes from 90° the degeneracy is broken, and one transition shifts to higher energy and the other to lower energy. Disulfide contributions to the near-UV CD spectrum can generally be distinguished from those of aromatic side chains because the former are much broader.[40,41] Little is known about higher energy disulfide transitions, although theoretical predictions indicate that they are likely to make only weak contributions to the CD spectrum.[42]

Proteins that have tightly bound cofactors, such as heme proteins, photosynthetic reaction centers and antenna proteins, flavoproteins, and pyridoxal phosphate- and NAD-dependent enzymes, provide a variety of chromophores which have absorption bands in the visible and UV region. The CD bands associated with the chromophoric groups are frequently quite intense, despite the fact that the isolated chromophores are achiral in many cases, and therefore have no CD, or are separated from the nearest chiral center by several bonds about which relatively free rotation can occur, and therefore have only weak CD. The extrinsic or induced CD observed in the visible and near-UV spectra of the proteins can provide useful information about the conformation and/or environment of the bound chromophore, which usually plays a critical role in the function of the protein.

Secondary Structure of Peptides

The common types of secondary structure adopted by peptides and proteins have distinctive CD spectra in the far-UV, as illustrated in Fig. 1.[43,44] The strongest and most characteristic spectrum is that of the α helix,[45] which has two negative bands of comparable magnitude near 222

[39] R. W. Woody, *Tetrahedron* **29**, 1273 (1973).
[40] E. H. Strickland, *Crit. Rev. Biochem.* **2**, 113 (1974).
[41] P. C. Kahn, this series, Vol. 61, p. 339.
[42] A. Rauk, *J. Am. Chem. Soc.* **106**, 6517 (1984).
[43] W. C. Johnson, Jr., and I. Tinoco, Jr., *J. Am. Chem. Soc.* **94**, 4389 (1972).
[44] S. Brahms and J. Brahms, *J. Mol. Biol.* **138**, 149 (1980).
[45] G. Holzwarth and P. Doty, *J. Am. Chem. Soc.* **87**, 218 (1965).

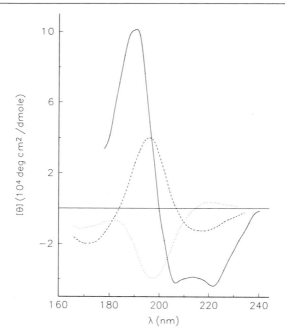

FIG. 1. CD spectra of three types of protein secondary structure: α helix (—), poly(Glu), pH 4.5[43]; β sheet (----), poly(Lys-Leu), 0.5 M NaF, pH 7[44]; unordered ($\cdots$), poly(Lys-Leu) in salt-free aqueous solution.[44]

and 208 nm and a stronger positive band near 190 nm. A wide range of synthetic peptides and polypeptides give rise to this type of spectrum.[46] Except for peptides with a large fraction of aromatic residues such as Tyr and Trp, for example, homopolymers of these amino acids, the α-helix CD spectrum is nearly independent of the nature of the side chains. It is also largely independent of solvent, so long as changes in solvent do not alter the extent of helix formation.

Peptides adopting the β-sheet conformation have a less intense and simpler CD spectrum, with a negative band near 217 nm, a positive band near 195 nm, and another negative band near 180 nm.[47] There is much greater variability among the CD spectra reported for model β structures than for α-helical models.[46] The amplitudes of the two long-wavelength bands, their ratios, and the wavelength of the positive band all show considerable variation with side chains, solvent, and other environmental

[46] R. W. Woody, *J. Polym. Sci. Macromol. Rev.* **12**, 181 (1977).
[47] S. Brahms, J. Brahms, G. Spach, and A. Brack, *Proc. Natl. Acad. Sci. U.S.A.* **74**, 3208 (1977).

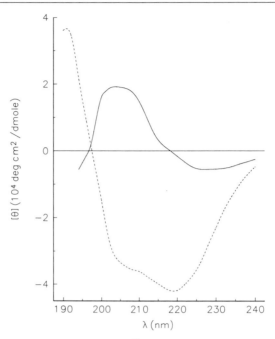

FIG. 2. CD spectra of two model β turns[49]: type I turn, cyclo(L-Ala-L-Ala-Aca) (Aca is ε-aminocaproyl) in water at 22° (—); type II turn, cyclo(L-Ala-D-Ala-Aca) in water at 22° (---). The CD is reported per residue (i.e., one-half the molar ellipticity).

factors. This wider range of CD parameters is probably attributable to the fact that β sheets can be antiparallel, parallel, or mixed; intra- or intermolecular; and are twisted to varying extents. Theoretical studies[48] suggest that the extent of twisting of β sheets is probably more important than is the distinction between antiparallel and parallel sheets.

The β turns occur over an even wider range of conformations, and, not surprisingly, there is no single CD pattern which is characteristic of β turns (Fig. 2).[49] Early theoretical studies[50] indicated that β turns could give rise to a variety of CD spectra but that most β turns should give a CD spectrum which resembles that of a β sheet, but shifted 5–10 nm to longer wavelengths. This type of spectrum, called class B, has a weak

[48] M. C. Manning, M. Illangasekare, and R. W. Woody, *Biophys. Chem.* **31,** 77 (1988).

[49] J. Bandekar, D. J. Evans, S. Krimm, S. J. Leach, S. Lee, J. R. McQuie, E. Minasian, G. Némethy, M. S. Pottle, H. A. Scherage, E. R. Stinson, and R. W. Woody, *Int. J. Pept. Protein Res.* **19,** 187 (1982).

[50] R. W. Woody, *in* "Peptides, Polypeptides, and Proteins" (E. R. Blout, F. A. Bovey, M. Goodman, and N. Lotan, eds.), p. 338. Wiley, New York, 1974.

negative band at 220–230 nm and a stronger positive band between 200 and 210 nm. Subsequent experimental work[49,51] has shown that the class B spectrum is indeed commonly observed for type II β turns. However type I (and the closely related type III) β turns typically have a spectrum which qualitatively resembles that of an α helix, but with the two long-wavelength negative bands having about half the amplitude of the α helix, and the positive band near 190 nm about one-fourth to one-third that of the helix (a class C spectrum in the nomenclature of Woody[50]).

The homopolymer poly(Pro) can adopt two very different conformations. Poly(Pro) I is a right-handed helix with *cis*-peptide bonds throughout, and it is stable only in relatively nonpolar solvents such as *n*-propanol. Poly(Pro) II is a left-handed 3-fold helix with trans residues, is the stable form in water, and has the conformation adopted by each of the three intertwined strands of collagen. The two types of poly(Pro) helix are readily distinguishable by CD,[52] as shown in Fig. 3.

The CD spectra of unordered polypeptides are characterized by a strong negative band near 200 nm but also have a weak band at longer wavelengths, although the latter band may be either positive or negative. Model systems in which the CD band near 220 nm is positive include such widely used models of unordered peptides as poly(Lys) and poly(Glu) at neutral pH. The CD spectra of those polypeptides resemble that of poly (Pro) II, and, to account for this resemblance, Tiffany and Krimm[53] proposed that the ionized homopolymers have a substantial amount of left-handed 3-fold helix. Although not widely accepted at first, this proposal has been supported by a number of other observations, especially VCD[54,55] and variable-temperature CD data,[56] as described in a review.[57] Based on CD spectral data, "unordered" peptides with a positive band near 220 nm appear to consist of short and frequently interrupted regions of left-handed poly(Pro) II-like helix at ordinary temperatures. "Unordered" peptides with negative CD near 220 nm have local conformations which are largely in the α-helical and β-structure regions of the Ramachandran map.

CD has been used extensively to follow the equilibrium between helix and unordered conformation. Peptides undergoing helix–coil transitions typically exhibit an isodichroic point near 203 nm as the temperature, solvent, pH, or ionic strength is varied. The observation of an isodichroic

[51] J. A. Smith and L. G. Pease, *Crit. Rev. Biochem.* **8**, 315 (1980).

[52] F. A. Bovey and F. P. Hood, *Biopolymers* **5**, 325 (1967).

[53] M. L. Tiffany and S. Krimm, *Biopolymers* **6**, 1379 (1968).

[54] S. C. Yasui and T. A. Keiderling, *J. Am. Chem. Soc.* **108**, 5576 (1986).

[55] M. G. Paterlini, T. B. Freedman, and L. A. Nafie, *Biopolymers* **25**, 1751 (1986).

[56] A. F. Drake, G. Siligardi, and W. A. Gibbons, *Biophys. Chem.* **31**, 143 (1988).

[57] R. W. Woody, *Adv. Biophys. Chem.* **2**, 37 (1992).

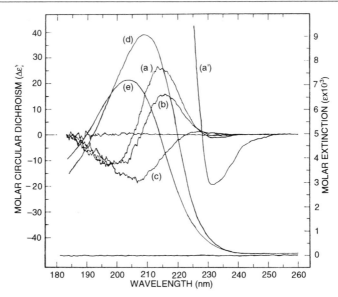

FIG. 3. CD and absorption spectra of poly(Pro) in trifluoroethanol.[52] Curve a represents a close approximation to the poly(Pro) I CD spectrum with the maximum recorded less than 20 min after dissolution. Curve b was recorded about 1 hr after dissolution. Curve c, representing poly(Pro) II, was recorded 19 hr after dissolution. Curve a′ was run at 10 min with a 10-fold larger path length. Curves d and e are the absorption spectra, run at 15 min (form I) and 19 hr (form II), respectively. [Reprinted with permission from Bovey and Hood (1967).[52] Copyright 1967 John Wiley & Sons, Inc.]

point indicates a two-state equilibrium, which can be described by a single equilibrium constant. The molar ellipticity at 220 nm is commonly measured in these studies. If the end points can be determined reliably, ΔH and ΔS for the transition can be obtained from a van't Hoff analysis of the thermal transition data. Such data have provided important information about the effects of amino acid substitution, charges on end groups, salt bridge formation, solvent composition, etc., on helix stability. These studies have been reviewed.[58] The value of $[\theta]_{220}$ for 100% helix and 100% coil is somewhat uncertain because of end effects in the helix and variations in the CD of the unordered conformation, respectively. However, the CD of the unordered conformation is weak at 220 nm, and there is now a basis for estimating the effects of helix length on the CD.[59]

[58] J. M. Scholtz and R. L. Baldwin, *Annu. Rev. Biophys. Biomol. Struct.* **21,** 95 (1992).
[59] P. J. Gans, P. C. Lyu, M. C. Manning, R. W. Woody, and N. R. Kallenbach, *Biopolymers* **31,** 1605 (1991).

In contrast to the extensive studies of helix–coil transitions, measurements of β–coil transitions[60] have been rare, although CD has proved useful in these cases also. The limited solubility of many β-sheet models has undoubtedly been a major factor inhibiting these studies, and the search continues for good models with which the effects of amino acid substitution on the stability of β structures can be characterized.

Secondary Structure of Proteins

The most extensive application of CD to biochemistry has been for estimating the secondary structure of proteins. Although CD cannot compete with X-ray diffraction and two-, three-, and four-dimensional (2D, 3D, and 4D) nuclear magnetic resonance (NMR) in providing a detailed three-dimensional structure of a protein, it can give estimates of the fraction of residues in α-helical, β-sheet, β-turn, and unordered conformations. This type of information is useful in the absence of high-resolution data, for example, in the initial stages of investigating a protein, or in cases where the protein cannot be crystallized or is too large for multidimensional NMR methods to be applicable. Even if a detailed three-dimensional structure is available, CD analysis of secondary structure is valuable because it is much less labor- and instrument-intensive than X-ray diffraction or multidimensional NMR, and it can also be used in rapid-mixing experiments. Thus, the effects of mutations, denaturants, and temperature can be studied, and the kinetics of protein folding and unfolding can be investigated much more efficiently by CD than by the high-resolution methods.

The analysis of CD spectra to obtain information about protein secondary structure has been reviewed extensively.[36,61-64] It is convenient to represent the CD spectrum of a protein as a column vector, **c**, where the elements of the matrix are the values of the CD spectrum at a set of wavelengths, generally spaced at equal intervals, for example, at every nanometer over a range from 180 to 260 nm. Similarly, we can represent the secondary structure of a protein by a column matrix, **f**, with m rows, each element of which is the fraction of residues in the protein in one of the secondary structural elements under consideration. The problem of determining the protein secondary structure can then be formulated as a matrix equation:

[60] W. L. Mattice, *Annu. Rev. Biophys. Biophys. Chem.* **18,** 93 (1989).
[61] R. W. Woody, *in* "The Peptides" (V. Hruby, ed.), Vol. 7, p. 15. Academic Press, New York, 1985.
[62] J. T. Yang, C.-S. C. Wu, and H. M. Martinez, this series, Vol. 130, p. 208.
[63] W. C. Johnson, Jr., *Annu. Rev. Biophys. Biophys. Chem.* **17,** 145 (1988).
[64] M. C. Manning, *J. Pharm. Biomed. Anal.* **7,** 1103 (1989).

$$\mathbf{f} = \mathbf{Xc} \qquad (13)$$

where $\mathbf{X}$ is a matrix which transforms the CD vector into the structural vector for a given protein. Thus, the problem is to find the matrix $\mathbf{X}$.

In nearly all currently used methods, $\mathbf{X}$ is solved for by utilizing the CD spectra of proteins for which the secondary structure has been determined by X-ray diffraction. Thus, the matrix equation [Eq. (13)] is generalized to

$$\mathbf{F} = \mathbf{XC} \qquad (14)$$

where $\mathbf{F}$ now contains n columns which are the secondary structural vectors of n proteins, determined from X-ray diffraction. $\mathbf{C}$ also consists of n columns, each of which is the CD spectrum of the corresponding protein. The matrix $\mathbf{X}$ then can be determined by a least-squares method and subsequently used to calculate the structure vectors for other proteins from the CD spectra, using Eq. (13). The problem with this method is that if one uses the minimal number of proteins required to specify $\mathbf{X}$ (i.e., the number of conformations considered, m), the resulting matrix is quite sensitive to the choice of proteins. This problem could be surmounted by using a larger set of proteins, but then the least-squares procedure becomes unstable. In general, the number of proteins considered is substantially greater than the number of secondary structures to be determined. Stable solutions for $\mathbf{X}$ have been obtained by a damped least-squares method,[65] by singular value decomposition,[66] or by a convex constraint transformation.[67]

Table I shows a comparison of the results for a number of methods.[68–73] The various methods have been tested by taking a set of proteins of known structure (from X-ray diffraction) and analyzing, in turn, each protein using the remaining proteins as the basis set. The methods are compared in terms of two performance criteria: the root-mean-square (rms) difference of fractions deduced from CD in comparison with the X-ray results, and the Pearson correlation coefficient. The correlation coefficient is defined so that perfect agreement between CD and X-ray results gives $r =$

[65] S. W. Provencher and J. Glöckner, *Biochemistry* **20**, 33 (1981).

[66] J. P. Hennessey, Jr., and W. C. Johnson, Jr., *Biochemistry* **20**, 1085 (1981).

[67] A. Perczel, M. Hollósi, G. Tusnády, and G. D. Fasman, *Protein Eng.* **4**, 669 (1991).

[68] C. T. Chang, C.-S. C. Wu, and J. T. Yang, *Anal. Biochem.* **91**, 13 (1978).

[69] I. A. Bolotina and V. Yu. Lugauskas, *Mol. Biol. (Engl. Transl.)* **19**, 1154 (1984).

[70] P. Manavalan and W. C. Johnson, Jr., *Anal. Biochem.* **167**, 76 (1987).

[71] A. Toumadje, S. W. Alcorn, and W. C. Johnson, Jr., *Anal. Biochem.* **200**, 321 (1992).

[72] I. H. M. van Stokkum, H. J. W. Spoelder, M. Bloemendal, R. van Grondelle, and F. C. A. Groen, *Anal. Biochem.* **191**, 110 (1990).

[73] N. Sreerama and R. W. Woody, *Anal. Biochem.* **209**, 32 (1993).

TABLE I

PERFORMANCE OF METHODS FOR SECONDARY STRUCTURE DETERMINATION BY
CIRCULAR DICHROISM

Method	α^a r^b	δ^b	β^a r^b	δ^b	T^a r^b	δ^b	U^a r^b	δ^b
CWY[68,c]	0.85	0.11	0.25	0.21	−0.31	0.15	0.46	0.15
BB[44]	0.92	0.02	0.93	0.12	0.33	0.07	0.65	0.12
BL[69]	0.97	0.07	0.48	0.06	0.33	0.06	0.72	0.06
PG[65]	0.96	0.05	0.94	0.06	0.31	0.10	0.49	0.11
HJ[66]	0.96	0.06	0.81	0.08	0.76	0.04	0.82	0.07
VS[70,d]	0.97	0.06	0.76	0.10	0.49	0.07	0.86	0.07
VS[71,e]	0.99	0.04	0.88	0.07	0.56	0.05	0.76	0.09
LL[72,f]	0.96	0.07	0.85	0.07	0.80	0.05	0.79	0.05
CCA[67,g]	0.93	0.11	0.71	0.10	0.73	0.20	0.35	0.09
							0.48	0.17
SC[73,h]	0.95	0.09	0.84	0.08	0.77	0.05	0.72	0.06

[a] The types of secondary structure are designated as follows: α, α helix; β, β sheet; T, β turn; U, unordered.

[b] Secondary structure fractions x_i from X-ray diffraction and y_i from CD are compared by two different criteria. r is the Pearson product-moment correlation, defined as

$$r = \frac{\Sigma x_i y_i - (1/n) \Sigma x_i \Sigma y_i}{\{\Sigma x_i^2 - [(\Sigma x_i)^2/n]\}^{1/2} \{\Sigma y_i^2 - [(\Sigma y_i)^2/n]\}^{1/2}}$$

where n is the number of cases considered. δ is the root-mean-square difference, defined as $\delta = \{\Sigma (x_i - y_i)^2/n\}^{1/2}$.

[c] r and δ were calculated by Provencher and Glöckner[65] by omitting the protein being analyzed from the reference set, to avoid bias. With the exception of the entry for SC[h], this procedure has been used for all entries.

[d] VS, Variable selection method, using data from 178 to 260 nm.

[e] Using data from 168 to 260 nm.

[f] LL, Locally linearized model. Results are from Sreerama and Woody,[73] using the method of van Stokkum et al.[72] but averaging over acceptable solutions rather than choosing the one with the sum of fractions closest to 1.

[g] CCA, Convex constraint analysis. No X-ray data are required. The two sets of values for the unordered conformation are due to two different classifications of secondary structure.

[h] In the self-consistent method, the CD spectrum of the protein being analyzed is retained in the matrix C and an initial guess for the secondary structure fractions is included in the matrix F. The guess is refined by an iterative process. Here, the initial guess is the secondary structure content of the protein in the basis set which has a CD spectrum most similar to that of the test protein.

1, whereas $r = 0$ implies no better agreement than random assignments would yield and negative r values imply even poorer agreement. It can be seen that most of the newer methods have r values exceeding 0.95 for α helix, 0.85 for β sheet, 0.75 for β turns, and 0.7 for unordered conformations. The rms deviations are less than 0.09 for α helix and β sheet, and 0.06 or less for β turn and unordered conformations. The reader should consult the original papers describing these methods for details.

Flexibility of the basis set is important and is a feature of the method of Provencher and Glöckner,[65] Bolotina and Lugauskas,[69] and the variable selection method[70,71] and its derivatives.[72,73] Given the deviations of helical and sheet structures from the ideal, as well as the idiosyncratic contributions of aromatic side chains, it is not surprising that fixed basis sets have limited ability to provide accurate analyses of protein CD spectra. It is also important to include data at shorter wavelengths. Hennessey and Johnson[66] showed that whereas CD spectra carried to 178 nm yield five significant basis spectra, when spectra to 190 nm were used, the β-strand and "other" structural contents were altered significantly, although the α-helix content was only slightly affected. Extension of data to 168 nm gave some improvement in performance relative to data terminated at 178 nm.[71] The need for deep UV data has been disputed by Venyaminov *et al.*,[74] who found no systematic deterioration in the quality of results with increasing truncation, using the Provencher and Glöckner method.[65] Nevertheless, it is advisable to use data to at least 185 nm and preferably to 178 nm, the practical lower limit with water as the solvent and fixed path length cells.

Because nearly all currently used methods for secondary structural analysis by CD use crystal structure data, two concerns need to be addressed. Are protein structures the same in solution as they are in the crystal? In a few specific cases, significant differences have been proposed on the basis of CD[75] or VCD.[76] The most definitive comparisons of solution versus crystal structure are provided by multidimensional NMR studies, which can yield solution structures for small proteins ($< \sim 20$ kDa), comparable in accuracy to those from X-ray diffraction. One review[77] concludes that, in most cases, the solution and crystal structures are very similar, allowing for some differences in the local conformation and dynamics of surface residues. In a few cases, distinct differences have been demon-

[74] S. Yu. Venyaminov, I. A. Baikalov, C.-S. C. Wu, and J. T. Yang, *Anal. Biochem.* **198,** 250 (1991).

[75] P. Bayley, S. Martin, and G. Jones, *FEBS Lett.* **238,** 61 (1988).

[76] M. Urbanova, R. K. Dukor, P. Pancoska, V. P. Gupta, and T. A. Keiderling, *Biochemistry* **30,** 10479 (1991).

[77] G. Wagner, S. G. Hyberts, and T. F. Havel, *Annu. Rev. Biophys. Biomol. Struct.* **21,** 167 (1992).

strated. Thus, the practice of using crystal structures to calibrate spectroscopic methods for determining solution structures is generally satisfactory, but anomalous cases may be encountered.

A second problem is that the crystal structure must be interpreted by assigning residues to various types of secondary structure. This is especially difficult at helix and β-strand termini. There are significant differences among the secondary structure fractions reported by crystallographers and those derived by "objective" analyses such as those of Kabsch and Sander[78] or Levitt and Greer,[79] as well as between the objective analyses. These differences have been discussed.[62,80]

Pancoska et al.[80] have shown that there is a linear correlation between the fraction of β sheet, β turn, and other structures and the fraction of α helix in a set of proteins. Furthermore, the deviations from the regression line become smaller as f_α increases. They have suggested that, to a significant extent, the success of far-UV CD in determining the fractions of conformations other than α-helix may be due to this correlation.

Infrared absorption and Raman frequencies are sensitive to protein secondary structure and have been used for the analysis of protein structure.[81,82] Relative to these methods, VCD spectroscopy has the advantage of using the greater dispersion of frequencies for different secondary structures characteristic of vibrational spectroscopy vis-à-vis electronic spectroscopy, plus the advantage of sign differences characteristic of chiroptical spectroscopy. VCD has been shown capable of providing secondary structure estimates for proteins[83–85] that are comparable, or in some cases somewhat superior, to those provided by far-UV CD when implemented at the same level, that is, without variable selection in either case. Further development of this technique is very promising. The principal disadvantages are the present limited availability of VCD instrumentation and the requirement for much higher concentrations than those needed in the far-UV. Relative to far-UV CD, VCD has the advantages of involving more localized transitions and much lower sensitivity to side-chain, especially aromatic, contributions.

Sarver and Krueger[86] have combined the far-UV CD and infrared absorption spectra (amide I) of a set of proteins and used the singular

[78] W. Kabsch and C. Sander, *Biopolymers* **22**, 2577 (1983).
[79] M. Levitt and J. Greer, *J. Mol. Biol.* **114**, 181 (1977).
[80] P. Pancoska, M. Blazek, and T. A. Keiderling, *Biochemistry* **31**, 10250 (1992).
[81] W. K. Surewicz, H. H. Mantsch, and D. Chapman, *Biochemistry* **32**, 389 (1993).
[82] R. W. Williams, this series, Vol. 130, p. 311.
[83] P. Pancoska, S. C. Yasui, and T. A. Keiderling, *Biochemistry* **28**, 5917 (1989).
[84] P. Pancoska, S. C. Yasui, and T. A. Keiderling, *Biochemistry* **30**, 5089 (1991).
[85] P. Pancoska and T. A. Keiderling, *Biochemistry* **30**, 6885 (1991).
[86] R. W. Sarver and W. C. Krueger, *Anal. Biochem.* **199**, 61 (1991).

value decomposition method to determine the secondary structures. The general rationale for such an approach is that combining the two kinds of spectra should lead to a significant increase in information content. More specifically, there is good reason to believe that CD and IR spectra should complement one another. CD is at its best in determining α-helix content, whereas the amide I bands of α helix and unordered conformations show significant overlap. Conversely, CD is less accurate for β sheets, but the amide I band of β sheets is at a distinctly different frequency from that of α helix or unordered conformations. Other combinations such as far-UV CD and VCD are of potential value.

Thus, developments in secondary structural analysis from CD have substantially improved the accuracy of the method. For α helix, there has been little improvement, but even the early methods gave rather good estimates of helix content. Estimates of β-sheet, β-turn, and other structural contents have shown the greatest improvement.

Several potential sources of error must be recognized in secondary structure analysis by CD.[64] First, α helices, and especially β sheets and β turns, can deviate significantly from standard or average structures. Second, α helices and β sheets of finite size will have CD spectra which depend on their length or their length and width, respectively. Third, adjacent elements of secondary structure may interact and lead to nonadditivity. Finally, aromatic and disulfide side chains may make significant contributions in the far-UV. Of these possible complications, the third is the least serious, as indicated by calculations on α helices in coiled coils[87] and β sheets in β-sandwich proteins.[88] The other factors can certainly be significant. However, the potential errors arising from these factors are decreased by the variable selection methods[70,72,73] which are replacing the methods using fixed basis sets. Of course, if a protein has CD contributions markedly different from any member(s) of the basis set, the analysis will be more subject to error. However, many such cases will be revealed by failure to meet the tests for a successful analysis. In addition, expansion of the basis set by adding more proteins with a wide range of conformations will increase the likelihood of a successful and reliable analysis.

Tertiary Structure of Proteins

The analysis of protein CD spectral effects arising from side chains has been discussed in two excellent reviews.[40,41] Side-chain CD bands can provide sensitive probes of protein conformational changes and of ligand binding, as illustrated in the following examples. Aromatic side-chain

[87] T. M. Cooper and R. W. Woody, *Biopolymers* **30**, 657 (1990).
[88] M. C. Manning and R. W. Woody, *Biopolymers* **26**, 1731 (1987).

CD bands have provided a useful marker for the R → T transition in hemoglobin.[89-91] The near-UV CD spectrum of heme proteins is generally dominated by a broad, intense band near 260 nm, which is due to a heme transition. Hemoglobin, however, also shows sharper bands in the 270–300 nm region which are due to Tyr and/or Trp. Deoxyhemoglobin has a marked negative band at 287 nm, whereas oxyhemoglobin has two weak negative features at 283 and 290 nm. Studies on chemically modified and mutant hemoglobins, in which the R → T transition is blocked, and results of binding the potent allosteric effector inositol hexaphosphate have shown that a strong 287 nm band is characteristic of the T (deoxy) conformation, whereas the R (liganded) conformation has weak positive or negative CD near 287 nm. The strong band in the T conformation was assigned to α-Tyr-42, which is hydrogen-bonded to β-Arg-40 in the $\alpha\beta$ interface.

Near-UV CD has also been useful in monitoring conformational transitions among insulin conformers in solution. Three hexameric conformers, T_6, T_3R_3, and R_6, have been observed in crystals.[92] Extensive studies by Wollmer and co-workers[93-95] have shown that the T → R transition is marked by an increase in magnitude of the negative CD between 250 and 255 nm. This transition can be induced by binding of polarizable anions, with the effectiveness following the Hofmeister series, and by phenol derivatives. The transition in the crystal involves conversion of residues 1–8 of the β chain from an extended conformation to α helix. In agreement with this, the far-UV CD also undergoes a significant increase in amplitude.

Protein–Ligand Interactions

CD changes provide a convenient spectroscopic method for following protein–ligand interactions. CD bands arising from the ligand can be conveniently monitored in the visible and near-UV. Ligand-induced changes in tertiary and quaternary structure can be detected in the near-UV CD,

[89] M. F. Perutz, J. E. Ladner, S. R. Simon, and C. Ho, *Biochemistry* **13**, 2163 (1974).
[90] M. F. Perutz, A. R. Fersht, S. R. Simon, and G. C. K. Roberts, *Biochemistry* **13**, 2174 (1974).
[91] M. F. Perutz, E. J. Heidner, J. E. Ladner, J. G. Beetlestone, C. Ho, and E. F. Slade, *Biochemistry* **13**, 2187 (1974).
[92] U. Derewenda, Z. Derewenda, E. J. Dodson, G. G. Dodson, C. D. Reynolds, G. D. Smith, C. Sparks, and D. Swenson, *Nature (London)* **338**, 594 (1989).
[93] H. Renscheidt, W. Strassburger, U. Glatter, A. Wollmer, G. G. Dodson, and D. A. Mercola, *Eur. J. Biochem.* **142**, 7 (1984).
[94] A. Wollmer, B. Rannefeld, B. R. Johansen, K. R. Hejnaes, P. Balschmidt, and F. B. Hansen, *Biol. Chem. Hoppe–Seyler* **368**, 903 (1987).
[95] P. Krüger, G. Gilge, Y. Cahuk, and A. Wollmer, *Biol. Chem. Hoppe–Seyler* **371**, 669 (1990).

and changes in protein secondary structure may be observable in the far-UV.

The use of CD to determine dissociation constants of protein–ligand complexes has been reviewed.[96–98] This application is relatively straightforward and of great utility. The concentration of the species being detected must be comparable to the dissociation constant of the complex, generally ranging from around $K_d/10$ to $10 K_d$. This assures that both the free and the complexed analyte concentrations can be measured with adequate accuracy. Typically, dissociation constants must be about 10^{-6} M or larger to be accurately measured by CD, as is also the case with isotropic absorption measurements. Fluorescence-detected CD (FDCD)[33] may permit the characterization of more tightly bound complexes, perhaps down to the 10 nM range for K_d, but this method has not been widely applied.

As with any spectroscopic probe, the wavelength to be used in CD measurements of binding constants must be chosen carefully. It should be a wavelength at which the difference in CD between the free and complexed analyte is at a maximum, but wavelengths near either the short- or long-range limit of the instrument should be avoided because of deterioration in the S/N ratio.

CD has a special advantage in the case of achiral ligands with absorption bands above 300 nm. In such a case, the complex will exhibit CD at wavelengths where neither partner has a CD band in the free form. Transition metal ions, hemes, pyridoxal phosphate, thiamin pyrophosphate, dyes, and many drugs are achiral, and complexes with proteins can be readily studied by CD. Many other interesting ligands exhibit only weak CD free in solution because the chromophores are separated by several single bonds from chiral centers. Such systems, which include flavin, NADH, and folic acid, generally have much stronger CD bands in the bound form than when free, and are therefore favorable objects for study by CD.

Many protein–ligand interactions are so strong that, under physiological conditions, ligand binding is irreversible, that is, dissociation constants are in nanomolars or less. Such systems include heme proteins and flavoproteins, for example. In such conjugated proteins, the extrinsic CD bands due to ligand transitions serve as a sensitive probe of the chromophore environment. For example, conformational changes induced by various effectors are readily detectable through changes in the extrinsic CD bands. It is generally difficult or impossible to interpret these conformational

[96] G. Blauer, *Struct. Bonding (Berlin)* **18,** 69 (1974).
[97] N. J. Greenfield, *CRC Crit. Rev. Biochem.* **3,** 71 (1975).
[98] M. Hatano, *Adv. Polym. Sci.* **77,** 1 (1986).

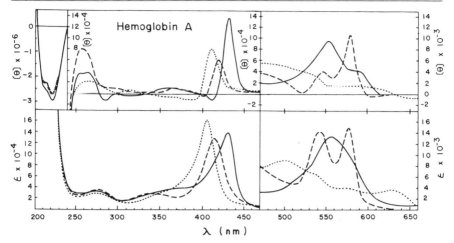

FIG. 4. CD and absorption spectra of native human hemoglobin (Hb)[104]: oxyHb (---); deoxyHb (—); metHb (·· ·). Ellipticities are per mol of heme. [Reprinted with permission from Y. Sugita, M. Nagai, and Y. Yoneyama, *J. Biol. Chem.* **246**, 383 (1971).[104] Copyright 1971, American Society of Biological Chemists.]

changes in quantitative structural terms, but qualitative inferences can be made in some cases. The heme group has a number of electronic transitions in the near-UV, visible, and (in some cases) near-IR. Therefore, heme proteins are excellent subjects for the study of extrinsic CD bands, and such studies have been reviewed.[99-103] In mammalian hemoglobin, strong positive CD bands are associated with the visible bands, the Soret band, and the δ or N band[104] (Fig. 4). Although these bands are shifted somewhat by changes in the oxidation state of the iron and replacement of the sixth ligand, the signs remain the same and the magnitude is not highly sensitive to these changes. Theoretical calculations[105-108] have shown that coupled-

[99] Y. P. Myer, this series, Vol. 54, p. 249.
[100] Y. P. Myer and A. Pande, *in* "The Porphyrins" (D. Dolphin, ed.), Vol. 3, p. 271. Academic Press, New York, 1978.
[101] R. W. Woody, *in* "Biochemical and Clinical Aspects of Hemoglobin Abnormalities" (W. S. Caughey, ed.), p. 279. Academic Press, New York, 1978.
[102] G. Geraci and L. J. Parkhurst, this series, Vol. 76, p. 262.
[103] Y. P. Myer, *Curr. Top. Bioenerg.* **14**, 149 (1985).
[104] Y. Sugita, M. Nagai, and Y. Yoneyama, *J. Biol. Chem.* **246**, 383 (1971).
[105] M.-C. Hsu and R. W. Woody, *J. Am. Chem. Soc.* **93**, 3515 (1971).
[106] J. Fleischhauer and A. Wollmer, *Z. Naturforsch. B: Anorg. Chem. Org. Chem. Biochem. Biophys. Biol.* **27B**, 530 (1972).
[107] R. W. Woody, *in* "Protein–Ligand Interactions" (H. Sund and G. Blauer, eds.), p. 60. de Gruyter, Berlin, 1975.
[108] W. Strassburger, A. Wollmer, H. Thiele, J. Fleischhauer, W. Steigemann, and E. Weber, *Z. Naturforsch. C: Biosci.* **33C**, 908 (1978).

oscillator interactions between the heme $\pi\pi^*$ transitions and far-UV $\pi\pi^*$ transitions in aromatic side chains near the heme account for the sign and approximate magnitude of all four observable $\pi\pi^*$ transitions in the CD spectrum of sperm whale myoglobin and both vertebrate and invertebrate hemoglobins.

CD can be used to monitor the phenomenon of heme isomerism in heme proteins.[109–111] The isomerism, discovered by LaMar and co-workers,[112,113] results from the presence of two forms of myoglobin which differ by 180° in the heme orientation about the α,γ-methine carbon axis. The two isomers are present in a 9:1 ratio at equilibrium in carbonmonoxymyoglobin (MbCO) but are in a 1:1 ratio in a sample freshly reconstituted from apomyoglobin and heme. Freshly reconstituted MbCO has a Soret CD band with only about half the amplitude of the native form,[109–111] suggesting that the minor isomer (at equilibrium) has only a weak Soret CD spectrum. The major form has a strong positive Soret CD band, whereas the minor form has a weakly negative Soret CD band.[111] These results and the CD of the heme undecapeptide from cytochrome[114] indicate that the origin of the Soret CD band must be more complex than proposed by Hsu and Woody.[105]

Membrane Proteins

CD spectroscopy has been used to study the conformation of a number of membrane proteins. The use of CD for determining the secondary structure of membrane proteins has been reviewed.[115] Membrane proteins which have been solubilized by detergents generally present no special difficulties for CD analysis, but the conformation is likely to be different from that in the native environment. The study of membrane proteins *in situ* by CD has been the subject of considerable controversy. Two kinds of artifacts which may afflict such studies have been identified.

Membrane fragments may be comparable in size to the wavelength of far-UV light and thus scatter the light strongly. To the extent that rcpl and lcpl are scattered differently by the fragments (CIDS), the differential

[109] W. R. Light, R. J. Rohlf, G. Palmer, and J. S. Olson, *J. Biol. Chem.* **262**, 46 (1987).

[110] H. S. Aojula, M. T. Wilson, and I. E. G. Morrison, *Biochem. J.* **243**, 205 (1987).

[111] H. S. Aojula, M. T. Wilson, G. R. Moore, and D. J. Williamson, *Biochem. J.* **250**, 853 (1988).

[112] G. N. LaMar, D. L. Budd, B. Viscio, K. M. Smith, and K. C. Langry, *Proc. Natl. Acad. Sci. U.S.A.* **75**, 5755 (1978).

[113] G. N. LaMar, N. L. Davis, D. W. Parish, and K. M. Smith, *J. Mol. Biol.* **168**, 887 (1983).

[114] G. Blauer, N. Sreerama, and R. W. Woody, *Biochemistry* **32**, 6674 (1993).

[115] M. P. Heyn, this series, Vol. 172, p. 575.

scattering will appear as CD.[14,116] There are two characteristic features of CIDS which often (but not always) reveal its presence. Scattering occurs at wavelengths where no absorption is observed and so does CIDS. CD spectra of samples which exhibit CIDS often have a long tail extending to wavelengths well above the longest wavelength absorption band (e.g., above 300 nm for proteins and nucleic acids). The CD spectra of samples exhibiting CIDS will also, in many cases, depend on the distance between the sample cell and the detector, which determines the fraction of scattered light detected by the photomultiplier. More complete collection of scattered light can be provided by a specially designed cell,[117] called the fluorscat cell, in which the sample cell is surrounded (except for the direction facing the light source) by a fluorophore solution. Effectively complete collection of the scattered light can be accomplished by adding an inert, achiral fluorophore to the sample and measuring the CD by FDCD.[33] With these methods,[118,119] the problem of differential scattering of membrane fragments can be solved, and the true CD can be obtained.

A second type of artifact which may occur in membrane protein preparations is caused by the inhomogeneous distribution of the protein in such systems. The protein is concentrated in the membrane fragments and absent from the surrounding solvent. This inhomogeneous distribution gives rise to an apparent reduction of the CD in regions of high absorption. The absorption analog of this phenomenon was discovered by Duysens[120] in studies of the absorption spectra of chloroplasts. Bustamante and Maestre[121] have shown that the flattening correction for CD is twice as large as that for absorption, in contrast to earlier analyses[122,123] which used the same correction factor for both CD and absorption. Eliminating the Duysens flattening effect is much more difficult than eliminating the CIDS contribution. Mao and Wallace[124] have shown that the flattening effect can be eliminated by incorporating the membrane protein in small unilamellar vesicles at a lipid/protein ratio sufficiently high that, on average, there is only one protein molecule per vesicle. However, this proce-

[116] I. Tinoco, Jr., M. F. Maestre, and C. Bustamante, *Trends Biochem. Sci.* **8**, 41 (1983).
[117] B. P. Dorman, J. E. Hearst, and M. F. Maestre, this series, Vol. 27, p. 767.
[118] C. Reich, M. F. Maestre, S. Edmondson, and D. M. Gray, *Biochemistry* **19**, 5208 (1980).
[119] I. Tinoco, Jr., and A. L. Williams, Jr., *Annu. Rev. Phys. Chem.* **35**, 329 (1984).
[120] L. N. M. Duysens, *Biochim. Biophys. Acta* **19**, 1 (1956).
[121] C. Bustamante and M. F. Maestre, *Proc. Natl. Acad. Sci. U.S.A.* **85**, 8482 (1988).
[122] D. W. Urry, *in* "Modern Physical Methods in Biochemistry" (A. Neuberger and L. L. M. van Deenen, eds.), p. 275. Elsevier, Amsterdam, 1985.
[123] B. A. Wallace and D. Mao, *Anal. Biochem.* **142**, 317 (1984).
[124] D. Mao and B. A. Wallace, *Biochemistry* **23**, 2667 (1984).

dure requires that the protein be removed from its native environment, which could cause irreversible conformational changes. In addition, the strong curvature of the lipid bilayer in small vesicles will affect the lipid distribution between the two leaflets of the bilayer and could lead to differences in protein conformation relative to the native membrane. The best solution would be to use photoacoustic detection of CD which, as Bustamante and Maestre[121] pointed out, will register the true CD spectrum and eliminate both types of artifacts. Although a photoacoustic-detected CD instrument has been built,[35] measurements on membranes by such an instrument have not been reported.

The membrane protein which has been most extensively studied by CD is bacteriorhodopsin (bR) from *Halobacterium halobium*. Two aspects of the CD spectrum of bR have been controversial, both of which have been reviewed by the principal protagonists of the opposing viewpoints: the analysis of the far-UV CD for secondary structure[125,126] and the interpretation of the visible CD spectrum arising from the retinal chromophore.[127,128]

The far-UV CD spectrum of bR is low in amplitude and has an unusually weak 208 nm band for a protein with around 75% α helix, as indicated by electron diffraction. Swords and Wallace[125] have attributed these anomalies entirely to CIDS and Duysens flattening, whereas Jap, Glaeser, and co-workers[126,129] have argued that the artifacts are minimal and that the unusual CD features reflect real conformational differences. In support of the latter, FDCD measurements[129] have shown that the CIDS contribution is negligible, and anomalies in the amide I frequency of bR[130] also support deviations[131,132] of the helical segments from normal α helix. For the visible region, the controversy concerns whether the couplet centered on the absorption maximum for the retinal chromophore is due to exciton coupling[127] among the bR molecules, or, as Wu and El-Sayed argue,[128] the couplet is due to some kind of protein heterogeneity. Cassim[127] has argued convincingly for the exciton interpretation, and the arguments are strengthened by theoretical calculations by Buss[133] which account

[125] N. A. Swords and B. A. Wallace, *Biochem. J.* **289,** 215 (1993).

[126] R. M. Glaeser, K. R. Downing, and B. K. Jap, *Biophys. J.* **59,** 934 (1991).

[127] J. Y. Cassim, *Biophys. J.* **63,** 1432 (1992).

[128] S. Wu and M. El-Sayed, *Biophys. J.* **60,** 190 (1991).

[129] B. K. Jap, M. F. Maestre, S. B. Hayward, and R. M. Glaeser, *Biophys. J.* **43,** 81 (1983).

[130] K. J. Rothschild and N. A. Clark, *Science* **204,** 311 (1979).

[131] S. Krimm and A. M. Dwivedi, *Science* **216,** 407 (1982).

[132] N. J. Gibson and J. Y. Cassim, *Biochemistry* **28,** 2134 (1989).

[133] V. Buss, private communication (1993).

for the observed couplet using the experimental purple membrane geometry.

Protein Folding

CD has found many applications in the study of protein folding.[134] Its ability to monitor both the overall backbone conformation (far-UV) and the environment of one or more specific side-chain chromophores (near-UV) has made CD a favorite method for following protein folding and unfolding (denaturation).

Much interest has centered on an intermediate state called the "molten globule"[135] observed in the denaturation of some proteins. CD was instrumental in first detecting this intermediate and remains an essential criterion for its identification. When the pH of an α-lactalbumin solution is decreased from neutral to a value near 2, the CD spectrum in the far-UV undergoes some relatively small changes, but the near-UV CD nearly vanishes.[136] These CD changes, together with observations of small-angle X-ray scattering, viscosity, fluorescence polarization, and calorimetry, led Dolgikh et al.[137] to postulate a structure for the protein which is slightly less compact than the native globule, retaining a substantial amount of secondary structure, but with little or no well-defined tertiary structure. This model has become known as the molten globule, the name of which derives from the disordered side chains (molten) and the somewhat expanded but still compact overall structure (globule). Such molten globule intermediates have been detected as equilibrium intermediates in the denaturation of a number of proteins. There is also evidence that molten globules occur as kinetic intermediates in protein folding, but it has not been established that they are directly on the folding pathway.

Improvements in CD instrumentation have made kinetic CD measurements possible in the far-UV. Stopped-flow measurements with a dead time of the order of a few milliseconds are now possible,[138] monitoring the ellipticity at a series of fixed wavelengths to follow the backbone conformation. Such measurements have been reported for folding of sev-

[134] C. R. Matthews, *Annu. Rev. Biochem.* **62**, 653 (1993).

[135] O. B. Ptitsyn, *J. Protein Chem.* **6**, 272 (1987).

[136] K. Kuwajima, K. Nitta, M. Yoneyama, and S. Sugai, *J. Mol. Biol.* **106**, 359 (1976).

[137] D. A Dolgikh, L. V. Abaturov, I. A. Bolotina, E. V. Brazhnikov, V. E. Bychkova, R. I. Gilmanshin, Yu. O. Lebedev, G. V. Semisotnov, E. I. Tiktopulo, and O. B. Ptitsyn, *Eur. J. Biophys.* **13**, 109 (1985).

[138] K. Kuwajima, H. Yamaya, S. Miwa, S. Sugai, and T. Nagamura, *FEBS Lett.* **221**, 115 (1987).

eral proteins.[139–141] It has been found that, within the dead time, a substantial fraction of the CD amplitude of the native proteins is recovered, implying that significant secondary structure formation occurs on a time scale of less than 10 msec.

Nucleic Acids

Chromophores

The chromophores which are responsible for the intrinsic CD of nucleic acids are the purine and pyrimidine bases. The absorption spectra of the heterocyclic molecules show only a few relatively broad bands. Numerous theoretical and experimental studies[142] have attempted to ascertain the number of electronically excited states, their energies, and the transition moment magnitudes and directions. Polarized absorption spectra of single crystals of the bases have greatly clarified the situation with respect to the $\pi\pi^*$ transitions in purines and pyrimidines, as summarized by Ho *et al.*[143] Theoretical calculations[144] have shown that electrostatic interactions among the bases in crystals can lead to mixing of excited states and significant changes in transition moment directions.

The situation with respect to $n\pi^*$ transitions is uncertain. Only a few $n\pi^*$ transitions have been tentatively identified in the biologically important bases. The paucity of concrete information on $n\pi^*$ transitions in these systems results from the difficulty of detecting these inherently weak transitions against the intense background of the allowed $\pi\pi^*$ transitions.

Secondary Structure

CD has been used extensively for following conformational transitions in DNA, such as denaturation and transitions from B $\rightarrow$ A and B $\rightarrow$ Z DNA. In addition, the CD of DNA triplexes has features which are distinct from those of the duplex. Some of these important applications of CD to the study of DNA have been described by Gray *et al.*[145] in this volume.

[139] K. Kuwajima, *Biochemistry* **27,** 7419 (1988).
[140] K. Kuwajima, E. P. Garvey, B. E. Finn, C. R. Matthews, and S. Sugai, *Biochemistry* **30,** 7693 (1991).
[141] A. F. Chaffotte, C. Cadieux, Y. Guillou, and M. E. Goldberg, *Biochemistry* **31,** 4303 (1992).
[142] P. R. Callis, *Annu. Rev. Phys. Chem.* **34,** 329 (1983).
[143] P. S. Ho, G. Zhou, and L. B. Clark, *Biopolymers* **30,** 151 (1990).
[144] D. Theiste, P. R. Callis, and R. W. Woody, *J. Am. Chem. Soc.* **113,** 3260 (1991).
[145] D. M. Gray, S.-H. Hung, and K. H. Johnson, this volume [3].

In addition, another volume of this series carried a chapter by Gray *et al.*[146] reviewing the CD of DNA. Other chapters in the latter volume described the use of CD in characterizing A-DNA[147] and Z-DNA.[148] Earlier reviews of the CD of DNA include those of Johnson,[4] Tinoco *et al.*,[15] and Woody.[46] Therefore, only a few specific aspects of the CD of DNA are discussed in this chapter.

The CD of DNA depends on the ionic strength and, to some extent, on the nature of the salt.[149–151] In general, increases in salt concentration lead to decreases in the positive 275 nm band with little change in the 245 nm band. These changes were initially interpreted as a B → C conformational change.[149] Subsequently, the salt dependence of DNA CD was correlated with changes in the helix winding angle (the angle between neighboring bases) deduced from measurements of superhelical density in closed circular DNA.[152] The CD at 275 nm is highly sensitive to the winding angle.[150,151] On going from dilute salt to 3.0 M CsCl, for example, the 275 nm band is nearly abolished. The number of base pairs per turn changes by -0.22, as compared with the change of $9.3 - 10 = -0.7$ expected for the B → C transition from X-ray studies. In general, it seems preferable to describe the salt effects as resulting from a change in the winding angle or number of base pairs per turn, rather than a B → C transition.

The effects of temperature on DNA conformation have also been extensively studied by CD. The largest effect of temperature is that associated with the denaturation or melting of DNA which occurs at high temperatures and can conveniently be followed either by CD or by absorption changes. The change in the CD associated with melting of double-stranded DNA is easily measured,[145,146] but it is not so large as one might expect in the 260 nm region. There is a decrease in the amplitude of both the positive 275 nm band and negative 245 nm bands, with a shift of the crossover point by approximately 4 nm to the red. Much larger changes are observed in the vacuum ultraviolet bands on denaturation.[153]

A less striking but very interesting effect of temperature is the premelting phenomenon,[154,155] that is, the changes in CD observed at temperatures

[146] D. B. Gray, R. L. Ratliff, and M. R. Vaughan, this series, Vol. 211, p. 389.
[147] V. I. Ivanov and D. Yu. Krylov, this series, Vol. 211, p. 111.
[148] B. H. Johnston, this series, Vol. 211, p. 127.
[149] S. Hanlon, S. Brudno, T. T. Wu, and B. Wolf, *Biochemistry* **14**, 1648 (1975).
[150] A. Chan, R. Kilkuskie, and S. Hanlon, *Biochemistry* **18**, 84 (1979).
[151] W. A. Baase and W. C. Johnson, Jr., *Nucleic Acids Res.* **6**, 797 (1979).
[152] P. Anderson and W. Bauer, *Biochemistry* **17**, 594 (1978).
[153] D. G. Lewis and W. C. Johnson, Jr., *J. Mol. Biol.* **86**, 91 (1974).
[154] E. Paleček, *Prog. Nucleic Acid Res. Mol. Biol.* **10**, 151 (1976).
[155] K. J. Breslauer, *Curr. Opin. Struct. Biol.* **1**, 416 (1991).

well below the melting point. Premelting is quite distinct from melting, because the CD changes in the long-wavelength band occur in the opposite direction; in premelting the magnitude of the CD at 275 nm increases with increasing temperature, whereas the 245 nm band shows a generally smaller decrease in magnitude. In addition, the absorbance at 260 nm does not change significantly at temperatures where premelting is observable in CD.[156] As described by Breslauer,[155] this transition may be associated with a helix–helix transition in which the low-temperature form has the unusual structural features observed in X-ray diffraction studies of oligomers with oligo(dA)·oligo(dT) tracts,[157,158] whereas the high-temperature form is normal B-DNA. Some natural DNAs, especially kinetoplast DNA from trypanosomes, have tracts of four or five A residues on one strand, repeating in phase with the DNA periodicity. These DNAs are strongly bent, as demonstrated by anomalously small electrophoretic mobility and by electron microscopy. The properties of kinetoplast DNA approach those of ordinary DNA over the premelting range, and these changes are accompanied by the typical premelting CD changes.[159]

Double-stranded RNA adopts a conformation like that of A-DNA, with 11 base pairs/turn and with the base pairs tilted by approximately 13° from perpendicular to the helix axis. Nominally single-stranded RNA such as tRNA, rRNA, and mRNA has a complex conformation. Thus, RNA is generally in the A-conformation and, except for homopolymers, has extensive secondary and tertiary structure. In agreement with this picture, both double- and single-stranded RNA have CD spectra which closely resemble the A-DNA CD spectrum.[160,161] The only significant difference is that RNA generally has a weak negative CD band on the long-wavelength side of the positive 270 nm band, whereas A-DNA does not.

RNA has never been observed in the B-conformation, but Z-form RNA has been demonstrated in some alternating G-C sequences.[162,163] Hall *et al.*[163] demonstrated that Z-RNA can also be obtained with an RNA contain-

[156] R. B. Gennis and C. R. Cantor, *J. Mol. Biol.* **65**, 381 (1972).

[157] M. Coll, C. A. Frederick, A. H. J. Wang, and A. Rich, *Proc. Natl. Acad. Sci. U.S.A.* **84**, 8385 (1987).

[158] H. C. M. Nelson, J. T. Finch, F. L. Bonaventura, and A. Klug, *Nature (London)* **330**, 221 (1987).

[159] J. C. Marini, P. N. Effron, T. C. Goodman, C. K. Singleton, R. D. Wells, R. M. Wartell, and P. T. Englund, *J. Biol. Chem.* **259**, 8974 (1984).

[160] B. D. Wells and J. T. Yang, *Biochemistry* **13**, 1311 (1974).

[161] B. D. Wells and J. T. Yang, *Biochemistry* **13**, 1317 (1974).

[162] S. Uesugi, M. Ohkubo, H. Urata, M. Ikehara, Y. Kobayashi, and Y. Kyogoku, *J. Am. Chem. Soc.* **106**, 3675 (1984).

[163] K. Hall, P. Cruz, I. Tinoco, Jr., T. M. Jovin, and J. H. van de Sande, *Nature (London)* **311**, 584 (1984).

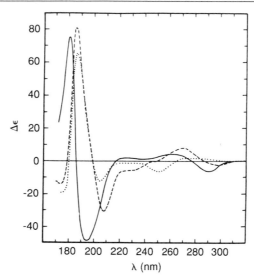

FIG. 5. CD spectra of poly[d(G-C)] at 22° in three different forms[164]: B ($\cdots$) in 10 mM phosphate buffer; A (---) in 0.67 mM phosphate, 80% trifluoroethanol; Z (—) in 10 mM phosphate, 2 M NaClO$_4$. [Reprinted with permission from J. H. Riazance, W. A. Baase, W. C. Johnson, Jr., K. Hall, P. Cruz, and I. Tinoco, Jr., *Nucleic Acids Res.* **13**, 4983 (1985).[164] Copyright 1985, IRL Press Ltd.]

ing only naturally occurring bases, poly[r(G-C)], although this requires both a high temperature (45°) and high salt concentration. CD evidence for the A → Z transformation consisted of the inversion of sign in the long-wavelength band, and this was also corroborated by NMR evidence for the Z-conformation. It has subsequently been demonstrated[164] that the vacuum-UV CD spectrum of poly[r(G-C)] at high salt and high temperatures is like that of Z-DNA and opposite in sign to that of A-RNA and -DNA and of B-DNA (Fig. 5).

The vacuum-UV CD spectrum of nucleic acids appears to be a reliable diagnostic of the sense of the helix, with right-handed helices (A- and B-forms) giving a positive couplet centered near 180 nm and left-handed helices (Z-form) giving a negative couplet centered near 190 nm.[164-167] By

[164] J. H. Riazance, W. A. Baase, W. C. Johnson, Jr., K. Hall, P. Cruz, and I. Tinoco, Jr., *Nucleic Acids Res.* **13**, 4983 (1985).
[165] C. A. Sprecher, W. A. Baase, and W. C. Johnson, Jr., *Biopolymers* **18**, 1009 (1979).
[166] J. C. Sutherland, K. P. Griffin, P. C. Keck, and P. Z. Takacs, *Proc. Natl. Acad. Sci. U.S.A.* **78**, 4801 (1981).
[167] S. Brahms, J. Vergne, J. G. Brahms, E. DiCapua, Ph. Bucher, and Th. Koller, *J. Mol. Biol.* **162,** 473 (1982).

contrast, the near-UV CD is not a reliable indicator of helix sense. Although most right-handed RNA and DNA duplexes have a positive long-wavelength band, some do not; for example, poly[d(I-C)][168] and poly[r(G-C)],[169] both right-handed, have negative bands at long wavelength. (Of course, as noted previously, many A-RNA molecules give a weak negative feature on the long-wavelength edge of the spectrum.)

Extensive studies of oligonucleotides and of simple-sequence polynucleotides have shown that the CD spectra depend on sequence and on base pairing.[169,170] Johnson and Gray[171] have described a method for using the information on sequence and base pairing contained in CD and absorption spectra to determine the fraction of A · U, G · C, and G · U base pairs; the fraction of A, U, G, and C in single-stranded regions; and the fraction of nearest-neighbor base pairs in an RNA. The method utilizes an extensive basis set of CD and absorption spectra for mono- and dinucleotides, a few higher oligomers, and polynucleotides (homopolymers and repeating di- and trinucleotides). The measured CD and absorption spectra of the unknown RNA, together with its known base composition, are fit by a linear combination of the basis spectra. A method of guided selection was used, in which basis spectra which contribute to negative fractions of base pairs, etc., are deleted. This method is closely related to the variable selection method[70] used in analyzing protein secondary structure.

Johnson and Gray[172] have applied their method of analysis to the secondary structure of *Escherichia coli* 5 S RNA, an RNA for which several secondary structural models have been proposed, based on extensive sequence data. The results were in good agreement with two of the proposed models but showed substantial differences with two other models. Thus, this method of secondary structure analysis for RNA can provide a test for proposed models based on sequence analysis.

Another application[173] of the Johnson and Gray method[171] was to an RNA (PK5) that had been shown by NMR to contain a pseudoknot at low temperatures and a hairpin at higher temperatures.[174] Fitting the basis spectra to the experimental CD and absorption spectra of the PK5 RNA gave base-pair, single-strand, and nearest-neighbor base-pair frequencies which agreed well with the pseudoknot structure at low temperatures

[168] Y. Mitsui, R. Langridge, R. C. Grant, M. Kodama, R. D. Wells, B. E. Shortle, and C. R. Cantor, *Nature (London)* **228,** 1166 (1970).
[169] D. M. Gray, J.-J. Liu, R. L. Ratliff, and F. S. Allen, *Biopolymers* **20,** 1337 (1981).
[170] R. A. Cox, W. Hirst, E. Godwin, and I. Kaiser, *Biochem. J.* **155,** 279 (1976).
[171] K. H. Johnson and D. M. Gray, *Biopolymers* **31,** 373 (1991).
[172] K. H. Johnson and D. M. Gray, *Biopolymers* **31,** 385 (1991).
[173] K. H. Johnson and D. M. Gray, *J. Biomol. Struct. Dyn.* **9,** 733 (1992).
[174] J. R. Wyatt, J. D. Puglisi, and I. Tinoco, Jr., *J. Mol. Biol.* **214,** 455 (1990).

and with the hairpin structure near room temperature. The successful application of this method to the pseudoknot structure implies that contributions from this type of tertiary structure do not significantly perturb the A-form CD spectrum and supports the applicability of this approach to other small RNAs.

Ligand Binding

CD has been applied extensively to investigations of the binding of small ligands such as dyes and antibiotics to nucleic acids. The kind of information obtainable from such studies is analogous to that for protein–ligand interactions: binding constants, kinetics of binding, and conformational changes in the ligand and/or the nucleic acid. Zimmer and Luck[175] have reviewed applications of CD to the study of drug binding to DNA.

Ethidium bromide (Etd[+]), a classic intercalating agent, is achiral. Its interactions with DNA and RNA have been extensively studied by CD.[176,177] The binding of Etd[+] has been found to inhibit the B → Z transition in DNA.[178] Binding of Etd[+] to B-DNA is noncooperative, whereas binding to Z-DNA is highly cooperative.[179] Similarly, CD studies of Etd[+] binding to double-stranded RNA show no cooperativity in the A-form and strong cooperativity in the Z-form.[180] These observations have been interpreted as evidence that Etd[+] has a strong preference for binding to right-handed duplexes of DNA and RNA, and that binding to Z-DNA induces a local conversion of the left-handed helix to a right-handed helix. Shafer et al.[181] interpreted the CD results as indicating that Etd[+] binds to left-handed helical regions at low dye/phosphate (D/P) ratios. Lamos et al.[182] have used FDCD to resolve this issue. Because only the Etd[+] is fluorescent, the large background of uncomplexed DNA at low D/P ratios is avoided in FDCD measurements. The FDCD results clearly supported a right-handed helical environment for bound Etd[+] even in DNA under Z-forming conditions at low P/D ratios. Moreover, the bound Etd[+] molecules are randomly positioned when bound to B-DNA but cluster strongly when bound to Z-DNA, consistent with small local regions of right-handed helix.

[175] C. Zimmer and G. Luck, *Adv. DNA Sequence Specific Agents* **1,** 51 (1992).
[176] C. Houssier, B. Hardy, and E. Fredericq, *Biopolymers* **13,** 1141 (1974).
[177] K. S. Dahl, A. Pardi, and I. Tinoco, Jr., *Biochemistry* **21,** 2730 (1982).
[178] F. Pohl and T. M. Jovin, *J. Mol. Biol.* **67,** 375 (1972).
[179] G. T. Walker, M. P. Stone, and T. R. Krugh, *Biochemistry* **24,** 7462 (1985).
[180] C. C. Hardin, G. T. Walker, and I. Tinoco, Jr., *Biochemistry* **27,** 4178 (1988).
[181] R. H. Shafer, S. C. Brown, A. Delbarre, and D. Wade, *Nucleic Acids Res.* **12,** 4679 (1984).
[182] M. L. Lamos, G. T. Walker, T. R. Krugh, and D. H. Turner, *Biochemistry* **25,** 687 (1986).

Many antibiotics which bind specifically to nucleic acids use a different mode for interaction. These agents bind in one of the grooves of B-DNA, rather than intercalating between base pairs. Netropsin (Nt) and distamycin (Dst), pyrrole-amidine antibiotics, are two classic examples of groove-binding drugs. The physical and DNA-binding properties of the drugs have been reviewed.[183] CD difference spectra between Nt–DNA complexes and free DNA show positive bands at 315 and 245 nm and negative bands at 275 and 220 nm. CD studies have provided evidence that Nt and Dst strongly prefer binding at A-T base pairs over G-C base pairs. Moreover, the binding is specific for B-DNA, as Nt and Dst binding to DNA in ethanol–water mixtures causes an A → B transition in the DNA. CD titrations of poly(dA) · poly(dT) with Nt demonstrate saturation at 0.1 Nt/base pair, corresponding to a site size of 5 base pairs. This agrees reasonably well with the site size of 4 base pairs inferred from other types of measurements, including CD titrations of DNA oligomers.[184]

Protein–Nucleic Acid Interactions

CD spectroscopy has been used extensively in the study of protein–DNA and protein–RNA interactions. Conformational changes in the protein and/or nucleic acid conformation can be detected readily by CD and in favorable cases can be assigned. Nonspecific interactions between proteins and nucleic acids are often characterized by dissociation constants in the micromolar range, so CD measurements can provide equilibrium constants. Specific interactions commonly have much smaller dissociation constants, but in some cases are in the micromolar range or can be shifted into this range by increasing the ionic strength. CD can be a valuable tool also for measuring the kinetics of protein–nucleic acid interactions.

The CD spectra of proteins and nucleic acids overlap extensively, but the spectra are complementary in a sense. In the region above 240 nm DNA and RNA have strong CD bands, whereas those of the aromatic side chains of the protein are relatively weak. Conversely, in the 210–230 nm region, protein CD is strong and nucleic acid CD, especially for B-DNA, is weak. These two windows therefore permit one to obtain information on the conformation of each of the two partners in nucleoprotein complexes.

Single-strand binding proteins (SSBP), also called helix-destabilizing proteins, are proteins which bind much more tightly to single-stranded polynucleotides than to double-stranded ones, thus lowering the melting

---

[183] Ch. Zimmer and U. Wähnert, *Prog. Biophys. Mol. Biol.* **47**, 31 (1986).
[184] Ch. Zimmer, G. Luck, and I. Frič, *Nucleic Acids. Res.* **3**, 1521 (1976).

point of the double-stranded polynucleotides. This class of proteins includes the gene 5 protein (g5p) of fd and other filamentous bacteriophages, gene 32 protein (gp32) of bacteriophage T4, *E. coli* single-strand binding protein, and adenovirus DNA-binding protein. The properties of these proteins and complexes with nucleic acids have been reviewed.[185,186]

The binding of SSBP to homopolynucleotides and to single-stranded DNA has been studied extensively by CD.[187-191] In general, the long-wavelength positive band of the DNA undergoes a substantial decrease in intensity, and, overall, the 260 nm band becomes significantly more negative. The CD signal as a function of the DNA/protein ratio gives information about the stoichiometry of the complexes. In several cases, there is evidence for two or more different modes of DNA binding. For example, gene 5 protein from fd binds to poly(dA) to form complexes with $n = 4$ and $n = 3$, where n is the number of nucleotides bound per protein molecule.[188]

There is disagreement concerning the conformation of the nucleic acid in the saturated complex. Scheerhagen *et al.*[189] emphasized that the CD of poly(rA) complexed with g5p from fd in the 260 nm region is distinct from that of poly(rA) at high temperatures. They calculated the CD and absorption spectra of poly(rA) with the DeVoe method, varying the geometry over a wide range, assuming a regular helical arrangement of the bases. They found that only geometries with strongly tilted bases and with a small twist between the bases (~20°) could account for the observed CD and absorption properties of the complex.

Sang and Gray[190] reported that the CD of the fd g5p–poly(rA) complex above 250 nm is equivalent to that of poly(rA) at 88°. They derived the spectrum of g5p in the complex, which agreed very well with the CD spectra of the free protein, except in the 230 nm region. The 230 nm region of the fd protein is known to have significant tyrosine contributions. Thus Sang and Gray concluded that there are no global conformational changes in the protein on complex formation, but that one or more tyrosine side chains are significantly perturbed. This is consistent with NMR and fluorescence evidence for tyrosine–base interactions in the g5p–ssDNA complex.

[185] J. E. Coleman and J. L. Oakley, *Crit. Rev. Biochem.* **7**, 247 (1980).
[186] J. W. Chase and K. R. Williams, *Annu. Rev. Biochem.* **55**, 103 (1986).
[187] L. A. Day, *Biochemistry* **12**, 5329 (1973).
[188] J. W. Kansy, B. A. Clack, and D. M. Gray, *J. Biomol. Struct. Dyn.* **3**, 1079 (1986).
[189] M. A. Scheerhagen, J. T. Bokma, C. A. Vlaanderen, J. Blok, and R. van Grondelle, *Biopolymers* **25**, 1419 (1986).
[190] B. C. Sang and D. M. Gray, *Biochemistry* **28**, 9502 (1989).
[191] M. L. Carpenter and G. G. Kneale, *J. Mol. Biol.* **217**, 681 (1991).

The differences between the two studies[189,190] may be due to different solution conditions. The same stoichiometry (4 nucleotides per g5p) was used, but there were significant differences in ionic strength and pH (7.7 versus 7.0), as well as the presence or absence of EDTA. The binding of SSBP, like other protein–DNA interactions, is known to be highly sensitive to ionic strength.

CD has been very useful in the study of several classes of specific DNA-binding proteins, especially the leucine zipper or bZIP proteins and helix–loop–helix proteins. Both classes involve dimeric proteins with dimerization domains and basic regions which actually recognize and contact the DNA. In the bZIP proteins, the leucine zipper region forms a coiled coil, with Leu residues spaced every seven residues forming part of the hydrophobic interface. In the helix–loop–helix proteins, the dimerization domain is more complex. CD provided much of the initial evidence for the coiled-coil structure of the leucine zipper class of proteins. O'Shea et al.[192] showed that a 33-peptide residue containing four of the leucine-containing heptet repeats from the protein GCN4 has a strong α-helix-like CD signal and is nearly 100% α helix at 0°.

Subsequently, CD has been used to study the formation of homo- and heterodimers of bZIP proteins, a process which is believed to be significant in the role of these proteins as regulators of transcription. Patel et al.[193] showed that a fragment of the bZIP protein Jun (residues 224–334) containing the leucine zipper and basic regions has a CD spectrum with a double minimum and an amplitude consistent with around 50% α helix. The corresponding fragment of Fos (116–211) has only a weak spectrum corresponding to about 23% helix. When the Fos and Jun fragments are mixed in equal proportions, however, the spectrum is like that of Jun, with around 50% helix. These results are consistent with other evidence which indicates that the Jun fragment can dimerize, forming homodimers, whereas the Fos fragment does not dimerize by itself. However, Fos and Jun can form heterodimers which are more stable than the Jun homodimers. The CD results also indicate that the coiled-coil formation substantially stabilizes the α helix relative to the monomeric state. When the Fos and/or Jun fragments are mixed with DNA containing the appropriate recognition site, a higher percentage of α helix is observed than in the absence of DNA. The increased helix formation is attributable to the basic regions also being converted to α helix. This is supported by experiments in which fragments lacking the basic region are used. These show the same behavior as the more complete fragments in the absence of DNA,

[192] E. K. O'Shea, R. Rutkowski, and P. S. Kim, Science 243, 538 (1988).
[193] L. Patel, C. Abate, and T. Curran, Nature (London) 347, 572 (1990).

but the behavior is not altered on the addition of DNA. Similar results have been reported for the basic region adjacent to the leucine zipper of GCN4.[194]

The basic region can bind to DNA in an α-helical conformation if it is stabilized in dimeric form by disulfide cross-linking. This was established by CD studies of Talanian et al.[195] using the basic region from GCN4 to which a Gly-Gly-Cys sequence was attached. Under oxidizing conditions, dimers are formed and some α-helix is present. On adding the specific DNA to which GCN4 binds, a large increase in helix content is observed.

Acknowledgment

This work was supported by National Institutes of Health Grant GM 22994.

[194] M. A. Weiss, T. Ellenberger, C. R. Wobbe, J. P. Lee, S. C. Harrison, and K. Struhl, *Nature (London)* **347**, 575 (1990).
[195] R. U. Talanian, C. J. McKnight, and P. S. Kim, *Science* **249**, 769 (1990).

[5] Bioinorganic Spectroscopy

By Edward I. Solomon, Martin L. Kirk, Daniel R. Gamelin, and Sabine Pulver

Introduction

Metalloproteins are spectroscopically attractive systems because metal ion active sites often have an open subshell electronic configuration (i.e., d^n with $0 < n < 10$) and thus exhibit transitions in many energy regions, each providing complementary insight into the enzyme active site. Key spectroscopic methods employed in bioinorganic chemistry are listed in Table I along with an indication of the information content. There are, of course, many uses of spectroscopy in bioinorganic chemistry which include analytical and kinetic applications; however, two goals should be emphasized as these directly relate to function: (1) Spectroscopic features can be used to probe the metal ion geometric and electronic structure and how these change with substrate and small molecule interactions with the protein. This provides molecular level insight into reaction mecha-

TABLE I

BIOINORGANIC SPECTROSCOPIC METHODS

Method	Parameters	Information content	Refs.
Ground state methods:			
Magnetic susceptibility	Molecular g value (g), axial (D) and rhombic (E) zero field splitting (ZFS) parameters, exchange interaction (J)	Number of unpaired electrons; probes ground state wave function at low resolution; defines antiferromagnetic and ferromagnetic interactions between metal centers; quantitates ground sublevel splittings on the order of kT	a
Mössbauer	Quadrupole splitting (ΔE_Q), isomer shift (δ), J, g, D, E, metal hyperfine (A_i^m)	For iron sites: oxidation and spin state; chemical environment; electric field gradient; orbital occupation of d levels; degree of valence delocalization in mixed-valence systems	b
Electron paramagnetic resonance (EPR)	g_i ($i = x, y, z$), A_i^m, ligand superhyperfine (A_i^l), D, E, J	For odd-electron metal sites: probes ground state wave function at high resolution; determination of atomic orbitals on metal and ligands contributing to MO containing unpaired e^- from e^-–nuclear hyperfine coupling; ligand field splittings via anisotropic g and ZFS tensors; determination of ligand bound to metal site from superhyperfine coupling; probes exchange interaction in coupled systems by resonance line position ($J < kT$) or variable temperature and relaxation studies ($J > kT$)	c
Electron nuclear double resonance (ENDOR)	A_i^m, A_i^l, quadrupole tensor (P_i), nuclear Zeeman splitting ($g_n\beta_n H$)	Combines sensitivity of EPR and high resolution of NMR to probe in detail ligand superhyperfine interactions with metal center and to identify specific type of ligand	d
Electron spin echo envelope modulation (ESEEM)	A_i^l, P_i^l	Complementary technique to ENDOR for measuring very small electron–nuclear hyperfine couplings	e
Vibrational (Raman and IR)	Energies, intensities, and polarizations	Identification of ligands coordinated to metal center; determination of metal–ligand and intraligand vibrational modes; bond strengths from force constants	f

72

Valence excited state methods:

Electronic absorption (ABS)	Energies, intensities, and band shapes	Direct probe of ligand field (LF) and charge transfer (CT) excited states; energies and intensities of LF transitions are powerful probe of metal site geometry; CT intensity and energies directly relate to metal–ligand orbital overlap and are thus sensitive probe of metal–ligand bonding	1
Polarized single crystal electronic absorption (linear dichroism)	Same as ABS plus polarization information	Polarization information provides a direct determination of selection rules and allows for rigorous assignments based on group theory; allows correlation of spectral features with geometric structure providing detailed insight into electronic structure of metal ion active site	1
Magnetic circular dichroism (MCD)	Same as ABS plus circular polarization induced by applied magnetic field and magnetic susceptibility	Greater sensitivity than ABS in observing weak transitions and greater resolution owing to differences in circular polarization; complementary selection rules aiding in assignment of electronic transitions; variable temperature–variable field (VTVH) MCD probes ground state sublevel splittings and associated wave function	2, g
Circular dichroism (CD)	Same as ABS plus circular polarization owing to dissymmetric nature of metal site	CD dispersion is signed quantity as in MCD, leading to enhanced resolution over ABS; complementary selection rule involving magnetic dipole character of transition	3, h
Resonance Raman	A-term (F_A) and B-term (F_B), intensity profiles, depolarization ratios ($\rho = I_\perp / I_\parallel$)	Excitation into electronic absorption results in intensity enhancement of normal modes of vibration which are coupled to electronic transition by either Franck–Condon or Herzberg–Teller coupling; allows for study of chromophoric active sites in biological molecules at low concentrations; allows assignment of CT (and in some cases LF) transitions based on nature of excited state distortion; can provide information on metal–ligand bonding as described above for vibrational spectroscopy	i

(continued)

73

TABLE I

BIOINORGANIC SPECTROSCOPIC METHODS

Method	Parameters	Information content	Refs.
Core excited state methods: X-Ray absorption spectroscopy (XAS)	Energies, intensities, and polarizations	Atom specific and allows for study of closed shell systems which are inaccessible via valence excited state methods; extended X-ray absorption fine structure (EXAFS) involves ionized scattered e^- and provides structural information (bond lengths and number of scatterers); X-ray absorption near-edge structure (XANES) involves transitions to bound states and is dependent on type of edge: *Metal K edge*: $1s$–$4p$ transitions are electric dipole allowed; energy and shape of X-ray edge correlates with oxidation state and geometry; $1s$–$3d$ is electric quadrupole allowed and thus has some absorption intensity, and noncentrosymmetric distortions can mix $1s$–$4p$ character into electric dipole forbidden $1s$–$3d,4s$ transitions; $3d/4p$ mixing probes potential contribution to metal hyperfine *Metal L edge*: 3- to 4-fold higher resolution than metal K edge; allowed $2p$–$3d$ transitions are observed; metal L edge contains information on spin state, oxidation state, and LF splitting of d orbitals; intensity probes metal–ligand covalency; L edge MCD possesses similar information content to MCD described above *Ligand K edge*: $1s$–$2p,3p$ transitions are electric dipole allowed; covalency mixes ligand p character into partially occupied metal d orbitals; intensity thus quantitates mixing (as with superhyperfine described above), and transition energy defines energies of ligand field states	1, *j, k*

| Photoelectron spectroscopy (PES) | Energies, polarizations, photoionization cross sections (intensity dependence on photon energy) | Measures kinetic energy and number of electrons ejected from sample, and therefore is surface sensitive and mostly applicable to active site model complexes (note that in XAS one detects photons rather than electrons, and method therefore does not possess surface sensitivity)

X-Ray photoelectron spectroscopy (XPS): involves core ionization which shows chemical shifts used to determine oxidation state and bonding information; probes exchange interactions between metal *d* and core electrons which directly relate to Fermi contact contribution to hyperfine; metal–ligand bonding information from satellite structure

Ultraviolet photoelectron spectroscopy (UPS): involves ionization of valence electrons which probes metal–ligand bonding and change with ionization (from resonance effects as photon energy is scanned through metal M edge), thus directly studying redox processes in metal complexes; allows for study of closed shell systems which are inaccessible via electronic absorption spectroscopy | 1, *l* |

[a] R. L. Carlin, "Magnetochemistry." Springer-Verlag, Berlin, 1986.

[b] P. Gutlich, R. Link, and A. Trautwein, "Mössbauer Spectroscopy and Transition Metal Chemistry." Springer-Verlag, Berlin, 1978.

[c] B. R. McGarvey, *Transition Met. Chem. (N.Y.)* **3**, 89 (1967).

[d] A. Schweiger, *Struct. Bonding (Berlin)* **51**, 1 (1982).

[e] W. B. Mims and J. Peisach, *in* "Advanced EPR" (A. J. Hoff, ed.), p. 1. Elsevier, Amsterdam, 1989.

[f] K. Nakamoto, "Infrared Spectra of Inorganic and Coordination Compounds," 2nd Ed. Wiley, New York, 1970.

[g] P. J. Stevens, *Annu. Rev. Phys. Chem.* **25**, 201 (1974).

[h] S. F. Mason, "Molecular Optical Activity and the Chiral Discriminations." Cambridge Univ. Press, New York, 1982.

[i] R. J. H. Clark and B. Stewart, *Struct. Bonding (Berlin)* **36**, 1 (1979).

[j] L. S. Kau, D. J. Spira-Solomon, J. E. Penner-Hahn, K. O. Hodgson, and E. I. Solomon. *J. Am. Chem. Soc.* **109**, 288 (1987).

[k] S. E. Shadle, J. E. Penner-Hahn, H. Schugar, B. Hedman, K. O. Hodgson, and E. I. Solomon, *J. Am. Chem. Soc.* **115**, 767 (1993).

[l] S. V. Didziulis, S. L. Cohen, A. A. Gerwirth, and E. I. Solomon, *J. Am. Chem. Soc.* **110**, 250 (1988).

nisms.[1,4] Although spectroscopy is often used to anticipate the results of X-ray crystallography, it should be viewed as providing complementary information related to bonding interactions and their contributions to reactivity. (2) Active sites often exhibit unique spectral features compared to small molecule inorganic complexes of the same metal ion. These reflect novel active site electronic structures which can make dominant contributions to the reactivity of the active sites in biology. Such spectral features often derive from intense low energy ligand-to-metal or metal-to-ligand charge transfer transitions which reflect highly covalent active sites. This topic has been discussed in some detail, and the reader is referred to recent reviews.[5–7]

In this chapter we consider the former goal, which focuses on the $d \rightarrow d$ (also called ligand field) transitions to excited states and their relationship to ground state properties. The unifying concept that allows one to extract geometric and electronic structural information from ligand field spectral features is ligand field theory, which is the theme of this chapter. We specifically use as an example nonheme high-spin ferrous (d^6) active sites. These are reduced sites so they do not exhibit low-energy ligand-to-metal charge transfer transitions. They are not particularly covalent, often do not exhibit electron paramagnetic resonance (EPR) signals (i.e., integer spin, $S = 2$ ions), and do not have the intense intraligand $\pi \rightarrow \pi^*$ transitions that dominate the spectra and electronic structure of ferrous heme proteins. The only spectral probes from Table I that have been generally employed for these enzymes are Mössbauer and metal K-edge X-ray absorption spectroscopy, and thus there has been only limited information available on their active sites. High-spin d^6 ferrous centers are open shell metal ions which have a number of ligand field transitions in the near infrared and visible regions of the absorption spectrum. These transitions can be interpreted in terms of crystal field theory and ligand field theory to obtain geometric and electronic structure information and can additionally be used to obtain ground state spin Hamiltonian parameters (even in the absence of an EPR signal) which provide further insight into the covalency of the active site.

[1] E. I. Solomon, *Comments Inorg. Chem.* **3**, 225 (1984).

[2] S. B. Piepho and P. N. Schatz, "Group Theory in Spectroscopy." Wiley (Interscience), New York, 1983.

[3] R. D. Gillard, *in* "Physical Methods in Advanced Inorganic Chemistry" (H. A. O. Hill and P. Day, eds.), p. 167. Interscience, New York, 1968.

[4] E. I. Solomon and Y. Zhang, *Acc. Chem. Res.* **25**, 343 (1992).

[5] E. I. Solomon and M. D. Lowery, *Science* **259**, 1575 (1993).

[6] E. I. Solomon, M. D. Lowery, L. B. LaCroix, and D. E. Root, this series, Vol. 226, p. 1.

[7] E. I. Solomon, M. J. Baldwin, and M. D. Lowery, *Chem. Rev.* **92**, 521 (1992).

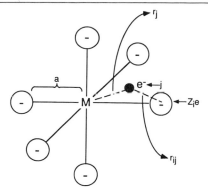

FIG. 1. One-electron metal ion in an octahedral crystal field of negatively charged ligands (indicated by circles).

Crystal Field Theory

Figure 1 depicts a one-electron metal ion in an octahedral field of negatively charged ligands with no overlap between the metal orbitals and the ligands. The real combinations of the five d orbitals are given in Table II, where d_{xy}, d_{xz}, and d_{yz} transform as the t_{2g} representation of the octahedral molecular symmetry group, and $d_{x^2-y^2}$ and d_{z^2} transform together as the e_g set. Crystal field theory considers the electrostatic repulsive interaction of electrons in these metal d orbitals with the negatively

TABLE II
REAL d ORBITALS

Octahedral symmetry	Real and complex wave functions	Orbital number
t_{2g}	$d_{xy} = \dfrac{1}{i(2)^{1/2}}(d_2 - d_{-2}) = R_{3d}\dfrac{15^{1/2}}{2(\pi)^{1/2}}xy$	5
t_{2g}	$d_{yz} = \dfrac{-1}{i(2)^{1/2}}(d_1 + d_{-1}) = R_{3d}\dfrac{15^{1/2}}{2(\pi)^{1/2}}yz$	4
t_{2g}	$d_{xz} = \dfrac{1}{-(2)^{1/2}}(d_1 - d_{-1}) = R_{3d}\dfrac{15^{1/2}}{2(\pi)^{1/2}}xz$	2
e_g	$d_{z^2} = d_0 = R_{3d}\dfrac{5^{1/2}}{4(\pi)^{1/2}}(2z^2 - x^2 - y^2)$	3
e_g	$d_{x^2-y^2} = \dfrac{1}{2^{1/2}}(d_2 + d_{-2}) = R_{3d}\dfrac{5^{1/2}}{4(\pi)^{1/2}}x^2 - y^2$	1

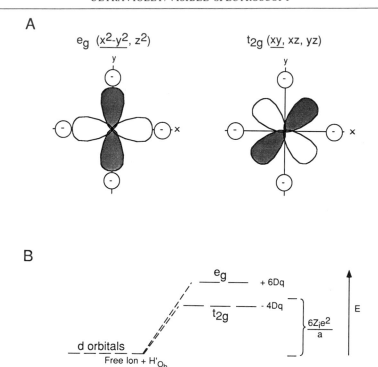

FIG. 2. (A) The $d_{x^2-y^2}$ and d_{xy} orbital interactions with negatively charged ligands in the xy molecular plane and (B) energy increase and splitting of the five d orbitals by an octahedral crystal field.

charged ligands.[8,9] Qualitatively, an electron in a $d_{x^2-y^2}$ orbital (lobes pointed directly at the ligands) has a greater repulsive interaction with the ligands than an electron in the d_{xy} orbital (lobes bisecting the ligand–metal bonds), and hence the e_g set of orbitals is raised more in energy than the t_{2g} set, resulting in a crystal field splitting of the d orbitals as shown in Fig. 2B.

Quantitatively, the crystal field potential energy term, $H'_{\text{crystal field}}$, may be evaluated as a perturbation over the real d orbitals in Table II using Eq. (1):

$$H'_{\text{crystal field}} = \sum_{i=\text{ligands}} \frac{Z_i e^2}{r_{ij}} \tag{1}$$

[8] H. Bethe, *Ann. Phys. (Leipzig)* **3**, 133 (1929).
[9] J. H. Van Vleck, *Phys. Rev.* **41**, 208 (1932).

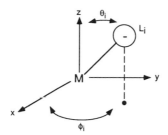

FIG. 3. Spherical polar coordinates in relationship to Cartesian coordinates.

where $Z_i e$ is the charge on ligand i, e is the charge on the electron, and r_{ij} is the distance between electron j and ligand i. This is accomplished by expanding $1/r_{ij}$ in terms of a product of spherical harmonics centered on the metal ion [Eq. (2)]:

$$\frac{1}{r_{ij}} = \sum_{l=0}^{\infty} \sum_{m=-l}^{+l} \frac{4\pi}{2l+1} \left(\frac{r_<^l}{r_>^{l+1}}\right) Y_l^m{}^*(\theta_i, \phi_i) Y_l^m(\theta_j, \phi_j) \qquad (2)$$

The polar coordinates θ and ϕ are defined in Fig. 3, where (θ_i, ϕ_i) refers to the coordinates of ligand i and (θ_j, ϕ_j) to the electron coordinates. The shorter radial distance to the origin, $r_<$, corresponds to the electron distance r_j in Fig. 1, and the larger radial distance to the origin, $r_>$, is the metal–ligand bond length, a. Thus, as given in Eq. (3),

$$H'_{\text{crystal field}} = \sum_{l=0}^{\infty} \sum_{m=-l}^{+l} A_l^m r_j^l Y_l^m(\theta_j, \phi_j) \qquad (3)$$

where

$$A_l^m = Z_i e^2 \sum_{i=\text{ligands}} \left(\frac{4\pi}{2l+1}\right)\left(\frac{1}{a^{l+1}}\right) Y_l^m{}^*(\theta_i, \phi_i)$$

the crystal field potential energy term involves a set of constants A_l^m, only dependent on the specific ligand charge arrangement around the metal ion, and $r_j^l Y_l^m(\theta_j, \phi_j)$, which operate only on the electron coordinates.

Although Eq. (3) involves an infinite sum of spherical harmonics, only $l = 0$, 2, and 4 contribute nonzero values when integrated over the d orbitals [Eq. (4)]:

$$H_{mn} = \int d_m H'_{\text{crystal field}} d_n \, d\tau \qquad (4)$$

Substituting the specific (θ_i, ϕ_i) values for the six ligands of an octahedral complex into Eq. (3) gives Eq. (5), where $r \equiv r_j$:

TABLE III
ONE-ELECTRON INTEGRALS IN TERMS OF DIRECTION COSINES[a]

Integral type (X^l, Y^m, Z^n)	Example	Integrated value[b]
(2)	$\int_0^{2\pi} \int_0^{\pi} X^2 \sin\theta \, d\theta \, d\phi$	5
(4)	X^4	3
(2, 2)	X^2Y^2	1
(6)	X^6	15/7
(4, 2)	X^4Y^2	3/7
(2, 2, 2)	$X^2Y^2Z^2$	1/7
(8)	X^8	5/3
(6, 2)	X^6Y^2	5/21
(4, 4)	X^4Y^4	1/7
(4, 2, 2)	$X^4Y^2Z^2$	1/21

[a] $X = x/r$, $Y = y/r$, $Z = z/r$.
[b] The normalization factor squared for the d wave functions, $[\frac{1}{2}(15/\pi)^{1/2}]^2$, has been included.

$$H'_{O_h} = \frac{6Z_i e^2}{a} + \left(\frac{49}{18}\right)^{1/2} (2\pi)^{1/2} \frac{Z_i e^2}{a^5} r^4 \left[Y_4^0 + \left(\frac{5}{14}\right)^{1/2} (Y_4^4 + Y_4^{-4}) \right] \quad (5)$$

To evaluate the octahedral field potential in Eq. (5) over the real d orbitals in Table II, it is convenient to transform Eq. (5) into Cartesian coordinates (using a table in Ballhausen[10]), which gives Eq. (6):

$$H'_{O_h} = \frac{6Z_i e^2}{a} + \frac{35 Z_i e^2}{4a^5} \left(x^4 + y^4 + z^4 - \frac{3}{5} r^4 \right) \quad (6)$$

The energies of the t_{2g} and e_g sets of d orbitals are obtained by substituting Eq. (6) and the real d orbitals in Table II into Eq. (4), which is evaluated using the appropriate integral values over angular coordinates from Table III. This results in Eqs. (7a) and (7b) with Dq defined in Eq. (7c) and

$$E_{t_{2g}} = \int (d_{xy}) H'_{O_h}(d_{xy}) r^2 \sin\theta \, dr \, d\theta \, d\phi = \frac{6Z_i e^2}{a} - 4Dq \quad (7a)$$

$$E_{e_g} = \int (d_{x^2-y^2}) H'_{O_h}(d_{x^2-y^2}) r^2 \sin\theta \, dr \, d\theta \, d\phi = \frac{6Z_i e^2}{a} + 6Dq \quad (7b)$$

$$Dq = \frac{Z_i e^2}{6a^5} \overline{r^4} \quad (7c)$$

[10] C. J. Ballhausen, "Introduction to Ligand Field Theory." McGraw-Hill, New York, 1962.

$\overline{r^4} \equiv \int_0^\infty [R_{3d}(r)]^2 r^4 r^2\, dr$. $R_{3d}(r)$ is the $3d$ radial wave function, and $\overline{r^4}$ corresponds to the fourth power of the radial distance of an electron from the nucleus averaged over this orbital. Thus, as shown in Fig. 2, all five of the d orbitals are raised in energy by an amount $6Z_i e^2/a$, which is on the order of 10 eV. The t_{2g}–e_g splitting is $10Dq$, which is on the order of 1–2 eV. An equivalent treatment for a tetrahedral crystal field gives Eq. (8):

$$H'_{T_d} = \frac{4Z_i e^2}{a} + \frac{35Z_i e^2}{9a^5}\left(x^4 + y^4 + z^4 - \frac{3}{5}r^4\right) \tag{8}$$

which, when evaluated over the d orbitals, leads to the generally used correlation that $Dq_{T_d} = -(4/9)Dq_{O_h}$.

Note that in the above treatment it was necessary to evaluate only two integrals of the crystal field potential, one over the d_{xy} orbital [Eq. (7a)] and the second over the $d_{x^2-y^2}$ orbital [Eq. (7b)]. This derives from the group theory of high symmetry crystal fields, where for an octahedral or tetrahedral complex each of the d orbital expressions is a good wave function for the molecular Hamiltonian and gives an energy appropriate for all members of the t_{2g} and e_g sets, respectively.[11] For a lower symmetry complex Eq. (4) must be expanded into a perturbation secular determination [Eq. (9)] with 25 matrix elements, H_{mn}, involving all pairwise combina-

$$|H_{mn} - E\delta_{mn}| = 0 \tag{9}$$

tions of the real d orbitals in Table II. The matrix elements for the secular determinant are given in Table IV in terms of $D_{lm} = \Sigma_{i=\text{ligands}} D^i_{lm}$, and $G_{lm} = \Sigma_{i=\text{ligands}} G^i_{lm}$, where the individual ligand position functions are given in Table IV with $\alpha^i_l = \dfrac{Z_i e^2}{a_i^{l+1}} \int_0^\infty (R_{3d})^2 r^l r^2\, dr$. Note that orbital numbers 1–5 are defined in Table II.

After choosing a coordinate system, inputting the polar coordinates of each ligand into Table IV, and obtaining the matrix elements for Eq. (9), the energies of the d orbitals are given in terms of a set of parameters α_0, α_2, and α_4 for each ligand.[12] The α_0 radial integral affects all the d orbitals by the same amount without changing the d orbital splittings and can therefore be eliminated. Thus, for a complex with low symmetry one needs two parameters (α_2, α_4) for each ligand to describe the crystal field splitting of the d orbitals. As the symmetry increases the number of parameters decreases, and in the octahedral and tetrahedral limit only $Dq(=\frac{1}{6}\alpha_4)$ is required, as derived in Eq. (7). Estimation of the Dq (or α_2, α_4) value is addressed in Section III.

[11] B. N. Figgis, "Introduction to Ligand Fields." Interscience, New York, 1967.
[12] A. L. Companion and M. A. Komarynsky, *J. Chem. Ed.* **41**, 257 (1964).

TABLE IV
MATRIX ELEMENTS AND LIGAND POSITION FUNCTIONS[a]

Integrals H_{mn} in terms of D_{lm} and G_{lm}	Ligand position functions
$H_{11} = D_{00} - \frac{1}{7}D_{20} + \frac{1}{56}D_{40} + \frac{5}{24}D_{44}$	$D_{00}{}^i = \alpha_0{}^i$
$H_{22} = D_{00} + \frac{1}{14}D_{20} - \frac{1}{14}D_{40} + \frac{3}{14}D_{22} + \frac{5}{42}D_{44}$	$D_{20}{}^i = \alpha_2{}^i (3\cos^2\theta_i - 1)$
$H_{33} = D_{00} + \frac{1}{7}D_{20} + \frac{3}{28}D_{40}$	$D_{40}{}^i = \alpha_4{}^i \left(\frac{35}{3}\cos^4\theta_i - 10\cos^2\theta_i + 1\right)$
$H_{44} = D_{00} - \frac{1}{14}D_{20} - \frac{1}{14}D_{40} - \frac{3}{14}D_{22} - \frac{5}{42}D_{44}$	$D_{21}{}^i = \alpha_2{}^i \sin^2\theta_i \cos\theta_i \cos\phi_i$
$H_{55} = D_{00} - \frac{1}{7}D_{20} + \frac{1}{56}D_{40} - \frac{5}{24}D_{44}$	$D_{22}{}^i = \alpha_2{}^i \sin^2\theta_i \cos 2\phi_i$
$H_{12} = \frac{3}{7}D_{21} - \frac{5}{28}D_{41} + \frac{5}{12}D_{43}$	$D_{41}{}^i = \alpha_4{}^i \sin^2\theta_i \cos\theta_i \left(\frac{7}{3}\cos^2\theta_i - 1\right)\cos\phi_i$
$H_{13} = \frac{-(3)^{1/2}}{7}D_{22} + \frac{5(3)^{1/2}}{84}D_{42}$	$D_{42}{}^i = \alpha_4{}^i \sin^2\theta_i (7\cos^2\theta_i - 1)\cos 2\phi_i$
$H_{14} = \frac{-3}{7}G_{21} + \frac{5}{28}G_{41} + \frac{5}{12}G_{43}$	$D_{43}{}^i = \alpha_4{}^i \sin^2\theta_i \cos^2\theta_i \cos 3\phi_i$
$H_{15} = \frac{5}{24}G_{44}$	$D_{44}{}^i = \alpha_4{}^i \sin^4\theta_i \cos 4\phi_i$
$H_{23} = \frac{(3)^{1/2}}{7}D_{21} + \frac{5(3)^{1/2}}{14}D_{41}$	$G_{21}{}^i = \alpha_2{}^i \sin\theta_i \cos\theta_i \sin\phi_i$
$H_{24} = \frac{3}{14}G_{22} + \frac{5}{42}G_{42}$	$G_{22}{}^i = \alpha_2{}^i \sin^2\theta_i \sin 2\phi_i$
$H_{25} = \frac{3}{7}G_{21} - \frac{5}{28}G_{41} + \frac{5}{12}G_{43}$	$G_{41}{}^i = \alpha_4{}^i \sin\theta_i \cos\theta_i \left(\frac{7}{3}\cos^2\theta_i - 1\right)\sin\phi_i$
$H_{34} = \frac{3^{1/2}}{7}G_{21} + \frac{5(3)^{1/2}}{14}G_{41}$	$G_{42}{}^i = \alpha_4{}^i \sin^2\theta_i (7\cos^2\theta_i - 1)\sin 2\phi_i$
$H_{35} = \frac{-(3)^{1/2}}{7}G_{22} + \frac{5(3)^{1/2}}{84}G_{42}$	$G_{43}{}^i = \alpha_4{}^i \sin^2\theta_i \cos\theta_i \sin 3\phi_i$
$H_{45} = \frac{3}{7}D_{21} - \frac{5}{28}D_{41} - \frac{5}{12}D_{43}$	$G_{44}{}^i = \alpha_4{}^i \sin^4\theta_i \sin 4\phi_i$

[a] Adapted from Companion and Komarynsky.[12]

For d^n metal sites with more than one electron or one hole one must add the n electrons to the crystal field split d orbitals, generate all possible many-electron wave functions, and allow for the additional perturbations caused by electron–electron repulsion within and between these states. The electron repulsion integrals evaluated for the real d orbitals are given in Table V. Note that three parameters (known as Racah parameters) are required to describe electron–electron repulsion: A, B, and C. Racah parameter A affects all states by the same amount and can be neglected

TABLE V
ELECTRON REPULSION MATRIX ELEMENTS OVER REAL d ORBITALS^a

Coulomb (J) and exchange (K) integrals in terms of Racah parameters A, B, and C

$J(xy, xy) = J(xz, xz) = J(yz, yz) = J(x^2 - y^2, x^2 - y^2) = J(z^2, z^2) = A + 4B + 3C$

$J(xy, yz) = J(xy, xz) = J(xz, yz) = J(x^2 - y^2, yz) = J(x^2 - y^2, xz) = A - 2B + C$

$J(z^2, xz) = J(z^2, yz) = A + 2B + C$

$J(z^2, xy) = J(z^2, x^2 - y^2) = A - 4B + C$

$J(x^2 - y^2, xy) = A + 4B + C$

$K(x^2 - y^2, xz) = K(x^2 - y^2, yz) = K(xz, yz) = K(xy, xz) = K(xy, yz) = 3B + C$

$K(z^2, x^2 - y^2) = K(z^2, xy) = 4B + C$

$K(z^2, yz) = K(z^2, xz) = B + C$

$K(x^2 - y^2, xy) = C$

Nonzero d orbital matrix elements				
a	b	c	d	$\left\langle ab \dfrac{1}{r_{12}} cd \right\rangle$
xz	z^2	xz	$x^2 - y^2$	$-2(3)^{1/2}B$
yz	z^2	yz	$x^2 - y^2$	$2(3)^{1/2}B$
z^2	yz	z^2	$x^2 - y^2$	$-(3)^{1/2}B$
z^2	xz	z^2	$x^2 - y^2$	$3^{1/2}B$
z^2	xy	xz	yz	$3^{1/2}B$
z^2	xz	xy	yz	$-2(3)^{1/2}B$
z^2	xy	yz	xz	$3^{1/2}B$
$x^2 - y^2$	xy	xz	yz	$3B$
$x^2 - y^2$	xy	yz	xz	$-3B$

^{a} Adapted from A. P. B. Lever, "Inorganic Electronic Spectroscopy." Elsevier, New York, 1984, and Ballhausen.[10]

with respect to energy splittings; B and C involve radial integrals and are given in Eq. (10):

$$B = \frac{e^2}{49} \int_0^\infty \int_0^\infty \frac{r_<^2}{r_>^3} [R_{3d}(r_1)]^2 [R_{3d}(r_2)]^2 r_1^2 \, dr_1 \, r_2^2 \, dr_2 - \frac{C}{7} \qquad (10a)$$

$$C = \frac{35e^2}{441} \int_0^\infty \int_0^\infty \frac{r_<^4}{r_>^5} [R_{3d}(r_1)]^2 [R_{3d}(r_2)]^2 r_1^2 \, dr_1 \, r_2^2 \, dr_2 \qquad (10b)$$

where the subscripts 1 and 2 refer to the two interacting electrons and $r_<$ and $r_>$ are the distances of these electrons closer to and farther from the nucleus. Based on theory the C/B ratio[13,14] is fixed at the experimental value of the free metal ion (between 4 and 5)[15] and all the electron repulsion in many-electron transition metal complexes is described in terms of only one Racah parameter, B.

A complete degenerate perturbation treatment including crystal field configurational energies and electron–electron repulsion for octahedral and tetrahedral complexes has been developed by Tanabe and Sugano.[13,14] Their matrices can be found in Ref. 14 and lead to the Tanabe–Sugano diagrams also given in that reference. The Tanabe–Sugano diagram appropriate for an octahedral d^6 complex is given in Fig. 4. Note that although there are many states contained in the diagram, all the energy splittings are defined by only the two parameters of crystal field theory for an octahedral complex, Dq and B. Every $d^6 \, O_h$ complex has specific Dq and B values, and the ratio defines all the excited state energies in terms of B. For the high-spin ferrous sites considered in this chapter the ground state is a quintet; these complexes therefore must have $Dq/B < 2.0$.

Ligand Field Theory

As presented in Section II, the observation of at least two excited states using electronic absorption spectroscopy allows experimental determination of Dq and B values for a given complex using Tanabe–Sugano diagrams and matrices. These values can then be compared to theory. Using $[Fe(H_2O)_6]^{2+}$ as an example, the experimental values of Dq and B are 1065 and 720 cm^{-1}, respectively.[16,17] Inserting a Slater type $3d$ radial

[13] Y. Tanabe and S. Sugano, *J. Phys. Soc. Jpn.* **9**, 753 (1954).
[14] S. Sugano, Y. Tanabe, and H. Kamimura, "Multiplets of Transition Metal Ions in Crystals." Academic Press, New York, 1970.
[15] E. U. Condon and G. H. Shortley, "The Theory of Atomic Spectra." Cambridge Univ. Press, London, 1953.
[16] F. A. Cotton and M. D. Meyers, *J. Am. Chem. Soc.* **82**, 5023 (1960).
[17] P. J. McCarthy and R. D. Bereman, *J. Coord. Chem.* **6**, 129 (1977).

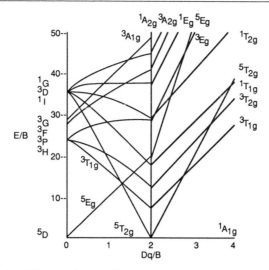

FIG. 4. Energy level diagram for the d^6 configuration in octahedral symmetry. (Adapted from Tanabe and Sugano.[13])

wave function[18] into Eq. (7c) and evaluating the fourth power of the radial distance of the electron from the metal over this $3d$ orbital yields a crystal field theory $10Dq$ value of approximately 100 cm^{-1}, which is an order of magnitude too small compared to experiment. The value of 1160 cm^{-1} for the free ferrous ion electron–electron repulsion parameter, B, calculated[19] by inputting radial functions into Eq. (10) is in reasonable agreement with the value of 917 cm^{-1} obtained from atomic spectroscopy,[13,20] but both are significantly larger than the value observed for the hexaaquoferrous complex. The differences between these values derive from the presence of covalency in $[\text{Fe}(\text{H}_2\text{O})_6]^{2+}$, which complicates the original assumption of crystal field theory that there is no overlap between the ligand valence and metal d orbitals.

A more complete description of the energy level diagram for metal complexes is given in Fig. 5, which corresponds to including ligand valence orbitals in Fig. 2. From the molecular orbital diagram in Fig. 5, the splitting of the t_{2g}/e_g sets of d orbitals derives from differences in σ versus π

[18] J. C. Slater, *Phys. Rev.* **34**, 1293 (1929).
[19] J. Ferguson, *Prog. Inorg. Chem.* **12**, 159 (1970).
[20] C. E. Moore, *Nat. Bur. Stand. Circular,* 467 (1952).

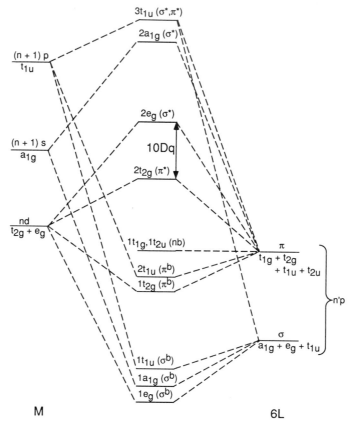

FIG. 5. Molecular orbital diagram for an octahedral metal complex having monoatomic σ donor and π donor ligands, where n and n' are the principal quantum numbers of the metal and ligand valence orbitals, respectively.

antibonding interactions with the ligands. In molecular orbital theory,[21,22] the d orbitals gain ligand character ψ_L owing to bonding [Eq. (11)]:

$$\psi_{(e_g\sigma^*)} = N[\psi_{nd}(e_g\sigma) - c\psi_L(e_g\sigma)] \tag{11a}$$

$$\psi_{(t_{2g}\pi^*)} = N'[\psi_{nd}(t_{2g}\pi) - c'\psi_L(t_{2g}\pi)] \tag{11b}$$

[21] C. J. Ballhausen and H. B. Gray, "Molecular Orbital Theory." Benjamin Cummings, Reading, Massachusetts, 1964.

[22] C. J. Ballhausen, "Molecular Electronic Structures of Transition Metal Complexes." McGraw-Hill, New York, 1979.

where c (c') is the orbital mixing coefficient and N (N') is a normalization factor. Van Vleck[23] has emphasized that molecular orbital theory corresponds in the limit of no overlap ($c = 0$) to crystal field theory and in the limit of complete covalent bonding ($c = 1$) to valence bond theory.

Molecular orbital theory is the most general approach for describing the bonding in metal complexes. Sophisticated calculations such as Hartree–Fock–self-consistent field–linear combination of atomic orbitals–molecular orbital plus configurational interaction (HF-SCF-LCAO-MO + CI)[24] or self-consistent field–Xα–scattered wave (SCF-Xα-SW)[25,26] are required for good agreement with experiment, however, and further require fairly high symmetry sites. Approximate methods (in particular the extended Hückel method[27,28] often employed), although providing qualitative insight, can give rather misleading results. Thus, whereas the energy splitting of the d levels and the repulsion between electrons in these orbitals clearly involve molecular orbital interactions with the ligands, high-level bonding descriptions are often not accessible.

There are many spectroscopic approaches[1] for experimentally estimating c in Eq. (11), and c^2 is generally found to be less than approximately 0.2 for many transition metal complexes. Thus, metal sites are much closer to the crystal field than the covalent limit. The approach that has developed to describe the electronic structure of transition metal complexes, generally referred to as ligand field theory,[10,11,14,29,30] starts from the crystal field model described in Section II and allows the crystal field parameters Dq and B to adjust to fit the experimental data. Trends in the experimental values of these parameters with ligand and metal variation are interpreted in terms of differences in covalency [i.e., c, $c' > 0$ in Eq. (11)]. These trends are summarized in Table VI.

The trend in $10Dq$ is known as the spectrochemical series and derives from the different types of bonding interactions a ligand can have with the metal center. Figure 5 depicts the interaction of the metal with six monoatomic anionic ligands such as fluoride, which is a σ and π donor.

[23] J. H. Van Vleck, *J. Chem. Phys.* **3**, 803 (1935).

[24] H. F. Schaefer, "The Electronic Structure of Atoms and Molecules: A Survey of Rigorous Quantum-Mechanical Results." Addison-Wesley, Reading, Massachusetts, 1972.

[25] K. H. Johnson, *Adv. Quantum Chem.* **7**, 143 (1973).

[26] J. W. D. Connolly, *in* "Modern Theoretical Chemistry" (G. A. Segal, eds.), Vol. 7, p. 105. Plenum, New York, 1977.

[27] M. Wolfsberg and J. Helmholz, *J. Chem. Phys.* **20**, 837 (1952).

[28] R. Hoffman, *J. Chem. Phys.* **39**, 1397 (1963).

[29] D. S. McClure, "Electronic Spectra of Molecules and Ions in Crystals." Academic Press, New York, 1959.

[30] J. S. Griffith, "The Theory of Transition Metal Ions." Cambridge Univ. Press, London, 1964.

TABLE VI
Spectrochemical and Nephelauxetic Series[a]

Spectrochemical series (increasing $10Dq$)	Nephelauxetic series (decreasing β)
Fixed metal system:	Fixed metal system:
$I^- < Br^- < Cl^- \approx {*}SCN^- \approx N_3^- < F^- <$ Urea $< OH^- < CH_3COO^- < ox < H_2O <$ ${*}NCS^- < EDTA < NH_3 \approx py < en \approx tren <$ o-phen $< {*}NO_2^- < H^- \approx CH_3^- < {*}CN^- <$ CO	$F^- > H_2O >$ urea $> NH_3 >$ en $\approx$ ox $>$ SCN${*}^- > Cl^- \approx CN^- >$ $Br^- > S^{2-} \approx I^-$
Fixed ligand system:	Fixed ligand system:
$Mn^{2+} < Co^{2+} \approx Ni^{2+} \approx Fe^{2+} < V^{2+} < Fe^{3+} <$ $Cr^{3+} < V^{3+} < Co^{3+} < Mn^{4+} < Mo^{3+} < Rh^{3+}$ $< Ru^{3+} < Ir^{3+} < Re^{4+} < Pt^{4+}$	$Mn^{2+} > Ni^{2+} \approx Co^{2+} \approx Mo^{3+} >$ $Cr^{3+} > Fe^{3+} > Rh^{3+} \approx Ir^{3+} >$ $Co^{3+} > Mn^{4+} > Pt^{4+}$

[a] Abbreviations: ox, oxalate; py, pyridine; en, ethylenediamine; tren, tris-aminoethylamine; o-phen, o-phenanthroline.

As shown in Fig. 6, σ donor ligands such as amines, which have no π antibonding interactions with the metal (i.e., nonbonding, nb), will have a larger splitting of the d orbitals than the σ donor, π donor ligands. Finally, ligands such as CN^- and CO, which have low-lying unoccupied π^* orbitals, have bonding interactions with the t_{2g} set of metal d orbitals (often referred to as π-backbonding). The latter act as σ donor, π acceptor ligands and have the highest values of $10Dq$. For a fixed ligand system, Dq increases by 40–80% when increasing the charge on the metal ion ($+2 \rightarrow +3 \rightarrow +4$) and by 30–40% on proceeding to the second and third transition metal series ($3d \rightarrow 4d \rightarrow 5d$).[31] These trends also follow differences in covalency, which derive from differences in valence orbital energy and overlap.

The effects of electron–electron repulsion are given by the parameter β, defined as $B_{complex}/B_{free\ ion}$. β is always found to be less than 1, and the trend in this value, given in Table VI, is referred to as the nephelauxetic (i.e., cloud expanding) series.[32] The value of B is generally found to decrease from the free ion value with increasing covalency of the ligand–metal bond. The dominant contribution to this nephelauxetic reduction of electron–electron repulsion is called central field covalency. All ligands are at least σ donors and contribute charge to the metal ion, thus decreasing the effective nuclear charge, Z_{eff}, of the metal ion in the com-

[31] H. L. Schläfer and G. Gliemann, "Ligand Field Parameters." Cambridge Univ. Press, New York, 1973.
[32] C. K. Jørgensen, "Absorption Spectra and Chemical Bonding in Complexes." Pergamon, New York, 1962.

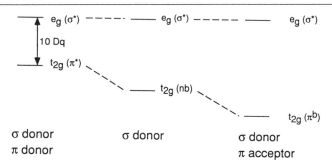

FIG. 6. Effect of ligand properties on magnitude of $10Dq$.

plex from that of the free ion. Substitution of a general Slater radial wave function into Eq. (10a) for B gives Eq. (12),[10] which shows that B is reduced as the effective nuclear charge on the metal ion is reduced. This

$$B = [204Z_{eff} - (5)14.7Z_{eff}] \text{ cm}^{-1} \qquad (12)$$

derives from the resultant expansion of the radial wave function, which lowers the repulsive interaction between electrons in the orbitals.

In the next section electronic absorption spectroscopy is used to observe and assign the ligand field excited states of a given ferrous site, which defines Dq/B in the Tanabe–Sugano diagram (Fig. 4) and provides an experimental estimate for these parameters.

Electronic Absorption Spectroscopy

The electronic absorption spectrum of $[Fe(H_2O)_6]^{2+}$ is given in Fig. 7.[16,17] The goal is to relate the relatively strong absorption bands at 9700 and 11,600 cm^{-1} and the weak band at 14,700 cm^{-1} to $d \rightarrow d$ transitions between the ground state, ψ_G, and specific excited states ψ_E on the high-spin side ($Dq/B < 2.0$) of the Tanabe–Sugano diagram in Fig. 4. To observe a transition in the absorption spectrum the oscillator strength, f, must be greater than 0. From Eq. (13a), f is experimentally given by the integrated

$$f_{exp} = 4.33 \times 10^{-9} \int \varepsilon(\nu) \, d\nu \qquad (13a)$$

area under an absorption band, where ν is the transition energy (in cm^{-1}). The theoretical expression for the oscillator strength is given in Eq. (13b):

$$f_{theory} = 1.085 \times 10^{-11} \, \nu \, [\int \psi_G \hat{M} \psi_E \, d\tau]^2 \qquad (13b)$$

where f_{theory} is proportional to the square of the transition moment integral between the ground and a specific excited state, and the transition moment integral is given in centimeters. If the wavelength of light is larger than

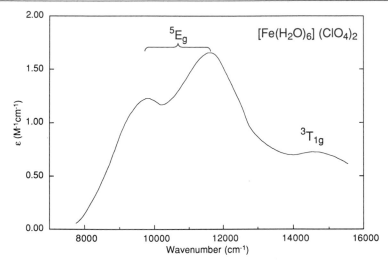

FIG. 7. Low temperature absorption spectrum of $[Fe(H_2O)_6](ClO_4)_2$. (Spectrum adapted from Cotton and Meyers[16] and McCarthy and Bereman.[17])

the dimensions of the electronic orbital on the metal center, the transition moment operator, $\hat{M}$, is given by a multiple expansion with the dominant term being the electric dipole operator, $\sum_j e\hat{r}_j \cdot \vec{E}$. Because this operator involves only the orbital coordinates of the j electrons, it does not affect spin. The wave functions ψ_G and ψ_E must therefore have the same spin quantum number for the integral in Eq. (13b) to be nonzero and for the transition to have some intensity in the absorption spectrum. This leads to the selection rule that $\Delta S = 0$ for *spin allowed* transitions. Furthermore, the electric dipole operator is a one-electron operator so only transitions to excited states which involve a one-electron orbital change are allowed. Finally, because $\hat{r}$ is a vector which is antisymmetric (i.e., ungerade) to the inversion symmetry operator, and the d orbitals are symmetric (i.e., gerade), all $d \rightarrow d$ transitions are parity (also referred to as Laporté) forbidden.

Active sites in proteins generally overcome this forbiddenness by a distortion of the ligand field around the metal ion such that there is no longer inversion symmetry. The effect of this distortion is to introduce a new term, $H'_{\text{ligand field}}$, into the potential energy expression [Eq. (6)], which mixes some higher energy Laporté allowed state ψ_U into the ligand field excited state. This is given by a first-order perturbation correction to ψ_E [Eq. (14)]:

$$\psi'_E = \psi_E + \left[\frac{\int \psi_E H'_{\text{ligand field}} \psi_U \, d\tau}{E_E - E_U}\right] \psi_U \qquad (14)$$

Substitution of the perturbed wave function into Eq. (13b) gives Eq. (15) for the intensity of a $\psi_G \rightarrow \psi_E$ transition:

$$f_{G \rightarrow E} = \frac{\nu_E}{\nu_U} \left[\frac{\int \psi_E H'_{\text{ligand field}} \psi_U \, d\tau}{E_E - E_U}\right]^2 f_U \qquad (15)$$

where ν_U is the energy (in cm^{-1}) of the Laporté allowed transition having an oscillator strength f_U. The intensity of a $d \rightarrow d$ transition is therefore proportional to the amount of ψ_U character mixed into the ligand field excited state. The Laporté allowed states, ψ_U, are charge transfer transitions, and the mixing coefficient in Eq. (15) is related to the covalent mixing of ligand character into the metal d orbitals. Because charge transfer transitions often have ε values around $10,000 \, M^{-1} \, cm^{-1}$ and the mixing of specific charge transfer transitions into ligand field excited states is on the order of a few percent, $d \rightarrow d$ transitions may have ε values around $100 \, M^{-1} \, cm^{-1}$, which will decrease in intensity as the site becomes less distorted.[33]

Additionally, the spin forbiddenness of ligand field transitions may be overcome by spin–orbit coupling, which can be described as a first-order perturbation on the wave function using the same approach as above. This mixing of spin allowed transitions into a spin forbidden transition is also expected to be small; therefore, spin forbidden ligand field transitions should have several orders of magnitude lower absorption intensity than spin allowed ligand field transitions. For a protein with an absorption ε of $100 \, M^{-1} \, cm^{-1}$, a 1 mM solution in a 1-cm cell gives an absorbance of 0.1, and thus only spin allowed ligand field transitions are generally observed in metalloproteins.

The ground state of an octahedral high-spin ferrous site is $^5T_{2g}$ (Fig. 4), and the only spin allowed ligand field excited state is ^5E_g. Therefore, the relatively intense bands in the absorption spectrum of $[Fe(H_2O)_6]^{2+}$ (Fig. 7) at 9700 and 11,600 cm^{-1} are assigned as two components of the $^5T_{2g} \rightarrow {}^5E_g$ transition. The approximately 2000 cm^{-1} splitting of these bands results from the combined effects of an excited state Jahn–Teller

---

[33] Note that the ε for the bands at 9700 and 11,600 cm^{-1} (Fig. 7) are significantly lower owing to the fact that this complex has inversion symmetry and must overcome the Laporté forbiddenness of the ligand field transitions by vibronic coupling.

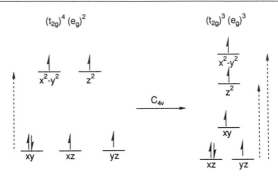

SCHEME I

distortion and a lowering of the octahedral site symmetry in the complex (*vide infra*). As shown in Fig. 4 the energy of the ^5E_g state is at $10Dq$ and independent of B, providing an experimental value for Dq of 10,650 cm^{-1}. The weak band at 14,700 cm^{-1} in Fig. 7 is assigned[17] as the lowest energy spin forbidden $d \rightarrow d$ transition, $^5T_{2g} \rightarrow {}^3T_{1g}$, the energy of which is dependent on B as given by the d^6 Tanabe–Sugano matrix.[14] The observed transition energy gives a B value of 720 cm^{-1}, corresponding to a β of 0.79, which is reduced from 1.0 owing to the covalency of the Fe^{2+}–OH_2 bond. The Dq/B value of $[Fe(H_2O)_6]^{2+}$ is 1.48, which determines the energies of the remaining ligand field excited states in Fig. 4.

The $^5T_{2g} \rightarrow {}^5E_g$ transition in Fig. 7 gives the $10Dq$ value for a ferrous center with six water ligands. The $[Fe(imidazole)_6]^{2+}$ complex is used to obtain a $10Dq$ estimate of approximately 11,400 cm^{-1} for six nitrogen ligands.[34] Using lower symmetry model complexes, experimental values for α_2 and α_4 (*vide supra*) can be estimated for oxygen and nitrogen ligands with normal ferrous–ligand bond lengths (O: $\alpha_2 = 17,300$ cm^{-1}, $\alpha_4 = 5760$ cm^{-1}; N: $\alpha_2 = 18,520$ cm^{-1}, $\alpha_4 = 6170$ cm^{-1}). Substitution of the experimental values into Eq. (9) using the matrix elements in Table IV allows determination of the d orbital splitting as a function of geometry. Because the $^5T_{2g} \rightarrow {}^5E_g$ transition corresponds to the $(t_{2g})^4(e_g)^2 \rightarrow (t_{2g})^3(e_g)^3$ one-electron promotion in Scheme I, the splitting of the d_{z^2} and $d_{x^2-y^2}$ orbitals reflects the splitting of the ^5E_g excited state and its energy dependence on active site distortion. This is summarized in Fig. 8, which shows that six-coordinate complexes (structure A) generally exhibit two spin allowed ligand field transitions in the 10,000 cm^{-1} region split by approximately 2000 cm^{-1}; five-coordinate complexes (B, C) show one band in the 10,000 cm^{-1} region and a second no more than about 5000 cm^{-1}. Four-

[34] J. Reedijk, *Recl. Trav. Chim. Pays-Bas* **88**, 1451 (1969).

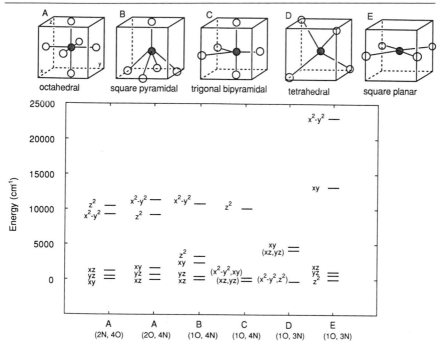

FIG. 8. Calculated d orbital energy level diagram for Fe(II) in a series of ligand fields. Iron is represented as a filled sphere and ligands as open spheres.

coordinate square planar complexes show a characteristic transition at approximately 20,000 cm^{-1} (E), and tetrahedral complexes (D) exhibit ligand field transitions in the region around 5000 cm^{-1} [a manifestation of the fact that $10Dq_{T_d} = -(4/9)10Dq_{O_h}$]. The $^5T_{2g}$ state also splits in energy owing to changes in ligand field geometry. This is discussed in Section VI.

As indicated above, the spin allowed ligand field bands are weak in the absorption spectrum of protein solutions and often occur in the near-infrared spectral region, which also has contributions from water vibrations. Therefore, $d \rightarrow d$ transitions are difficult to observe directly using electronic absorption spectroscopy. Circular dichroism (CD) (see [4] in this volume) and magnetic circular dichroism (MCD) (see [6] in this volume) spectroscopies involve different intensity mechanisms and are useful for observing transitions to ligand field excited states.[2,35] Intensity in the CD spectrum is given by the rotational strength, R [Eq. (16)]:

[35] P. J. Stephens, *Adv. Chem. Phys.* **35**, 197 (1976).

$$R_{exp} = (22.9 \times 10^{-40}) \left(\int \frac{\Delta\varepsilon}{\nu} d\nu \right) \tag{16a}$$

$$R_{theory} = (4.7 \times 10^{-24}) \text{Im}(\int \psi_G \vec{M}_{electric\ dipole} \psi_E\ d\tau \int \psi_G \vec{M}_{magnetic\ dipole} \psi_E\ d\tau) \tag{16b}$$

where $\vec{M}_{magnetic\ dipole} = \beta(\hat{L} + 2\hat{S})\vec{H}$ and $\vec{H}$ is the magnetic component of light. R is dependent on the projection of the electric dipole transition moment onto the magnetic dipole transition moment for a transition to a specific excited state. The projection will be nonzero only if the metal ion site is optically active, as is the case for metalloprotein centers. From Eq. (16b), the rotational strength is proportional to the amount of magnetic dipole character in a ligand field transition, which can be obtained from ligand field theory by evaluation of the orbital angular momentum operator, $\hat{L}$, between ψ_G and ψ_E. The electric dipole moment can be obtained from the absorption spectrum. However, the projection of the two transition moments is generally only accessible for the simplest structurally defined transition metal complexes.[36] CD spectroscopy of metalloproteins is therefore primarily used simply to observe ligand field transitions which may be unobservable in the absorption spectrum. It should be noted that CD intensities of vibrations are much smaller than those of electronic transitions,[37,38] and that ligand field transitions generally have some magnetic as well as electric dipole character[3]; therefore, the near-IR region of the CD spectrum is usually dominated by ligand field transitions.

The MCD intensity of paramagnetic transition metal complexes, which increases with decreasing temperature, involves a C-term mechanism. The simplest case is a spin $\frac{1}{2}$ system where the ground and excited states split linearly in energy with increasing magnetic field (Fig. 9). Two transitions of equal magnitude but opposite circular polarization are predicted for a transition between the sublevels of the ground and the specific excited state. The Zeeman splitting with experimentally accessible fields (6–7 Tesla) is on the order of 10 cm^{-1}, whereas absorption bands are generally a few thousand cm^{-1} broad. As a result the transitions will mostly cancel, giving a broad, weak, derivative shaped MCD signal as shown in Fig. 9B (top). This is a MCD A-term, which is observed at high temperature when both components of the ground state are equally thermally populated. As the temperature is decreased the $M_S = -\frac{1}{2}$ component of the ground state increases in relative population, and cancellation of the transitions no

[36] F. S. Richardson, *Chem. Rev.* **79**, 17 (1979).
[37] G. A. Osborne, J. C. Cheng, and P. J. Stephens, *Rev. Sci. Instrum.* **44**, 10 (1973).
[38] T. A. Keiderling, *Appl. Spectrosc. Rev.* **17**, 189 (1981).

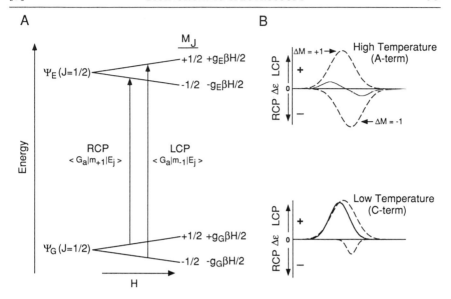

FIG. 9. Magnetic circular dichroism selection rules. (A) Transitions between Zeeman-split J (or S) = $\frac{1}{2}$ paramagnetic ground (G) and excited (E) states. Right circularly polarized (RCP) and left circularly polarized (LCP) transitions are indicated. (B) MCD spectra associated with the energy scheme in (A). The dashed bands correspond to individual left and right polarized transitions, whereas the solid line spectra give the dichroism. The C-term spectrum corresponds to a Gaussian band shape, which peaks at the absorption maximum, whereas an A-term reflects its first derivative which crosses zero at the absorption maximum.

longer occurs, allowing one component to increase in intensity (the left circularly polarized component in the example in Fig. 9B, bottom). This is a MCD C-term, and its magnitude is given quantitatively by the difference between left and right circularly polarized absorption intensity to a given excited state, which is Boltzmann averaged over the Zeeman split components of a degenerate ground state. Equation (17a) describes C-term MCD intensity, ΔA, in the limit of $kT \gg g\beta H_z$:

$$\frac{\Delta A}{E} = \frac{\gamma C_0 \beta H f(E)}{kT} \tag{17a}$$

where γ depends on constants such as the dielectric constant and refractive index, β is the Bohr magneton, k is the Boltzmann constant, H_z is the external magnetic field along the molecular z axis, and $f(E)$ is the energy-dependent band shape. The MCD C-term coefficient, C_0, is given by Eq. (17b):

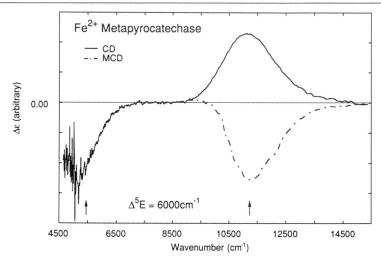

FIG. 10. Low temperature MCD and room temperature CD spectra of the nonheme Fe^{2+} active site in resting metapyrocatechase. The MCD spectrum was recorded at 4.2 K and 5.9 T. Note that MCD data were not recorded below 9500 cm^{-1}.

$$C_0 = \frac{-1}{d} \sum_{a,j} \langle Ga|L_z + 2S_z|Ga \rangle \times (|\langle Ga|m_{-1}|Ej \rangle|^2 - |\langle Ga|m_{+1}|Ej \rangle|^2) \quad (17b)$$

where a and j are sublevels of the ground state, G, and excited state, E, respectively, d is the degeneracy of the ground state, and $\langle Ga|m_{\pm 1}|Ej \rangle$ is the electric dipole transition moment for absorption of left ($+$) or right ($-$) circularly polarized light. Equation (17) gives the sign (left or right circular polarization) and magnitude (dependent on the ground state Zeeman splitting) of the MCD band, and the C-term intensity is proportional to $1/T$. The low temperature C-term MCD intensity will be two orders of magnitude higher than the high temperature limit for a paramagnetic ground state; therefore, spin allowed ligand field transitions can usually be observed in the low temperature MCD spectrum.

Using the nonheme ferrous enzyme metapyrocatechase as an example, although no ligand field transitions with ε below 10 M^{-1} cm^{-1} are observed in the absorption spectrum, the CD and low temperature MCD spectra (Fig. 10) clearly show two bands: one at 11,240 cm^{-1} and the second at 5220 cm^{-1}.[39] From Fig. 8 this allows one to estimate the effective geometry of this active site as a five-coordinate distorted square pyramidal structure.

[39] P. A. Mabrouk, A. M. Orville, J. D. Lipscomb, and E. I. Solomon, *J. Am. Chem. Soc.* **113**, 4053 (1991).

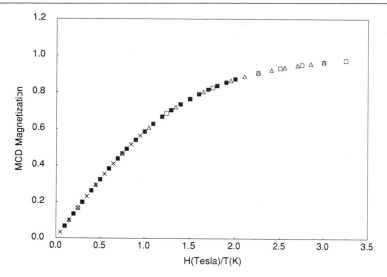

FIG. 11. Saturation-magnetization behavior of an isolated Kramers doublet ($S = \frac{1}{2}$) ground state. A series of fields are given from 0 to 7 T for four fixed temperatures: (□) 2, (△) 3.33, (■) 5, and (×) 10 K.

Excited State Probes of Ground State Spin Hamiltonian Parameters

As discussed in the previous section, the sign of the MCD band for a specific transition relates to the nature of the excited state. The variable temperature–variable field MCD (VTVH MCD) behavior of the transition can be further used to obtain the linear polarization of a transition and in particular provides an estimate of the ground state spin Hamiltonian parameters of the metal site, even in the absence of an EPR signal.[40–42] Equation (17a), which gives the linear dependence of MCD amplitude with H/T, requires that the thermal energy be much greater than the Zeeman splitting of the ground state sublevels in Fig. 9 ($kT \gg g\beta H$). This behavior is no longer observed as the low temperature, high field limit is approached. Eventually only the lowest sublevel of the ground state is populated and the MCD signal becomes independent of temperature (and magnetic field) and is said to be saturated. Variable temperature MCD can be used to obtain a saturation magnetization curve, as shown, for example, in Fig. 11, where saturation occurs at large values of H/T. The saturation magneti-

[40] J. W. Whittaker and E. I. Solomon, *J. Am. Chem. Soc.* **110,** 5329 (1988).
[41] P. N. Schatz, R. L. Mowery, and E. R. Krausz, *Mol. Phys.* **35,** 1537 (1978).
[42] A. J. Thomson and M. K. Johnson, *Biochem. J.* **191,** 411 (1980).

zation MCD intensity is Boltzmann averaged over the Zeeman split ground state sublevels as in Eq. (18):

$$\Delta A \propto \frac{\sum_{a,j} ([\langle Ga|m_{-1}|Ej\rangle]^2 - [\langle Ga|m_{+1}|Ej\rangle]^2) \exp\left[(-\langle Ga|L + 2S|Ga\rangle)\frac{\beta H}{kT}\right]}{\sum_{a} \exp\left[(-\langle Ga|L + 2S|Ga\rangle)\frac{\beta H}{kT}\right]}$$

(18)

The specific case of a Kramers doublet (i.e., $S = \frac{1}{2}$) ground state with orientational averaging over all space gives Eq. (19)[43,44]:

$$\Delta A \propto M_{xy}^2 \left[\int_0^{\pi/2} \frac{\cos^2 \theta \sin \theta}{\Gamma} g_{\parallel} \tanh\left(\frac{\Gamma \beta H}{2kT}\right) d\theta \right.$$
$$\left. - 2^{1/2} \frac{M_z}{M_{xy}} \int_0^{\pi/2} \frac{\sin^3 \theta}{\Gamma} g_{\perp} \tanh\left(\frac{\Gamma \beta H}{2kT}\right) d\theta \right]$$

(19)

where $\Gamma = (\sin^2 \theta g_{\perp} + \cos^2 \theta g_{\parallel})^{1/2}$, M_{xy} and M_z are the transition moment integrals in the directions indicated, and θ is the orientation of the molecular z axis relative to the applied magnetic field. The data in Fig. 11 are for an $S = \frac{1}{2}$ ion and can be fit to Eq. (19), which allows determination of the ground state g values and the linear polarization of the transition to a given excited state, providing further insight into the assignment of that excited state.

The saturation magnetization behavior of an $S = \frac{1}{2}$ ion is dependent only on H/T [Eq. (19)]. If the field is varied at a series of fixed temperatures and the MCD amplitude is plotted as a function of H/T, all points fall on the same curve, as illustrated in Fig. 11. This is not the case for a non-Kramers (integer spin) ion such as the ferrous center in metapyrocatechase ($S = 2$).[39] The saturation magnetization curves for this site do not superimpose (Fig. 12A). Insight into the origin of the nesting behavior in Fig. 12A is given by Fig. 12B, which displays the isofield data as a function of temperature. The fact that the MCD signal intensity increases with decreasing temperature indicates that the paramagnetic $M_S = \pm 2$ components of the zero field split $S = 2$ ground state are lowest in energy (Fig. 13, center) with $D < 0$ (see Section VI for a description of zero field splitting and the origin of D). The data in Fig. 12B show that the low temperature saturated MCD amplitude increases nonlinearly with increasing magnetic field and eventually becomes constant at high magnetic field.

[43] D. E. Bennet and M. K. Johnson, *Biochim. Biophys. Acta* **911**, 71 (1987).
[44] Y. Zhang, M. S. Gebhard, and E. I. Solomon, *J. Am. Chem. Soc.* **113**, 5162 (1991).

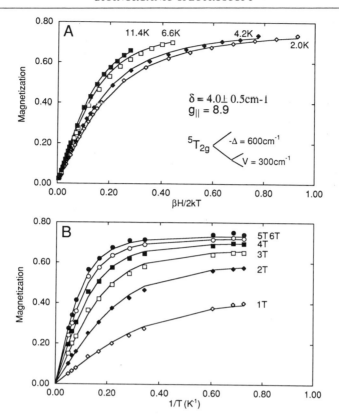

FIG. 12. Magnetization behavior for resting metapyrocatechase. (A) MCD amplitude at 890 nm for a range of applied magnetic field strengths (0–5.9 T) at a series of fixed temperatures plotted as a function of $\beta H/2kT$. (B) Replot of saturation data shown in (A) as a function of temperature at fixed fields. Solid curves represent the fit to the data for an isolated non-Kramers doublet with rhombic zero field splitting [Eq. (20)].

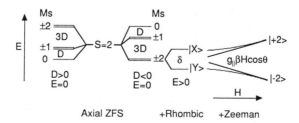

FIG. 13. Energy level diagram for a $S = 2$ state. Left to right: Effects of positive and negative axial zero field splitting, rhombic splitting of the non-Kramers ± 2 doublet, and Zeeman effect on this doublet.

This indicates that the wave function of the lowest component of the ground state changes with the applied magnetic field. This behavior is expected for a non-Kramers doublet where the $\pm$ components are split in zero magnetic field owing to rhombic distortion ($|E| > 0$) of the site. As shown in Fig. 13 (right), the rhombic wave functions are real at zero magnetic field and become pure complex wave functions at high magnetic field which are MCD active.[40] The model in Fig. 13 can be used in Eq. (18) to obtain an expression for MCD amplitude as a function of magnetic field and temperature for a non-Kramers doublet, as given in Eq. (20) for an x, y polarized transition (an additional term in $g_\perp$ is present for a transition with z polarization):

$$\Delta A \propto M_{xy}^2 \int_0^{\pi/2} \frac{\cos^2 \theta \sin \theta}{\Lambda} g_\| \beta H \tanh \left(\frac{\Lambda}{2kT} \right) d\theta \tag{20}$$

where $\Lambda = [\delta^2 + (g_\| \beta H \cos \theta)^2]^{1/2}$ and δ is the zero field splitting of the ground state doublet.

Fitting this non-Kramers doublet model to the data in Fig. 12 provides an experimental estimate of δ and $g_\|$ for the ground state doublet in Fig. 13 of 4 cm^{-1} and 8.9, respectively. The ground state parameters can now be interpreted in terms of the geometric and electronic structure of the metal site using ligand field theory as described in the next section.

Ligand Field Origin of Spin Hamiltonian Parameters

As described in the last section, VTVH MCD can provide estimates of ground state spin Hamiltonian parameters, in particular the components of the g tensor and the zero field splitting. In this section we consider the origin of these parameters from ligand field theory and show that these ground state parameters derive from spin–orbit interactions with ligand field excited states. Only when these excited states have been defined by electronic absorption spectroscopies can the information content of the ground state spin Hamiltonian parameters be maximized.

The ground state Zeeman splitting of an $S = \frac{1}{2}$ ion (Fig. 9) is given in Eq. (21):

$$E_{\text{Zeeman}} = \beta \langle G \pm \tfrac{1}{2} | \hat{L} + 2\hat{S} | G \pm \tfrac{1}{2} \rangle \vec{H} \tag{21}$$

For a pure $S = \frac{1}{2}$ ground state, the orbital angular momentum operator contribution to E_{Zeeman} is zero, and the g value equals 2.00 and is isotropic (i.e., the same for the magnetic field oriented along the x, y, and z directions of the complex). This would be the case for an unpaired electron in a nondegenerate d orbital in a low symmetry distorted site without the inclusion of spin–orbit coupling. Spin–orbit coupling may be fairly large in transition metal ions, however, and this mixes ligand field excited states into the nondegenerate ground state, which leads to nonzero values of the orbital angular momentum operator in Eq. (21) and to anisotropic g values that differ from 2.00. Spin–orbit mixing of ligand field excited states into the ground state combined with the Zeeman effect gives the first-order perturbation expression for g values [Eq. (22)]:

$$g_i = 2 \left(1 - k^2 \lambda \sum_{E \neq 0} \frac{\langle \psi_G | L_i | \psi_E \rangle \langle \psi_E | L_i | \psi_G \rangle}{E_E - E_G} \right) \qquad (22)$$

where $i = x$, y, and z, ψ_G is the ground state, and ψ_E are the ligand field excited states. The spin–orbit coupling constant, λ, equals $\pm \zeta_{3d} / 2S$ where ζ_{3d} is the one-electron radial integral for spin–orbit coupling given in Table VII, S is the total spin of the ground state, and $+ \; (-)$ is appropriate for a less than (greater than) half-filled subshell. k^2 is the Stevens orbital reduction factor (vide infra).

The values for the orbital angular momentum operator evaluated over the specific d orbitals needed in Eq. (22) are given in Table VIII. The energy denominator in Eq. (22) is equal to the transition energy to a specific ligand field excited state, which can be obtained experimentally from electronic absorption spectroscopy. When experimental ligand field transition energies are available, ligand field calculated g values are generally in qualitative agreement with experiment but deviate too much from 2.00.[45] This is because covalency must still be included, which mixes the d orbitals with the ligands and thus reduces the orbital angular momentum. The Stevens orbital reduction factor, k^2, in Eq. (22) accounts for covalency and decreases from 1.0 with increasing delocalization of the d electron onto the ligands.[46] This factor can be related to specific covalent interactions between d orbitals and the ligands[47] and provides an experimental estimate for c and c' in Eq. (11).

Ground states with $S > \frac{1}{2}$ have an additional term in the spin Hamiltonian which allows for the $2S + 1$ spin degeneracy in M_S to split in energy even in the absence of a magnetic field. This is called zero field splitting (ZFS), which is given by Eq. (23):

$$\hat{H} = D(\hat{S}_z^2 - \tfrac{1}{3}S(S + 1)) + E(\hat{S}_x^2 - \hat{S}_y^2) \qquad (23)$$

[45] E. I. Solomon, A. A. Gewirth, and S. L. Cohen, in "Understanding Molecular Properties" (A. E. Hansen, J. Avery, and J. P. Dahl, eds.), p. 27. Reidel, Dordrecht, The Netherlands, 1987.

[46] K. W. H. Stevens, Proc. R. Soc. London A **226**, 96 (1954).

[47] M. Gerloch and J. R. Miller, Prog. Inorg. Chem. **10**, 1 (1968).

TABLE VII
FREE ION SINGLE-ELECTRON SPIN–ORBIT COUPLING PARAMETERS (ζ_{nd}) FOR TRANSITION METAL ELEMENTS[a]

Group	Metal	Oxidation number						
		0	1+	2+	3+	4+	5+	6+
4	Ti	70	90	123	155			
	Zr		(300)	(400)	(500)			
	Hf							
5	V	95	135	170	210	250		
	Nb		(420)	(610)	(800)			
	Ta				(1400)			
6	Cr	135	185	230	275	355	380	
	Mo			(670)	800	(850)	(900)	
	W			(1500)	(1800)	(2300)	(2700)	
7	Mn	190	255	300	355	415	475	540
	Tc			(950)	(1200)	(1300)	(1500)	(1700)
	Re			(2100)	(2500)	(3300)	(3700)	(4200)
8	Fe	275	335	400	460	520	590	665
	Ru				(1250)	(1400)	(1500)	(1700)
	Os				(3000)	(4000)	(4500)	(5000)
9	Co	390	455	515	580	650	715	790
	Rh					(1700)	(1850)	(2100)
	Ir					(5000)	(5500)	(6000)
10	Ni		565	630	705	790	865	950
	Pd		(1300)	(1600)				
	Pt		(3400)					
11	Cu			830	890	960	1030	1130
	Ag			(1800)				
	Au			(5000)				

[a] Data in cm^{-1}. Values in parentheses are only estimates. Adapted from Figgis.[11]

TABLE VIII
EFFECT OF L_i ON REAL d ORBITALS[a]

$\hat{L}_x$	$\hat{L}_y$	$\hat{L}_z$
$\hat{L}_x d_{xz} = -id_{xy}$	$\hat{L}_y d_{xz} = -id_{x^2-y^2} - i(3)^{1/2}d_{z^2}$	$\hat{L}_z d_{xz} = id_{yz}$
$\hat{L}_x d_{yz} = i(3)^{1/2}d_{z^2} + id_{x^2-y^2}$	$\hat{L}_y d_{yz} = -id_{xy}$	$\hat{L}_z d_{yz} = id_{xz}$
$\hat{L}_x d_{xy} = id_{xz}$	$\hat{L}_y d_{xy} = -id_{yz}$	$\hat{L}_z d_{xy} = -2id_{x^2-y^2}$
$\hat{L}_x d_{x^2-y^2} = -id_{yz}$	$\hat{L}_y d_{x^2-y^2} = -id_{xz}$	$\hat{L}_z d_{x^2-y^2} = 2id_{xy}$
$\hat{L}_x d_{z^2} = -i(3)^{1/2}d_{yz}$	$\hat{L}_y d_{z^2} = i(3)^{1/2}d_{xz}$	$\hat{L}_z d_{z^2} = 0$

[a] Adapted from Ballhausen.[22]

where D is the axial ($z \neq x, y$) and E the rhombic ($x \neq y$) ZFS contribution. Figure 14 shows as an example the ground state splitting of a high-spin d^5 $S = \frac{5}{2}$ ion in an axially distorted site [$D \neq 0$, $E = 0$ in Eq. (23)]. For an orbitally nondegenerate ground state, pure spin wave functions cannot split in energy owing to a distortion (from octahedral or tetrahedral) of the metal site. In ligand field theory this splitting occurs through second order spin–orbit coupling of the ground state with specific ligand field excited states. These excited states are orbitally degenerate in the octahedral (or tetrahedral) limit but split in energy owing to axial distortion of the metal site. In the case of high-spin d^5 complexes this involves the lowest energy ligand field excited state, $^4T_{1g}$, which splits into two components at energies E_z and $E_{x,y}$ with axial distortion. Griffith[30] has derived an expression for the zero field splitting of the $^6A_{1g}$ ground state shown in Fig. 14. This expression [Eq. (24), where $\zeta_{Fe^{3+}} = 430$ cm^{-1}] involves spin–orbit coupling ($H_{S.O.}$) of the ground state with the axially split $^4T_{1g}$ excited state:

$$D = \frac{1}{180}\left[\frac{\langle ^4T_1(z)|H_{S.O.}|^6A_1\rangle^2}{E_z} - \frac{\langle ^4T_1(x,y)|H_{S.O.}|^6A_1\rangle^2}{E_{x,y}}\right] \quad (24a)$$

$$D = \frac{(\zeta_{Fe^{3+}})^2}{5}\left[\frac{1}{E_z} - \frac{1}{E_{x,y}}\right] \quad (24b)$$

Substitution of experimental ligand field excited state energies into the energy denominators in Eq. (24) results in poor agreement with experimen-

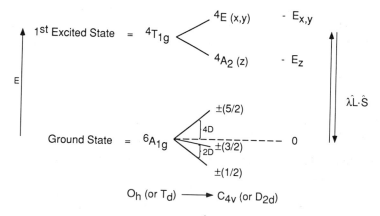

FIG. 14. Ligand field theory for a high-spin d^5 ($S = \frac{5}{2}$) ground and first excited state in an axially distorted site.

tal D values and in some cases predicts the wrong sign of D.[48] Covalent delocalization of the d orbitals must be included in Eq. (24) for a quantitative analysis of D (and E), resulting in Eq. (25):

$$D = \frac{(\zeta_{Fe^{3+}})^2}{5} \left[\frac{k_z^2}{E_z} - \frac{k_{x,y}^2}{E_{x,y}} \right] \tag{25}$$

If D is less than zero, $k_{x,y}^2$ must be greater than k_z^2, where the k_i^2 terms allow for anisotropic covalent reduction of the spin–orbit coupling. The k_i values directly relate to differences in covalent delocalization of the different d orbitals due to formation of metal–ligand bonds.[48]

The $^5T_{2g}$ ground state of the high-spin ferrous example considered throughout this chapter is complicated by orbital degeneracy, which results in in-state orbital angular momentum. This can directly couple to the spin angular momentum and produce in-state zero field splitting. The T_{2g} state has an effective angular momentum $L' = -1$ which couples to the $S = 2$ causing the $^5T_{2g}$ state to split, even in the octahedral limit (Fig. 15, bottom left). A weak axial ligand field distortion splits the $^5T_{2g}$ by an amount $-\Delta$, resulting in the ^5E_g component of the $^5T_{2g}$ state lowest in energy. This again has in-state spin–orbit splitting (Fig. 15, bottom center). Finally, a rhombic ligand field distortion of the ferrous site splits the ^5E_g component of the $^5T_{2g}$ by an amount v. Spin–orbit coupling over the rhombically split $^5T_{2g}$ state leads to an energy diagram (Fig. 15, bottom) which is different from that given by the spin Hamiltonian in Eq. (23) for an $S = 2$ ion (Fig. 13, $D < 0$, $|E| > 0$). Ligand field theory can therefore be used to calculate the $g_\parallel$ and δ of the lowest non-Kramers doublet in Fig. 15 as a function of $-\Delta$ and v, which are given in Fig. 16.[40] In summary, the VTVH MCD data in Section V (Fig. 12 for metapyrocatechase) give $g_\parallel$ and δ which are interpreted in terms of ligand field theory (Fig. 16) to give $-\Delta$ and v, the axial and rhombic splittings of the t_{2g} set of d orbitals in the ferrous active site.[39]

Structure/Function Insight from Electronic Absorption Spectroscopy and Ligand Field Theory: An Example

We complete this review with a specific example[39] of the use of spectroscopy to obtain insight into the mechanism of an active site on a molecular level. Metapyrocatechase is representative of a class of nonheme iron enzymes which in the ferrous oxidation state react with dioxygen to cata-

[48] M. S. Gebhard, J. C. Deaton, S. A. Koch, M. Millar, and E. I. Solomon, *J. Am. Chem. Soc.* **112,** 2217 (1990).

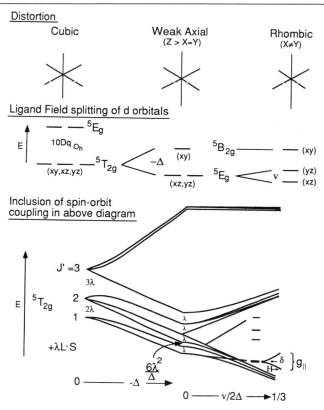

FIG. 15. Ligand field theory for the $^5T_{2g}$ ground state. Under a weak axial distortion an orbital doublet ground state is stabilized $(-\Delta)$ and undergoes rhombic splitting (v) (middle). The effective $L' = 1$ orbital angular momentum of the $^5T_{2g}$ leads to a multiplet splitting within the $^5T_{2g}$ state with an inverted Hund's rule ordering (bottom, left). The axial perturbation leaves orbital angular momentum in the ^5E component, leading to a ladder of doublets (bottom, middle). A rhombic distortion leaves a non-Kramers doublet lowest (bottom, right). The inset expands the bottom part of the diagram and shows the splitting of the lowest pair of sublevels in a magnetic field. The splitting pattern converges with the predictions of the $S = 2$ spin Hamiltonian in the extreme rhombic limit $v/2\Delta = \frac{1}{3}$.

lyze the extradiol dioxygenation of catechol.[49,50] The splitting of the ^5E_g excited state obtained from CD and MCD spectroscopies (Fig. 10) puts the $d_{x^2-y^2}$ level at 11,240 cm^{-1} and the d_{z^2} at 5220 cm^{-1}. The VTVH MCD data for the 11,240 cm^{-1} excited state (Fig. 12) allows determination of

[49] M. Nozaki, *Top. Curr. Chem.* **78,** 145 (1979).

[50] J. D. Lipscomb, J. W. Whittaker, and D. M. Arciero, *in* "Oxygenases and Oxygen Metabolism—4: Symposium in Honor of Osamu Hayaishi" (M. Nozaki, S. Yamamoto, M. J. Coon, L. Ernster, and R. W. Estabrook, eds.), p. 27. Academic Press, New York, 1982.

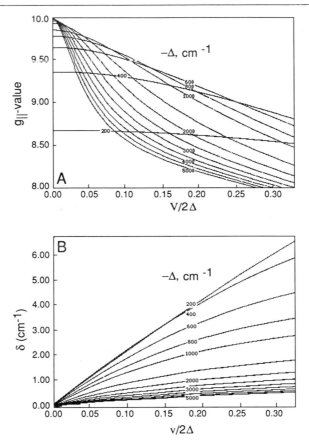

FIG. 16. Ground-state parameters $g_{\parallel}$ (A) and rhombic splitting δ (B) for a $^5T_{2g}$ ground state as a function of the axial (Δ) and rhombic (v) splitting of the t_{2g} set of d orbitals, calculated from a ligand field Hamiltonian with $\lambda = -80$ cm^{-1}.

δ and $g_{\parallel}$ for the ground state non-Kramers doublet described in Section VI. This provides an estimate of the ligand field splitting of the $^5T_{2g}$ state in the ferrous site, allowing calculation of the d_{xy} and d_{xz} orbital energies at 750 and 300 cm^{-1}, respectively, above the lowest energy d_{yz} orbital. Thus electronic spectroscopy defines the experimental ligand field splitting of the d orbitals of the active site, which provides geometric and electronic structural insight. In particular, this can be used to study small molecule and substrate interactions with the ferrous center which relate to the reaction mechanism.

Addition of azide to the resting enzyme results in no change in either the excited or ground state spectral features (Fig. 17A), indicating that this inhibitor does not bind to the active site of the resting enzyme. Alternatively, addition of substrate (Fig. 17B) causes a large increase in energy of the transition to the $d_{x^2-y^2}$ level and causes the ground state to become more easily saturated, indicating that δ has decreased, thus implying that the $^5T_{2g}$ splitting has increased (Fig. 16B). This strongly supports results

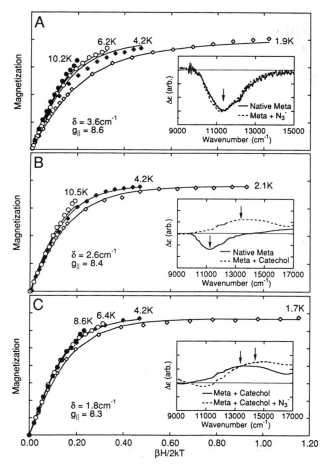

FIG. 17. Saturation–magnetization behavior for (A) native metapyrocatechase at 890 nm, (B) anaerobic metapyrocatechase–catechol complex at 750 nm, and (C) anaerobic metapyrocatechase–catechol–azide complex at 680 nm. The MCD intensity amplitude for a range of magnetic fields (0–5.9 T) at a series of fixed temperatures is plotted as a function of $\beta H/2kT$. The insets show MCD spectra at 4.2 K and 5.9 T of resting metapyrocatechase, its substrate complex, and the enzyme–substrate–azide complex.

from EPR studies on the ground state of the NO complex of the enzyme, which show that substrate binds directly to the iron active site.[51,52] The data in Fig. 17 provide the first evidence for substrate binding to the EPR inactive catalytically relevant ferrous site. (No change was observed in Mössbauer studies of the ferrous enzyme.[53]) Anaerobic addition of azide to the substrate-bound ferrous active site results in a further shift of the $d_{x^2-y^2}$ ligand field transition and an additional change in the ground state splitting. Thus, substrate binding activates the ferrous site for small molecule binding, directly reflecting the catalytic bi–uni mechanism, which requires substrate binding for O_2 reactivity.[54]

The excited and ground state data on metapyrocatechase described above provide an experimental ligand field energy level diagram for the ferrous active site (Fig. 18). Correlating the d orbital energy splitting of the resting enzyme with the ligand field splittings associated with different ferrous geometries (Fig. 8) indicates that the effective symmetry of the resting active site is five-coordinate distorted square pyramidal. Substrate binding causes the $d_{x^2-y^2}$ and d_{xy} orbitals to go up in energy. Because catechol is a weak ligand for Fe(II), this indicates that the catechol binds axially, causing the ferrous center to shift into the equatorial plane. The d_{xz}/d_{yz} orbital splitting increases, indicating that catechol binds in a bidentate fashion with the second oxygen in the equatorial plane. Substrate binding activates the ferrous site for small molecule binding, and the energy diagram in Fig. 18 shows that there are both geometric and electronic contributions to this activation. As discussed above, substrate binding results in a shift of the ferrous ion into the equatorial plane and a large increase in rhombic splitting of the t_{2g} set of d orbitals. This localizes the extra electron of the d^6 configuration in the d_{yz} orbital, producing anisotropic electron density in the equatorial plane. Localization weakens the interaction of the ferrous center with electron donating ligands along the y direction and activates this equatorial position for reaction with electron accepting ligands, in particular the dioxygen involved in catalysis.

Concluding Comments

A few points should be emphasized from this review: (1) The information content of electronic absorption spectroscopy on metalloproteins is extremely high and is accessible through ligand field theory. (2) Ligand

[51] D. M. Arciero, A. M. Orville, and J. D. Lipscomb, *J. Biol. Chem.* **260,** 14035 (1985).
[52] D. M. Arciero and J. D. Lipscomb, *J. Biol. Chem.* **261,** 2170 (1986).
[53] Y. Tatsuno, Y. Saeki, M. Nozaki, S. Otsuka, and Y. Maeda, *FEBS Lett.* **112,** 83 (1983).
[54] K. Hori, T. Hashimoto, and M. Nozaki, *J. Biochem. (Tokyo)* **74,** 375 (1973).

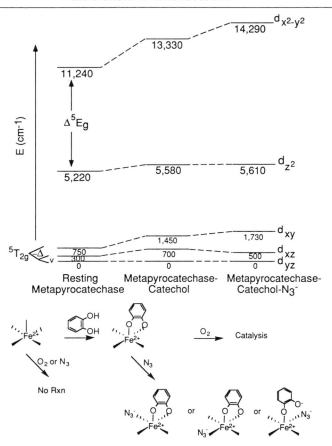

FIG. 18. Experimental d orbital energy level diagram for resting metapyrocatechase, its substrate complex, and the enzyme–substrate–azide ternary system (top). The spectroscopically effective structural mechanism derived from this energy diagram for the Fe(II) active site in metapyrocatechase is also shown (bottom).

field transitions are generally weak, and for paramagnetic metalloenzymes they are most readily observed using low temperature MCD spectroscopy. (3) Excited states can be used to obtain ground state information on a metal ion active site through VTVH MCD spectroscopy. (4) Ligand field theory defines the information content of the ground state spin Hamiltonian parameters. This includes both excited state energies and ligand–metal bond covalencies. These contributions to g values and ZFS parameters are interdependent but can be uncoupled if the excited state energies are obtained directly by electronic spectroscopy. (5) Bioinorganic spectroscopy provides insight into electronic as well as geometric structures of metal ion active sites, which make significant contributions to biological reactivity.

Acknowledgments

This work was supported by the National Science Foundation Biophysics Program (MCB 9019752) and the National Institute of Health (GM 40392). E.I.S. expresses sincere appreciation to all the students and collaborators listed as coauthors in the literature cited for their commitment and contributions to this research. Special thanks to Professor John Lipscomb for the collaborative studies on metapyrocatechase.

[6] Magnetic Circular Dichroism

By John C. Sutherland

Introduction

Magnetic circular dichroism (MCD) is the differential absorption of left and right circularly polarized light induced by an externally applied magnetic field. MCD is one of the several forms of "higher order" spectroscopy that can augment the information provided by the measurement of the absorption of a sample using unpolarized electromagnetic radiation. The additional information provided by MCD can aid in resolving complex absorption spectra. Some biologically important molecules exhibit particularly strong MCD signals and can thus be detected and quantified even when present in complex mixtures. The most sophisticated applications of MCD involve the extraction of properties of the system under study by comparison of experimental spectral parameters with theoretical models. The theory and some of the biological applications of MCD have been reviewed elsewhere.[1-9]

[1] P. J. Stephens, *Annu. Rev. Phys. Chem.* **25,** 201 (1974).
[2] P. J. Stephens, *Adv. Chem. Phys.* **35,** 197 (1976).
[3] B. Holmquist and B. L. Vallee, this series, Vol. 49 Part G, p. 149.
[4] J. C. Sutherland, *in* "The Porphyrins" (D. Dolphin, ed.), p. 225. Academic Press, New York, 1978.
[5] A. J. Thomson, M. R. Cheesman, and S. J. George, this series, Vol. 226, p. 199. Academic Press, New York, 1993.
[6] J. C. Sutherland and B. Holmquist, *Annu. Rev. Biophys. Bioeng.* **9,** 293 (1980).
[7] J. H. Dawson and D. M. Dooley, *in* "Iron Porphyrins" (A. B. P. Lever and H. B. Gray, eds.), p. 1. VCH, New York, 1986.

MCD is closely related to some types of spectroscopy but must be clearly differentiated from others. The relationship between MCD and "natural" circular dichroism[10] (CD) is particularly important. CD is the differential absorption of left and right circularly polarized light resulting from the asymmetry of the absorbing moiety or its immediate environment. CD usually tells about the physical conformation of the absorbing species, whereas MCD usually tells about its electronic state. Although the information provided is quite different, the instrumentation required to perform MCD measurements is, almost always, a superset of that required for CD, hence distributing the cost of a spectrometer over a broader range of applications. In contrast, magnetic optical rotatory dispersion (MORD), the difference between the refractive indices of right and left circularly polarized light induced by an external magnetic field, provides information equivalent to that obtained from MCD, but requires different instrumentation. MCD and MORD are linked by Kramers–Kronig integral transforms,[11] as are CD and ORD. The absorption based spectra are usually easier to interpret, so MCD and CD are the methods of choice.

MCD measurements are performed on samples containing randomly oriented absorbing species and with the magnetic field oriented parallel (or antiparallel) to the direction of the probing beam of radiation, so that the field does not introduce a preferred direction compared to the polarization of the beam. A magnetic field can generate sufficient torque to partially orient large macromolecules and macromolecular arrays.[12,13] Observations made on samples that are not randomly oriented or that can be aligned by the magnetic field can generate magnetic field-dependent signals that do not fit the definition of MCD as discussed here, but which may provide other forms of useful information. Magnetic fields oriented perpendicular to the direction of the probing beam can produce magnetic linear dichroism that, as a result of the methods usually employed in CD and MCD measurements, can produce apparent MCD signals, the origin and interpretation of which are different from what we call MCD.

[8] M. J. Stillman, in "Phthalocyanines, Properties and Applications" (C. C. Leznoff and A. P. B. Lever, eds.), p. 227. VCH, New York, 1993.

[9] Y. A. Sharonov, in "Soviet Scientific Reviews Section D: Physicochemical Biology Reviews" (V. P. Skulachev, ed.), p. 1. Harwood Academic Publ., London, 1991.

[10] R. W. Woody, this volume [4].

[11] C. R. Cantor and P. R. Schimmel, "Biophysical Chemistry." Freeman, New York, 1980.

[12] G. Maret and K. Dransfeld, *Physica* **86–88B,** 1077 (1977).

[13] G. Garab, A. Faludi-Daniel, J. C. Sutherland, and G. Hind, *Biochem.* **27,** 2425 (1988).

Signals directly analogous to MCD have been observed in both fluo-rescence[14,15] and Raman scattering,[16,17] but the use of these measurements for the study of biological materials has been limited. MCD is usually detected by observing light that has been transmitted through a sample. However, much the same information can be obtained from monitoring fluorescence emitted by a sample, an experiment referred to as fluores-cence-detected MCD[18] (FDMCD), which can detect the MCD of both fluorescent and nonfluorescent molecules. Although it has not been used frequently in the ultraviolet (UV), visible, or infrared (IR), FDMCD may be the detection method of choice for the measurement of MCD in the X-ray region.[19] Special techniques have been developed to record transient MCD spectra with a temporal resolution of nanoseconds.[20]

Basic Principles

Measurement of Magnetic Circular Dichroism

Integral versus Differential MCD. The definition of MCD presented above specifies that an external magnetic field causes the absorption of left circularly polarized light to differ from that of right circularly polarized light. In principle, one could measure the absorption spectrum of a sample using first one and then the other circular polarization. MCD is that portion of the difference between the two spectra that is induced by application of an external magnetic field. Such measurements are performed only very rarely in the visible, UV, and IR regions of the spectrum because the field-induced effects are usually too small to be detected in this manner. Most MCD is measured as the difference between the absorption of left and right circularly polarized light using the polarization modulation method described below. The relationship between the CD, MCD, the intensities of left and right circularly polarized light, and other observable parameters are given by the Beer–Lambert law. Suppose that a beam of radiation of wavelength λ and intensity $I_0(\lambda)$ is incident on a sample of thickness l,

[14] R. A. Shatwell and A. J. McCaffery, J. Chem. Soc., Chem. Commun., 546 (1973).
[15] R. A. Shatwell, R. Gale, R. A. McCaffery, and K. Sichel, J. Am. Chem. Soc. 97, 7015 (1975).
[16] L. D. Barron, Nature (London) 257, 372 (1975).
[17] L. D. Barron, Chem. Phys. Lett. 46, 579 (1977).
[18] J. C. Sutherland and H. Low, Proc. Natl. Acad. Sci. U.S.A. 73, 276 (1976).
[19] J. van Elp, S. J. George, J. Chen, G. Peng, C. T. Chen, L. H. Tjeng, G. Meigs, H.-J. Lin, Z. H. Zhou, M. W. W. Adams, B. G. Searle, and S. P. Cramer, Proc. Natl. Acad. Sci. U.S.A. 90, 9664 (1993).
[20] R. A. Goldbeck and D. S. Kliger, this series, Vol. 226, p. 147.

and that the intensity of the beam which emerges from the far side of the sample is of intensity $I(\lambda)$, as shown schematically in Fig. 1. The absorbance of the sample at that wavelength, $A(\lambda)$, is defined in terms of $I_0(\lambda)$ and $I(\lambda)$, and it is related to the molar concentration, c, and the molar extinction coefficient, $\varepsilon(\lambda)$, of the absorbing species at wavelength λ as shown by Eq. (1).[11] We could perform the same measurement using first

$$A(\lambda) \equiv \log I_0(\lambda)/I(\lambda) = c\varepsilon(\lambda)l \tag{1}$$

left and then right circularly polarized light, hence obtaining $A_L(\lambda)$ and $A_R(\lambda)$, respectively. The net circular dichroism, $\Delta A(\lambda)$, is defined by the equation $\Delta A = A_L - A_R$ and can be written as the sum of the contributions of the CD and MCD according to Eq. (2). Equation (2) presumes (in

$$\Delta A(\lambda, H) = \Delta A_{CD}(\lambda) + H\Delta A_{MCD}(\lambda) \tag{2}$$

agreement with both theory and experiments) that the contributions of CD and MCD to the total observed differential absorption are independent, and that the observed effect is a linear function of the magnetic field strength, H. In the SI system of units, magnetic field strengths are specified in teslas (1 T = 10^4 Gauss). As A and ΔA_{CD} are dimensionless, it follows that ΔA_{MCD} is expressed in units of T^{-1}. Dividing each of the differential absorbances by the product of concentration and path length gives the corresponding differential molar extinction coefficient: $\Delta\varepsilon(\lambda)$, $\Delta\varepsilon_{CD}(\lambda)$, and $\Delta\varepsilon_{MCD}(\lambda)$. The units of $\Delta\varepsilon$ and $\Delta\varepsilon_{CD}$ are M^{-1} cm^{-1}, whereas the units of $\Delta\varepsilon_{MCD}$ are M^{-1} cm^{-1} T^{-1}. In practice, $\Delta\varepsilon_{MCD}(\lambda)$, ΔA_{MCD}, $\Delta\varepsilon$, and ΔA are all used to report MCD spectra. If in doubt, check the units! CD is also reported in units of ellipticity ($\theta = 33 \Delta A$) and molar ellipticity ($[\theta] = 3300 \Delta\varepsilon$). Although θ and $[\theta]$ are used less frequently in MCD studies, they are still used extensively in the biochemical CD literature.

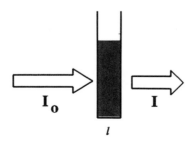

FIG. 1. Transmission of an optical beam by a sample of thickness l. The incident beam is monochromatic and characterized by wavelength λ and intensity I_0, whereas the intensity of the transmitted beam is I.

Experimental spectra reported in ellipticity units can be separated into the contributions of CD and MCD by an equation equivalent to Eq. (2).

Instruments for Measuring Magnetic Circular Dichroism

MCD Spectrometers for Ultraviolet, Visible, and Near-Infrared Spectral Regions. Most MCD spectroscopy is performed in the spectral region with wavelengths from roughly 120 nm in the vacuum ultraviolet (VUV) to 1200 nm (1.2 μm) in the near-infrared (IR) where light can be detected by a photomultiplier. Figure 2 shows a schematic diagram of such an instrument, which is basically a CD spectrometer fitted with a magnet to generate a field at the position of the sample. A key component of such an instrument is the photoelastic modulator (PEM) that converts the polarization of a linearly polarized monochromatic light beam into first left and then right circularly polarized components while keeping the total intensity

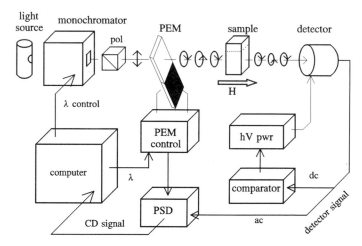

FIG. 2. Schematic diagram of a CD/MCD spectrometer using a photomultiplier as the detector. A separate polarizer (pol) is needed if, as is usually the case with conventional broad spectrum sources, the light emerging from the monochromator is not linearly polarized. The photoelastic modulator (PEM) must be programmed to provide the correct phase retardation to render the light alternately left and right circularly polarized at each wavelength. The static magnetic field, H, is usually supplied by either an electromagnet or a superconducting magnet. Care is required in modifying a CD spectrometer to measure MCD. In addition to physically fitting a magnet in the sample compartment, it is necessary to prevent the magnetic field from interfering with the operation of the instrument. For example, most photomultipliers will not operate in even a moderate magnetic field. Collection of spectra in digital form, usually with a computer that controls the operation of the spectrometer, is highly desirable, as it greatly facilitates the separation of CD and MCD signals [see Eq. (2)] as well as the subtraction of any instrumental baseline.

of the light constant.[21] The polarization is modulated at a frequency of (typically) 50 kHz. Differential absorption of left and right circularly polarized light by the sample results in an amplitude modulation of the beam reaching the detector. If $I_0(\lambda)$ is the intensity of the incident beam for both polarizations, $I_{L/R}(\lambda)$ is the intensity of the transmitted beam for the two circular polarizations, $A(\lambda)$ is the average absorption of the sample at wavelength λ, and $\Delta A(\lambda) = A_L(\lambda) - A_R(\lambda)$ is the net CD [see Eq. (2)], then $\Delta I(\lambda) \equiv I_L(\lambda) - I_R(\lambda) = I_0 10^{-(A + \Delta A/2)} - I_0 10^{-(A - \Delta A/2)}$. Series expansion of the exponential terms, discarding of higher order differentials (assuming that $\Delta A \ll A$), and solving for the CD gives

$$\Delta A(\lambda) = \frac{-\Delta I(\lambda)}{\ln(10)\, I(\lambda)} \tag{3}$$

Thus, to measure the CD, one must obtain the ratio of ΔI, that small portion of the detected signal that is modulated with the same frequency and phase as the polarization changes introduced by the PEM, divided by I, the far larger unmodulated component of the detected signal. This is conveniently accomplished in instruments that use photomultipliers as detectors by controlling the voltage applied to the detector, and hence its gain, to maintain the average signal, I, at a fixed value. The small modulated signal, ΔI, is amplified with the same gain as the unmodulated signal, and hence it is a direct measure of the net CD. ΔI is measured with a phase-sensitive detector (PSD, sometimes called a "lock-in amplifier"), which can extract a small periodic signal of known frequency and phase from a large background of incoherent noise. The proportionality constant relating the measured magnitude of the modulated signal with CD is obtained empirically using a material such as camphorsulfonic acid.[22] Once the CD is calibrated, the strength of the magnetic field at the sample, H, can be determined from the magnitude of MCD signal of a standard compound, or H can be determined independently using, for example, the Hall effect.

MCD Spectrometers for the Infrared. There are two distinct types of instruments used for IR CD/MCD. "Dispersive" instruments are similar to the design shown in Fig. 2, except that the detector is a solid state device, the gain of which cannot be controlled as can the gain of a photomultiplier. The ratio of $\Delta I/I$ in such instruments is obtained with some form of electronic divider circuit. The second class of IR CD/MCD instruments use modified Fourier transform spectrometers.[23] The long-wave-

[21] J. C. Kemp, *J. Opt. Soc. Am.* **59**, 950 (1969).
[22] G. C. Chen and J. T. Yang, *Anal. Lett.* **10**, 1195 (1977).
[23] L. A. Nafie and M. Diem, *Appl. Spectrosc.* **33**, 130 (1979).

length limit of IR CD/MCD spectrometers is greater than 10 μm (<1000 cm^{-1}). CD/MCD spectrometers that operate in the near-IR ($\lambda < \sim 2$ μm) are available from commercial sources, whereas mid-IR instruments must still be built in the laboratory.

Magnets. Most MCD spectrometers use either electromagnets or superconducting magnets. Electromagnets typically produce magnetic fields in the range from 1 to 2 T, whereas superconducting magnets used for MCD typically generate a maximum field ranging from 4 to 10 T. As the magnitude of the signal observed in an MCD experiment scales with the strength of the magnetic field [see Eq. (2)], higher fields are desirable. However, electromagnets are far easier and less expensive to operate than superconducting magnets. The choice of magnets for studies in which *C*-type MCD spectra (*vide infra*) are frequently observed (e.g., metalloproteins) presents a paradox. At low temperatures, $\Delta\varepsilon_{MCD}$ will tend to be large, hence reducing the need for the higher fields that are usually generated by a superconducting magnet. However, superconducting magnets are already fitted with cryogenic plumbing capable of reaching temperatures near absolute zero, and hence are frequently used in such studies. The ability to cool samples to near absolute zero is also used in the study of small molecules that can be trapped in a matrix consisting of an inert "gas" such as nitrogen or argon.[24]

Magnetic Circular Dichroism Spectra

MCD spectra are classified as *A*-, *B*-, or *C*-type, nomenclature that is singularly uninformative. *A*- and *B*-type MCD result from the perturbation of the states of a molecule involved in the observed photon-induced transition by the external magnetic field. As the change in energy produced by an external magnetic field is small compared to the intrinsic energy of the electrons that define the energy of the absorbing species, the problem can be treated using quantum mechanical perturbation theory. Typically, *B*-type spectra are the least intense type of MCD while *C*-type are the most intense, and I shall discuss them in that order.

B-Type MCD Spectra. First we consider *B*-type MCD spectra that are the only contribution to the MCD when both the initial and final states responsible for a transition are unique in energy. If in the absence of an external magnetic field the energies of the initial and final states involved in the absorption of a photon are different from those of any other states in the molecule (or, more generally, the absorbing species), we say that the states are nondegenerate, as is the photon-induced transition between

[24] B. E. Williamson, T. C. VanCott, M. E. Boyle, G. C. Misener, M. Stillman, and P. N. Schatz, *J. Am. Chem. Soc.* **114**, 2412 (1992).

them. The magnetic field modifies or perturbs these states slightly. Each perturbed state of the absorbing species can be represented as the sum of the unperturbed state plus small contributions from other states of the unperturbed absorber that are "mixed" with the unperturbed state by the external magnetic field. The symmetry of the energy perturbation introduced by an external magnetic field is such that only states with transition dipoles that are nonparallel are mixed together by the magnetic field. Thus a transition that is linearly polarized in the absence of an external magnetic field will become slightly elliptically polarized in the presence of the field, provided there are one or more transitions that have an absorption dipole that is not exactly parallel. We can decompose the elliptical transition dipole into a sum of the original linear component plus a small circular component which represents the stronger absorption of one circular polarization compared to the other circular polarization, that is, the MCD. The perturbations introduced by a magnetic field affect the pairs of states coupled by the magnetic field in a fashion that is identical in magnitude but opposite in sign, so, in the event that most of the observed MCD is due to the mixing of just two states, the MCD of the corresponding absorption bands will have the same (integrated) magnitude but opposite signs. Thus, B-type MCD spectra are similar in shape to the absorption spectrum for the same transition, can be of either sign, and tend to occur in pairs with opposite signs. An energy level diagram for a system that will exhibit B-type MCD is shown in Fig. 3, and the corresponding absorption and MCD spectra are shown in Fig. 4.

A-Type MCD Spectra. A-type spectra occur if either the initial or the final state involved in a transition is degenerate, that is, if there are two or more states with exactly the same energy in the absence of an applied magnetic field. Mathematically, energy degeneracy can occur only if the absorbing species is sufficiently symmetric, possessing at least a 3-fold symmetry axis. An applied magnetic field can remove the degeneracy by shifting the formerly unresolved transitions to slightly higher and lower energies. This problem is treated using degenerate perturbation theory. The component shifted to higher energy will absorb light of the opposite circular polarization compared to the component shifted to lower energy, hence producing MCD, as shown in Fig. 5. The splitting of degenerate levels by an applied magnetic field is well known from atomic spectroscopy as the Zeeman effect. (Thus, it might have been called Z-type MCD, rather than A-type, but was not.) In the case of atomic spectra, however, the widths of the absorption lines are less than the energy splitting that can be induced by a reasonable laboratory magnet, so the effect can be detected without recourse to measurement of circular polarization. Most systems of biological interest, however, involve polyatomic molecules in condensed

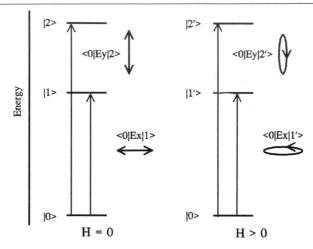

FIG. 3. Energy level diagram illustrating B-type MCD. Two nondegenerate excited states, $|1\rangle$ and $|2\rangle$, are mixed by the external magnetic field, H. Transitions from the ground state of the system, $|0\rangle$, to the perturbed excited states, $|1'\rangle$ and $|2'\rangle$, are elliptically polarized in opposite senses.

media, so spectral bandwidths are usually much greater than the energy splitting induced by laboratory magnets and are very difficult to observe in absorption spectra measured using unpolarized light. Such splitting may be resolved by MCD. As the components of the transition that are shifted in opposite spectral directions absorb opposite circular polarizations, the MCD resembles the derivative of the absorption spectrum, as shown in Fig. 6. Thus, a quantitative comparison of the MCD and the derivative of the absorption band can be used to determine the magnitude of the field-induced splitting. The magnitude of the splitting per unit magnetic field is related to the angular momentum of the degenerate state and hence can be used to determine this parameter.[1,2]

C-Type MCD Spectra. If the initial state involved in a transition is degenerate, a different mechanism can contribute to the differential absorption of left and right circularly polarized light. For a normal absorption measurement, the initial state will be the ground state of the system. The multiple energy levels of the magnetic field-resolved degenerate states will be populated according to a Boltzmann probability distribution. As the lower energy states are more likely to be populated, they will account for a greater share of the absorption by the system, and, if the more populated states absorb one circular polarization preferentially, an MCD signal will be observed. An energy level diagram depicting this situation

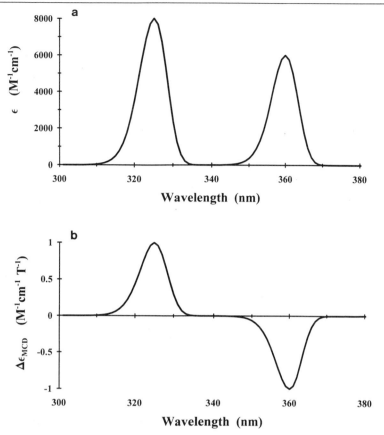

FIG. 4. (a) Absorption spectrum corresponding to the energy level diagram shown in Fig. 3. Within the resolution of a typical spectrophotometer, exactly the same absorption spectrum would be recorded in the presence of a magnetic field as in its absence. (b) MCD spectrum for the energy levels shown in Fig. 3. The two B-type MCD spectra have shapes similar to the corresponding absorption bands, but opposite signs.

is shown in Fig. 7. In contrast to A- and B-type spectra, the C-type MCD will be strongly temperature dependent, particularly at lower temperatures, as illustrated in Fig. 8b. The corresponding absorption spectrum is, however, far less sensitive to temperature than the MCD, as illustrated in Fig. 8a. C-type MCD results from a degenerate state of the absorbing system, and hence it will always be combined with an A-type MCD spectrum. For the case of a simple 2-fold degeneracy, the observed MCD spectrum will approach the A-type spectrum, which resembles the deriva-

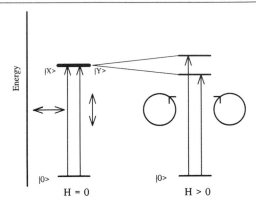

FIG. 5. Energy level diagram illustrating A-type MCD. The first excited state is composed of two states that we label $|X\rangle$ and $|Y\rangle$. The transition dipoles from the ground state $|0\rangle$ to $|X\rangle$ and $|Y\rangle$ are perpendicular to one another, but in the absence of an external field, the energies of the two states are equal. Linear combinations of $|X\rangle$ and $|Y\rangle$ will also be valid descriptions of the system in the absence of an external field, so the system can be described equally well as $|\pm\rangle = (|X\rangle \pm i|Y\rangle)/2^{1/2}$. The transitions from $|0\rangle$ to $|+\rangle$ and $|-\rangle$ will be circularly polarized in opposite senses to one another, but, as they have the same energy, there will be no net polarization observed. The magnetic field shifts one component to a slightly higher energy and the other component to a lower energy (the amplitudes of the shifts being equal). For most molecules of biological interest, the width of the absorption bands is greater than the field-induced splitting of the energy levels; hence, the splitting is difficult to observe in the unpolarized absorption spectrum, even if easily observable in the MCD.

tive of the absorption spectrum, at high temperatures, because the two states resolved by the magnetic field will be populated equally. As the temperature approaches absolute zero, however, the lower energy component of the initial state will be preferentially populated, and the MCD spectrum will approach the shape of the absorption spectrum.

Summary of Spectral Types. Assuming that the energies of field-induced splittings are small compared to kT, the MCD spectrum displayed as a function of photon energy ν (in units of wave numbers, cm^{-1}), can be expanded as a sum of A, B, and C terms according to the following expression:

$$\frac{\Delta\varepsilon_{MCD}(\nu)}{\nu} \propto \sum_i \left[A_i \frac{df_i(\nu)}{d\nu} + \left(B_i + \frac{C_i}{kT} \right) f_i(\nu) \right] \tag{4}$$

where the summation is over all transitions in the designated spectral domain, and $f_i(\nu)$ and $df_i(\nu)/d\nu$ are the shape and first derivatives of the

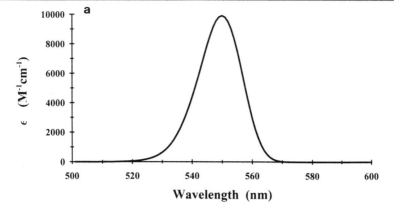

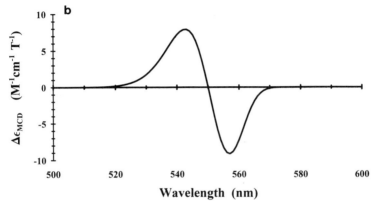

Fig. 6. (a) Absorption and (b) MCD spectra corresponding to the energy level diagram shown in Fig. 5. The slight broadening of the absorption spectrum induced by a magnetic field is usually too small to be observed. However, the corresponding A-type MCD of the magnitude shown would be easily observable using the type of spectrometer shown in Fig. 2.

shape of the absorption of the ith transition, normalized such that $\int f_i(\nu)\, d\nu = 1$. The corresponding expression for the unpolarized absorption spectrum is

$$\frac{\varepsilon(\nu)}{\nu} \propto \sum_i D_i f_i(\nu) \tag{5}$$

The coefficients A, B, C, and D for each transition can be expressed in terms of quantum mechanical matrix elements, and hence related to molecular parameters.[1,2]

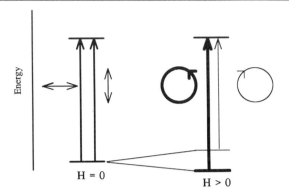

FIG. 7. Energy level diagram for a system in which the ground state is 2-fold degenerate. The quantum mechanical treatment of the degenerate states is similar to the case shown in Fig. 5; however, differential population of the states resolved in energy by application of a magnetic field results in greater absorption of the circular polarization absorbed by the state shifted to lower energy. Note that this will be the absorption band that appears shifted to higher energy. The result is a temperature-dependent, C-type MCD spectrum.

Information from Magnetic Circular Dichroism Spectra

MCD can provide several types of useful information, but I must also mention some of the potential traps in the interpretation of MCD spectra. The theory of MCD depends on the degeneracy of the initial and/or final states of the transition(s) involved. Because exactly degenerate transitions can exist only if the absorbing species possesses at least a 3-fold axis of symmetry, it might appear that experimental MCD could be used to determine something about the symmetry, or lack thereof, of the absorber. This does not follow, however, for two reasons. First, symmetric molecules (benzene is a good example) have both degenerate and nondegenerate transitions, so the existence of a B-type MCD spectrum does not imply that the absorbing species is not symmetric. Second, an MCD spectrum that resembles the derivative of the absorption spectrum may either be a true A-type spectrum or result from two closely spaced B-type spectra of opposite sign. For example, the MCD of both the 260 and 200 nm bands of adenine and hypoxanthine resemble the derivative of the corresponding absorption bands,[25,26] as illustrated in Fig. 9, but the known symmetry of the molecules proscribes their classification as true A-type spectra. Such spectra have been described as pseudo-A-type. Similarly, the shape of

[25] W. Voelter, R. Record, R. Bunnenberg, and C. Djerassi, J. Am. Chem. Soc. 90, 6163 (1968).
[26] J. C. Sutherland and K. Griffin, Biopolymers 23, 2715 (1984).

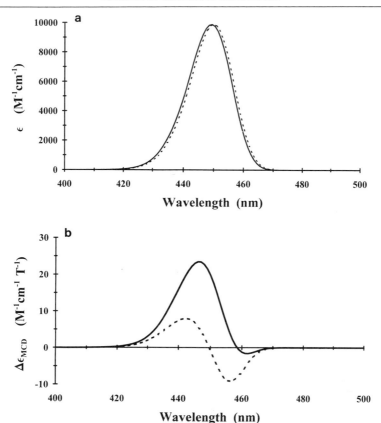

FIG. 8. (a) Absorption and (b) MCD spectra corresponding to the energy level diagram shown in Fig. 7. For temperatures at which kT is much greater than the difference in energy of the magnetic field-resolved degenerate states, the MCD will be dominated by an A-type spectrum and will resemble the derivative of the absorption spectrum, as in the case illustrated by the dashed curves. For temperatures at which kT is comparable to the difference in energies of the resolved states (as shown by the solid curves), however, the higher probability of occupation of the lower energy state will result in the preferential absorption of one circular polarization over the other. This produces a dramatic effect on the MCD spectrum but only a minor effect in the net absorption spectrum.

the MCD spectrum alone cannot distinguish A- and C-type spectra, because two closely spaced C-type spectra of opposite sign may maintain the shape of the derivative of the absorption band even as the temperature is lowered, as illustrated in Fig. 10. The change, or lack of change, in the magnitude of the MCD signal at low temperatures is the definitive test for C-type MCD.

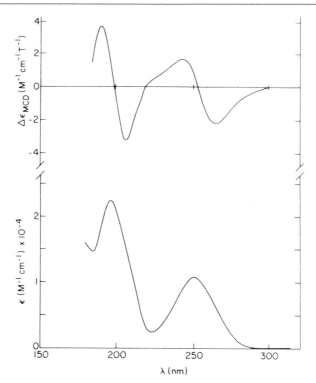

FIG. 9. Absorption spectrum (bottom) and MCD spectrum (top), of hypoxanthine in dilute phosphate buffer at pH 6 from 180 to 300 nm. The MCD spectrum is similar in shape to the derivative of the absorption spectrum. The spectra demonstrate the ability of MCD to resolve transitions that are not easily detected in absorption spectroscopy. They also demonstrate pseudo-A-type MCD spectra, since the symmetry of hypoxanthine forbids the existence of a true A-type MCD. This illustrates that information about the symmetry of a chromophore cannot be deduced from a comparison of the shape of the absorption and MCD spectra. (Adapted from Sutherland and Griffin,[26] with permission.)

MCD is, however, an exellent method of detecting the existence of multiple transitions that are not resolved in the absorption spectrum. Thus, if the shape of the MCD spectrum differs from that of the corresponding absorption spectrum, we know that the observed absorption spectrum is the sum of two or more unresolved transitions, as illustrated for riboflavin in Fig. 11. The converse is usually, but not always, true. In other words, if the MCD has the same shape as the corresponding absorption spectrum, there is likely to be exactly one transition involved. The slight reservation expressed in the previous sentence is required because of the possibility that an absorption envelope contains two or more transitions, all of which

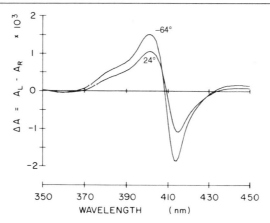

FIG. 10. MCD spectra of ferricytochrome c at $+24°$ and $-64°$, illustrating that closely spaced C-type MCD bands can resemble the derivative of the absorption spectrum. Thus, temperature dependence, and not the shape of the spectrum, is the proper hallmark for C-type MCD. The data also show that temperatures near absolute zero are not needed to demonstrate the temperature dependence of MCD spectra. [Adapted from J. C. Sutherland, *Anal. Biochem.* **113,** 108 (1981), with permission.]

have transition dipoles directed parallel to one another. Such transitions will not be coupled to one another by a magnetic field, and could all give B-type spectra (owing to coupling to transitions at different energies). The resulting sum of the B-type MCD spectra would resemble the shape of the absorption spectrum.

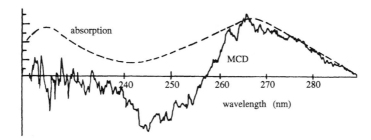

FIG. 11. Absorption (---) and MCD spectra (—) of FMN in 0.1 M phosphate buffer at pH 7. The negative MCD in the region from 240 to 257 nm in this early MCD spectrum illustrated the detection of electronic transitions not revealed in the absorption spectrum. The existence of a distinct electronic transition centered near 245 nm in the UV spectrum of compounds containing the isoalloxazine ring was critical to the proper interpretation of the UV resonance Raman spectra of these compounds [R. A. Copeland and T. G. Spiro, *J. Phys. Chem.* **90,** 6648 (1986)]. [Adapted from G. Tollin, *Biochemistry* **7,** 1720 (1968), with permission. Note that the sign of the MCD is inverted compared to the original presentation of the spectrum.]

MCD, in combination with the unpolarized absorption spectrum, is a powerful tool for characterizing a chromophore. The diversity of possible MCD spectral shapes provides more qualitative information than is available from absorption alone. The ratio of the magnitudes of the MCD and absorption at defined wavelengths provides an even more informative quantitative criterion. Because both MCD and absorption are proportional to the concentration of the absorbing species, their ratio is independent of concentration and, hence, is an intrinsic property of the material; that is,

$$\frac{\Delta A_{MCD}(\lambda_1)}{A(\lambda_2)} = \frac{\Delta\varepsilon_{MCD}(\lambda_1)}{\varepsilon(\lambda_2)} \tag{6}$$

This ratio, the MCD/absorption anisotropy, thus provides an intrinsic characterization of the material being studied from spectral measurements alone and without knowing its concentration. The wide range of values of this ratio enhances the utility of this type of characterization. The two wavelengths specified in Eq. (6) would normally locate the peak magnitudes of the MCD and absorption, respectively. One cannot simply specify the measurement of both MCD and absorption at the peak of the absorption band because, in the case of A-type spectra, the MCD may be zero at this wavelength. Alternatively, we could plot the anisotropy ratio as a function of wavelength (or wave number) and use this "spectrum" as the signature of a chromophore.

Experimental MCD can be compared with theoretical predictions to determine details of the physical and/or electronic structure of molecules. In the case of cyclic π-electron systems, the sign of the longest wavelength A- or B-type MCD spectrum can be related to the differences in energy of the two occupied orbitals that are highest in energy and the two unoccupied orbitals that are lowest in energy (see Michl[27] and related papers in the same issue). Detailed comparisons can be made between experimental absorption, MCD, and electron paramagnetic resonance (EPR) spectra and theoretical predictions for metal ions in heme and other quasi-symmetric environments using ligand field theory, particularly for transitions that occur in the near-IR (see, e.g., Refs. 28 and 29). The same may be true of MCD studies using soft X-rays.[19] Compared to EPR, MCD has the useful property that signals are observed for systems with no unpaired spins. The near-infrared MCD of heme proteins has been particularly

[27] J. Michl, *J. Am. Chem. Soc.* **100,** 6801 (1978).
[28] P. A. Gadsby and A. J. Thomson, *J. Am. Chem. Soc.* **112,** 5003 (1990).
[29] Q. Peng, R. Timkovich, P. C. Loewen, and J. Peterson, *FEBS Letters* **309,** 157 (1992).

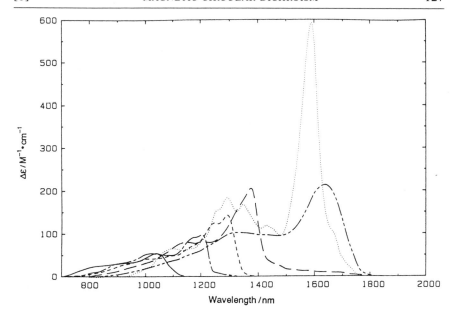

FIG. 12. Near-infrared MCD spectrum recorded at 4.2 K for six low-spin heme proteins, each containing protoheme IX with histidine as one ligand but with different second ligands. The spectra illustrate the sensitivity of the near-IR MCD to the ligation of the heme ring and demonstrate that MCD in this region can be used to assign axial ligands. The near-IR MCD spectra, in combination with EPR, were also used to determine the electronic structure factors for the proteins. (From Gadsby and Thomson,[28] with permission.)

useful in such quantitative studies because the spectra in this region are due to $d \rightarrow d$ and charge transfer transitions and, thus, are more sensitive to the axial ligands of the heme group, as illustrated in Fig. 12.

The ability to quantify a species with a distinct MCD spectrum and large MCD/absorption anisotropy in the presence of other absorbing species with lower anisotropies has also been used to quantify the amino acid tryptophan in proteins containing tyrosine residues that absorb in the same spectral region.[30–32] Tryptophan exhibits a strong positive signal at 297 nm that is easily distinguished from the MCD of tyrosine at that wavelength. The large anisotropy and distinctive shape of the MCD spec-

[30] G. Barth, R. Records, E. Bunnerberg, C. Djerassi, and W. Voelter, *J. Am. Chem. Soc.* **93**, 2545 (1971).
[31] G. Barth, E. Bunnerberg, and C. Djerassi, *Anal. Biochem.* **48**, 471 (1972).
[32] B. Holmquist and B. L. Vallee, *Biochemistry* **12**, 4409 (1973).

trum of porphyrin rings have been used to quantify these materials in urine[33] and to search for them in samples of lunar dust.[34]

The wide range of values of the ratio of MCD to absorption forms the basis of another practical, albeit rarely used, application in biophysical spectroscopy. Molecules with a characteristic MCD spectrum and high MCD/absorption anisotropy in a limited spectral region (heme proteins are a good example) can be detected and quantified in complex mixtures that exhibit continuous absorption and scattering of light across the spectrum, such as crude cell or tissue extracts.[35] This use of MCD is similar to one-beam, two-wavelength absorption difference spectroscopy, which is widely used in studies of metabolism and the biochemistry of heme proteins.[36]

Magnetic Circular Dichroism in the Extreme Ultraviolet and X-Ray Regions: Wavelengths Less Than 100 Nanometers

Although scarcely used at present, the previously inaccessible extreme-UV and X-ray region of the spectrum, along with the infrared, represents the frontier for MCD spectroscopy. Unlike the case for IR wavelengths, however, the instrumentation required for wavelengths less than 100 nm is radically different from that used in any other spectral region. The development of synchrotron radiation sources has made possible the measurement of MCD in the extreme-UV (XUV, $\lambda < 100$ nm) and X-ray regions of the spectrum. Three properties of synchrotron radiation are critical in this application: broad, continuous spectral distribution (extending from the IR all the way to X-rays), high intensity compared to conventional XUV and X-ray sources, and the polarization properties of the beam.

As in the visible and near-IR regions, much of the interest in biological MCD will involve proteins, and other systems, that contain metals. This conclusion is based on the fact that there are so many carbon, nitrogen, and oxygen atoms in most biologically interesting molecules that one can derive little information from the MCD spectra of these elements. In contrast, there are usually no more than a few metal ions in a given protein. X-Ray absorption spectroscopy of these materials has already

[33] S. M. Kalman, G. Barth, R. E. Linder, E. Bunnerberg, and C. Djerassi, *Anal. Biochem.* **52,** 83 (1973).

[34] G. W. Hodgson, E. Peterson, K. A. Kvenvolden, E. Bunnenberg, B. Halpern, and C. Ponnamperuma, *Science* **167,** 763 (1970).

[35] P. M. Dolinger, M. Kielczewski, J. R. Trudell, G. Barth, R. E. Linder, E. Bunnenberg, and C. Djerassi, *Proc. Natl. Acad. Sci. U.S.A.* **71,** 399 (1974).

[36] B. Chance, *Rev. Sci. Instrum.* **22,** 634 (1951).

proved useful as a probe of the environment of the metal component of proteins, and MCD should enhance this information, as it does in other spectral regions. The particular advantage of the XUV and X-ray regions, compared to the UV, visible, and near-IR, is that the range of metals that can be studied is increased to include those that do not absorb at the longer wavelengths.

Trends and Predictions

MCD is useful in probing the electronic structure of small molecules containing conjugated π-electron systems. The variety of molecules in this class is limited, however, because biological systems tend to use the same molecules over and over in different combinations, as exemplified by proteins and nucleic acids. Natural CD is useful in probing the huge variety of materials that can be constructed from a defined set of molecular building blocks, whereas MCD is useful in characterizing the "blocks" themselves, a drastically more limited enterprise. This limits the extent to which MCD will be used in the ultraviolet, the region where amino acids and nucleic acid bases absorb. The major opportunity for biological studies in the UV is the extension of MCD studies of such molecules into the vacuum UV ($\lambda < {\sim}200$ nm). Such studies complement CD studies of proteins, nucleic acids, and other macromolecules that can now be extended into the VUV.

The opportunities for MCD spectroscopy in the visible region are far more extensive. The MCD of the visible absorption bands of porphyrins have been probed by MCD since the beginning of the field. The sensitivity of the MCD spectra to the environment of porphyrin derivatives, in solution as well as embedded in proteins, provides information on both electronic structure and physical conformation. As an example, Dawson and colleagues have used MCD to probe the identity of the heme-type prosthetic group of myeloperoxidase.[37] Such capabilities combined with the seemingly infinite number of metalloproteins available in nature provide ample opportunities for the productive use of MCD.

Prospects for MCD spectroscopy in the near-IR are also bright. The weak bands of heme and other metalloproteins are usually either $d \rightarrow d$ transitions of the electrons on the metal ion or charge transfer transitions involving orbitals of both the metal and its ligands. Thus, this spectral region provides direct information on the state of the metal ion, which usually is of critical importance in the function of the protein. The work

[37] M. Sono, A. M. Bracete, A. M. Huff, M. Ikeda-Sato, and J. H. Dawson, *Proc. Natl. Acad. Sci. U.S.A.* **88,** 11148 (1991).

of Thomson and colleagues was noted above,[28] but numerous examples could be given of interesting near-IR MCD originating from several laboratories. The introduction of commercial CD/MCD spectrometers that operate up to wavelengths of about 2 μm has made it easier to perform such experiments. Worldwide, there are roughly a dozen well-equipped laboratories that study metalloproteins using MCD spectroscopy in the visible and near-IR.

MCD in the mid-IR, where absorption is due to changes in vibrational, as opposed to electronic, energy levels, has lagged far behind the development of natural CD in this region. The work of Keiderling and colleagues[38] demonstrated that MCD can indeed be measured in the mid-IR. As has happened in other regions of the spectrum, porphyrins were among the first materials studied and produced a large easily measurable signal. The surprisingly large magnitude of the vibrational MCD of highly symmetric molecules appears to reflect the influence of vibronic coupling.[38–40] The mid-IR is one of the frontiers of MCD spectroscopy.

The other frontier for MCD spectroscopy is at the short wavelength end of the spectrum ($\lambda \leq 100$ nm), where facilities for measuring MCD in the XUV, soft X-ray, and X-ray regions have relatively recently become available or are being developed. As with other spectral regions, metalloproteins are likely to be the major focus, because the L edge absorptions of transition metals and K edge absorptions of lanthanides should show large MCD anisotropies. A particular advantage of X-ray absorption spectroscopy is that metals which are "spectroscopically silent" in the visible and near-IR should be easy to study.

Because of the cost and difficulty of operation of advanced CD spectrometers and superconducting magnets, MCD has developed as a field dominated by a relatively small number of specialists. The extension of MCD technology into the X-ray domain may continue this trend, as a "beam line" (experimental station) suitable for MCD spectroscopy may cost several million dollars—exclusive of the cost of the synchrotron itself. However, because of the high cost of such instruments, they must be operated as "user facilities" and hence available for use by many scientists. Freed from the burden of developing and maintaining specialized instrumentation, scientists can focus on the preparation of interesting samples and the interpretation of experimental data. Thus, the move toward "big science" may have the paradoxical effect of opening the measurement of MCD to a wider range of scientists.

[38] P. V. Croatto and T. A. Keiderling, *Chem. Phys. Lett.* **144,** 455 (1988).
[39] M. Pawlikowski and O. S. Mortensen, *Chem. Phys. Lett.* **168,** 140 (1990).
[40] M. Pawlikowski, M. Pilch, and O. S. Mortensen, *J. Chem. Phys.* **96,** 4982 (1992).

Acknowledgments

The preparation of this chapter was supported by the Office of Health and Environmental Research, U.S. Department of Energy. I thank Stephen Cramer, University of California, Davis, John Dawson, University of South Caroline, Louise Heusinkveld, Brookhaven National Laboratory, Timothy Keiderling, University of Illinois at Chicago, Yu. Sharonov, Engelhardt Institute of Molecular Biology, Moscow, Peter Siddons, Brookhaven National Laboratory, Philip Stephens, University of Southern California, Martin Stillman, University of Western Ontario, and Andrew Thomson, University of East Anglia, for assistance in obtaining preprints, reprints, and references to the current MCD literature.

[7] Low-Temperature Spectroscopy

By ROBERT H. AUSTIN and SHYAMSUNDER ERRAMILLI

Introduction

This chapter is divided into three parts: some practical advice on how to do low-temperature spectroscopy, a discussion of the "internal" structure of visible absorption spectra and their temperature dependence, and a brief discussion of the utilization of low-temperature spectroscopy to probe protein dynamics. We would also like to go beyond the traditional use of low temperatures to trap intermediate chemical states and to speculate about the insight that low-temperature spectroscopy provides into the role of protein dynamics in controlling chemical reactions.

Probably the main utilization of low-temperature spectroscopic studies in biophysics has been to trap short-lived intermediate chemical states of a protein. Typically the chromophore alone is the object of study, and the protein is viewed basically as a structure which simply holds the chromophore in place or may contribute some critical amino acids in close proximity to the chromophore. The protein is not believed to play an overtly dynamic role, a dynamic role being one where the reaction process is steered by time-dependent protein conformational changes. Temperature is used to slow down or arrest the reaction kinetics occurring at the chromophore and perhaps to discover additional chemical states which have too short a lifetime to be observed at room temperature. Such kinds of kinetic arrest studies have been successfully used in studies of heme proteins,[1,2] the photocycle of rhodopsin,[3] bacteriorhodopsin,[4] and the photosynthetic reaction center.[5] The success of these studies has already justified the usefulness of low-temperature spectroscopy in biomolecular physics and has answered criticisms which suggest that low-temperature studies are "not physiologically relevant."

Although low-temperature studies have been used to measure transient chemical species extremely well, there is another aspect of low-temperature studies that condensed matter physicists have long used but biophysicists have not exploited much (with a few exceptions): low-temperature studies can be used to probe and characterize the dynamics and energetics of interactions in condensed matter systems. Biological macromolecules are in fact condensed matter systems in which low-temperature spectroscopy can be used as a probe of the dynamics of the system as it relaxes to equilibrium. The connection between biopolymer dynamics and the transport physics of semiconductors and insulators is tempting but fraught with traps, and more than one eminent physicist has gone astray assuming that biomolecules are in fact some sort of supersemiconductors. In fact, biomolecules are a mix of what appears to be both a glassy and a highly ordered physical system,[6] and the chromophores in the protein are molecules rather than simple defect sites so that one cannot write down the wave function for these systems as can be done for a color center.

However, an analogy with condensed matter physics is useful. If one simply measures the resistivity of some material at room temperature, it is not possible to understand the mechanism responsible for impeding the flow of electrons because there are many contributions to the resistivity, from phonons, defects, and other electrons. However, studies as a function of a temperature can separate the effects of phonons, defects, and interaction with other electrons because the temperature dependence of each of these contributions is characteristically different. We believe that the approach of condensed matter physicists has much to teach us in biophysics, particularly in the study of proteins in which the reaction

[1] T. Yonetani, T. Iizuka, H. Yamamoto, and B. Chance, in "Oxidases and Related Redox Systems" (T. E. King, H. S. Mason, and M. Morrison, eds.), p. 401. Univ. Park Press, Baltimore, Maryland, 1973.
[2] T. Iizuka, H. Yamamoto, M. Kotani, and T. Yonetani, Biochim. Biophys. Acta 371, 1715 (1974).
[3] K. Bagley, G. Dollinger, L. Eisenstein, A. K. Singh, and L. Zimanyi, Proc. Natl. Acad. Sci. U.S.A. 79, 4972 (1982).
[4] T. N. Earnest, P. Roepe, K. J. Rothschild, S. K. Das Gupta, and J. Herzfeld, in "Biophysical Studies of Retinal Proteins" (T. Ebrey, H. Frauenfelder, B. Honig, and K. Nakanishi, eds.), p. 133. Univ. of Illinois Press, Urbana, 1987.
[5] A collection of articles may be found in "Antennas and Reaction Centers of Photosynthetic Bacteria" (M. E. Michel-Beyerle, ed.), Springer-Verlag, Berlin, 1985.
[6] D. Stein (ed.), "Spin Glasses and Biology." World Scientific, Singapore, 1992.

progress is steered by time-dependent protein conformational changes. The same approach that has been successful in condensed matter physics is valuable for biophysicists as well: temperature studies are an excellent way in which the contributions arising from different molecular interactions can be understood.

In the case of chemical reactions within a biomolecule the situation is of course quite different from that of charge transport in a semiconductor or metal. First, unlike semiconductors biomolecules are designed to do a specific task, and their mechanisms are not readily understood. Second, as mentioned above, the biopolymer is typically an insulator and locally highly disordered, so the usual band picture of semiconductors does not work. Third, there usually are a few isolated sites (perhaps just one) where the actual reaction occurs. Thus, typically an organic chemist would simply view the protein as a scaffold which holds ligands at convenient distances to facilitate the reaction. However, proteins are quite large, considering the fact that the reaction occurs only over a very small fractional volume of the molecule. We can ask how all the other parts of the protein play a role in protein reactivity. Finally, proteins are polymers which are covalently bonded only along the backbone, not throughout the structure as is true for most conventional materials. The actual three-dimensional structure of the protein is quite fragile, held together by weak hydrogen bonds and salt bridges. Although one can work with analogy with condensed matter systems, many aspects of the protein environment are quite unique.

Probably the main reason for the neglect of low-temperature spectroscopy in biophysics is the unstated opinion of many biophysicists that these kinds of studies are irrelevant to the biological aspects of the reaction. Another reason, connected to the first claim of irrelevance, is the fact that any such study of low-temperature dynamics necessarily involves construction of theoretical models, and in complex molecules this is a notoriously difficult and speculative thing to do. It is not our intention to provide the reader with a comprehensive review of all low-temperature studies on biological systems; such a review would be impossible in the limited space of this chapter. We would like to stress applications that go beyond the traditional use of low temperatures to trap intermediate chemical states and to speculate about the insight that low-temperature spectroscopy provides into the role of protein dynamics in controlling chemical reactions.

Historical Overview

One of the earliest studies of spectroscopy of biological samples at temperatures around 4.2 K was done by Yonetani and co-workers,[1,2] who measured the optical absorption spectrum of sperm whale myoglobin and other heme proteins. This work also exploited the capability of low temperatures to trap photoexcited states in which the oxygen ligand is dissociated from the heme proteins. Frauenfelder and co-workers have honed low-temperature studies on the dynamics of photodissociated carbonmonoxymyoglobin (MbCO) to an art and have revealed the remarkable complexity that even "simple" proteins can exhibit.[7,8] One of the first extensions of low temperature techniques to the infrared regime was reported by Alben et al.[9] The ability to ratio photodissociated spectra with respect to "dark" spectra in the same sample has led to studies in which absorbance changes as small as approximately 10^{-5} can be routinely assayed. This technique was perfected by Rothschild and co-workers[4] to address the important role of specific residues in the photocycle of bacteriorhodopsin. Low-temperature studies on electron transfer proteins date back to the early classic work of DeVault and Chance[10] who measured the electron transfer rate from cytochrome c of the bacteriochlorophyll dimer $(BChl)_2$ as a function of temperature and found that electrons "tunnel" across surprisingly large distances. Studies of charge transfer in the reaction center of *Rhodopseudomonas viridis* by Michel and co-workers[11] have suggested the importance of low-frequency modes in determining the spectrum of this important protein. Low-temperature studies on the perturbation of the function of this protein by the presence of far-infrared radiation[12] and studies by Martin and co-workers using femtosecond laser spectroscopy[13] all point not only to the usefulness of low-temperature studies, but also to exciting new ideas that can emerge if experiments are combined with simple models of how proteins function.

[7] A. Ansari, J. Berendzen, S. F. Bowne, H. Frauenfelder, I. E. T. Iben, T. B. Sauke, E. Shyamsunder, and R. D. Young, *Proc. Natl. Acad. Sci. U.S.A.* **85,** 5000 (1985).

[8] R. H. Austin, K. W. Beeson, L. Eisenstein, H. Frauenfelder, and I. C. Gunsalus, *Biochemistry* **14,** 5355 (1975).

[9] J. O. Alben, D. Beece, S. F. Bowne, L. Eisenstein, H. Frauenfelder, D. Good, M. C. Marden, P. P. Moh, L. Reinisch, A. H. Reynolds, and K. T. Yue, *Phys. Rev. Lett.* **44,** 1157 (1980).

[10] D. DeVault and B. Chance, *Biophys. J.* **6,** 825 (1966).

[11] P. O. J. Scherer, S. F. Fischer, J. K. H. Horber, M. E. Michel-Beyerle, and H. Michel, *in* "Antennas and Reaction Centers of Photosynthetic Bacteria (Proceedings of the Feldafing Workshop)" (M. E. Michel-Beyerle, ed.), p. 131. Springer-Verlag, Berlin, 1985.

[12] R. H. Austin, M. K. Hong, C. Moser, and J. Plombon, *Chem. Phys.* **158,** 473 (1991).

[13] M. H. Vos, F. Rappaport, J.-C. Lambry, J. Breton, and J.-L. Martin, *Nature (London)* **363,** 320 (1993).

Techniques in Low-Temperature Spectroscopy

Cryostats

Most of the experiments discussed below require the ability to vary the temperature continuously from a little above room temperature down to about 4 K. The simplest Dewar flask of course is an immersion Dewar where the sample sits in the cryogenic fluid. They are inexpensive and easy to use but permit studies only at the boiling points of commonly available cryogenic liquids, and thus their utility is limited in this field.

The next step from simple immersion cryostats is the variable-temperature cryostat, where the cryogenic fluid is transported from a storage vessel under controlled flow rates and used to cool the sample. The first kind of variable-temperature cryostat, exemplified by the reliable Janis Vari-Temp cryostats (Janis Research Corp., Wilmington, MA), has an internal liquid helium reservoir with a fine needle valve at the reservoir base that admits the liquid helium into a small diameter tube. The tube introduces helium flow into the sample chamber, where it cools the sample (at atmospheric pressure typically) and eventually vents via a long vacuum insulated tube which passes axially through the helium reservoir. The vented gas can be collected for recovery of the helium, which is a precious and dwindling resource. By filling the sample region with liquid helium and pumping on the helium one can attain temperatures below 4.2 K. For experiments that do not have to go below 77 K, liquid nitrogen can be used as the cryogenic liquid.

This kind of cryostat has the advantage of maintaining the sample at atmospheric pressure (so thermal contact is excellent), liquids can be studied up to or above room temperature, and samples can be readily changed during a run. There are disadvantages. The Dewar flask is bulky and must be mounted upright, and the transfer of liquid helium from a bulk storage is both awkward and inefficient. Thus, these Dewar flasks are not convenient to use in confined places. The internal cryogen storage volume of the Dewar flask in the most common Vari-Temp model (Model 4-DT) is about 4 liters, it must be refilled about every 10 hr, and it requires constant attention during a long run. The central sample region is separated by a set of windows from a vacuum-filled insulating region, which in turn has to be separated from the ambient air by an additional set of windows. The inner set of windows is usually bonded with an epoxy adhesive directly to metal; the glass–epoxy–metal seals can fail with repeated cycling.

Because the sample itself has to be contained in a cell equipped with transparent windows, an absorption measurement has to pass light through no fewer than six windows. Although this is not a problem with studies in the visible region of the spectrum, infrared studies can be difficult because of the need to use an infrared transmitting material with the desired thermal properties.

The second type of variable-temperature cryostat involves the use of a relatively long (6 feet or so) vacuum insulated transfer line which continuously feeds cryogenic fluid from a large storage vessel supplied by a local cryogen supplier. Such flow cryostats are available from many manufacturers, including Janis Research, Cryo Industies (Atkinson, NH), and APD Cryogenics (Allentown, PA). The cryogenic fluid is used to cool a copper block which is in a vacuum enclosure. The copper block is in intimate thermal contact with the sample, which is mounted in vacuum. In principle the cryogenic fluid could be used to cool the sample directly, as in the Vari-Temp, by flowing into a chamber containing the sample; however, the inefficiencies involved with the long transfer line and the long thermal lag between cooldown of the line and transfer of the fluid make it much more efficient to cool the block and simultaneously apply electric heat to the sample block to maintain the desired temperature. The advantage of this mode is that the large storage Dewar can contain enough helium for several days of experimentation with little crisis intervention by the experimenter. Another advantage to the flow cryostat is the relatively small size of the cryostat head which allows the convenient mounting of a cooled sample within an apparatus such as a Fourier transform interferometer (FTIR) or similar spectroscopic apparatus. Further, the lack of an inner and outer chamber means that two fewer windows have to be used. There are disadvantages. In flow cryostats the sample is mounted in vacuum: this makes it imperative that the sample have excellent thermal contact with the cooled block. Temperature studies of liquids require vacuum seals between the fluid and the vacuum, and sample changing may be a traumatic enterprise if the head is cold.

A third class of cryostats uses a separately housed gas compressor and an expander connected to the cryogenic head to generate cryogenic temperatures directly at the sample. Complete systems are available from many manufacturers, including APD Cryogenics, Cryomech (Syracuse, NY), and Janis Research. This kind of cryostat also requires vacuum mounting of the sample in contact with the cooled metal block because of the relatively low cooling power of the compressor (a few watts). The big advantages are the fact that the sample can be maintained at low temperatures virtually indefinitely and the relatively compact size of the cryogenic head. The disadvantages are the relatively high cost of the compressor and expansion head and the possibility of significant vibration arising from the expander. If experiments are done where cryogenic fluids like liquid helium or liquid nitrogen are very difficult to obtain, the compressor/expander is very useful.

Solvents

The book by Douzou on cryobiochemistry provides an excellent discussion of the types of solvents that are suitable for low-temperature studies.[14] The need for investigating the effects of low temperatures on various solvents other than water was for cryoprotection of the dissolved proteins and the requirement for optically transparent samples. Samples even as thin as around 0.1 cm polycrystalline ice do not transmit light well; however, if glycerol is added to about 60–75% (v/v), a clear glass is formed at about 200 K.

Much of the early low-temperature spectroscopic studies assumed that the addition of glycerol did not change the protein structure. It is now clear that the solvent can influence spectra in a strong way by (perhaps) slaving the protein dynamics to the solvent glass transition[15] and influencing the static structure of the protein. Cordone et al.[16] have shown that the charge transfer band at approximately 760 nm observed in heme proteins such as myoglobin (Mb) is solvent dependent. Because the charge transfer band in Mb is known to be sensitive to protein conformation,[17] this is somewhat disconcerting. However, as in evolution one can take advantage of mistakes, and there have been a number of experiments which have exploited the charge transfer band. In particular, the charge transfer band sensitivity to protein conformation has been used to infer that in the case of myoglobin the protein dynamics are "slaved" to the viscosity of the solvent.[17-19] The realization that the dynamics of proteins is strongly influenced by the solvent has led to the view that no protein studies can be conducted without explicitly considering the nature of the solvent.

Douzou[14] has described the effect of temperature on not only viscosity but also pH and other properties of commonly used cryosolvents. The chemical nature of the buffers used can become important, particularly

[14] P. Douzou, 'Cryobiochemistry.'' Academic Press, London, 1977.
[15] I. E. T. Iben, D. Braunstein, W. Doster, H. Frauenfelder, M. K. Hong, J. B. Johnson, S. Luck, P. Ormos, A. Schulte, P. J. Steinbach, A. H. Xie, and R. D. Young, Phys. Rev. Lett. 62, 1916 (1989).
[16] L. Cordone, L. Cupane, M. Leone, and E. Vitrano, Biopolymers 29, 639 (1990).
[17] P. J. Steinbach, A. Ansari, J. Berendzen, D. Braunstein, K. Chu, B. R. Cowen, D. Ehrenstein, H. Frauenfelder, J. B. Johnson, D. C. Lamb, S. Luck, J. R. Mourant, O. P. Nienhaus, R. Philipp, A. Xie, and R. D. Young, Biochemistry 30, 3988 (1991).
[18] N. Agmon, Biochemistry 27, 3507 (1988).
[19] B. F. Campbell, M. R. Chance, and J. M. Friedman, Science 238, 373 (1987).

because of possible variations induced in the pH of the sample by temperature. At present, it is not possible to offer generalizations on what is the best solvent to use for low-temperature studies, and we refer the reader to the book by Douzou.[14] In general, we can state that complex buffers such as Tris are to be avoided, since they show a strong pH dependence with temperature, whereas simple buffers such as sodium phosphate show little temperature dependence.

Thin Films

We discuss later why infrared spectroscopy is very important if the coupling between the protein and the active site is to be probed. We have already mentioned the charge transfer band in heme proteins seen in the near-infrared at approximately 760 nm which is sensitive to protein conformation, probably indirectly via coupling between the iron atom and the proximal histidine linkage. The mid-infrared region, from 1.5 to about 20 μm, contains a large number of absorption lines not related to the chromophore and is an obvious place to look for protein conformation changes.

Unfortunately, water has very strong absorption in this spectral range, exhibiting extremely wide bands. Most proteins require water as a solvent. To do precision IR spectra of proteins, the optical path of the water must be made very small, on the order of 10 μm or so. If the path length is that small, a very high protein concentration is required, on the order of 10 mM typically. There are two ways to achieve spectra at these protein concentrations, either by forming thin films or by using a cell where the IR beam probes via the evanescent wave of total internal reflection.

The total internal reflection cell (Harrick Scientific, Ossining, NY) has the advantage that the laborious formation of good thin films is avoided. The cell has the disadvantage of being susceptible to breakage if the solvent freezes; it is also very bulky, costly, and has small acceptance angles, so that the light flux is not high.

Most low-temperature infrared spectroscopy has been performed using thin films. Some membrane proteins, such as bacteriorhodopsin, lend themselves easily to thin film use. The use of thin films affords some advantages, such as small volumes (tens of microliters), good compatibility with vacuum cryostat mounting, and excellent thermal contact with the cold stage.

Typically the thin films are made by deposition of a known amount of protein dissolved in a relatively high vapor pressure solvent such as water on a circular window of infrared-transmitting material (see below). Frequently, the next step is to remove some fraction of the solvent in order to achieve the desired thinness of the water or similar solvent. We have experience only in the removal of water, which is most conveniently done by placing the sample in a vacuum desiccator and reducing the pressure slowly to avoid bubbling of the solvent. In the case of bacteriorhodopsin and other such membrane-bound proteins, we have found that the hydrophobic membrane suspension often does not "stick" well to the substrate, and a very inhomogeneous concave sample is formed consisting of a thin center and a thick outer ring. Professor Sol Gruner from the Physics Department at Princeton suggested that charging of the substrate by passing a tesla coil radio frequency (RF) field generator (the brown cylindrical device used to leak-check glass vacuum systems) over the surface might help. We have verified this by charging the substrate, applying the liquid quickly in a few seconds, then immediately placing the sample in the vacuum desiccator, and drying the sample down in vacuum, which resulted in greatly improved, highly uniform thin films. This procedure was of greatest benefit when plastic substrates such as polymethylpentane (TPX) were used for far-infrared work (see section on optical substrates). Once a thin sample film is obtained, it should be mounted in a sample holder, made preferably of OFHC (oxygen-free high conductivity) copper, and the holder should be bolted to the cold finger, taking care to put a thin indium foil, or a conductive thermal grease (Wakefield Engineering, Wakefield, MA), between the mating surfaces.

We cannot overemphasize the importance of paying attention to good thermal connection between the copper block and the sample holder in a vacuum environment, particularly at very low temperatures where the thermal conductivity of most materials is not good. In an experiment in which a sample was mounted carelessly, on cooling to 10 K and illuminating the sample, a gap of about 50 μm opened between the sample and the copper block because of differential thermal expansion. The sample yielded an infrared spectrum characteristic of 320 K, while the copper block remained at 10 K!

Temperature Control

Although we have built a number of temperature controllers over the years for cryogenic work, sophisticated commercial temperature controllers are readily available. The basic concept is simple: the actual temperature, sensed as a voltage, is compared by a difference amplifier to a reference voltage representing the desired temperature. The difference is amplified by a variable gain amplifier to represent the error voltage. This is then summed with a fixed voltage representing the quiescent heater voltage needed to provide the basic offset heat to counteract the removal of heat from the cold tip. The summed output is then input to a linear power amplifier which is connected to the sample heater. This kind of circuit can be easily built and works as long as one is willing to adjust the DC offset voltage manually. More sophisticated temperature controllers which integrate the error voltage and thus automatically adjust the DC offset voltage are very useful, as are adjustable time-constant filters which adjust the system response to compensate for the great decrease in the specific heat that occurs below approximately 10 K. This kind of a control technique goes under the name of PID (proportional integral derivative) and is now very well understood.

The simplest and most inexpensive temperature monitors are thermocouples, such as copper–constantin (type T). Although the temperature coefficient of the thermoelectric effect is low (39 μV/K at room temperature for type T), there exist excellent tables for common thermocouples which make them extremely reproducible (see, e.g., the helpful handbook on temperature measurements from Omega Engineering, Stamford, CT). Unfortunately, below about 30 K the thermoelectric coefficient decreases by at least a factor of 10 and thermocouples become not useful (unless one uses exotic and expensive couples like gold/iron). Further, because the voltages measured are so low, artifacts such as ground loops and stray return currents from the typical 1–5 A current being applied to the heater can lead to serious errors.

The forward voltage across a semiconductor diode such as a silicon diode or a GaAs diode is a convenient temperature sensor. Usually a 10 μA current from a precision current source (which is easily made from a single field-effect transistor and an operational amplifier)[20] is used as the exciting current, and the forward voltage is a convenient 1–2 V. Further, because the temperature coefficient of the diode, about 1 mV/K at room temperature, increases by approximately a factor of 10 below about 20 K, the silicon diodes have excellent low-temperature sensitivity. The main drawback is the variation in temperature coefficient and absolute voltage even for a given diode type. Fortunately, there are now available diodes from various cryogenic companies which are highly reproducible, to within ±0.5 K. Unless one needs highly precise measurements, diodes are the most robust temperature monitors. Silicon diodes with excellent reproducible calibration curves are available from Lake Shore Cryotronics (Columbus, OH).

High precision work and reproducible measurements are best made via precision metal resistors which have a highly stable and linear variation

---

[20] P. Horowitz and W. Hill, "The Art of Electronics." Cambridge Univ. Press, Cambridge, 1989.

of resistance with temperature. These resistors, called RTDs, usually are made using platinum and are somewhat expensive and fragile. Calibrated meters for both silicon diodes and RTDs are available which will linearize the output and provide a direct readout in terms of kelvins. Such RTDs are available from a number of vendors, including Omega Engineering and Lake Shore Cryotronics. Other low-temperature sensors like capacitance sensors are of interest primarily to condensed matter physicists interested in low temperatures and high magnetic fields.

Optical Substrates

Window materials are important if spectroscopy is to be done in the visible and IR, and some compromises must be made regarding costs. Common materials such as quartz do not transmit at wavelengths longer than about 3 μm. The next material in breadth of transmission is sapphire, which has excellent hardness and thermal conductivity but does not transmit above about 5 μm. Calcium fluoride transmits to about 10 μm, is reasonably hard and virtually insoluble in water, and is an attractive material. If one is willing to forgo good visible transmission, materials like KRS-5 (potassium thallium bromide) offer excellent transmission to over 40 μm, but KRS-5 is quite soft and does not transmit at wavelengths shorter than about 0.6 μm. Finally, there is always diamond, an extremely hard material with twice the thermal conductivity of copper at room temperature and transmission virtually flat from 0.2 μm to the far-infrared (hundreds of micrometers). Actually, diamond is not so expensive now that vapor-deposited diamond has arrived. Thin diamond windows with thickness of about 10 μm with a 5 mm diameter can be obtained fairly reasonably from one of the more complete sources of optical substrates such as Janos Technology (Townshend, VT).

The low-energy collective modes (200–20 cm^{-1}), in the far-infrared (FIR), may play an important role in protein dynamics. To study directly the low-temperature spectroscopy of these modes via absorption spectroscopy, some different materials must be used. One of the most useful substrates in this region is the plastic TPX, which is also optically transparent. Optically thick materials such as Teflon and black polyethylene also are transparent in the FIR but are not usable in the visible, making alignment using HeNe laser beams difficult. Another useful material is Z-cut quartz, crystalline quartz cut along an axis so that the material is not birefringent. This material is transparent from about 100 cm^{-1} on down. Diamond, of course, is very transparent in this region, but use of diamond for anything other than windows with diameters of 3 mm on down is very expensive.

Much of the mechanical complexity in cryogenic studies is due to the need for optical access to the sample. It would seem natural that fiber optic probes, which are beginning to be available even in the infrared, are going to simplify cryogenics considerably. An interesting development is the availability of small, low-cost fiber-based spectrometers. The ability to make observations on small volumes of samples means that cryogenic systems will not require great cooling power. A further advantage is that the fiber can be mounted very close to, or even in, the matrix, which alleviates many of the problems of beam scatter and wander common to glassy or highly viscous fluids (see Dinkel and Lytle[21]).

Optical Absorption and Emission at Low Temperatures

To utilize the effect of temperature on the absorption of biomolecules, it is necessary to understand the physical origin of the optical absorption spectra and the origin of temperature effects. We present a simple, brief picture of the underlying physics of broad absorption bands.

Line Width in Spectra

Let us discuss the origins of the optical spectra observed at low temperatures in chromophore-containing biomolecules. Absorption studies in the UV/visible region of the spectrum are by far the most commonly used form of spectroscopy in biological applications. If a molecular transition has some effective frequency-dependent cross section $\sigma(\nu)$ for photon absorption, then the intensity of a light beam $I(z)$ passing through a solution containing N molecules/cm^3 decreases exponentially as a function of depth z:

$$I(z) = I(0)e^{-\sigma(\nu)Nz} = I(0)10^{-A} \tag{1}$$

where $I(0)$ is the incident intensity. The absorption coefficient A is commonly given by

$$A = \varepsilon(\nu)Cz \tag{2}$$

where C is the concentration of the molecular species in mM/cm^3 and $\varepsilon(\nu)$ is the frequency-dependent extinction coefficient, measured in cm^2/mM. Typically the optical absorption spectra is given as a plot $\varepsilon(\lambda)$ versus the wavelength λ. Before attempting a quantitative analysis, it is important to convert the data to plots of ε as a function of the frequency $\nu = c/\lambda$, where c is the speed of light, or in terms of inverse wavelength $1/\lambda$,

[21] D. M. Dinkel and F. E. Lytle, *Appl. Spectrosc.* **46,** 1732 (1992).

expressed in units of wavenumbers (cm^{-1}). It is also customary to attempt to fit optical spectra to Gaussian line shapes and to extract peak positions, peak widths, and integrated areas from such fits. We would like to point out that there is no *a priori* reason why peaks should have Gaussian shapes.

One way to parameterize the absorbance spectrum $\varepsilon(\nu)$ is by calculation of the moments of the millimolar extinction coefficient distribution. The nth moment of a distribution is given by

$$M_n = \frac{\int_0^\infty \varepsilon(\nu)\nu^n \, d\nu}{M_0} \tag{3}$$

The zeroth moment M_0 is simply the area under the curve, represents the normalization constant, and is related to the oscillator strength of the transition. Because $\varepsilon(\nu)$ is effectively a probability distribution, it is clear that the first moment for symmetric spectra gives the average frequency $\bar{\nu}$. The second moment for symmetric spectra is related to the width of the transition by a variance $m^2 = M_2/M_0 - \bar{\nu}^2$. Microscopic models usually are expected to calculate at least some of these moments for comparison with experiments. In computing the moments numerically, integrations are somewhat arbitrarily limited to some finite region spanning the band.

We now turn to the question of line shapes. Absorption bands in proteins are determined by both homogeneous and inhomogeneous effects. First consider homogeneous effects. Simple atomic two-level transitions usually are approximated by a Lorentzian line shape of the form

$$\sigma(\nu) = \frac{Ne^2}{4\pi\varepsilon_0 mc} \left[\frac{(1/2\pi\tau)}{(\nu - \nu_0)^2 + (1/2\pi\tau)^2} \right] \tag{4}$$

where ν_0 is the peak frequency of the spectrum, N is the number of chromophores per unit volume, m is the mass of the electron, e is the charge of the electron, and τ is the natural lifetime of the excited state.

The full width at half-maximum (fwhm) of a Lorentzian transition is a direct measure of the lifetime of the excited state and is equal to $1/\pi\tau$. The natural lifetime τ of an electronic excited state with electric dipole allowed relaxation in a two-level system is

$$\tau = \frac{3\hbar c^3}{32\pi^3 D\nu^3} \tag{5}$$

where D is the transition dipole moment connecting the two states. The maximum transition dipole moment for an atom is approximately the electron charge e times the size of an atom, roughly 1 Å. If we calculate

τ for light of wavelength 500 nm (blue-green) we find that in the optical regime excited state lifetimes should be about 1 nsec long, and the natural line widths of the transitions should be approximately 0.02 cm^{-1} or 0.01 Å wide. This is a very sharp, narrow-line transition! This line width, called the natural line width, is, however, rarely realized in condensed matter systems. The observed optical line widths in condensed matter systems in general, and biological systems in particular, are very wide, on the order of 1000 cm^{-1} or 500 Å wide. Why are the bands so wide? What are we missing? Moreover, is what we are missing of importance for understanding how low-temperature spectroscopy can aid in understanding protein function?

Homogeneous versus Heterogeneous Broadening

We have ignored four basic mechanisms that give rise to line broadening of optical transitions in condensed systems over and above the natural line width. The first mechanism, an obvious extension of the previous section, is simply the fact that in a condensed phase lifetimes of excited states can be greatly decreased by some different de-excitation channel opening up and removing the energy in a nonradiative manner. The second mechanism comes from the vibrational manifold of states that underlie the excited electronic states. These vibrational states can either be highly localized or collective modes. The third mechanism arises from the possibility of a heterogeneous distribution of solvent environments embedding the chromophore. The fourth mechanism is heterogeneous broadening owing to distributions of structurally different molecules, even in the case of homogeneous solvent environments. Intuitively, we expect that the large biological macromolecules might have significantly different ways to fold into their three-dimensional (3-D) structures, that these molecules might exhibit the largest degree of heterogeneous broadening, and that it might even be biologically relevant.

Taking the easy case of line broadening first, one way for the lines to be so wide is if the excited state lifetimes are very short, on the order of femtoseconds. Fourier analysis indicates that the spectral line width $\Delta \nu$ of a transition of lifetime τ (ignoring factors of π) is approximately $1/\tau$, and so the fractional line width of the line $\Delta \nu / \nu_0 \approx 1/\tau \nu_0$. It is more convenient to use energy units of inverse wavelength ($1/\lambda$, or cm^{-1}). Let the energy of the center frequency be given as E_0 (in cm^{-1}) with lifetime τ. The fractional line width then is

$$\frac{\Delta \nu}{\nu_0} = \frac{1}{\tau E_0 c} \tag{6}$$

where c is the speed of light. This broadening mechanism is a homogeneous line broadening since each molecule will exhibit such a broadened transition spectrum. Because optical transitions in condensed systems typically have fractional line widths of 0.05, with E_0 around 2×10^4 cm^{-1}, if the line width is due to lifetime broadening then the excited state lifetimes must be on the order of 4×10^{-14} sec. In the case of heme (iron-containing) proteins extremely short excited state lifetimes such as the above are possible owing to very fast relaxation of the excited state into a large manifold of other states.

However, as mentioned above simple lifetime shortening cannot be responsible for the wide lines seen in other molecules. For example, in many fluorescent molecules the observed lifetimes are near the lifetime calculated from Eq. (5), yet the observed width of the optical transitions remain far broader than the natural width. A common dye molecule such as rhodamine 6G has very broad optical absorption bands with widths on the order of 1000 Å, but it has a measured radiative lifetime on the order of 10 nsec. Homogeneous lifetime broadening is clearly not the reason that the transitions are broad. The line width of the dye electronic transition must be the result of some combination of heterogeneous environment, coupling to vibrational transitions, and large-scale collective modes. An important goal of some low-temperature studies has been to separate these effects, because each is affected rather differently by temperature.

Vibrons versus Phonons versus Conformational States

Next, consider the manner in which vibrational modes couple into the electronic transitions and give rise to spectral broadening. In the simple harmonic oscillator picture the angular frequency of the mid-IR transitions come from the beating of local atoms against one another. The frequency is given by $\nu = 1/2\pi(\kappa/m)^{1/2}$, where κ is the spring constant and m is the mass of the vibrating entity. If m is small, on the order of a carbon atom (i.e., interatom vibration), ν will be typically in the mid-IR region, from 500 to 3000 cm^{-1}. These localized, high-energy states are called "vibrons."

The low-frequency collective modes are due to the matrix that the chromophore is embedded in and may originate from either the solvent if the chromophore is simply dissolved in a solvent or the protein if the chromophore is embedded in a protein. We call these low-frequency modes "phonons" to distinguish them from vibrons. (Note that in condensed matter literature, the high-frequency modes are also called phonons; we avoid this description.)

The phonon modes have a large effective "mass" since the motion is collective among many atoms, and the characteristic frequencies of the motion will be small: in the case of an infinite solid the energies of the states will range from zero to a maximum value determined by the speed of sound in the material and the interatomic spacing of the atoms. As a rough estimate we can assume that the speed of sound v_s in the material is about 10^5 cm/sec and that the interatomic spacing d is about 2 Å; hence, the maximum frequency ω_{max} of the phonons is $10^5/2 \times 10^{-8}$, equal to a maximum energy E_{max} of 18 meV, and can be excited by a photon of energy 170 cm^{-1}. This maximum frequency is sometimes called the "Debye frequency."

The phonons may or may not couple to the electronic transition in the chromophore, and the extent to which they couple to the electronic transition is difficult to predict. The phonons serve also to define the temperature of the system. However, the phonons may also serve a more direct role in controlling protein function. As physicists interested in how the protein controls protein reactivity in a global, dynamic sense, it is this phonon band that is of the greatest interest to us since it represents the collective modes of the protein molecule.

Consider an electronic transition coupled to both the relatively sharp vibron transitions and the broad spectrum of phonons. The vibrons will split the single-electron transitions characterized by the quantum number n into a series of lines characterized by the vibrational quantum number j. Each vibron level is split further into closely spaced phonon lines, which we represent by a set of quantum numbers l, representing the phonon modes. Thus, in our simple representation a state can be given by the three quantum numbers (in order of decreasing energy) (n, j, l). However, the number of phonon modes is so large that experimentally, at high temperatures, one measures only an unresolved band.

At sufficiently low temperatures, it is possible to imagine an electronic transition between a ground state ψ_g and an excited state ψ_e where the ground state has no thermally occupied phonon levels. The lowest possible frequency of absorption involves an electronic transition from $(n = 0, j = 0, l = 0)$ to $(n = 1, j = 0, l = 0)$. (Note that the usual selection rules that describe atomic and molecular transitions in vacuum are violated in condensed phases.) Within the absorption band corresponding to this electronic transition, excitations of vibrons will lead to a number of sharp lines since (at least for the lowest values for j) the vibrational levels consist of a series of relatively sharp lines of approximately 1 cm^{-1} line width, whose width is the convolution of the excited state vibron lifetime and the electronic state lifetime.

Such an absorption spectrum containing only $(0, 0, 0) \rightarrow (n, l, 0)$

transitions will consist of sharp lines, called the zero phonon lines (ZPL). Adding in phonon interactions will give rise to an additional quasi-continuum of levels to each vibrational level, as we said before. It should now be clear that what is actually observed in an absorption or an emission experiment depends on the temperature, as well as the strength of the coupling between electrons, phonons, and vibrons.

Note that there is one zero phonon line for each vibronic state. If there is strong electron–phonon interaction then in addition to the ZPL there will be additional electronic transitions associated with the various phonon excited states. In absorption spectra, all such additional transitions give rise to·a broad band at energies higher than the ZPL. The absorption spectrum consists of a sharp line at a frequency ν_0 for the ZPL, and a broad band at higher frequencies (toward the blue region of the spectrum). This band is called the phonon wing (PW). The shape of the band is determined by the density of states and the strength of the coupling. At very low phonon frequencies, the density of phonon states is very small, and the contribution of the phonon wing to the absorption near ν_0 is very small. As the phonon frequencies rise, the density of states also increases, and the absorption arising from the phonon wing also increases to a peak, set roughly by the Debye frequency ν_D. The peak of the phonon wing occurs roughly at $\nu_0 + \nu_D$. In nearly all experiments, there is considerable absorption beyond this maximum, owing to multiple phonon processes. It is rather difficult experimentally to distinguish between multiple phonon processes and single phonon processes. For this reason, one must be careful in trying to extract the density of states directly from the shape of the phonon wing.

The most important question concerning the above discussion is whether the absorption bands of "real proteins" in frozen solutions in fact consist of sharp ZPL with broad phonon wings or not. Before addressing this issue, we discuss briefly the emission spectrum that may be expected to result from electron–vibron–phonon coupling. The excitation spectrum of a molecule is typically higher in energy than the emission spectrum, the fluorescence spectrum. The origin of this shift is crucial to determining the electron–phonon coupling constant, as we will show. In a complex system approximations must always be made. In our case we use the Born–Oppenheimer approximation which states that the total wave function of the system can be written as the product of the electronic wave function ψ_e and the nuclear ψ_N since the masses of the electrons and the nuclei are so different. Further, the electronic excitation time is "instantaneous" compared to the time it takes nuclei to move, so that the transitions occur vertically between the two potential energy surfaces of the ground and excited states.

Imagine that one obtains an emission spectrum of a dye molecule which is free in vacuum. In the absence of a damping mechanism, the emission spectrum will exist as a series of sharp zero phonon lines, exactly overlaying the excitation lines. Now imagine what happens as the chromophore is put into some sort of extended solvent (it could be just the protein alone). Depending on the extent to which the solvent interacts with the protein the emission lines will be shifted from the vacuum values, owing to both a vertical shift in absolute energy and a ''lateral'' shift because the excited state atomic positions find that a new value lowers the energy of the system. These shifts are really a measure of the electron–phonon interaction since, if the solvent did not interact with the environment, there would be no shift possible.

It is usual to assume that in the electronic excited state (n, j, l) the nuclear coordinates have an equilibrium position displaced some distance q_0 from the ground $(0, j, l)$ state owing to the change in the spatial extent of the electronic state. However, as stated above, during the transition the nuclear coordinates cannot change, and hence the electron must make vertical transitions between fixed values of q. In such a transition the energy change $\Delta E(q)$ for a given q is

$$\Delta E(q) = \frac{1}{2} K q^2 - \frac{1}{2} K (q - q_0)^2 + E_0 = \frac{K}{2} (2 q q_0 - a_0^2) + E_0 \qquad (7)$$

where E_0 is the energy gap between the $(0, 0, 0)$ state and the $(n, 0, 0)$ state. The vibron excitation transition at $T = 0$ will thus be shifted an energy distance $K q_0^2$ from the expected emission energy in the absence of potential energy surface shift. Further, the emission of the photons from the excited state $(m, 0, 0)$ is red-shifted to longer wavelengths owing to this Franck–Condon factor. Typically, one observes about a 0.05 eV, or 500 cm^{-1} shift between the minimum of the ground $(0, 0, 0)$ and $(1, 0, 0)$ excited state. This shift is known as the Stokes shift. As far as condensed matter systems go, this is a huge electron–phonon interaction and simply says that the electrons are very highly localized in these systems: they are insulators.

Figure 1 shows a simplistic and idealized sketch of the energy levels and what one would expect for the absorption spectrum at $T = 0$ K if a very narrow line width laser is used as the probing source in the absence of inhomogeneous broadening. Because $T = 0$ K, the ground state $(0, 0, 0)$ is the only occupied state. At the extreme red edge of the absorption spectrum only the $(1, 0, 0)$ excited state can be reached, and it has no excited phonon. However, the $(1, 0, 0)$ state can relax either to the no excited phonon state $(0, 0, 0)$ (giving rise to the ZPL) or into the continuum of excited phonon states $(0, 0, j)$, giving rise to a phonon wing. If the

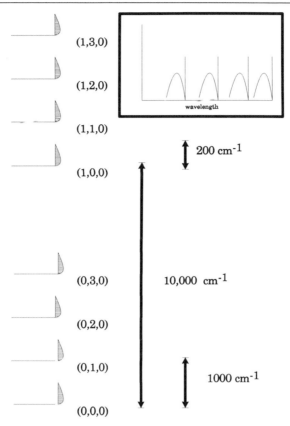

FIG. 1. Simplistic view of a homogeneously broadened chromophore optical absorption spectrum at 0 K. The left-hand side shows the vibrational and phonon splitting of two electronic levels. The typical vibrational levels are split by 1000 cm^{-1}, and the phonons form a quasi-continuous distribution of width approximately equal to 200 cm^{-1}. The upper right-hand corner shows the absorption spectrum at 0 K with the sharp zero phonon lines of approximately 1 cm^{-1} width and the phonon wing with width approximately 200 cm^{-1}.

system relaxes by photon emission (fluorescence) these transitions must emit lower energy photons than the ZPL in our simple situation, and phase-space considerations convolved into the idealized emission spectrum result in an asymmetrical smear known as the phonon wing (PW), of width about 200 cm^{-1}. The extent of this smear is a function of the extent to which the phonon density of states $f(\nu, T)$ couples to the chromophore (see [10] in this volume).

In reality, such a beautiful set of narrow lines exists only if the chromophores are put in certain special solvents where the electron–phonon interaction is particularly weak. Water in particular is a very poor solvent for resolving these spectral lines since it forms an inherently disordered solid. The structural heterogeneity of solid ice is such that the chromophore sees a heterogeneous environment and the lines are heterogeneously broadened by several hundred cm^{-1}, enough to essentially make the trees become a continuous forest. The extent of this broadening is such that it is difficult to measure the effects of the last two mechanisms of line broadening, the heterogeneous ones, solvent and structural heterogeneity. They act as perturbants on the large broadening caused by vibrational manifolds, and it is difficult to predict the extent to which the lines are further broadened by this distribution of conformational states.

The art of the experimenter lies in finding ways to extract the various elements that give rise to the widths of the lines. A physicist will always try to cool the sample down first to see what happens at low temperatures.

Temperature Effects

Consider the relationship between the ZPL and the PW (the Debye–Waller factor). Note that there is an interesting analogy between the ZPL and the Mössbauer effect: in the ZPL no phonon is exchanged with the medium and the transition line width is the natural line width, as in the Mössbauer effect where the decaying nucleus exchanges effectively no phonon with the matrix and the nuclear transition line width is again the natural line width.[22] A more common analogy is with X-ray diffraction: if the X-ray photon scatters elastically with the atom, then a coherent diffraction spot can form; if the X-ray photon scatters inelastically, coherence is lost and a diffuse background of scattered photons occurs. The Debye–Waller factor α in X-ray diffraction is defined as the ratio of the integrated intensity of the diffraction spot to the sum of the integrated intensities of the diffuse background and the diffraction spot. In a similar manner, α is defined in low-temperature fluorescence spectroscopy as the ratio of the integrated intensity I_{ZPL} of the zero phonon line (elastic scattering as it were) to the sum of I_{ZPL} and the integrated phonon wing I_{PW} (inelastic scattering in our analogy)[23]:

$$\alpha = \frac{I_{ZPL}}{I_{ZPL} + I_{PW}} \tag{8}$$

[22] H. Frauenfelder, "The Mössbauer Effect." Benjamin, New York, 1962.

[23] R. I. Personov, in "Spectroscopy and Excitation Dynamics of Condensed Molecular Systems" (V. M. Agranovich and R. M. Hochstrasser, eds.), p. 555. North-Holland Publ., Amsterdam, 1983.

It is possible to get an expression for α in terms of the temperature-dependent integrated phonon density of states $f(T)$ which is formally determined by including both the temperature-dependent occupation of the allowed states (Bose–Einstein statistics) and the strength of the electron–phonon coupling. Just to be specific we shall write the expressions, although they are idealized and not calculable at present.

First, we note that $f(T)$, the integrated phonon density of states, is given by

$$f(T) = \int_0^\infty f(\nu, T)\, d\nu \tag{9}$$

where $f(\nu, T)$ is the frequency- and temperature-dependent phonon density of states. Next, we note that $f(\nu, T)$ can be expressed as

$$f(\nu, T) = [n(\nu) + 1]f_0(\nu) + n(-\nu)f_0(-\nu) \tag{10}$$

where $n(\nu) = [\exp(\nu/kT)-1]^{-1}$ is the Bose–Einstein density of states and $f_0(\nu)$ is a measure of the electron–phonon coupling constant:

$$f_0(\nu) = 6Nq^2(\nu)\rho(\nu) \tag{11}$$

Here at last we have direct physical quantities: N is the number of atoms in the crystal, $q^2(\nu)$ is the root mean square (RMS) average displacement of oscillators of frequency ν (the same q that we discussed in the section on the Stokes shift), and $\rho(\nu)$ is the Debye density of states for a solid. It is also possible to express $f_0(\nu)$ as a measure of the probability of creating an excited phonon state at $T = 0$ K, and can be expanded as a power series in terms of one-phonon, two-phonon, etc., excited state creation.

Putting everything into the expression for the Debye–Waller factor we finally get

$$\alpha(T) = \exp\left\{ -\int_0^\infty f(\nu)\left[\frac{2}{(e^{h\nu/kT} - 1)} + 1\right] d\nu\right\} \tag{12}$$

where k is the Boltzmann constant. Figure 2 shows a plot of α versus T, where as a simple illustrative example we assumed that $f(\nu)$ is constant from 0 to 100 cm^{-1}, is zero at higher frequencies, and is normalized to unity quantum yield for single-phonon excited state production. Note that even with these optimistic numbers α is appreciably different from 0 only at temperatures below 20 K.

Because, as we discuss below, heterogeneous broadening is important in biological molecules, techniques such as low-temperature fluorescence and spectral hole burning at low temperatures (this volume [10]) are important probes of heterogeneity. In fluorescence techniques the excited state

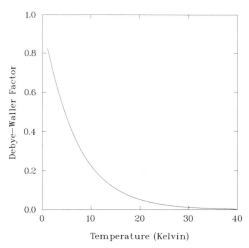

FIG. 2. Plot of the Debye–Waller factor α as described in the text versus temperature.

decays both through the ZPL and via a phonon cascade into the excited phonon levels of the ground states, which are not thermally occupied. However, note that in the excitation process the PW will not be visible since excitation can occur only from thermally populated levels to those levels which have the exciting photon energy. In certain cases, the act of photoexcitation if carried on for an extended period of time can bleach, or destroy, the excited state. In that case, within the (broadened) excitation absorption profile a narrow "hole" will appear owing to the loss of molecules with the heterogeneous profile. The hole in principle will have the fwhm of the ZPL, which as we have mentioned can be less than 0.1 cm^{-1} wide.

In the presence of an environment which adds a random amount of shift to the (presumably sharp) $(0, 0, 0) \rightarrow (1, j, 0)$ transitions that underlie the absorption profile and at temperatures where the Debye–Waller factor is small, we might then expect to find that the absorption profile is now changed into essentially a Gaussian line width by the central limit theorem. The zero-temperature fwhm σ of this line is not due to any lifetime effects but instead is due to the sum of the heterogeneous effects and the phonon spectrum. At 0 K the system is excited from the ground state $(0, 0, 0)$ into the various states $(1, j, l)$. Because the vibronic states (j) have a minimum excitation of about 200 cm^{-1} or so, for temperatures up to the normal biological range of 300 K they cannot contribute to the temperature dependence of the line width. However, the phonons are quasi-continuous from 0 cm^{-1} up to approximately 200 cm^{-1}, and as one increases the temperature the ground excitation levels will become increasingly spread among phonon excited states. This convolution of an increasing ground state phase space with the (fixed) excited state phase space gives rise to the increase of the fwhm of the optical transition with increasing tempera ture.

Although in principle detailed knowledge of the actual phonon phase space is necessary in order to predict accurately the temperature dependence of the line width, a simple approximation can suffice. We simply assume a variation of the Einstein model for the specific heat dependence of solids, namely, that there exists a single phonon mode of energy $\hbar\omega_0$. It is then straightforward to show that the fwhm $\sigma(T)$ of the band must vary with temperature as

$$\sigma(T) = \sigma_0 \coth(\hbar\omega_0/k_B T) \tag{13}$$

Although an approximation, it works pretty well.

As an illustration, consider some low-temperature absorption spectra we have done on thin-film samples of purified *Rhodobacter sphaeroides* reaction centers.[12] The three absorption bands of Fig. 3a are plotted as a function of temperature. Note that the band due to the special pair shows a strong temperature-dependent frequency shift (Fig. 3b) of the absorption maximum in addition to a temperature-dependent line width. Figure 3c shows a plot of the line width σ versus temperature, done by fitting three Gaussians (for lack of a theory) to the three peaks observed, and shows that the "average" value of the phonon field interacting with the chromophore has an energy of 150 cm^{-1}.

We thus see that the homogeneous causes of line broadening typically are continuous as a function of temperature and can at least be partially understood. The heterogeneous effects are more complicated. Consider first the effects of the local solvent environment on the line width. As long as the solvent is frozen, that heterogeneous environment is independent of temperature to first order anyway and can be probed by some of the frequency-selective effects we discuss later. However, once the solvent melts then the issue becomes a dynamic one: at time scales less than the relaxation time of the solvent molecules around the chromophore, the environment is heterogeneous, but at times long compared to solvent relaxation time, the environment is time-averaged and homogeneous. In a similar manner, if macromolecules have a distribution of conformations, we must still consider the rate at which the conformations can relax. If we probe at short time scales compared to this relaxation rate we will see a heterogeneous set of states, whereas if we probe at long time scales we

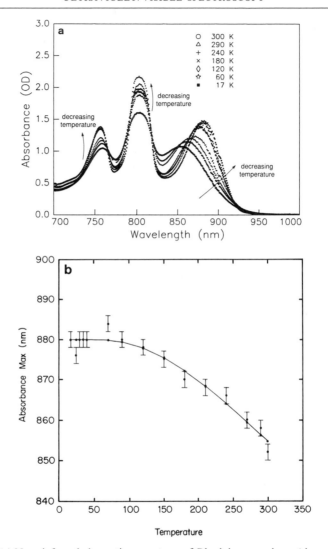

FIG. 3. (a) Near-infrared absorption spectrum of *Rhodobacter sphaeroides* reaction centers as a function of temperature. (b) Absorption maximum of the special pair band at 860 nm plotted versus temperature. (c) Line width of the special pair band at 860 nm versus temperature. (From Austin *et al.*[12])

will see a time-averaged homogeneous state. Perhaps the reader can see that temperature is a variable even more crucial to untangling the state complexity for the heterogeneous mechanisms of broadening than it is in the homogeneous cases.

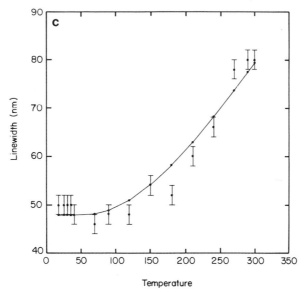

FIG. 3. (*continued*)

Spectral Heterogeneity in Biomolecules

Myoglobin Recombination Kinetics

To this point we have been primarily concerned with the connection between spectral broadening and the basic physics underlying the broadening. However, a connection must be made ultimately to the function of the molecule being studied. In the case of a protein, that connection must be with the rate at which the protein performs its biological function. If we were to find that proteins have a homogeneous, exponential rate of reaction, then most of the considerations above would be of academic interest only.

As an example of the direct connection between the analysis we have done above, we cite the work of the Frauenfelder group,[8] which leads one to expect heterogeneous line broadening in biomolecules as determined from studies of reaction rates at low temperatures. It is our belief that mapping of the time-dependent heterogeneity of the protein structures will lead to important information about biologically important protein dynamics.

Very briefly, Frauenfelder and colleagues have studied the recombination of carbon monoxide (and other ligands) with the iron atom in the heme-containing protein myoglobin (among other heme proteins). It is extremely easy to study carbon monoxide (CO) recombination at low temperatures because the bond between CO and the iron atom can be broken with near unity quantum yield by a photon, and the optical absorption spectrum of the heme is perturbed substantially by the absence of CO. We note here that it would be wonderful to try to do high-resolution low-temperature fluorescence spectroscopy on the heme, and perhaps also try hole-burning experiments, but unfortunately the excited state lifetime of the iron-containing heme group is on the order of femtoseconds owing to rapid relaxation into the iron orbitals.

It was observed that at temperatures below about 200 K the recombination kinetics of the CO changed from predominantly a bimolecular process to a geminate, unimolecular process. This was not too surprising since, if the solvent freezes or becomes extremely viscous, one could imagine that the photolyzed ligand would not be able to get out of the protein and hence the recombination would be unimolecular.

What was surprising, however, was the observation that the kinetics at low temperatures were not exponential in time as would be expected for a simple unimolecular process but instead were very nonexponential, to a rough approximation resembling a power law:

$$N(t) \approx N(0) \times (1 + t/\tau)^{-n} \tag{14}$$

where n is the slope of the decay kinetics when plotted on a log $N(t)$ versus log t plot and τ is some characteristic time. A simple experiment, where the sample was repeatedly photolyzed at some low temperature, revealed that the recombination could not be "pumped" into long-lived states, effectively ruling out diffusional models for the power-law kinetics. It was concluded that the nonexponential kinetics was due to a heterogeneous distribution of conformations of the myoglobin molecule; this distribution of conformations was characterized by a distribution of reaction rates which resulted in the observed nonexponential kinetics. At temperatures above the "glass transition" at about 200 K, it was posited that the myoglobin could rapidly relax among the various conformations which gave rise to the low-temperature nonexponential kinetics. As long as the CO could escape from protein fast enough and return slowly enough to allow a sufficient time for the protein to diffuse among all the allowed conformations, the recombination was bimolecular and characterized by an average rate constant.

The idea of a dynamic exchange among protein conformations that Frauenfelder proposed was at the time radical.[8] Clearly, since the reaction rates at low temperatures spanned many order of magnitude in time, it would be important to identify spectroscopically some signature that

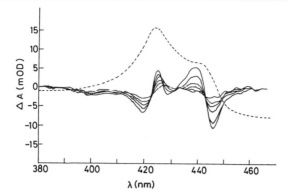

FIG. 4. Difference changes in the Soret absorbance region with recombination at 40 K. If the MbCO recombined with no variation in rate with spectral position, then the plot should be flat. The two double derivative peaks are due to Mb (deoxy) on the right-hand side and MbCO on the left. The dashed line shows a spectrum of partially photolyzed MbCO to show the position of the Mb and MbCO peaks. (From Ormos et al.[25])

differentiates slowly reacting molecules from quickly reacting ones, and ultimately get at the structural differences between the species. However, the work really just involved using low temperatures to slow relaxation processes and did not address how the conformational heterogeneity could be seen in the low-temperature spectroscopy.

In later experiments Frauenfelder et al. have observed that within the broadened Soret absorption profile there is an underlying heterogeneity, and they have examined the relationship between the "functionally important" inhomogeneous broadening of the rates of rebinding.[24] Because, as we mentioned above, there is a distribution of conformational substrates in the protein which have different rates of recombination, we might expect that these substates might also have different absorption profiles. We present in Fig. 4 an illustration of exactly this effect.[25] Figure 4 shows the time-dependent spectral changes that occur in the photolyzed deoxyMb spectrum as the ligand rebinds at 5 K, determined by normalizing the initial deoxyMb spectrum and doing in effect a double derivative of the spectral changes on recombination. Note that the bleaching observed at the longer wavelengths and the absorption increase observed at shorter wavelengths is indicative of a correlation between shorter wavelengths in absorption and slower recombination rate. One can estimate from the observed derivative curve that the inhomogeneous broadening is approxi-

[24] V. Srajer, K. T. Schomacker, and P. M. Champion, Phys. Rev. Lett. **57**, 1267 (1986).
[25] P. Ormos, A. Ansari, D. Braunstein, B. R. Cowen, H. F. Frauenfelder, M. K. Hong, I. E. T. Iben, T. B. Sauke, and R. D. Young, Biophys. J. **57**, 191 (1990).

mately 2 nm wide in a band with center wavelength of approximately 440 nm, or (in energy units) $2/440 \times 20{,}000$ cm^{-1} or 100 cm^{-1} full width. Although this is of the same magnitude as the Debye broadening expected from the phonon wing, Debye broadening should give rise to homogeneous effects, not the heterogeneous effects seen.

Charge Transfer Bands

A charge transfer band occurs when an electron donor and an electron acceptor become close enough together for there to be appreciable probability for charge to brought from the donor to acceptor via photoexcitation. Charge transfer bands are usually not observed in the absorption spectra of the individual molecules. One can generalize this concept a bit to include internal electron transfer between different functional groups in a macromolecule. Charge transfer bands are of particular interest in temperature studies because they involve intimate coupling between electronic energy levels and phonon modes, since the strength of a charge transfer band is a strong function of the distance between the donor and acceptor groups and this distance is modulated by the collective phonon modes. This coupling means that charge transfer bands are among the spectral features known to be the most sensitive to temperature changes. Experiments in two different protein systems have yielded remarkably similar looking models for analyzing the observed data. The first involves measurements on the so-called band III centered around 760 nm in deoxy-myoglobin and hemoglobin,[16] and the second involves studies of the transient changes in the charge transfer bands of the photosynthetic reaction center by Vos and co-workers.[13]

The theory for understanding the effects of temperature on charge transfer bands was derived from models developed by condensed matter physicists a long time ago for understanding the optical spectra of color centers in solids. Although the theory has been well developed for the case of an electronic transition coupled to a large number of phonon modes, all the observed data on proteins are analyzed assuming a very simplified picture in which there is just one low-frequency mode that is strongly coupled to the transition. All other phonon modes are weakly coupled and serve only to define the temperature of the specimen. Such a model is similar in spirit to the Einstein model of specific heat in solids, although in order to explain the observed shift in the peak position with temperature, it is necessary to invoke nonlinear or anharmonic effects, in which the (angular) frequency of the mode coupled to the electronic transition has a different value ω_g in the ground state than in the excited state (ω_u). Within the Einstein model, it is straightforward to show that the average frequency is given by

$$\bar{\nu} = M_1/M_0 = \nu_0 + \frac{1}{2}q_0^2\omega_u^2 + \frac{1}{8\pi}\frac{(\omega_u^2 - \omega_g^2)}{\omega_g}\coth(\hbar\omega/2k_BT) \quad (15)$$

where ν_0 is the frequency of the translation in the absence of any coupling (related to the energy difference between the minima of the ground and excited states) and q_0 is the shift in the equilibrium position of the normal mode (as discussed earlier). Note that the temperature dependence enters only in the third term, and vanishes unless ω_u is different from ω_g. The width can be deduced from the variance m, which is related to the second moment

$$m^2 = m_0^2 + \frac{q_0^2\omega_u^4}{4\pi\omega_g}\coth(\hbar\omega_g/2kT) + \cdots \quad (16)$$

where m_0^2 is the intrinsic width of the band. The dots indicate that higher order quadratic terms in $\coth(\hbar\omega_g/2kT)$ have been omitted; the full expression can be found in Scherer et al.[11]

How well do observed charge transfer bands fit this simple model? The data suggest that the photosynthetic reaction center fits the model reasonably well and allow an extraction of ω_g of approximately 200 cm^{-1}. Cupane et al. also find (for hemoglobin) that the data for band III may be fit, yielding a value of ω_g of about 240 cm^{-1}. However, they also find that the case of deoxymyoglobin is more complicated and is not consistent with the simple model. They find in addition that the solvent composition has a strong influence on the nature of the charge transfer band.

One possible explanation for the observed discrepancies is that the simple model ignores the effect of heterogeneity in the protein environment. Although the traditional picture is that heterogeneous effects are independent of temperature, we demonstrated above that Frauenfelder et al.[8] have shown that heterogeneous effects can lead to profound additional temperature dependence in observed spectra in the Soret region. In their analysis of the Soret region of both deoxyMb and MbCO, Srajer et al.[24] also conclude that there is substantial inhomogeneous broadening. We can conclude from such analysis that, at the very least, any attempts at a detailed analysis of absorption bands in proteins has to take into account broadening owing to heterogeneity. The question is whether one can use low-temperature spectroscopy to separate out inhomogeneous effects.

In fact, a series of workers[17-19] have shown that there is a direct connection between the rate of CO recombination and the center frequency of the underlying homogeneously broadened distribution of states. Friedman and co-workers[19] and Agmon[18] were the first to notice that the apparent shift in the center frequency of the band III transition with time during recombination was really due to a correlated shift in the center of gravity of the underlying distribution caused by a connection between the center wavelength $\bar{\nu}$ and the rebinding rate k. The general relationship is that the red-shifted wavelengths in the charge transfer band have the faster rate of rebinding. Work from Nienhaus *et al.*[26] has shown conclusively that it is the protein that relaxes after photolysis, leading to an increased enthalpic barrier. Unfortunately, the homogeneous line width of the states underlying the heterogeneous line seems to be approximately 150 cm^{-1}, compared to the total line width of 300 cm^{-1}. This strong overlap makes any sort of high-resolution low-temperature spectroscopy difficult at best.

Low-Temperature Fluorescence Line Narrowing

There have been attempts to address the issue of low spectral resolution at low temperatures. Vanderkooi and colleagues have done some experiments using low-temperature narrow-line fluorescence spectroscopy to do truly high-resolution spectroscopy of the ZPLs in proteins and to observe conformational heterogeneity in proteins.[27]

The essentials of the idea were worked out by Personov and co-workers[23] in glassy systems. They showed that, if one uses a tunable, highly monochromatic laser than it is possible to extract sharp features in the emission spectra in glassy systems. To appreciate the idea recall that, if the absorption transition between two electronic energy levels is studied at very low temperatures, only the ZPL in principle can be excited. When the excited state relaxes back to the ground state by the emison of a photon, the system can go into the entire manifold of phonon states. Thus the emission spectrum will show not only the ZPL but also the phonon wing.

The true natural line width of the states without the complication of the phonon wing can be studied by low-temperature absorption hole-burning experiments. Because Friedrich ([10] in this volume) presents a discussion of hole burning in protein spectroscopy, we confine ourselves to a brief review of the results of Vanderkooi using low-temperature fluroescence techniques.

If a ZPL $(0, 0, 0) \rightarrow (1, j, 0)$ excitation transition is performed, the relaxation of the excited state back to the ground state can go through

[26] G. U. Nienhaus, J. R. Mourant, K. Chu, and H. Frauenfelder, *Biochemistry* submitted (1994).

[27] A. D. Kaposi and J. M. Vanderkooi, *Proc. Natl. Acad. Sci. U.S.A.* **89,** 11371 (1992).

the ground state phonon states. Thus, the fluorescence emission spectrum can in principle show both the ZPL and the phonon wing, and it can give the extent of the electron–phonon coupling.

Such an experiment cannot unfortunately be done with an iron-containing heme group since there is virtually no fluorescence yield and a very short excited state lifetime. Line-narrowing procedures must use either iron-free porphyrins or porphyrins containing atoms such as magnesium or zinc. Because these substitutions must be made by replacing the heme there is always the danger that the proteins have structures different from those of the native species. In particular, because the proximal iron–histidine-33 bond is believed to play a rather important role in the conformational changes that occur on photolysis, and this bond is missing in the modified proteins, the experiments described here can give an indication only of the protein–chromophore coupling; they are blind to the ligand-induced conformational changes that are at the heart of enzyme action.

In any event, let us examine low-temperature fluorescence results from a magnesium-porphyrin myoglobin (Fig. 5).[27,28] If the reader actually consults the cited works, he/she may be confused by the term population distribution function (PDF) used by the authors, which is easily confused with the phonon wing (PW) discussed up to this point. Vanderkooi et al. use the term PDF as an acronym for the conformational distribution that Frauenfelder et al. have used to explain low-temperature reaction kinetic distributions, and they measure this distribution function by picking out the variation in a ZPL intensity with excitation wavelength.

Vanderkooi et al. demonstrated the inhomogeneous distribution arising from protein substates. The lowest ZPL line for Mg-porphyrin lies at approximately 16,700 cm^{-1}. Although a ZPL is always narrow, there will be an inhomogeneous distribution of ZPLs if there exists a distribution of conformational states. Excitation at a particular wavelength λ will pick out a particular subset of conformational states with ZPLs resonant with the excitation wavelength. Scanning the excitation line and observing the dependence of the ZPL line with excitation then gives a measure of the width of the distribution of conformational states. On the other hand, the PW emission is homogeneous, so the PW emission spectrum should be independent of the excitation energy, when corrected for the difference between the absorption and emission wavelength. The ZPL intensity, although sharp at any particular excitation wavelength, will show a intensity variation with excitation energy spread over the range of substate inhomogeneous broadening. Vanderkooi and co-workers found that the

[28] A. D. Kaposi, J. Fidy, S. S. Stavrov, and J. M. Vanderkooi, J. Phys. Chem. 97, 6319 (1993).

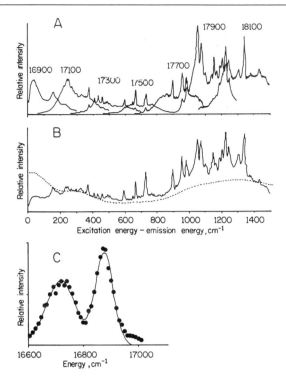

FIG. 5. (A) Temperature-narrowed fluorescence emission spectrum of Mg-substituted myoglobin (MgMb) at 4.2 K as a function of excitation wavelength. (B) Reconstructed emission spectrum of MgMb at 4.2 K, created by summing the spectral components in (A). (C) Measured inhomogeneous ZPL of the line downshifted 375 cm^{-1} from the excitation wavelength. (From Kaposi and Vanderkooi.[27])

fluorescence-narrowed emission spectrum in the (1, 0, 0) → (0, 0, 0) ZPL region did indeed show a range of inhomogeneous broadening, as we show in Fig. 5C. If we compare the Soret result observed by Ormos et al.[25] with observations by Vanderkooi for heterogeneous line broadening in fluorescence, we find that, although Vanderkooi obtains a bimodal distribution, the width of the homogeneous broadening is similar to the Soret work.

As Ormos et al. point out,[25] their methods actually ascribe a functional relevance to the inhomogeneous line broadening, namely, ligand recombination rates. Unfortunately, because the fluorescent analogs do not bind a ligand, it is not possible at present to move up in temperature from the 5 K region and determine how rapidly the inhomogeneous distribution can interconvert above the glass transition. This is a crucial issue in protein dynamics and is highly controversial. Many of those engaged in the debate about the room temperature kinetics perhaps have not read and understood the points by Frauenfelder concerning the low-temperature distribution of protein substates, and/or the dynamic analysis of the relaxation among these substates by Agmon. The point is that the protein via vertical Franck–Condon terms at low temperature is "excited" into a strained set of conformational substates, and at some temperature it will be able to relax among these substates.

However, low-temperature spectroscopy, in particular the fluorescence-narrowed emission, can in principle tell one further thing about the influence of proteins on chromophores. There can be a quasi-continuous tail at energies below the ZPL peak, and it represents coupling of the dipole transition to the low-frequency modes (the phonons) of the protein or the solvent. The relative strength of the ZPL to the phonon tail is the Debye–Waller factor and is a measure of the strength of the phonon–chromophore coupling. The phonon band is expected to be homogeneously broadened. Unfortunately, at this point no one has been able to use low-temperature spectroscopy to resolve the ZPL and the Debye spectrum in a protein, or observe the Debye factor as a function of temperature. We look forward to these important measurements.

The Far-Infrared: Uncharted and Shaky Ground of Collective Modes

There are differences between chemistry and physics. Chemistry is concerned with the local configuration of chemical bonds, whereas physics is concerned with the general aspects of how things work. We have avoided the mid-IR region of the spectrum because, unless one believes in solitons, one is again mainly involved with local conformation changes in the mid-IR. As we implied with the soliton comment, modes like the amide I band at 1665 cm^{-1} are part of a hydrogen-bonded network in α helices, and long-range effects are possibly present in this spectral region.

However, the far-infrared, from about 500 cm^{-1} down, is surely the province of the collective, delocalized "modes" of the protein dynamics. Because the characteristic energy of these modes is comparable to kT, they might actually play a role in directing enzyme action, although this is by no means obvious from what we know.

There is some evidence that the collective modes are coupled to the electronic transitions in proteins, although these measurements are somewhat indirect since no one has yet seen direct Debye–Waller factors in the ZPL fluorescence work. The main evidence has come from the work of Vos et al. involving electron transfer studies in photosynthetic reaction centers.[13] Martin and collaborators have engineered a mutant of the photosynthetic reaction center which contains only the "h" α-helix chain spanning the membrane, so that the electron transfer rate was slowed from the usual rate of 2 psec to hundreds of picoseconds after photoexcitation. The mere fact that the electron travels down only one side of a system with two nearly identical paths would indicate that simple distance alone and site recognition energies are not the sole determinants in the rate of electron transfer, but that protein dynamics and/or intermediate states may play an equally important role in directing charge flow.

From the low-temperature fluorescence work of Vanderkooi, we would expect that if the red edge of the absorption spectrum of the complex is excited, then only a few of the collective (Debye) modes will be excited. Work of Vanderkooi was steady state, and thus only the time-averaged distribution of collective modes is seen; in a time-resolved experiment with a subpicosecond time scale resolution the actual occupation of the collective mode can be observed directly, since the fact that the mode is coupled to the chromophore makes the chromophore absorbance a function of the collective mode occupation. If the coherence time of the collective mode is sufficiently long, one might actually expect to find a periodic modulation of the chromophore absorbance as collective mode energy oscillates between the few modes excited. If we assume that the Debye spectrum has a maximum at about 100 cm^{-1}, then these modes should show up with a frequency of about 3×10^{12} Hz, or a period of about 0.4 psec.

Figure 6 is a somewhat complex time-resolved transient absorbance plot of the oscillations seen as a function of temperature and wavelength.[13] The wavelength dependence is as we would have anticipated: at the red edge two clear modes are seen at 50 and 100 cm^{-1}, whereas as the excitation wavelength is moved further toward the central frequency the size of the oscillations seems to become blurred, perhaps owing to the large number of modes excited. Also, at low temperatures the signal is clear and strong with a dephasing time of about 2 picosec, whereas with increasing temperature the signal decreases in amplitude and the dephasing time decreases, presumably owing to the thermal activation of very low frequency modes which give rise to inelastic scattering. However, even at room temperature a clear signal was seen. Note that the temperature dependence of the charge transfer band line width mentioned earlier could also be fit to a phonon of energy approximately equal to 150 cm^{-1}.

The startling aspect of these experiments is that there are only two low-frequency modes seen of rather narrow line width. One would expect from our discussion of the quasi-continuous Debye spectrum that any Fourier transform of the modes would be quasi-continuous; instead only two are present, indicating a very close coupling to a very narrow subset

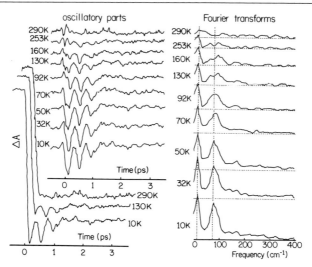

FIG. 6. Temperature dependence of emission oscillations detected at 945 nm in a mutant of *Rhodobacter capsulatus*. (From Vos et al.[13])

of the phonon spectrum. Although these experiments are fascinating, no clear connection with the actual biological process of electron transfer has been shown. Warshel et al.[29] have proposed specific mechanisms by which the fluctuating distances between the transfer sites as determined by excitation of the phonons could influence the electron transfer rates via the exponential dependence of the tunneling rate with distance. However, there have been few attempts (other than the work by Schulten[30]) to do a specific calculation using the known structure of the reaction center.

We end this review by discussing briefly some experiments that one of us (R.H.A.) has been involved in for a number of years using the pulsed output of the University of California, Santa Barbara, free-electron laser to do more direct studies of the influence of the continuum modes of the protein on the control of reactivity. Most of the experiments that we have cited to this point were done at low temperature out of the suspicion that temperature plays a major role in controlling protein conformational fluctuations. Because kT at $T = 300$ K is on the order of 400 cm^{-1}, there may be important protein collective vibrational modes with energies well below 400 cm^{-1} that are important in controlling the reaction rates of proteins. Taken a step further, by pumping a large amount of energy into a FIR collective mode you might see a change in the reaction rate of a

[29] A. Warshel, Z. T. Chu, and W. W. Parson, *Science* **246**, 112 (1989).
[30] K. Schulten and M. Tesch, *Chem. Phys.* **158**, 421 (1991).

protein. The results of Vos *et al.*[13] would seem to indicate that there is some foundation to doing FIR pump experiments.

We have attempted such low-temperature FIR pump experiments. Although we have carried out experiments on carbon monoxide recombination,[31] bacteriorhodopsin,[32] and reaction centers,[12] we just briefly discuss our experiment on reaction centers because we have previously discussed the temperature dependence of the reaction center charge transfer bands and the time-resolved work of Vos *et al.* above.

In our experiment we made thin films of the reaction centers using the methods discussed in the techniques section, and we mounted the sample in a flow cryostat. All the optics used in the experiments had to be transparent in the far-infrared, so the Dewar windows were Z-cut quartz and the sample was placed on TPX slides. A strong light at 860 nm illuminated the sample as an actinic source of photons. Its purpose was to initiate constantly the charge transfer loop—namely, fast picosecond electron transfer owing to excitation of the special pair creating an oxidized special pair and subsequent back-tunneling of the electron on a time scale of microseconds. In the presence of a strong actinic light the ratio of oxidized to reduced states R is set at some equilibrium between the photon-assisted rate k_{pump} and the back-tunneling rate k_{relax}: $R = N_{reduced}/N_{oxidized} = k_{relax}/k_{pump}$. If excitation into the collective modes changes either of the two rates, then the ratio R will be perturbed and a change in the absorbance will be observed.

We will not go into the details of the experiment nor the reasons why we believe the signal is due to nonthermal effects of pumping the collective modes, except to replot in Fig. 7 the size of the prompt transient signal seen with a 20-μsec FIR pulse at 85 cm^{-1} as a function of temperature. Note that the signal has a complex temperature dependence, changing sign at approximately 180 K. Without further information concerning the response of the collective modes to saturating levels of FIR radiation, it is difficult at present to make any quantitative analysis of these results, other than to point out that not only do optical experiments indicate coupling between the phonons and the chromophore, but so do direct excitation experiments.

[31] R. H. Austin, M. W. Roberson, and P. Mansky, *Phys. Rev. Lett.* **62**, 1912 (1989).

[32] R. H. Austin, M. K. Hong, and J. Plumbom, *in* "Princeton Lectures in Biophysics" (W. Bialek, ed.). World Scientific, Singapore, 1992.

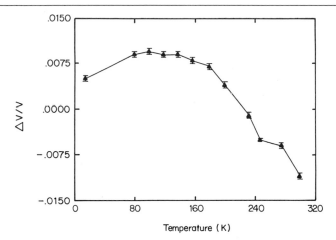

FIG. 7. Prompt transmission change observed in a thin *Rhodobacter sphaeroides* film as a function of temperature. The signal plotted is the prompt change during the 20 μsec wide 85 cm^{-1} FIR pulse. (From Austin *et al.*[12])

Extent of Biological Relevance

Low-temperature spectroscopy, when used to trap or characterize intermediate states, is of clear importance in biology and biophysics. One finds a far less receptive audience when it comes to claims that the low-temperature dynamics we discussed, accessible through low-temperature spectroscopy, is of biological relevance. If biology occurs at about 293 K, why worry about the way the system relaxes at 5 K? Perhaps the processes that can be resolved at low temperatures still occur at higher temperatures but at much faster speeds. By mapping the time and amplitude course of a kinetic component versus temperature, one can gain a great deal of understanding concerning the origin of the process.

However, low-temperature spectroscopy reveals that the chromophore absorption spectrum is heterogeneously broadened, and Frauenfelder has shown that there is a functional link between the spectral line position and the recombination rate. Agmon and Hopfield[33] pointed out that the protein myoglobin is photolyzed into a strained configuration and must undergo conformational relaxation to reach the true "deoxy" configuration, and that the recombination rate is slowest in the fully relaxed deoxy

[33] N. Agmon and J. J. Hopfield, *J. Chem. Phys.* **79**, 2042 (1983).

state. Agmon and co-workers have pointed out that myoglobin actually uses the conformational relaxation of the protein to control the binding of the ligand: the time-dependent decrease in the recombination rate serves to control the dynamics of ligand release and recombination. The protein seems to use conformational relaxation to control (on the nanosecond time scale) an important biological process. Low-temperature spectroscopy has been critical in the development of this fundamental model of protein reactivity, and we are confident that applications of this technique to unraveling the fundamentals of protein action are just beginning.

Acknowledgments

We thank Prof. Ken Sauer for extraordinary patience as we prepared this chapter, Prof. Jane Vanderkooi for helpful conversations concerning experiments, and Prof. Hans Frauenfelder for reviewing the manuscript.

[8] Rapid-Scanning Ultraviolet/Visible Spectroscopy Applied in Stopped-Flow Studies

By Peter S. Brzović and Michael F. Dunn

Introduction

If the investigation of the chemical mechanisms of enzyme-catalyzed reactions is to be successful, then detection and identification of transient intermediates along the reaction path are essential components. The application of rapid-scanning stopped-flow (RSSF) UV/visible spectroscopy to the study of enzyme catalysis may be of considerable significance in certain instances as a tool for both the direct detection and the identification of transient chemical intermediates.

Formation and decay of transient intermediates constitute a phenomenon that is ubiquitous to enzyme catalysis. Time-resolved UV/visible optical spectra which document the absorbance spectral changes that accompany the conversion of chromophoric reactants to final products by way of transient intermediates have the potential for identification of these transient species. The RSSF experiment provides three-dimensional data sets consisting of optical absorbance as a function of both wavelength and time. Owing to the temporal nature of the acquired data, an RSSF experiment can be a powerful tool for sorting out the timing of intermediate formation and decay with respect to other events that take place during the catalytic cycle.[1] Both intrinsic chromophores and appropriate tailored chromophoric probes can give information, not only about the structure of an intermediate, but also about the nature of the microenvironment and bonding interactions at the site. Because enzyme-bound intermediates are formed in the unique, nonaqueous environments of catalytic sites, the spectra of transient species may reflect not only the covalent bonding changes, but also the weak bonding interactions between the site and the reacting substrate and the constellation of electrostatic force fields present at the site. The effects arising from the unusual microenvironment of the site sometimes can be quite large and can result in the shifting of the spectral bands by as much as 100 nm in the visible region of the spectrum in comparison to the chromophore (or a suitable model) in an aqueous milieu.[2-6] Because such perturbations are not at present, *a priori,* very predictable, and may only be detectable as properties associated with a transient species, RSSF spectroscopy is uniquely capable of characterizing such systems.[1,7,8]

Enzyme-catalyzed processes occur on fast time scales (microseconds to seconds). Until the 1980s, the acquisition of time-resolved spectra under the conditions of a single turnover (i.e., transient or pre-steady-state conditions) in rapid mixing experiments had been technically difficult to achieve and, therefore, not much used. Almost all of the published rapid-scanning UV/visible spectroscopy carried out to investigate enzyme catalysis has relied on home-built instruments or instruments assembled from hardware components designed for other applications.[1] Until comparatively recently, the constraints imposed both by the availability of instrumentation and by the large volume of data generated in a single rapid-scanning experiment have greatly impeded the development of rapid-scanning stopped-flow instrumentation for UV/visible spectroscopy.

Two technological developments have begun to change this situation. Low-cost personal computers with sufficient memory and speed to accommodate the data acquisition and storage requirements are now available, and solid state detector arrays with near single-photon counting capabili-

[1] P. S. Brzović and M. F. Dunn, *in* "Bioanalytical Instrumentation," (C. H. Suelter, ed.), Vol. 37, p. 191. J. Wiley & Sons, Inc., New York, 1993.
[2] S. A. Bernhard, S.-J. Lau, and H. F. Noller, *Biochemistry* **4,** 1108 (1965).
[3] O. P. Malhotra and S. A. Bernard, *J. Biol. Chem.* **243,** 1243 (1968).
[4] P. M. Hinkle and J. F. Kirsch, *Biochemistry* **9,** 4633 (1970).
[5] M. F. Dunn and J. S. Hutchison, *Biochemistry* **12,** 4882 (1973).
[6] P. D. Buckley and M. F. Dunn, *in* "Enzymology of Carbonyl Metabolism: Aldehyde Dehydrogenase and Aldo/Keto Reductase" (H. Weiner, ed.), p. 23. Alan R. Liss, New York, 1982.
[7] K. H. Dahl and M. F. Dunn, *Biochemistry* **23,** 4094 (1984).
[8] K. H. Dahl and M. F. Dunn, *Biochemistry* **23,** 6829 (1984).

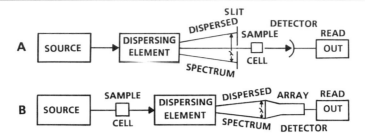

FIG. 1. Schematic diagrams of (A) conventional and (B) diode array scanning spectrophotometers operated in the dispersion mode. (A) The spectrum is dispersed by a rotating or moving grating or prism (the dispersing element) and scanned across an exit slit. A single-element detector (usually a photomultiplier tube) is used to measure the intensity of a monochromatic beam after it has passed through the sample. (B) The sample is illuminated with white light prior to dispersion. The dispersed spectrum is imaged on a linear array detector, and the signals from individual elements provide the information necessary to generate spectral information. (Redrawn from Santini *et al.*[9] with permission.) Copyright 1973 American Chemical Society.

ties and high-speed analog to digital (A/D) converters are commercially available. Consequently, in 1993 at least five companies introduced commercial rapid-scanning stopped-flow (RSSF) UV/visible spectrometer systems which employ either silicon photodiode array detectors (four of five) or a single-element (phototube) detector suitable for use in the study of enzyme catalytic mechanism (see Brzović and Dunn[1] for a review of the available instrumentation). The instruments are capable of achieving repetitive scanning rates on a millisecond time scale with at least 1–2 nm wavelength resolution and high sensitivity. Sophisticated data analysis software packages also are becoming available. These no doubt will provide increased capabilities for RSSF data analysis.

Instrumentation

Two approaches are in use for the design of instrumentation for RSSF spectroscopy.[1,9] In both, the basic apparatus consists of a stopped-flow rapid-mixing system integrated into a scanning UV/visible spectrophotometer operated in the dispersion mode. The classic instrument (Fig. 1A) employs an optical system that transmits "monochromatic" light from a monochromator through the sample onto a single-element detector (usually a phototube). Wavelength-dependent spectra data are obtained by mechanically scanning the dispersed spectrum across the exit slit of the

[9] R. E. Santini, M. J. Milano, and H. L. Pardue, *Anal. Chem.* **45**, 915A (1973).

monochromator.[10,11] The alternative method employs an optical system with no moving parts. The sample is illuminated with "white" light prior to dispersion in a grating polychromator (Fig. 1B). The dispersed spectrum is imaged onto an array detector (usually a linear photodiode array).[12,13] Spectra as a function of time are reconstructed from the individual diode signals.

In the single-element detector system, spectral information is generated by sampling the phototube signal at a rate that is rapid relative to the scan rate using a high-speed, high-resolution A/D converter and computerized data acquisition system. In the array detector system, spectral information is provided by the individual signals from the individual elements of the array. These signals also are digitized via a high-speed, high-resolution A/D converter and stored in computer memory. Both types of instruments are commercially available. Because the technology for RSSF applications has been changing relatively rapidly, it is not clear which approach is more advantageous. Most of the enzyme mechanism work published since the early 1980s has been carried out using array detector systems.[1] Until relatively recently, the available array detector technology limited the repetitive scanning rate to approximately 5 msec/scan for a 250–300 nm wavelength range. The improved A/D converter rates and diode arrays that have become available possess scanning rates with a lower limit of about 1 msec/scan. Owing to the properties of the photomultiplier tube, the single-element detector system is inherently capable of faster scan rates. A commercial apparatus which has just appeared on the market claims a scanning speed of 1 msec/scan (over a 220 nm wavelength range), and scan rates of 0.3 to 0.1 msec/scan are advertised. No published work from this instrument has appeared. Unless there are substantial advantages to be achieved from signal averaging, scan rates greater than 1 msec/scan would seem to be of little advantage for RSSF applications.

Phototube technology provides a wide linear dynamic range, rapid (instantaneous) response, and photon counting capability. Wavelength resolution is determined by the characteristics of the grating and the slit widths used. Hence, it is easy to achieve a wavelength resolution below 1.0 nm. Simple silicon photodiodes (SPD) arrays are characterized by rapid response times, wide spectral sensitivity, linearity of responses and

[10] R. B. Coolen, N. Papadakis, J. Avery, C. G. Enke, and J. L. Dye, *Anal. Chem.* **47,** 1649 (1975).

[11] N. Papadakis, R. B. Coolen, and J. L. Dye, *Anal. Chem.* **47,** 1644 (1975).

[12] Y. Talmi, *in* "Multichannel Image Detectors" (Y. Talmi, ed.), Vol. 2, p. 332. American Chemical Society, Washington, D.C., 1983.

[13] D. G. Jones, *Anal. Chem.* **54,** 1057A (1985).

a fairly large dynamic range, low noise, and temporal stability, and they approach a photon counting capability.[12] Most diode array systems achieve a wavelength resolution of 1 to 2 nm. In contrast to a phototube, an SPD is an integrating device. The signal level is specified by the light intensity and the dark current integrated over the time of exposure. Consequently the signal level can be modulated by controlling the exposure time. Intensified arrays also are available. These devices achieve a single-photon counting capability.[12] For most RSSF work, the simple SPD array has proved to be sufficient. Intensified arrays may be useful in extending RSSF spectral measurements into the 200 to 300 nm region of the spectrum or in applications where the signal intensity is low.

Experimental Design

Qualitative Analysis

At the most basic and functional level, RSSF spectroscopy has proved to be a very useful tool for qualitative interrogation of new systems to determine if there are detectable amounts of chromophoric species that accumulate in an enzyme-catalyzed reaction. The design of such experiments is usually dictated by a combination of factors that includes consideration of the rates of the processes under study and the spectral properties of reactants, products, and predicted intermediates.

Enzyme-catalyzed reactions occur with turnover numbers (k_{cat}) that range from 10^6 to 1 sec^{-1} or slower. Many enzymes fall into the 10^3 to 1 sec^{-1} range. Because the parameter k_{cat} is determined by the process (or processes) that limits the rate at which turnover occurs, the rate constants for all the individual, non-rate-limiting processes along the reaction path will be larger (sometimes much larger) than k_{cat}. There appears to be a practical limit of about 100 to 200 μsec for the time required to mix two aqueous solutions in a stopped-flow apparatus so that the ensuing changes in absorbance that accompany reaction can be measured.[14] Most commercial stopped-flow rapid-mixing UV/visible absorbance spectrometers claim mixing dead times between 500 μsec and 3 msec (depending on path length, flow rates, and observation cuvette volumes). The combined rapid-mixing rapid-scanning experiment dead time places practical limitations on the magnitudes of the rates of enzyme-mediated processes that can be observed with stopped-flow technology. With large changes in molar extinction coefficients and an approximately 1 msec experiment dead time, it is possible to measure first-order (or pseudo-first-order) processes with

[14] R. L. Berger, B. Balko, and H. F. Chapman, *Rev. Sci. Instrum.* **39**, 493 (1968).

rate constants between 500 and 2000 sec^{-1}. Consequently, the use of rapid-mixing technology extends the dynamic range of UV/visible absorbance measurements to the millisecond range, and therefore some (and in favorable cases most) of the events that occur in a single turnover for enzymes with k_{cat} values no greater than 10^3 sec^{-1} may be amenable to investigation provided there are suitable changes in absorbance that accompany reaction.

If intermediate detection is the primary objective, it is likely that a simple inspection and comparison of the spectra contained in the data set will be sufficient to determine whether an intermediate accumulates in detectable amounts. For this purpose, it is useful to superimpose some or all of the spectra on the same absorbance versus wavelength plot so that the positions and shapes of bands can be easily compared throughout the time interval of interest (see Figs. 5–8 and 11–13 in the section on significant applications). It often turns out that the rates of intermediate formation and decay are such that the amount of intermediate that accumulates is sufficiently large to be manifest in transient spectral bands derived from the intermediate that are obvious to the investigator. In other instances, the spectral changes may be more subtle, and the evidence for the appearance of an intermediate must be inferred from the perturbed spectra of reactants and products. For example, the lack of isosbestic points in the set of time-resolved spectra is good evidence that a transient intermediate is formed during reaction (see Tinoco[15] and van Amerongen and van Grondelle[16]).

Quantitative Analysis

Rapid-scanning stopped-flow spectroscopy usually is carried out by initiating the reaction under conditions far from equilibrium. Enzyme-catalyzed reactions frequently involve complicated reaction time courses containing multiple relaxations derived from binding events, conformational transitions, and intermediate formation and decay. Therefore, the central underlying assumption critical to relaxation kinetic theory,[17] that the perturbations are small compared to the equilibrium concentrations of the species present, generally does not apply. For the purpose of quantitative analysis of the kinetic data, this potential complication is usually circumvented by selecting experimental conditions of concentrations such that steps in the reaction with a molecularity greater than one are carried out under pseudo-first-order conditions, that is, conditions where the con-

[15] I. Tinoco, Jr., this volume [2].
[16] H. van Amerongen and R. van Grondelle, this volume [9].
[17] C. F. Bernasconi, "Relaxation Kinetics." Academic Press, New York, 1976.

centration of one reactant (usually either the enzyme or one substrate) is kept small with respect to all others. In the case of a bimolecular process, such as A + B → C with forward rate k, the reaction time course is rendered pseudo-first-order by placing one reactant in large excess over the other (i.e., [B] ≫ [A] so that $d[C]/dt = k[A][B] \approx k[B]_0[A] = k'[A]$, where k' is the pseudo-first-order rate constant equal to $k[B]$. Buffering can be used as an alternative means for achieving a decrease in the apparent kinetic order of a reaction, thereby reducing the complexity of the time course. Buffering, of course, is widely used to maintain the concentrations of H^+ or OH^- constant during reaction. In some instances where the reaction under examination involves metal ions, chelators can be used to buffer the concentration of the metal ion.

If experimental conditions are carefully selected so that all individual steps are either first-order or pseudo-first-order processes, then, under transient or pre-steady-state conditions, the reaction time courses will take the form of a sum of exponentials (i.e., a linear combination of the individual rate equations for each relaxation), such that the observable, time-dependent changes in absorbance, ΔX, are given by the relationship

$$\Delta X = \sum_{j=1}^{n} a_i e^{-t/\tau_i} \tag{1}$$

where a_i is the amplitude factor for each relaxation, τ_i. Therefore, the same relaxation rate expressions derived under the assumption of a small perturbation[17] are valid and can be applied to situations where the perturbation is large (as when enzyme and substrates contained in separate syringes are mixed in a stopped-flow apparatus).

If the conditions of concentration are inappropriately chosen, then the observed time courses may contain relaxations that are not exponential in form. Analytical methods for extracting meaningful rate and amplitude data would necessarily be designed for each specific case. Such time courses can easily be mistaken for biphasic exponential processes, a conclusion that would lead to inappropriate mechanistic ideas.

If the relaxation spectrum of the kinetic time course consists entirely of exponential processes, then the data can be analyzed with standard software programs for the decomposition of the time course into its component exponentials. In practice, the analysis of systems consisting of more than two or three exponentials is difficult unless the relaxation rates have similar amplitudes and are well separated in magnitude.[17] As will be shown, the RSSF experiment can be a very useful qualitative and sometimes semiquantitative tool for the interrogation of multiphasic reactions. By inspection, it is usually possible to identify special wavelengths for single-wavelength, stopped-flow (SWSF) kinetic analysis where the complexity of the time course is reduced as a consequence of the presence of false isosbestic points which persist during one phase of the reaction. The single-wavelength time courses measured at the special wavelengths will contain one fewer relaxation, thereby rendering the time course less complex.

Observation of optical absorbance as a function of both wavelength and time, as is the case in a single RSSF experiment, generates a three-dimensional surface which requires the accumulation of a large amount of spectroscopic data. The ready availability of high-speed laboratory computers capable of handling large amounts of data allow application of modern methods of data analysis to RSSF spectral profiles. Two approaches which appear to have particular value for analysis of RSSF spectra in the investigation of enzyme mechanism are methods based on global analysis[18] and singular value decomposition[19] (also discussed by Ross and Leurgans, this volume [27]). Detailed and quantitative information may be derived from these methods including (1) robust determination of relaxation rates from the experimental data with a high degree of confidence, (2) determination of the minimal number of intermediates and spectral components necessary to describe the observed relaxations and spectral changes in the three-dimensional (3-D) data matrix, and (3) analysis of the spectral shapes of individual intermediates. Data analysis software incorporating these analytic approaches is currently available from a variety of sources and is part of the commercial RSSF spectrometer packages for several systems.

In our experience, the following protocol has been helpful when embarking on the investigation of a new enzyme system with the objective of detecting and characterizing intermediates in the catalytic mechanism. Once a suitable set of chromophoric probes has been incorporated into the system, a preliminary set of RSSF experiments is carried out to survey the conditions of concentration and pH available to optimize the accumulation and detection of intermediates. These conditions usually require that the concentration of enzyme sites be relatively high (high micromolar, and usually in the 10 to 50 μM range). As a general rule, the facile detection of an intermediate by direct spectrophotometric observation of the UV/visible spectrum requires that the rate of formation be comparable to, or greater than, the rate of decay. If a spectral band of the intermediate has a large ε_{max}, then it may be possible to detect an intermediate even though

[18] J. M. Beechem, this series, Vol. 210, p. 37.
[19] E. R. Henry and J. Hofrichter, this series, Vol. 210, p. 129.

the rate of decay is severalfold greater than the rate of formation (provided that other absorption bands do not obscure the spectrum of the intermediate). (For an example, see MacGibbon et al.[20])

The two most important features of the RSSF data sets that implicate the presence of a transient intermediate are the following: (1) The transient appearance and decay of new spectral bands not associated with reactant or product species. Of course, to qualify as a possible reaction intermediate, the transient species must satisfy tests of kinetic competence. (2) The absence of true isosbestic points in the set of time-resolved spectra imply the accumulation of transient species.

Design of Chromophoric Probes

The successful application of RSSF spectroscopy to the investigation of enzyme catalytic mechanism is critically dependent on the availability of chromophoric probes that undergo spectral changes which are directly coupled to the physical and chemical events under study. Many enzymes possess suitable chromophoric cofactors and coenzymes that directly participate in catalysis either as a reactant (e.g., a coenzyme such as NADH in dehydrogenase reactions) or as a reactive functional group (e.g., a cofactor such as PLP in transamination reactions). Examples of naturally occurring chromophoric cofactors and coenzymes include NAD(P)H, flavins, pyridoxal phosphate (PLP), heme, coenzyme B_{12}, folates, type I blue copper centers, and Ni(II) centers. These coenzymes and cofactors are notable because the chromophore in each instance participates in catalysis either via covalent bonding or, in the case of metal ion cofactors, via inner sphere coordination interactions that are integral components of the catalytic reaction. The bonding interactions alter the spectrum of the chromophore, and the resulting spectral changes occur at wavelengths above 300 nm and, therefore, are well separated from the enzyme polypeptide chromophores (Phe, Tyr, Trp, Cys, and His). The naturally occurring chromophoric probes have been extensively exploited as a means for the investigation of catalytic mechanism in both static and kinetic UV/visible spectrophotometric studies, and the RSSF literature is dominated by work on a subset of these enzymes (see Brzović and Dunn[1] for a comprehensive review).

Although a few enzymes have naturally occurring chromophoric substrates, most enzymes do not require a chromophoric cofactor, coenzyme, or substrate for catalysis. If RSSF spectroscopy is to be of use in the study of enzyme systems that lack a natural chromophoric signal, then

[20] A. K. H. MacGibbon, K. Peace, S. C. Koerber, and M. F. Dunn, *Biochemistry* **26,** 3058 (1987).

some innovation is needed to introduce chromophoric probes that can be used to investigate the catalytic mechanisms of these enzymes. Where substrate, cofactor, or coenzyme specificity is sufficiently broad, it has been possible to design chromophoric analogs of the natural substrate, cofactor, or coenzyme. This approach has been particularly useful in instances where highly conjugated aromatic compounds can be substituted for the natural aliphatic substrates. For example, artificial substrates containing a chromophoric aromatic leaving group have been widely used to study the catalytic mechanisms of all classes of proteases, as well as many esterases, phosphatases, polysaccharide lyases, and sulfatases. The substrate specificities of some dehydrogenases are sufficiently broad to allow substitution of chromophoric aromatic substrates for the natural aliphatic substrates (e.g., alcohol dehydrogenase, aldehyde dehydrogenase, glyceraldehyde-3-phosphate dehydrogenase, and acyl-CoA dehydrogenase).

The best chromophoric probes (from a spectroscopic point of view) are compounds with large extinction coefficients and that undergo large spectral shifts during the course of the reaction at wavelength well separated from the intrinsic chromophores contributed by the aromatic side-chain residues of the protein. In the most informative examples, the chromophore spectrum is altered both by the covalent changes that accompany reaction and by the special microenvironment of the enzyme site. Such probes can be used to determine both the nature of the covalent chemical steps in catalysis and the nature of the weak bonding interactions and electrostatic fields at the site which are responsible for catalysis. The arylacryloyl chromophore has been a particularly successful probe in this respect. It was first used in the study of serine proteases,[2,21] then with glyceraldehyde-3-phosphate dehydrogenase,[3] alcohol dehydrogenase,[5] aldehyde dehydrogenase,[6] and, more recently, with acyl-CoA dehydrogenase[22,23] and enolyl-CoA hydratase (crotonase).[24]

The work with liver alcohol dehydrogenase (LADH) is particularly illustrative of the information that can be obtained from a suitable chromophoric probe. Dunn and Hutchison[5] discovered that the LADH-catalyzed reduction of 4-*trans*-(N,N-dimethylamino)cinnamaldehyde (DACA) occurs via a transient intermediate wherein the carbonyl oxygen of DACA forms an inner sphere coordination complex with the active site zinc ion. The function of the bonding interaction is to activate the aldehyde carbonyl

[21] M. L. Bender, G. R. Schonbaum, and B. Zerner, *J. Am. Chem. Soc.* **84**, 2540 (1962).
[22] J. T. McFarland, M. Lee, J. Reinsch, and W. Raven, *Biochemistry* **21**, 1224 (1982).
[23] J. K. Johnson and D. K. Srivastava, *Biochemistry* **32**, 8004 (1993).
[24] P. Brzović, R. D'Ordine, V. Anderson, and M. F. Dunn, unpublished results (1993).

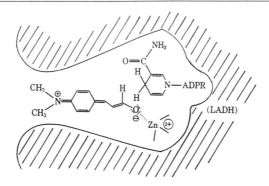

Fig. 2. Diagram depicting the active site of horse liver alcohol dehydrogenase (LADH) with 4-*trans*-(*N,N*-dimethylamino)cinnamaldehyde coordinated to the active site zinc ion. (From Dunn and Hutchison[5] with permission.) Copyright 1973 American Chemical Society.

for hydride ion attack. The catalytic interaction causes the long-wavelength $\pi \rightarrow \pi^*$ transition of DACA to shift from 398 nm (in water) to 464 nm when coordinated to the active site zinc ion of the E(NADH) complex, where E represents the enzyme (Fig. 2).[5,25] Subsequent investigations showed that the magnitude of the shift is dependent on the active site metal ion ($Co^{2+} > Zn^{2+} \approx Ni^{2+} > Cd^{2+}$),[26] the oxidation state of the coenzyme (the abortive NAD^+ complex gives a λ_{max} of 495 nm),[7,8] and the net charge at the active site (introduction of a negative charge via carboxymethylation of one of the Cys sulfhydryl groups liganded to the active site zinc ion shifts the spectrum of inner sphere coordinated DACA to shorter wavelengths).[8] As a consequence of these studies with DACA, it was established that the active site zinc ion plays a Lewis acid catalytic role and that the enzyme site provides a highly polar nonaqueous milieu which promotes formation of the "carbocation-" and "alkoxide ion-like" species (Fig. 3) that interconvert during hydride transfer. The electrostatic interactions between zinc ion, coenzyme, the protein matrix, and the reacting substrate comprise an electrostatic strain-distortion effect that "solvates" the carbocation and alkoxide ion-like species (Fig. 3) and lowers the activation energy for the transfer of hydride ion.[27]

Chromophoric cofactor analogs have been used to excellent advantage in some systems. For example, a wide variety of biological reactions are

[25] E. Cedergren-Zeppezauer, J.-P. Samama, and H. Eklund, *Biochemistry* **21**, 4895 (1982).

[26] M. F. Dunn, H. Dietrich, A. K. H. MacGibbon, S. C. Koerber, and M. Zeppezauer, *Biochemistry* **21**, 354 (1982).

[27] M. F. Dunn, *in* "Handbook on Metal–Ligand Interactions in Biological Fluids" (G. Berthon, ed.), in press. Dekker, New York, 1994.

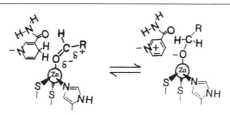

FIG. 3. Carbocation- and alkoxide ion-like structures of the reactive species postulated to be involved in the hydride transfer step of the horse liver alcohol dehydrogenase catalytic mechanism.

catalyzed by zinc metalloproteins. In most of the enzymes, the Zn^{2+} cofactor is directly involved in catalysis. Zn^{2+} is a spectroscopically "silent" metal ion. However, it is usually possible to substitute Co(II) for Zn(II), and the resulting Co(II)-substituted enzymes usually exhibit catalytic activities comparable to that of the native Zn(II)-containing enzyme. Because the metal ion coordination sphere almost invariably has a distorted tetrahedral or pentacoordinate geometry, Co(II) substitution introduces a chromophoric probe with relatively intense $d \rightarrow d$ and ligand-to-metal charge transfer (LMCT) electronic transitions that may be sensitive to the substrate interactions and protein conformation changes that occur in the catalytic cycle. The Co(II) $d \rightarrow d$ transitions of tetrahedral complexes generally exhibit extinction coefficients in the range 200 to 500 M^{-1} cm^{-1} with λ_{max} in the 520–650 nm range, pentacoordinate complexes give slightly lower values (50 to 200 M^{-1} cm^{-1}) with λ_{max} in the 500 to 600 nm range, whereas octahedral complexes give considerably weaker transitions in the 500 nm range.[28] Depending on the nature of the donor ligand(s), the Co(II) LMCT bands can be quite intense. Because the $d \rightarrow d$ transitions of tetrahedral Co(II) complexes occur at relatively low energies, the positions of these bands are very sensitive to the chemical composition of the ligand field, to distortions of bond lengths and bond angles, and to electrostatic fields in the vicinity of the metal ion. RSSF studies of Co(II)-substituted LADH (see below) have established that the Co(II) $d \rightarrow d$ and LMCT bands are sensitive probes of the chemical and conformational events that occur during reaction.[29]

The substitution of Ni(II) or Cu(II) for Zn(II) also has some potential for the introduction of a chromophoric probe. However, the Ni(II) and Cu(II) $d \rightarrow d$ and LMCT bands generally are not as useful as are the

[28] F. A. Cotton and G. Wilkinson, in "Advanced Inorganic Chemistry," 3rd Ed., p. 652 and 893. (Wiley) Interscience, New York, 1972.
[29] C. Sartorius, M. Gerber, M. Zeppezauer, and M. F. Dunn, *Biochemistry* **26**, 871 (1987).

corresponding Co(II) transitions. Therefore, Co(II) substitution has been the preferred route for introducing a chromophoric metal center into zinc metalloproteins.

Chemical modification of organic cofactors and coenzymes to introduce improved chromophoric properties has been of limited use. Replacement of the amide moiety of the $NAD(P)^+/NAD(P)H$ nicotinamide ring by the thioamide, the carboxaldehyde, or the methyl ketone functionalities shifts the $\pi \rightarrow \pi^*$ transition of the 1,4-dihydronicotinamide ring of the reduced coenzyme from 340 to 375 nm [3-thionicotinamide-NAD(P)H], to 358 nm [3-pyridinealdehyde-NAD(P)H], or to 363 nm [3-acetylpyridine-NAD(P)H].[30] These coenzyme derivatives are chemically functional analogs for most NAD(P)H-requiring dehydrogenases. The red-shifted spectra of the analogs can be used to good advantage in situations where the 340 nm region of the spectrum is obscured by overlapping spectral bands from other components of the system (see Buckley and Dunn[6] for an RSSF example).

The use of analogs to investigate reaction mechanisms has been of considerable importance to the field of enzymology. In most instances, it has been found that the analogues follow reaction pathways with ground states and transition states qualitatively similar to those of the natural substrates. However, the highly conjugated π systems of chromophoric analogs usually alter the free energies of ground states and activated complexes. This may have the advantage of causing certain intermediates that, for kinetic reasons, do not accumulate with the natural substrate to become detectable species (sometimes even quasi-stable species) in the analog reaction. This happenstance can make it possible to acquire information about the mechanism of catalysis that otherwise would be difficult to obtain. However, the investigator must be alert to the possibility that the analog might follow a pathway that diverges at some point from that of the natural substrate. This could result in an alteration of the roles played by site functional groups in catalysis and/or the generation of an unexpected reaction product or the accumulation of an unexpected chemical intermediate. Although such events can raise confusing issues about mechanism, an altered pathway can give useful insights about the properties of catalytic residues even though the reaction catalyzed is bogus.

Significant Applications: Use of Rapid-Scanning Stopped-Flow Spectroscopy to Investigate Enzyme Structure–Function Relationships

Rapid-scanning stopped-flow techniques have been used to investigate ligand binding and enzyme catalysis for a wide variety of proteins that

[30] Pabst Laboratories Circular OR-18, 3rd printing, Milwaukee, Wisconsin, July, 1965.

require naturally occurring chromophoric cofactors. Changes in the UV/visible spectroscopic properties of the cofactor which accompany ligand binding and/or catalysis provides invaluable information for the elucidation of enzyme mechanism and protein structure–function relationships. These systems include dehydrogenases (see below), pyridoxal phosphate-dependent enzymes (see below), flavoproteins, and a large number of hemoproteins, including myoglobin and hemoglobin, cytochrome oxidase, and a number of catalases and peroxidases (see Brzović and Dunn[1] for a comprehensive review of the RSSF literature).

One emphasis of research in our laboratory has been the investigation of the catalytic mechanisms of NAD^+-requiring dehydrogenases.[5–8,27] A second area of research has focused on PLP-dependent enzymes that catalyze either β or γ elimination/replacement reactions.[1,31,32] RSSF spectroscopy has been particularly useful for the characterization of the reaction mechanism, allosteric interactions, and structure–function relationships in the tryptophan synthase multienzyme complex from enteric bacteria.

To illustrate the use of RSSF spectroscopy to study enzyme catalysis, we give two systems detailed treatment herein, namely, horse liver alcohol dehydrogenase and *Salmonella typhimurium* tryptophan synthase. For an inclusive review of RSSF spectroscopy applications in the field of enzymology, see Brzović and Dunn.[1]

Investigations of Catalytic Mechanism of Active-Site Co(II)-Substituted Liver Alcohol Dehydrogenase (NADH)

The specific substitution of Co^{2+} for the active site Zn^{2+} of horse liver alcohol dehydrogenase (LADH) achieved by the Zeppezauer group[33,34] opened the way for mechanistic studies of LADH that have greatly enriched our knowledge of the catalytic mechanism. Through the use of RSSF UV/visible spectroscopy and Co(II)-substituted LADH, new understanding concerning the roles played by (1) protein conformation, (2) the active site metal ion, (3) the ionization of coordinated alcohol, and (4) electrostatic force fields in catalysis has been obtained.[26,27,29]

The spectrum of active-site-substituted Co(II)LADH contains LMCT bands in the 300–450 nm region and $d \rightarrow d$ transitions characteristic of tetrahedral Co(II) in the 500–700 nm region.[33] The positions and intensities

[31] W. F. Drewe, Jr., and M. F. Dunn, *Biochemistry* **24**, 3977 (1985).
[32] W. F. Drewe, Jr., and M. F. Dunn, *Biochemistry* **25**, 2494 (1986).
[33] W. Maret, I. Anderson, H. Dietrich, H. Schneider-Berhlöur, R. Einarsson, and M. Zeppezauer, *Eur. J. Biochem.* **98**, 501 (1979).

of the spectral bands are sensitive to (a) both the presence of and the oxidation state of bound coenzyme, (b) the nature of the fourth (exogenous) ligand coordinated to the active site metal ion, and (c) the pH (Fig. 4).[29] The three-dimensional structures of the native enzyme and of binary and ternary complexes establish that monodentate ligands form pseudotetrahedral $Zn(II)NS_2X$ complexes with the active site metal ion where X is an exogenous monodentate ligand, N is His-67, and S is Cys-46 and Cys-174. Co(II) substitution give crystalline complexes which are isomorphous

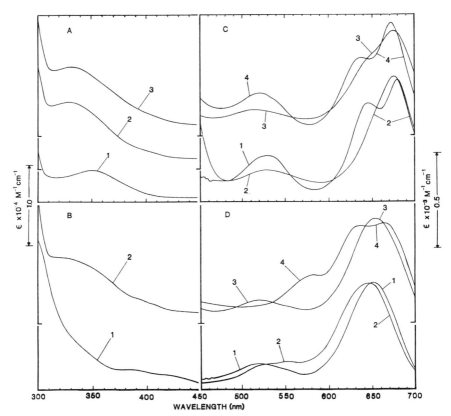

FIG. 4. Spectra of Co(II)E complexes at 25° and (A, B) 300–450 nm or (C and D) 450–700 nm. Spectrum A-1, Co(II)E, pH 5.6; A-2, Co(II)E(NADH), pH 9.0; A-3, Co(II)E(NADH, Pyr), pH 9.0; B-1, Co(II)E(NAD$^+$–Pyr); B-2, Co(II)E(NADH,IBA), pH 9.0; C-1, Co(II)E (NAD$^+$–Pyr), pH 9.0; C-2, Co(II)E(NADH,IBA), pH 9.0; C-3, Co(II)E(NADH), pH 9.0; C-4, Co(II)E(NADH,Pyr), pH 9.0; D-1, Co(II)E, pH 6.3; D-2, Co(II)E, pH 9.0; D-3, Co(II)E (NAD$^+$), pH 5.6; D-4, Co(II)E(NAD$^+$), pH 9.0. For clarity in viewing the spectra, each graph is separated by vertical offsets indicated on the ordinates. Spectra were measured with the RSSF spectrophotometer to take advantage of the time-resolving capabilities of the instrument. (From Sartorius et al.[29] with permission.) Copyright 1987 American Chemical Society.

(by X-ray diffraction methods) with the corresponding Zn(II)–enzyme complexes.[34] Figure 4 shows that the formation of binary and ternary complexes causes characteristic perturbations of the positions and intensities of the LMCT and $d \rightarrow d$ transitions.

Strategies for Transient Kinetic Measurements. To simplify the kinetic behavior of the system, two sets of experimental conditions were selected. In the direction of aldehyde reduction, reaction was carried out under conditions where both aldehyde and NADH were present in large excess relative to the initial enzyme concentration, $[E]_0$. To achieve the limitation of reaction to a single turnover of sites, the reaction was carried out in the presence of the potent inhibitor pyrazole (Pyr). Pyrazole reacts rapidly and quasi-irreversibly to trap enzyme-bound NAD^+ product in the form of a covalent adduct at the 4-position of the nicotinamide ring [Eq. (2)].

$$E(NADH) + RCHO + H^+ \rightleftharpoons \cdots \rightleftharpoons E(NAD^+, RCH_2OH)$$
$$\rightleftharpoons E(NAD^+) + RCH_2OH \qquad (2)$$
$$E(NAD^+) + Pyr \rightarrow E(NAD-Pyr)_{adduct}$$

Because Pyr binds only weakly to the E(NADH) complex, conditions were chosen such that the Pyr-NAD^+ reaction limits the enzyme to a single turnover of sites without interference with the preceding steps.[35,36]

In the direction of alcohol oxidation, no potent quasi-irreversible inhibitor was available for trapping the E(NADH) complex. However, NADH dissociation from the binary complex is rate-determining for turnover in this direction. Therefore, the binding of isobutyramide (IBA) to the E(NADH) complex decreases both the NADH dissociation constant (K_D) and the rate of NADH dissociation [Eq. (3)].[37,38] Because IBA binds tightly only to the E(NADH) complex, the presence of IBA greatly slows the steady-state rate of alcohol oxidation but does not perturb the pre-steady-state phase of the reaction. Therefore, the RSSF spectral changes that occur during a single turnover cycle of the Co(II)E catalyzed reactions of aldehydes with NADH were determined in the presence of Pyr; the pre-steady-state RSSF spectral changes in the Co(II)E catalyzed reactions of alcohols with NAD^+ were determined in the presence of IBA.[29] To investigate the pH dependence of the time courses, RSSF studies were

[34] G. Schneider, H. Eklund, E. Cedergren-Zeppezauer, and M. Zeppezauer, *Proc. Natl. Acad. Sci. U.S.A.* **80,** 5289 (1983).

[35] J. T. McFarland and S. A. Bernhard, *Biochemistry* **11,** 1486 (1972).

[36] N. H. Becker and J. D. Roberts, *Biochemistry* **23,** 3336 (1984).

[37] P. L. Luisi and E. Bignetti, *J. Mol. Biol.* **88,** 653 (1974).

[38] J. Kvassman and G. Pettersson, *Eur. J. Biochem.* **87,** 417 (1978).

carried out over the pH range 4.8 to 9.0 using a ''pH-jump'' method[39] to examine pH regions where enzyme and/or coenzyme is unstable.

Studies of Aldehyde Reduction. The results of the work of Sartorius *et al.*[29] are summarized in the RSSF data sets shown in Figs. 5 through 9. The data in Fig. 5 and 6 are typical of the spectral changes that occur under conditions of high and low pH during aldehyde reduction. The data shown are for the reduction of benzaldehyde at pH 9.0 (Fig. 5A,B) and at pH 5.6 (Fig. 5C,D). Single-wavelength time courses are shown in the insets to the right of each panel in Fig. 5. In Fig. 5A,B, the complete set of time-resolved spectra are shown in A-1 and B-1, respectively. Casual inspection of the RSSF data reveals that at pH 9 the reaction involves the formation and decay of an intermediate. This also is obvious from the single-wavelength time courses measured at 397 nm (Fig. 5Ad and 5Ba, c, and d). In Fig. 5A,B, the spectral changes that occur in each phase are shown as subsets A-2 and B-2 (the fast phase), and A-3 and B-3 (the slow phase). The changes which occur in the two phases are further delineated in the difference spectra subsets shown in Fig. 6A,B. From close inspection of the spectra, it is clear that intermediate formation, as measured by the appearance of a new $d \rightarrow d$ band centered at approximately 570 nm (Figs. 5B and 6C), is accompanied by NADH oxidation (Figs. 5A and 6A). On the other hand, disappearance of the intermediate is concomitant with formation of the Co(II)E(NAD–Pyr) adduct (Fig. 6B,D). These observations are consistent with the assignment of the intermediate as a ternary complex of enzyme, NAD^+, and some form of benzyl alcohol. As the pH is decreased below 7.5, the amplitude of the transient 570 nm band decreases, and at pH 5.6 (Fig. 5C,D), the time-resolved spectra are characterized by sharp isosbestic points, indicating that reaction intermediates do not accumulate under these conditions.

Studies of Alcohol Oxidation. When the reaction was investigated from the direction of alcohol oxidation under pre-steady-state conditions in the presence of IBA, the time-resolved spectra obtained from RSSF measurements again show evidence for the formation of a transient intermediate in the NAD^+-mediated oxidation of benzyl alcohol.[29] Data collected at pH values of 9.0, 5.6, and 4.8 are shown in Figs. 7 and 8. In the wavelength region 300 to 450 nm and at pH 9.0, the time-resolved spectra are characterized by a fast, pre-steady-state (exponential) phase dominated by the appearance of bound NADH. This process is followed by an approximately zeroth-order (steady-state) phase in which free NADH is generated by multiple turnovers. The difference spectra in Fig. 7C,D compare the changes which occur in the pre-steady-state phase with those in the steady

[39] R. G. Morris, G. Saliman, and M. F. Dunn, *Biochemistry* **19,** 725 (1980).

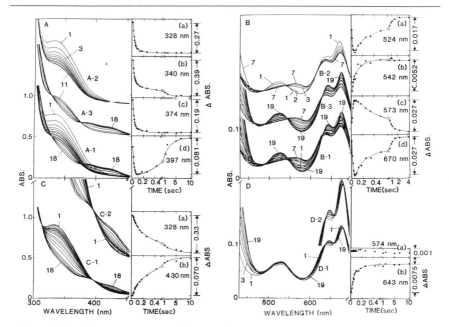

FIG. 5. Rapid-scanning stopped-flow UV/visible spectra (numbered in chronological order) showing the spectral changes (300–700 nm) that occur during the reduction of benzaldehyde by NADH catalyzed by Co(II)E at pH 9 (A, B) and pH 5.6 (C, D) and 25°. The repetitive scan rate was 8.605 msec/scan. At longer times, delays were introduced between scans to give the desired spacing. The actual timing patterns used are indicated by the data points shown in the single-wavelength time courses. The complete sets of time-resolved spectra for a single turnover are shown respectively in A-1, B-1, C-1, and D-1. To aid in visualizing the spectral changes at pH 9, subsets of spectra for the fast phase (A-2 and B-2) and for the slow phase (A-3 and B-3) are shown offset above the complete sets. In (C) and (D), the apparent isosbestic points are shown on expanded scales in C-2 and D-2. Single-wavelength time courses at selected wavelengths are shown on the right-hand side of each graph. Conditions after mixing were as follows: (A) [Co(II)E], 35 μN; [NADH], 66.6 μM; [benzaldehyde], 2 mM; [Pyr], 25 mM; 0.1 M sodium pyrophosphate and 50 mM Bis–Tris, final pH 9.0. (B) [Co(II)E], 105 μN; [NADH], 172 μM; [benzaldehyde], 2 mM; [Pyr], 25 mM; 50 mM glycine hydrochloride and 50 mM Bis–Tris, final pH 9.0. (C) [Co(II)E], 35 μN; [NADH], 66.6 μM; [benzaldehyde], 2 mM; [Pyr], 75 mM; 50 mM sodium citrate and 50 mM Bis–Tris, final pH 5.6. (D) [Co(II)E], 90 μN; [NADH], 110 μM; [benzaldehyde], 2 mM; [Pyr], 75 mM; 0.1 M MES and 33 mM Bis–Tris, final pH 5.6. In each experiment, the final pH was achieved by mixing Co(II)E in pH 6.5 Bis–Tris buffer with NADH, benzaldehyde, and Pyr preincubated in a second buffer system (i.e., sodium citrate, MES, glycine, or sodium pyrophosphate as indicated above). (From Sartorius et al.[29] with permission.) Copyright 1987 American Chemical Society.

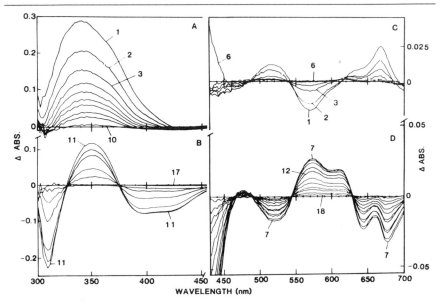

FIG. 6. Difference spectra (numbered in chronological order) calculated from the data given in Fig. 5A,B showing the changes which occur during the fast (A, C) and slow (B, D) relaxations. Difference spectra in (A) were calculated by subtracting spectrum 11 of Fig. 5A from spectra 1–10; difference spectra in (C) were calculated by subtracting spectrum 7 of Fig. 5B from spectra 1–6. The difference spectra in (B) were calculated by subtracting spectrum 18 of Fig. 5A from spectra 11–17; difference spectra in (D) were calculated by subtracting spectrum 18 of Fig. 5B from spectra 7–17. (From Sartorius et al.[29] with permission.) Copyright 1987 American Chemical Society.

state. The pre-steady-state changes are consistent with the rapid oxidation of bound alcohol, the concomitant formation of bound NADH, and the perturbation of LMCT bands from the Co(II) center. In the 450 to 700 nm region (Fig. 8A), the time-resolved spectra show that a transient species with a λ_{max} of 570 nm is formed very rapidly and decays with a rate that is similar to the rate of appearance of enzyme-bound NADH (compare the insets to Figs. 7Aa and 8Aa). It is notable that the $d \rightarrow d$ transitions of the species formed in the very fast initial phase of the benzyl alcohol reaction are highly similar both to those of the stable $Co(II)E(NAD^+,$ trifluoroethanol) complex (Fig. 9) and to the transient species seen in the reduction of benzaldehyde. Below pH 7.5, the amount of intermediate that accumulates decreases. Traces of the intermediate are detectable in the spectra collected at pH 5.6 (Fig. 8C). At pH 4.8, this species could not be detected (Fig. 8B,D). At that pH, the spectral changes indicate

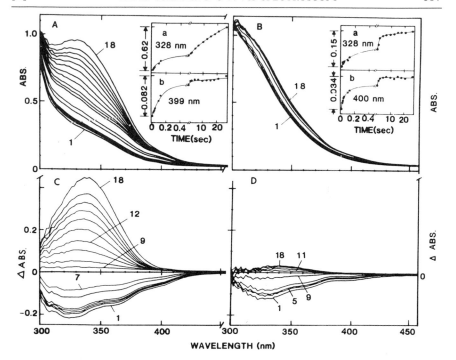

FIG. 7. Time-resolved spectra and difference spectra for the pre-steady state and steady-state phases of the Co(II)E-catalyzed oxidation of benzyl alcohol by NAD$^+$ at pH 9 (A, C) and pH 4.8 (B, D) and 25° for the wavelength range 300–450 nm. Scanning was carried out as described in the caption to Fig. 5. Difference spectra in (C) and (D) were calculated by subtracting the last spectrum collected in the pre-steady-state phase [respectively spectrum 8 of (A) and spectrum 10 of (B)] from all other spectra in the set. Single-wavelength time courses are shown in insets a and b of (A) and (B). Conditions after mixing were as follows: (A, C) [Co(II)E], 29 μN; [NAD$^+$], 1.4 mM; [benzyl alcohol], 2 mM; [IBA], 50 mM; 50 mM glycine and 50 mM Bis–Tris, final pH 9.0; (B, D) [Co(II)E], 29 μN; [NAD$^+$], 3.5 mM; [benzyl alcohol], 4 mM; [IBA], 50 mM; 10 mM H$_2$SO$_4$ and 50 mM Bis–Tris, final pH 4.8. Reaction was initiated by mixing enzyme in Bis–Tris buffer (pH 6.5) with NAD$^+$, benzyl alcohol, and IBA preincubated in the above-indicated solutions. (From Sartorius et al.[29] with permission.) Copyright 1987 American Chemical Society.

that reaction proceeds from a species which resembles the Co(II)E(NAD$^+$) complex to the Co(II)E(NADH,IBA) complex.

Because no intermediate was detected at low pH, two explanations were considered. (1) The binding of substrate is pH-dependent, and, at low pH, insufficient substrate was used to saturate binding, therefore, preventing undetectable amounts of intermediate from accumulating under these conditions. (2) Alternatively, the spectrum of the intermediate is pH-

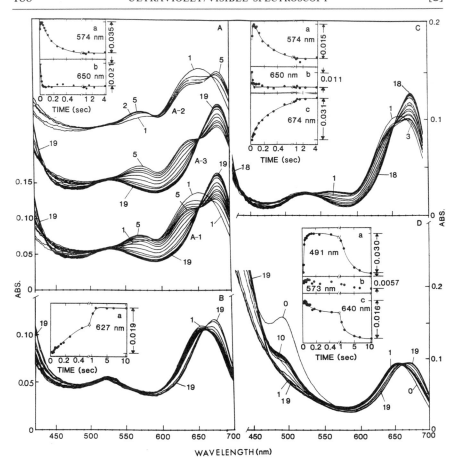

FIG. 8. Rapid-scanning visible spectra showing the changes in the Co(II)E spectral bands during the pre-steady-state and steady-state phases of benzyl alcohol oxidation by NAD$^+$ at pH 9 (A), pH 4.8 (B, D), and pH 5.6 (C) and 25°. Scanning was carried out as described in the caption to Fig. 5. The pre-steady-state reaction at pH 9.0 (A) occurs in two kinetic relaxations; the spectra in A-1 show the changes in both relaxations as the steady state is approached; the changes which occur in each relaxation are shown offset in A-2 (the fast relaxation) and A-3 (the slow relaxation), respectively. The time-resolved spectra in (B) and (D) show the changes which occur in the presence (D) and absence (B) of a trace of DACA. Spectra 1–19 in (D) are the time-resolved spectra for the reaction of Co(II)E with NAD$^+$ and benzyl alcohol in the presence of IBA and DACA. In (D), spectrum 0 is the spectrum of the Co(II)E(NAD$^+$,DACA) complex in the absence of benzyl alcohol (but with IBA present) measured in a separate experiment. Conditions after mixing were as follows: (A) [Co(II)E], 101 μN; [NAD$^+$], 1.58 mM; [benzyl alcohol], 24.1 mM; [IBA], 45 mM; 50 mM glycine–50 mM Bis and Tris, final pH 9.0; (B) [Co(II)E], 89 μN; [NAD$^+$], 3.5 mM; [benzyl alcohol], 4 mM; [IBA], 50 mM; 10 mM H$_2$SO$_4$ and 50 mM Bis and Tris, final pH 4.8; (C) [Co(II)E], 103 μN; [NAD$^+$], 3.95 mM; [benzyl alcohol], 1.8 mM; [IBA], 45 mM; 7.5 mM

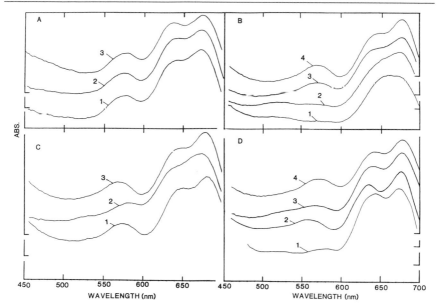

FIG. 9. Comparison of the spectra of transient intermediates formed during alcohol oxidation (B, C) and aldehyde reduction (D) with the spectrum of the Co(II)E(NAD⁺,trifluoroethanol) complex (A) in the 450–700 nm region at 25°. (A) Spectra of the Co(II)E (NAD⁺,trifluoroethanol) complex: 1, pH 4.3; 2, pH 6.23; 3, pH 9.0. The spectra shown in (B), (C), and (D) have been selected from RSSF data sets (such as those presented in Figs. 5 and 8) to show the maximum amount of 570 nm intermediate. (B) Oxidation of benzyl alcohol: 1, pH 4.8; 2, pH 5.6; 3, pH 6.8; 4, pH 9. (C) Comparison of the transient intermediates formed during the oxidation of (1) 4-nitrobenzyl alcohol, (2) ethanol, and (3) benzyl alcohol, all at pH 9. (D) Comparison of the transient intermediates formed during the oxidation of (1) acetaldehyde, (2) anisaldehyde, (3) benzaldehyde, and (4) 4-nitrobenzaldehyde, all at pH 9. The amplitudes of the spectra have been normalized to the same concentration of Co(II)E. (From Sartorius et al.[29] with permission.) Copyright 1987 American Chemical Society.

dependent, and therefore a protonated transient species with a spectrum closely resembling the spectrum of the Co(II)E(NAD⁺) complex is formed as an intermediate. These possibilities were tested by using 4-*trans-p-* (*N,N*-dimethylamino)cinnamaldehyde (DACA) as an indicator of ligand binding to the substrate site of the Co(II)E(NAD⁺) complex.

H₂SO₄ and 50 mM Bis–Tris, final pH 5.6; (D) spectra 1–19, [Co(II)E], 108 μN; [NAD⁺], 3.95 mM; [benzyl alcohol], 4 mM; [DACA], 4.14 μM; [IBA], 50 mM; 10 mM H₂SO₄ and 50 mM Bis–Tris, final pH 4.8; (D) spectrum 0, same as for spectra 1–19 except that no benzyl alcohol was present. Premixing conditions were similar to those described in Figs. 5 and 7. (From Sartorius et al.[29] with permission.) Copyright 1987 American Chemical Society.

The experiments of Dahl and Dunn[7] established that DACA binds very rapidly and reversibly to give an abortive ternary E(NAD$^+$,DACA) complex wherein the carbonyl oxygen of DACA is coordinated to the active site metal ion. This bonding interaction shifts the $\pi \to \pi^*$ transition of DACA from a λ_{max} of 398 nm (in water) to 495 nm in the zinc–enzyme ternary complex. In the Co(II)-substituted enzyme system, the DACA λ_{max} is shifted to 499 nm. Because the binding of DACA occurs with association and dissociation rate constants that are rapid relative to the oxidation of benzyl alcohohl, the 499 nm band of the Co(II)E(NAD$^+$, DACA) ternary complex can be used to monitor the binding and reaction of this substrate. Figure 8D compares the spectrum of the Co(II)E(NAD$^+$, DACA) complex (spectrum 0) with the set of RSSF spectra collected under conditions where the Co(II)E enzyme is reacted with DACA, NAD$^+$, benzyl alcohol, and IBA at pH 4.8 (spectra 1–19). The spectra establish that during the experiment dead time, the amount of the 499 nm DACA ternary complex formed (spectra 1–10, Fig. 8D) is decreased in amplitude by about two-thirds, indicating that benzyl alcohol and DACA compete for the substrate site under these conditions. As oxidation progresses in the first turnover (spectra 10–19, Fig. 8D), the residual DACA complex disappears as all pre-steady-state enzyme species are converted to the Co(II)E(NADH,IBA) complex. The experiment shows that benzyl alcohol binds sufficiently tightly at pH 4.8 to displace DACA, and therefore an unfavorable pH dependence for benzyl alcohol binding can be eliminated as the explanation for why the 570 nm intermediate is not detected in the RSSF time course.

Structural Assignment for 570 nm Intermediate. The RSSF studies presented in Fig. 8 establish that an intermediate with some form of benzyl alcohol bound to the Co(II)E(NAD$^+$) complex accumulates during the catalytic cycle at alkaline pH. Comparison of the $d \to d$ transitions of the complex with the spectra of stable Co(II)E(NAD$^+$,ligand) ternary complexes (Fig. 9) shows a strong correlation between the spectrum of the intermediate and the spectra of ternary complexes containing NAD$^+$ and monovalent anions and also to the Co(II)E(NAD$^+$,trifluoroethanol) complex. These similarities provide strong evidence in support of the assignment of the 570 nm absorbing intermediate as the ternary complex of NAD$^+$ and the coordinated alkoxide ion of benzyl alcohol. RSSF studies at low pH suggest that the pH dependence of intermediate formation is a consequence of the ionization of the inner sphere coordinated alcohol. Therefore, these studies imply an apparent pK_a between 6 and 7 for the ternary Co(II)E(NAD$^+$,benzyl alcohol) complex. The pH dependence of the Co(II)E(NAD$^+$) binary complex is consistent with the ionization of coordinated water to coordinated hydroxide ion with an apparent pK_a between 7 and 8.[38]

Mechanistic Implications. The native Zn(II)-substituted enzyme has been shown to undergo a conformational change from an "open" structure, where the active site clefts are solvent accessible, to a "closed" structure on formation of ternary complexes where the substrate binding cleft becomes a narrow hydrophobic tunnel from which solvent is excluded.[25] The $d \to d$ transitions of the Co(II)-substituted enzyme are sensitive both to the change from an open to a closed conformation and to the electronic properties of inner sphere coordinated ligands. It is likely that in the enzyme–NAD$^+$ complexes, the ionization of the coordinated ligand (water, benzyl alcohol, or trifluoroethanol) is accompanied by a change in protein conformation from the open conformation to the closed conformation.[29]

If the pH dependencies of the spectra of Co(II)E ternary complexes are correctly interpreted as arising from the ionization of coordinated ligands in the benzyl alcohol, trifluoroethanol, and water systems, then the pK_a values of these coordinated species (between 4 to 8) are much lower than the pK_a values of the free species. In aqueous solution, the pK_a of benzyl alcohol is 15.1, that of trifluooethanol is 12.3, and that of water is 15.6.[40] Therefore, the electrostatic interactions provided by the metal ion together with NAD$^+$ and the protein matrix must perturb the pK_a of the ligand by about 10 orders of magnitude (in comparison to an aqueous milieu). Coordination to the active site metal ion must contribute a significant portion of this perturbation. Water pK_a values in the range of 8.7 to 9.3 have been reported for various small-molecule zinc complexes containing a coordinated water molecule.[27] The binding of NAD$^+$ introduces an additional electrostatic component from the positively charged nicotinamide ring.[7] Thus, the apparent pK_a of the coordinated water molecule is decreased from 9.2 to 7.6.[38] Dipolar residues from the protein also could contribute to the electrostatic field to make the site more electropositive. Taken together, these findings support a mechanism for LADH in which catalysis of hydride transfer occurs within the closed conformation of the enzyme. This closed conformation provides a highly polar, nonaqueous environment.[25,27]

In summary, the RSSF studies of Co(II)-substituted LADH detect and characterize a transient intermediate that is proposed to be the Co(II)E (NAD$^+$,alkoxide ion) complex. The spectral changes in the Co(II) $d \to$ d transitions imply that the pK_a of the coordinated alcohol must be perturbed

[40] P. Ballinger and F. A. Long, *J. Am. Chem. Soc.* **82,** 795 (1960).

(α -REACTION)

GLU 49
ASP 60
H+

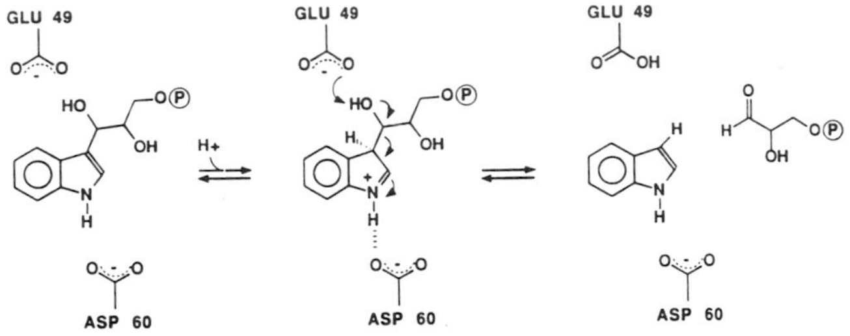

(β -REACTION)

Stage I

ENZ
H+
E(Ain)
410nm
E(GD)
310-340nm
E(Aex1)
422nm
E(Q1)
460nm?
-OH
E(A-A)
350nm
shoulder 380-530nm

Stage II

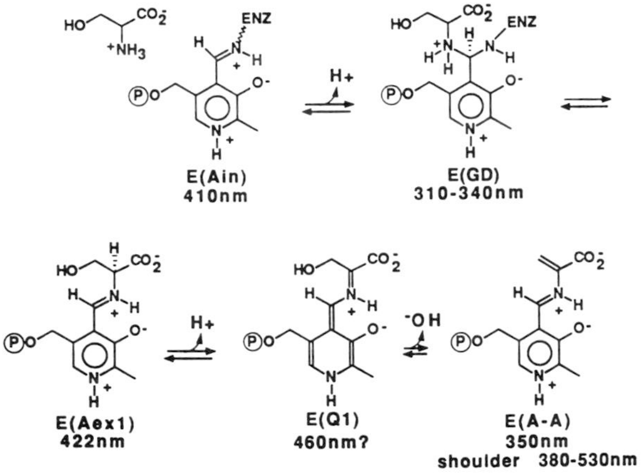

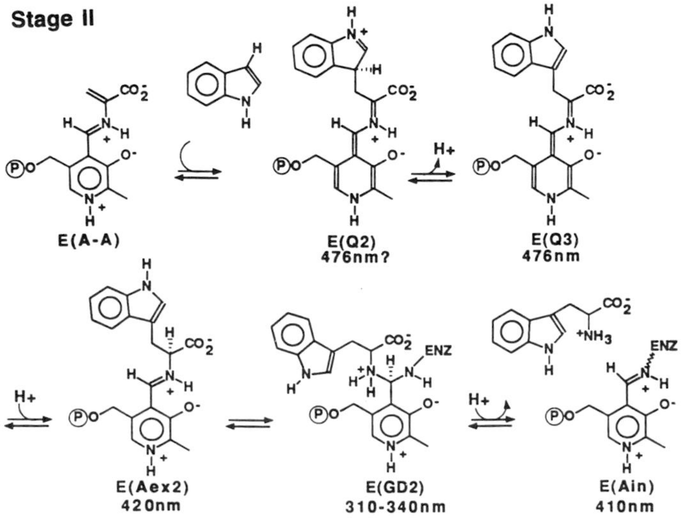

E(A-A)
H+
E(Q2)
476nm?
E(Q3)
476nm
H+
E(Aex2)
420nm
ENZ
E(GD2)
310-340nm
H+
E(Ain)
410nm

to a lower value by about 10 orders of magnitude in comparison to the pK_a of the free alcohol. Furthermore, changes in the $d \rightarrow d$ transitions that accompany ternary complex formation are consistent with a change in protein conformation from an open to a closed structure. These data are consistent with a catalytic mechanism wherein the highly electropositive LADH site stabilizes the highly polar hydride transfer transition state via an electrostatic strain–distortion effect.[27]

Investigations of Reactions Catalyzed by $\alpha_2\beta_2$ Tryptophan Synthase Bienzyme Complex

The tryptophan synthase bienzyme complex from enteric bacteria is composed of heterologous α and β subunits arranged in a nearly linear $\alpha-\beta-\beta-\alpha$ array.[41] The α subunit catalyzes the aldolytic cleavage of 1-(indol-3-yl)glycerol 3-phosphate (IGP) to indole and glyceraldehyde 3-phosphate (G3P) (α-reaction, Fig. 10), whereas the β subunit catalyzes the PLP-dependent condensation of L-Ser and indole to yield L-Trp (β reaction, Fig. 10). The overall $\alpha\beta$ reaction catalyzed by the bienzyme complex is essentially the sum of the individual α and β reactions. Indole, the common intermediate produced at the α site, is channeled to the β site via a tunnel located in the interior of the protein complex that directly interconnects the α and β catalytic centers.[41,42] Although the individual subunits may be isolated and are functional, formation of the bienzyme complex not only increases the catalytic activities of the separate subunits by nearly 100-fold, but also alters the thermodynamic stability of β site reaction intermediates. Thus, the tryptophan synthase system is considered to be a prototype multienzyme complex in which metabolites are directly channeled between successive metabolic enzymes. Allosteric interactions between the α and β subunits are essential for channeling in this system and serve to coordinate catalytic events between the heterologous

[41] C. C. Hyde, A. Ahmed, E. A. Padlan, E. W. Miles, and D. R. Davies, *J. Biol. Chem.* **263**, 17857 (1988).
[42] M. F. Dunn, V. Aguilar, P. S. Brzović, W. F. Drewe, Jr., K. F. Houben, C. A. Leja, and M. Roy, *Biochemistry* **29**, 8598 (1990).

FIG. 10. Mechanism of the reactions catalyzed by both the α and β subunits of the tryptophan synthase bienzyme complex. The α reaction involves cleavage of IGP. The β reaction involves two stages: stage I is reaction of L-Ser with E(Ain) to give E(A-A); stage II is reaction of indole with E(A-A) to give L-Trp. Indole, the common intermediate, is directly channeled between the two catalytic centers via a 25-Å-long tunnel that interconnects the α and β active sites through the interior of the protein. See text for details.

active sites in the complex.[42,43] RSSF spectroscopy has been an invaluable tool for the investigation of the reaction mechanism catalyzed by the PLP-dependent β subunit of the $\alpha_2\beta_2$ complex. RSSF spectroscopy is also a powerful method for studying both allosteric interactions between the heterologous α and β subunits of the bienzyme complex and the effects of site-specific mutants on both catalysis and allostery in the bienzyme complex.

Spectroscopic Characterization of Reaction of L-Serine at β Active Site of $\alpha_2\beta_2$ Tryptophan Synthase Multienzyme Complex. The reactions catalyzed at the PLP-dependent β site may be divided into two separate stages (Fig. 10). Stage I involves the reaction of L-Ser with the PLP cofactor to form the electrophilic α-aminoacrylate intermediate, E(A-A). During stage II, indole is activated as a nucleophile which rapidly reacts with the E(A-A) intermediate to ultimately yield L-Trp. The RSSF time-resolved spectral changes that occur during both stage I and stage II of the wild-type enzyme-catalyzed β reaction are shown in Fig. 11.[31,32] Examination of Fig. 11 shows the transient formation of several reaction intermediates at the β subunit active site during the course of the β reaction. The sequence of catalytic events and the positions and band shapes of the spectra of intermediates formed give important clues regarding the chemical structure of the observed species. Spectroscopic assignments are further facilitated by the wealth of information in the literature regarding the structure both of PLP-like model compounds in solution and of stable enzyme–substrate complexes observed in other PLP-dependent enzyme systems.[44] In certain cases, isotopically labeled substrates and substrate analogs have been used to influence preferentially the accumulation of specific reaction intermediates and aid in the identification of species. Furthermore, utilization of calculated difference spectra and derivative spectra has been helpful in resolving overlapping spectral bands and has resulted in the identification of most of the expected intermediates involved in the synthesis of L-Trp catalyzed by the β subunit.

In the native enzyme, the PLP cofactor is covalently bound at the β site via a Schiff base linkage [the internal aldimine, E(Ain)] to the ε-amino group of an active site lysine residue. This configuration of the cofactor gives rise to an absorption band centered at 410 nm with a shoulder at shorter wavelengths (spectrum 0, Fig. 11A). On mixing with L-Ser (Fig. 11A), there is a rapid formation of a new spectral band $(1/\tau_1)$ with a λ_{max}

[43] P. S. Brzović, K. Ngo, and M. F. Dunn, *Biochemistry* **31**, 3831 (1992).

[44] R. G. Kallen, T. Korpella, A. E. Martell, Y. Matsushima, D. E. Metzler, T. V. Morozov, I. M. Ralston, F. A. Savin, Y. M. Torchinshky, and H. Ueno, *in* "Transaminases" (P. Christen and D. E. Metzler, eds.), p. 37. Wiley, New York, 1985.

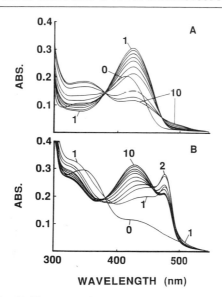

FIG. 11. Time-resolved RSSF spectra for both stage I (A) and stage II (B) of the β-reaction catalyzed by the tryptophan synthase $\alpha_2\beta_2$ bienzyme complex from *S. typhimurium* at pH 7.8 and 25°. (A) RSSF spectra for the reaction of 40 m*M* L-Ser with 10 μM $\alpha_2\beta_2$ bienzyme complex. (B) RSSF spectra for the reaction of 1 m*M* indole with 13 μM $\alpha_2\beta_2$ bienzyme complex preequilibrated with 40 m*M* L-Ser. Spectra for both (A) and (B) were collected 8.53, 17.1, 25.6, 34.1, 42.6, 59.7, 85.3, 170.6, 255.8, and 341.1 msec after flow had stopped and mixing was complete. All concentrations refer to reaction conditions immediately after mixing.

of 422 nm that accumulates within the mixing dead time of the stopped-flow instrument (typically 3 to 4 msec.) This intermediate has been shown to correspond to the external aldimine, E(Aex1), (Fig. 10) formed between L-Ser and the PLP cofactor.[31] When L-α-[^{2}H]Ser is substituted for isotopically normal L-Ser, the transient species with λ_{max} 422 nm also is rapidly formed after mixing (data not shown). However, the relative absorbance at 422 nm is larger, the spectral bandwidth is narrower than that observed with the isotopically normal compound, and the decay of this species to form the E(A-A) intermediate is correspondingly slower. These findings indicate that α-proton abstraction is partially rate-determining in the formation of the E(A-A) complex. Second, comparison of the transient (E(Aex1) spectral bands observed with isotopically normal L-Ser and L-α-[^{2}H]Ser provides evidence that the L-Ser quinonoid with a λ_{max} around 460 nm, E(Q1), is present as a minor species which contributes to the spectral changes observed during stage I of the β reaction.

The 422 nm species subsequently decays in a biphasic process ($1/\tau_2 =$

$10 \text{ sec}^{-1} > 1/\tau_3 = 1 \text{ sec}^{-1}$) to give the spectrum of the E(A-A) complex (Fig. 11A, spectrum 8). Because the reaction of L-Ser to form E(A-A) is reversible, the final spectrum at equilibrium is composed of a distribution of stage I reaction intermediates.

Characterization of Reaction of Indole with E(A-A). The RSSF spectra for stage II of the β reaction are shown in Fig. 11B. The reaction of the wild-type E(A-A) complex with indole is dominated by the rapid accumulation ($1/\tau = 200 \text{ sec}^{-1}$) of a new species characterized by a spectral band with a λ_{max} of 476 nm (Fig. 11B). This spectral band corresponds to the L-Trp quinonoidal species, E(Q3),[32] that accumulates after the formation of the C–C bond between the β carbon of E(A-A) and indole (Fig. 10). Formation of E(Q3) is followed by the accumulation of another species with a spectral band centered at 420 nm. The position and shape of the 420 nm spectral band and the observed sequence of catalytic events identify this species as the L-Trp external aldimine, E(Aex2).[32] As the reaction approaches the steady state, the spectrum of E(Q3) is evident as a shoulder at 476 nm adjacent to the larger (E(Aex2) band and is present only as a minor component in the spectrum.

Identification and characterization of the transient spectral intermediates involved in the β reaction provided an important foundation for the investigation of reciprocal allosteric interactions between the heterologous α and β subunits and enzyme structure–function relationships. Site-specific mutations introduced within an enzyme active site may affect both the formation and decay of intermediates, as well as influence the environment and/or the conformation of the enzyme active site. In certain cases, the spectra of PLP-derived intermediates have been shown to be a sensitive indicator of the integrity of the active site environment.[45] Second, in enzyme reactions that involve a number of catalytic intermediates it is particularly important to be able to determine which step (or steps) in a catalytic pathway have been altered as a result of a particular mutation. Without such information, classic comparisons of steady-state rate constants may be misleading.[46] For example, changes in the transient accumulation of intermediates may be obscured in steady-state measurements by other, partially rate-limiting steps in a different part of the catalytic pathway. The observation of intermediates via RSSF spectroscopy provides a means for directly examining the changes induced by a mutation on elementary catalytic steps in a reaction pathway.

One example of the use of RSSF spectroscopy to explore enzyme structure–function relationships in the tryptophan synthase system in-

---

[45] P. S. Brzović, A. M. Kayastha, E. W. Miles, and M. F. Dunn, *Biochemistry* **31**, 1180 (1992).
[46] J. R. Knowles, *Science* **236**, 1252 (1987).

volves the substitution of Arg-179 by Leu in a flexible loop adjacent to the α active site which connects strand 6 with helix 6 in the α/β barrel topology of the α subunit.[47] The substitution does not affect the steady-state kinetics of the β reaction and causes only a modest change in the turnover rate of the overall $\alpha\beta$ reaction from IGP and L-Ser. The K_m and k_{cat} values for the $\alpha\beta$ reaction catalyzed by the αR179L mutant are altered by only 4-fold and 5-fold, respectively.[48] The fact that the αR179L mutant retains significant catalytic activity in reactions catalyzed by the α subunit permits characterization of the effect of the mutation in loop 6 on the overall catalytic cycle of the $\alpha\beta$ reaction. Moreover, the reciprocal allosteric interactions observed between the α and β subunits of the bienzyme complex and the close coordination of α- and β-catalytic activities suggest that mutations which affect the catalytic activity of one enzyme may be monitored by studying the associated changes in activity at the heterologous active site nearly 30 Å away. Because amino acid replacements in the α subunit do not affect the primary amino acid sequence of the β subunit, alterations in the reactivity of the $\alpha_2\beta_2$ complex must be due primarily to differences in the reactivity of the α subunit and/or $\alpha\beta$ subunit interactions.

Consistent with the results from steady-state kinetic studies, RSSF experiments show that the mutation in the α subunit does not alter the series of spectral changes observed for either stage I or stage II of the β reaction. The positions and shapes of the observed spectral bands, and the kinetics of the transient accumulation of intermediate during the αR179L-catalyzed β reactions, are nearly identical to those found for the wild-type catalyzed reaction shown in Fig. 11. These findings give strong evidence that the αR179L mutation does not alter the conformation of the β subunit, the environment of the β active site, or the interaction between α and β subunits in the bienzyme complex.

The RSSF spectral changes for the reaction of IGP with the wild-type E(A-A) complex are shown in Fig. 12A. The reaction is characterized by a broad decrease in absorbance below 385 nm and a corresponding increase in absorbance between 385 and 500 nm. There is an apparent isoabsorptive point at 385 nm. Difference spectra for the reaction (Fig. 13A) reveal that the spectral changes are dominated by the accumulation of a quinonoidal absorbance band at 476 nm that corresponds to the formation of the E(Q3) intermediate (Fig. 10). However, under identical experimental conditions, only very small spectral changes occur in the $\alpha\beta$ reaction catalyzed by

[47] P. S. Brzović, C. C. Hyde, E. W. Miles, and M. F. Dunn, *Biochemistry* **32**, 10404 (1993).
[48] H. Kawasaki, R. Bauerle, G. Zon, S. Ahmed, and E. W. Miles, *J. Biol. Chem.* **262**, 10678 (1987).

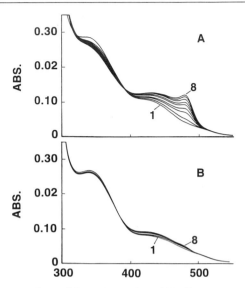

FIG. 12. RSSF spectra of the wild-type (A) and αR179L (B) enzyme-catalyzed $\alpha\beta$ reaction. In each case, 0.2 mM IGP was mixed with either 10 mM wild-type of 8.1 mM αR179L $\alpha_2\beta_2$ preequilibrated with 40 mM L-Ser. Spectra were collected at 8.53, 17.1, 34.1, 42.6, 76.8, 255.8, 426.4, and 639.6 msec after flow had stopped. (From Brzović et al.[43] with permission.) Copyright 1994 American Chemical Society.

the αR179L mutant (Fig. 12B). It is not clear from the spectra whether the changes represent the formation of stage II intermediates or result from allosteric effects on the distribution of stage I intermediates at the β site caused by the binding of IGP to the α site. However, comparison of difference spectra for both the β and $\alpha\beta$ reactions indicate that stage II intermediates are present in the αR179L enzyme-catalyzed reaction (see Fig. 13), although the accumulation of these species has been greatly reduced by the mutation in the α subunit. The calculated difference spectral changes of the αR179L $\alpha\beta$ reaction are qualitatively similar to the αR179L enzyme-catalyzed β reaction in the absence of a ligand bound at the α site.

The observed changes both in the rates of formation and in the extent of accumulation of reaction intermediates at the β active site of the αR179L mutant have been shown to be predominantly due to changes in the affinity of ligands for the α active site. Comparison of the dependence of the rate of E(Q3) formation, as followed by the increase in absorbance at 476 nm, on the concentration of IGP between the wild-type and αR179L enzyme-catalyzed $\alpha\beta$ reactions (data not shown) reveal that the binding affinity

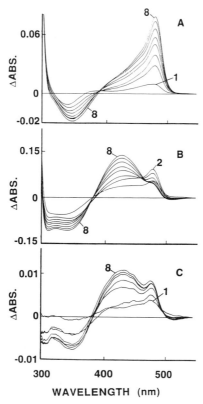

FIG. 13. Calculated difference spectra for wild-type and αR179L enzyme-catalyzed β and $\alpha\beta$ reactions described in Fig. 12A,B. (A) Difference spectra calculated from the wild-type enzyme-catalyzed $\alpha\beta$ reaction shown in Fig. 12A. Spectrum 1 of Fig. 12A was subtracted from subsequent spectra. (B) Difference spectra of the αR179L enzyme-catalyzed $\alpha\beta$ reaction shown in Fig. 12B. Spectrum 1, which is nearly identical to the αR179L E(A-A) spectrum, was subtracted from subsequent spectra in the data set. (C) αR179L enzyme-catalyzed β reaction. Spectrum 1 of Fig. 12B was subtracted from subsequent spectra. The spectral changes in the reaction are identical to the wild-type enzyme-catalyzed β reaction. (From Brzović et al.[43] with permission.) Copyright 1994 American Chemical Society.

of IGP has been reduced by at least 15-fold as a result of the αR179L mutation. Equilibrium binding studies show substantial reductions in the affinity for other α-subunit-specific ligands, including the product of the α reaction, glyceraldehyde 3-phosphate.[47] These results indicate that loop 6 is necessary for the tight binding of substrates and ligands at the α site.

Because of the extensive structural homology between the α subunit of tryptophan synthase and triose phosphate isomerase,[41] it is likely that loop 6 is involved in the conformational transition from an open to a closed structure that occurs on ligand binding to the α active site.[43,49,50] Movement of loop 6 on ligand binding to the α site serves to close off the active site from the solvent, thus sequestering the substrate and initial products of the α reaction within the confines of the multienzyme complex. Closure of the α subunit has the essential role of promoting the diffusion of indole, produced from the cleavage of IGP at the α site, through the interconnecting tunnel to the β active site and preventing the escape of indole directly into the bulk solvent. Substitution of Arg by Leu at this position probably alters either the mobility or the structural integrity of the flexible loop, thereby adversely affecting loop function.

Concluding Remarks

As evidenced in the preceding paragraphs, RSSF spectroscopy provides a means for the direct detection and characterization of catalytic intermediates. Not only does RSSF UV/visible spectroscopy allow a determination of the number of different absorbing species which accumulate and decay, but the spectra of intermediates also contain considerable information regarding chemical structure and bonding interactions. This type of information can be extractly only indirectly from the number of observed relaxations and the concentration dependence of the calculated relaxation rates obtained from single-wavelength measurements. Furthermore, it is difficult to extract reliable rate data from multiexponential single-wavelength stopped-flow time courses with relaxations that are poorly resolved. RSSF spectroscopy may guide the choice of appropriate wavelengths for SWSF measurements that will help to simplify kinetic analysis and aid study of the accumulation and decay of intermediates whose presence might otherwise be obscured owing to measurement at an ill-suited wavelength. The eluciation of catalytic mechanism and spectral characterization of certain, or in favorable cases all, reaction intermediates can provide a much needed foundation for examining protein structure–function relationships. Especially for enzymes with multiintermediate catalytic pathways, RSSF is a useful tool for detecting changes induced by site-specific amino acid substitutions on either the accumulation and decay of intermediates or the spectral properties of specific intermediates which may arise from alterations in the active site environment.

[49] D. L. Pompliano, P. Anusch, and J. R. Knowles, Biochemistry 29, 3186 (1990).
[50] P. S. Brzović, Y. Sawa, C. C. Hyde, E. W. Miles, and M. F. Dunn, J. Biol. Chem. 267, 13028 (1992).

A large number of enzyme systems exist with intrinsic chromophoric cofactors which are well suited for investigation with RSSF techniques. With some ingenuity chromophoric substrates or cofactors may be developed to study enzyme mechanism in other systems. Herein, we have given the reader brief examples that illustrate the usefulness of RSSF spectroscopy. We hope, with the increasing availability of low-cost computer hardware and a number of different commercial RSSF spectrometers, that RSSF methods will become more widely used by enzymologists for the study of enzyme mechanism and protein structure–function relationships.

Acknowledgment

We thank the National Science Foundation for support of this work and the instrumentation described in this chapter.

[9] Transient Absorption Spectroscopy in Study of Processes and Dynamics in Biology

By HERBERT VAN AMERONGEN and RIENK VAN GRONDELLE

Introduction

Many processes in nature are accompanied by color changes. For example, when the visual pigment rhodopsin is excited by light, it undergoes a complex fast isomerization reaction, whereby the wavelength of maximum absorption shifts from about 500 nm to approximately 550 nm: a dramatic color change! This is just one example of a (bio)chemical reaction for which the reaction product has quite different absorption characteristics than the original species and the resulting changes in color might even be visible with the naked eye. However, color changes can also take place that are not (easily) detected by the eye, for instance, when they occur in the ultraviolet (UV) or infrared (IR) or when they are small. Transient absorption measurements aim at the accurate detection of these absorption changes in order to unravel the identity of the intermediates and the kinetics of the processes under study.

The term "transient absorption" is commonly used for techniques that detect relatively fast (<1 sec) changes in absorption arising from electronic transitions in molecules in the wavelength region between 200 and 1200 nm, that is, from the far-UV to the near-IR. Transient IR absorption spectra (vibrational transitions) can be studied, but the required techniques are not discussed here. A variety of different processes can be investigated using the technique of transient absorption: (bio)chemical reactions induced by mixing of reactants, reactions in biological materials induced by ionizing radiation or light (electronic excitation transfer between photosynthetic pigments, light-induced electron transfer), and changes in conformation. Some processes are more easily detected if the polarization of the absorption or emission is probed, and this subject is treated elsewhere in this volume (H. van Amerongen and W. S. Struve [11]).

Kinetic studies require that a process is started on a time scale which is fast with respect to the time scale of the kinetics of interest. For this reason, the study of subnanosecond phenomena is limited to those that can be induced by a short light pulse. Pulses of several femtoseconds can now be produced; for instance, the primary events in bacteriorhodopsin have been investigated with pulses of 6 fsec, the shortet light pulses available today.[1]

In this chapter we mainly limit ourselves to the time region from several femtoseconds to milliseconds and to processes that are induced by illumination with light. One could roughly divide this time region into two parts, from femtoseconds to nanoseconds and from nanoseconds to milliseconds. The latter time scale can be investigated with electronic detection, and the described methods can also be used for processes that are initiated in other ways, for example, in stopped-flow experiments where reactants are rapidly mixed or pulse-radiolysis experiments where a short electron pulse induces a chemical reaction. After an introduction to the basic principles of transient absorption, we first treat "slow" transient absorption measurements (nanoseconds–milliseconds).

The largest part of the chapter deals with "fast" pump–probe measurements (femtoseconds–nanoseconds). Although fast processes can also be studied "electronically" with the use of a streak camera (see, e.g., Knox and Mourou[2]), there are only few examples of this method in biology or biochemistry, and we do not consider that possibility here. In the "fast" experiments a short pulse is used to excite (pump) the sample, and a short pulse, which has been delayed with respect to the pump pulse, is used to probe absorption changes. The time between the arrival of pump and probe pulses in the sample can be controlled mechanically by slowly varying the path length that one of the pulses traverses before it hits the sample. A delay of 30 cm in the optical pathway leads to a time delay of 1 nsec owing to the finite speed of light (3×10^8 msec^{-1}). One should note that the detection electronics need not be fast because a series of steady-state experiments is performed with a quasi-fixed time delay be-

[1] R. A. Mathies, C. H. Brito Cruz, W. T. Pollard, and C. V. Shank, *Science* **240**, 777 (1988).
[2] W. Knox and G. Mourou, *Opt. Commun.* **37**, 203 (1981).

tween the pump and probe pulses. Excellent reviews on ultrafast spectroscopy with applications in biology exist,[3–8] and we refer to those for additional information.

Principles of Transient Absorption

Optical Density

Transient absorption measurements should be performed in wavelength regions where the molecules that one wishes to excite or the "reaction products" show absorption or changes in absorption. We remark that in general biological molecules show broad, sometimes multiple absorption bands and that they undergo absorption changes over large parts of the accessible wavelength region. The measurements are based on the well-known Beer–Lambert absorption law:

$$I(\lambda) = I_0(\lambda)\ 10^{-C\varepsilon(\lambda)l} \tag{1}$$

where I_0 is the intensity of the light of a particular wavelength λ that impinges on the sample and I is the intensity of the transmitted light. C is the concentration of absorbing molecules (expressed in M), ε is the molar extinction coefficient of these molecules (expressed in $M^{-1}\,cm^{-1}$), and l is the path length traversed through the sample (expressed in cm). The OD (optical density) or alternatively A (absorbance) is defined as $-\log(I/I_0)$, and therefore

$$OD(\lambda) = C\varepsilon(\lambda)l \tag{2}$$

When a mixture of absorbing molecules is present, the following expression can simply be used:

$$OD(\lambda) = \Sigma\ C_i\varepsilon_i(\lambda)l \tag{3}$$

In transient absorption measurements changes in absorption (ΔOD) are measured. The change in absorption is observed by a difference in detected intensity ΔI between I_1 and I_2 before and after the start of the induced processes. In many experiments this change is small ($\Delta I/I \ll 1$). In that case the following useful approximation can be made:

[3] C. V. Shank, in "Ultrashort Laser Pulses and Applications" (W. Kaiser, ed.), p. 5. Springer-Verlag, Berlin, 1988.

[4] G. R. Fleming, Annu. Rev. Phys. Chem. 37, 81 (1986).

[5] R. M. Hochstrasser and J. K. Johnson, in "Ultrashort Laser Pulses and Applications" (W. Kaiser, ed.), p. 357. Springer-Verlag, Berlin, 1988.

[6] A. R. Holzwarth, Q. Rev. Biophys. 22, 239 (1989).

[7] J.-L. Martin and M. H. Vos, Annu. Rev. Biophys. Struct. 21, 199 (1992).

[8] S. V. Chekalin, Yu. A. Matveets, and A. P. Yartsev, Rev. Phys. Appl. 22, 1761 (1987).

$$\Delta OD = -\log(I_2/I_0) - [-\log(I_1/I_0] = -\log(I_2/I_1) = -\log(1 + \Delta I/I_1)$$
$$= -0.434 \ln(1 + \Delta I/I_1) \approx -0.434 \Delta I/I_1 \tag{4}$$

that is, the change in intensity is directly proportional to the change in absorption. In case ΔI becomes too large, ΔOD should be calculated according to $\Delta OD = -\log(1 + \Delta I/I_1)$, but in order to be exact a homogeneous illumination throughout the sample should be obtained, which can be realized only approximately. It is important to choose an appropriate concentration at which to perform optimal experiments. The change in optical density increases when the concentration is increased, but this limits the amount of detected light. Excessive concentrations should therefore be avoided, and as a rule of thumb one should look for an optimal OD around 0.5 at the excitation/detection wavelength in a one-color pump–probe experiment. When an entire wavelength region is probed, however, OD values of up to 1 or even higher in the maximum are preferred in order to get good signal/noise ratios over the entire wavelength region.

Ground State Absorption, Excited State Absorption, and Stimulated Emission

In transient absorption measurements different absorption changes play a role: first is the bleaching of the absorbing molecules which lose their ground state absorption when they are promoted to an excited state or when they undergo a reaction; concomitantly, a different species is created with a different absorption spectrum, for instance, a reaction product, a molecule that has accepted or donated an electron, or a molecule that is in the excited state and shows excited state absorption (ESA); third, a molecule in the excited state can show stimulated emission (SE) induced by the probe light, which leads to an increased intensity on the detector, giving rise to an apparent additional bleaching. This SE has a spectrum similar to the fluorescence spectrum, and its intensity can be of the same order of magnitude as the bleaching of the absorption from the ground state to the lowest excited state. In Fig. 1 a schematic energy level system of a chromophore (colored molecule) is given.

Interactions between Chromophores

When a chromophore is excited, the absorption spectrum of a nearby chromophore may also be influenced owing to an interaction with the excited molecule, which is observed, for instance, in the photosynthetic bacterial reaction center of purple bacteria.[9,10] In the same complex the

---

[9] C. Kirmaier and D. Holten, *Photosynth. Res.* **13**, 225 (1987).
[10] M. H. Vos, J.-C. Lambry, S. J. Robles, D. C. Youvan, J. Breton, and J.-L. Martin, *Proc. Natl. Acad. Sci. U.S.A.* **89**, 613 (1992).

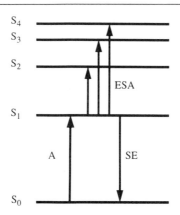

FIG. 1. Schematic view of the energy levels of a chromophore. Molecules in the ground state S_0 can be promoted to an excited state (in this case the first excited state S_1) by absorption (A) of light. Being in the first excited state, the molecule can then absorb a photon (ESA, excited state absorption) to go to a higher excited state. On the other hand, it can be stimulated to return to the ground state (SE) by photons from the probe beam which have an energy that corresponds to the energy difference between S_1 and S_0.

transfer of an electron from one electron carrier to the next not only changes the absorption spectra of the reduced and oxidized molecules but also influences those of the neighboring pigments through electrostatic interactions (electrochromic shift)[9] (see also Fig. 6[10a]).

Intensity

In transient absorption measurements the number of probing photons in the time window of interest should be sufficiently large to get a satisfactory signal to noise (S/N) ratio. For instance, in a nanosecond experiment the number of photons during 1 nsec should preferably be similar to the number of photons during a millisecond experiment. For this reason a continuous lamp can be used as a probe source in the latter experiment, whereas a flash lamp is preferred in the former.

Both the pump and probe beam intensities should be as stable as possible. It is advisable to measure the pump and probe beam intensities (I_{pump} and I_{probe}) independently and normalize the signals obtained to these intensities. In a first-order approximation the observed signal is proportional to the product $I_{pump} \times I_{probe}$. In practice $\langle I_{pump} \rangle$ and $\langle I_{probe} \rangle$ are measured, and normalization is performed with $\langle I_{pump} \rangle \langle I_{probe} \rangle$ which is a good approximation as long as I_{pump} and I_{probe} do not vary much. Here,

[10a] C. Kirmaier and D. Holten, *Isr. J. Chem.* **28**, 79 (1988).

the angular brackets denote averaging of the intensity over a certain period of time. I_{probe} should be small with respect to I_{pump} in order to minimize secondary excitation effects by the probe light. As shown below, double modulation techniques offer a way to circumvent these unwanted actinic effects of the probe beam, allowing high probe intensities.

In transient absorption experiments often many traces are averaged, and one should avoid the buildup of (temporary) reaction products. This effect can be minimized by lowering the pump and probe intensities or the repetition frequency, but this will concomitantly lead to a decrease of the S/N ratio. In the ideal case the sample is replaced between each pair of pump and probe shots, for instance, by flowing the sample or placing it in a rotating cell; however, complete replacement is often not feasible at high repetition rates. A combination of rotation and translation further reduces the disturbing effects of reaction product buildup.[11] Obviously, these requirements are difficult to meet for experiments at low temperatures (e.g., 4 or 77 K).

An important quantity that is often presented with an experimental curve is the number of photons per pulse per square centimeter. For instance, a pulse energy of 1 μJ at 1000 nm corresponds to 5×10^{12} photons, and the excitation density depends on how strong the light beams are focused into the sample. For a collimated laser beam with a diameter d_0 in the lens (which focuses the beam into the sample at the focal distance f) the diameter d in the sample is approximately $d = f\lambda/\pi d_0$. Focusing pump and probe beams into a small intersecting volume leads to larger signals, but too high excitation densities should be avoided as explained above.

An additional complication of high densities arises in aggregates of dyes or in photosynthetic pigment–protein complexes containing a large number of pigments between which extensive electronic excitation transfer takes place. In these "aggregates" excitation annihilation can occur when an excitation is transferred between pigments and meets another singlet or triplet excitation. This annihilation process can be quite efficient and leads to an accelerated depopulation of the excited state (see, e.g., Sundström et al.[12]). To determine the contribution of these annihilation processes one should check the intensity dependence of the decay curves.

One can get a rough approximation of the fraction of excited molecules from the number of incident photons per pulse per square centimeter. The relation OD = $\log(I/I_0)$ yields directly the percentage of the light

[11] P. I. van Noort, D. A. Gormin, T. J. Aartsma, and J. Amesz, *Biochim. Biophys. Acta* **1140**, 15 (1992).
[12] V. Sundström, T. Gillbro, R. A. Gadonas, and A. Piskarsas, *J. Chem. Phys.* **89**, 2754 (1988).

absorbed on passing the sample and thus how many molecules (N_1) have been excited. The total number (N_2) of molecules present in the same volume is calculated from the optical density and the molar extinction coefficient, and from that the excited fraction N_1/N_2 can be determined.

The path length through the sample after the probe beam has crossed the path of the pump beam should be minimized in order to avoid unnecessary intensity losses. To reduce the amount of pump light on the detector, the pump and probe beam hit the sample with a certain angle with respect to one another. It may be difficult to avoid some scattered pump light reaching the detector. In real-time experiments with electronic detection this makes a measurement immediately after the pulse difficult or impossible. In pump–probe experiments where the pump and probe pulses are delayed with respect to one another this scattering gives only a constant background, but this can be a significant source of noise. Modulation of both pump and probe beam can minimize this disturbing effect (see below). In one-color experiments the coherent coupling artifact may lead to enhanced scattering when the pump and probe beams pass the sample at the same time (see also below).

Analysis of Results

Instrument Function

Infinitely short light pulses of one particular wavelength would be ideal for pump–probe measurements. However, from the Heisenberg uncertainty principle ($\Delta E \Delta t \geq h/2$) it is immediately clear that these cannot be obtained. For instance, in the wavelength region from 600 to 700 nm a 100 fsec pulse has a full width at half-maximum (fwhm) of at least 5–10 nm, and, owing to imperfections in the laser, additional broadening to 20 nm can take place (see, e.g., Durrant et al.[13] and Hastings et al.[14]). Especially for processes where the inverse rate corresponds to a similar time scale as the pulse width, it is important to realize that the measured trace is a convolution of the studied kinetics and the instrument function. This instrument function in "slow" measurements is due to the finite response time of the detection system and the width of the exciting pulse. In "fast" experiments this instrument function is determined entirely by the cross-correlation function $G(t)$ of the pump and probe pulse:

[13] J. R. Durrant, G. Hastings, Q. Hong, J. Barber, G. Porter, and D. R. Klug, *Chem. Phys. Lett.* **188**, 54 (1992).

[14] G. Hastings, J. R. Durrant, J. Barber, G. Porter, and D. R. Klug, *Biochemistry* **31**, 7638 (1992).

$$G(t) = \int_{-\infty}^{\infty} I_{\text{pump}}(t')I_{\text{probe}}(t - t') \, dt \tag{5}$$

In fitting the experimental results, the instrument function is considered as a set of closely spaced "delta functions" (extremely short pulses), all giving rise to the same kinetic curve but with different starting points and different amplitudes. The results are then fitted to a sum of decay curves with the same decay parameters and with different starting points and starting heights, according to the shape of the instrument function. Formally this can be expressed as

$$\Delta A(t) = \int_{0}^{t} G(t - t') \, \Delta A'(t') \, dt' \tag{6}$$

where the measured change in absorption $\Delta A(t)$ is a convolution of $G(t)$ and the real absorption change $\Delta A'(t')$ that has to be recovered from the measured signal. An accurate determination of the instrument function is required (see below). For "slow" real-time experiments (slower than 1 nsec) this convolution is usually not necessary because the pulses are either too short on the relevant time scales or measurements immediately after the excitation are not possible at all owing to a burst of scattering (or fluorescence) during and immediately after the pulse. Whereas scattering is an "instantaneous process," the fluorescence can have a lifetime of several nanoseconds. This burst of light can even "knock out" the detector for longer times (microseconds).

Analysis of Kinetics

Transient absorption traces are most often fitted with a sum of exponentials, and the decay rates are related to the processes taking place. When a range of wavelengths is probed, all the decay curves can be fitted simultaneously to a minimum number of decay parameters in a "global analysis" according to

$$\Delta A(\lambda, t) = \Sigma \, a_{\text{m}}(\lambda) \exp(-\tau/\tau_{\text{m}}) \tag{7}$$

(see, e.g., Holzworth et al.[15]). The change in absorption ΔA at wavelength λ and time t is written as a sum of exponentials with decay times τ_{m} and corresponding preexponentials a_{m}. The decay times in such an analysis are taken to be the same at all wavelengths, whereas the a_{m} can vary with wavelength. The $a_{\text{m}}(\lambda)$ constitute a spectrum associated with decay time τ_{m}. This is known as a decay-associated spectrum (DAS). Such a spectrum can give further information on the participating molecules and processes

[15] A. R. Holzwarth, J. Wendler, and G. W. Suter, *Biophys. J.* **51**, 1 (1987).

occurring (an example of a global analysis fit is given in Fig. 2[15a]). With the use of target analysis one can fit the observed kinetics to an explicit kinetic model.[16,17] In the case of an inhomogeneous distribution of closely spaced decay rates, the decay can be fitted with a phenomenological "stretched exponential" equation: $a \exp\{-(kt)^\alpha\} + b$, where k reflects some average decay rate and α gives information on the broadness of the distribution (see, e.g., Narajan et al.[18]).

In general one can decide how many independent decay components are needed to describe the observed kinetics by closely inspecting the residuals (difference between measured and fitted curves). The singular value decomposition (SVD) method[19] is a statistical tool to determine the maximum number of components that can be extracted with confidence from the experimental traces.

"Slow" Transient Absorption Measurements: Nanoseconds to Milliseconds

For "slow" transient absorption measurements the exciting light (or actinic light) is generally provided by a short flash, for instance, from a pulsed laser at an angle of 90° from the measuring (probe) light. The pump pulses have a high intensity at the excitation wavelength as compared to the measuring light, which in many cases is provided by a (pulsed) lamp with a broad spectral range. When depolarization effects play a role on the time scale of interest, polarizers should be placed in the pump and probe branch, just before the sample. The "excitation polarizer" should be oriented vertically and the "detection polarizer" at an angle of 54.7° with respect to the vertical axis. Note that the omission of polarizers can lead to erroneous results, even if the pump and probe beams are nonpolarized.

The excitation light can be provided by a mode-locked (50–100 psec) or Q-switched (several nanoseconds) frequency-doubled Nd-YAG (neodymium–yttrium-aluminum-garnet) laser (532 nm) or a dye laser pumped by one of these, providing pulses of less intensity at longer wavelengths.

[15a] P. J. M. van Kan, M. L. Groot, I. H. M. van Stokkum, S. L. S. Kwa, R. van Grondelle, and J. P. Dekker, in "Research in Photosynthesis, Volume I" (N. Murata, ed.), p. 271. Kluwer Academic Press, Dordrecht, The Netherlands, 1992.

[16] J. M. Beechem, M. Ameloot, and L. Brand, Chem. Phys. Lett. 120, 466 (1985).

[17] J. M. Beechem, M. Ameloot, and L. Brand, Anal. Instrum. 14, 379 (1985).

[18] V. Narajan, W. W. Parson, D. Gaul, and C. Schenck, Proc. Natl. Acad. Sci. U.S.A. 87, 7888 (1990).

[19] J. Hofrichter, E. R. Henry, J. H. Sommer, R. Deutsch, M. Ikeda-Saito, T. Yonetani, and W. A. Eaton, Biochemistry 24, 2667 (1985).

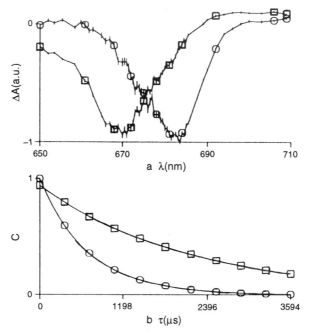

FIG. 2. (a) The photosynthetic pigment–protein complex CP47 of photosystem II from spinach has an absorption peak around 680 nm, owing to chlorophyll a. On excitation a significant fraction of the excited molecules goes from the singlet excited state via the process of intersystem crossing into a triplet state which has a much longer lifetime than the singlet excited state, and this can be observed as long-lived bleaching near the absorption peak. The absorption difference spectrum is called a triplet minus singlet spectrum. For CP47 the decay of the long-lived states was described [S. G. Boxer, R. A. Goldstein, D. J. Lockhurst, T. R. Middendorf, and L. Takiff, *J. Phys. Chem.* **93**, 8280 (1989)] by two exponentials (b) in the wavelength region from 650 to 700 nm, according to $A_1(\lambda)\, e^{-t/\tau_1} + A_2(\lambda)\, e^{-t/\tau_2}$, where $\tau_1 = 0.7$ msec (○) and $\tau_2 = 2.2$ msec (□). The corresponding decay-associated spectra $A_1(\lambda)$ and $A_2(\lambda)$ have the spectral characteristics of bleached absorption bands [centered at 683 (○) and 670 (□) nm]. The latter is due to "unconnected" chlorophyll a, which cannot transfer its excitation energy to lower energy pigments in CP47. The measurements were performed at 77 K. [With permission from P. J. M. van Kan, M. L. Groot, I. H. M. van Stokkum, S. L. S. Kwa, R. van Grondelle, and J. P. Dekker, *in* "Research in Photosynthesis, Volume 1" (N. Murata, ed.), p. 271. Kluwer Academic Press, Dordrecht, The Netherlands, 1992.] Better preparations can now be made that lack the 670 nm component in the triplet minus singlet spectrum, after it turned out to be due to free chlorophyll (M. L. Groot and J. P. Dekker, private communication).

In the case of a mode-locked laser a cavity dumper can be used to reduce the repetition rate to below the time scale on which the processes of interest take place. For shorter excitation wavelengths, for instance, a nitrogen laser (337 nm) can be used, which can pump a dye laser.

To record processes between 1 nsec and 1 μsec a high intensity for the probe light in a short time is required which can be produced by a pulsed xenon flash lamp. For slower processes (>1 μsec) continuous lamps (xenon arc, deuterium or tungsten lamp) are available. To minimize heating and actinic effects, a mechanical shutter is placed between the probe lamp and the sample. It opens shortly before the excitation pulse arrives and closes after the time interval of interest. The probe wavelength can be selected by a monochromator or an interference filter. A monochromator placed in front of the sample reduces the amount of unwanted heating and actinic effects by the probe beam. A monochromator placed behind the sample reduces the amount of scattered (elastic or Raman) and fluorescence light on the detector. The transmitted light is detected by a photomultiplier (PM) or a photodiode (PD). Photomultipliers are commonly used in the UV, whereas both PM and PD detectors (the latter are significantly less expensive but less sensitive) are used at longer wavelengths and have a time resolution of a few nanoseconds. Most information can be obtained in a short measuring time with the use of an optical multichannel analyzer (OMA) in combination with a polychromator. The transmitted light of the lamp spectrum after passing the sample is dispersed, and the spectrum is recorded with a diode array as a function of time.

One of the reasons that scattered light and fluorescence light can be disturbing is that the probe light is provided in many cases by a lamp with a relatively large spot size and a "strongly" diverging beam. In gathering all the probe light after the sample and focusing in onto the detector, much fluorescence and scattered light is collected from the "wide" space angle. The use of a laser in the probe branch both enhances the amount of probe light at a certain wavelength and limits the space angle over which the scattered and fluorescence light is collected. For instance, a dye laser or a titanium–sapphire laser can be used to enable one to tune the detection wavelength. However, a laser usually offers less tunability than the combination of a lamp and a monochromator.

The output signal of the PD/PM can be fed into a transient recorder, which can record and store traces with a resolution of 1 nsec per channel or slower. Because the changes in intensity are usually small with respect to the total intensity, it is desirable to apply an electronic offset which compensates for the steady-state intensity level. In this way the dynamic range is enlarged. Recorders are available that have a built-in compensation possibility.

To relate intensity changes to absorption changes one should measure the transmitted probe intensity before and after the excitation pulse and/ or the transmitted probe intensity with and without the excitation pulse.

In the case of pulsed lamps, the pulse profile should be reproducible or should be recorded simultaneously.

"Fast" Transient Absorption Measurements: Femtoseconds to Nanoseconds

This section deals with measurements on a femtosecond to nanosecond time scale where the pump and probe pulses have been delayed with respect to one another as was explained above. Three different types of methods are treated below which all make use of this principle: (1) one-color experiments in which both the pump and probe pulses are obtained from the same laser and have the same color (these experiments can be performed with a high repetition frequency of approximately 100 MHz), (2) two-color experiments with independently tunable pump and probe wavelengths in which two dye lasers (driven by the same pump laser) produce pulses with the same repetition rate as in one-color experiments, and (3) continuum experiments in which amplified short pulses of selected wavelength excite the sample and a continuum of light is used to probe the subsequent changes over a broad spectral range (the repetition rate of these experiments is typically between 1 Hz and 1 kHz, although laser amplifiers with repetition rates of several hundreds of kilohertz have been designed).

Ultrafast One-Color Pump–Probe Experiments

Experimental Setup. The first example of a one-color pump–probe experiment on a picosecond time scale was given by Shank *et al.*,[20] where hemoglobin was studied, and dates back to the first mode-locked Ar^+ synchronously pumped dye laser systems developed by Shank and Ippen. Since then many different groups have made use of the one-color pump–probe technique.[11,21–26] A simplified scheme of such a setup is given in Fig. 3. A high repetition rate (76 MHz) frequency-doubled (532 nm) mode-locked Nd-YAG laser (or Ar^+ ion laser) is used to pump a dye laser with 50–100 psec pulses. In the dye laser one dye jet functions as a lasing medium leading to pulses which can be shorter than 10 psec. An additional

[20] C. V. Shank, E. P. Ippen, and R. Bersohn, *Science* **193,** 50 (1976).
[21] R. A. Engh, J. W. Petrich, and G. R. Fleming, *J. Phys. Chem.* **89,** 618 (1985).
[22] T. P. Causgrove, S. Yang, and W. S. Struve, *J. Phys. Chem.* **93,** 6844 (1989).
[23] S. Lin, H. van Amerongen, and W. S. Struve, *Biochim. Biophys. Acta* **1060,** 13 (1991).
[24] E. Åkesson, V. Sundström, and T. Gillbro, *Chem. Phys. Lett.* **121,** 513 (1985).
[25] F. Zhang, T. Gillbro, R. van Grondelle, and V. Sundström, *Biophys. J.* **61,** 694 (1992).
[26] W. F. Beck and K. Sauer, *J. Phys. Chem.* **96,** 4658 (1992).

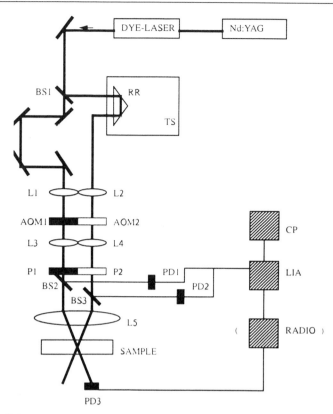

FIG. 3. Schematic representation of a one-color pump–probe setup. The setup is a simplified version of the setup in the laboratory of Dr. W. S. Struve (Iowa State University, Chemistry Department). The high repetition rate Nd-YAG laser pumps the dye laser, which delivers a train of short light pulses. The light is split into two beams (pump and probe beams) by beam splitter 1 (BS1). Via the computer-controlled optical delay line, consisting of a retroreflector (RR) on a variable translation stage (TS), the moment of arrival in the sample of the pump pulse relative to that of the probe pulse can be continuously adjusted. The lenses L1 and L2 focus the pump and probe beams through the acoustooptic modulators. The polarizers P1 and P2 control the polarization of the pump and probe beams with respect to one another. Photodiodes PD1 and PD2 record the stability of both beams. The final signal can be normalized to the pump and probe beam intensities. Lens L5 focuses both beams into the sample, where they overlap. PD3 records the probe beam intensity. A radio detects signals at the difference frequency between the frequencies of AOM1 and AOM2, and via an additional lock-in amplifier (LIA) the signal is registered by a computer (CP) which also collects the signals from PD1 and PD2 via analog to digital converters of the LIA.

(saturable absorber) jet within the cavity of the dye laser can further shorten the pulses to approximately 1 psec. The use of intracavity prisms can compress these pulses to approximately 100 fsec. The repetition rate can be decreased by a factor of approximately 1000 with the use of a cavity dumper. Specific combinations of lasing and saturable absorber dyes can provide light in a wavelength region of many tens of nanometers. Many dyes are available that allow access to a variety of wavelength regions from 550 to 1000 nm. An intracavity birefringent filter can select the appropriate wavelength, and a maximum output power of several hundreds of milliwatts is achievable.

Alternatively, a colliding pulse modelocking (CPM) laser (620 nm, 100 MHz, 50 fsec) can be used to pump the dye laser. In the wavelength region 700–1000 nm a Ti–sapphire laser (76 MHz), pumped by a continuous wave (cw) Ar^+ laser, is especially suitable to provide pulses shorter than 100 fsec, whereas the wavelength (for a specific intracavity mirror set) is easily changed within a wavelength region of 100 nm.

In the experiment the train of short pulses is split by a beam splitter into a pump and a probe beam. The path length traversed by either the pump or probe beam can be changed with the use of an optical delay line as explained above. In this specific example the path traversed by the pump light is varied. With the lens L5 (Fig. 3) both beams are focused into the sample where they spatially overlap. The probe light is detected with a photodiode (PD3). The delay line consists of a retroreflector on a translation stage which can be displaced in a quasi-continuous way in steps of several "femtoseconds." Attention should be paid to the fact that the retroreflector moves parallel to the incoming pump beam. This can easily be checked by imaging the pump spot on a distant "screen" and looking whether the spot remains in the same place on moving the retroreflector. A retroreflector is preferred over a rooftop prism because the former always reflects the pump light perfectly parallel to the incoming beam, which guarantees that it is focused by the lens L3 on the same position in the sample. Therefore, the overlap region is essentially the same at all delay times. Behind the sample the spot might be slightly displaced. As the probe beam does not traverse a variable distance (as the pump beam does) the position where the beam hits PD3 remains exactly the same. This leads to maximum stability.

Intensity Modulation. The S/N ratio is improved by modulating the pump-beam intensity either by the use of a chopper or an acoustooptic modulator (with a frequency f_1). By tuning the detection to this frequency with the use of a lock-in amplifier (LIA), one can selectively look at variations in the detected probe-beam intensity, caused by variations in the pump-beam intensity (leading to variations in the amount of excitation with frequency f_1). High modulation frequencies (~ 1 MHz) reject low-frequency noise arising from the pump laser or dye laser. A complication arises when pump light scatters from the sample into the detector. This effect becomes more important when the size of the particles increases, but it occurs also in low-temperature measurements where light is scattered from imperfections in the glass. Because the scattered light shows the same modulation frequency as the induced modulation of the transmitted probe beam, this increases the measured signal. Although the scattered light ideally leads to an offset which is independent of the delay time, it will still increase the noise. Additional modulation of the probe-beam intensity with a frequency f_2 selects for changes in the probe-beam intensity caused by the pump-beam excitations by detection of either the sum $(f_1 + f_2)$ or difference $(f_1 - f_2)$ frequency. In some studies[22,23] this is done, for instance, with the use of a radio. The signal-bearing output of the radio with a frequency of 50 kHz is subsequently fed into a lock-in amplifier (LIA).

Coherent Coupling Artifacts. For the detection of ultrashort events (<1 psec) in one-color experiments the coherent coupling artifact which shows up as a subpicosecond "spike" on the pump–probe trace can be a considerable nuisance. The "coherence spike" is ascribed to scattering processes which occur owing to the coherence of the pump and probe beams when these are derived from the same laser. The coherence spike is essentially symmetric in time and is further discussed by Fleming.[4] One could simply try to "cut off" the spike from the signal by computer, which is probably justified if the spike is narrow with respect to the total pulse. However, when the pulse spike becomes nearly transform limited (i.e., it is as short as the Heisenberg uncertainty principle allows), the shape of the coherence spike starts to resemble the total pulse autocorrelation shape, and such an operation is not easy. Another correction method is called antisymmetrization[11,21] and makes use of the fact that the coherence spike is (nearly) symmetric. By subtracting the total trace from its mirror image (mirrored around the center of the spike) the coherence spike disappears. It is however, not always obvious where the center is, and very fast "real" processes might still go undetected.

Intensity and Polarization. To correct for changes in the pump and probe intensities a fraction of the incident pump and probe beams is measured simultaneously with the transient absorption signal using additional photodiodes (PD1 and PD2, Fig. 3). To avoid unwanted polarization effects, polarizers should be placed in both the pump and probe beams with an angle between the polarization directions of 54.7° (the magic angle).

The focal length of L3 determines the overlap volume of the pump and probe beams (as shown above). A short focal length leads to a small overlap volume and consequently to a higher intensity in the overlap region giving rise to larger signals.[22,23] One should realize that for specific applications the intensity in the overlap region may be too high as explained above. Special attention is paid[24,25] to get very low excitation densities. Concomitantly a larger focal length is taken. In case of double modulation it is not necessary to keep the intensity of the probe beam low with respect to that of the pump beam. One should check for an optimal combination of pump- and probe-beam intensities.

Autocorrelation Function. The instrument function can be obtained with the use of a doubling crystal in place of the sample. For ease of use the doubling crystal can also be placed in a separate but identical branch. If oriented properly, the doubling crystal produces light with a frequency that is the sum of the frequencies of the pump and probe lights. Thus the frequency-doubled light gives the instrument function because this frequency doubling occurs only when both the pump and probe pulses temporally overlap.

Despite its obvious disadvantage of limited spectral information, a one-color pump–probe experiment has the advantage of being performed at a high repetition rate. In effect, a large number of experiments are performed in a short time, thereby increasing the signal to noise ratio. An example of results from a one-color experiment is given in Fig. 4.[26a]

Ultrafast Two-Color Pump–Probe Experiments

Experimental Setup. An obvious extension of the one-color pump–probe experiments is the application of two-color experiments in which two independently tunable dye lasers share the same pump laser. One can use the same high repetition rate and obtain spectral evolutions on excitation at selected wavelengths. The measurements are performed in essentially the same way as one-color experiments.[11,25–28] A disadvantage is the broadened instrument function (cross-correlation function) caused by time jitter between the two pulses, since they are not obtained from the same dye laser. This leads to a full-width half-maximum (fwhm) value of the instrument function of approximately 5–10 psec.

Although in principle this broadening can be avoided, the experiment becomes much more complicated. The instrument function can be determined in the same way as in a one-color experiment. Alternatively, use

[26a] L. M. P. Beekman, F. van Mourik, M. R. Jones, H. M. Visser, C. N. Hunter, and R. van Grondelle, *Biochemistry* **33,** 3413 (1994).

[27] S. Lin, H. van Amerongen, and W. S. Struve, *Biochim. Biophys. Acta* **1140,** 6 (1992).

[28] S. L. S. Kwa, H. van Amerongen, S. Lin, J. P. Dekker, R. van Grondelle, and W. S. Struve, *Biochim. Biophys. Acta* **1102,** 202 (1991).

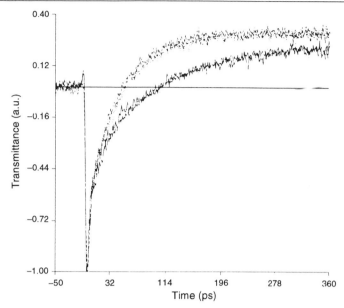

FIG. 4. Results from one-color experiments on chromatophores from the photosynthetic bacterium *Rhodobacter sphaeroides*. The chromatophores consist of a reaction center surrounded by a pool of pigments which dominate the absorption at 861 nm, at which wavelength the measurements were performed. The full width at half-maximum of the autocorrelation function of the laser pulses was 5 psec. Shown are the results for chromatophores from the wild-type bacterium and three mutants in which the tyrosine at the M210 position in the reaction center has been replaced by a histidine, phenylalanine, and leucine, respectively. The mutation supposedly changes the rate of primary charge separation in the reaction center. The curves have been normalized in the negative peak. The wild-type and His mutant traces are almost identical (top two traces), as are the other two. At the start of the pulses a bleaching of the excited antenna pigments leads to an increased transmittance. Very quickly (still within the excitation pulse) the excitation is transferred to pigments which absorb at longer wavelengths and show excited state absorption at 861 nm, which leads to the decrease in transmittance. From the antenna pigments the excitation is transferred to the special pair of the reaction center, which can subsequently donate an electron to the primary acceptor, leading to a bleaching signal again (of the primary donor) at long times. Whereas the spectral equilibration within the antenna is fast (at most a few picoseconds), the total charge separation time ranges from 45 to 60 psec for the four different species. (With permission from Beekman *et al.*[26a])

can be made of the fact that the cross-correlation function has a nearly Gaussian shape. Replacing the sample by a long-lived dye leads to a slowly decaying pump–probe signal, convoluted with the instrument function. By fitting the pump–probe trace with a Gaussian function and a (one-exponential) decay, the instrument function can be determined rather accurately (H. van Amerongen and S. Lin, unpublished results, 1990).

The above-treated (sub)picosecond techniques have been widely applied in the study of electronic excitation energy transfer in photosynthetic pigment–protein antenna complexes. Energy transfer between pigments with slightly different absorption spectra can be studied. Also the process of electron transfer in reaction centers of photosynthetic bacteria, algae, and green plants have been studied in detail. To obtain more extensive spectral details, continuum pump–probe experiments have been used quite extensively (see below). In addition, conformational changes within bacteriorhodopsin have been studied with this technique. In all these experiments one can probe the bleaching, excited state absorption, stimulated emission, and absorption differences arising from changes in conformation (isomerization) or in the identity of molecules (oxidation, reduction, protonation, etc.).

Ultrafast Continuum Pump–Probe Experiments

Several laboratories now make use of the ultrafast continuum pump–probe technique in the study of ultrafast processes in biological molecules or molecular complexes.[1,7–10,13,14,29–43] Notable molecules and complexes under study are the photosynthetic reaction centers of purple bacteria, the reaction centers of photosystems I and II of green plants,

[29] W. Holzapfel, U. Finkele, W. Kaiser, D. Oesterhelt, H. Scheer, H. U. Stilz, and W. Zinth, *Chem. Phys. Lett.* **160,** 1 (1989).

[30] M. R. Wasielewski and D. M. Tiede, *FEBS Lett.* **204,** 368 (1986).

[31] J. L. Martin, J. Breton, A. J. Hoff, A. Migus, and A. Antonetti, *Proc. Natl. Acad. Sci. U.S.A.* **83,** 957 (1986).

[32] G. R. Fleming, J.-L. Martin, and J. Breton, *Nature (London)* **333,** 190 (1988).

[33] M. H. Vos, J.-C. Lambry, S. J. Robles, D. C. Youvan, J. Breton, and J.-L. Martin, *Proc. Natl. Acad. Sci. U.S.A.* **88,** 8885 (1991).

[34] J. Breton, J.-L. Martin, G. R. Fleming, and J.-C. Lambry, *Biochemistry* **27,** 8276 (1988).

[35] R. A. Mathies, S. W. Lin, J. B. Ames, and W. T. Pollard, *Annu. Rev. Biophys. Biophys. Chem.* **20,** 491 (1991).

[36] R. L. Fork, C. H. Brito Cruz, P. C. Becker, and C. V. Shank, *Opt. Lett.* **12,** 483 (1987).

[37] N. W. Woodbury, M. Becker, D. Middendorf, and W. W. Parson, *Biochemistry* **24,** 7516 (1985).

[38] V. Nagarajan, W. W. Parson, D. Gaul, and C. Schenck, *Proc. Natl. Acad. Sci. U.S.A.* **87,** 7888 (1990).

[39] C. Kirmaier and D. Holten, *Isr. J. Chem.* **28,** 79 (1988).

[40] C. Kirmaier and D. Holten, *Biochemistry* **30,** 609 (1991).

[41] C.-K. Chan, L. X.-Q. Chen, T. J. Dimagno, D. K. Hanson, S. L. Nance, M. Schiffer, J. R. Norris, and G. R. Fleming, *Chem. Phys. Lett.* **176,** 366 (1991).

[42] C.-K. Chan, T. J. Dimagno, L. X.-Q. Chen, J. R. Norris, and G. R. Fleming, *Proc. Natl. Acad. Sci. U.S.A.* **88,** 11202 (1992).

[43] A. K. W. Taguchi, J. W. Stocker, R. G. Alden, T. P. Causgrove, J. M. Peloquin, S. G. Boxer, and N. W. Woodbury, *Biochemistry* **31,** 10345 (1992).

bacteriorhodopsin, visual pigments, and heme proteins. The basic setup is discussed below as well as some problems that one might or will encounter. Some results on reaction centers from purple bacteria and photosystem II (PSII) are discussed. A review has appeared by Martin and Vos[7] which, besides applications of the technique to the study of the reaction centers of purple bacteria and heme proteins, gives a global description of an ultrafast continuum pump–probe set-up. Therefore, we give only a short overview of such a system with some discussion; for further information, the reader is referred to the above-mentioned references.

Experimental Setup. In general, at the heart of a continuum pump–probe experiment is a high repetition rate laser. This can be a passively mode-locked CPM laser which is pumped by a cw Ar^+ laser and which produces pulses typically shorter than 100 fsec around 620 nm with a pulse energy of approximately 0.1 nJ. Alternatively, a mode-locked Nd-YAG laser (532 nm) pumps a dye laser. In that case pulses at somewhat shorter wavelengths can be obtained, which are higher in energy when the same repetition rate is used (typically 0.5 nJ), but the pulses are usually longer, even after pulse compression (>100 fsec).

The pulses are then amplified by leading them through media which have been excited by high-intensity nanosecond pulses at a low repetition rate, for instance, by a Q-switched Nd-YAG laser (10 Hz, 532 nm, 5 nsec pulses). When the subpicosecond pulse passes the excited medium it induces stimulated emission, and the pulse wil be amplified. After several stages (usually 3–5) of amplification the pulses are amplified to energies of hundreds of microjoules. The largest amplification is obtained during the first step and decreases in subsequent steps. Note that the repetition rate is reduced to that of the pumping amplification laser (e.g., 76 MHz → 10 Hz). The high density of excited states also leads to unwanted spontaneous stimulated emission, also called amplified spontaneous emission (ASE), and its intensity is much higher than that of the subpicosecond amplified pulses. This ASE is distributed much wider in time and space, and if not eliminated to a large extent it will make the measurements with high time resolution largely impossible. One can reduce the amount of disturbing ASE by (1) spatial filtering (SF) using pinholes between and after amplification steps or (2) a saturable absorber (SA) placed in between and behind the amplification steps. Because the desired ultrafast pulses have a highly concentrated intensity distribution in time, they will bleach to a very large extent (saturate) the ground-state absorption of the SA after which it is transparent for the remaining major part of the pulse. As the ASE is spread more in time and space, a lower percentage will pass the SA. Moreover, the ASE is usually peaked at a different wavelength than the pulses which have to be amplified, and one can choose the SA or a mixture of SAs such that it is more difficult for the ASE to pass than for the ultrashort pulses. Thus, it is possible to decrease the contribution of ASE to the amplified beam to less than 5%. The amplified pulses should be intense enough to create a white light continuum in a water cell owing to nonlinear processes (see below).

Before we continue to explain how the amplified pulses are further used, we want to stress that the continuum measurements are performed at a much lower repetition rate than the previously discussed one- and two-color experiments. Therefore, much larger averaging times are needed for continuum measurements, and it could be tempting to use high pulse intensities to obtain an increased S/N ratio with the hazard of building up a population of metastable states (see above). In some studies,[13,14] a copper-vapor laser is used to pump the amplifying medium at a repetition rate of 6.5 kHz with lower pulse energies, and a higher S/N ratio can be obtained. In another study,[43] a regenerative amplifier operates at a repetition rate of 540 Hz. This amplifier is driven by the same Nd-YAG laser that also produces the pulses that have to be amplified.

It has become possible to obtain amplified pulses in the wavelength region from approximately 700 to 1000 nm at a repetition rate of several hundreds of kilohertz from a Ti–sapphire laser, but to our knowledge no applications to biological molecules have been reported so far.

Part (or all) of the light from the amplified pulses is used for continuum generation. The light impinges on a cell with water or ethylene glycol. Owing to nonlinear processes a broad continuum spectrum is produced with a maximum centered at the pump wavelength and ranging from approximately 400 to 1200 nm. The pulse length is hardly changed. The use of water leads to a brighter white light source, but ethylene glycol gives better pulses.

Part (or all) of the white light is used to probe the sample. To this end the probe beam is split further into two parts, one of which crosses the pump beam in the sample, whereas the other (the reference beam) passes the sample without intersecting the pump beam. The probe and reference beams are then used to calculate the change in absorption. One can select a probing wavelength at will by using a monochromator or filter. To avoid pulse-broadening effects caused by the monochromator, one should place the monochromator behind the sample, where only the amount of light is important and not the pulse profile. The latter can be significantly distorted by a monochromator. We would like to stress that the Heisenberg uncertainty relation cannot be circumvented by placing the monochromator behind the sample. Therefore, one should always be aware that the product of wavelength and time resolution has a lower limit according to this relation (see above).

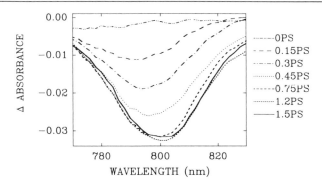

FIG. 5. Spectra recorded with a dual diode array detecting system in a continuum experiment. Membranes of the heliobacterium *Heliobacillus mobilis* were excited at 590 nm with a repetition rate of 540 Hz (full width at half-maximum 200 fsec). Spectra were taken at the indicated times after excitation. The authors conclude that this early time spectral evolution is probably due to the excitation distribution among different spectral forms. [With permission from S. Lin, H.-C. Chiou, and R. E. Blankenship, *in* "Research in Photosynthesis, Volume 1" (N. Murata, ed.), p. 417. Kluwer Academic Press, Dordrecht, The Netherlands, 1992.]

Most spectral information is obtained if a dispersive grating (polychromator) disperses the probe light on a linear diode array, such that an entire spectrum can be measured at once. A second diode array should then be used to record the reference beam (an example is given in Fig. 5[43a]). Pulse compression is needed in both the pump and probe beams to compensate for group velocity dispersion (GVD) in the entire setup.

A completely different method was developed and applied to the study of ultrafast processes in bacteriorhodopsin.[1,35,36,44] Pulses from a CPM laser were amplified, and then the light was chirped in an optical fiber. With the use of pulse compression ultrashort pulses of 6–12 fsec were obtained. The ultrashort pulses correspond to a broad spectral range because of the Heisenberg uncertainty principle. We refer the reader elsewhere[36,44] for further details.

Depending on the desired excitation wavelength, different ways can be used to pump the sample. Part of the amplified pulses can be used to excite the sample directly (after intensity reduction with neutral density filters) as is done, for instance, by Woodbury *et al.*[37] One can also amplify a selected wavelength region of the generated continuum light in a one- or two-stage amplifier. In this way a broadly tunable excitation source is

[43a] S. Lin, H.-C. Chiou, and R. E. Blankenship, *in* "Research in Photosynthesis, Volume 1" (N. Murata, ed.), p. 417. Kluwer Academic Press, Dordrecht, The Netherlands, 1992.
[44] S. L. Dexheimer, Q. Wang, L. A. Peteanu, W. T. Pollard, R. A. Mathies, and C. V. Shank, *Chem. Phys. Lett.* **188,** 61 (1992).

obtained. Examples are available in the literature.[7,31,32,14,40,41] Wasielewski and Tiede conducted experiments in which the amplified 611 nm light is led through CH_4 where the second Stokes Raman line at 950 nm is selected to obtain a high-intensity pump pulse. In a similar way, Kirmaier and Holten[40] obtained light at 870 nm with the use of C_6H_{12}.

The basic scheme of Fig. 3 also applies to the continuum measurements. Either the pump or the probe beam is delayed. For the same reason as explained above, it is better to delay the pump beam. In addition, the light of the pump and the probe beams should be polarized with an angle between the polarization directions of 54.7°.

Ultrafast continuum pump–probe spectroscopy has led to the discovery of subpicosecond processes in bacteriorhodopsin.[1,35,44,45] This protein contains a retinal chromophore and acts as a light-induced proton pump that creates a potential difference across a membrane in *Halobacterium halobium,* which can be used as a free-energy source.[35] Changes in the bleaching, SE, and ESA were observed and were used in concert to describe, assign, and model the observed spectral changes and kinetics.[1,35,44,45]

Reaction Centers from Photosynthetic Purple Bacteria. Numerous studies have been performed on the photosynthetic reaction centers of the purple bacteria *Rhodobacter sphaeroides* and *Rhodopseudomonas viridis*. The solution of the crystal structure of these reaction center proteins[46,47] has directed a large amount of attention of experimentalists and theoreticians toward their function. In nature the reaction centers are surrounded by an array of antenna pigment–protein complexes. When light is absorbed by one of the pigments the excitation is rapidly transferred from pigment to pigment until it ends up in the reaction center, where the excitation energy is used to transfer an electron from a donor to an acceptor. After several secondary electron transfer steps this finally leads to a potential difference across the membrane in which the reaction center is embedded. This potential difference can be used for further chemical processes. Understanding the very effective mechanism of charge separation is a challenge, and many experiments have been directed at determining and understanding the transfer rates and related spectral changes (see also Fig. 6[39]).

The reaction center contains four bacteriochlorophyll (BChl) and two bacteriopheophytin (BPheo) molecules. Two bacteriochlorophyll mole-

[45] W. Zinth, J. Dobler, K. Dressler, and W. Kaiser, *in* "Ultrafast Phenomena VI" (T. Yajima, K. Yoshihara, C. B. Harris, and S. Shionoyad, eds.), p. 581. Springer-Verlag, Berlin and Heidelberg, 1988.
[46] J. Deisenhofer, O. Epp, K. Miki, R. Huber, and H. Michel, *J. Mol. Biol.* **180**, 385 (1984).
[47] J. P. Allen and G. Feher, *Proc. Natl. Acad. Sci. U.S.A.* **81**, 4795 (1984).

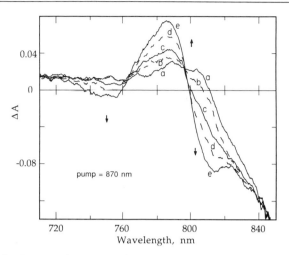

FIG. 6. Results from continuum experiments on the reaction center from the photosynthetic purple bacterium *Rhodobacter capsulatus*. Excitation was performed at 870 nm, with 350 fsec pulses. The absorption difference spectra are shown after 600 fsec (a), 1.6 psec (b), 3.2 psec (c), 6.0 psec (d), and 16.8 psec (e). The arrows depict how the absorption changes with increasing time. At early times the primary donor is excited, leading to bleaching and stimulated emission at long wavelengths and excited state absorption at shorter wavelengths. Subsequently, an electron transfer takes place to one of the bacteriopheophytins, leading to an additional bleaching of the bacteriopheophytin band around 750 nm. Owing to the charge separation the absorption spectrum of the accessory bacteriochlorophylls, absorbing around 800 nm, shifts from longer to shorter wavelengths (electrochromic shift), leading to the derivative-like difference spectrum around 800 nm. The presence of two isosbestic points was interpreted as a strong indication by the authors for direct transfer of an electron from the primary donor to BPheo since no intermediate state is observed. [With permission from C. Kirmaier and D. Holten, *Isr. J. Chem.* **28,** 79 (1988).]

cules are close together and form the so-called special pair (primary donor) which donates an electron to a BPheo in the L-branch within a few picoseconds after excitation (2–3 psec) which then in approximately 200 psec is transferred to a quinone molecule (see, e.g., Holten *et al.*[48]). The protein has a quasi-C_2 symmetry axis around which the L-branch is mapped onto the M-branch after rotation over 180°. This holds not only for the pigments but also for a significant part of the protein. Although there is general agreement about these two transfer times, there are important remaining questions. The first transfer step is extremely fast and would rather be expected to occur on a nanosecond time scale if only a BChl and a BPheo molecule would be present at a distance similar to that in the reaction

[48] D. Holten, M. W. Windsor, W. W. Parson, and J. P. Thornber, *Biochim. Biophys. Acta* **501,** 112 (1978).

center. Despite the structural quasi-C_2 symmetry the electron appears to go down only one branch (the L-branch).

Major questions concern the role of the protein medium in electron transfer, and mutated reaction centers are being produced where "important" amino acids have been selectively replaced. The role of the accessory bacteriochlorophyll molecules has also been studied, with research focusing on the question of whether the accessory BChl in the L-branch acts as a true intermediate in the electron transfer.[29,49] Furthermore, one would like to understand what the specific role of the dimer nature of the primary donor is (see, e.g., Boxer et al.[50]).

Most of our knowledge about the ultrafast kinetics in the reaction centers stems from transient absorption measurements.[8–10,29,31–33,37–42] An example of the kinetics that are associated with charge separation in the reaction center of R. sphaeroides is shown in Fig. 6. The appearance of the spectral changes in the region 750 to 900 nm represents the formation of the first charge separation in about 3 psec.

Important in the assignment of these transfer steps is obviously the identification of the absorption bands in the spectrum of the reaction center. In the near-infrared absorption region there are three bands visible for the R. sphaeroides reaction center.[51] With the use of spectral differences induced by reduction or oxidation it was possible to assign the longest wavelength band in the Q_y region (870 nm) to the special pair, the middle band (800 nm) to the accessory chlorophylls, and the third (770 nm) to the pheophytins.

The special dimer shows strong excitonic interaction leading to two perpendicularly polarized absorption bands, where the high-energy band is only weak and is present as a shoulder on the accessory BChl band.[52] The BPheo pigments have almost identical Q_y bands, but the Q_x bands (around 540 nm) differ and this has led to the notion that only one BPheo is reduced (see, e.g., Kirmaier and Holten[9]). The difference is especially clear at cryogenic temperatures. The assignment of this band to the BPheo in the L-branch was based on the linear dichroism spectrum of the reaction center of Rhodopseudomonas viridis.[53] Later, this assignment was confirmed with the (M)L214H mutant in which one BPheo was replaced by a BChl molecule.[54] Several mutants have now been studied with fast

[49] C. Kirmaier and D. Holten, Biochemistry 30, 609 (1991).
[50] S. G. Boxer, R. A. Goldstein, D. J. Lockhurst, T. R. Middendorf, and L. Takiff, J. Phys. Chem. 93, 8280 (1989).
[51] C. Kirmaier, D. Holten, and W. W. Parson, Biochim. Biophys. Acta 810, 49 (1985).
[52] A. Vermeglio and G. Paillotin, Biochim. Biophys. Acta 681, 32 (1982).
[53] W. Zinth, W. Kaiser, and H. Michel, Biochim. Biophys. Acta 723, 128 (1983).
[54] C. Kirmaier, D. Gaul, R. De Berg, D. Holten, and C. C. Schenck, Science 251, 922 (1991).

spectroscopy. Most studies are directed toward the Q_y region, but one should always bear in mind that the unidirectionality cannot *a priori* be taken for granted in these mutants and should be tested.

Controversy remains whether the electron transfer occurs via the accessory BChl in the L-branch. Transient absorption measurements which showed an intermediate reduction of such a monomeric BChl[29] have been disputed by Kirmaier and Holten.[49]

The assignment of absorption bands to specific pigments and spectral changes to specific transfer steps is complicated by several factors. (1) All the pigments are close together, and therefore the absorption bands might show an excitonic character; in other words, a specific absorption band is not entirely due to a single chromophore but has a collective nature (although one pigment might dominate). Ultrafast experiments have shown that directly after excitation of the special pair a very complicated difference spectrum was created, which could indicate significant excitonic interactions.[10] (2) Also a change in the oxidation/reduction state can alter the spectroscopic properties of nearby chromophores (see also Fig. 6), which is also the case in the bacterial reaction center (electrochromic shift) (see, e.g., Kirmaier and Holten[9]). (3) On excitation with ultrashort pulses, damped oscillations were present in the transient curves, which were suggested to be due to coherent motion of a vibrational wave packet, thereby complicating the whole ultrafast kinetics enormously.[33] (4) Finally, not all reaction centers are exactly alike, as they may all be in slightly different conformations which have specific spectroscopic and transfer properties. As long as this inhomogeneity is small it will not seriously influence the kinetics, but it is by no means obvious that one can neglect it *a priori,* especially at low temperatures or at ultrashort times.

Reaction Centers from Photosystem II of Green Plants. As a final example we mention the ultrafast research on the photosynthetic reaction center from photosystem II (PSII) of green plants. The reaction center contains six chlorophylls (Chl) and two pheophytins (Pheo). The absorption bands of all these pigments overlap extensively.[55] Therefore, it is even more difficult to assign the different steps in the electron transfer process. Large uncertainty exists about the transfer time from the primary donor to the pheophytin. For instance, the transfer time was reported to be 3 psec.[56] However, a time constant of 21 psec was reported for the same electron transfer event.[13,14] A difference in the experimental condi-

[55] K. Gounaris, D. J. Chapman, P. Booth, B. Crystall, L. B. Giorgi, D. R. Klug, G. Porter, and J. Barber, *FEBS Lett.* **265,** 88 (1990).

[56] M. R. Wasielewski, D. G. Johnson, M. Seibert, and Govindjee, *Proc. Natl. Acad. Sci. U.S.A.* **86,** 524 (1989).

tions used by the two groups concerns the applied excitation intensity, which was a factor of ten higher in the first study.[56]

An unanswered question is whether the primary donor is also a dimer in PSII. From the transient measurements it was tentatively suggested that a dimer is bleached first (corresponding to a large bleaching) and after that the excitation localizes on one of the pigments constituting the dimer, leading to a diminished bleaching.[13,14] However, these experiments were performed with the pump and probe beam polarized parallel, and this could seriously effect the apparent magnitude of the bleaching. A setup in which the directions of the polarization make an angle of 54.7° is preferable.

We have given a brief overview of the techniques used in transient absorption on a femtosecond to millisecond time scale. It is by no means exhaustive and is to a certain extent superficial; it is meant to serve only as a guideline to think about possible applications of the technique in the study of biological systems. We refer to the cited references and literature given therein for more detailed experimental and theoretical information.

Acknowledgments

The authors thank Mrs. M. L. Groot for providing Fig. 2, Mr. L. M. P. Beekman for providing Fig. 4, Dr. S. Lin for providing Fig. 5, and Dr. C. Kirmaier for providing Fig. 6. This work was supported by the Dutch Foundation for Biophysics (SvB), financed by the Netherlands Organization for Scientific Research (NWO).

[10] Hole Burning Spectroscopy and Physics of Proteins

By JOSEF FRIEDRICH

Introduction

Spectral hole burning is an optical analog of the radio frequency saturation experiments in nuclear magnetic resonance (NMR) as introduced in the famous work by Bloembergen, Purcell, and Pound in 1948.[1] The NMR saturation experiments are dynamic in nature, that is, the hole or the saturation dip relaxes with a rate constant as given by spin–lattice relaxation. The analogous experiment in the optical domain was performed for the first time by Szabo.[2] In the optical domain, the relaxation processes

[1] N. Bloembergen, E. M. Purcell, and R. V. Pound, *Phys. Rev.* **73,** 679 (1948).
[2] A. Szabo, *Phys. Rev.* **B11,** 4512 (1975).

are orders of magnitude faster than in the NMR domain; hence, dynamic saturation of optical transitions persists just for nanoseconds to microseconds. In 1974, the situation changed drastically when two groups simultaneously succeeded in burning persistent holes either by exploiting frequency selective photochemistry or persistent photophysical transformations.[3,4] It is the persistency which makes spectral hole burning so attractive, because it enables the experimentalist to use this high-resolution optical technique to investigate electronic ground state phenomena, namely, phenomena which are connected with the nonequilibrium nature of structurally disordered materials and which show up, for instance, in extremely slow relaxation processes.[5] Glasses and polymers belong to this class of materials. However, proteins, too, fall into this category. They are at the edge between order and disorder. It is this in-between nature of proteins which gives rise to unusual and unexpected physical behavior.

Apart from ground state spectroscopy, hole burning is also an excellent tool for investigating properties of the electronic excited states. For instance, it has been extensively used to investigate the primary processes in photosynthetic assemblies, such as energy transfer processes, electron–phonon coupling, and charge separation processes.[6,7] As a matter of fact, hole burning was applied to biomaterials for the first time[8] using a chromoprotein of photosynthesis.

In addition to the relation of hole burning techniques to dynamic saturation methods, it is also important to stress the close relation to another rather different technique, namely, Mössbauer spectroscopy. The hole burning technique has been called the optical analog of the Mössbauer effect.[9] The relation is very close, indeed. On the one hand, both techniques work at the ultimate limit of resolution given by the natural line width. It is not the lasers which limit the resolution in hole burning; it is the fast excited state lifetimes which set the resolution limit. For organic dye molecules these lifetimes are in the nanosecond time regime. Conse-

---

[3] A. A. Gorokhovskii, R. K. Kaarli, and L. A. Rebane, *JETP Lett.* (*Engl. Trans.*) **20**, 216 (1974).

[4] B. M. Kharlamov, R. I. Personov, and L. A. Bykovskaya, *Opt. Commun.* **12**, 191 (1974).

[5] W. Breinl, J. Friedrich, and D. Haarer, *Chem. Phys. Lett.* **106**, 487 (1984); W. Breinl, J. Friedrich, and D. Haarer, *J. Chem. Phys.* **81**, 3916 (1984).

[6] R. Jankowiak, J. M. Hayes, and G. J. Small, *Chem. Rev.* **93**, 1471 (1993).

[7] N. S. Reddy, P. A. Lyle, and G. J. Small, *Photosynth. Res.* **31**, 167 (1992).

[8] J. Friedrich, H. Scheer, B. Zickendraht-Wendelstadt, and D. Haarer, *J. Am. Chem. Soc.* **103**, 1030 (1981); J. Friedrich, H. Scheer, B. Zickendraht-Wendelstadt, and D. Haarer, *J. Chem. Phys.* **74**, 2260 (1981).

[9] K. K. Rebane and L. A. Rebane, *in* "Persistent Spectral Hole Burning: Science and Applications" (W. E. Moerner, ed.), p. 17. Springer-Verlag, Berlin, 1988.

quently, typical widths of burnt-in holes are in the mega- to gigahertz (MHz to GHz) regime. On the other hand, it is the symmetry between space and momentum in the Hamiltonian of the harmonic oscillator which leads to almost identical spectral features in both techniques: a very sharp zero-phonon line superimposed on a background of lattice vibrations.

Since its discovery in 1974, persistent spectral hole burning has been successfully applied to an ever-increasing manifold of problems in various fields. Disorder phenomena in glasses and polymers, which show up in characteristic line-broadening effects, comprise an extensively studied area.[6,10,11] Because of the high resolution of the technique the influence of external fields such as electric fields or strain fields has been investigated in great detail in all kind of materials: crystals, glasses, polymers, and proteins. In some systems, hole burning was used to measure hyperfine and nuclear quadrupole levels.[12] The technique can also be used with advantage to measure nuclear spin conversion processes as they occur, for instance, in methyl groups as a consequence of the Pauli principle.[13,14] Hole burning also has practical applications: the most spectacular application was proposed in 1978 in a famous IBM patent.[15] It was suggested that the narrow spectral dips be used as information bits in the frequency domain. Optical memories with storage capacities far beyond the so-called diffraction limit seemed possible. Although this has never been achieved, there has been great progress in holographic image storage and pattern recognition.[16]

In protein physics, hole burning experiments have contributed deep insights into the photophysics of photosynthesis and disorder phenomena in the ground state of proteins. As a consequence, this chapter has three main sections: the first deals with the general aspects of the technique and the physics involved; the second highlights the specific contributions from hole burning experiments to our understanding of photosynthesis; and the third focuses on the power of the hole burning technique in investigating disorder phenomena in proteins. By comparing proteins with

[10] J. Friedrich and D. Haarer, in "Optical Spectroscopy of Glasses" (I. Zschokke, ed.), p. 149. Reidel, Dordrecht, The Netherlands, 1986.

[11] L. R. Narasimhan, K. A. Littau, D. W. Pack, Y. S. Bai, A. Elschner, and M. D. Fayer, *Chem. Rev.* **90,** 439 (1990).

[12] R. M. Macfarlane and R. M. Shelby, in "Persistent Spectral Hole Burning: Science and Applications" (W. E. Moerner, ed.), p. 127. Springer-Verlag, Berlin, 1988.

[13] C. v. Borczyskovski, A. Oppenländer, H.-P. Trommsdorff, and J. C. Vial, *Phys. Rev. Lett.* **65,** 3277 (1990).

[14] K. Orth, P. Schellenberg, and J. Friedrich, *J. Chem. Phys.* **99,** 1 (1993).

[15] G. Castro, D. Haarer, R. M. Macfarlane, and H.-P. Trommsdorff, "Frequency Selective Optical Data Storage Systems." U.S. Patent No. 4,101,976 (1978).

[16] U. P. Wild and A. Renn, *J. Mol. Electron.* **7,** 1 (1991).

crystals and glasses, I shall try to characterize the specific features of the protein state of matter.

Basic Aspects of Spectral Hole Burning

Perfect Crystals

Hole burning is a technique[17] whose application is intimately related to disorder in the system considered. The disorder may arise from a variety of possibilities. Holes can be burnt into Doppler-broadened gas phase spectra where disorder arises from the Maxwell–Boltzmann velocity distribution. In NMR spectroscopy, disorder is frequently dominated by field inhomogeneities. In the spectroscopy of matrix-isolated probe molecules, disorder is due to structural imperfections. Figure 1a represents an ordered lattice structure doped with dye probe molecules. If the concentration is sufficiently low, the probe molecules are isolated from one another. Because the environment of each probe molecule is identical, the spectral features will be identical. Figure 1b shows the corresponding absorption band of the S_1 state. There is a sharp zero-phonon transition with a width γ, which is accompanied at the high-energy side by lattice phonons. This line shape reflects nothing more than the Franck–Condon principle applied to lattice vibrations. The relative intensity α in the zero-phonon line, that is, in the pure electronic transition, is called the Debye–Waller factor:

$$\alpha = \frac{\text{intensity in zero-phonon line}}{\text{total intensity}} \tag{1}$$

α is determined by the strength of the electron–phonon coupling. The electrons of the probe molecule couple to the lattice phonons because, in the excited state, the charge distribution is different and the lattice molecules feel this difference and adjust positions accordingly. This adjustment process reflects lattice motions (i.e., phonons).

At low temperatures the width of the pure electronic transition is much narrower than that of the associated phonon transitions because electronic relaxation is much slower than vibronic relaxation. This is the reason why the zero-phonon line is so prominent in the spectrum. With increasing temperature the line shape will lose its characteristic features because (1) the Debye–Waller factor drops rapidly with increasing temperature, so that for many systems the intensity in the zero-phonon line is close to zero above 50 K; and (2) the width γ of the transition increases strongly with temperature. As a consequence of this thermal broadening, the peak

[17] J. Friedrich and D. Haarer, *Angew Chem., Int. Ed. Engl.* **23**, 113 (1984).

a

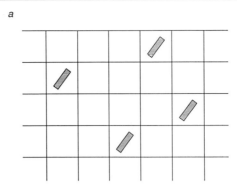

b

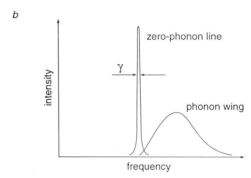

FIG. 1. (a) Perfectly ordered lattice doped with guest molecules in low concentration. (b) Absorption line shape of a guest molecule in a perfect crystal, where γ is the homogeneous width.

intensity drops. Both effects lead to a broad unstructured line shape at higher temperatures.

According to the Heisenberg principle, the width γ is associated with a lifetime T_2:

$$\gamma = \frac{1}{2\pi T_2} \qquad (2)$$

T_2 is the so-called dephasing time, and γ is called the homogeneous line width. T_2 is the lifetime of the excited state coherence, and its physical meaning is easily understood. Suppose the ensemble of probe molecules is excited with a short pulse; then, the associated transition dipole moments will oscillate in phase for a while. Lattice vibrations as well as decay processes to the ground state will destroy the coherence. After time T_2 has elapsed, only a fraction $(1/e)$ of the initial ensemble is still coherently oscillating. Along these lines of reasoning, it is clear that the width γ (or the time constant T_2) is composed of two physically rather different contributions, namely, contributions from decay processes of the excited state population leading to a finite lifetime T_1 (so-called T_1 processes) and contributions arising from pure phase-destroying processes (commonly referred to as T_2 processes):

$$\frac{1}{T_2} = \frac{1}{T_1} + \frac{2}{T_2^*} \tag{3}$$

Pure phase-destroying processes originate from the thermal motion of the lattice. Hence, as the temperature T approaches zero, T_2^* becomes infinitely long so that the width is solely determined through the excited state lifetime. In this limit γ is called the natural line width. The conclusion is that in some limiting cases, that is, whenever dephasing processes can be neglected, the lifetime of the excited state can be determined from the associated line width. Note that the factor 2 in Eq. (3) arises from the fact that, with respect to optical energies, the lifetime of the ground state is infinitely long.

Influence of Disorder and How to Get Around It

Figure 2a represents a structurally disordered host lattice. In contrast to the well-ordered structure (Fig. 1a), the individual guest molecules see different environments. As a consequence the respective absorption energies experience a dispersion. The associated band is rather broad. The respective band shape is built from a convolution of the single molecule line shape (Fig. 1b) with the respective distribution of absorption frequencies, as shown in Fig. 2b. The broadening due to static disorder is called inhomogeneous broadening. The respective width Γ can be quite large. For organic dye molecules in organic glasses, Γ can comprise several hundred wave numbers. For proteins, a typical order of magnitude value is 100 cm^{-1}. Even good quality crystals show inhomogeneous broadening. There, a typical value is about 1 cm^{-1}.

It is instructive to compare the homogeneous width γ with the inhomogeneous width Γ. At sufficiently low temperatures γ is lifetime limited. For organic dye probes, a typical lifetime is a few tens of nanoseconds. Consequently, the associated width γ is on the order of a few tens of megahertz (1 cm^{-1} = 30,000 megahertz). Hence, there can be as much as 5 to 6 orders of magnitude between the homogeneous and the inhomogeneous width. The homogeneous line shape is, even for good quality crystals, completely buried beneath the disorder-dominated inhomogeneous band.

a

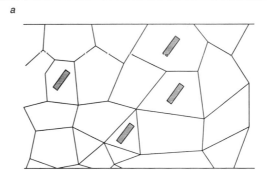

b

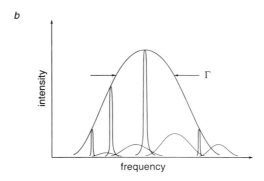

FIG. 2. (a) Disordered lattice doped with guest molecules. (b) Absorption line shape of guest molecules in a disordered lattice, where Γ is the inhomogeneous width.

Spectral hole burning is one way to get around inhomogeneous broadening as shown in Fig. 3. One basic requirement is that the dye probe has some kind of a long-lived reservoir state. For instance, the reservoir state could be a photochemical state, an ionized state, or a long-lived spin state. In the following, we call the reservoir state the photoproduct state. A second requirement is that the spectral properties of the photoproduct state are different from those of the ground state, so that both states can be distinguished in the spectrum.

Then, hole burning works in the following way (Fig. 3). The radiation from a narrow bandwidth laser excites a subensemble of probe molecules which absorb in a range around the laser frequency as roughly given by the homogeneous width γ. This subensemble is transferred to the photoproduct state and, hence, is removed from its original position in the spectrum. As a consequence, a hole appears in the spectrum whose contours are essentially determined by the homogeneous line shape function (Fig. 1b) as long as the irradiation dose is kept sufficiently low. That

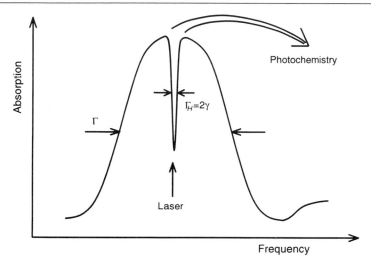

FIG. 3. Principle of photochemical hole burning.

is, there is a sharp zero-phonon hole accompanied by a phonon side hole. The phonon side hole appears, as a rule, on the low-energy side of the zero-phonon hole in contrast to what is shown in Fig. 1b.

This phenomenon comes from the fact that the spectrum of the hole is actually a convolution of the single molecule line shape (Fig. 1b) with itself, reflecting the two procedures associated with the registration of a hole burning spectrum, namely, burning and reading. That the overall hole shape is asymmetric with respect to the zero-phonon hole is a saturation phenomenon. It should also be stressed that, because of this convolution procedure, the width of the zero-phonon hole Γ_H is actually twice the homogeneous line width:

$$\Gamma_H = 2\gamma \tag{4}$$

Equation (4) holds only in case of vanishing irradiation dose. Hence, the homogeneous line width of a probe molecule follows from the measured hole width through an extrapolation procedure to irradiation dose zero. If the dose is too high, saturation phenomena occur: the transformation process to the photoproduct state ceases in the center of the hole because most of the molecules have been transformed after a while. It is, however, still occurring in the wings of the line because there are still many molecules which can be excited via their Lorentzian tails which overlap the laser frequency, producing an artificial broadening.

Advantages of Hole Burning Technique

Hole burning has wide application. According to Eqs. (2) and (4), it can be used to measure electronic relaxation times, dephasing times, and lifetimes. Although dephasing processes in chromoproteins were studied using the hole burning technique,[18,19] it is in measuring very fast lifetimes as they occur, for instance, in energy and electron transfer systems such as the photosynthetic assemblies, where hole burning offers an experimentally easy access. Note that the shorter the lifetime the easier it is to measure it with hole burning. The holes become very broad and can be measured without difficulty. However, the inhomogeneous width sets an upper limit. If the lifetime becomes so short that the associated natural line broadening approaches the inhomogeneous width, it becomes more difficult to separate both contributions in a precise way. Nevertheless, lifetimes on the order of 50 fsec have been measured.[20,21]

In measuring very short lifetimes hole burning competes with time-domain techniques. However, there is another application where the hole burning method is unique: the spectroscopy of the ground state. Here, it is the sharpness of the hole on which the application is based. Once a hole is burned, it remains almost forever. Because it can be orders of magnitude narrower than the inhomogeneous width, it can serve as a highly sensitive detector for small perturbations imposed on the system either through internal changes of the lattice as they occur during structural relaxation processes[10] or through external perturbations induced by electric,[22] magnetic,[23] or pressure fields.[24] As long as the perturbations bring about changes of the hole (line shifts, line broadening) on the order of the homogeneous width, they can easily be measured. Figure 4 shows a hole burnt into the absorption spectrum of mesoporphyrin IX-substituted horseradish peroxidase as it changes under isotropic pressure.

Basic Phototransformation Processes in Hole Burning

As has been pointed out above, there is a large variety of reservoir states where populations can be stored long enough so that the population

[18] S. G. Boxer, D. S. Gottfried, D. J. Lockart, and T. R. Middendorf, *J. Chem. Phys.* **86,** 2439 (1987).

[19] W. Köhler, J. Friedrich, R. Fischer, and H. Scheer, *Chem. Phys. Lett.* **146,** 280 (1988).

[20] S. G. Johnson and G. J. Small, *Chem. Phys. Lett.* **155,** 371 (1989).

[21] J. Pahapill and L. A. Rebane, *Chem. Phys. Lett.* **158,** 283 (1989).

[22] M. Maier, *Appl. Phys.* **B41,** 73 (1986).

[23] A. I. M. Dicker, M. Noort, S. Voelker, and J. H. Van Der Waals, *Chem. Phys. Lett.* **73,** 1 (1980).

[24] Th. Sesselmann, W. Richter, D. Haarer, and H. Morawitz, *Phys. Rev.* **B36,** 7601 (1987).

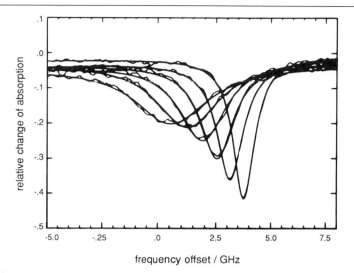

FIG. 4. Behavior of a spectral hole under isotropic pressure variations (0–1.5 MPa). Sample: Mesoporphyrin IX-substituted horseradish peroxidase. Temperature: 1.5 K. Burn wavelength: 6131 Å [From J. Zollfrank, J. Friedrich, J. Fidy, and J. M. Vanderkooi, *J. Chem. Phys.* **94**, 8600 (1991)].

depletion at the laser frequency, that is, the hole, can be scanned. However, there are two major categories of hole burning processes, namely, photochemical transformation processes and photophysical transformation processes.

The photochemical transformation processes involve almost any kind of classic photoreaction which does not couple too strongly to the lattice so that zero-phonon lines can still be observed. Examples are bond-breaking reactions, isomerization reactions, proton transfer reactions, and electron transfer reactions. Very often, the spectral shift between educt and photoproduct state is larger than the inhomogeneous width. If so, then the transformation process can be characterized as being photochemical in nature. However, this is not always the case, and it can happen that the spectral shifts are less than the inhomogeneous widths. Generally speaking, it is not always a straightforward procedure to elucidate the nature of the transformation processes.

The second category of reactions, the so-called photophysical transformations, are not understood in great detail. Their characteristic feature is that the chromophore or the probe molecule is not affected at all. It is usually assumed that light irradiation changes the mutual configuration between chromophore and lattice, thereby inducing a frequency change which manifests itself in hole burning. Such a change in the local configuration could, for instance, be brought about by nonradiative energy release from the excited chromophore which leads to a relative configurational change of the solvent cage and the respective chromophore molecule. Because the reaction is a rearrangement of the selected molecules in the host lattice, it is clear that the photoproduct is confined to the spectral range covered by the inhomogeneous band of the educt. Photophysical hole burning systems were among the first hole burning systems discovered.[4] Examples are perylene or tetracene in alcohol glasses. Photophysical reactions of the type described above can occur in proteins as well and even in crystalline host materials.

Hole Burning Techniques

Hole burning is conceptually a simple technique, namely, a straightforward absorption experiment. This simple concept allows for a diversity of technical variations which offer specific advantages, either in the sensitivity of the detection or in the elucidation of specific physical phenomena.

For many questions the straightforward transmission detection of the burnt-in hole is sufficient. After burning, the laser is scanned over the spectral range where burning was performed, and the transmitted intensity is measured. This procedure is good enough for optical densities on the order of 1 and for a relative hole depth of a few percent. Problems may arise when the optical density is rather low or when the sample has a bad glass quality. In this case, it is better to work in the fluorescence detection mode: Instead of measuring the transmission of the laser radiation one looks at the fluorescence as one scans the burned region in the absorption. A simple calculation shows that below an OD (optical density) of 0.5, fluorescence detection is better; above 0.5, the transmission mode should be preferred.

If the signal-to-noise (S/N) ratio is a problem, zero-background techniques should be used. Several such techniques have been developed. Holographic and polarization methods seem to be the most important ones. In the holographic technique,[16] the hole is burnt via two interfering beams. The photoreaction transforms the interference pattern into an absorption or refractive index grating. Detection of the hole is performed by observing the beam diffracted from the grating as a function of the respective wavelength.

In the polarization technique[25] one uses the fact that a hole burnt with polarized light is optically anisotropic. Hence, if the sample is put between two crossed polarizers, the transmission increases as the wavelength is tuned into the spectral range of the hole. Owing to the macroscopic aniso-

[25] W. Köhler, W. Breinl, and J. Friedrich, *J. Phys. Chem.* **89**, 2473 (1985).

tropy, the hole changes the polarization direction of the light. Hence, light can pass through the couple of crossed polarizers but is still blocked in the isotropic ranges of the spectrum.

Apart from improving the signal-to-noise ratio, hole burning is widely used in combination with external fields, namely, with electric fields and pressure fields, as mentioned above, and offers a tremendous advantage. In glasses, but also in proteins and even in crystals, the inhomogeneous width is so large that huge electric or pressure fields are needed to induce any significant variation in the line shape from which microscopic information could be obtained. Holes are orders of magnitude narrower than the inhomogeneous width; hence, significant changes can be obtained with extremely low fields. From electric field experiments (Stark effect) information on the difference of the dipole moments in the ground and excited state is obtained. Pressure experiments yield information on the chromophore–solvent interaction as well as on elastic properties of the sample, such as compressibility.[26]

Finally, an important and useful variant in hole burning spectroscopy is the temperature cycling technique.[27] A hole is burnt at low temperatures, then the temperature is varied in a cyclic fashion. The change in the hole width is measured as a function of $T_b - T_{ex}$, where T_b is the burn temperature and T_{ex} the so-called excursion temperature, which is the maximum temperature in the cycle. Note that in this type of experiment, it is the changes which the hole experiences during a temperature cycle that are measured. These changes are indicative of slight structural alterations in the system which may occur in disordered materials as a consequence of increasing temperature.

As a rule, disordered systems (and proteins are a special class among them) are nonequilibrium systems. Hence, the structural alterations are irreversible with temperature and can be seen as a residual broadening of the hole. Instead of measuring the hole width, it is also useful to measure the change of the hole area as a function of excursion temperature. Such measures yield information on the distribution of barrier heights between photoproduct and educt.[28,29]

Hole Burning and Photosynthesis

The pigments of cyanobacterial photosynthesis, phycoerythrin and phycocyanin, were among the first proteins on which spectral hole burning

[26] G. Gradl, J. Zollfrank, W. Breinl, and J. Friedrich, *J. Chem. Phys.* **94**, 7619 (1991).

[27] W. Köhler, J. Zollfrank, and J. Friedrich, *Phys. Rev.* **B39**, 5414 (1989).

[28] W. Köhler and J. Friedrich, *Phys. Rev. Lett.* **59**, 2199 (1987).

[29] J. Fidy, J. M. Vanderkooi, J. Zollfrank, and J. Friedrich, *Biophys. J.* **61**, 381 (1992).

experiments were performed.[8,30] Since then, a series of fundamental topics regarding the photophysics of the photosynthetic assembly has been addressed with hole burning experiments: energy transfer time in antennas,[31] excitonic versus incoherent coupling,[6,7,20] charge separation kinetics in the reaction center,[6,7] electron–phonon coupling and thermalization processes,[6,7] and dispersive excited state kinetics and disorder effects.[32] In the following, I sketch the relation of these issues to the basic functioning of the photosynthetic process.

Photosynthetic Apparatus: How Does It Work and What Are the Basic Photophysical Processes Involved?

Figure 5 represents a simplified energy level scheme of the photosynthetic assembly of blue-green and red algae. It consists of a highly organized antenna system and the reaction center (RC). The antenna pigments involved are phycoerythrocyanin (PEC), phycocyanin (PC), and allophycocyanin (APC). The assembly is stabilized and optimized through the various proteins involved and through the thylakoid membrane. The antenna system, in this case the so-called phycobilisome, is a highly organized supramolecular array of pigment complexes. It is located on top of the reaction center, so that effective energy transfer is ensured (Fig. 6).[33] The reaction center is a membrane chromoprotein; its active electron transporting chain is shown in Fig. 7.

There is a symmetry axis, which indicates that there is a second branch (not shown in Fig. 7) symmetrically arranged, yet inactive in the charge transport. The antenna system sits on top of the reaction center. Note that there are differences between the photosynthetic assemblies of green plants and bacteria, but the basic physical processes are the same, namely, light absorption in the antennas, energy transfer to the reaction center, and charge separation. Part of the photon energy is finally stored in an electric field which builds up across the membrane as a consequence of charge separation. With the aid of special enzymes the energy of the field is used to synthesize hydrocarbons.[34]

Aspects of Physics of Energy Transport in Antennas

As stressed above, hole burning provides a rather easy access to ultrashort relaxation times via the homogeneous line width [Eqs. (2) and (3)].

[30] J. Friedrich, *Mol. Cryst. Liq. Cryst.* **183,** 91 (1990).
[31] W. Köhler, J. Friedrich, R. Fischer, and H. Scheer, *J. Chem. Phys.* **89,** 871 (1988).
[32] G. J. Small, J. M. Hayes, and R. J. Silbey, *J. Phys. Chem.* **96,** 7499 (1992).
[33] E. Mörschel and E. Rhiel, *in* "Electron Microscopy of Proteins" (J. R. Harris and W. Home, eds.), Vol. 6, p. 209. Academic Press, London, 1987.
[34] J. Deisenhofer and H. Michel, *Angew. Chem.* **101,** 872 (1989).

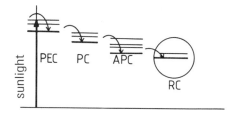

FIG. 5. Energy level scheme of the antenna system of blue-green and red algae.

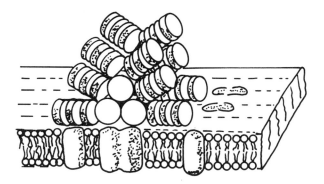

FIG. 6. Antennas (phycobilisomes), membrane, and reaction center.

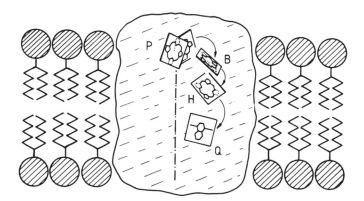

FIG. 7. Reaction center with the active branch. P, Special pair; B, bacteriochlorophyll; H, bacteriopheophytin; Q, quinone. The inactive branch is symmetrically arranged.

In case there is a significant contribution to the homogeneous width from fast energy transfer times, they can be determined or at least estimated from hole burning data. This looks as if hole burning is nothing but a supplemental technique for time domain experiments; however, this is by no means the case. As a frequency domain technique, hole burning is much more sensitive to the specific features of the transport mechanism involved. So far, all information on coherent energy transport in antenna systems stems from hole burning experiments.[6,7]

From Fig. 5, it is obvious that energy transport in photosynthetic assemblies is quite a complicated process. There are transport processes which occur within one pigment, say, PEC in Fig. 5. In addition, there are processes which occur between different pigments. The transport energy may be resonant for a single transfer step, yet over the whole chain it is nonresonant, that is, energy is lost during the process. The transport is not only energetically (downward) directed but also spatially directed, namely, from the antennas to the reaction center.

It has long been questioned whether energy transport in photosynthesis is coherent (exciton-like) or stochastic (wave packet-like). Before going into detail, I address the characteristic physical features of these two limiting cases and the problem of how they can experimentally be distinguished.

If energy transport over N chromophores is coherent, the excited state wave function is delocalized over the N molecules and has a common phase. This means that the excitation is wavelike. Because wavelike excitations carry momentum, new selection rules come in. In addition the spectrum may change. The simplest case is when the wave function is delocalized over at least two chromophores. One then has a dimer. Depending on the mutual orientation, one of the dimer states may be forbidden or only partially allowed, with the other carrying the rest of the total oscillator strength. The polarization of the transitions involved with respect to the individual molecular frame changes. Yet they are still fully polarized, namely, perpendicular to one another. As to transport properties, it is evident that an individual energy transfer time can no longer be defined. However, new relaxation processes come into play. In the simple two chromophore dimer "miniexciton," the upper Davidov component can relax to the lower one. These are processes comparable to internal conversion in a N-center supermolecule.[35,36]

If energy transport is stochastic, excitation is always localized at an individual chromophore. Energy transfer can be resonant or nonresonant.

[35] A. S. Davydov, "Theory of Molecular Excitons." McGraw-Hill, New York, 1962.
[36] N. R. S. Reddy and G. J. Small, *J. Phys. Chem.* **94**, 7545 (1991).

If it is nonresonant, the transfer is accompanied by vibrational excitations in order to ensure energy conservation. A transfer time is clearly defined. Most importantly, the spectrum of the individual chromophore is not changed. In disordered systems, stochastic energy transport is always accompanied by fluorescence depolarization.

Coherence in the excited state of an N-chromophore system means that the individual chromophores are basically indistinguishable. This is a symmetry requirement inherent to the problem. Hence, whether transport is coherent or not depends on the relative coupling strength (e.g., dipole–dipole coupling) between like chromophores as compared to all interactions which break the above symmetry requirement so that individual chromophores can be distinguished (through their energy, orientation, etc.). Symmetry breaking interactions are connected with dynamic or static disorder. Dynamic disorder stems from thermal energy and can be frozen out. As a consequence coherent transport is more likely to occur at low temperatures. A measure for static disorder is inhomogeneous line broadening. Owing to the interplay between these various interactions, it is *a priori* not clear whether the excitation is localized or delocalized over several like chromophores.

The first systematic hole burning study of energy transfer in photosynthetic pigments was performed with phycobilisomes from the alga *Mastigocladus laminosus*.[31] Figure 8 shows the absorption together with a ΔOD spectrum. The ΔOD spectrum represents the difference spectrum between the initial absorption and the absorption modified by laser irradiation at

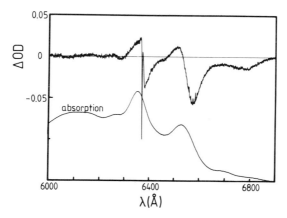

Fig. 8. Low-temperature absorption (4 K) and ΔOD spectra of phycobilisomes of *Mastigocladus laminosus*. Laser excitation was carried out at 6380 Å, as documented by the sharp zero-phonon hole [From W. Köhler, J. Friedrich, R. Fischer, and H. Scheer, *J. Chem. Phys.* **89,** 871 (1988)].

6380 Å. The two main bands in the absorption spectrum correspond to phycocyanin (6315 Å) and allophycocyanin (6530 Å). The spectrum shows several noteworthy features. First, there is a sharp zero-phonon line at the burn frequency within the phycocyanin band. Its width is about 0.3 cm^{-1}. Hence, we conclude from Eq. (1) that energy transfer occurs on time scales longer than 16 psec. That there is indeed efficient energy transfer is also clearly seen from Fig. 8. When hole burning is, for example, carried out in the phycocyanin band, we observe that the allophycocyanin band around 6530 Å is simultaneously bleached, because a broad hole occurs in this frequency range, too. This can be understood as follows: For some chromophores the phototransformation occurs in the resonantly excited state leading to the sharp hole. However, most of the resonantly excited phycocyanin chromophores undergo energy transfer to allophycocyanin, and phototransformation occurs there. During such a transfer the energy selectivity is usually lost, and hence the nonresonant hole is broad.

The experiments on phycobilisomes did not give evidence for exciton-like energy transport. In particular, no specific dependence of the zero-phonon line width as a function of burn frequency could be found, a result which was considered as strong support for stochastic transport processes.

The first unambiguous evidence for the formation of coherent, exciton-like states came from experiments on antennas of *Prosthecochloris aestuari*.[20] Figure 9 shows the inhomogeneous absorption together with the ΔOD spectrum. The arrow in Fig. 9 indicates the position where hole burning was performed. Clearly, a sharp zero-phonon hole appears at the laser frequency; some hundred wave numbers to the blue, a very broad side hole appears simultaneously. This side hole has a width on the order of 50 cm^{-1}. From these observations it followed immediately that the bands at 12,300 and 12,100 cm^{-1} are connected. If they originated from independent weakly coupled chromophores they could not be simultaneously bleached. When hole burning was performed directly in the 12,300 cm^{-1} band, again a broad hole appeared but this time resonantly with the laser frequency. This experiment gave evidence that the large width of the hole in the 12,300 cm^{-1} band was due to an ultrafast homogeneous line-broadening mechanism.

From these and additional experiments, the following picture emerged. The S_1 absorption (Fig. 9) represent delocalized exciton levels of a trimer of subunits arranged around a 3-fold symmetry axis. Each subunit contains 7 symmetry-inequivalent bacteriochlorophyll *a* molecules. The narrow and the broad hole represent lower and upper exciton levels. The large width of the upper component arises from the fact that this level has an

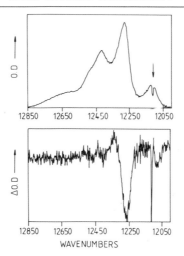

FIG. 9. Absorption and ΔOD spectra of the antenna complex of *Prosthecochloris aestuari*. The arrow indicates where hole burning was performed [Data from S. G. Johnson and G. J. Small, *Chem. Phys. Lett.* **155**, 371 (1989)].

extremely short lifetime on the order of 100 fsec owing to the possibility of relaxing to the lower component via phonon emission. Because the lower state does not have this possibility it is much sharper.

Meanwhile it was found that coherent energy transport seems to be quite common in antennas of the photosynthetic assembly. Not only in *Prosthecochloris aestuari* but also in the light harvesting complexes of *Rhodobacter sphaeroides* (B800–B850 and B875) is fast interexciton phonon scattering present.[6,7,37]

From Fig. 5 it is evident that energy is funneled from the high-energy states of the antennas to the reaction center. The overall process is associated with energy loss via phonon emission.

The transport mechanism—coherent versus incoherent—can have a strong influence on the overall probability that a photon absorbed somewhere in the antenna will feed the reaction center with energy. Suppose the transfer is stochastic and occurs with a probability p per individual step. If there are n transfer steps, the overall probability that energy is fed into the reaction center is p^n and can be rather low although p may be close to 1. On the other hand, if transport is coherent, the process of feeding the reaction center can be viewed as kind of an internal conversion process in a supermolecule comprising the various chromophores. As

[37] N. R. S. Reddy, S. V. Kolaczkowski, and G. J. Small, *Science* **260**, 68 (1993).

seen from the broad holes, these processes are extremely fast, and as a consequence the overall probability that energy is fed into the reaction center can be very high.

Electron Transport in Reaction Center

The central chromophore in the reaction center is the so-called special pair (P), a pair of chlorophyll molecules (Fig. 7). It is there that charge separation occurs. The special pair is excited, either via energy transfer from the antennas or directly via light absorption. There is much evidence for a considerable charge transfer character in the excited state P* of the special pair, including a strong electron–phonon coupling, a considerable Stark shift, as well as support from electronic structure calculations. The initial P* state decays into a charge separated state: $P*BH \rightarrow P^+B^-H \rightarrow P^+BH^-$, where B stands for bacteriochlorophyll and H for bacteriopheophytin. In Fig. 7, Q denotes the final electron acceptor, namely, quinone. While the electron is on its way down the chain to the quinone acceptor, the special pair is in an oxidized state, which has different spectroscopic properties. Hence, if P* is excited with a narrow laser, a hole will appear in the respective inhomogeneously broadened absorption band as long as the special pair remains in the photooxidized state. This is the case for a few microseconds. Then the special pair is reduced again, the ground state is restored, and the reaction center is ready for the next absorption event. Hence, hole burning in the special pair state P* is transient in nature, contrary to the case for the antennas, where it is persistent. Whereas in the antennas there is strong evidence that the reaction is based on a photophysical transformation, as outlined above, it is a transient photooxidation process in the reaction center. The transient nature of the reaction together with strong electron–phonon coupling makes hole burning in reaction centers experimentally more challenging. (Note that it has been discovered that persistent hole burning in reaction centers of *Rhodopseudomonas viridis* is possible, too.[37])

Figure 10 shows a hole burnt into the perdeuterated reaction center of wild-type *Rhodobacter sphaeroides*. There are some noteworthy features. First, there is a rather sharp, although small, zero-phonon hole with a width of 6 cm^{-1}. According to Eqs. (3) and (4), the associated lifetime is 0.9 psec at 4.2 K. In other words, within 0.9 psec, a charge-separated state is formed from the initially excited P* state. The same result is obtained from time domain experiments.[38] However, the time domain experiments essentially populated electronic–vibrational states, whereas

[38] G. R. Fleming, J.-L. Martin, and J. Breton, *Nature (London)* **96**, 7499 (1988).

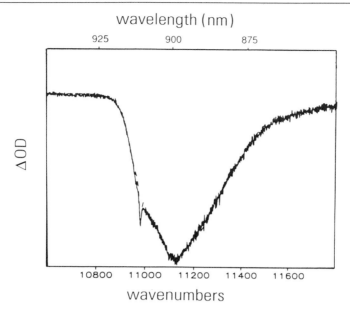

FIG. 10. ΔOD spectrum of the deuterated reaction center P870 of wild-type *Rhodobacter sphaeroides*. Note the sharp zero-phonon hole at the laser wavelength 910 Å [Data from N. R. S. Reddy, P. A. Lyle, and G. J. Small, *Photosynth. Res.* **31,** 167 (1992)].

the low-temperature hole burning experiment directly probed the decay of the purely electronic state. Hence, the conclusion is that thermalization of phonon and vibrational modes occurs prior to the primary charge separation process. The exact nature of the primary charge-separated state and its associated dynamics is still under discussion. The problems addressed concern the role of the bacteriochlorophyll (B, Fig. 7), namely, whether it enters the transfer chain as a real or virtual intermediate. This question touches the nature of the charge transfer process itself. If the bacteriochlorophyll states enter as virtual states into the transfer process, the transfer process is classified as a superexchange mechanism, because the electron is directly transferred to the bacteriopheophytin.

Second, apart from the sharp zero-phonon hole in Fig. 10, there is a broad side hole whose maximum is shifted to the blue compared to the laser frequency. The side hole represents molecular and lattice vibrational transitions which are simultaneously bleached with the pure electronic line. From the relative magnitude of the area under the zero-phonon hole and the side hole, we see that the coupling to vibrational transitions is rather strong, as is expected for states with a considerable amount of charge transfer character.

The phonon coupling scheme which emerged from various hole burning experiments on reaction centers of *Rhodobacter sphaeroides* and *Rhodopseudomonas viridis* is as follows. The transition to the P* state couples moderately strongly to protein phonon modes with a mean frequency of 20–30 cm^{-1}. In addition to the protein phonons, there is also a moderately strong coupling to a so-called special pair marker mode with typical energies of 100 cm^{-1}. The pair modes involve intermolecular motions of the monomers of the pair. They cannot always be seen in the spectra, because their relaxation can be in the femtosecond range. Nevertheless, they contribute to the intensity of the rather smooth side band.

The strong electron–phonon coupling yields a significant contribution to the broadband absorption spectra. Owing to the strong phonon contribution, the hole-burnt spectra depend in a specific fashion on the burn frequency. For example, the sharp zero-phonon feature in Fig. 9 is observed only for burning the red edge of the inhomogeneously broadened absorption band. Above a certain excess energy, the holes lose their sharply structured features.

Apart from bacterial reaction centers, hole burning was also performed in reaction centers of green plants (P680, P700). Although there are distinct differences, it seems that the essential features of the electron transfer kinetics are very similar.

An interesting observation has been made in emission studies from P* of some bacterial reaction centers[39]: associated decay patterns are nonexponential. It is not yet clear from where this nonexponentiality comes, but there seems to be agreement that it must be strongly related to the structural heterogeneity of the proteins, which could lead to a dispersion of the parameters which govern the electron transfer process and, hence, to nonexponential features in the decay pattern of P*.[32] In the next section, I address the problem of structural heterogeneity of proteins and its possible role in function performance in more detail.

Disorder in Ordered Structures: A Characteristic Feature of Proteins

Structural disorder has severe implications for all kinds of physical phenomena. We mentioned above the nonexponential relaxation features of the P* state in the reaction center. Nonexponential behavior in all kinds of relaxation and transport processes is the rule rather than the exception whenever disorder plays a significant role. In addition to nonexponential behavior, the specific features of the microscopic processes involved, such as their temperature dependence, may be severely influenced by

---

[39] A. Holzwarth, *in* "Ultrafast VIII." Springer-Verlag, Berlin, 1992 (abstract).

disorder. Disorder is also responsible for nonergodic behavior as it is reflected, for example, in thermally irreversible features of frozen proteins. As stressed above, hole burning is extremely sensitive in probing disorder effects. In this context, I address the problems of spectral diffusion and pressure phenomena in hole burning.

Basic Aspects of Protein State of Matter

Whereas in the preceding section we dealt with excited states of chromophores, we deal with ground state phenomena of the protein in this section. The solid-state physics of proteins is very intriguing because proteins are at the edge between order and disorder. They can be crystallized so that quite highly resolved X-ray diffraction patterns can be obtained, very much like those of small molecular crystals. Yet an analysis of the Debye–Waller factors of the diffracted intensity pattern shows that the mean square displacements $\langle x^2 \rangle$ of the protein building blocks behave in a nonuniform fashion and are, for $T \rightarrow 0$, much larger than typical values obtained for small molecules.[40] From this it was concluded that a protein can exist in a huge variety of substructures despite the overall order. Because of the inherent structural indeterminism, proteins are often compared with glasses. However, the picture for proteins is distinctly different. It has been suggested that the substructures occur on rather well-separated hierarchy levels, a shown in Fig. 11.[41]

As an example, myoglobin, which is the best studied protein in this respect, exists in two conformations, which are called the "tensed" and "relaxed" conformations, referring to the two states in which the ligand is bound or released. We associate a zero-order hierarchy level with these two conformations. The free enthalpy G as a function of a relevant configurational coordinate seems to have a well-defined minimum.

A closer look at the energy surface, however, reveals that it is rugged. In one definite conformation, the protein can exist in at least three different conformational substates differing by the binding angle of the ligand with respect to the heme plane.[42] These states are called the "taxonomic" states. Myoglobin has different properties in the three conformational substates, for instance, different kinetic behaviors and different IR frequencies.[43] A further hierarchy level is immediately obvious from the fact

[40] H. Frauenfelder, F. Parak, and R. D. Young, *Annu. Rev. Biophys. Biophys. Chem.* **17**, 451 (1988).

[41] A. Ansari, J. Berendzen, S. F. Bowne, H. Frauenfelder, I. E. T. Iben, T. S. Sauke, E. Shyamsunder, and R. D. Young, *Proc. Natl. Acad. Sci. U.S.A.* **82**, 5000 (1985).

[42] H. Frauenfelder, S. G. Sligar, and P. Wolynes, *Science* **254**, 1598 (1991).

[43] H. Frauenfelder, G. U. Nienhaus, and J. B. Johnson, *Ber. Bunsen-Ges. Phys. Chem.* **95**, 272 (1991).

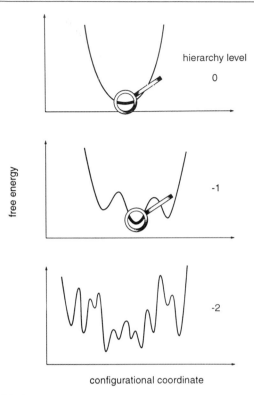

FIG. 11. Hierarchical structural disorder in proteins. Progressively lower hierarchy levels are characterized by negative numbers. The magnifying glass symbolizes the increasingly higher resolution.

that the spectral bands are inhomogeneously broadened. It is assumed (and there is strong experimental evidence) that there are further levels of structural disorder with progressively lower hierarchy levels. Yet it becomes increasingly difficult to address them separately.

Conformational substates with low hierarchy, that is, those with small separating barriers, may be in equilibrium at a given temperature. However, some of the conformational substates are definitely not. As a consequence, a protein behaves in a nonergodic fashion. It is unable to sample its total structural phase space within practical time scales. We shall show how hole burning experiments can support this model of the protein state of matter. In these kind of investigations one problem arises that has to be considered as carefully as possible. Because the protein under investigation is, as a rule, dissolved in a glass-forming solvent which can induce disorder effects in the optical transitions of the chromophore by itself,

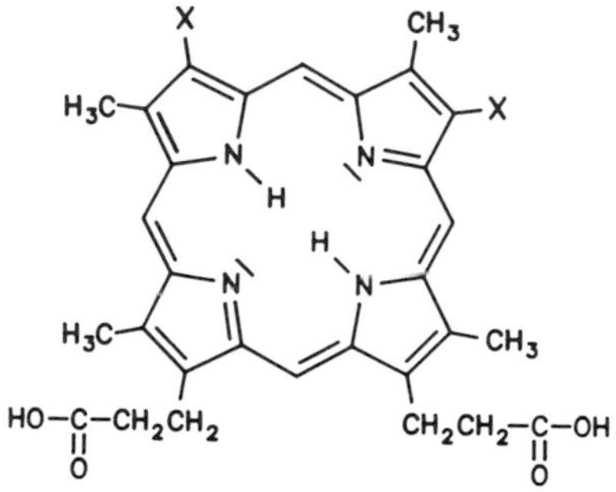

Protoporphyrin IX: X = CH CH$_2$

Mesoporphyrin IX: X = CH$_2$CH$_3$

FIG. 12. Protoporphyrin IX and mesoprophyrin IX, substitutes for the heme group in myoglobin and horseradish peroxidase in order to enable high-resolution hole burning.

the influence of the solvent has to be separated as well as possible. For this reason we have chosen heme proteins, namely, myoglobin and horseradish peroxidase, for our experiments. In these proteins, the chromophore is rather well shielded from the solvent because it is buried within the protein pocket. However, the natural heme cannot be used for narrow bandwidth hole burning, because its lifetime is so short that the respective absorption bands are homogeneously broadened to a large extent.[21] To overcome this disadvantage, the natural heme group was substituted by free base porphyrin analogs: protoporphyrin IX for myoglobin and mesoporphyrin IX for horseradish peroxidase (Fig. 12). In some cases it was possible to bring the chromophores directly into a host glass, so that a 1 : 1 comparison between protein solution and dye solution was possible. Free base porphyrins are well-known hole burning systems.[44,45] The photoreaction is based on a phototautomerism of the two inner ring protons. Note that a series of tautomer states is possible.[29,46]

Thermocyclic Hole Burning Experiments and Spectral Diffusion Phenomena

There is direct evidence of conformational substates in a protein, and some of the substates have extremely low energy barriers so that the

[44] S. Voelker and J. H. van der Waals, *Mol. Phys.* **32**, 1703 (1976).

[45] S. Voelker, *in* "Relaxation Processes in Molecular Excited States" (J. Fuenfschilling, ed.), p. 113. Kluwer Academic Publishers, Dordrecht, The Netherlands, 1989.

[46] J. Zollfrank, J. Friedrich, J. Fidy, and J. M. Vanderkooi, *Biophys. J.* **59**, 305 (1991).

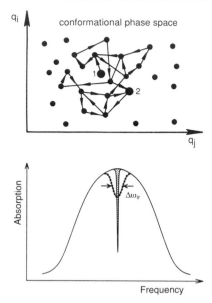

FIG. 13. A spectral hole as a phase space marker. The sharp hole is burnt at the lowest temperature, when the protein is in substate 1. As a consequence of cyclic temperature variation, the protein is trapped in substate 2. The process of structural detrapping and retrapping is reflected in a broadening of the hole ($\Delta\omega_{ir}$).

protein can escape from a structural trap even at temperatures close to 4 K.[46–49] Figure 13 reflects the principle of a thermocyclic hole burning experiment. The top part of Fig. 13 represents the structural phase space of a protein. Each point is meant to represent a certain conformational substate, where q_i and q_j are conformational coordinates. Suppose we cool the system close to 0 K. Then, all dynamics ceases and the system occupies one substate out of a huge manifold. Assume that it is substate 1 in Fig. 13. To measure the dynamics associated with structural relaxation processes, we have to label the protein in the very state it occupies. We do this by burning a narrow hole into the absorption spectrum.

After burning, the temperature is increased to a value T_{ex}, called the excursion temperature. Because the protein has now thermal energy, it starts to explore its structural phase space (a trajectory is sketched in Fig. 13). After a while, the temperature is reduced again to the burn temperature which was, in our case, close to 0 K. The protein is again frozen in a

[47] W. Köhler and J. Friedrich, *J. Chem. Phys.* **90,** 1270 (1989).
[48] W. Köhler and J. Friedrich, *Europhys. Lett.* **7,** 517 (1988).
[49] J. Zollfrank, J. Friedrich, J. Fidy, and J. M. Vanderkooi, *J. Chem. Phys.* **95,** 3134 (1991).

definite state, yet in a different one (substate 2 in Fig. 13). In this state the microscopic fields are different compared to state 1. The chromophore feels these differences and adjusts its absorption energy accordingly. As a consequence, the hole broadens. This broadening is solely based on the structural relaxation processes induced by the temperature cycle, because the line width is always measured at the same temperature, namely, the burn temperature. We call this type of broadening thermally induced spectral diffusion broadening.[48]

In Fig. 14[49] the change of the hole width $\Delta\omega_{ir}$ through a thermal cycle is plotted as a function of the excursion temperature. The two data series represent the behavior of the protein in a glycerol/water glass and the chromophore directly dissolved in a glass. Spectral diffusion broadening in the glass is smooth. It follows a power law with an exponent of 3/2. In the protein, however, there is a distinct step around 12 K. In this temperature range the protein obviously undergoes a structural change which is reflected in the broadening. We conclude that there must be conformational substates with very low barriers.

This experiment shows that, although proteins are similar to glasses, they are also very different. The smoothness of the spectral diffusion broadening in the glass reflects the fact that there are no distinguishable hierarchy levels and that it is, under the conditions of our experiment, essentially an infinite system. The protein is finite. There is no smooth distribution of structural states. Within the temperature range accessible

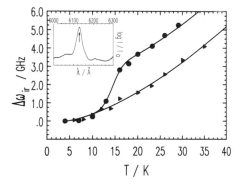

FIG. 14. Thermally induced spectral diffusion broadening in mesoporphyrin IX-substituted horseradish peroxidase (●) as compared to mesoporphyrin IX in a glass (▲). Inset shows the long-wavelength range of the protein absorption. The arrow indicates the hole burning frequency [From J. Zollfrank, J. Friedrich, J. Fidy, and J. M. Vanderkooi, *Biophys. J.* **59**, 305 (1991)].

to the experiment, there is one rather discrete structural barrier which can be crossed. The fact that spectral diffusion broadening does occur is a strong hint of an additional hierarchy level buried below the inhomogeneously broadened band.

Spectral Holes under Pressure: What Do We Learn about Proteins?

In this section we show that hole burning can be used to measure the compressibility of a protein molecule by purely spectroscopic means.[50] The compressibility, in turn, is a structure-sensitive parameter. It determines the respective volume fluctuations. Our results demonstrate that rather small structural changes are enough to induce significant changes in the compressibility. In other words, conformational substates on a sufficiently large hierarchy level have significantly different elastic properties. In the following I present two experiments. The first compares pressure phenomena in spectral holes burnt into the absorption of a protein, namely, protoporphyrin IX-substituted myoglobin, with the respective behavior in a host glass.[51] From the comparison we learn about the specific features which characterize the ground state of a protein as compared to that of a glass. The second experiment concerns the enzyme horseradish peroxidase.[52] Horseradish peroxidase is a medium size heme protein of 34 kDa. Its function is the oxidation of small aromatic substrate molecules in plant roots. To this end, the protein incorporates the respective substrate molecules into the heme pocket. The incorporation of a substrate molecule leads to a dramatic softening of the package density of the protein obviously because of a structural reorganization. For hole burning the heme group is replaced by mesoporphyrin IX. It should be stressed that this substitution does not alter the substrate binding affinity.

When spectral holes are exposed to isotropic pressure changes, the holes will shift and broaden. We start with a simple pressure shift–solvent shift model[24] assuming that the pressure-induced shift s is proportional to the solvent shift ν_s. If the solvent shift originates from a solute–solvent interaction proportional to $1/R^n$, then it is straightforward to show that

$$s = \frac{n}{3}\kappa\nu_s\Delta p \tag{5}$$

The most reasonable choice is $n = 6$ (dispersion interaction). The point is that, in disordered materials, ν_s is not a fixed quantity but instead varies

[50] J. Zollfrank, J. Friedrich, J. Fidy, and J. M. Vanderkooi, *J. Chem. Phys.* **94**, 8600 (1991).
[51] J. Gafert, J. Friedrich, and F. Parak, *J. Chem. Phys.* **99**, 2478 (1993).
[52] J. Fidy, J. M. Vanderkooi, J. Zollfrank, and J. Friedrich, *Biophys. J.* **63**, 1605 (1992).

over the inhomogeneous band. Because our experiment is frequency selective, we can select chromophores with different ν_s by burning holes at various positions in the inhomogeneous band. According to the definition of the solvent shift ν_s, we can write

$$\nu_s = \nu_b - \nu_{vac} \tag{6}$$

with ν_b being the burn frequency (i.e., the laser frequency) and ν_{vac} being the absorption frequency of the isolated chromophore.[26,53,54] We call this frequency the vacuum absorption frequency. Then, from a plot of $s/\Delta p$ versus ν_b, we can directly determine the compressibility κ. Note that κ is, in our case, a local compressibility. The chromophore just feels the compression of the lattice in a range determined by the dispersion interaction. Because this interaction is of sufficiently short range and does not exceed the dimensions of the protein, it is the compressibility of the protein molecule which we measure and not that of the host lattice.

If Eqs. (5) and (6) describe the physics reasonably well, the pressure shift s should vary in a linear fashion with burn frequency ν_b, and there should be a frequency, namely, ν_{vac}, where this shift vanishes. At ν_{vac}, the hole should broaden, only. We tested these predictions in a series of glasses with a variety of dye probes and found perfect agreement with the model.[53,55] This encouraged us to apply the technique to proteins. We also stress that the conditions under which Eq. (5) holds can be derived from a microscopic model.[56] Figure 4 shows how a hole burnt into the inhomogeneously broadened absorption spectrum of a protein (mesoporphyrin IX-substituted horseradish peroxidase) deforms under pressure variations up to 1.5 MPa.

Consider the pressure tuning hole burning experiments of protoporphyrin IX dissolved directly in a glass (Fig. 15). As a glass matrix we have chosen glycerol doped with dimethylformamide (30%, w/w) in order to ensure a proper solubility. Figure 15a shows the inhomogeneously broadened absorption spectrum; Fig. 15b shows the shift of the holes with pressure and the broadening of the holes with pressure over the burn frequency. The results agree perfectly with the pressure shift–solvent shift model described above. The shift with pressure is linear in burn frequency (linear color effect) in agreement with Eqs. (5) and (6). There is a well-defined frequency (arrow in Fig. 15a) where the pressure shift vanishes, also in agreement with Eqs. (5) and (6). From Eq. (6) it is clear

[53] J. Zollfrank and J. Friedrich, *J. Phys. Chem.* **96**, 7889 (1992).
[54] J. Zollfrank and J. Friedrich, *J. Opt. Soc. Am.* **B9**, 956 (1992).
[55] H. Pschierer, J. Friedrich, H. Falk, and W. Schmitzberger, *J. Phys. Chem.* **97**, 6902 (1993).
[56] B. B. Laird and J. L. Skinner, *J. Chem. Phys.* **90**, 3274 (1989).

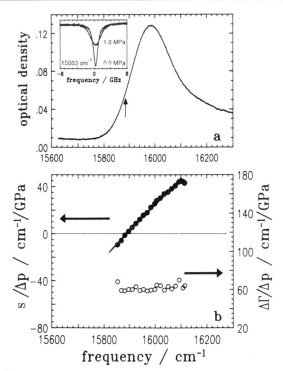

FIG. 15. Pressure tuning hole burning spectroscopy of protoporphyrin IX in a glass. (a) Absorption spectrum. The arrow indicates the frequency where the pressure shift vanishes. The corresponding hole is shown in the inset. (b) Shifts and broadening $\Delta\Gamma$ of the hole with pressure as a function of burn frequency [From J. Gafert, J. Friedrich, and F. Parak, *J. Chem. Phys.* **99**, 2478 (1993)].

that this frequency has to be identified with the vacuum frequency of the dye probe. The inset in Fig. 15a shows that a hole at this frequency does not shift under pressure; however, it broadens. The slope of the straight line in Fig. 15b determines the compressibility κ of the host glass. We obtained a value of 0.11 GPa^{-1}, which is the correct order of magnitude.

To make the description complete we also show the broadening of the hole under pressure, which is independent of frequency. This is also in agreement with more general models of glasses.[56] In simple words, it reflects the fact that, in a glass, there is no correlation between local environment and excitation frequency within the inhomogeneous band. At any frequency, the whole ensemble of environments is probed. The degeneracy is lifted under pressure and leads to the observed broadening.

The protein shows a markedly different behavior (Fig. 16). The color effect in the shift with pressure varies in a nonlinear fashion over the inhomogeneous band. In the red and blue edge, we measure a linear behavior, yet the corresponding slopes differ by a factor of almost 3. Because we associated the slope with the compressibility, we have to conclude that the compressibility of the protein varies within a factor of 3.

What is the physics behind this finding? Let us discuss this question by comparing the structural phase space of a glass with that of a protein. Figure 17 represents an artist's view of the phase spaces. We focus on the phase space of the glass first. Each point is meant to represent a global structural state of the respective systems. We assume that the ensemble of dye molecules probes the possible structural states in a uniform way. Each dye molecule sees a different environment and adjusts energy levels accordingly. However, the spectral adjustment in the excited state is rather independent of that of the ground state, and, as a consequence, there is no correlation between absorption frequency and environment. This means that if we select an ensemble of absorbers with a narrow bandwidth laser, we probe all structural states in the glass (Fig. 17a), that is, the whole occupied phase space. Hence, the compressibility parameters which we measure are bulk compressibilities.

In a protein (Fig. 17b), there are states distinguished according to the hierarchical model discussed above. In myoglobin, for example, the so-called taxonomic states fall into this category.[42] Conformational substates with smaller structural variations group around the distinguished states. In other words, the phase space is no longer homogeneous. Instead, it decomposes into islands.

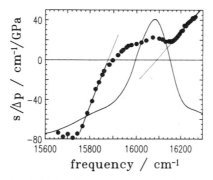

FIG. 16. Pressure tuning hole burning spectroscopy of protoporphyrin IX-substituted myoglobin. Plotted is the shift of the hole with pressure as a function of burn frequency in the inhomogeneously broadened absorption. Note the different slopes in the red and blue edges [Data from J. Gafert, J. Friedrich, and F. Parak, *J. Chem. Phys.* **99**, 2478 (1993)].

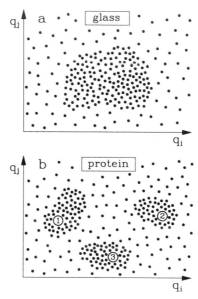

FIG. 17. Structural phase space of a glass (a) and a protein (b). Each point is meant to represent a global structure, where q_i and q_j are coordinates.

The point of our model, which we call the "correlated phase space model," is that the various islands correspond to specific tautomer states. To put it differently, a certain protein structure as reflected, for example, through island 1, supports a specific tautomer structure (or vice versa). If so, protein structures can be optically selected by irradiating into a specific tautomer band. In this case, it is evident that an experiment averages over the selected structures only. That is, averages do not comprise the whole phase space but are confined to the islands. Because the islands reflect different structures, structure-sensitive parameters such as the compressibility are expected to be different. Protoporphyrin IX, as well as mesoporphyrin IX, can exist in a variety of tautomeric states[46] which differ by the arrangement of inner ring protons. In myoglobin, these tautomeric states are rather close in energy. There are indications that the absorption band (Fig. 16) is actually a superposition of several tautomers. Hence, as the burn frequency is tuned across the absorption band, one selects different structures with different properties. In the band center, however, the various tautomers strongly overlap. Because of this overlap, the measured pressure shift represents an average over the behavior of all the tautomers, and, in fact, the quantity of interest, namely, the slope of the shift per unit pressure, can attain almost all possible values. Therefore, in the band center the data are not conclusive. However, as one tunes the burn frequency into the wings of the inhomogeneous band, specific tautomers eventually dominate, and the data can be interpreted on the basis of the model outlined above. This is clearly seen in Fig. 16. In the wings straight lines are measured in agreement with Eq. (5). From the difference in the slope we have to conclude that we select protein substructures in the wings (islands in Fig. 17) whose compressibilities differ by as much as a factor of 3.

The results of our experiments on horseradish peroxidase are summarized in Fig. 18. We performed pressure tuning experiments on the protein in the presence of a substrate molecule (HRP–NHA complex) as well as on the free enzyme. A sketch of the prosthetic group together with some adjacent amino acids and the substrate molecule is shown in the inset to Fig. 18a. The substrate molecule is naphthohydroxamic acid. The obvious result from this experiment is very clear and fits into the correlated phase space model outlined above. The inclusion of a substrate molecule is linked to a structural change of the apoprotein. As is obvious from the large compressibility in the complexed form (as expressed through the steep slope), this structural change leads to a rather dramatic softening of the packing density of the protein around the chromophore. So the general result is similiar to the case for myoglobin: rather small perturbations, as obvious from small changes in the spectra, lead to rather dramatic changes in structure-sensitive parameters, such as the compressibility.

A few noteworthy remarks should be added. First, all our results fit on an absolute scale quite well into the scenario of what is known on compressibilities of proteins. They are compatible with the data on specific heat and thermal expansion.[51] Second, we stress that within the pressure range of our experiments, proteins are indeed fully elastic. This can, for instance, be seen from the inset of Fig. 18b, where several pressure tuning cycles are superimposed. The traces clearly show that there is full reversibility. No plastic deformations occur. Third, pressure experiments on a series of discrete sites in organic crystals show that strong fluctuations in microcompressibilities can indeed occur.[57]

We conclude this section by addressing the problem of the scale of the structural changes associated with the large compressibility variations. We specifically ask whether these changes could be observed in X-ray diffraction patterns. To answer this question we ask how much do we, on average, have to change the structure in order to observe a band shift on the order of the inhomogeneous width. This is also the order of the tautomer splitting which is the proper energy scale to be considered. We

---

[57] P. Schellenberg, J. Friedrich, and J. Kikas, *J. Chem. Phys.* **100**, 5501 (1994).

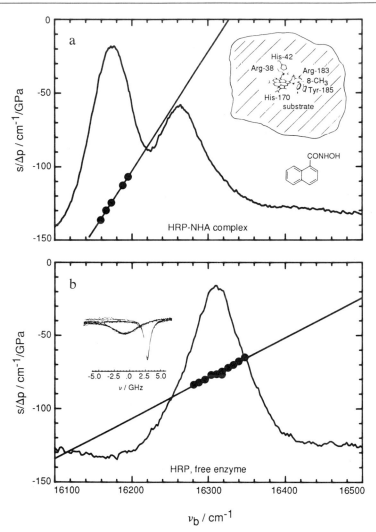

FIG. 18. Pressure tuning hole burning spectroscopy of mesoporphyrin IX-substituted horseradish peroxidase. (a) Spectrum of the enzyme complexed with a substrate molecule, namely, naphthohydroxamic acid. Inset shows the arrangement of the prosthetic group, the substrate molecule, and some amino acids. (b) Spectrum of the free enzyme. Inset shows the behavior of a hole under pressure cycles (0–2 MPa). Several traces are superimposed. Note the full reversibility [From J. Fidy, J. M. Vanderkooi, J. Zollfrank, and J. Friedrich, *Biophys. J.* **63,** 1605 (1992)].

decompose the whole protein volume in elementary units occupied by just one atom. Then, the change ΔV of the protein volume is given by

$$\Delta V = 3V \frac{\Delta a}{a}$$

where a^3 is the volume of the elementary unit. a is on the order of an interatomic distance. On the other hand,

$$\Delta V = \kappa V \Delta p$$

Because a pressure change on the order of 1 MPa induced a shift on the order of 0.1 cm^{-1}, we need something like 1 GPa to induce a shift on the order of the inhomogeneous band. Hence, $\kappa \Delta p = 0.1$ and we have

$$\frac{\Delta a}{a} = 0.03$$

The conclusion is that even if such pronounced optical changes as the shift of a peak on the order of the inhomogeneous width are induced, the associated average structural changes are just at the resolution limit of X-rays as obtained from an analysis of the associated Debye–Waller factors.[51]

Acknowledgments

I thank J. Fidy, H. Frauenfelder, J. Gafert, F. Parak, and H. Scheer for many stimulating discussions. I thank Prof. G. J. Small for sending me his publications on photosynthesis research. Financial support came from the DFG (Fr 456/17-1, SFB 213, Graduiertenkolleg "Nichtlineare Dynamik"), the Fonds der Chemischen Industrie, and the WTZ program with Hungary.

[11] Polarized Optical Spectroscopy of Chromoproteins

By HERBERT VAN AMERONGEN and WALTER S. STRUVE

Introduction

The spectroscopic techniques in linear dichroism, polarized fluorescence, and polarized pump–probe experiments differ from those in the corresponding unpolarized experiments principally through the addition of linear polarizers in the excitation light beam and (when applicable) in the fluorescence or absorbance probe beam emerging from the sample. Hence, the spectroscopic methodology of dynamic linear dichroism experiments is essentially identical to that described in the chapters on absorption difference spectroscopy (H. van Amerongen and R. van Grondelle, this volume [9]), and time-resolved fluorescence spectroscopy (A. R. Holzwarth, this volume [14]). An extensive review of steady-state linear dichroism in nucleic acids and DNA–protein complexes has been published by Nordén et al.[1] This chapter focuses instead on steady-state and time-resolved studies of chromoproteins such as phytochrome and photosynthetic antennae. Major current issues in polarized optical spectroscopy include (1) relationships of optical anisotropies to coherence phenomena, molecular motion, electronic energy transfer, and chromophore architecture; and (2) preparation and characterization of partially ordered protein samples. Coherence phenomena in assemblies of strongly interacting chromophores (e.g., in photosynthetic antennae and in synthetic aromatic homodimers such as binaphthyl) are a subject of intense current interest. Although single crystals of the vast majority of chromoproteins remain inaccessible for linear dichroism studies, greatly improved techniques have been developed for the preparation of partially ordered proteins. These topics form the main emphases of this chapter. Readers who are primarily interested in experimental aspects of the subject may wish to read Section II of this chapter before addressing the theoretical material in Section I.

I. Randomly Oriented Proteins

The polarizations $\mathbf{e}_i$, $\mathbf{e}_f$ of incident and/or emerging light beams will be specified with respect to a laboratory-fixed coordinate system (xyz). Because chromoproteins in a homogeneous sample usually exhibit well-defined molecular structure, it is more useful to specify the chromophore transition moments (which are denoted in this chapter with Greek vectors such as $\boldsymbol{\mu}$ and $\boldsymbol{\nu}$) with reference to protein-fixed coordinates (abc). It is assumed throughout this chapter that protein orientational diffusion is negligible during the experimental time scale; the transformation between laboratory- and protein-fixed coordinates is then fixed in time. Depolarization accompanying rotational diffusion of molecules and proteins in solution has been extensively treated elsewhere[2,3] (see also D. M. Jameson and W. H. Sawyer, this volume [12]).

A. Preparation of Excited States and Coherence

The outcome of a polarized optical experiment may be profoundly influenced by the presence of exciton couplings and coherence in many-

[1] B. Nordén, M. Kubista, and T. Kurucsev, Q. Rev. Biophys. 25, 51 (1992).
[2] T. Tao, Biopolymers 8, 609 (1969).
[3] T. J. Chuang and K. B. Eisenthal, J. Chem. Phys. 57, 5094 (1972).

chromophore systems. The simplest prototype is a pair of identical chromophores, whose electronic excited states may be described using either a site representation (in which the basis states $|1\rangle$, $|2\rangle$ represent excitations localized on chromophores 1 and 2, respectively) or a delocalized representation [in which the basis states $|+\rangle = 2^{-1/2}(|1\rangle + |2\rangle)$ and $|-\rangle = 2^{-1/2}(|1\rangle - |2\rangle)$ are exciton states]. In the presence of an intermolecular interaction with energy H_{12}, the original (degenerate) energy $E_{1(2)}$ of the lowest excited singlet state of the dimer becomes split into exciton components with energies E_+, E_-. Because the delocalized states $|+\rangle$, $|-\rangle$ are linear combinations of the site states $|1\rangle$, $|2\rangle$, the transition moment orientations μ_+, μ_- of the delocalized states differ from those (μ_1, μ_2) in the site representation.

All of the information concerning excited state populations in an ensemble of such dimers residues in the four elements ρ_{kl} of a time-dependent density matrix ρ,[4] where the indices k, l are 1 or 2 in the site representation, and $+$ or $-$ in the delocalized representation. The initial values of the density matrix elements in the delocalized representation, immediately after electric dipole-allowed absorption of a photon with polarization $\mathbf{e_i}$,[5] are given by[6]

$$\rho_{++}(0) = K_+ B_+ (\hat{e}_i \cdot \hat{\mu}_+)^2 \qquad \rho_{--}(0) = K_- B_- (\hat{e}_i \cdot \hat{\mu}_-)^2$$
$$\rho_{+-}(0) = \rho_{-+}(0) = (K_+ K_- B_+ B_-)^{1/2} (\hat{e}_i \cdot \hat{\mu}_+)(\hat{e}_i \cdot \hat{\mu}_-) \tag{1}$$

Here B_+, B_- are the Einstein coefficients[7] for absorptive transitions to the two exciton states, and K_+, K_- are the incident light intensities at the respective transition wavelengths. (The Einstein coefficients are proportional to the extinction coefficients of the absorption bands). Throughout this chapter, the notation $\hat{x}$ indicates a unit vector that points in the same direction as $\mathbf{x}$.

The subsequent time evolution of the density matrix elements $\rho_{kl}(t)$ is governed by the stochastic Liouville equations (Section I,B).[4] In the electric dipole approximation, the intensity of fluorescence subsequently emitted by such a dimer with polarization $\mathbf{e_f}$ will be

$$I(t) = A_+ \rho_{++}(t)(\hat{e}_f \cdot \hat{\mu}_+)^2 + A_- \rho_{--}(t)(\hat{e}_f \cdot \hat{\mu}_-)^2$$
$$+ (A_+ A_-)^{1/2}(\rho_{+-} + \rho_{-+})(\hat{e}_f \cdot \hat{\mu}_+)(\hat{e}_f \cdot \hat{\mu}_-) \tag{2}$$

[4] C. Cohen-Tannoudji, B. Diu, and F. Laloë, "Quantum Mechanics," Vol. 1. Wiley(Interscience), New York, 1977; J. J. Sakurai, "Advanced Quantum Mechanics." Addison-Wesley, Reading, Massachusetts, 1967.
[5] W. S. Struve, "Fundamentals of Molecular Spectroscopy." Wiley(Interscience), New York, 1989.
[6] T. S. Rahman, R. S. Knox, and V. M. Kenkre, Chem. Phys. 44, 197 (1979).
[7] J. B. Birks, "Photophysics of Aromatic Molecules." Wiley(Interscience), London, 1970.

where A_+, A_- are the Einstein coefficients for spontaneous emission from the exciton states. If one of the exciton components is selectively excited (e.g., $K_+ = 0$), the initial values of the off-diagonal density matrix elements ρ_{+-}, ρ_{-+} [Eq. (1)] are zero. Because the diagonal and off-diagonal density matrix elements are uncoupled in the stochastic Liouville equations for the delocalized representation (see Section I,B), the cross terms in the fluorescence intensities will then vanish for all subsequent times, with the result that the fluorescence spectrum and polarization become simply sums of the respective quantities for the two exciton components. In the limit of spectrally broad excitation ($K_+ = K_-$), both of the off-diagonal elements initialize to large, nonzero values. The resulting interference terms in Eq. (2) may then drastically modulate the fluorescence polarization, with a time dependence that reflects processes responsible for the decay in ρ_{+-} and ρ_{-+}. This situation arises in the ultrafast spectroscopy of strongly coupled chromophores: femtosecond excitation pulses are inherently spectrally broad, and such time resolution affords direct observation of coherence decays in ρ_{+-}, ρ_{-+}. In Section IB, we concentrate discussion on narrow band excitation of single, nondegenerate electronic states, with measurements performed on such a coarse time scale that coherence effects [embodied in cross terms analogous to those in Eqs. (1) and (2)] are unimportant. The more general case is considered in Section I,C.

B. Incoherently Excited Systems

In a static linear dichroism experiment, one measures the sample absorbance using light propagating along the laboratory z axis and polarized along either the x or y axis. For a single electric-dipole electronic transition from state 0 to state n, the parallel and perpendicular absorbance components are then proportional to

$$A_{\parallel} = \sum_n B_{0n} \langle (\hat{x} \cdot \hat{\mu}_n)^2 \rangle \tag{3a}$$

$$A_{\perp} = \sum_n B_{0n} \langle (\hat{y} \cdot \hat{\mu}_n)^2 \rangle \tag{3b}$$

where μ_n and B_{0n} are the pertinent transition moment and Einstein coefficient,[7] respectively, for absorption. For economy of notation, the single index n runs over both chromophores and excited states, so that a given n denotes a particular excited state on a particular chromophore. The brackets $\langle \rangle$ indicate averaging over the chromoprotein orientational distribution. For chromoproteins exhibiting multiple absorption bands, these expressions must be summed over n (i.e., over absorbing species and electronic transitions). We define the reduced linear dichroism as

$$LD = \frac{A_\parallel - A_\perp}{A_\parallel + 2A_\perp} = \frac{\sum_n B_{0n}[\langle(\hat{x} \cdot \hat{\mu}_n)^2\rangle - \langle(\hat{y} \cdot \hat{\mu}_n)^2\rangle]}{\sum_n B_{0n}} = \frac{\sum_n B_{0n}(LD)_n}{\sum_n B_{0n}} \quad (4)$$

Defined in this way, the reduced linear dichroism for a given absorption band depends only on the transition moment directions μ; furthermore, the total LD observed at any wavelength is the sum of reduced linear dichroisms for the individual bands, weighted by their extinction coefficients.

Another quantity that characterizes optical anisotropy is the dichroic ratio $A_\parallel/A_\perp$; a frequently used alternative definition of the reduced linear dichroism is $LD = (A_\parallel - A_\perp)/(A_\parallel + A_\perp)$. In isotropic (randomly oriented) samples, there is nothing to distinguish physically between the laboratory x and y directions, so that such samples exhibit zero linear dichroism ($A_\parallel = A_\perp$). Hence, static linear dichroism studies are applicable only to oriented samples.

In a polarized fluorescence experiment, the time-dependent fluorescence intensity $I(t)$ observed using excitation and analyzing polarizations $\mathbf{e_i}$, $\mathbf{e_f}$ is

$$I(t) = \sum_{mn} K_n p_m(t)\langle(\hat{e}_i \cdot \hat{\mu}_n)^2(\hat{e}_f \cdot \hat{\nu}_m)^2\rangle B_{0n} A_{m0} \quad (5)$$

Here μ_n, ν_m are the absorption and emission transition moments, $p_m(t)$ is the excited state population for emitting electronic state m, A_{m0} is its Einstein coefficient for spontaneous emission,[7] and K_n is the laser intensity at the wavelength for the $0 \to n$ transition. The use of independent indices n, m allows for the possibility that the absorbing and emitting states may be different electronic states on the same chromophore (e.g., owing to internal conversion). Alternatively, they may represent excitations on different chromophores, as a consequence of electronic energy transfer. The absence in Eq. (5) of cross terms analogous to those in Eqs. (1) and (2) stems from the incoherent nature of the monitored excited state population(s). In a conventional (but by no means universal) experimental geometry, the excitation beam is polarized along the x axis and propagates along the y axis; sample fluorescence is viewed along the z axis. The fluorescence components polarized parallel and perpendicular to the excitation polarization are then

$$I_\parallel(t) = \sum_{mn} K_n p_m(t)\langle(\hat{x}\cdot\hat{\mu}_n)^2(\hat{x}\cdot\hat{v}_m)^2\rangle B_{0n}A_{m0} \tag{6a}$$

$$I_\perp(t) = \sum_{mn} K_n p_m(t)\langle(\hat{x}\cdot\hat{\mu}_n)^2(\hat{y}\cdot\hat{v}_m)^2\rangle B_{0n}A_{m0} \tag{6b}$$

For this case the fluorescence anisotropy function $r(t)$ is defined by

$$r(t) = \frac{I_\parallel(t) - I_\perp(t)}{I_\parallel(t) + 2I_\perp(t)} \tag{7}$$

Unlike the static linear dichroism, the anisotropy function $r(t)$ exhibits nonzero values even in randomly oriented samples, because polarized excitation creates an anisotropic excited state population through the electric dipole orientational factor $(\mathbf{x}\cdot\boldsymbol{\mu})^2$. However, polarized fluorescence experiments in oriented samples are potentially far more informative than in isotropically random samples (Section II,A).

In a pump–probe experiment, one probes the absorbance changes $\Delta A(t)$ after exciting the sample at time $t = 0$ with a short laser pulse. For collinear pump and probe beams, the polarized absorption difference signals detected using probe polarizations parallel and perpendicular to the pump polarization are proportional to

$$\Delta A_\parallel(t) = \sum_{mn} K_n p_m(t)B_{0n}\langle(\hat{x}\cdot\hat{\mu}_n)^2[B_{ms}(\hat{x}\cdot\hat{v}_{ms})^2$$

$$- B_{0m}(\hat{x}\cdot\hat{v}_{0m})^2 - B_{m0}(\hat{x}\cdot\hat{v}_{m0})^2]\rangle \tag{8a}$$

$$\Delta A_\perp(t) = \sum_{mn} K_n p_m(t)B_{0n}\langle(\hat{x}\cdot\hat{\mu}_n)^2[B_{ms}(\hat{y}\cdot\hat{v}_{ms})^2$$

$$- B_{0m}(\hat{y}\cdot\hat{v}_{0m})^2 - B_{m0}(\hat{y}\cdot\hat{v}_{m0})^2]\rangle \tag{8b}$$

Here the first, second, and third sets of terms inside the square brackets represent contributions arising from excited state absorption ($m \to s$), ground-state photobleaching ($0 \to m$), and stimulated emission ($m \to 0$), respectively.[5] It is important to recognize that the photobleaching terms depend on B_{0m} and $\boldsymbol{v}_{0m}$ (rather than on B_{0n} and $\boldsymbol{v}_{0n}$, since photobleaching can "migrate" from the initially excited chromophore(s) because of electronic energy transfer. Equation (8) has been oversimplified for clarity. Excitation of state m on a particular chromophore bleaches all ground state absorption bands on that chromophore (not just $0 \to m$). The pump–probe anisotropy function $r(t)$ can be defined in analogy to the fluorescence anisotropy [Eq. (7)].

The protein orientational averaging in Eqs. (3), (5), (6), and (8) requires a transformation between laboratory-fixed and protein-fixed coordinates. The protein-fixed axes (abc) can be reoriented into the laboratory-fixed axes (xyz) by applying a sequence of counterclockwise rotations through the independent Euler angles ($\chi\theta\phi$). The first rotation is made through the angle χ about the c axis and reorients the directions (abc) into ($x'y'z'$) with $z' = c$. The second rotation is made through the angle θ about the y axis, yielding new coordinate axes ($x''y''z''$) with $y'' = y'$. The final rotation is made through the angle ϕ about the $z''(=z)$ axis. (This sequence of rotations is illustrated in perspective by Marion in *Classical Dynamics of Particles and Systems*.[8] Many alternative definitions of Euler angles exist in the literature, leading to contrasting expressions for the protein to laboratory transformation.) The rotation matrix λ for this transformation,

$$
\begin{bmatrix} x \\ y \\ z \end{bmatrix} = \begin{bmatrix} \lambda_{xa} & \lambda_{xb} & \lambda_{xc} \\ \lambda_{ya} & \lambda_{yb} & \lambda_{yc} \\ \lambda_{za} & \lambda_{zb} & \lambda_{zc} \end{bmatrix} \begin{bmatrix} a \\ b \\ c \end{bmatrix} \tag{9}
$$

is easily shown to be[8]

$$
\lambda = \begin{bmatrix} \cos\chi\cos\theta\cos\phi - \sin\chi\sin\phi & \sin\chi\cos\theta\cos\phi + \cos\chi\sin\phi & \sin\theta\cos\phi \\ -\cos\chi\cos\theta\sin\phi - \sin\chi\cos\phi & -\sin\chi\cos\theta\sin\phi + \cos\chi\cos\phi & -\sin\theta\sin\phi \\ -\cos\chi\sin\theta & -\sin\chi\sin\theta & \cos\theta \end{bmatrix}
$$

For illustration, we evaluate the orientational averages required in the calculation of polarized fluorescence intensities [Eq. (6)] emitted by a randomly oriented protein sample. For a protein with orientational distribution $f(\chi\theta\phi) \equiv f(\Omega)$, the orientational average required in $I_\perp(t)$ is

$$
\langle (\hat{x} \cdot \hat{\mu})^2 (\hat{y} \cdot \hat{\nu})^2 \rangle = \int_0^{2\pi} d\chi \int_0^\pi \sin\theta \, d\theta \int_0^{2\pi} d\phi (\hat{x} \cdot \hat{\mu})^2 (\hat{y} \cdot \hat{\nu})^2 f(\Omega) \tag{10}
$$

If fluorescence arising from a particular transition $m \to 0$ is monitored after exciting a particular absorption band $0 \to n$, the protein-fixed absorption and emission transition moment directions can be specified without loss of generality as $\mu = (001)$, $\nu = (abc)$. In this case, Eq. (9) yields

$$
\hat{x} \cdot \hat{\mu} = \lambda_{xc} \qquad \hat{y} \cdot \hat{\nu} = \lambda_{ya}a + \lambda_{yb}b + \lambda_{yc}c \tag{11}
$$

[8] J. B. Marion, "Classical Dynamics of Particles and Systems." Academic Press, New York, 1965. In that reference, the second Euler rotation is made about the x' rather than y' axis; the rotation matrix correspondingly differs from ours.

In the special case of a randomly oriented protein $[f(\Omega) = 1/8\pi^2]$, the only nonvanishing contributions to the orientational average are then

$$\langle(\hat{x}\cdot\hat{\mu})^2(\hat{y}\cdot\hat{\nu})^2\rangle = a^2\langle\lambda_{xc}^2\lambda_{ya}^2\rangle + b^2\langle\lambda_{xc}^2\lambda_{yb}^2\rangle + c^2\langle\lambda_{xc}^2\lambda_{yc}^2\rangle$$

$$= \frac{2}{15}(a^2 + b^2) + \frac{1}{15}c^2$$

which equals $(2 - c^2)/15$ because $a^2 + b^2 + c^2 = 1$. Similarly, the orientational average required for $I_\parallel(t)$ is

$$\langle(\hat{x}\cdot\hat{\mu})^2(\hat{x}\cdot\hat{\nu})^2\rangle = a^2\langle\lambda_{xc}^2\lambda_{xa}^2\rangle + b^2\langle\lambda_{xc}^2\lambda_{xb}^2\rangle + c^2\langle\lambda_{xc}^4\rangle = \frac{1}{15}(a^2 + b^2) + \frac{1}{5}c^2$$

which is $(2c^2 + 1)/15$. Substitution of these averages into the expressions for the polarized fluorescence intensities [Eq. (6)] then leads to a concise expression for the fluorescence anisotropy:

$$r = \frac{3c^2 - 1}{5} \tag{12}$$

The possible fluorescence anisotropies r thus range from -0.2 (for perpendicular absorption and emission transition moments, $c = 0$) to a maximum of $+0.4$ (for parallel transition moments, $c = 1$). Complete depolarization ($r = 0$) occurs only when the absorption and emission moments are separated by the magic angle 54.7°, for which $(3c^2 - 1) = 0$. This example illustrates the sensitivity of the anisotropy to chromophore architecture in the protein; this emerges as a consequence of the well-defined chromoprotein structure.

Pump–probe anisotropies can differ considerably from fluorescence anisotropies in the same chromoprotein, because stimulated emission (SE), excited state absorption (ESA), and photobleaching (PB) can all exhibit independent transition moment orientations. By analogy to Eq. (12), the anisotropy observed in a pump–probe experiment in which excited state m has been populated following excitation of state n is

$$r = \frac{1}{5}\left[\frac{3(c_{ms}^2 B_{ms} - c_{0m}^2 B_{0m} - c_{m0}^2 B_{m0})}{B_{ms} - B_{0m} - B_{m0}} - 1\right] \tag{13}$$

which shows that the total anisotropy is a weighted average of anisotropies for the overlapping electronic transitions. In the special case where

PB, SE, and ESA all exhibit the same transition moment orientation, the pump–probe and fluorescence anisotropies coincide. In a contrasting case in which mutually parallel PB and SE moments are both perpendicular to the ESA moment, it is easily shown that the pump–probe anisotropy is

$$r = \frac{\Delta A_{\parallel} - \Delta A_{\perp}}{\Delta A_{\parallel} + 2\Delta A_{\perp}} = \frac{x + 2}{5(1 - x)} \tag{14}$$

where $x = B_{ms}/(B_{0m} + B_{m0})$ is the ratio of cross sections for ESA and PB/SE at the probe laser wavelength. The possible anisotropies r are then at least $+0.4$ (when PB/SE dominate the absorption difference signal) or at most -0.2 (when ESA dominates). These domains complement the range of anisotropies generally observed in fluorescence depolarization [Eq. (12)].

Beck and Sauer[9] studied pump–probe anisotropy decays in allophyco-cyanin trimers from a mutant of the cyanobacterium *Synechococcus* PCC 6301. The antenna contains three, rotationally equivalent pairs of strongly coupled phycocyanobilin chromophores, which give rise to exciton levels $|+\rangle, |-\rangle$ separated by about $210 \, \text{cm}^{-1}$. Excitation at wavelengths between 650 and 670 nm prepares one of three (nearly degenerate) states $|+\rangle$. For trimers excited and probed at 670 nm, the anisotropy function r(t) decays from $r(0) = 0.36$ to $r(\infty) = 0.06$, with a lifetime of 71 psec. At this wavelength, the absorption difference is apparently dominated by PB/SE; the corresponding isotropic PB/SE decay only exhibits a very small 71 psec component. Hence, this 71 psec decay component reflects energy equilibration among three nearly degenerate phycocyanobilin pairs. Owing to the 3-fold symmetry in the chromophore architecture, this corresponds to an approximately 210 psec time constant for individual transfer steps between chromophore pairs.[10] This example illustrates, with unusual clarity, the utility of anisotropy studies in highlighting energy transfer kinetics involving spectrally equivalent molecules.

A fundamental limitation of the approach described in this section is the assumption that the occupation probabilities $p_m(t)$ and transition moments completely define the nature of the excited states. As pointed out in Section I,A, the excitation of multiple exciton components [cf. Eqs. (1) and (2)] leads to the presence of interference terms, which must be considered in a more general treatment of fluorescence and absorption difference experiments.

[9] W. F. Beck and K. Sauer, *J. Phys. Chem.* **96,** 4658 (1992).
[10] T. P. Causgrove, S. Yang, and W. S. Struve, *J. Phys. Chem.* **92,** 6790 (1988).

C. Coherently Excited Systems: Stochastic Liouville Equation

To our knowledge, the only case for which depolarization has been treated in terms of coherent excited state dynamics is the system of two identical, strongly interacting chromophores.[6,11,12] Rahman, Knox, and Kenkre (hereafter called RKK[6]) considered the time evolution of the ensemble-averaged density matrix[4] for a homodimer in which the angle between the chromophore transition moments μ_1, μ_2 is θ. (This density matrix, introduced at the beginning of Section I,A, lends itself directly to calculations of time-resolved fluorescence and absorption difference anisotropies.) RKK showed that when detailed balancing is incorporated along with the effects of radiative and nonradiative decay, three of the equations of motion for the time-dependent homodimer density matrix elements ρ_{++}, ρ_{--}, ρ_{+-}, ρ_{-+} in the delocalized representation are given by

$$\frac{d\rho_{++}}{dt} = -\gamma\rho_{++} + \gamma'\rho_{--} - \frac{(1 + \cos\theta)\rho_{++}}{\tau_0} - \left(\frac{1}{\tau} - \frac{1}{\tau_0}\right)\rho_{++} \quad (15a)$$

$$\frac{d\rho_{--}}{dt} = -\gamma'\rho_{--} + \gamma\rho_{++} - \frac{(1 - \cos\theta)\rho_{--}}{\tau_0} - \left(\frac{1}{\tau} - \frac{1}{\tau_0}\right)\rho_{--} \quad (15b)$$

$$\frac{d\rho_{+-}}{dt} = -2iJ\rho_{+-} - \left(2A + \frac{1}{\tau}\right)\rho_{+-} - (\Gamma + B - A)(\rho_{+-} - \rho_{-+}) \quad (15c)$$

A fourth equation is generated from Eq. (15c) by interchanging ρ_{+-} and ρ_{-+} everywhere, and reversing the sign of the first term on the right-hand side. Here γ, γ' are related to local fluctuations in $E_{1(2)}$ and control the rates of downward and upward transitions between the exciton components, Γ is their mean, $1/\tau_0$ and $1/\tau$ are the radiative and total excited state decay rates, and J is related to the interaction matrix element H_{12}. The constants A, B are related to the fluctuation in the interaction, and therefore to H_{12}^2. The Förster rate for excitation transfer between the two chromophores is given by $F = A + 2J^2/(2\Gamma + 2B + 1/\tau)$ in the weak coupling regime for this formalism.

The stochastic Liouville equations are readily solved for the time-dependent density matrix elements ρ_{kl} (e.g., through Laplace transforms); the latter may then be used in turn to develop expressions for the polarized fluorescence or absorption difference signals. The initial values of the density matrix elements under δ-function pulsed excitation are given by

[11] F. Zhu, C. Galli, and R. M. Hochstrasser, *J. Chem. Phys.* **98**, 1042 (1993).
[12] K. Wynne and R. M. Hochstrasser, *Chem. Phys.* **171**, 179 (1993).

Eq. (1), whereas the time-dependent fluorescence intensities are given by Eq. (2). If the homodimer is excited with a laser pulse at zero time, the density matrix elements will evolve as

$$\rho_{++}(t) = a_+(t)\rho_{++}(0) + b_+(t)\rho_{--}(0)$$
$$\rho_{--}(t) = a_-(t)\rho_{--}(0) + b_-(t)\rho_{++}(0)$$
$$\rho_{+-}(t) = c_+(t)\rho_{+-}(0) + d_+(t)\rho_{-+}(0)$$
$$\rho_{-+}(t) = c_-(t)\rho_{-+}(0) + d_-(t)\rho_{+-}(0)$$

because no couplings exist between diagonal and off-diagonal density matrix elements in the delocalized representation [Eq. (15)]. We necessarily have $a_+(0) = a_-(0) = c_+(0) = c_-(0) = 1$; the other four coefficients initialize to zero. Combining the latter expressions with Eq. (1) and substitution into Eq. (2) yield the dimer fluorescence intensities for x-polarized laser excitation:

$$
\begin{aligned}
I_\parallel(t) = {} & K_+A_+B_+a_+(t)\langle(\hat{x}\cdot\hat{\mu}_+)^4\rangle + K_-A_+B_-b_+(t)\langle(\hat{x}\cdot\hat{\mu}_-)^2(\hat{x}\cdot\hat{\mu}_+)^2\rangle \\
& + K_-A_-B_-a_-(t)\langle(\hat{x}\cdot\hat{\mu}_-)^4\rangle + K_+A_-B_+b_-(t)\langle(\hat{x}\cdot\hat{\mu}_+)^2(\hat{x}\cdot\hat{\mu}_-)^2\rangle \\
& + (K_+K_-A_+A_-B_+B_-)^{1/2}[c_+(t) + d_+(t) + c_-(t) \\
& + d_-(t)]\langle(\hat{x}\cdot\hat{\mu}_+)^2(\hat{x}\cdot\hat{\mu}_-)^2 \quad\quad\quad\quad\quad\quad\quad\quad\text{(16a)}
\end{aligned}
$$

and

$$
\begin{aligned}
I_\perp(t) = {} & K_+A_+B_+a_+(t)\langle(\hat{x}\cdot\hat{\mu}_+)^2(\hat{y}\cdot\hat{\mu}_+)^2\rangle \\
& + K_-A_+B_-b_+(t)\langle(\hat{x}\cdot\hat{\mu}_-)^2(\hat{y}\cdot\hat{\mu}_+)^2\rangle \\
& + K_-A_-B_-a_-(t)\langle(\hat{x}\cdot\hat{\mu}_-)^2(\hat{y}\cdot\hat{\mu}_-)^2\rangle \\
& + K_+A_-B_+b_-(t)\langle(\hat{x}\cdot\hat{\mu}_+)^2(\hat{y}\cdot\hat{\mu}_-)^2\rangle \\
& + (K_+K_-A_+A_-B_+B_-)^{1/2}[c_+(t) + d_+(t) + c_-(t) + d_-(t)] \\
& \langle(\hat{x}\cdot\hat{\mu}_+)(\hat{x}\cdot\hat{\mu}_-)(\hat{y}\cdot\hat{\mu}_+)(\hat{y}\cdot\hat{\mu}_-)\rangle \quad\quad\quad\quad\text{(16b)}
\end{aligned}
$$

The transition moment orientations μ_+, μ_- in a symmetric dimer are orthogonal in the delocalized representation, irrespective of the transition moment directions on the individual chromophores[13]; their unit vectors can thus be specified without loss of generality as (001) and (ab0) in the protein reference frame. Orientational averaging of the fluorescence intensities in Eq. (16) then leads to

$$
I_\parallel(t) = \frac{3}{15}K_+A_+B_+a_+(t) + \frac{3}{15}K_-A_-B_-a_-(t) + \frac{1}{15}K_-B_-A_+b_+(t)
$$

$$
+ \frac{1}{15}K_+B_+A_-b_-(t) + \frac{1}{15}(K_+K_-A_+A_-B_+B_-)^{1/2}
$$

$$
[c_+(t) + d_+(t) + c_-(t) + d_-(t)] \quad\quad\quad\quad\quad\quad\quad\text{(17a)}
$$

[13] R. W. Chambers, T. Kajiwara, and D. R. Kearns, *J. Phys. Chem.* **78**, 380 (1974).

and

$$I_{\|}(t) = \frac{1}{15} K_+ A_+ B_+ a_+(t) + \frac{1}{15} K_- A_- B_- a_-(t) + \frac{2}{15} K_- A_+ B_- b_+(t)$$

$$+ \frac{2}{15} K_+ A_- B_+ b_-(t) - \frac{1}{30} (K_+ K_- A_+ A_- B_+ B_-)^{1/2}$$

$$[c_+(t) + d_+(t) + c_-(t) + d_-(t)] \tag{17b}$$

Applying the initial conditions for the time-dependent coefficients $a_\pm(t)$, $b_\pm(t)$, etc., to Eq. (17) and then substitution into Eq. (7) lead to the initial anisotropy:

$$r(0) = \frac{2K_+ A_+ B_+ + 2K_- A_- B_- + 3(K_+ K_- A_+ A_- B_+ B_-)^{1/2}}{5(K_+ A_+ B_+ + K_- A_- B_-)} \tag{18}$$

We next make use of the fact that in a homodimer exhibiting an angle of θ between the monomer transition moments, the Einstein coefficients B_+ and B_- for absorptive transitions to the exciton states $|+\rangle$, $|-\rangle$ are proportional to $(1 \pm \cos \theta)^6$; similar relationships hold for the spontaneous emission coefficients A_+, A_-. In the limit of spectrally broad excitation ($K_+ = K_-$), the initial anisotropy is then

$$r(0) = \frac{7 + \cos^2 \theta}{10(1 + \cos^2 \theta)} \tag{19}$$

Hence, for broadband excitation of a symmetric dimer, the initial anisotropy may fall anywhere from 0.4 to 0.7, depending on the transition moment geometry.[14] This expression concurs with one derived by Wynne and Hochstrasser[12] using Redfield theory.[15] In the special case where $\theta = 90°$, the time-dependent anisotropy function $r(t)$ has the form[12,14]

$$r_{90}(t) = \frac{1}{10} [1 + 3e^{-2\Gamma t} + 3\xi(t)] \tag{20}$$

For general θ (in the limit where the total excited state decay rate $1/\tau$ is negligible compared to A, B, Γ, and J), it can be shown using Eq. (17) that

[14] R. S. Knox and D. Gülen, *Photochem. Photobiol.* **57**, 40 (1993) [Proceedings of the Second International Workshop on Photosynthetic Antenna Systems, Freising, Germany, March 30–April 3, 1992].
[15] A. G. Redfield, *Adv. Magn. Reson.* **1**, 1 (1965).

$$r(t) = \frac{r_{90}(t) + \frac{1}{10}[3 + e^{-2\Gamma t} - 3\xi(t)]\cos^2\theta + \frac{2}{5}(1 - e^{-2\Gamma t})\frac{(\gamma' - \gamma)}{2\Gamma}\cos\theta}{1 + e^{-2\Gamma t}\cos^2\theta + (1 - e^{-2\Gamma t})\frac{(\gamma' - \gamma)}{2\Gamma}\cos\theta}$$

(21)

Here the function $\xi(t)$, which initializes to unity at $t = 0$, is

$$\xi(t) = e^{-(A+B+\Gamma)t}\left[(B + \Gamma - A)\frac{\sin\sqrt{4J^2 - (B + \Gamma - A)^2}\,t}{\sqrt{4J^2 - (B + \Gamma - A)^2}}\right.$$

$$\left. + \cos\sqrt{4J^2 - (B + \Gamma - A)^2}\,t\right]$$

(22)

Because the rates γ, γ' for upward and downward transitions between exciton components separated by energy Δ must satisfy detailed balancing, the magnitude of the factor $(\gamma - \gamma')/2\Gamma$ that appears in Eq. (21) is $\tanh(\Delta/2kT)$ at temperature T. For exciton splittings Δ of several hundred wave numbers (common in photosynthetic antennae), this factor is significant even at 300 K. Finally, we note that in the weak coupling limit[6] (in which $J \approx 0$ and $(B + \Gamma - A) \approx \Gamma$), $\xi(t)$ no longer contains sinusoidal terms. The third term on the right-hand side of Eq. (20) then assumes the form $\exp(-2Ft)$, so that this decay component in the anisotropy function reflects ordinary Förster excitation transfer kinetics between the two chromophores.

Equations (20) and (21) predict that the anisotropy should initially decay rapidly from $r(0) \geq 0.4$ with lifetime 2Γ, then decay more slowly from $r = 0.4$ with dynamics that resemble Förster kinetics in the limit of weak coupling. No anisotropies $r(0) > 0.4$ have been reported in numerous pump–probe studies[16] of energy transport in bacterial and green plant antennae, conducted at time resolutions varying from around 10 psec to about 300 fsec. For these systems, $(2\Gamma)^{-1}$ therefore appears to be bounded from above by (conservatively) approximately 1 psec. Anisotropies $r(0) \approx 0.7$ have been observed in magnesium tetraphenylporphyrin,[17] a system to which the homodimer model appears to be directly applicable.

[16] T. Gillbro, V. Sundström, A. Sandström, M. Spangfort, and B. Andersson, *FEBS Lett.* **193**, 267 (1985); S. L. S. Kwa, H. van Amerongen, S. Lin, J. P. Dekker, R. van Grondelle, and W. S. Struve, *Biochim. Biophys. Acta* **1102**, 202 (1992); S. Lin, H. van Amerongen, and W. S. Struve, *Biochim. Biophys. Acta* **1060**, 13 (1991); and references therein.

[17] C. Galli, K. Wynne, S. M. LeCours, M. J. Therien, and R. M. Hochstrasser, *Chem. Phys. Lett.* **206**, 493 (1993).

II. Partially Oriented Proteins

A. Relationship between Observables and Protein Properties

Physically diverse techniques exist for preparation of partially oriented proteins, including alignment in electric and magnetic fields, embedding of proteins in stretched polymer films, and expansion/compression of protein-containing gels. Most techniques establish a unique, laboratory-fixed symmetry axis (called the director **n**): the resulting nonrandom orientational distribution $f(\Omega)$ is independent of the azimuthal angle about the director. In the discussion that follows, we use the example of a gel compressed between parallel optical flats. For this case, **n** is perpendicular to the cell windows. Many of the conclusions reached herein will apply to the other orientational methods as well. Because of the azimuthal symmetry associated with the director, no linear dichroism will be exhibited if the incident light propagates along **n**.

Two arrangements [termed cases (a) and (b)] that do show linear dichroism are illustrated in Fig. 1. In case (a), the propagation axis is tilted by an angle ω from the director. In case (b), which may be viewed as the limit of case (a) when $\omega = 90°$, the light beam traverses the gel along a path normal to the director. When applied to the polarized absorbance

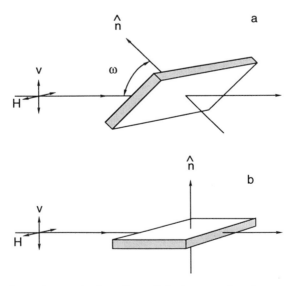

FIG. 1. Experimental geometry for a linear dichroism experiment on a partially oriented sample with director **n**. In case (a), **n** is displaced through the angle ω from the light propagation vector, about an axis that is parallel to $\mathbf{e_H}$. In case (b), $\omega = \pi/2$.

components for the horizontal and vertical polarizations shown in Fig. 1, Eq. (3) becomes

$$A_{\mathrm{H}} = B_{01}\langle(\hat{e}_{\mathrm{H}} \cdot \hat{\mu})^2\rangle \equiv \frac{B_{01}}{3}[\langle 2P_2(\hat{e}_{\mathrm{H}} \cdot \hat{\mu})\rangle + 1]$$

$$A_{\mathrm{V}} = B_{01}\langle(\hat{e}_{\mathrm{V}} \cdot \hat{\mu})^2\rangle \equiv \frac{B_{01}}{3}[\langle 2P_2(\hat{e}_{\mathrm{V}} \cdot \hat{\mu})\rangle + 1]$$

(23)

where $P_2(x)$ denotes the second-order Legendre polynomial in x, and the other notation is similar to that in Section I. The orientational averaging required here is far more complicated than that for a random distribution. If the proteins are embedded in membranes, the membranes will exhibit a certain orientational distribution with respect to **n**, the proteins will exhibit a certain distribution with respect to the membranes, and finally the chromophores (presumably) exhibit well-characterized orientations with respect to the proteins. (Elucidating the latter chromophore–protein architecture is a central objective.) This entails a series of rotational transformations, whose algebra is greatly simplified by exploiting the transformation properties of second-rank tensors[18] that are extensively used in studies of quantum mechanical angular momentum.[19] [Readers uninitiated in these mathematical topics have the option of proceeding directly to Eq. (28), which gives the final result for the linear dichroism.]

The angular part of the hydrogenlike atom wave functions are given by the associated Legendre polynomials $P_{jm}(\theta, \phi)$, in which θ, ϕ are referenced to an arbitrary coordinate system (x, y, z). If the latter system is displaced into a new coordinate system by successive rotations through the Euler angles (α, β, γ), the original wave functions can always be expressed as linear combinations of a rotated set of associated Legendre polynomials $P_{jm}(\theta', \phi')$, in which θ', ϕ' are referenced to the new coordinate system (x', y', z'), via[18]

$$P_{jm}(\theta, \phi) = \sum_{m'=-j}^{j} D_{mm'}^{(j)*}(\alpha\beta\gamma)P_{jm'}(\theta', \phi')$$

(24)

The complex-valued rotation matrix elements $D_{mm'}^{(j)} = \exp(-im\alpha)$ $d_{mm'}(\beta)\exp(-im'\gamma)$ (which are also known as Wigner polynomials) reduce into associated Legendre polynomials $P_{jm}(\alpha, \beta)$ for $m' = 0$, and into ordinary Legendre polynomials $P_j(\cos\beta)$ for $m = m' = 0$. They have been extensively tabulated for other combinations of j, m, and m'.[19]

[18] K. Gottfried, "Quantum Mechanics." Benjamin, New York, 1966.
[19] D. M. Brink and G. R. Satchler, "Angular Momentum," 2nd Ed. Oxford Univ. Press, London and New York, 1971.

The orientational averages required in Eq. (23) are facilitated by a well-known addition theorem (also called the closure relationship) for the Wigner polynomials[18]:

$$P_2(\hat{e}_i \cdot \hat{\mu}) = \sum_{m=-2}^{2} D_{m0}^{(2)*}(\phi_i \theta_i 0) D_{m0}^{(2)}(\phi_\mu \theta_\mu 0) \tag{25}$$

where $(\phi_i \theta_i 0)$ and $(\phi_\mu \theta_\mu 0)$ give the orientations of $\mathbf{e}_i$ and $\boldsymbol{\mu}$ in terms of Euler angles in the laboratory-fixed system. [The orientation (000) in laboratory-fixed coordinates denotes a vector parallel to the z axis; only two independent Euler angles are required to specify the orientation of a cylindrically symmetric object such as a vector.] The rotational transformation [Eq. (24)] can be used to express the orientational factors in Eq. (23) in terms of the transition moment orientations $(\alpha_c \beta_c 0)$ in cell-fixed coordinates:

$$D_{m0}^{(2)}(\phi_\mu \theta_\mu 0) = \sum_{m'=-2}^{2} D_{mm'}^{(2)*}(\alpha_{cl} \beta_{cl} \gamma_{cl}) D_{m'0}^{(2)}(\alpha_c \beta_c 0) \tag{26}$$

In this context, the angles $(\alpha_{cl} \beta_{cl} \gamma_{cl})$ that rotate the cell frame into the laboratory frame are $(0, \pi/2 - \omega, 0)$ and (000) for cases (a) and (b) in Fig. 1. Because the polarizations $(\phi_i \theta_i 0)$ are given by (000) and $(\pi/2, \pi/2, 0)$ for $\mathbf{e}_V$ and $\mathbf{e}_H$ respectively (Fig. 1), Eqs. (25) and (26) can be combined with explicit expressions for $d_{mm'}^{(2)}(\beta)$[19] to yield

$$P_2(\hat{e}_H \cdot \hat{\mu}) = -\frac{1}{2} P_2(\cos \beta_c)$$
$$P_2(\hat{e}_H \cdot \hat{\mu}) = P_2(\sin \omega) P_2(\cos \beta_c) \tag{27}$$

Here β_c is the angle between the transition moment and the director $\mathbf{n}$. Using S_μ to denote the orientationally averaged quantity $\langle P_2(\cos \beta_c) \rangle$, the reduced linear dichroism [Eq. (4)] becomes

$$LD = \frac{S_\mu \sin^2 \omega}{1 - S_\mu \cos^2 \omega} \tag{28}$$

In the special case (b) for $\omega = \pi/2$ (Fig. 1), the reduced linear dichroism reduces to $LD = S_\mu$. By concatenating a series of rotational transformations (protein → membrane → cell) and applying the closure relationship, it can be shown[20] that

$$S_\mu = \langle P_2(\cos \beta_{mc}) \rangle \langle P_2(\cos \beta_{pm}) \rangle P_2(\cos \beta_\mu) \tag{29}$$

where β_{mc} is the angle between $\mathbf{n}$ and the membrane plane, β_{pm} is the angle between the latter and the protein symmetry axis, and β_μ is the

[20] M. van Gurp, Ph.D. Thesis, University of Utrecht, The Netherlands (1988).

(well-defined) transition moment orientation with respect to the protein symmetry axis. For proteins that have been solubilized from the membrane, the factor $\langle P_2(\cos \beta_{mc})\rangle\langle P_2(\cos \beta_{pm})\rangle$ (which is known as the order parameter $\langle P_2 \rangle$) coalesces into a single average $\langle P_2(\cos \beta_{cp})\rangle$. The value of $\langle P_2 \rangle$ measures the extent of sample orientation: it ranges from 0 in randomly ordered proteins to 1 in a perfectly ordered sample.

Equation (29) indicates that since the protein-fixed moment direction and the order parameter both influence the linear dichroism, the latter technique alone cannot determine the projection $\cos \beta_\mu$ of the moment along the protein symmetry axis unless $\langle P_2 \rangle$ is independently known. Even when $\langle P_2 \rangle$ is given, linear dichroism studies give no information about the azimuthal moment orientation α_μ of the chromophore in the protein frame; the situation is further complicated if several electronic transitions or absorbing species are present. Finally, we note that if the chromophore orientation is well characterized, linear dichroism experiments do not reveal the full protein orientational distribution $f(\Omega)$, but only its moment with respect to $P_2(\cos \beta_{cp})$.

Hemelrijk et al.[21] studied the linear dichroism spectrum of oblate LHC-II trimers from the chlorophyll (Chl) a/b peripheral antenna of photosystem II, using two-dimensional gel compression in polyacrylamide (PAA). There are at least six spectral forms of Chl pigments in the antenna, whose Q_y spectrum extends from 630 to 685 nm; the orientations of the porphyrin planes are known[22] to be approximately normal to the trimer plane. The low-energy 676 nm Chl a component exhibits a large positive LD, indicating that its average Q_y moment is nearly parallel to the trimer plane. The Q_x transition for the same chromophores must therefore be essentially perpendicular to the trimer plane. A question then arises as to whether LCH-II absorption at 640 nm is dominated by Q_y transitions in 640 nm Chl b spectral form(s), or by Q_x transitions in the 676 nm Chl a component. The ocurrence of positive linear dichroism at this wavelength rules out the latter possibility. At least two of the LHC-II Chl b bands (at 648, 657 nm) exhibit negative linear dichroism, indicating that the Q_y moments of these species are oriented at an angle β_μ less than 54.7° with respect to the trimer symmetry axis [cf. Eq. (29)].

Kooyman et al.[23,24] devised a polarized fluorescence technique that extracts the maximum possible information about chromophore architec-

[21] P. W. Hemelrijk, S. L. S. Kwa, R. van Grondelle, and J. P. Dekker, Biochim. Biophys. Acta 1098, 159 (1992).
[22] W. Kühlbrandt and D. N. Wang, Nature (London) 350, 130 (1991).
[23] R. P. H. Kooyman, Y. K. Levine, and B. W. van der Meer, Chem. Phys. 60, 317 (1981).
[24] R. P. H. Kooyman, M. H. Vos, and Y. K. Levine, Chem. Phys. 81, 461 (1981).

ture in a partially oriented protein (Fig. 2). A compressed gel lies parallel to the xy plane with its director $\mathbf{n}$ along the z axis. The fluorescence excitation and analyzing beam paths lie in the xz plane, making angles of θ and ϕ, respectively, with the z axis. Vertical polarization in either beam is defined as parallel to the x axis, and it is expressed in terms of Euler angles via $\mathbf{e}_{i(f)} = (0, \pi/2, 0)$. The horizontal polarizations are defined as $\mathbf{e}_i = (\pi/2, \pi/2 + \theta, 0)$ and $\mathbf{e}_f = (\pi/2, \pi - \phi, 0)$. The absorption and fluorescence transition moment orientations are $\boldsymbol{\mu} = (\alpha_\mu, \beta_\mu, 0)$ and $\boldsymbol{\nu} = [\alpha_\nu(t), \beta_\nu(t), 0]$. The polarized fluorescence intensities [Eq. (5)] may be recast in terms of second-order Legendre polynomials:

$$I(t) = \frac{1}{9} p(t) \langle [1 + P_2(\hat{e}_i \cdot \hat{\mu})]\{1 + P_2[\hat{e}_f \cdot \hat{\nu}(t)]\} \rangle \tag{30}$$

A treatment fully analogous to the foregoing analysis of linear dichroism then leads to the following expressions for four independent fluorescence intensities:

$$
\begin{aligned}
I_{VV} &= p(t)[1 - S_\mu - S_\nu(t) + G_0(t) + 3G_2(t)] \\
I_{VH} &= p(t)\{1 - S_\mu - S_\nu(t) + G_0(t) - 3G_2(t) \\
&\quad + [3S_\nu(t) - 3G_0(t) + 3G_2(t)] \sin^2 \phi\} \\
I_{HV} &= p(t)\{1 - S_\mu - S_\nu(t) + G_0(t) - 3G_2(t) \\
&\quad + [3S_\mu - 3G_0(t) + 3G_2(t)] \sin^2 \theta\} \\
I_{HH} &= p(t)\{1 - S_\mu - S_\nu(t) + G_0(t) - 3G_2(t) \\
&\quad + [3S_\nu(t) - 3G_0(t) - 3G_2(t)] \sin^2 \phi \\
&\quad + [3S_\mu - 3G_0(t) - 3G_2(t)] \sin^2 \theta \\
&\quad + [9G_0(t) + 3G_2(t)] \sin^2 \phi \sin^2 \theta + 3G_1(t) \sin 2\phi \sin 2\theta\}
\end{aligned}
\tag{31}
$$

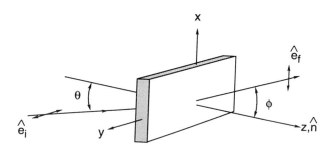

FIG. 2. Experimental geometry for a fluorescence depolarization experiment. The director $\mathbf{n}$ is parallel to the z axis, and the cell windows are parallel to the xy plane. The excitation and detected emission bands propagate in the xz plane, along directions that make angles of θ and ϕ, respectively, with the z axis. Horizontal and vertical beam polarizations are illustrated for $\mathbf{e}_i$ and $\mathbf{e}_f$, respectively.

These fluorescence intensities contain the five anisotropic parameters:

$$S_\mu = \langle P_2 \rangle \langle P_2(\cos \beta_\mu) \rangle$$
$$S_\nu(t) = \langle P_2 \rangle \langle P_2(\cos \beta_\nu) \rangle \tag{32}$$
$$G_k(t) = G_{k0} g(t) + 3 G_{k1} h(t) + \tfrac{3}{4} G_{k2} j(t)$$

where k is 0, 1, or 2; the transition moment correlation functions[25] are

$$g(t) = \langle P_2(\cos \beta_\mu) P_2 \cos[\beta_\nu(t)] \rangle$$
$$h(t) = \langle \sin \beta_\mu \cos \beta_\mu \sin \beta_\nu(t) \cos \beta_\nu(t) \cos[\alpha_\mu - \alpha_\nu(t)] \rangle \tag{33}$$
$$j(t) = \langle \sin^2 \beta_\mu \sin^2 \beta_\nu(t) \cos 2[\alpha_\mu - \alpha_\nu(t)] \rangle$$

The nine parameters G_{kl} ($k, l = 0, 1, 2$) are known linear combinations[20] of the two protein order parameters $\langle P_2 \rangle$ and $\langle P_4 \rangle = \langle P_4(\cos \beta_{cp}) \rangle$. The latter are moments of the protein orientational distribution in the gel and depend on the extent of gel compression. In highly oriented samples, $\langle P_2 \rangle$ is positive and negative, respectively, for unidirectional compression of oblate and prolate proteins (which have implicitly been assumed to be cylindrically symmetric). A frequent origin of time dependence in the observed fluorescence transition moment orientation $\nu(t)$ is electronic energy transfer between pigments in photosynthetic antennae. It is now known that the LHC-II light-harvesting antenna in green plants,[22] the photosystem I core antenna–reaction center complex in green plants and cyanobacteria,[26] C-phycocyanin antennae in cyanobacteria,[27] and base-plate bacteriochlorophyll (BChl) a antennae in green bacteria[28] all exist as trimers of identical subunits. In a uniaxial protein with rotationally equivalent protein subunits, the correlation function $g(t)$ reflects only energy transfer between rotationally inequivalent pigments (i.e., between chromophores with different moment projections along the trimer symmetry axis). The other two correlation functions $h(t)$ and $j(t)$ reveal the dynamics of intersubunit energy transfers between rotationally equivalent pigments.

B. Protein Orientation Techniques

One of the most promising techniques for preparation of oriented proteins appears to be expansion or compression in gels. Introduced by

[25] H. van Amerongen and W. S. Struve, *J. Lumin.* **51**, 29 (1992).

[26] H. T. Witt, N. Krauss, W. Hinrichs, I. Witt, P. Fromme, and W. Saenger, *Res. Photosynth.* **I.4**, 521 (1992).

[27] T. Schirmer, R. Huber, M. Schneider, W. Bode, M. Miller, and M. L. Hackert, *J. Mol. Biol.* **188**, 651 (1986).

[28] D. E. Tronrud, M. F. Schmid, and B. W. Matthews, *J. Mol. Biol.* **188**, 443 (1986).

Abdourakhmanov *et al.* in 1979,[29] the method can be applied over a wide range of pH and salt concentrations, and it appears to minimize protein degradation and light-scattering artifacts at visible wavelengths.[30] Such artifacts become more pronounced in the near-UV, where suitable background corrections must be applied.[31]

In a typical polyacrylamide (PAA) gel preparation, 10% (w/v) acrylamide, 0.3% (w/v) bisacrylamide cross-linker, and 0.03% (w/v) N,N,N',N'-tetramethylethyldiamine are added to a solution of molecules in a buffer containing 50–60% (w/v) glycerol. (The glycerol is required only in low-temperature experiments.) Ammonium persulfate [APS, 0.05% (w/v)] is finally added to initiate the polymerization reaction; the amount of APS is kept as low as possible in order to minimize the concentration of free radicals. The PAA gels are essentially transparent for wavelengths below 300 nm. Gelatin gels may be used as an alternative gel medium. Gelatin is fluid at temperatures above 40°, but cools into gels of high optical quality; its greenish yellow color is easily corrected for in absorption spectra of oriented proteins. The absorption and/or circular dichroism spectra of some particles become distorted in PAA but are unaffected in gelatin; the CP47 antenna protein complex associated with photosystem II in green plants exhibits this behavior.[32]

Ganago *et al.*[33] formulated a theory for predicting the order parameter $\langle P_2 \rangle$ in gels that are expanded or compressed without change in volume. If a gel is compressed along the x and y axes (or expanded along the z axis), the director **n** is parallel to the z axis. The deformed gel dimensions are related to the initial dimensions l_x, l_y, l_z by

$$l'_x = \frac{l_x}{n^{1/2}} \qquad l'_y = \frac{l_y}{n^{1/2}} \qquad l'_z = nl_z \qquad (34)$$

where n is the deformation parameter; by definition, $n > 1$. In effect, this theory evaluates the change in orientation of an imaginary straight line drawn in the gel, when the latter is compressed from $n = 1$ to an arbitrary final value. The result is averaged over the random initial orientations in the uncompressed gel.

A rodlike (highly prolate) particle is assumed to reorient in the same way. For disklike particles, the reorientation of a plane is considered

[29] I. A. Abdourakhmanov, A. O. Ganago, Yu. E. Erokhin, A. A. Solov'ev, and V. A. Chugunov, *Biochim. Biophys. Acta* **546,** 183 (1979).

[30] P. Haworth, C. J. Arntzen, P. Tapie, and J. Breton, *Biochim. Biophys. Acta* **679,** 428 (1982).

[31] H. van Amerongen and R. van Grondelle, *J. Mol. Biol.* **209,** 433 (1989).

[32] H. van Amerongen, unpublished results.

[33] A. O. Ganago, M. V. Fok, I. A. Abdourakhmanov, A. A. Solov'ev, and Yu. E. Erokhin, *Mol. Biol.* (*Moscow*) **14,** 381 (1980).

instead. The theoretical order parameters $\langle P_2 \rangle$ so obtained may be expressed in terms of the functions

$$T(n) = \frac{n^3}{n^3 - 1} \left[1 - \frac{\tan^{-1}(n^3 - 1)^{1/2}}{(n^3 - 1)^{1/2}} \right]$$

$$L(n) = \frac{1}{n^3 - 1} \left\{ \left(\frac{n^3}{n^3 - 1} \right)^{1/2} \ln[n^{3/2} + (n^3 - 1)^{1/2} - 1] \right\}$$

(35)

For rod-shaped molecules, the predicted order parameter is $\langle P_2 \rangle = [3T(n) - 1]/2$; for disk-shaped particles, it is $\langle P_2 \rangle = [3L(n) - 1]/2$. Both of the functions $T(n)$ and $L(n)$ approach the value $\frac{1}{3}$ for an uncompressed gel ($n = 1$), corresponding to $\langle P_2 \rangle = 0$. In the high compression limit ($n \to \infty$), $T(n)$ and $L(n)$ approach 1 and 0, respectively. Hence, for a gel compressed in two directions (or expanded in one direction), the limiting order parameter $\langle P_2 \rangle$ for large n is 1.0 for rods and -0.5 for disks. If the gel is instead compressed along one direction (e.g., the z axis, which becomes the director **n**) and allowed to expand in the other two directions, the functions $T(n)$ and $L(n)$ become interchanged for disks and rods. For given order parameter $\langle P_2 \rangle$, the reduced linear dichroism obtained in the experimental geometry of Fig. 1b is $LD = \langle P_2 \rangle P_2(\cos \beta_\mu)$. Linear dichroism measurements thus offer information about the orientation of chromophore transition moments with respect to the protein symmetry axis (or with respect to the membrane plane, if it is present).

In a third compression scheme, the gel is compressed by the factor n in one direction, maintaining the sample dimension constant along the beam propagation direction and allowing the gel to expand by the same factor n in only one perpendicular direction.[34] For this scheme (which is more easily accomplished than biaxial pressing), the theoretical linear dichroism is $LD = \Phi(n)\langle P_2(\cos \beta_\mu) \rangle$, where

$$\Phi(n) = \frac{n^4 + 1}{n^4 - 1} \left[1 - \frac{n(n^2 - 1)^2}{(n^4 + 1)|n^4 - 1|} \int_0^1 \frac{x^2 + 2[n/(n^2 - 1)]^2}{\{x^4 - x^2 - [n/(n^2 - 1)]^2\}^{1/2}} \, dx \right]$$

(36)

Because rotational symmetry is not present in this compression scheme, the order parameter $\langle P_2 \rangle$ is undefined here.

These theories provide only upper bounds to the amount of orientation $|\langle P_2 \rangle|$ or $|\Phi(n)|$ that can be expected for a given degree of compression n. The absolute upper limit that can be accomplished for n in PAA gels is around 2, since cracking occurs at higher (and frequently lower) n. For

[34] H. van Amerongen, H. Vasmel, and R. van Grondelle, *Biophys. J.* **54**, 65 (1988).

$n < 2$, $\Phi(n)$ in Eq. (36) is well approximated by $\pm 4(n^2 - 1)/5(n^2 + 1)$, where the upper and lower signs refer to prolate and oblate proteins. Very high degrees of compression can distort the linear dichroism spectra (probably through deformation of the proteins themselves). Linear dichroism experiments have been routinely performed at $n \approx 1.55$. For this compression, Eq. (35) predicts $\langle P_2 \rangle$ values of 0.28 and -0.23 for rods and disks, respectively (compression in two directions); these values would be interchanged for compression in one direction and expansion in two directions. Van Amerongen et al.[35] studied the linear dichroism of oriented chlorosomes and B800–850 complexes (which may be treated to first order as rods and disks, respectively). For gels compressed in two directions, they characterized order parameters $\langle P_2 \rangle$ of 0.28 and -0.17. Some proteins orient less well than others; compression of bovine γ-crystallins in PAA gels yields $\Phi(n)$ values on the order of 0.01.[36]

Molecules may be oriented in solution by the presence of a strong, uniform external magnetic field **H**. One may define a protein-fixed coordinate system (abc) which diagonalizes the diamagnetic susceptibility tensor χ of the protein in such a way that its nonzero elements along the diagonal are χ_a, χ_b, χ_c.[37] In the special case where $\chi_a = \chi_b$, the magnetic energy of the molecule in the field **H** becomes[38]

$$
\begin{aligned}
U &= -\frac{1}{2} V H^2 (\chi_a a^2 + \chi_b b^2 + \chi_c c^2) \\
&= -\frac{1}{2} V H^2 (\Delta \chi \cos^2 \theta + \chi_a)
\end{aligned}
\tag{37}
$$

where a, b, and $c \equiv \cos \theta$ are the direction cosines of the protein-fixed axes with respect to the magnetic field, V is the effective molecular volume, and $\Delta \chi = \chi_c - \chi_a$ is the anisotropy in the diamagnetic susceptibility. [No magnetic field-induced reorientation can occur in the absence of such an anisotropy. In such a case ($\Delta \chi = 0$), the induced magnetic moment would always be parallel to the external field **H**, and the magnetic energy U would be independent of θ, so that the orientational distribution would remain random even in the presence of the field.]

Appropriate weighting with Boltzmann factors then leads to an expression for the order parameter at thermal equilibrium:

[35] H. van Amerongen, B. van Haeringen, M. van Gurp, and R. van Grondelle, *Biophys. J.* **59**, 992 (1991).

[36] M. Bloemendal, J. A. M. Leunissen, H. van Amerongen, and R. van Grondelle, *J. Mol. Biol.* **216**, 181 (1990).

[37] N. E. Geacintov, F. van Nostrand, J. F. Becker, and J. B. Tinkel, *Biochim. Biophys. Acta* **267**, 65 (1972).

[38] J. D. Jackson, "Classical Electrodynamics." Wiley, New York, 1962.

$$\langle P_2 \rangle = \frac{\displaystyle\int_0^\pi P_2(\cos\theta)\, e^{VH^2\Delta\chi\,\cos^2\theta/2kT} \sin\theta\, d\theta}{\displaystyle\int_0^\pi e^{VH^2\Delta\chi\,\cos^2\theta/2kT} \sin\theta\, d\theta} \tag{38}$$

As expected, $\langle P_2 \rangle$ approaches 0 and 1 in the limits of low and high fields ($H \rightarrow 0$ and large, respectively). Polarized fluorescence studies of aqueous suspensions of *Chlorella, Scenedesmus, Euglena,* and spinach chloroplasts have indicated that the experimental field-dependent order parameters conform closely to Eq. (38), with saturation behavior ($\langle P_2 \rangle \approx 1$) setting in at fields over 10 kG.[37] Hence, the technique can afford significant extents of orientation. However, it cannot be applied to microscopic particles (e.g., chromophores or chromoproteins) owing to the minuscule magnetic susceptibilities.[30] The particle size must be at least mesoscopic; whether the technique can be applied to membranes probably depends on the size of the membrane fragments.

Molecules may similarly be oriented using external electric fields **E**.[39] For a uniaxial molecule with diagonalized electric polarizability tensor components $\alpha_a = \alpha_b \neq \alpha_c$, the polarizability contribution to the electrostatic energy U in the **E** field is

$$U_p = -\tfrac{1}{2}(\alpha_c - \alpha_a)E^2 \cos^2\theta \tag{39}$$

The contribution of the permanent molecular dipole moment μ to U is

$$U_d = -\mu EC \cos\theta \tag{40}$$

where C is an internal field correction that depends on the solvent and protein dielectric constants.[40] In analogy to Eq. (38) for magnetic field orientation, the order parameter achieved in the electric field becomes

$$\langle P_2 \rangle = \frac{\displaystyle\int_0^\pi P_2(\cos\theta)\, e^{\beta\cos\theta + \gamma\cos^2\theta} \sin\theta\, d\theta}{\displaystyle\int_0^\pi e^{\beta\cos\theta + \gamma\cos^2\theta} \sin\theta\, d\theta} \tag{41}$$

with $\beta = \mu EC/kT$ and $\gamma = (\alpha_c - \alpha_a)E^2/2kT$. In the limit of large **E**, $\langle P_2 \rangle$ approaches 1, irrespective of the relative contributions of U_p and U_d. In

[39] E. Fredericq and C. Houssier, "Electric Dichroism and Electric Birefringence." Oxford Univ. Press (Clarendon), Oxford, 1973.
[40] C. T. O'Konski, K. Yoshioka, and W. H. Orttung, *J. Phys. Chem.* **63,** 1558 (1959).

the limit where the permanent dipole contribution dominates ($U_d \gg U_p$), the order parameter assumes the analytical form[40]

$$\langle P_2 \rangle = 1 - \frac{3}{\beta}\left(\coth \beta - \frac{1}{\beta}\right) \tag{42}$$

In the opposite limit where $U_p \gg U_d$, $\langle P_2 \rangle$ becomes

$$\langle P_2 \rangle = \frac{3}{4}\left(\frac{e^\gamma}{\gamma^{1/2}E(\gamma^{1/2})} - \frac{1}{\gamma}\right) - \frac{1}{2} \tag{43}$$

where $E(x) = \int_0^x \exp(t^2)\, dt$. High electric fields (several kV/cm) are normally required to orient macromolecules. Van Amerongen and Van Grondelle[31] used 8 kV/cm to achieve $\langle P_2 \rangle = 0.7$ in chlorosomes from *Chloroflexus aurantiacus;* significantly less orientation is obtained in many molecules under similar conditions. The technique is limited to solutions with low salt concentrations (typically <10 mM NaCl) and must be implemented with pulsed (rather than static) electric fields.

An example of orientation in stretched films was described by Bolt and Sauer,[41] who mixed a 35% solution of low molecular weight polyvinyl alcohol (PVA) in Tris-HCl (pH 7.0) with a buffer suspension of dialyzed BChl *a* protein from *Rhodobacter sphaeroides*. The mixture was spread over a microscope slide and dried in N_2 in the dark; the resulting PVA film was peeled off and equilibrated at 100% humidity for 1 hr. Stretching such films in one dimension preferentially orients the longest dimension of the embedded macromolecules along the stretching axis. The technique has been used in linear dichroism studies of bacterial reaction centers[42] and thylakoids and isolated pigment–protein complexes from green plants.[43] The drying procedure sometimes introduces unwanted effects, however, such as blue shifts in absorption spectra of pea thylakoids due to protein degradation.[30] Unlike the situation in gel expansion/compression techniques, the film dimensions perpendicular to the stretching direction (thickness and width) are not necessarily reduced by the same factor. Hence, the macromolecule orientational distributions in stretched films may not be axially symmetric. Thulstrup and Michl showed, for example, that the orientational distributions of tropolone anion and 1,10-phenanthroline in PVA stretched films are not uniaxial.[44]

Space limitations preclude discussion of several other orientation tech-

[41] J. Bolt and K. Sauer, *Biochim. Biophys. Acta* **546,** 54 (1979).
[42] C. N. Rafferty and R. K. Clayton, *Biochim. Biophys. Acta* **502,** 51 (1978).
[43] J. Biggins and J. Svejkovsky, *Biochim. Biophys. Acta* **592,** 565 (1980).
[44] E. W. Thulstrup and J. Michl, *J. Phys. Chem.* **84,** 82 (1980).

niques which have been covered in depth in an excellent review by Nordén et al.[1] Flow fields in Couette cells are an effective technique for aligning long molecules such as DNA. Like polymer film stretching techniques, they have the drawback that the distributions so produced are not uniaxial. However, this is not a major disadvantage for macromolecules that exhibit near-cylindrical symmetry. Alignment in liquid crystals works well for small polynuclear aromatic hydrocarbons. Electrophoretic orientation in gels has, to our knowledge, been applied only to polynucleotides.

Acknowledgments

The Ames Laboratory is operated for the U.S. Department of Energy by Iowa State University under Contract No. W-7405-Eng-82. This work was supported by the division of Chemical Sciences, Office of Basic Energy Sciences. We thank Dr. R. S. Knox for commenting on a draft of the manuscript. One of us (W.S.S.) is indebted to Dr. G. van Ginkel for sending copies of the Ph.D. theses of M. van Gurp and H. van Langen.

[12] Fluorescence Anisotropy Applied to Biomolecular Interactions

By DAVID M. JAMESON and WILLIAM H. SAWYER

Introduction

The appeal of fluorescence spectroscopy in the study of biomolecular systems lies in the characteristic time scale of the emission process, the sensitivity of the technique, and its ability to accommodate rapid and facile changes in the solvent milieu under conditions corresponding to thermodynamic equilibrium. The time scale of the emission process invites exploitation in two related manners. First, information on hydrodynamic aspects of the system is available from steady-state or time-resolved measurements. Second, detailed information on local dynamic processes within the biomolecular matrix may be derived. Information on hydrodynamic aspects of a macromolecular system may be used to study binding processes, that is, the association of small ligands with macromolecules or macromolecule–macromolecule interactions. In this chapter we focus on the latter applications of polarization or anisotropy data. We shall also try to clarify aspects of this area that our experience has shown to be occasionally misunderstood by initiates.

Basic Principles

Polarized Emission and Rotational Diffusion

Further exposition on this topic requires that we clarify the definitions of polarization and anisotropy and describe the molecular events which determine their values. An excellent treatment of the origins of the polarization of fluorescence has been given by Weber.[1] Our treatment here will not enter into the same rigorous detail, and readers with a sustaining interest are referred to that article[1] and the references therein. On excitation of a fluorescent solution and observation of the emission at right angles both to the direction of propagation of the exciting light and to the direction of the electric vector, the polarization of the emission is defined as

$$P = (I_\parallel - I_\perp)/(I_\parallel + I_\perp) \qquad (1)$$

where we understand $I_\parallel$ and $I_\perp$ to be the intensities of the emission viewed through parallel and perpendicularly arranged polarizers, respectively. Anisotropy (r), on the other hand, is defined as

$$r = (I_\parallel - I_\perp)/(I_\parallel + 2I_\perp) \qquad (2)$$

We then note that

$$r = 2P/(3 - P) \qquad \text{or} \qquad r = 2/3(1/P - 1/3)^{-1} \qquad (3)$$

The relations between polarization, fluorescence lifetime, and rotational diffusion were explored by F. Perrin,[2] no doubt inspired by the work of his father, J. Perrin, on the translational diffusion of macroscopic particles. The Perrin equation may be written

$$(1/P - 1/3) = (1/P_0 - 1/3)(1 + 3\tau/\rho) \qquad (4)$$

where P_0 is the limiting or intrinsic polarization in the absence of depolarizing influences such as rotation or energy transfer,[3] τ is the excited state

---

[1] G. Weber, *in* "Fluorescence and Phosphorescence Analysis" (D. M. Hercules, ed.), p. 217. Wiley, New York, 1966.

[2] F. Perrin, *J. Phys. Radium* **7**, 390 (1926).

[3] The value of P_0 is determined by the angle (θ) between the absorption oscillator and the emission oscillator. For fluorophores randomly oriented in solution this value can, in principle, range between $+\frac{1}{2}$ and $-\frac{1}{3}$, its magnitude being given by the expression

$$(1/P_0 - 1/3) = (2/3)[(3 \cos^2 \theta - 1/2)/2]^{-1}$$

The positive limit, 0.5 (or 0.4 for anisotropy), is often assumed appropriate and, in fact, was a reasonable approximation in the days when only a few fluorophores based on naphthalene and fluorescein were utilized along with arc lamps/monochromator systems for excitation. Given the immense proliferation of probes currently available (see, e.g., a catalog from Molecular Probes, Eugene, OR) and the increasing use of lasers with

lifetime, and ρ the Debye rotational relaxation time. For a spherical molecule,

$$\rho = 3\eta V/RT \tag{5}$$

where η is the viscosity of the medium, V the molar volume monitored by the probe, R the gas constant, and T the absolute temperature. At this point we should draw attention to the fact that one often encounters in the literature the term τ_c, the rotational correlation time, and that[4]

$$\rho = 3\tau_c \tag{6}$$

Hence, when approaching the literature one must be very clear on this difference and on which definition is being utilized.

---

various fixed outputs (such as 325 nm for helium–cadmium lasers), the fluorescence practitioner would do well to verify the P_0 value for the probe of choice given the particular excitation and emission wavelengths utilized. Pyrene, for example, exhibits a P_0 which is generally low (<0.2) and varies with both excitation and emission wavelength. This consideration also applies to intrinsic protein fluorescence: one often notices excitation in these cases at 295 nm, to minimize excitation of tyrosine, yet the highest P_0 of tryptophan (0.4) is not attained at excitation wavelengths below 300 nm.

[4] The rotational relaxation time was originally defined by Debye in relation to studies on dielectric phenomena. Dipolar molecules orient themselves in an electric field, and this preferential orientation decays and disappears on removal of the field. Because the orientation in which the molecular dipoles align antiparallel to the electric field is not equivalent to the orientation in which they lie parallel to it, the orientational distribution is a function of cos θ, the angle between the direction of the electric field and that of the dipole axis. The decay of the orientational polarizability at time t is then given by cos $\theta(t) =$ exp$(-t/\rho)$ where ρ is the rotational relaxation time. For a spherical molecule Einstein's theory of molecular rotations gives $\rho = 3\eta v/kT$ where η is the solvent viscosity and v the molecular volume. In work on fluorescence depolarization by rotational diffusion Perrin used this definition by Debye as giving the characteristic time for the molecular rotations of spheres. Photoselection takes place because absorption of light is proportional to the cosine square of the angle between the transition moment in absorption and the electric vector of the exciting light, and molecules lying 180° apart in orientation have the same absorption probability. As a result the average cosine square of the angle of random molecular orientation and the electric vector of the light is not zero, as in the dielectric experiment, but rather 1/3. If instead of defining the decay of the orientation of the illumination according to the average angle of a preferred orientation, we were to do it according to the decay of the observable changes in polarization, the characteristic time for disorientation would appear to be three times shorter than for the previous case. This later point of view was adopted by Bloch in computing the decay in the nuclear polarization and was named by him the rotational correlation time. Various authors have followed Bloch and write the Perrin equation for the depolarization owing to molecular rotations as

$$r = r_0/(1 + \tau/\tau_c)$$

where $\tau_c = \rho/3$ (the authors thank Gregorio Weber for this historical insight).

The method of Perrin was later extended by Weber[5] to the case of nonspherical molecules such as ellipsoids of revolution. In addition to depolarization owing to overall or "global" rotational motion of the target macromolecule one must often take account of "local" probe motion. By "local" motion we understand any mobility of the probe in excess of that expected by the rotation of a rigid body to which it is attached. Hence, "local" motion includes internal or domain motions as well as specific movement of the probe molecule about its point of attachment to the macromolecule. The latter motion is dramatically reduced in cases of noncovalently bound probes, such as ANS, porphyrins, and NADH to name a few. In these cases the attainment of significant binding energies usually entails several points of interaction, thus reducing the probe mobility. In contradistinction covalent probes, including the intrinsic protein fluorophores such as tryptophan and tyrosine, often experience considerable movement limited only by the structural barriers imposed by the macromolecular framework.

Additivity of Polarization/Anisotropy

As stated earlier our goal here is not to expound on hydrodynamic information per se but rather to ascertain how the hydrodynamic (and dynamic) considerations may be utilized to study biomolecular interactions. The use of polarization/anisotropy to elucidate binding isotherms arose from the observations of Weber[5] on the additivity properties of polarization. Specifically, he demonstrated that (for polarized excitation)

$$(1/P_{obs} - 1/3)^{-1} = \Sigma f_i (1/P_i - 1/3)^{-1} \tag{7}$$

where P_{obs} is the actual polarization observed arising from i components and f_i represents the fractional contribution of the ith component to the total emission intensity. In terms of anisotropies then

$$r_{obs} = \Sigma f_i r_i \tag{8}$$

The motivation for considering the additivity function arose from the realization that a population of nonspherical proteins could give rise to a distribution of rotational rates depending on the orientation of the probe along the respective rotational axes. Hence, a clear understanding of the ways in which the individual contributions sum to the total signal was important. These considerations led immediately to the work of Laurence[6] who first described the application of polarization methods to follow the binding of small ligands to proteins. Dandliker and co-workers later applied

[5] G. Weber, *Biochem. J.* **51**, 145 (1952).
[6] D. J. R. Laurence, *Biochem. J.* **51**, 168 (1952).

these principles explicitly to the study of antibody–antigen interactions.[7,8] We may note at this point that Jablonski[9] introduced the term anisotropy, as defined above, and drew attention to the additive nature of the function and to the fact that the sum $(I_\parallel + 2I_\perp)$ represents the total contribution to an emitting system viewed at right angle to the excitation. One must not, however, lose sight of the fact that anisotropy and polarization have the same information content, and the use of one or the other term is a matter of convenience not of substance.

Although conceptually straightforward, the utilization of polarization/ anisotropy data to obtain binding isotherms can be subject to experimental subtleties. Once the f_i terms are determined the problem would seem virtually solved at least so far as having the information necessary to construct a binding isotherm (such as a Bjerrum or Scatchard plot[10]). The difficulty lies in the determination and significance of the f_i terms. One must realize that these terms represent, prima facia, the contribution of the ith component to the photocurrent. To convert the data to molar quantities assumes a thorough appreciation of the instrument response characteristics, including the detector and monochromator response to parallel and perpendicularly polarized light as a function of wavelength,[11] as well as the relative quantum yields of the free and bound probes.

The appeal of optical spectroscopy in the study of binding processes owes in large measure, as alluded to earlier, to the fact that these methods permit acquisition of data under conditions of thermodynamic equilibrium. Most other methods in fact rely on separation of free and bound materials, for example, by ultrafiltration, chromatography, or sedimentation. One must then clearly recognize the potential limits of these methods as regards the rapidity with which a new equilibrium may be established after the separation procedure and the possible perturbing effect of the method on the initial equilibrium (e.g., pressure effects during centrifugation). With spectroscopic methods one strives to obtain directly the concentrations of free and bound material by virtue of a variation in some observable parameter on complex formation. In the case of fluorescence the simplest case would be that in which the spectral maximum or quantum yield changes sensibly on binding. It may happen, however, that such changes are too slight to be experimentally resolvable (*vida infra*). In almost all such cases, though, be they small molecule–macromolecule or macromol-

[7] W. B. Dandliker and G. A. Feijen, *Biochem. Biophys. Res. Commun.* **5**, 299 (1961).

[8] W. B. Dandliker, M.-L. Hsu, J. Levin, and B. R. Rao, this series, Vol. 74, p. 3.

[9] A. Jablonski, *Acta Physiol. Pol.* **16**, 471 (1957).

[10] G. Weber, "Protein Interactions." Chapman & Hall, New York and London, 1992.

[11] D. M. Jameson, *in* "Fluorescein Hapten: An Immunological Probe" (E. Voss, Jr., ed.), p. 23. CRC Press, Boca Raton, Florida, 1984.

ecule–macromolecule interactions, changes in rotational rates do occur on complex formation and hence indicate the utility of a method based on molecular rotational rates, such as polarization/anisotropy methods. Another limitation with intensity measurements may be encountered if the binding is sufficiently weak that appreciable concentrations of the molecular species involved must be utilized to assure complex formation. In such cases the optical densities of the solution may be unavoidably high (even allowing for smaller cuvette or front-face geometries), and the requisite inner filter corrections may lower confidence in the binding data. Polarization/anisotropy measurements, like lifetime measurements, are intensive quantities and as such are not subject to first-order inner filter corrections.

Applications

Small Molecule–Protein Interactions

Kasprzak and Villafranca[12] characterized the binding of the fluorescent nucleotide 1,N^6-ethenoadenosine 5'-monophosphate (εAMP) to *Escherichia coli* carbamoyl-phosphate synthase (CPS) using fluorescence anisotropy titrations. This approach was particularly relevant in this system since the binding of εAMP to CPS is not accompanied by alterations in the quantum yield or emission maximum of the probe. To ascertain the anisotropy of the bound εAMP (r_b) they first titrated a small concentration of probe with excess enzyme until the anisotropy approached a limiting value (Fig. 1a). Next, CPS, at a constant concentration, was titrated with εAMP, and a Scatchard plot was constructed as shown in Fig. 1b. The fraction of bound εAMP, f_b, at any given probe/CPS ratio, was calculated from the observed anisotropy, r_{obs}, by

$$f_b = (r_{obs} - r_{in})/(r_b - r_{in}) \tag{9}$$

where r_{in} is the initial anisotropy. In this case, because the quantum yield does not change on binding, additional corrections as discussed earlier are not required.

The above procedure applies well to situations in which the ligand is the source of the fluorescence signal. If the acceptor is the source of the fluorescence signal then a slightly different procedure is recommended as illustrated in the section on intrinsic protein fluorescence in protein–DNA interactions. An extensive treatment of the application of these methods to antigen–antibody interactions has been given by Dandliker et al.[8] Other

[12] A. A. Kasprzak and J. F. Villafranca, *Biochemistry* **27**, 8050 (1988).

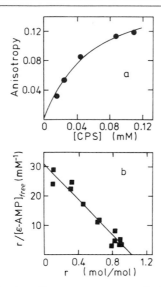

FIG. 1. (a) Anisotropy titration of a constant amount of εATP (15 μM) with carbamoyl-phosphate synthase. The solid line corresponds to $r_b = 0.176$ and $K_d = 0.04$ mM. (b) Scatchard plot calculated from data obtained from the titration of a constant amount of enzyme (51.5 μM) with εATP. The solid line corresponds to $K_d = 0.033$ mM and $n = 1$. (From Kasprzak and Villafranca.[12])

authors who have explicitly considered the application of polarization methods to the characterization of small molecule–protein interactions include Rajkowski and Cittanova,[13] Teichberg and Shinitsky,[14] and Deranleau et al.[15]

Peptide–Protein Interactions

Malencik and Anderson have described a series of studies, using various probe strategies, on the interaction of peptides with calmodulin.[16] Because calmodulin does not contain tryptophan, they were able to observe the increase in the anisotropy of tryptophan in various peptides on binding. For example, the anisotropy from the tryptophan of *Polistes* mastparan, a toxic peptide from the social wasp, on excitation at 294 nm, increased from around 0.02 to 0.13 on binding to calmodulin. In another study Malencik and Anderson[17] characterized the association of melittin,

[13] K. M. Rajkowski and N. Cittanova, *J. Theor. Biol.* **93**, 691 (1981).
[14] V. I. Teichberg and M. Shinitsky, *J. Mol. Biol.* **74**, 519 (1973).
[15] D. A. Deranleau, T. Binkert, and P. Bally, *J. Theor. Biol.* **86**, 477 (1980).
[16] D. A. Malencik and S. R. Anderson, *Biochem. Biophys. Res. Commun.* **114**, 50 (1983).
[17] D. A. Malencik and S. R. Anderson, *Biochemistry* **27**, 1941 (1988).

a tryptophan-containing 26-residue peptide, with calmodulin and three types of myosin light chain kinases. The authors make the point that, in such protein studies, the choice of excitation wavelength represents a compromise between the limiting anisotropy value, the sensitivity, and the minimization of tyrosine fluorescence; in their studies 294 nm excitation generally optimized these considerations. They also explicitly extended the additivity of the anisotropy principle to the case of three fluorescent species, namely, the unbound melittin, the unbound light chain, and the complex.

Protein–DNA Interactions

Intrinsic Protein Fluorescence. As in the case of protein–ligand interactions, the largest change in anisotropy on binding will occur when the fluorescence anisotropy of the smallest molecular weight species is being followed. Thus, for the binding of a protein to long DNA the fluorescence of the protein should be observed. On the other hand, for the binding of a protein to a short recognition sequence of DNA, such as a 20 base pair (bp) operator sequence recognized by a repressor protein, it may be more beneficial to label the DNA.

It is important to appreciate the interplay between steady-state anisotropy and the lifetime of the fluorophore. The short lifetime (1–6 nsec) of tryptophan in proteins does not necessarily preclude its use for monitoring protein–DNA interactions. The correlation time for the global tumbling of the protein may be long relative to the lifetime of tryptophan such that the anisotropy of the protein in free solution is high and may increase little on binding DNA. However, it is often observed that the indole side chains of tryptophan have substantial mobility owing to rotation about the C_α–C_β bond such that the anisotropy of the protein in free solution is small. In these cases there might be a useful increase in anisotropy on binding to DNA. For example, the tryptophan anisotropy of the Ada protein, a DNA repair enzyme from *E. coli*, increases on binding nonspecifically to DNA,[18] as shown in Fig. 2. In this case, the anisotropy of the free Ada protein is relatively large, indicating restricted motion of the tryptophan residues. Nevertheless, the small increase in anisotropy (~0.02 units) was sufficient for analysis of the binding equilibria. Corrections for the inner filter effect were not necessary. The stoichiometry was confirmed by a reverse titration in which the Ada protein was added to a constant amount of DNA.

[18] M. Takahashi, K. Sakumi, and M. Sekiguchi, *Biochemistry* **29**, 3431 (1990).

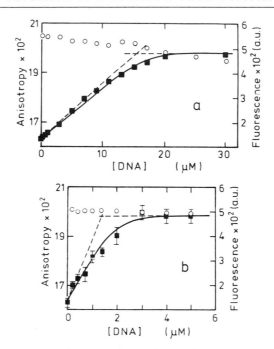

FIG. 2. The tryptophan anisotropy of the Ada protein (■) on interaction with DNA at two protein concentrations: (a) 2.0 M; (b) 0.2 M. No change was observed in the steady-state fluorescence signal (○). The dashed lines are extrapolations to estimate the site size. The solid line is the theoretical curve for an association constant of $4 \times 10^7 \ M^{-1}$ and a site size of 7 bp assuming no site–site interactions. (From Takahashi et al.[18])

The measurement of the tryptophan anisotropy has the additional advantage that it can be used to establish the polymeric state of the protein. In the case of the Ada protein, the anisotropy was independent of protein concentration, consistent with the observation that the protein remains in a monomeric state over a wide range of protein concentrations. An increase in the steady-state tryptophan anisotropy of the trp repressor on binding a 26 bp operator sequence has also been observed.[19]

Besides quantitatively reflecting the binding of a protein to DNA, studies of time-resolved anisotropy provide information about the dynamics of the tryptophan chromophore. In favorable circumstances, the fluorescence emitted from two tryptophan residues in a protein can be resolved.[20] However, it is desirable that studies be carried out on proteins

[19] T. Fernando and C. A. Royer, Biochemistry 31, 3429 (1992).
[20] C. A. Royer, J. A. Gardner, J. M. Beechem, J.-C. Brochon, and K. S. Mathews, Biophys. J. 58, 363 (1990).

which contain a single tryptophan so that the anisotropy decays can be attributed to a specific fluorophore at a defined position in the amino acid sequence. To this end, several DNA-binding proteins have been genetically engineered to contain a single tryptophan residue for fluorescence analysis.[21,22] We may also note here that although one expects and usually does observe an increase in the anisotropy of the tryptophan emission of a protein on binding to another macromolecule, a decrease in this function is also possible if the local motion of the tryptophan increases appreciably on binding. Such a situation occurs in the binding of elongation factor Tu to elongation factor Ts.[23]

Use of Extrinsic Probes. The intrinsic fluorescence intensity of some proteins does not change on binding to DNA. For example, the fluorescence of the single tryptophan in the avian retroviral nucleocapsid protein does not change on binding to nucleic acid.[24] The tryptophan fluorescence of the TyrR repressor does not change on binding to its operator sequence (M. Bailey and P. Hagmar, personal communication, 1993). To increase the versatility of the fluorescence technique, extrinsic probes may be covalently conjugated to either the DNA-binding protein or to the DNA itself. The coupling of probes to proteins has been reviewed elsewhere.[25]

DNA has no measurable intrinsic fluorescence except at low temperatures,[26] and for many years the use of fluorescence techniques to monitor DNA–protein interactions has not been possible owing to the absence of methods to attach fluorescent probes covalently at particular points in the base sequence. Intercalating probes such as ethidium bromide and the acridine dyes have been used extensively but have the disadvantage that they bind almost nonspecifically along the DNA helix. Moreover, the lengthening of the DNA helix that results from the intercalation may severely interfere with the binding of proteins. An alternative approach is to incorporate fluorescent nucleotide analogs into synthetic oligonucleotides. The fluorescence of poly($1,N^6$-ethenoadenylic acid) has been used to examine the nonspecific binding of recA protein and eukaryotic polypeptide chain initiation factors,[27,28] and the adenine analog 2-aminopu-

[21] T. Hard, M. H. Sayre, E. P. Geiduschek, and D. R. Kearns, *Biochemistry* **28**, 2813 (1989).
[22] C. R. Guest, R. A. Hochstrasser, C. G. Dupuy, D. J. Allen, S. J. Benkovic, and D. P. Millar, *Biochemistry* **30**, 8759 (1991).
[23] D. M. Jameson, E. Gratton, and J. F. Eccleston, *Biochemistry* **26**, 3894 (1987).
[24] J. Secnik, Q. Wang, C.-M. Chang, and J. E. Jentoft, *Biochemistry* **29**, 7991 (1990).
[25] R. P. Haugland, *in* "Excited States of Biopolymers" (R. F. Steiner, ed.), p. 29. Plenum, New York, 1983.
[26] J. Eisinger and A. A. Lamola, *in* "Excited States of Proteins and Nucleic Acids" (R. F. Steiner and I. Weinryb, eds.), p. 107. Plenum, New York, 1971.
[27] M. Chabbert, C. Cazenave, and C. Helene, *Biochemistry* **26**, 2218 (1987).
[28] D. J. Goss, C. L. Woodley, and A. J. Wahba, *Biochemistry* **26**, 1551 (1987).

rine has been used to examine the dynamics of mismatched base pairs in DNA.[29]

Royer *et al.*[30] studied the interaction of histone H2A-H2B with 200 bp DNA fragments from chicken erythrocytes by labeling the histones with dansyl chloride (DNS). In this case labeling of the histones was accomplished by incubation with a 10-fold molar excess of dansyl chloride for 10 min at pH 8.0; free dye was separated from labeled protein by gel filtration chromatography. In such cases of random covalent labeling it is often of interest to ascertain the degree of labeling. To this end one typically utilizes the best estimate of the molar extinction coefficient of the dye bound to the protein. Because this value may depend, to a greater or lesser extent, on the environment of the probe, it is often only an approximate method. In the study by Royer *et al.*[30] a labeling ratio of approximately 1 DNS per H2A-H2B dimer was determined (average labeling ratios typically reflect a distribution of labels among the target population; if truly random labeling is achieved then one may assume a Poisson distribution of labels). The polarization titrations indicated stoichiometries of between 4 and 16 histone octamers per DNA 200-mer with affinities in the nanomolar range.

Labeling DNA. The availability of methods to conjugate fluorophores to oligonucleotides has greatly widened the scope for fluorescence analysis of DNA–protein interactions. Strategies based on fluorescence quenching and resonance energy transfer, as well as anisotropy, are now possible but as yet have been infrequently applied. A number of methods of specifically labeling synthetic oligonucleotides is now available. The most common derivatives are as follows (Fig. 3).

1. *Terminal labeling:* A primary aliphatic amine can be conjugated to the 5'-phosphate group of an oligonucleotide using standard phosphoramidite chemistry. Labeled products are commercially available from Applied Biosystems (Foster City, CA) and Clontech (Palo Alto, CA) and come with spacers of 2, 3, 6, or 12 carbon atoms between the 5' oxygen and the protected amino group. 5'-Thiol modifiers are also available. Labeling at the 3' terminus is now possible with an amino-modified controlled pore glass suitable for use in oligonucleotide synthesizers (Applied Biosystems and Clontech). After cleavage from the column and deprotection, a primary amino group is available for conjugation at the 3' terminus of the oligonucleotide. An alternative strategy for labeling the 5' terminus is to prepare a 5'-phosphorylated derivative of the oligonucleotide with T4

[29] C. R. Guest, R. A. Hochstrasser, L. C. Sowers, and D. P. Millar, *Biochemistry* **30**, 3271 (1991).
[30] C. A. Royer, T. Ropp, and S. F. Scarlata, *Biophys. Chem.* **43**, 197 (1992).

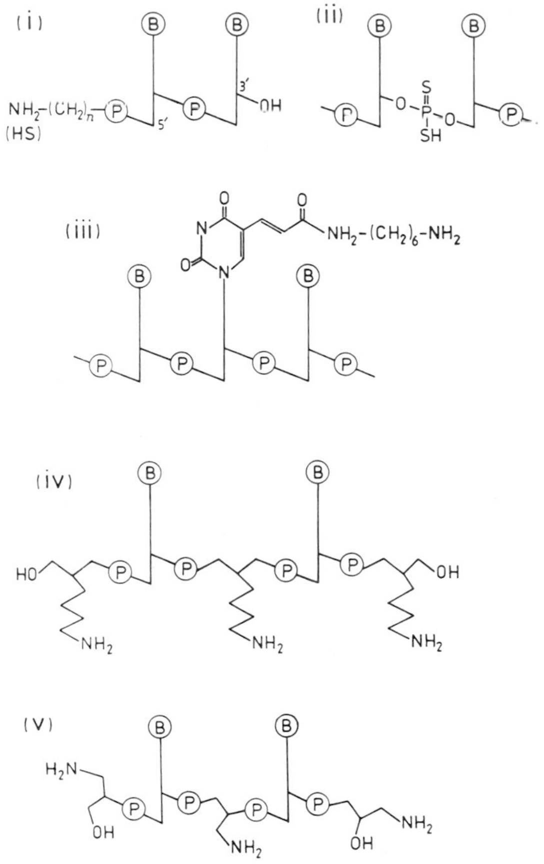

FIG. 3. Modified oligonucleotides for the conjugation of amino or thiol reactive fluorescent probes to DNA. (i) Labeling at the 5′ terminus (Applied Biosystems, Clontech); (ii) labeling at the internucleotide phosphate using a phosphorodithioate derivative; (iii) labeling of a modified base: an amino-modified thymidine (Applied Biosystems); (iv) internucleotide labeling using a 3-carbon bridge with an aminobutyl side arm [the reagent can also be placed at the 3′ and 5′ ends of the oligonucleotide (Clontech)]; and (v) internucleotide labeling with a 2-carbon bridge having an aminomethyl side arm; [the reagent can also be placed at the 3′ and 5′ ends of the oligonucleotide (Clontech)].

polynucleotide kinase and to couple cystamine using a carbodiimide reaction: reduction of the disulfide bond with dithiothreitol provides a thiol group that can be linked to a maleimide or iodoacetate probe derivative.[31]

[31] T. Heyduk and J. C. Lee, Proc. Natl. Acad. Sci. U.S.A. 87, 1744 (1990).

2. *Labeling at internucleotide phosphorus:* Synthesis of deoxynucleo-tide 3'-phosphorothioamidites, incorporation into DNA using commer-cially available oligonucleotide synthesizers, and subsequent alkylation have been reviewed by Caruthers *et al.*[32]

3. *Labeling modified base:* Deoxyuridines substituted at C-5 carrying a blocked primary aliphatic amino group have been prepared by Haralam-bidis *et al.*[33-35] The chemistry allows for the incorporation of spacer arms of various lengths. Fluorescent derivatives of 5-(aminopropyl)deoxyuridine have been reported by Gibson and Benkovic[36] and used to study the interaction of the Klenow fragment of *E. coli* DNA polymerase I with DNA.[36-38] An amino-modified thymidine phosphoamidite with an 11 atom spacer is now available from Applied Biosystems.

4. *Labeling between nucleotides:* Replacement of a nucleoside with a 3 carbon bridge between the 3'-phosphate of one nucleotide and the 5'-phosphate of the next is said to conserve the interphosphate distance. The bridge carries an aminobutyl side arm which can be conjugated to a fluorescent probe. This method has the disadvantage that a nucleotide is replaced by a foreign bridging group. Standard phosphoamidite chemistry can be used, and a reagent known as Uni-Link AminoModifier is available from Clontech.

In our experience, labeling the 5' terminus with the dansyl fluorophore connected via a 6 carbon spacer does not affect the melting point of the duplex and the degree of hypochromicity observed. Nor does the melting affect the fluorescence characteristics of the dansyl group (M. Bailey and W. H. Sawyer, unpublished observations, 1993). These observations suggest that end labeling of oligonucleotides does not interfere with duplex formation and that the dansyl label on its spacer is oriented away from the helix and does not fold back to interact with the helical barrel. Labeling of short oligonucleotides at the 3' or 5' terminus may be sufficient to permit the monitoring of the binding equilibria using anisotropy and has the advantage that the fluorophore cannot interfere directly with the pro-tein–DNA interaction site. The length of a spacer arm between the DNA helix and the probe may be crucial. If the fluorophore is attached via a

[32] M. H. Caruthers, G. Beaton, J. V. Wu, and W. Wiesler, this series, Vol. 211, p. 3.

[33] J. Haralambidis, M. Chai, and G. W. Tregear, *Nucleic Acids Res.* **15,** 4857 (1987).

[34] J. Haralambidis, L. Duncan, K. Angus, and G. W. Tregear, *Nucleic Acids Res.* **18,** 493 (1990).

[35] J. Haralambidis, K. Angus, S. Pownall, L. Duncan, M. Chai, and G. W. Tregear, *Nucleic Acids Res.* **18,** 501 (1990).

[36] K. J. Gibson and S. J. Benkovic, *Nucleic Acids Res.* **15,** 6455 (1987).

[37] D. J. Allen, P. L. Darke, and S. J. Benkovic, *Biochemistry* **28,** 4601 (1989).

[38] D. J. Allen and S. J. Benkovic, *Biochemistry* **28,** 9586 (1989).

flexible extension, it may retain its flexibility on binding to the protein such that the anisotropy is unaffected. Thus short spacer arms are desirable in that rapid depolarizing motions of the probe at the point of attachment are limited and the anisotropy will more accurately reflect the global motion of the DNA and the DNA–protein complex. The possibility that the DNA-binding protein interacts with the fluorophore as well as or instead of the double helix should be carefully examined. Determining whether the protein binds the fluorescent probe or a low molecular weight fluorescent conjugate (e.g., dansyl lysine) is one way to test this possibility.

The binding of the *E. coli* cAMP receptor protein to a 32 bp fragment of the *lac* promoter provides another example of this strategy.[31] The DNA was labeled with a coumarin derivative, and the binding stoichiometry was determined by titration of the ligand into a fixed concentration of acceptor under conditions of stoichiometric binding, that is, when all added ligand is bound. Stoichiometric binding can be realized if the acceptor concentration, or more precisely the concentration of binding sites, is at least 10-fold greater than the K_d. A plot of the anisotropy versus the molar ratio of ligand to acceptor should then have a clear inflection point (as illustrated in Fig. 4a), indicating the binding stoichiometry. A second titration, carried out under conditions of equilibrium binding in which the acceptor concentration is approximately equal to the K_d, then yields values

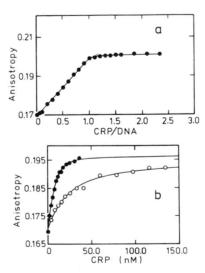

Fig. 4. (a) Stoichiometric titration of a coumarin-labeled 32 bp DNA fragment of the *lac* promoter (32-bp-CPM) with the cAMP receptor protein (CRP). The concentration of 32-bp-CPM was fixed at 98.5 n*M*. (b) Titration of 32-bp-CPM (11.1 n*M*) with CRP in the presence of 0.5 μ*M* cAMP (○) and 500 μ*M* cAMP (●). (From Heyduk and Lee.[31])

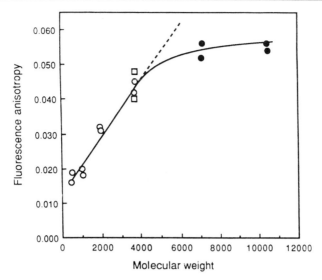

Fɪɢ. 5. Fluorescence anisotropy of oligonucleotides covalently labeled with fluorescein at the internucleotide phosphate. The open symbols are for oligothymidine molecules containing 2, 5, or 11 bases. The filled symbols are for 11-mers complexed with the complementary dA strand; [fluorescent probe] = [oligonucleotide] = 0.1 M; 0.1 M phosphate buffer (pH 7.0), 20°. (From Murakami *et al.*[39])

of the fractional saturation which, together with the site number determined from the first titration, provides the information necessary for construction of a binding plot (Fig. 4b).

An example of steady-state anisotropy measurements of a series of fluorescein-labeled oligothymidylates is shown in Fig. 5.[39] The anisotropy increases linearly with chain length (2–11 nucleotides) and therefore with molecular weight. Increasing the molecular weight by formation of a duplex is less effective in increasing the anisotropy compared to increasing the chain length. In this system, the steady-state anisotropy is determined by the angles between the absorption and emission transition moments of the dye and the axis of rotation and by three depolarizing rotations: (i) end-to-end tumbling, (ii) rotation about the long axis of the molecule or helix, and (iii) motion of the probe at the point of attachment. Time-resolved anisotropy measurements (*vida infra*) are useful for resolving some of these contributions. For example, for a 5′-mansyl-labeled 42-mer the independent motion of the probe at the point of attachment has a

[39] A. Murakami, M. Nakaura, Y. Nakatsuji, S. Nagahara, Q. Tran-Cong, and K. Makino, *Nucleic Acids Res.* **19,** 4097 (1991).

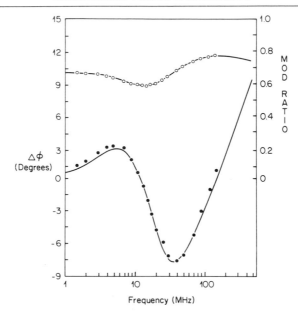

Fig. 6. Example of anomalous phase delay resulting from associative decay. Differential phase (●) and modulation (○) data and a two-component fit (solid lines) are shown for a system containing free ethidium bromide and ethidium bromide bound to Phe-tRNA[Phe]. The lifetimes of free and bound ethidium bromide were 1.86 and 26 nsec, respectively, and the associated rotational relaxation times were 0.54 and 136 nsec, respectively.

correlation time of 300–400 psec compared to global motion of the chain at 4–5 nsec (D. P. Millar, M. Bailey, and W. H. Sawyer, unpublished observations, 1992). Correlation times of 240–590 psec for a dansyl probe attached to a C-5-modified uridine base within an oligonucleotide have also been ascribed to independent motion of the probe. A longer correlation time, attributed to the global tumbling of the DNA helix, was shown to increase substantially on binding of the Klenow fragment of DNA polymerase I.[22] In this case, changes in the steady-state fluorescence intensity were used to measure the binding of the protein to the duplex.

Time-Resolved Methods

Time-resolved fluorescence methods are discussed elsewhere in this volume, and we shall not enter into lengthy descriptions of the technique but rather will point out some interesting aspects of time-resolved anisotropy methods as they apply to binding studies. The two principal time-resolved methodologies presently utilized are the impulse–response

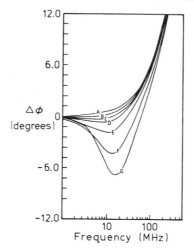

FIG. 7. Same system as Fig. 6 but starting with only free probe (A) and additions of bound probe (B–G). The molar contributions for the bound probes are as follows: (A) 0.0%, (B) 0.18%, (C) 0.36%, (D) 0.54%, (E) 1.1%, (F) 2.1%, and (G) 3.2%.

method and the harmonic response method. In the first technique[40] the direct time decay of the emission, following a brief excitation pulse, is recorded, whereas in the latter technique the excited state decay parameters are deduced from the response of the emitting system to sinusoidally modulated excitation.[41] In the harmonic response method one can resolve rotational modalities of the emitting system by monitoring the phase delay between the perpendicular and parallel components of the emission, on excitation by parallel polarized light.[42] One is thus able, in favorable circumstances, to separate "global" from "local" motions of bound fluorophores.

In principle, one can also resolve free from bound probes, and, in fact, an interesting phenomenon occurs in these types of measurements which can be used to determine binding isotherms. Namely, if the lifetime of the fluorophore increases on binding, in addition to its rotational rate, then one may observe, at particular modulation frequencies, anomalous phase delays, that is, negative values for the phase delay between the parallel and perpendicular components, which provide a sensitive measure

[40] J. Yguerabide and E. E. Yguerabide, in "Optical Techniques in Biological Research" (D. L. Rousseau, ed.), p. 181. Academic Press, Orlando, Florida, 1984.
[41] T. L. Hazlett and D. M. Jameson, Proc. Soc. Photo-Opt. Instrum. Eng. **909**, 412 (1988).
[42] D. M. Jameson and T. L. Hazlett, in "Biophysical and Biochemical Aspects of Fluorescence Spectroscopy" (G. Dewey, ed.), p. 105. Plenum, New York, 1991.

of the extent of binding. This effect (known informally as "Chip dip") is illustrated in Figs. 6 and 7 which concern the ethidium bromide/tRNA system.[41] In this case the free ethidium bromide has a lifetime of 1.86 nsec and a rotational relaxation time of 0.54 nsec. On binding the Phe-tRNA^{Phe} the lifetime increases to 26 nsec and the rotational relaxation increases to 136 nsec. A single rotating species cannot give rise to a negative value for the differential phase delay (assuming excitation at a wavelength which gives a positive P_0), and the negative excursion for this function shown in Fig. 6 is due to the presence of free ethidium bromide in equilibrium with bound material. Figure 7 shows the calculated effects for increasing amounts of bound material.

We should note that, given the difference in quantum yield between the free and bound probe, the fractional intensities utilized in Fig. 7 actually represent small percentages of bound probe on a molar basis. In fact, considering the accuracy of the differential phase measurement (better than 0.1°) one can detect, in this system, on the order of 0.1% bound probe. This phenomenon also occurs in time-domain measurements. Specifically, if one monitors the anisotropy decay of a system which displays multiple lifetimes associated with multiple rotational diffusion rates then one may observe a decline at short times of the anisotropy followed by a rise at latter times and subsequent decrease. This "dip and rise" effect has been observed by Millar and co-workers[22] in studies on protein–DNA interactions, specifically in the case of the interaction of a fluorescent DNA duplex with the Klenow fragment of DNA polymerase.

Acknowledgments

D. M. J. is an Established Investigator of the American Heart Association and wishes to acknowledge support from National Science Foundation Grant DMB-9005195. W. H. S. acknowledges support from the Australian Research Council.

[13] Fluorescence Resonance Energy Transfer

By Paul R. Selvin

Introduction

Fluorescence resonance energy transfer (FRET) is a technique for measuring the distance between two points which are separated by approximately 10–75 Å. The technique is valuable because measurements can be made under physiological (or other) conditions with near angstrom resolution and with the exquisite sensitivity of fluorescence measurements. For these reasons FRET has found wide use in polymer science, biochemistry, and structural biology. Reviews have appeared on FRET applied to actin structure,[1] nucleic acids,[2] phycobiliproteins,[3-5] cell surfaces with an emphasis on protein interaction,[6] *in situ* imaging,[7] diffusion,[8] and microscopy.[9-12] Among systems already studied in the literature are DNAs such as oligonucleotides[13-16] and Holliday junctions[14,17,18]; proteins such as oligopeptides,[19] rhodopsin,[8,20] myosin,[21] various calcium binders,[22] and major histocompatibility complexes[23]; RNA[24,25]; and nucleic

[1] M. Miki, S. I. O'Donoghue, and C. G. Dos Remedios, *J. Muscle Res. Cell Motil.* **13,** 132 (1992).

[2] R. M. Clegg, this series, Vol. 211, p. 353.

[3] A. N. Glazer and L. Stryer, this series, Vol. 184, p. 188.

[4] A. N. Glazer, *J. Biol. Chem.* **264,** 1 (1989).

[5] R. Huber, *EMBO J.* **8,** 2125 (1989).

[6] L. Matyus, *J. Photochem. Photobiol. B* **12,** 323 (1992).

[7] G. Bottiroli, A. C. Croce, and R. Ramponi, *J. Photochem. Photobiol. B* **12,** 413 (1992).

[8] L. Stryer, D. D. Thomas, and C. F. Meares, *Annu. Rev. Biophys. Bioeng.* **11,** 203 (1982).

[9] B. Herman, *Methods Cell Biol.* **30,** 219 (1989).

[10] T. M. Jovin and D. J. Arndt-Jovin, *Annu. Rev. Biophys. Biophys. Chem.* **18,** 271 (1989).

[11] T. M. Jovin and D. J. Arndt-Jovin, in "Microspectrofluorimetry of Single Living Cells" (E. Kohen, J. S. Ploem, and J. G. Hirschberg, eds.), p. 99. Academic Press, Orlando, Florida, 1989.

[12] T. M. Jovin, D. J. Arndt-Jovin, G. Marriott, R. M. Clegg, M. Robert-Nicoud, and T. Schormann, in "Optical Microscopy for Biology" (B. Herman and K. Jacobson, eds.), p. 575. Wiley–Liss, New York, 1990.

[13] R. Cardullo, S. Agrawal, C. Flores, P. C. Zamecnik, and D. E. Wolf, *Proc. Natl. Acad. Sci. U.S.A.* **85,** 8780 (1988).

[14] J. P. Cooper and P. J. Hagerman, *Biochemistry* **29,** 9261 (1990).

[15] H. Ozaki and L. W. McLaughlin, *Nucleic Acids Res.* **20,** 5205 (1992).

[16] R. M. Clegg, A. I. Murchie, A. Zechel, and D. M. Lilley, *Proc. Natl. Acad. Sci. U.S.A.* **90,** 2994 (1993).

[17] A. I. Murchie, R. M. Clegg, E. von Kitzing, D. R. Duckett, S. Diekmann, and D. M. Lilley, *Nature (London)* **341,** 763 (1989).

[18] R. M. Clegg, A. I. H. Murchie, A. Zechel, C. Carlberg, S. Diekmann, and D. M. J. Lilley, *Biochemistry* **31,** 4846 (1992).

[19] L. Stryer and R. P. Haugland, *Proc. Natl. Acad. Sci. U.S.A.* **98,** 719 (1967).

[20] R. O. Leder, S. L. Helgerson, and D. D. Thomas, *J. Mol. Biol.* **209,** 683 (1989).

[21] H. C. Cheung, I. Gryczynski, H. Malak, G. Wiczk, M. L. Johnson, and J. R. Lakowicz, *Biophys. Chem.* **40,** 1 (1991).

[22] D. T. Cronce and W. D. Horrocks, Jr., *Biochemistry* 31, 7963 (1992).

[23] R. Tampé, B. R. Clark, and H. M. McConnell, *Science* **254,** 87 (1991).

[24] D. J. Robbins, O. W. J. Odom, J. Lynch, D. Dottavio-Martin, G. Kramer, and B. Hardesty, *Biochemistry* **20,** 5301 (1981).

[25] K. Beardsley and C. R. Cantor, *Proc. Natl. Acad. Sci. U.S.A.* **65,** 39 (1970).

acid–protein complexes such as nucleosomes,[26-29] chromatin[30,31] and protein–promoter interactions.[32] FRET has also been used to monitor dynamic processes such as actin assembly,[1,33] nucleosome assembly, and human immunodeficiency virus (HIV) protease activity.[34] Van der Meer and co-workers have recently written a book on FRET.[35] Excellent reviews of the basics of FRET, which complement this chapter, have appeared.[36-38]

The idea behind the technique is to label the two points of interest with different dyes; one, which must be fluorescent, is called the donor, and the other, which is not necessarily fluorescent but often is, is called the acceptor. By choosing dyes with the appropriate spectral characteristics, the donor, after being excited by light, can transfer energy to the acceptor. The efficiency of energy transfer depends on the inverse sixth power of the distance between the dyes. In general, the acceptor must be within 10–75 Å to get reasonable energy transfer, the exact range depending on the dyes chosen.

If one measures the amount of energy transfer, it is therefore possible to determine the distance between donor and acceptor. Qualitatively, the farther apart the donor and acceptor, the less energy transfer. The extent of energy transfer can be measured because the fluorescence of the donor (both intensity and lifetime) decreases, or is quenched, and the acceptor, if fluorescent, increases its fluorescence, or becomes "sensitized," with energy transfer. These changes in fluorescence can be measured by comparing a complex labeled with both donor and acceptor to ones labeled only with donor and only with acceptor. The experimental and theoretical details are presented below.

Other Uses

Measuring the (static) distance between two points, although the main use of FRET, is just one application. A number of workers have also used

[26] H. Eshaghpour, A. E. Dieterich, D. Crothers, and C. R. Cantor, *Biochemistry* **19,** 1797 (1980).
[27] D. G. Chung and P. N. Lewis, *Biochemistry* **25,** 5036 (1986).
[28] D. G. Chung and P. N. Lewis, *Biochemistry* **25,** 2048 (1986).
[29] J. Widom, in preparation (1994).
[30] M. Lerho, M. Favazza, and C. Houssier, *J. Biomol. Struct. Dyn.* **7,** 1301 (1990).
[31] G. Sarlet, S. Muller, and C. Houssier, *J. Biomol. Struct. Dyn.* **10,** 35 (1992).
[32] T. Heyduk and J. C. Lee, *Proc. Natl. Acad. Sci. U.S.A.* **87,** 1744 (1990).
[33] D. L. Taylor, J. Reidler, J. A. Spudich, and L. Stryer, *J. Cell Biol.* **89,** 363 (1981).
[34] E. D. Matayoshi, G. T. Wang, G. A. Krafft, and J. Erickson, *Science* **247,** 954 (1990).
[35] B. W. van der Meer, G. Coker III, and S. Y. Chen, "Resonance Energy Transfer." VCH, New York, 1994.
[36] C. R. Cantor and P. R. Schimmel, "Biophysical Chemistry," Vol. 2, pp. 448–455. Freeman, San Francisco, California, 1980.

FRET to measure dynamic processes, including enzymatic activity such as HIV proteolysis[34] or molecular assembly such as actin assembly.[1,33] In these cases the distance between two points changes as a function of some dynamic process; the ends of a polypeptide separate after cleavage by HIV, for example. As long as the dynamic process is not too fast (long enough to acquire a reasonable fluorescent signal–roughly, a few minutes with ease or, with effort, subsecond), the process can be monitored by FRET.

Yet another application of FRET is to measure rates of diffusion and distances of closest approach. In these cases, one uses a long-lived donor (lifetime on the order of a microsecond to millisecond). During the excited state lifetime of the donor, the donor and/or acceptor can diffuse. If the distance which they can diffuse during the donor lifetime is on the order of the average distance separating the donor and acceptor, the amount of energy transfer will depend on the diffusion coefficients, and hence measurements of energy transfer can shed light on this quantity. If the average distance is considerably less than the diffusional distance, the donor and acceptor approach each other many times during the donor lifetime, and energy transfer will depend mostly on the distance of closest approach.

A third and relatively new application of FRET is the generation of new compound dyes with spectral characteristics that combine the best of both dyes. The idea is to attach covalently a donor and acceptor together in close proximity to one another. In the simplest case, where the absorption or emission properties of the individual dyes do not change, the absorption characteristic of the compound dye is the sum of the two individual dyes. At the same time, the emission is dominated by the acceptor since almost all of the energy absorbed by the donor is transferred to the acceptor. This results in dyes having potentially large Stokes shifts (the sum of the donor and acceptor Stokes shifts) and excellent quantum yields. So far, this work has mainly been applied to phycobiliproteins and DNA dyes.[39–42]

[37] R. H. Fairclough and C. R. Cantor, this series, Vol. 48, p. 347.
[38] L. Stryer, *Annu. Rev. Biochem.* **47,** 819 (1978).
[39] A. N. Glazer and L. Stryer, *Biophys. J.* **43,** 383 (1983).
[40] S. C. Benson, P. Singh, and A. N. Glazer, *in* "Human Genome Program, Contractor–Grantee Workshop III," Santa Fe, New Mexico, U.S. Department of Energy, Publ. ORNL/M-2588, 1993.
[41] S. C. Benson, P. Singh, and A. N. Glazer, *Nucleic Acids Res.* **21,** 5727 (1993).
[42] R. P. Haugland, "Molecular Probes, Inc., Catalog" (K. D. Larison, ed.), 5th Ed., Molecular Probes, Eugene, Oregon, 1992–1994.

Choice of Technique

Alternative techniques which give some of the same information as FRET include X-ray crystallography, nuclear magnetic resonance (NMR), cryoelectron microscopy, and biochemical methods such as gel-shift assays and cross-linking studies. Briefly, X-ray crystallography and NMR both produce potentially complete structural information but require large quantities of material. For *in vitro* studies, FRET is often used to get initial structural information, with the complete solution coming later from X-ray crystallography or NMR. X-Ray crystallography and NMR are limited to *in vitro* measurements and can analyze only relatively small molecules, restrictions which do not apply to FRET. In addition, in the case of X-ray crystallography, one is also faced with the difficult problem of crystallization and isomorphous replacement.

Cryoelectron microscopy achieves the high resolution of electron microscopy while minimizing some of the sample preparation artifacts associated with drying, staining, and interaction with the solid support. The technique has been shown to reproduce structural features of viruses, some protein crystals, and even more flexible samples such as DNA.[43] The technique is limited to using thin (100 nm) samples, contrast is fairly low, and aggregation can occur during the cooling period. A number of reviews on cryoelectron microscopy have appeared.[43–47]

Gel-shift assays, in which the mobility of a sample in a polyacrylamide gel is a function of the structure of the molecule (e.g., bent or straight, compact or extended) can supply some of the same information as FRET. This is because alterations in molecular shape can lead to changes in gel mobility and also changes in the distance (and hence energy transfer) between two site-specifically placed dyes (e.g., in DNA[17,48]). Sample preparation in the gel-shift techniques is generally easier, since no labels need be attached, but the technique has the disadvantage that the complex must be stable over the course of hours and structural changes must be inferred from changes in mobility, which are not well understood theoretically.

Cross-linking studies, like gel-shift assays, have the advantage that site-specific labels need not be introduced into the macromolecule. On the

[43] J. Dubochet, M. Adrian, I. Dustin, P. Furrer, and A. Stasiak, this series, Vol. 211, p. 507.
[44] R. H. Wade and D. Chretien, *J. Struct. Biol.* **110,** 1 (1993).
[45] R. Schroder, W. Hofmann, J. Menetret, K. Holmes, and R. Goody, *Electron Microsc. Rev.* **5,** 171 (1992).
[46] K. Meller, *Electron Microsco. Rev.* **5,** 341 (1992).
[47] P. Flicker, R. Milligan, and D. Applegate, *Adv. Biophys.* **27,** 185 (1991).
[48] U. K. Snyder, J. F. Thompson, and A. Landy, *Nature (London)* **341,** 255 (1989).

other hand, it is not always clear where the cross-linking is taking place. Furthermore, if cross-linking does not occur, this may be because the sites are not in close proximity or because the sites are not chemically reactive. Finally, if flexible chemical linkers are used, the distance determination is limited to saying that the two points of interest are less than or equal to the maximal extension of the cross-linker. A number of reviews on cross-linking techniques have appeared.[49–52]

Problems

It is also important to understand the limitations of FRET. Although these are explored in some detail later, the most important drawback of FRET is its limited ability to measure absolute distances. It is quite good at measuring relative distances, namely, whether two points are closer together under condition A than condition B. The problem is that the efficiency of energy transfer depends not only on the distance between the donor and acceptor, but on the relative orientation of the dyes as well, a factor which is often not precisely known. Even when measuring relative distances one must take care to ensure that the orientation factor does not change between the two systems one is comparing. Unfortunately, this orientation factor can be significant, multiplying the fitted-distance anywhere from 0 to $4^{1/6} = 1.26$. Polarizaton measurements on the donor and acceptor can be made which constrain this factor, but rarely do they eliminate all uncertainty. In addition, FRET is limited in its ability to measure absolute distances because there is usually uncertainty in the exact position of the FRET dyes owing to flexibility in the linker arm used to attach the dyes. For these reasons FRET is more easily and reliably used as a measure of relative distance.

A second problem with FRET is the very sharp distance dependence. This has two drawbacks: (1) it is difficult to measure relatively long distances because the signal is very weak, and (2) the signal tends to be "all or none," that is, if the two points are less than a certain characteristic distance (this distance, known as R_0, is the distance at which 50% of the energy is transferred and is a function of the particular dyes chosen), almost all of the energy is transferred, but if greater than this distance, very little energy is transferred. It is therefore helpful to have some estimate of the distance of interest before a FRET measurement is undertaken.

[49] J. Brunner, *Annu. Rev. Biochem.* **62**, 483 (1993).
[50] J. Brunner, this series, Vol. 172, p. 628.
[51] M. Brinkley, *Bioconjugate Chem.* **3**, 2 (1992).
[52] S. S. Wong and L. J. C. Wong, *Enzyme Microb. Technol.* **14**, 866 (1992).

Underlying Principles: Förster's Theory of Dipole–Dipole Interaction

Here we present a brief review of the physical principles underlying fluorescence energy transfer. The theory was developed primarily by Förster and extended by Dexter.[53,54] Förster did some early experimental studies,[55] and Stryer and Haugland convincingly showed that fluorescence energy transfer could be used as a "molecular ruler" to measure distances.[19] Emphasis is on developing an intuitive feel for the important relevant parameters. Both a classical and a quantum mechanical approach are given.

In FRET, a fluorescent donor molecule transfers energy to an acceptor molecule, which is usually but not necessarily a fluorescent molecule. The mechanism is a nonradiative induced dipole–induced dipole interaction[36]: "nonradiative" because no photons are "passed" between the dye molecules; "dipole–dipole" because each dye molecule acts like a classical (or quantum mechanical) dipole antenna, emitting and absorbing energy; "induced" because the dipoles are not permanent but are a result of electric fields which create them. On energy transfer, the signal in a FRET experiment is a decrease in the fluorescence intensity and lifetime of the donor and, if the acceptor is also fluorescent, an increase in the fluorescence of the acceptor. The changes are measured by comparing the fluorescence of a complex containing both donor and acceptor to that of complexes containing only donor or only acceptor. The fluorescence of the donor decreases in the presence of acceptor because some of the energy goes to the acceptor instead of into the radiation (or photon) field. The lifetime of the donor also decreases because the energy transfer to the acceptor is another pathway for the excited state to decay to the ground state.

The efficiency of energy transfer (E), which is defined as the fraction of donor molecules de-excited via energy transfer to the acceptor, therefore equals

$$E = (1 - I_{D_A}/I_D) = 1 - \tau_{D_A}/\tau_D \tag{1}$$

where I_{D_A} and τ_{D_A} are the intensity and lifetime, respectively, of the donor in the presence of acceptor, and I_D and τ_D in the absence of acceptor. The efficiency of energy transfer can also be measured by looking at the increase in fluorescence of the acceptor:

$$E = (I_{A_D}/I_A - 1)(\varepsilon_A/\varepsilon_D) \tag{2}$$

[53] T. Förster, *Mod. Quantum. Chem. Lect. Istanbul Int. Summer Sch.* (1965).
[54] D. L. Dexter, *J. Chem. Phys.* **21**, 836 (1953).
[55] T. Förster, *Z. Naturforsch. A: Astrophys. Phys. Phys. Chem.* **4**, 321 (1949).

where I_{A_D} is the emission of the acceptor in the presence of the donor (consisting of fluorescence arising from energy transfer and from direct excitation of the acceptor) and I_A is the fluorescence of the acceptor-only labeled sample (consisting of fluorescence arising from direct excitation only); ε_A and ε_D are the molar extinction coefficients of the acceptor and donor, respectively, at the wavelength of excitation. Equations (1) and (2) assume complete labeling; in other words, the doubly labeled complex is completely labeled with donor and acceptor. They can be readily modified to account for incomplete labeling.[56]

To get distance information from these experimental parameters, one needs to know how the efficiency of energy transfer depends on distance. Förster showed that

$$E = 1/(1 + R^6/R_0^6) \tag{3}$$

where R is the distance between the donor and acceptor and R_0 is a characteristic distance, typically 10–50 Å, related to properties of the donor and acceptor.[36,53] From Eq. (3) it is easy to see that R_0 is the distance at which 50% of the energy is transferred:

$$R_0 = (8.79 \times 10^{-5} Jq_D n^{-4} \kappa^2)^{1/6} \quad \text{(in Å)} \tag{4}$$

$$J = \int \varepsilon_A(\lambda) f_D(\lambda) \lambda^4 \, d\lambda \bigg/ \int f_D(\lambda) \, d\lambda \quad \text{(in } M^{-1} \text{cm}^{-1} \text{nm}^4) \tag{5}$$

where J is the normalized spectral overlap of the donor emission (f_D) and acceptor absorption (ε_A), q_D is the quantum efficiency (or quantum yield) for donor emission in the absence of acceptor (q_D is the number of photons emitted divided by number of photons absorbed), n is the index of refraction (typically 1.3–1.4), and κ^2 is a geometric factor related to the relative angle of the two transition dipoles.

Figure 1 shows an example of the spectrum of a commonly used donor and acceptor and the intensity changes which occur with energy transfer. Figure 2 shows the experimental verification of the R^{-6} dependence of energy transfer using dansyl as a donor and naphthyl as an acceptor, separated by a series of rigid, polyproline linkers.

To generate an intuitive feeling for Eqs. (1)–(5), we first show how the form of Eq. (3) for the fraction of energy transferred is physically reasonable, then we show the sixth power dependence. Finally, we discuss the parameters which determine R_0 in Eqs. (4) and (5).

Form of Equation

We first show that the form of the energy transfer efficiency qualitatively looks like $1/[1 + f(r)]$ where $f(r)$ is some function of the distance,

---

[56] B. Epe, K. G. Steinhauser, and P. Woolley, *Proc. Natl. Acad. Sci. U.S.A.* **80**, 2579 (1983).

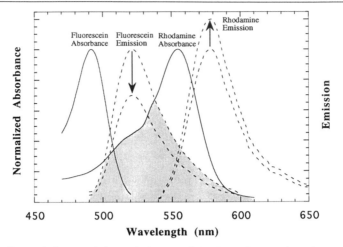

FIG. 1. Spectral characteristics and changes of the donor fluorescein and acceptor tetra-methylrhodamine undergoing energy transfer. The donor intensity decreases and the acceptor is sensitized with energy transfer. The spectral overlap which makes energy transfer possible is shown in gray. The absorbance and emission intensities are normalized for display purposes. The R_0 for the pair is approximately 45 Å.

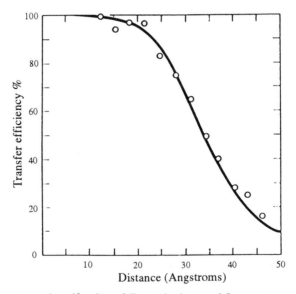

FIG. 2. Experimental verification of Förster's theory of fluorescence energy transfer. Energy transfer was studied with a series of end-labeled oligopeptides, dansyl-(polyproline)$_n$-naphthyl. The solid line is a fit to the data with Eq. (3) showing the R^{-6} dependence. (From Stryer and Haugland.[19]

r, between the donor and acceptor. Consider a donor which has been excited by light. The excited state will decay with rate $k_{nd} + k_{et}$, where k_{nd} is the sum of all distance-independent rates such as fluorescence (k_f) and heat (k_h) ($k_{nd} = k_f + k_h = \tau_D^{-1}$) and k_{et} is the distant-dependent rate of energy transfer to an acceptor.

The fraction of donor molecules giving energy to the acceptor is therefore just

$$E = k_{et}/(k_{et} + k_{nd}) = 1/(1 + k_{nd}/k_{et}) = 1/(1 + 1/\tau_D k_{et}) \qquad (6)$$

which is the form of Förster's equation [Eq. (3)] if k_{et} depends on R^{-6} and R_0 is related to constants in k_{et} and k_{nd}. In addition, the excited state donor lifetime decreases from $1/k_{nd}$ (τ_D) without the acceptor, to $1/(k_{nd} + k_{et})$ (τ_{D_A}) with the acceptor, directly leading to Eq. (1). When the acceptor is nearby, the energy transfer rate is fast compared to other decay pathways and most of the energy is transferred. If the acceptor is farther away this rate is less and the efficiency of energy transfer decreases. For later reference, we note that combining Eq. (3) and Eq. (6) yields

$$k_{et} = \tau_D^{-1}(R/R_0)^6 \qquad (7)$$

Why is the distance dependence of energy transfer R^{-6}? It comes about because the extent of energy transfer depends on the square of the electric field produced by the donor, and this field decays as R^{-3} at distances relevant in FRET. First we examine the donor electric field. The electric field of the donor arises because the incident excitation light induces electrons in the donor to oscillate (or, in quantum mechanics terms, induces transitions). This creates an induced, electric dipole moment in the donor, which creates its own, characteristic electric field. The dipole field of the donor has two parts, one which dies away like $1/R$, the other which decays like $1/R^3$. Far away from the molecule, the $1/R$ term dominates. This is called the radiation field and is what we see as fluorescent photons. (The energy carried away goes like the square of the electric field, and therefore drops off as $1/R^2$, as it must to conserve energy.) Close up, however, within a wavelength of light, the $1/R^3$ term dominates. This field is the one of interest in FRET. It does not carry away energy, that is, does not radiate, and so sometimes FRET is called nonradiative energy transfer. (In very rare instances, when using lanthanides as donors, for example, the donor may act partially like a magnetic dipole[57] or, in the solid phase, even like an electric quadrupole[58-60] instead of, or in addition

[57] J.-C. G. Bunzli and G. R. Choppin (eds.), "Lanthanide Probes in Life, Chemical and Earth Sciences: Theory and Practice," Elsevier, New York, 1989.
[58] R. Reisfeld and L. Boehm, *J. Solid State Chem.* **4**, 417 (1972).
[59] E. Nakazawa and S. Shionoya, *J. Chem. Phys.* **47**, 3211 (1967).
[60] R. Reisfeld, E. Greenberg, and R. Velapoldi, *J. Chem. Phys.* **56**, 1698 (1972).

to, acting like an electric dipole. Such cases are covered theoretically by Dexter in a seminal paper.[54])

To understand why energy transfer is proportional to the square of the R^{-3} field of the donor, we must understand how the acceptor interacts with (takes energy from) the donor electric field. If an acceptor molecule is in this close-up electric field, its electrons will be induced to oscillate, creating an induced dipole moment, $\mathbf{p}_A$, in the acceptor (just as an induced dipole was formed in the donor by the incident electric field of the exciting light). The size of the dipole is related to the size of the electric field, E_D, creating it: $p_A = \alpha_A E_D$, where α_A is the polarizability of the acceptor and is a measure of how easily the electrons can be made to oscillate.

What is the fraction (or efficiency) of energy transferred? It is simply the energy absorbed by the acceptor from the donor, divided by the energy absorbed by the donor from the excitation light. The latter term is independent of any distances of interest and so is irrelevant here. Moreover, the amount of energy absorbed by the acceptor is just $p_A \cdot E_D = \alpha_A E_D^2$. Consequently, because E_D decays as R^{-3}, the amount of energy absorbed by the acceptor is a function of R^{-6} between donor and acceptor.

Of course the proper treatment of energy transfer is via quantum mechanics. The analysis is very straightforward, and an excellent outline is presented by Cantor and Schimmel.[36] The excitation light induces transitions in the donor to an excited (singlet) state. This decays rapidly to the lowest excited state. The donor can then relax either via fluorescence, nonradiative procesess, or interaction with the acceptor via a dipole–dipole interaction (see Fig. 3). The Hamiltonian or energy of interaction between the donor and acceptor is

$$H = (\mu_D \cdot \mu_A)/R^3 + (\mu_D \cdot \mathbf{R})(\mu_A \cdot \mathbf{R})/R^5 \tag{8}$$

where $\mu_D(\mu_A)$ is the transition dipole moment of the donor (acceptor) and $\mathbf{R}$ is the vector separating their centers. According to Fermi's rule, the rate of inducing transitions is proportional to the square of the Hamiltonian matrix element between final and initial states:

$$k_{et} \propto [\langle D^*|\langle A|[(\mu_D \cdot \mu_A)/R^3 - 3(\mu_D \cdot \mathbf{R})(\mu_A \cdot \mathbf{R})/R^5 |D\rangle|A^*\rangle]^2 \tag{9}$$

where the initial state is the product of the excited state of the donor ($\langle D^*|$) and the ground state of the acceptor ($\langle A|$) and the final state is the product of the donor ground state ($|D\rangle$) and acceptor excited state ($|A^*\rangle$). We can write the wave function as this simple product because we assume the coupling between donor and acceptor is weak, and so the individual wave functions are not much perturbed. (If this is not the case, one gets into exciton coupling where the absorption spectra of the individual dyes change in the donor–acceptor complex). Only those donor and acceptor wave functions with nearly the same energy will significantly contribute to the rate; this is the resonance condition of FRET. The dot products can be explicitly written and Eq. (9) separated into those quantities depending on the donor wave functions and those on the acceptor wave functions, and those depending on the relative orientation:

$$k_{et} \, \alpha \, R^{-6} \, \langle D^*|\mu_D|D\rangle^2 \langle A|\mu_A|A^*\rangle^2 [\langle D^*\langle A|(\cos \theta_{DA} - 3 \cos \theta_D \cos \theta_A) \, |D\rangle|A^*\rangle]^2 \tag{10}$$

The rate is therefore proportional to the square of the transition dipole moments of the acceptor and donor, which can be related to the absorption and emission properties of each dye, respectively.[36] The rate also depends on a geometric factor.

Combining Eq. (10) and Eq. (7) yields Förster's equation [Eq. (3)] where R_0 is a function of the acceptor absorption cross section, the donor emission efficiency, and also the relative angles of the donor and acceptor dipoles [see Eqs. (4) and (5)].

Parameters in R_0

We can now understand the parameters which enter into R_0 [Eq. (4)]. R_0 is a measure of how well the donor and acceptor can transfer energy to one another, where a large R_0 indicates that the donor and acceptor

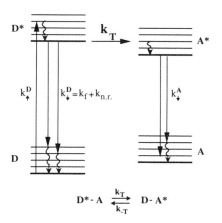

FIG. 3. Jablonski energy level diagram. The donor is excited and rapidly drops to the lowest vibrational level of the excited state, where it can radiatively (primarily via fluorescence) or nonradiatively decay to the group state, or transfer energy to the acceptor. Only those levels of the donor and acceptor with similar energies contribute significantly to the transfer rate. Once the acceptor is excited, rapid vibrational relaxation prevents back transfer. The acceptor then decays to the ground state via fluorescence or heat.

can transfer energy efficiently even if they are relatively far apart. To transfer energy from donor to acceptor, the electric field produced by the donor must be at a frequency (or wavelength) that can induce transitions in the acceptor (or, classically, a frequency which can efficiently drive the electrons in the acceptor). Consequently, R_0 depends on the spectral overlap term, J, which is a measure of how well the donor fluorescence frequency (or wavelength) and acceptor absorbance overlap in wavelength. Why is there a λ dependence in the J term? The answer is that the electric field dies off as $(\lambda/r)^3$. If λ is large (further to the red) the electric field drops off more slowly, and energy transfer can occur at farther distances. Mathematically, the λ^4 power comes about because μ_A^2 is proportional to $\varepsilon_A\lambda$ and μ_D^2 is proportional to $\lambda^3 q_D/\tau_D$.[36] Note that the dependence of μ_D^2 on τ_D^{-1} eliminates the donor lifetime dependence from R_0. This is expected because, although the rate depends on the lifetime, a longer lifetime means a slower rate integrated over a longer time. The index of refraction enters into R_0 because the size of the electric field produced by the donor is modified by the polarizability of the medium, which is directly related to the index of refraction. Finally, the quantum yield of the donor, q_D, enters because it is a measure of how well the donor converts the energy it has absorbed into an electric field (as opposed to converting the energy to phonons, or heat). (A donor with a high quantum yield efficiently creates a large electric near-field, which is relevant for FRET, as well as a large electric far-field, the latter being fluorescent photons.)

The last and most troubling term in R_0 is κ^2, which arises because the efficiency of energy transfer depends on the relative orientation of the two dyes (the $\mu_D \cdot \mu_A$ term) and the relative orientation in space [the $3(\mu_D \cdot \mathbf{R})(\mu_A \cdot \mathbf{R})$ term]. The expression for κ^2 is

$$\kappa^2 = (\cos \theta_{DA} - 3 \cos \theta_D \cos \theta_A)^2 \tag{11}$$

where θ_{DA} is the angle between the donor and acceptor transition dipole moments and θ_D (θ_A) is the angle between the donor (acceptor) transition dipole moment and the $\mathbf{R}$ vector joining the two dyes.

One can immediately see that κ^2 can vary from 0 if all angles are perpendicular to 4 if all angles are parallel. If the orientation of the dipoles is random, because they are moving rapidly (within the donor lifetime) then $\kappa^2 = 2/3$. This assumption is often made, even if not strictly true, and accounts for much of the uncertainty in FRET measurements. If just the donor or just the acceptor is randomized, then $1/3 < \kappa^2 < 4/3$. In this case the uncertainty in measured distance is approximately $\pm 11\%$.[38] In reality, what usually happens is that the dyes undergo fast, restricted motion such that κ^2 approaches 2/3, with some uncertainty remaining.

Dale and Eisinger have analyzed the effect of rotational mobility[61]; Stryer presents an analysis of the errors introduced by assuming $\kappa^2 = 2/3$[38]; and van der Meer *et al.* present a treatment on the effects of restricted rotational and translational diffusion.[62] Experimentally, one determines the rotational mobility of the dyes by a steady-state or time-resolved fluorescence depolarization experiment.[63]

In practice, R_0 is often measured from a model system (see, e.g., Fig. 2) and assumed to apply to the system of interest, or R_0 is calculated from Eqs. (4) and (5), assuming a value $\kappa^2 = 2/3$.

Labeling

Perhaps the most difficult aspect of FRET is the problem of labeling the sites of interest with the appropriate dyes. One must choose dyes that are spectrally compatible and that can be site specifically labeled without significantly perturbing the original structure of the molecule of interest. In general it is also important that the sample be completely labeled with both donor and acceptor, or at least that the extent of labeling be known. Analyses have appeared which take into account incomplete labeling.[2,64] Waggoner ([15] in this volume) reviews fluorescent labeling, and Brinkley has reviewed techniques for labeling proteins with dyes, haptens, and cross-linking reagents.[51]

Naturally Occurring Fluorophores

The ideal situation is if the biological molecule is fluorescent, or can be made so with slight modification. In proteins, tryptophan is a naturally occurring amino acid which makes a reasonable donor. Tyrosine is also fluorescent but is rarely a good donor to an external acceptor because it is often quenched by tryptophans in the protein.[36] Beardsley and Cantor have used the fluorescent Y-base associated with yeast phenylalanine transfer RNA.[25] For DNA and RNA there are a number of fluorescent nucleotide analogs.[38,42] Yet another possibility is if the protein binds a ligand which is either fluorescent or can be made so. In metal-binding proteins, zinc can be replaced with cobalt, which can act as a good acceptor because of wide visible absorbance.[22,65] Terbium, another metal, can act

[61] R. E. Dale, J. Eisinger, and W. E. Blumberg, *Biophys. J.* **26**, 161 (1979).

[62] B. W. van der Meer, M. A. Raymer, S. L. Wagoner, R. L. Hackney, J. M. Beechem, and E. Gratton, *Biophys. J.* **64**, 1243 (1993).

[63] J. R. Lakowicz, "Principles of Fluorescence Spectroscopy." Plenum, New York, 1983.

[64] J. R. Lakowicz, I. Gryczynski, W. Wiczk, J. Kusba, and M. L. Johnson, *Anal. Biochem.* **195**, 243 (1991).

[65] S. A. Latt, D. S. Auld, and B. L. Vallee, *Proc. Natl. Acad. Sci. U.S.A.* **67**, 1383 (1970).

as an excellent donor especially if there is a tryptophan nearby. (The tryptophan, in effect, increases the absorption cross section of the terbium by absorbing light and passing energy to the terbium, which by itself, has very weak absorbance.)[66] Terbium and europium have also been used as isomorphous replacements for calcium[22] and in metal-binding engineered proteins.[67,68]

Fluorescent Dyes

There is a large number of dyes to choose from, many of which are listed in the Molecular Probes (Eugene, OR) catalog.[42] Ideally, one picks a pair with R_0 equal to the distance to be measured, since small changes in distance around R_0 lead to large changes in signal. However, because the distance is generally unknown, or only approximately known, it is wise to pick a donor–acceptor pair with a large R_0, equal to or larger than the distance to be measured. If necessary, R_0 can be decreased by adding reagents such as iodine that reduce the donor quantum yield or by choosing an acceptor with poorer spectral overlap to the donor.

To achieve a large R_0 one wants a donor with a high quantum yield of long-wavelength emission and an acceptor with a large absorbance at the donor fluorescence wavelengths. Unfortunately, as the emission becomes more red, the quantum yield tends to drop, although there are some promising new dyes with reasonable quantum yields in the red, such as La Jolla Blue,[69] CY-5,[70,71] and Bodipy[42] as well as nucleic acid stains such as BOBO and POPO,[42] and also the phycobiliproteins.[42]

Increasing R_0 and thereby getting a large signal is only part of the strategy. Minimizing background is also important. When measuring donor quenching or sensitized emission, it is desirable to have little spectral overlap between donor and acceptor fluorescence. For sensitized emission, it is also desirable for the acceptor not to be excited directly by the excitation light. This requires the acceptor absorbance to be small where the donor absorbance is large. Finally, to maximize the sensitized emission signal, it is desirable for the acceptor to have a good quantum yield.

The ideal situation is therefore when both donor and acceptor have high quantum yields and large Stokes shifts. Unfortunately a large Stokes

[66] P. Cioni, G. B. Strambini, and P. Degan, *J. Photochem. Photobiol. B* **13**, 289 (1992).

[67] J. P. MacManus, C. W. Hogue, B. J. Marsden, M. Sikorska, and A. G. Szabo, *J. Biol. Chem.* **265**, 10358 (1990).

[68] I. D. Clark, J. P. MacManus, D. Banville, and A. G. Szabo, *Anal. Biochem.* **210**, 1 (1993).

[69] R. Devlin, R. M. Studholme, W. B. Dandliker, E. Fahy, K. Blumeyer, and S. S. Ghosh, *Clin. Chem.* **39**, 1939 (1993).

[70] H. Yu, L. Ernst, M. Wagner, and A. Waggoner, *Nucleic Acids Res.* **20**, 83 (1992).

[71] R. B. Mujumdar, L. A. Ernst, S. R. Mujumdar, C. J. Lewis, and A. S. Waggoner, *Bioconjugate Chem.* **4**, 105 (1993).

shift generally implies a small quantum yield, so there is a trade-off in these properties. (In the section on future directions, we discuss an energy transfer scheme using lanthanides as donors and organic chromophores as acceptors which gets around these problems.)

Perhaps the most popular donor is fluorescein, which often has a quantum yield exceeding 0.5. It can be used with eosin ($R_0 = 50$–54 Å[14]), chlorofluorescein (≈ 50Å), tetramethylrhodamine ($R_0 = 45$ Å[16]), tetraethyl-rhodamine ($R_0 = 40$ Å[37]), or Texas Red (R_0 not reported but $\ll 40$ Å). (All R_0 values are approximate because they depend on the donor quantum yield and emission spectra which can change depending on solvent conditions.) In this series, as R_0 decreases, the spectral separation increases. Greater spectral separation is especially helpful if one is measuring the sensitized emission of the acceptor (see measurement section below) or if only a relatively small percentage of molecules have both donor and acceptor bound or are capable of transferring energy. (See the FRET work of McConnell and co-workers on conformational changes in high mobility group (HMG) proteins for an example of such a case.[23]

Tetramethylrhodamine is perhaps the most popular acceptor with fluorescein because of its large R_0 and because the acceptor fluorescence is somewhat separated from donor fluorescence. Fluorescein fluorescence can be monitored from 500 to 525 nm with no contamination from acceptor fluorescence. A new carbocyanine dye, CY-3, is spectrally very similar to tetramethylrhodamine but is reported to have somewhat higher maximum absorbance.[71–73] Eosin as acceptor has a somewhat higher R_0 than tetramethylrhodamine, but its emission strongly overlaps that of fluorescein, such that there is not even a separable maximum. Fluorescein as donor does have the disadvantage that it may be quenched approximately 50% in proteins,[42] and it has a short, multiexponential (main component of 3 nsec) lifetime which makes donor lifetime measurements difficult, of limited accuracy, and especially problematic if R is significantly greater than R_0. The quantum yield of fluorescein is also a strong function of pH below pH 8 and decreases with increasing Na^+, Cl^-, and Mg^{2+}.[18]

Other dyes such as dansyl and AEDANS are popular donors in protein studies, in part because of the relatively long lifetimes (13–20 nsec), large (150 nm) Stokes shifts, and reasonable quantum yields (0.1–0.5).[42] If steric hindrance is not a problem and a large-sized donor or acceptor can be used, the multichromophoric phycobiliproteins (molecular weight of 104,000 for B- or R-phycoerythrin; 240,000 for allophycocyanin) make excellent donors or acceptors, having extinction coefficients which can exceed

[72] "Biological Detection Catalog." Biological Detections Inc., Pittsburgh, Pennsylvania, 1993.
[73] "Research Organics Inc., Catalog." Cleveland, Ohio, 1993.

2×10^6 and quantum yields of 0.68–0.98. They are also available with reactive groups.[42] Fairclough and Cantor present a useful list of donor–acceptor pairs, including spectra and R_0 values,[37] and van der Meer and co-workers[35] are compiling an extensive list.

Site-Specific Attachment

Fortunately there is a wide variety of dyes that are available with reactive groups. For attachment to amines, the most common reactive groups are succinimdyl esters and isothiocyanates. The former have generally better coupling efficiency if the coupling can be done in an organic phase. For attachment to synthetic DNA, an amine-modified base can be introduced via commercially available phosphoramidite, either internally or on the 3' or 5' ends.[74–76] Direct attachment to the DNA backbone has also been achieved.[15] Relatively short polypeptides can often be readily labeled, especially if they contain a unique cysteine or amine-containing amino acid.[19,34,77] For proteins, lysines (and to a significantly lesser extent arginines) are available for labeling.[78]

With proteins the problem is generally that there are too many reactive sites. Site-directed mutagenesis can sometimes be used to introduce a free sulfhydryl group which can then be coupled to a dye via an iodoacetamide or a maleimide. An extensive list of proteins which have been labeled with iodoacetamides has been tabulated.[42] Site-directed mutagenesis can also introduce a tryptophan (as donor) if one is not already present. ANS or AEDANS are good acceptors for tryptophan, both yielding an R_0 of 22 Å. Proteins with N-terminal serine or threonine can also be specifically labeled.[37,38] Objects which can be biotinylated can be labeled with fluorescent avidin or streptavidin. More generally, fluorescently labeled antibodies can be used to bind to a wide variety of biological substrates.

The reactive groups are often attached to the dyes via a n-carbon linker, where n typically ranges from 2 to 12. The linker often allows relatively free rotation of the dye, which minimizes uncertainty in κ^2. It can also minimize quenching of the dye, especially if the dye is quenched by a hydrophobic environment. The linker, however, has the disadvantage of adding uncertainty to the exact position of the dye. In general, the

[74] "Clonetech Catalog." Palo Alto, California, 1993.
[75] "Glen Research Catalog." Sterling, Virginia, 1993.
[76] "Peninsula Laboratories Catalog." Belmont, California, 1993.
[77] Y. Pouny, D. Rapaport, A. Mor, P. Nicolas, and Y. Shai, *Biochemistry* 31, 12416 (1992).
[78] T. E. Creighton, "Proteins: Structures and Molecular Properties," 2nd Ed., Freeman, New York, 1993.

minimal length that allows free rotation of the dye and does not cause quenching is desirable. A six-carbon linker is a good starting point.

Testing for Altered Structures

It is important to check whether the introduction of fluorescent labels alters the macromolecular structure. For DNA, testing the hybridization melting temperature is a crude measure. For proteins, comparing the function (enzymatic activity, binding constant, etc.) of the labeled versus unlabeled molecule is an excellent check.

How to Measure Energy Transfer

There are several ways to measure the amount of energy transfer: (1) decrease in donor intensity or quantum yield; (2) increase in intensity of acceptor emission (sensitized emission); (3) decrease in lifetime of donor; (4) decrease in photobleaching of donor; and (5) change in lifetime of sensitized emission (discussed in the section on future directions below).

Donor Intensity

The simplest way to measure energy transfer is to measure the decrease in fluorescence of the donor in the presence and absence of acceptor. The fractional decrease in the donor fluorescence with the acceptor present is equal to the efficiency of energy transfer [see Eq. (1)]. The only instrumentation necessary is a steady-state fluorimeter. Besides simplicity, the steady-state measurement has the advantage that even a relatively small amount of energy transfer can be measured. If care is taken in measuring concentrations and in the spectroscopy, a 5% decrease in fluorescence is measurable: this corresponds to a distance of 1.6 R_0. One caution is that the optical density (OD) of the sample must be kept sufficiently low that no appreciable absorption of the donor fluorescence takes place; less than 0.02 OD is recommended. Another caution is that the donor and acceptors should be chosen so that there is a region of donor-only fluorescence. Although it is not necessary to have an acceptor which is fluorescent in this experiment, it is important confirmation that energy transfer is taking place because an increase in acceptor emission can arise only from energy transfer (assuming the optical density is low enough), whereas donor quenching can arise from several trivial sources. With a fluorescent acceptor, one can also measure the polarization of the acceptor emission, which tells about its rigidity and hence limits the uncertainty in κ^2.

In addition to measuring the decrease in donor fluorescence, if one wants to calculate R_0, it is necessary to measure the quantum yield of the donor in the absence of acceptor, and also to measure the steady-state or time-resolved polarization of both donor and acceptor. Quantum yield measurements are generally done by comparing total fluorescence of the sample to a reference with known quantum yield.[36,79,80] Polarization measurements are important to limit uncertainty in κ^2, and they are covered by H. van Amerongen and W. S. Struve (this volume [11]) as well as in a book by Lakowicz on fluorescence.[63]

Sensitized Emission

With the same steady-state fluorimeter, one can also measure the sensitized emission of the acceptor. The efficiency of energy transferred calculated via donor quenching should agree with that calculated by sensitized emission. Sensitized emission can also be particularly useful when measuring long distances or when the sample is inhomogeneous and only a small fraction of the sample contributes to energy transfer.[23] In either case it is helpful if the fluorescence of the acceptor is well separated from that of the donor. In principle, one can measure the energy transfer via sensitized emission based on Eq. (2). In reality, there are experimental difficulties, discussed by Epe *et al.*[56] In particular, for distances much beyond R_0, the ratio I_{A_D}/I_A approaches unity, and even small errors in measurements lead to large errors in the calculated energy transfer.

A number of workers have attempted to make sensitized emission (and donor quenching) measurements more robust. In general these techniques attempt to reduce the number of independent samples (donor–acceptor, donor only, acceptor only) which must be compared. Fairclough and Cantor cover standard methods.[37] Epe and co-workers[56] developed a technique where a donor–acceptor labeled sample is measured and then enzymatically digested, thereby separating the donor from the acceptor and eliminating energy transfer. Donor quenching or sensitized emission can therefore be made on one sample. Clegg and co-workers have developed an analysis of acceptor emission which yields reproducible results, even when measuring samples with small energy transfer and under significantly different conditions.[2,17,18] They applied the analysis to DNA samples labeled with fluorescein and tetramethylrhodamine, but it should be generally applicable to other FRET pairs. Clegg has outlined the many advantages of the technique.[2] We present a brief outline of the technique.

[79] G. Weber and F. W. J. Teale, *Trans. Faraday Soc.* **53**, 646 (1957).
[80] J. B. Birks, "Photophysics of Aromatic Molecules." Wiley (Interscience), New York, 1970.

Steady-state emission spectra of a donor–acceptor labeled sample and a donor-only labeled sample are taken. The donor emission is removed from the donor–acceptor emission spectrum by subtracting the normalized donor-only emission spectrum. This leaves the fluorescence of the acceptor due to direct excitation and due to energy transfer (see Fig. 4). Clegg and co-workers call this the "extracted acceptor emission" spectrum, F_{em}^A. Note that this process does not require the concentration of donor-only sample to be the same as the donor–acceptor sample—only the shape of the donor spectrum is used. This spectrum is divided by a fluorescence value (often the maximum) of an emission spectrum taken on the donor–acceptor complex excited at a wavelength where only the acceptor absorbs (565 nm for fluorescein–tetramethylrhodamine). Alternatively, one can divide by the maximum of the excitation spectrum of the donor–acceptor complex (excitation at 400–590 nm, emission in the range 580–600 nm, for fluorescein–rhodamine). In either case, the resultant ratio spectrum, "$(ratio)_A$," is normalized for quantum yield of acceptor, for concentration of total molecules, and for incomplete acceptor labeling.

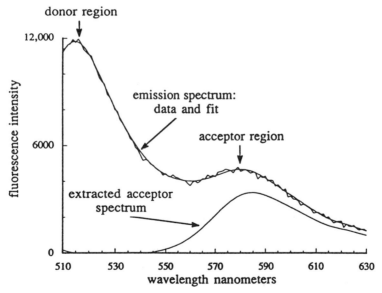

FIG. 4. Emission spectrum of fluorescein–tetramethylrhodamine labeled DNA oligomer. The region where only the donor emits (500–530 nm) is fit to a fluorescein-only spectrum and then subtracted from the entire spectrum, leaving the "extracted acceptor spectrum" consisting of acceptor fluorescence arising from energy transfer and from direct fluorescence. This spectrum is then fit using Eq. (4). The Raman background is subtracted from each fluorescence spectrum. (From R. M. Clegg, this series, Vol. 221, p. 372.)

Clegg and co-workers[2,18] have shown this ratio to be

$$(\text{ratio})_A = F_{em}{}^A(\nu_1,\nu')/F_{em}{}^A(\nu_2,\nu'')$$
$$= \{E \cdot \varepsilon^D(\nu')/\varepsilon^A(\nu'') + \varepsilon^A(\nu')/\varepsilon^A(\nu'')\}[\phi(\nu_1)/\phi(\nu_2)] \quad (12)$$

where superscripts D and A refer to donor and acceptor, ε is the molar extinction coefficient, E is the efficiency of energy transferred, ν' and ν_1 are the excitation and emission wavelengths, respectively, for the FRET measurement, ν'' is the excitation wavelength where the acceptor alone absorbs (560 nm for rhodamine), and ν_2 is the wavelength(s) where the acceptor emission is measured. $\phi(\nu_i)$ is an "emission spectrum shape function" of the acceptor, where the integral over ν is proportional to the quantum yield. It is simply the fluorescence of the acceptor-only sample at wavelength ν_i. [Because $\phi(\nu_i)$ enters only as a ratio, the concentration of acceptor is unimportant.] For the fluorescein–tetramethylrhodamine pair, ν_1 and ν_2 are both 585 nm. The second term within the braces, $\varepsilon^A(\nu')/\varepsilon^A(\nu'')$ can be measured on the acceptor-only sample via absorbance or via an excitation spectrum.

In the scheme developed by Clegg and co-workers,[2,17,18] it is not necessary to compare the intensities of a donor-only or an acceptor-only sample with the doubly-labeled sample; hence, errors in concentration between samples do not lead to errors in measured energy transfer. When comparing samples under different solvent conditions, it is necessary to measure the quantum yield of donor in each case since this parameter enters into R_0. More general equations can be found which include the effect of incomplete labeling, and for the case when the donor and acceptor absorbances and emissions overlap significantly.[2]

Donor Lifetime

Measurement of the donor lifetime, which typically is 2–25 nsec, requires adequate time resolution. Two techniques, time-correlated single-photon counting and frequency-domain fluorimetry modulation, can be used (see A. R. Holzwarth, this volume [14]). Excellent books have been written which include discussion of each technique,[63,81] and Lakowicz and co-workers have discussed advances in frequency-domain instrumentation and applications to FRET.[82] Donor lifetime measurements, unlike steady-state measurements, are capable of detecting multiple donor–acceptor transfer efficiencies in the sample. These lead to multiexponential decays. Donor lifetime measurements are also not affected by an inner-filter effect

[81] D. V. O'Connor and D. Phillips, "Time-Correlated Single Photon Counting." Academic Press, London, 1984.
[82] K. W. Berndt, J. R. Lakowicz, and I. Gryczynski, *Anal. Biochem.* **192,** 131 (1991).

where the donor fluorescence is absorbed by the acceptor. Even more significantly, donor lifetime is not sensitive to concentration; therefore, differences in concentration between the donor-only sample and the donor–acceptor sample do not lead to errors in the measurement of energy transfer.

Donor Photobleaching

A fourth way to measure energy transfer is based on changes in photobleachability of the donor. The technique, pioneered by the Jovin group, is particularly well suited to FRET in a microscope where the high light intensities necessary are readily accessible and where other FRET techniques have yielded only qualitative results.[10-12] (The discussion here is adopted from Ref. 10.) The idea is that the donor photobleaches more slowly if energy transfer to an acceptor is occurring since energy transfer is an alternative pathway for the excited state to give up energy. It can be shown that the fractional change in the photobleaching time constant caused by energy transfer is the same as the fractional change in the fluorescence lifetime of the donor. Thus, in the simplest case, the efficiency of energy transfer is just $1 - \tau_{bl}/\tau_{bl}'$ where τ_{bl} (τ_{bl}') is the bleaching time constant in the absence (presence) of acceptor. The rate of photobleaching can be easily measured in solution or in a microscope.

A second, related way of measuring energy transfer is based on changes in quantum yield. First a low-light level image is taken of a sample labeled only with donor. The fluorescent intensity at any spot is proportional to the quantum yield and total number of fluorophores. To normalize by the total number of fluorophores (in a manner independent of quantum yield), a high-light level image is taken, and all the fluorescent photons are counted (integrated) until complete photodestruction has occurred. The ratio of the low-light image to the integrated, high-light image is proportional to the quantum yield. The procedure is repeated with a donor–acceptor label sample, and the energy transfer is just the usual 1 minus the ratio of quantum yields [analogous to Eq. (1)]. If a linear camera [e.g., one based on a charge-coupled device (CCD) sensor] is used as the detector, a pixel-by-pixel energy transfer image can be obtained. A number of workers have applied photobleaching-FRET to epitope mapping in T cell lines,[83] visualizing receptor aggregation on the surface of living mast cells,[84] and

[83] G. Szaba, Jr., P. S. Pine, J. L. Weaver, M. Kasari, and A. Aszalos, *Biophys. J.* **61,** 661 (1992).
[84] U. Kubitscheck, R. Schweitzer-Stenner, D. J. Arndt-Jovin, T. M. Jovin, and I. Pecht, *Biophys. J.* **64,** 110 (1993).

studying the binding of haptens to monoclonal immunoglobulins on cell surface receptors.[85]

Controls

In all of the above techniques it is important to subtract background arising from Raman, specular, or other sources. The best way to do this is to prepare a sample identical to the fluorescence samples, but without the attached dyes. It is also important to assure that the energy transfer is arising only from intramolecular energy transfer, and not from diffusional contact or aggregation. To control for this, one can mix a donor-only labeled sample and an acceptor-only labeled sample under conditions where they will not form a donor–acceptor complex. Noncomplementary DNA strands, for example, can be mixed together, or a donor-only labeled protein can be mixed with an acceptor-only labeled protein, etc. FRET measurements should also be made with magic angle settings (analyzer set to 54°) to assure no polarization artifacts.

Examples and Application of Fluorescence Resonance Energy Transfer

Human Immunodeficiency Virus Protease

Matayoshi and co-workers have developed a simple and effective means for measuring HIV protease activity using FRET.[34] This is an example of measuring dynamics with FRET. The idea is to use a relatively short polypeptide that HIV can cleave into two and attach a donor on one end and an acceptor on the other end (see Fig. 5A). If the polypeptide is intact, the donor is highly quenched. On cleavage, the energy transfer is eliminated and the donor fluoresces. By making the polypeptide with an end-to-end distance less than R_0, the increase in signal on cleavage was 40-fold. They used EDANS as donor and DABCYL, a nonfluorescent dye, as acceptor (see Fig. 5B).

DNA Structure

One of the more recent applications of FRET is in the study of DNA structure. Clegg and co-workers have published much in this field[16–18] including a 1992 review.[2] They have measured the end-to-end distances of a series of DNA oligonucleotides, ranging from 8 to 20 base pairs (bp)

[85] U. Kubitscheck, M. Kircheis, R. Schweitzer-Stenner, W. Dreybrodt, T. M. Jovin, and I. Pecht, *Biophy. J.* **60,** 307 (1991).

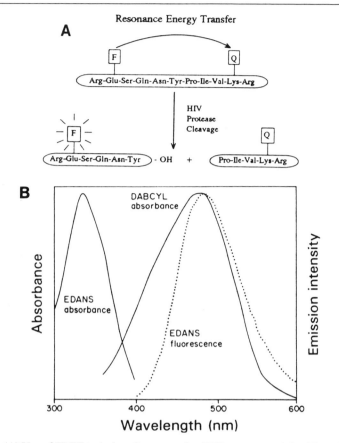

FIG. 5. (A) Use of FRET technique for measuring HIV protease activity. The polypeptide was Arg-Glu(EDANS)-Ser-Gln-Asn-Tyr-Pro-Ile-Val-Lys(DABCYL)-Arg. Because a simple system which could be commercialized was desired (fluorescence means HIV activity, no fluorescence means no HIV activity), the nonfluorescent acceptor was beneficial because it does not contribute signal after cleavage. This system also has the attribute that detailed information about the dyes and their energy transfer properties are not needed. (B) Good overlap between donor emission (maximum at 490 nm, excitation at 340 nm) and acceptor absorbance (maximum of 28,000 M^{-1} cm^{-1}). (From Molecular Probes Catalog.[42])

in length (see Fig. 6). Although the structure of short oligonucleotides is well understood, their detailed study showed conclusively that FRET can be used to study DNA structures, despite some early confusion.[14] The helical repeat of the DNA in solution was observed and the enthalpy of strand hybridization calculated. The helical repeat can be seen in the modulation of the R^{-6} energy transfer as a function of number of base pairs separating donor and acceptor.

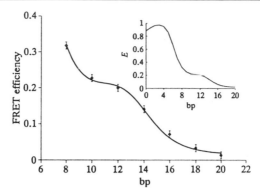

FIG. 6. Fluorescence energy transfer on a series of DNA oligomers differing in length from 8 to 20 bp. The modulation in the energy transfer decay curve (compare with Fig. 2) arises from the helical geometry of the DNA. Inset shows the theoretical FRET signal calculated from Eq. (12) for 0–20 bp. (From Clegg et al.[16])

Clegg and co-workers used fluorescein as donor and tetramethylrhodamine as acceptor (R_0 = 45 Å), attaching the dyes to the 5′ ends of complementary DNA strands. Others have also used fluorescein–eosin.[14,15] The ability to measure the relatively long distances of a 20-mer (end-to-end distance of approximately 72 Å plus linker lengths) required careful measurement and analysis of the sensitized emission (discussed above in the section on measurement). They took care to ensure that the local environment around the dyes was constant for all samples. They also adjusted the dye linker length to ensure that at least one of the dyes was rotationally mobile. (The fluorescein had a low steady-state anisotropy of 0.07; the rhodamine was less mobile, with an anisotropy of 0.25. In this case, the distance error should be less than 10%.)

Cardullo has also used FRET to study the hybridization of DNA oligonucleotides. Donor (fluorescein) and acceptor (tetramethylrhodamine) were placed on the 5′ ends of single-stranded DNA oligomers. On hybridization, energy transfer took place. (This is quite analogous to the HIV study cited above, where here the quantity of interest is hybridization rather than cleavage.) FRET has the advantage that hybridization can be measured at quite low concentration (<100 nM), in contrast to standard absorbance melting studies which require micromolar quantities. Measurement of hybridization is just a specific example of the more general problem of measuring binding constants. FRET can be useful in this regard because measurements can be made at low concentrations (in contrast to NMR), which is necessary for measuring large binding constants.

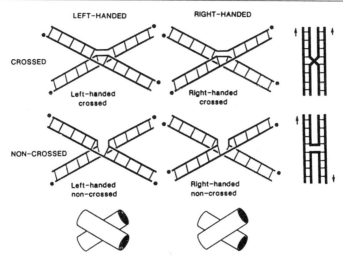

FIG. 7. Use of FRET to show that the overall geometry of the four-way DNA junctions is a right-handed noncrossed structure. The 5′ end of the DNA strands are labeled with filled circles. Noncrossed and crossed structures generate antiparallel or parallel alignment of DNA sequences, shown by the arrows at right. The six possible end-to-end distances were measured by labeling the appropriate 5′ ends with fluorescein (donor) or tetramethylrhodamine (acceptor) and monitoring energy transfer. (From Murchie et al.[17])

A more sophisticated example of FRET is its application to DNA four-way junctions, also called Holliday junctions. These are believed to be important intermediates in homologous recombination where genetic material is swapped between chromosomes.[86] The geometry of the junction presumably facilitates this swapping process, in combination with enzymes such as resolvase. Lilley and co-workers examined the three-dimensional structure of the junction by several methods, including gel electrophoresis and FRET.[17,18,87] Hagerman and co-workers have also studied the problem.[14,88] Crudely speaking, the junction looks like a nonplanar X (Fig. 7). Gel electrophoresis measurements indicated that the X structure involved strands which did not cross, but this conclusion was based on poorly understood assumptions about DNA mobility in gels. In contrast, the same information could be determined using FRET, where the assumptions are minimal. In addition, the junctions, because they are nonplanar, had a handedness to them (right- or left-handed). FRET could be used to distin-

[86] R. Holliday, Gen. Res. 5, 282 (1964).

[87] D. R. Duckett, A. I. H. Murchie, S. Dickmann, E. von Kitzing, B. Kemper, and D. M. J. Lilley, Cell (Cambridge, Mass.) 55, 79 (1988).

[88] J. P. Cooper and P. Hagerman, J. Mol. Biol. 198, 711 (1987).

guish the crossed from noncrossed structure and right-handed from left-handed because the different proposed structures had different end-to-end distances, which could be measured by end-labeling them with donor and acceptor dyes. Fortunately, measurement of absolute distances was not required to differentiate between the possible structures. Crossed versus noncrossed models, for example, gave different predictions about which two of the six end-to-end distances were closest. Handedness of a junction could be determined by changing the length of one arm of the X (in effect, walking the end-labeled dye around the DNA helix of the arm): left-handed versus right-handed models gave different predictions for when the acceptor and donor would be closest or farthest away from one another. Lilley and co-workers concluded that four-way DNA junctions are right-handed noncrossed structures.[17]

Diffusion-Enhanced Fluorescence Resonance Energy Transfer

FRET has been used to measure diffusional rates and to examine the accessibility of certain sites to collisional quenching. Stryer et al. have presented an excellent review,[8] and Thomas and co-workers put the technique on a firm experimental foundation.[89] Three time regimes can be distinguished in FRET. If D is the sum of the donor and acceptor diffusion coefficients, s is the average separation, and τ is the donor lifetime, then the static limit, where the donor and acceptor do not appreciably move during the donor lifetime, is when $6D\tau \ll s^2$. The intermediate region, which is useful for measuring diffusion constants, is when $6D\tau \approx s^2$. The rapid diffusion, where donor and acceptor collide many times and energy transfer is a sensitive function of the closest distance, is when $6D\tau \gg s^2$. If the acceptor (or donor) is free to diffuse and has a diffusion constant of approximately 10^{-6} cm^2/sec (a typical value for a small dye) and the donor has a lifetime of a millisecond, then the rapid diffusion regime can occur if the acceptor concentration is on the order of 1 μM or more. The intermediate regime, which is less often used, has been achieved using naphthalene, which has a lifetime of approximately 100 nsec.[90]

To achieve the rapid diffusion limit requires a very long-lived donor. Chelates of terbium, which typically have lifetimes of 1.5–2.2 msec, are frequently used.[8] Fluorescein and rhodamine make excellent acceptors for terbium. Chelates of europium also have long lifetimes, 0.5–2.3 msec,[91] although these appear not to have been used. (Cronce and Horrocks have used europium in a calcium-binding site as a donor in a rapid diffusion

[89] D. D. Thomas, W. F. Carlsen, and L. Stryer, *Proc. Natl. Acad. Sci. U.S.A.* **75**, 5746 (1979).

[90] Y. Elkana, J. Feitelson, and E. Katchalski, *J. Chem. Phys.* **48**, 2399 (1968).

[91] C. C. Bryden and C. N. Reilley, *Anal. Chem.* **54**, 610 (1982).

experiment.[22] The chelates are often polyaminocarboxylic, such as EDTA. By modifying the chelator to alter its charge and measuring distances of closest approach, one can deduce the local charge surrounding the acceptor. Replacing one or two carboxyl groups of $EDTA^{4-}$ with alcohol groups to make HED3A or BED2A, respectively, yields chelates with one or two fewer negative charges. In addition, by attaching an organic chromophore to the chelator, the extremely weak absorbance of the terbium or europium ($\sim 1\ M^{-1}\ cm^{-1}$ or less) can be increased several thousandfold.[8,92-94] By operating in D_2O, the quantum yield of the donor is often assumed to be unity (H_2O partially quenches lanthanide luminescence[95,96]), although the exact value is difficult to measure.

Energy transfer is typically measured by a decrease in the donor lifetime and, in the rapid diffusion limit, is a sensitive function of closest approach. Energy transfer based on a Förster dipole–dipole mechanism is proportional to a^{-4} in three dimensions and a^{-3} in two dimensions (e.g., on membranes), where a is the distance of closest approach.[8] If the distance of closest approach is less than 10 or 11 Å, energy transfer is dominated by a Dexter type exchange mechanism owing to overlapping wave functions between the donor and acceptor.[8,54] The exact distance of closest approach then becomes difficult to measure.

Lerho et al. have used FRET in the rapid diffusion limit to measure the accessibility of H5 histone in chromatin as a function of salt.[30] They labeled the H5 histone with fluorescein-labeled antibody and used TbHED3A and $TbEDTA^-$ as the freely diffusing donor. They found that the fluorescein becomes less accessible to the chelates as ionic strength increased. In the presence of DNA, they found H5 to be already folded at low ionic strength and the fluorescein inaccessible to the donor chelates. Thomas and co-workers have performed a number of experiments measuring the position of retinal in membranes by its ability to quench freely diffusing terbium chelates.[8,20,97] Retinal is the chromophore in rhodopsin, which acts as a signal transducer in vision, and in bacteriorhodopsin, which acts as a light-driven proton pump. They conclude that retinal in both bacteriorhodopsin and in bovine rhodopsin is buried with respect to both inner and outer membrane surfaces. In the case of rhodopsin, they measure a distance of 22 Å from the inner surface and 28 Å from the outer

[92] A. Canfi, M. P. Bailey, and B. F. Rocks, Analyst 114, 1405 (1989).
[93] T. Ando, T. Yamamoto, N. Kobayashi, and E. Munekata, Biochim. Biophys. Acta 1102, 186 (1992).
[94] A. K. Saha et al., J. Am. Chem. Soc. 115, 11032 (1993).
[95] W. D. Horrocks, Jr., and D. R. Sudnick, Acc. Chem. Res. 14, 384 (1981).
[96] W. D. Horrocks, Jr., and D. R. Sudnick, J. Am. Chem. Soc. 101, 334 (1979).
[97] D. D. Thomas and L. Stryer, J. Mol. Biol. 154, 145 (1982).

surface[97]; for bacteriorhodopsin they measure the retinal to be approximately 10 Å from the periplasmic surface.

Generating New Dyes

An application of FRET which is still in its infancy but shows great promise is the production of new heterodimeric dyes. One dye serves as the energy donor, the other as the energy acceptor. Over 10 years ago, Glazer and Stryer applied this technique to fluorescent phycobiliproteins,[39] covalently attaching phycoerythrin to allophycocyanin via a disulfide cross-link. The extent of energy transfer was 90%. The complex had the intense absorption of phycoerythrin around 545 nm, with the fluorescence maximum of allophycocyanin (660 nm), a Stokes shift of 115 nm. Glazer and co-workers have applied this technique to DNA dyes.[40,41] Again, the extent of energy transfer is approximately 90%, and they have created a series of dyes which have the same or similar absorption characteristics but with differing emission wavelengths (see Fig. 8). This allows simultaneous excitation with one light source and independent detection of the emission. In addition, the linker used to join the donor and acceptor is positively charged to enhance binding to the DNA and is of such a length to promote intercalation. Because the dyes intercalate and are therefore separated by a DNA base, there does not appear to be excitonic coupling, which would alter the spectral characteristics of the individual dye. The energy transfer mechanism is therefore presumed to be Förster type, and the spectral characteristics of the compound dye are therefore very similar to the sum of those of the individual dyes. Furthermore, a large number of such dyes is possible because the spectral overlap between donor and acceptor need not be large since they are so close together that efficient energy transfer takes place. By choosing one dye (donor) with large absorbance (the quantum yield need not be large) and the other dye (acceptor) with good quantum yield, unusually bright dyes should be possible (see A. Waggoner, this volume [15]). Molecular Probes has generated a series of DNA dyes based on homodimeric and heterodimeric compounds which span a wide wavelength range.[42] Application to molecules other than DNA will depend on the ability to link dyes without creating significant excitonic coupling.

Problems and Future Directions: Luminescence Resonance Energy Transfer Using Lanthanide Chelates

Despite the numerous successful applications of FRET, the technique has a number of drawbacks. First, the maximum distance which can be measured is less than optimal for many biological applications. Second,

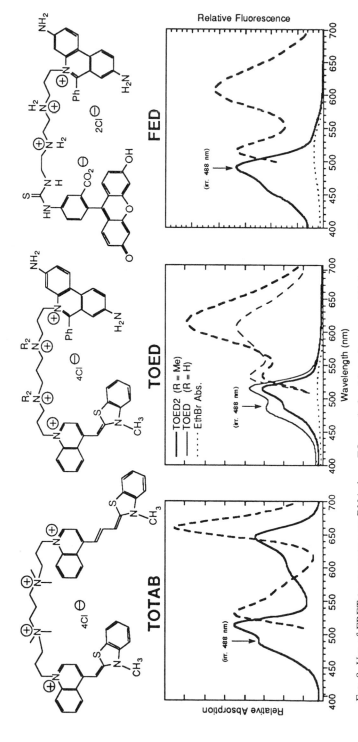

FIG. 8. Use of FRET to generate new DNA dyes. TO stands for thioazole orange, AB for Azure Blue, F for fluorescein, and ED for ethidium bromide. The dyes bind tightly to DNA such that they are stable even during electrophoresis. Solid lines represent absorption; dashed lines, emission. (From Benson et al.[40,41])

the lifetime of commonly used donor fluorophores are short (typically a few nanoseconds) and are often multiexponential, making lifetime measurements difficult and of limited accuracy. An accuracy of 10% limits measurements of quenching of more than 10% and hence less than $1.45R_0$. For R_0 values of 40–55 Å, the largest yet attained for small dyes, the maximum measurable distance via donor quenching is therefore 58–72 Å. Third, when measuring the sensitized emission of acceptor the signal-to-background ratio is poor, typically on the order of 1 : 1. The background arises from interfering fluorescence from the donor and from direct excitation of the acceptor by the laser or excitation light. The poor signal-to-background ratio limits the maximum measurable distance and also makes measurement of the lifetime of the sensitized emission not feasible. The large background also severely inhibits the use of FRET on biological systems that are impure (where, e.g., only a small percentage of donor–acceptor complexes form). Fourth, distances are difficult to determine precisely because of the uncertainty in κ^2.

We have developed an energy transfer system which overcomes these difficulties. We use a luminescent lanthanide chelate as donor and an organic dye such as fluorescein, rhodamine, or CY-5 as acceptor. A number of workers have noted that the luminescent lanthanide elements terbium and europium are attractive donors because they have multiple transition dipole moments such that they act as randomized donors even in the absence of any rotational motion. This limits κ^2 ($1/3 < \kappa^2 < 4/3$) even if the acceptor is stationary. Furthermore, the lifetimes are extremely long (0.6–2.3 msec) and single exponential, and thus are easy to measure. The quantum yields of the donors are also likely to be large (approaching 1 in D_2O), although the exact value is difficult to measure. The lanthanides can be relatively easily attached to macromolecules via the chelator, and, by covalently coupling an organic chromophore onto the chelator, the lanthanides can also be easily excited.[92–94,98,99] In addition, the spectral overlap is large when using terbium as donor and fluorescein or rhodamine as acceptor, or when using europium as donor and CY-5 as acceptor. The net result is unusually large R_0 values exceeding 50 Å (depending on whether the experiment is performed in H_2O or D_2O, what quantum yield is assumed, and which donor–acceptor pair is used). With terbium and rhodamine, for example, we calculate an R_0 of 65 Å, assuming the quantum yield of terbium is 1 in D_2O, and find that energy transfer experiments are consistent with this value[98] (see below). Using europium and CY-5 in D_2O we calculate an unusually large R_0 of 70 Å and, again, find that experiments are consistent with this value.[99] Using a europium cryptate

[98] P. R. Selvin and J. E. Hearst, *Proc. Natl. Acad. Sci. U.S.A.* **91,** 10024 (1994).
[99] P. R. Selvin, T. M. Rana, and J. E. Hearst, *J. Am. Chem. Soc.* **116,** 6029 (1994).

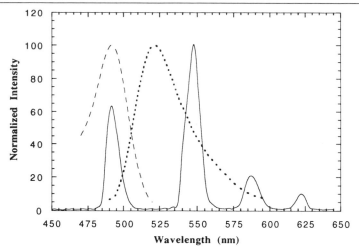

FIG. 9. Emission spectrum of a terbium chelate (structure shown in Fig. 10), used as a donor (solid line), along with absorbance (dotted line) and emission (dashed line) spectra of fluorescein, used as acceptor. Note that emission of the donor is silent around 520 nm, where the acceptor emission is maximal. Also note the excellent overlap between the 492 nm donor emission line and fluorescein absorbance, leading to a large R_0 ($\cong 52 Å$). By using a pulsed excitation source and monitoring at 520 nm, any signal arises only from sensitized emission, that is, fluorescein fluorescence due only to energy transfer (see Fig. 11).

as donor and allophycocyanin as acceptor, Mathis reports an exceptionally large R_0 of 90 Å.[100]

For all of these reasons, lanthanide donors and organic acceptors have been used in diffusion-enhanced energy transfer experiments. We find, however, that they can be used quite effectively in static FRET as well, as long as care is taken to ensure that no intermolecular energy transfer takes place. Furthermore, the most powerful aspect of these donor–acceptor pairs appears not to have been recognized until relatively recently: one can measure the sensitized emission of the acceptor without any interfering background.[98,99,100] This is in contrast to most donor–acceptor pairs, where the sensitized emission is much less than the background. Such a dark-background sensitized emission experiment has a number of advantages. First, like all sensitized emission experiments, it is less susceptible to artifacts than donor quenching. Second, because even a small amount of energy transfer yields fluorescence much above background, distances well beyond R_0 can be measured. Third, it is possible to measure the lifetime of the sensitized emission. (We measure not the nanosecond lifetime of each acceptor molecule, but the millisecond decay of the ensemble of acceptors.) By measuring the lifetime of the sensitized emission, studies are insensitive to concentration effects, to quantum

[100] G. Mathis, *Clin. Chem.* **39**, 1953 (1993).

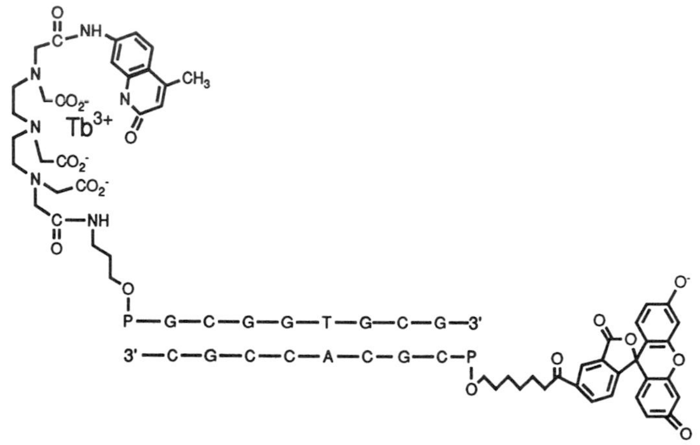

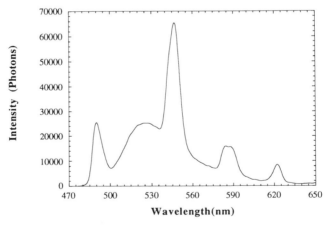

FIG. 10. Structure or terbium chelate, namely, terbium diethylenetriaminepentaacetic acid coupled to carbostyril 124. The carbostyril absorbs light (maximally at 327 nm) and transfers energy to the terbium. The net result is an increase in the effective absorbance of the terbium by several thousandfold. The DTPA chelate shields the terbium from the quenching effects of water and allows for easy attachment to macromolecules. Here the macromolecule is an 8-mer DNA oligomer modified with a primary amine on the 5' end. The acceptor, fluorescein, is attached to the 5' end of a complementary DNA oligomer.

FIG. 11. Emission spectrum showing energy transfer after excitation at 337 nm with a 2 nsec pulse, with the signal collected after a 80 μsec delay with a 7 msec gate. Each point (taken every 2 nm) is the average of 160 pulses. The sensitized emission signal-to-background ratio at 520 nm is approximately 400 : 1. The efficiency of energy transfer is 70%.

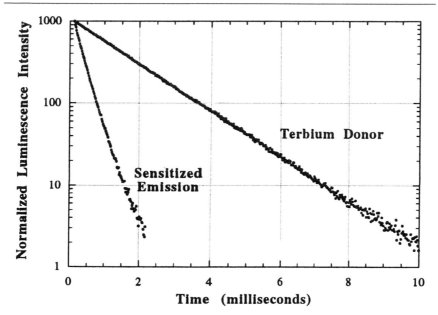

FIG. 12. Lifetime of 1.56 msec for unquenched terbium donor and 270 μsec for the fluorescein sensitized emission of donor–acceptor complex from Fig. 11. The sensitized emission lifetime indicates 83% quenching. Only completely labeled donor–acceptor complexes contribute to the sensitized emission signal. The somewhat higher energy transfer efficiency measured via lifetime (83%) versus intensities (70%, Fig. 11) is due to some donor-only species which contributes signal to intensity but not lifetime measurements.

yields (except as they affect R_0), and to incomplete labeling; only those species that are labeled with both donor and acceptor contribute to the signal.

We achieve a dark-background, sensitized emission experiment by eliminating the two usual sources of background. The donor fluorescence can be eliminated because the lanthanide luminescence is highly spiked and has regions of darkness (see Fig. 9). Terbium, for example, is silent at 520 nm. The fluorescence of the acceptor arising from direct excitation can be eliminated by using pulsed excitation and gating the detector off for a brief period, during which time the acceptor fluorescence dies away (lifetime typically a few nanoseconds) while the donor remains excited (lifetime of 1.5 msec in H_2O, 2.2 msec in D_2O for terbium). As a result, any fluorescence striking the detector in the donor dark region after a few microseconds is due only to energy transfer.

In Figures 11 and 12 we show an experiment where a terbium chelate (terbium diethylenetriaminepentaacetic acid coupled to carbostyril 124) transfers energy to fluorescein. The donor and acceptor are separated by an 8-mer DNA duplex oligomer (Fig. 10). The sensitized emission is measured with no background at around 520 nm (Fig. 11).

The 8-mer DNA oligomer is used to separate rigidly the donor and acceptor, and the complex is immersed in a viscous sucrose solution to eliminate intermolecular interactions. (We find that intermolecular interactions do not significantly contribute to energy transfer for this system even without the sucrose, presumably because of charge repulsion of the DNA oligomers and because the diffusional rate of energy transfer is much smaller than that energy transfer arising from the acceptor fixed to the same DNA oligomer as the donor.) The efficiency of energy transfer based on both donor quenching and the integrated sensitized emission area is 70%. Figure 12 shows measurement of the unquenched donor lifetime (1.5 msec) in the absence of acceptor and the sensitized emission lifetime in the donor–acceptor complex (lifetime 270 μsec). The lifetimes indicate a quenching of 83%. (The discrepancy between 70 and 83% appears to be due to a small fraction of terbiums which cannot transfer energy, presumably arising from some donor-only complex. This lessens the percent quenching as measured by donor intensity quenching but does not affect the percent quenching as measured by the sensitized emission lifetime.) Note that the sensitized emission lifetime measurement is completely insensitive to incomplete labeling and to absolute concentrations.

The ability to measure intensities and lifetimes of both donor and acceptor emission with high accuracy and excellent signal-to-background, coupled with the unusually large R_0s, makes luminescence resonance energy transfer a potentially powerful technique for measuring distances in biological systems.

Acknowledgments

P.R.S. acknowledges support from the National Institutes of Science, grant RO1GM41911, and from the Office of Energy Research, Office of Health and Environmental Research of the Department of Energy under contract DEACO3-76SF00098.

[14] Time-Resolved Fluorescence Spectroscopy

By ALFRED R. HOLZWARTH

Introduction

Fluorescence, that is, the phenomenon of light emission from an electronically excited state of a molecule, has found numerous and still rapidly growing applications for studies in the life sciences. The principal advantages of fluorescence techniques for probing molecular properties or environments in biological systems are evident from the high sensitivity, high selectivity, and nondestructive nature, and from the fact that the methods can also be applied easily to strongly scattering materials. Early applications were mainly concerned with probing the influence of molecular environments on spectra and intensities of the fluorescing states of either intrinsic fluorescing groups or species or, alternatively, artificial fluorescence labels introduced into the biological systems to be studied.

Since the rapid development of time-resolved fluorescence methods starting in the 1970s mainly in physical chemistry, increasing use has been made in life sciences of these technological and methodological advances. Although steady-state techniques in fluorescence will maintain important roles in the future, the main interest in the application of fluorescence techniques in biochemical and biological research has shifted to time-resolved techniques.[1-10] This is because the extension into the time dimension generally provides more detailed information while at the same time incorporating most of the features and possibilities of steady-state techniques. It is thus possible with time-resolved fluorescence techniques to probe the structural and dynamic features of macromolecules or supramolecular assemblies like proteins, membranes,[8] and nucleic acids,[11] to gain information on photobiological events,[2] and to acquire two-dimensional microscopic images of chemical distributions in cells,[12] to name just the most important applications. On the technical side the development of time-resolved fluorescence has benefited equally from dramatic technical advances in the field of pulsed light sources (mostly lasers), new types of

[1] A. R. Holzwarth, in "Chlorophylls" (H. Scheer, ed.), p. 1125. CRC Press, Boca Raton, Florida, 1991.

[2] A. R. Holzwarth, Q. Rev. Biophys. 22, 239 (1989).

[3] J. N. Demas, in "Molecular Luminescence Spectroscopy" (S. G. Schulman, ed.), p. 79. Wiley, New York, 1988.

[4] R. E. Dale, Stud. Biophys. 121, 5 (1987).

[5] N. E. Geacintov, Photochem. Photobiol. 45, 547 (1987).

[6] A. Van Hoek, K. Vos, and A. J. W. G. Visser, IEEE J. Quantum Electron. 23, 1812 (1987).

[7] J. M. Beechem and L. Brand, Annu. Rev. Biochem. 54, 43 (1985).

[8] L. Brand, J. R. Knutson, L. Davenport, J. M. Beechem, R. E. Dale, D. G. Walbridge, and A. A. Kowalczyk, in "Spectroscopy and the Dynamics of Molecular Biological Systems" (P. M. Bayley and R. E. Dale, eds.), p. 259. Academic Press, London, 1985.

[9] K. Kinosita, S. Kawato, and A. Ikegami, Adv. Biophys. 17, 147 (1984).

[10] F. Pellegrino and R. R. Alfano, in "Biological Events Probed by Ultrafast Laser Spectroscopy" (R. R. Alfano, ed.), p. 27. Academic Press, New York, 1982.

[11] M. Daniels, L. P. Hart, P. S. Ho, J.-P. Ballini, and P. Vigny, in "Time-Resolved Laser Spectroscopy in Biochemistry II" (J. R. Lakowicz, ed.), p. 304. SPIE, Bellingham, Washington, 1990.

[12] M. A. J. Rodgers and P. A. Firey, Photochem. Photobiol. 42, 613 (1985).

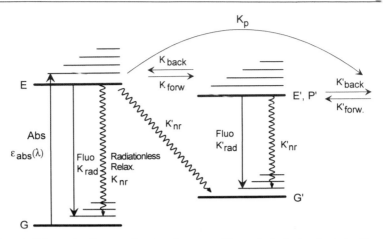

FIG. 1. Schematic diagram showing the states (E, excited electronic state; G, ground state; P, photochemical product) and the interconnecting processes with rate constants k_i relevant for understanding time-resolved fluorescence processes. Abs, Absorption; Fluo, fluorescence; k_{back} and k_{forw} are rate constants of excited state processes involving equilibria (energy transfer, proton transfer, etc.). The states indicated with a prime (E', P', G') are secondary states reached from the initially excited state E via processes k_i or k_i'.

detectors, and advances in methods of data analysis and computer technology.

It is the aim of this chapter to review the most widely used techniques for time-resolved fluorescence in the life sciences, to point out possible future developments, and to give examples of typical and important applications. The reader less familiar with fluorescence techniques is referred to the chapters on fluorescence energy transfer by P. R. Selvin (this volume [13]), fluorescence labeling by A. Waggoner ([15]), and polarization by H. van Amerongen and W. S. Struve ([11] in this volume). The information given in those chapters actually forms the basis for a better understanding of the usefulness of time-resolved techniques described here. Several general reviews and monographs on techniques have appeared, and the reader is referred to those for further details.[2,3,13-15]

Fluorescence techniques in the life sciences are concerned either with the excited state properties of a natural chromophore in a biological system and their dependence on the microenvironment or with the influence of the microenvironment on an artificial "fluorescence label" or probe molecule. A general schematic diagram illustrating the relevant processes and

[13] J. N. Demas, "Excited State Lifetime Measurements," p. 1. Academic Press, New York, 1983.

[14] J. R. Lakowicz, "Principles of Fluorescence Spectroscopy," p. 1. Plenum, New York, 1986.

[15] J. R. Lakowicz (ed.), "Topics in Fluorescence Spectroscopy." Plenum, New York, 1991.

states that can be probed by fluorescence techniques is given in Fig. 1. The principal parameters that can be measured are (i) fluorescence spectra, thus giving information on the energy or conformation of a chromophore and how it is influenced by environment; (ii) intensity of fluorescence, which is generally determined, although often in a nontransparent fashion, by the molecular rate constants k_i depopulating the excited state (see Fig. 1) and their dependence on the environment of the fluorescent chromophore; (iii) time dependence of the fluorescence intensity or spectrum after excitation with a brief light flash, which is determined directly by the rate constants of the depopulating processes (Fig. 1); or (iv) orientation and movement of a chromophore relative to the framework of a system of biological interest. At the focus of time-resolved spectroscopy are generally the latter two topics, but information on the others are often included; with carefully designed instrumentation information can be obtained at the same time on all of them. A collection of fluorescence life-time standards and common fluorescence lifetime labels has been published.[16,17]

Fluorescence Kinetics

In the simplest case of a single type of fluorescing chromophore being depopulated by various processes, the relationships determining the fluorescence decay kinetics and their relation to steady-state parameters are given by the following equations. The excited state decay is given by a first-order differential equation:

$$\frac{d[M(t)^*]}{dt} = -k_M[M(t)^*] \tag{1}$$

which on integration yields an exponential decay of the excited state of the form

$$[M(t)^*] = [M^*]_0 \exp^{-k_M t} \tag{2}$$

where $[M(t)^*]$ and $[M^*]_0$ represent the excited state concentrations at time t and $t = 0$, respectively. τ_M is the excited state decay time, which is related to the molecular rate constants (see Fig. 1) by

$$k_M = \frac{1}{\tau_M} = \Sigma(k_{rad} + k_{nr} + k_p + k_{ET} + k_Q[Q]) \tag{3}$$

where k_{rad} is the purely radiative rate constant, k_{nr} the rate constant of all radiationless processes like internal conversion or intersystem crossing, k_p the rate constant of photochemical processes, k_{ET} the rate constant for

[16] R. A. Lampert, L. A. Chewert, D. Phillips, D. V. O'Connor, A. J. Roberts, and S. R. Meech, Anal. Chem. 55, 68 (1983).
[17] R. F. Chen and C. H. Scott, Anal. Lett. 18A, 393 (1985).

energy transfer, and k_Q the rate constant for (static) quenching by a quencher Q with concentration [Q]. The time-resolved kinetics of a chromophore is related to the steady-state properties [spectrum $I_F(\lambda)$, quantum yield ϕ_F] via the following equations:

$$I_F(\lambda_{fl}) = \frac{\int_0^\infty I(t)_{fl}\, dt}{N_{abs}} = \frac{\int_0^\infty k_{rad}[M(t)^*]\, dt}{N_{abs}}$$

$$= \int_0^\infty \frac{k_{rad}[M^*]_0 \exp^{-k_M t}\, dt}{N_{abs}}$$

$$= \int_0^\infty k_{rad} \exp^{-k_M t}\, dt \tag{4}$$

$$\phi_F = \int I_F(\lambda_{fl})\, d\lambda = \frac{k_{rad}}{k_M} \tag{5}$$

Kinetic equations for more complex situations, involving multiple chromophores, multiple excited states, excited state equilibria, energy transfer, etc., may be found in Demas.[13] Each of the rate constants and thus the fluorescence lifetime, in principle, depends on the properties of the molecular environment of the fluorescent chromophore. Therefore, any particular environment or changes therein during a biochemically relevant process may be probed by its influence on the fluorescence decay kinetics. Thus, the often weak fluorescence not only reports on fluorescence processes of chromophores but equally so on nonfluorescent processes (energy transfer, photochemistry, quenching, conformational changes or rigidity, rotational relaxation, etc.).

Important Applications of Time-Resolved Fluorescence

A rough estimation of the number of papers that have appeared in the life sciences applying time-resolved fluorescence techniques gives a result that easily exceeds 10,000. A substantial number of more specialized reviews have appeared recently. Also several excellent monographs dealing with either techniques or applications are available.[14,15] Thus, we do not intend nor does available space allow a review of all these papers. Rather the choice of literature cited in the following will be dictated primarily by practical considerations concerning usefulness for the reader as general examples to illustrate the potential of the method(s) rather than indicating priority. A strong emphasis is given to literature that has appeared during the last decade.

Dynamics of Proteins and Membranes

A vast field for the application of time-resolved fluorescence methods is in the study of the internal and overall dynamics of proteins. Either intrinsic natural fluorescence probes or artifical fluorescence labels may be used. The natural probes are the aromatic amino acids Trp, Tyr, and Phe.[18-21] In particular Trp is strongly fluorescent and has been used for numerous time-resolved studies. Both the fluorescence decay itself of the aromatic amino acids as well as the time-resolved anisotropy decays contain information on the internal dynamics of proteins, the environment of the fluorescent amino acids, changes in conformation, etc. A review describing these studies in detail has been published by Beechem and Brand.[7] In contrast to common expectation, the fluorescence decay of proteins that contain only a single Trp is not a single exponential function. Rather it reflects the dynamic interconversion between a multitude of conformational substates of proteins.[22] The nonsingle-exponential decay of Trp fluorescence in proteins has been widely used to study conformational heterogeneity and dynamics of proteins,[7,8,23,24] and such studies are now being combined with site-directed mutagenesis.[25]

The segmental or global motion of a protein is often important for function. So far relatively little is known about such processes. It has been shown that the newly developed fluorescence instrumentation that allows the simultaneous lifetime measurement at several wavelengths (see Fig. 7) in a very short time (milliseconds) is a powerful method to study such motions, including protein folding and unfolding dynamics.[26] The method exploits the distance and orientation dependence of fluorescence energy transfer of probes attached to a protein. In applying powerful global analysis methods three different conformational distribution parameters can be obtained simultaneously from a single measurement: (i) average interprobe distance, (ii) width and shape of the interprobe distance distribution, and (iii) rate of change of the interprobe distance. From these

[18] H. Vogel, Q. Rev. Biophys. **25**, 433 (1992).

[19] E. Bucci and R. F. Steiner, Biophys. Chem. **30**, 199 (1988).

[20] A. P. Demchenko, Essays Biochem. **22**, 120 (1986).

[21] W. C. Galley, in "Biochemical Fluorescence Concepts" (R. F. Chen and H. Edelhoch, eds.), p. 409. Dekker, New York, 1976.

[22] H. Frauenfelder, F. Parak, and R. D. Young, Annu. Rev. Biophys. Biophys. Chem. **17**, 451 (1988).

[23] G. Careri, P. Fasella, and E. Gratton, Annu. Rev. Biophys. Bioeng. **8**, 69 (1979).

[24] J. R. Alcala, E. Gratton, and F. G. Prendergast, Biophys. J. **51**, 925 (1987).

[25] C. A. Royer, J. A. Gardner, J. M. Beechem, J.-C. Brochon, and K. S. Matthews, Biophys. J. **58**, 363 (1990).

[26] J. M. Beechem and E. Haas, Biophys. J. **55**, 1225 (1989).

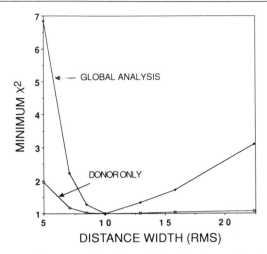

FIG. 2. Accuracy of determination of the width of an interprobe (donor–acceptor pair) distance distribution in a labeled protein from time-resolved fluorescence kinetics. The optimal value of the width is 10 Å. (From Beechem and Haas.[26])

parameters detailed information on the conformational dynamics can be deduced. An example of analysis of interprobe distance width is shown in Fig. 2. The advantage of simultaneous global analysis of donor and acceptor fluorescence kinetics versus analysis of the donor fluorescence alone is demonstrated.

Quenching studies of protein fluorescence provide answers regarding the accessibility of certain internal or external groups to quencher molecules.[27] Another application concerns the study of associative behavior and properties of proteins and membranes.[8] The rationale is that the fluorescence transition is polarized and this polarization can be exploited in time-resolved analysis and interpreted in terms of the rotation or tumbling motion which in turn is determined by the viscosity and structure of the environment of the fluorescing group. In particular, anisotropy decay studies have yielded a great deal of information on the mobility of natural and artificial membranes and/or the dynamics of proteins as well as small molecules in membranes.[8,28] For such studies fluorescence lifetime labels that can be attached to proteins or that dissolve in membranes have

[27] M. R. Eftink and C. A. Ghiron, *Anal. Biochem.* **114,** 199 (1981).
[28] L.B.-A. Johansson and G. Lindblom, *Q. Rev. Biophys.* **13,** 63 (1980).

often been used.[17,29-38] The observed phenomena in such studies may be rotational relaxation,[9] energy transfer between fixed or moving groups, changes in polarity of certain regions of membranes or proteins, drug binding to proteins,[39] etc. An example of protein–protein as well as protein–membrane interactions is given by a study of the inactivation of the Ca^{2+}-ATPase activity in sarcoplasmic reticulum membranes by time-resolved anisotropy of phosphorescence and fluorescence. The increase in time-resolved fluorescence anisotropy of the membrane probe diphenylhexatriene (DPH) caused by a decreased mobility of the membrane on melittin binding to the membrane protein Ca^{2+}-ATPase is shown in Fig. 3.

Photoreceptors

Another widely studied field where fluorescence techniques have been used successfully involves the properties of proteins that contain fluorescent groups other than the aromatic amino acids. There are numerous examples of such systems in biochemistry and biology. A very prominent group includes photosynthetic antenna pigments, namely, chlorophylls, phycobiliproteins, and carotenoids. A vast literature exists on the use of time-resolved spectroscopy to study energy transfer processes in antenna systems,[1,40,41] as well as charge separation and recombination in reaction centers.[1,41,42] In fact time-resolved fluorescence is probably the most widely used tool in the study of the primary processes of photosynthesis. Its advantages lie in the high sensitivity and noninvasive nature of the

[29] J. Voss, W. Birmachu, D. M. Hussey, and D. D. Thomas, *Biochemistry* **30**, 7498 (1991).

[30] J. Storch, N. M. Bass, and A. M. Kleinfeld, *J. Biol. Chem.* **264**, 8708 (1989).

[31] T. P. Burghardt and K. Ajtai, in "Time-Resolved Laser Spectroscopy in Biochemistry" (J. R. Lakowicz, ed.), p. 121. SPIE, Bellingham, Washington, 1988.

[32] P. Tompa, J. Bar, and J. Batke, in "Time-Resolved Laser Spectroscopy in Biochemistry" (J. R. Lakowicz, ed.), p. 147. SPIE, Bellingham, Washington, 1988.

[33] R. Fiorini, M. Valentino, S. Wang, M. Glaser, and E. Gratton, *Biochemistry* **26**, 3864 (1987).

[34] F. T. Greenaway and J. W. Ledbetter, *Biophys. Chem.* **28**, 265 (1987).

[35] D. L. Sackett and J. Wolff, *Anal. Biochem.* **167**, 228 (1987).

[36] J. Szollosi, S. Damjanovich, S. A. Mulhern, and L. Tron, *Prog. Biophys. Mol. Biol.* **49**, 65 (1987).

[37] P. B. O'hara and K. Gorski, *Fed. Proc.* **45**, 1601 (1986).

[38] B. E. Peerce and E. M. Wright, *Proc. Natl. Acad. Sci. U.S.A.* **83**, 8092 (1986).

[39] Y. Kato, T. Horie, M. Hayashi, and S. Awazu, *Biochem. Pharmacol.* **34**, 2555 (1985).

[40] A. R. Holzwarth, in "The Light Reactions" (J. Barber, ed.), p. 95. Elsevier, Amsterdam, 1987.

[41] K. K. Karukstis and K. Sauer, *J. Cell. Biochem.* **23**, 131 (1983).

[42] M. G. Müller, K. Griebenow, and A. R. Holzwarth, *Chem. Phys. Lett.* **199**, 465 (1992).

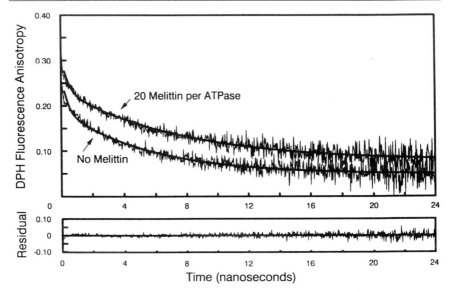

FIG. 3. Effect of melittin binding to Ca^{2+}-ATPase in sarcoplasmatic reticulum on the time-resolved anisotropy decay of the membrane probe diphenylhexatriene. (From Voss et al.[29])

technique. Thus isolated pigment–protein complexes as well as intact algae, chloroplasts, or intact leaves are suitable study objects.[40,43–45] Both structural and functional information can be obtained from such studies. Figure 4 shows the time-resolved equilibration of the excitons in the inhomogeneously broadened chlorophyll antenna of photosystem I from a cyanobacterium. The importance of sensitive time resolution techniques in the study of electron transfer processes in photosynthetic reaction centers of bacteria and higher plants has been demonstrated.[46–49] Often fluorescence techniques give important complementary information to other techniques like flash photolysis. Another example of chromo-

[43] I. Mukerji and K. Sauer, *Biochim. Biophys. Acta* **1142**, 311 (1993).

[44] H. Dau and K. Sauer, *Biochim. Biophys. Acta* **1102**, 91 (1992).

[45] I. Mukerji and K. Sauer, *in* "Photosynthesis" (W. R. Briggs, ed.), p. 105. Alan R. Liss, New York, 1989.

[46] M. G. Müller, K. Griebenow, and A. R. Holzwarth, *Biochim. Biophys. Acta* **1098**, 1 (1991).

[47] M. Du, X. L. Xie, Y. W. Jia, L. Mets, and G. R. Fleming, *Chem. Phys. Lett.* **201**, 535 (1993).

[48] Z. Wang, R. M. Pearlstein, Y. Jia, G. R. Fleming, and J. R. Norris, *Chem. Phys.* **176**, 421 (1993).

[49] P. Hamm, K. A. Gray, D. Oesterhelt, R. Feick, H. Scheer, and W. Zinth, *Biochim. Biophys. Acta* **1142**, 99 (1993).

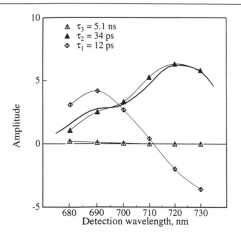

FIG. 4. Decay-associated spectra showing the exciton equilibration (12 psec component with positive and negative amplitudes) in the spectrally inhomogeneous antenna of photosystem I from the cyanobacterium *Synechococcus* sp. [From A. R. Holzwarth, G. Schatz, H. Brock, and E. Bittersmann, *Biophys. J.* **64,** 1813 (1993).]

phore–protein complexes that have been studied extensively by fluorescence is phytochrome, the receptor for photomorphogenic responses in higher plants.[50]

Time-Resolved Fluoroimmunoassay

An ever-increasing group of applications for time-resolved fluorescence methods in clinical chemistry has been developed which in many cases can replace the widely used but now problematic radioimmunoassays.[51,52] The method is based on the labeling of specific antibodies or antigens with highly fluorescent labels. Various kinds of fluorescence effects altering either the fluorescence lifetime or the fluorescence anisotropy can be used. The fluorescence assays are highly sensitive, specific, and often less complicated than other methods. The theory and applications of fluorescence immunoassays have been described extensively by Morrison.[52] Various kinds of labels involving, for example, rare earth chelators[53] or phycobiliproteins[52,54] and various other groups have been

[50] K. Schaffner, S. E. Braslavsky, and A. R. Holzwarth, *in* "Advances in Photochemistry" (D. H. Volman, G. S. Hammond, and K. Gollnick, eds.), p. 229. Wiley, New York (1990).

[51] I. A. Hemmila, *ISI Atlas Sci.: Immunol.* **1,** 231 (1988).

[52] L. E. Morrison, *Anal. Biochem.* **174,** 101 (1988).

[53] E. P. Diamandis, R. C. Morton, E. Reichstein, and M. J. Khosravi, *Anal. Chem.* **61,** 48 (1989).

[54] M. N. Kronick, *J. Immunol. Methods* **92,** 1 (1986).

Concentric
Distribution

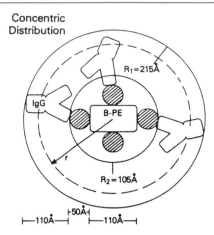

FIG. 5. Spatial model for antibody–antigen complexes involving B-phycoerythrin (B-PE) as energy acceptor and antibodies to human immunoglobulin G loaded with the energy donor pyrene butyrate. The shaded circles represent fluorescent antibody fragments covalently labeled with B-PE. (From Morrison.[52])

tested successfully and allow highly sensitive assays down to the level of tens of attomoles. An example employing phycobiliproteins as energy accepting labels and demonstrating the advantage of measuring simultaneously the donor fluorescence as well as the acceptor fluorescence is shown in Fig. 5. The antibody–antigen interaction brings donor and acceptor into close distance and thus makes the energy transfer highly efficient and specific. The range of applications spans hormone determination, drug level determination, antibody detection to test for certain diseases, etc. This field is rapidly developing, driven by the availability of more user-friendly fluorescence lifetime instrumentation that can be handled in clinical laboratories.

Nucleic Acids

Time-resolved fluorescence has been applied successfully to the study of the structure and dynamics of nucleic acids via the intrinsic fluorescence of the bases or via interchelating fluorophores.[11,55–60] Other important

[55] T. M. Nordlund, P. Wu, S. Andersson, L. Nilsson, R. Rigler, A. Graslund, L. W. McLaughlin, and B. Gildea, in "Time-Resolved Laser Spectroscopy in Biochemistry II" (J. R. Lakowicz, ed.), p. 344. SPIE, Bellingham, Washington, 1990.
[56] T. M. Nordlund, S. Andersson, L. Nilsson, R. Rigler, A. Graslund, and L. W. McLaughlin, *Biochemistry* **28**, 9095 (1989).
[57] S. Georghiou, T. M. Nordlund, and A. M. Saim, *Photochem. Photobiol.* **41**, 209 (1985).
[58] G. G. Kneale and R. W. van Resandt, *Eur. J. Biochem.* **149**, 85 (1985).
[59] M. L. Lamos and D. H. Turner, *Biochemistry* **24**, 2819 (1985).
[60] D. P. Millar, R. J. Robbins, and A. H. Zewail, *J. Chem. Phys.* **76**, 2080 (1982).

applications relate, for example, to nucleic acid–protein interactions and recognition.[61]

Two-Dimensional Imaging

A method that combines microscopic resolution and high-resolution lifetime measurement has been developed.[12] This powerful technique allows detailed study of the location and environment of fluorescent labels in cells and other microscopic studies. It is expected that two-dimensional imaging will find a number of interesting applications in the near future.

Techniques

The phenomena discussed above can be studied using various techniques, and not all methods are equally suitable in a particular case. Two principally different kinds of methods for measuring fluorescence lifetimes exist,[3] namely, pulse methods and modulation or phase-shift methods. Phase-shift methods, despite the fact that they have been known for a longer time, have not found widespread use during the last decade. However, important technical advances have been made in phase-shift methods which in fact have inspired many researchers to apply them more frequently.[62] Nevertheless, pulse methods are still the most widely used today, in particular for high time resolution. If carried out properly both types of methods must and will give the same result. Details of the measuring problem will determine which method is more appropriate in a particular case.

In pulse fluorometry the sample is excited with a brief flash of light whose duration [full width at half-maximum (fwhm)] ideally should be shorter than the time scale of the fluorescence decay. Because the longest fluorescence lifetimes of molecules are in the range of hundreds of nanoseconds and the fastest phenomena of interest in biology are on the subpicosecond time scale, this may require rather short light pulses that can be obtained only by special, usually mode-locked, laser systems. The fluorescence response of the system on a light pulse can then be measured using various techniques. The most widely used are single-photon counting, streak cameras, and fluorescence up-conversion. In contrast to pulse methods, the phase-shift technique usually employs a continuous light source whose intensity is modulated.

[61] D. P. Millar, D. J. Allen, and S. J. Benkovic, in "Time-Resolved Laser Spectroscopy in Biochemistry" (J. R. Lakowicz, ed.), p. 392. SPIE, Bellingham, Washington, 1990.

[62] J. R. Lakowicz, G. Laczko, H. Cherek, E. Gratton, and M. Limkeman, *Biophys. J.* **46**, 463 (1984).

The various techniques for measuring fluorescence decays are described below in more detail. All fluorescence techniques, to varying degrees, have the advantage that they are zero-background techniques and thus are usually more sensitive compared to other methods, like absorption.

Pulse Techniques and Data Analysis

A typical pulse experiment involves measurement of the fluorescence decay by one of the techniques described in the following. Quite typically the measured decay curve with either technique does not represent directly the time course of the fluorescent sample on a δ-shaped excitation pulse (impulse response in information theory). Rather the measured decay curve is distorted by the finite duration of the excitation pulse and by the final response time of the detectors employed. This problem can be overcome by measuring under exactly the same optical conditions the fluorescence decay of the unknown sample and subsequently the optical response of a purely scattering sample (nonfluorescent) that responds promptly, that is, without delay to the exciting light. Dilute suspensions of microscopic particles (silica, powdered milk, nondairy creamer, etc.) have been used as scatterers. An important requirement is that scatterers be nonfluorescent in the wavelength region of interest. The signal measured from such scatterers is usually called the "prompt response" of the apparatus or the "apparatus function." An exact mathematical relationship, the so-called convolution integral,[63] describes the relationship of the δ response of the fluorescing sample, the time course of the exciting light, and the time response of the apparatus. This relationship is usually employed to perform the reverse operation mathematically, that is, a so-called deconvolution during the data analysis, in order to extract the "true" undistorted fluorescence decay of the sample.[64,65]

Although being mathematically exact, this procedure has one principal difficulty, namely, the fact that the apparatus function in most cases cannot be measured at the same wavelength as the fluorescence decay of the sample. Rather it is normally measured at the excitation wavelength. For deconvolution in a strict sense the apparatus response must be the same for both sample and scatterer. This requires a wavelength independence of the detector response which is not always easy to achieve, in particular with the single-photon counting technique. Various methods have been proposed to correct for any wavelength dependence in the detector re-

[63] A. Grinvald and I. Z. Steinberg, *Anal. Biochem.* **59**, 583 (1974).
[64] U. P. Wild, A. R. Holzwarth, and H. P. Good, *Rev. Sci. Instrum.* **48**, 1621 (1977).
[65] W. R. Ware, L. J. Doemeny, and T. L. Nemzek, *J. Phys. Chem.* **77**, 2038 (1973).

sponse (the same problem occurs with the phase-shift technique).[66] Although such correction techniques can in principle correct for the wavelength response, they have not found widespread use. The reason is that the procedures are time-consuming, tedious, and not always reliable. Most use reference compounds that are assumed to have a fluorescence response that is independent of the wavelength. This assumption then shifts the problem to the purity of the reference sample and/or to the exact knowledge of the photophysics of the reference compound. The latter is complex, particularly in the very short time range below tens of picoseconds. By far the best solution thus is to design the instrument and the measurement in such a way that any wavelength dependence of the detectors either can be ignored or can be described by a linear term only, that is, a simple time shift. This time shift can usually be corrected for mathematically by the introduction of a shift parameter in the analysis.[66]

Although being two independent steps in principle, deconvolution (yielding the δ response of the fluorescing sample) and extraction of the kinetic parameters (lifetimes and amplitudes) of a pulse fluorescence experiment are usually performed in a combined deconvolution/data analysis step (the same holds for the phase-shift technique). Careful analysis of the problem reveals that the true δ response of the fluorescent sample is hardly ever required or used, and thus there is no need to attempt to calculate it from the data.[64,67] Rather the fluorescence decay is almost always described in terms of some physical or kinetic model, often given by a sum of a few exponentials. Then the task of the data analysis consists of determining either the kinetic parameters (lifetimes and amplitudes) and/or the actual physical parameters of interest (rate constants, activation energies, etc.). The former analysis, often called lifetime analysis, can be done rather independently of any actual physical or kinetic model by using numerical curve-fitting algorithms.[63–65,68] Limitations in the number of extractable kinetic components are determined mainly by the accuracy and signal/noise ratio in the data and the relative differences in lifetimes and amplitudes of the components.[69–75]

[66] M. Van den Zegel, N. Boens, D. Daems, and F. C. De Schryver, *Chem. Phys.* **101**, 311 (1986).

[67] A. E. W. Knight and B. K. Selinger, *Spectrochim. Acta* **27A**, 1223 (1971).

[68] A. E. W. Knight and B. K. Selinger, *Aust. J. Chem.* **26**, 1 (1973).

[69] B. K. Selinger and A. L. Hinde, *in* "Time-Resolved Fluorescence Spectroscopy in Biochemistry and Biology" (R. B. Cundall and R. E. Dale, eds.), p. 129. Plenum, New York, 1983.

[70] B. K. Selinger, C. M. Harris, and A. J. Kallir, *in* "Time-Resolved Fluorescence Spectroscopy in Biochemistry and Biology" (R. B. Cundall and R. E. Dale, eds.), p. 143. Plenum, New York, 1983.

The reliability of these procedures and the uncertainty in extracted parameters can and should be checked regularly by analyzing simulated data of comparable signal/noise ratio and complexity as the experimental data.[71]

In lifetime analysis the assignment and interpretation of the extracted parameters within a physical model of the problem are done after the analysis. Quite a different philosophy lies behind another procedure where the physical parameters of interest are directly extracted from the data. This procedure has been called "target analysis" or "compartment analysis."[72,75–78] The rationale behind this procedure is the idea that one is not really interested merely in a mathematically correct description of the fluorescence decay parameters. Rather the actual interest is in the physics of the problem and the parameters that determine it, that is, one should determine the physically relevant parameters directly from the data, rather than indirectly from some purely mathematical parameters (like lifetimes and amplitudes). This approach then requires a physical model of the problem which can be given in a mathematically defined form. Thus simple data fitting, as is the case in lifetime analysis procedures, is replaced by physical model testing. Among other advantages, this approach has the benefit of providing more information on the problem and of allowing one to extract more complex physical relationships from the data. For this reason the much more powerful model testing approach has become increasingly popular in fluorescence kinetics analysis and should replace mere data fitting wherever possible. There is not enough space here to discuss this problem in detail, and the reader is referred to more specialized in-depth treatments of fluorescence decay analysis.[75] Because the pulse response and the frequency response of a fluorescing system are directly

[71] B. K. Selinger and C. M. Harris, in "Time-Resolved Fluorescence Spectroscopy in Biochemistry and Biology" (R. B. Cundall and R. E. Dale, eds.), p. 155. Plenum, New York, 1983.

[72] J. M. Beechem, M. Ameloot, and L. Brand, *Anal. Instrum.* **14,** 379 (1985).

[73] J. M. Beechem and L. Brand, *Photochem. Photobiol.* **44,** 323 (1986).

[74] L. Brand, J. R. Knutson, L. Davenport, J. M. Beechem, R. E. Dale, D. G. Walbridge, and A. A. Kowalczyk, in "Spectroscopy and the Dynamics of Molecular Biological Systems" (P. M. Bayley and R. E. Dale, eds.), p. 259. Academic Press, London, 1985.

[75] J. M. Beechem, E. Gratton, M. Ameloot, J. R. Knutson, and L. Brand, in "Topics in Fluorescence Spectroscopy" (J. R. Lakowicz, ed.), p. 241. Plenum, New York, 1991.

[76] R. Weidner and S. Georghiou, in "Time-Resolved Laser Spectroscopy in Biochemistry II" (J. R. Lakowicz, ed.), p. 717. SPIE, Bellingham, Washington, 1990.

[77] F. J. Knorr and J. M. Harris, *Anal. Chem.* **53,** 272 (1981).

[78] J. R. Knutson, J. M. Beechem, and L. Brand, *Chem. Phys. Lett.* **102,** 501 (1983).

related via the Fourier transform, all these considerations apply equally to phase-shift methods.

Global Analysis Methods. One of the most important breakthroughs in time-resolved spectroscopy during the last decade has been the introduction and rigorous theoretical treatment of the so-called "global" or "simultaneous" data analysis methods. In this procedure fluorescence decay data taken at various conditions of emission wavelength, excitation wavelength, temperature, concentration, etc., are analyzed in a combined procedure. The underlying idea is that certain parameters in a fluorescence decay remain constant or change in a defined manner with other independent parameters. Then a combined analysis where several fluorescence decays are linked together becomes possible. In this way the amount of information that can be extracted from a data set is increased dramatically.[75] Global analysis procedures have found numerous applications and should be used whenever possible. It is not possible to cite all of the papers here that actually applied this technique, but examples are cited throughout this chapter. The first to apply this technique were Knorr and Harris.[77]

Single-Photon Timing. The term single-photon timing (SPT, also called time-correlated single-photon counting) relates to the fact that individual photons are being detected. This already points to the fact that it is a highly sensitive technique. The SPT technique belongs to the category of pulse techniques. A schematic diagram of an SPT apparatus is shown in Fig. 6. A weak, pulsed light source excites the sample at high repetition rate (kilo- to megahertz). The sample emits fluorescence photons which are detected by a sensitive detector, usually a photomultiplier, that is capable of responding to a single photon with a defined and measurable electric output pulse. A reference detector (photodiode) starts a clock (time-to-amplitude converter, TAC). The TAC after being started actually charges a capacitor with constant current. The output pulse caused by the first fluorescence photon is used to stop the clock (i.e., charging of the capacitor is stopped). Discriminators are used to shape the electric pulses from the detectors and to present a standard electric signal to the inputs of the TAC start and stop channels. This keeps the timing jitter low and helps to discard unwanted radio frequency (rf) interference signals or other noise that would distort the measurement. The accumulated voltage in the TAC is proportional to the time between start and stop pulses, and it is converted by an analog-to-digital converter (ADC) to a number. This number is then interpreted as a channel number in a multichannel analyzer (MCA) or computer memory, and the content of the corresponding channel is incremented by one.

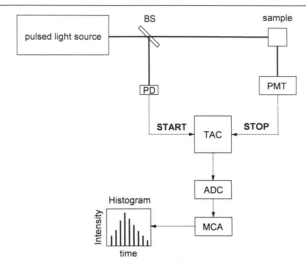

FIG. 6. Diagram of an SPT apparatus. BS, Beam splitter; PD, photodiode; PMT, photo-multiplier tube; TAC, time–amplitude converter; ADC, analog–digital converter; MCA, multichannel analyzer.

This type of measurement is also called a "delayed coincidence method" and was originally invented in solid-state physics to measure the lifetimes of radioactive nuclei and later was used with optical detection of scintillating materials.[79–83] In the SPT experiment the delayed coincidence between an exciting light pulse and the occurrence of the first fluorescence photon is being measured.[68] Of course, the measurement of a single photon does not contain any information on the time course of the fluorescence. However, if such an experiment is repeated many times at high repetition rate (typically 10^6 to 10^8 times depending on the accuracy required and the complexity of the kinetics), the computer memory will contain a histogram of the probability of the occurrence of a fluorescence photon as a function of delay time after excitation. It can be shown that the shape of the histogram matches exactly the time course of the fluorescing sample. Thus the SPT technique is a stochastic and also a digital technique. This is in fact one of the most important reasons for its

[79] P. K. Ludwig, Adv. Radiat. Chem. 3, 1 (1972).
[80] H. Bader, H. R. Gordon, and O. B. Brown, Rev. Sci. Instrum. 43, 1407 (1972).
[81] R. Z. Bachrach, Rev. Sci. Instrum. 43, 734 (1972).
[82] B. E. A. Saleh and B. K. Selinger, Appl. Opt. 16, 1408 (1977).
[83] P. Wahl, Chem. Phys. 22, 245 (1977).

high dynamic range and sensitivity. The SPT technique has been described extensively by O'Connor and Phillips.[84]

Light Sources for Single-Photon Timing and Other Pulsed Methods. Because the TAC can handle only one stop pulse after a start pulse, the probability for occurrence of a second photon must be kept negligible. This requires that the excitation intensity be kept at such a low level that, on average, only one fluorescence photon is being detected per hundred excitation pulses (i.e., the duty cycle is low). The low duty cycle in general is by no means a disadvantage in SPT, in particular with weakly fluorescent samples. However, it demands that a high repetition rate light source be used. A further requirement for the light source is that its pulse shape must remain constant during the time of the measurement of the fluorescence decay and the apparatus response function. Traditionally free-running or gated flash lamps, using hydrogen or nitrogen as gas filling, providing pulse widths in the range of a few nanoseconds have been used as excitation sources.[84] Modern flash lamps which can be obtained commercially have good stability and pulse widths down to about 1 nsec. The most pronounced disadvantage of flash lamps is the low intensity, particularly if selective excitation at particular wavelengths is required.

High repetition rate mode-locked lasers have been increasingly used as excitation sources. These are generally synchronously pumped dye lasers or solid-state titanium–sapphire lasers. The pump sources for the lasers are either mode-locked ion lasers (argon or krypton) or mode-locked neodymium–YAG (yttrium-aluminum-garnet) lasers. In addition to having all the advantages of lasers (high intensity, easy focusing, wavelength selectivity, etc.), these light sources allow repetition rates of up to 100 MHz and pulse widths to the subpicosecond or even the femtosecond range. In fact the repetition rates are often too high for many measuring problems and have to be reduced by additional devices like cavity dumpers or pulse pickers to more acceptable levels in the range of a few megahertz.[84] Solid-state diode lasers allowing rather narrow pulse widths in the picosecond time range have also been applied. Their principal advantages are compactness and relatively low cost.[85,86] Disadvantages are the often low intensity (in particular, if the second harmonic of the optical output frequency is being used) and the very limited choice of excitation wavelengths. Development of solid-state devices is very rapid, however, and one may expect that diode lasers will become much more versatile in the

[84] D. V. O'Connor and D. Phillips, "Time-Correlated Single Photon Counting," p. 1. Academic Press, London, 1984.

[85] Y. Silberberg and P. W. Smith, *IEEE J. Quantum Electron.* **22,** 759 (1986).

[86] D. L. Farrens and P.-S. Song, *Photochem. Photobiol.* **54,** 313 (1991).

coming years. A more exotic and less generally accessible light source, which gives the advantages of a large wavelength tunability and relatively intense short pulses, is a synchrotron storage ring.[87]

Using laser excitation sources in combination with fast detectors, the time resolution of the SPT technique has been extended into the range of a few picoseconds.[88] Such a high time resolution requires, however, the optimization of the apparatus at all levels, namely, high stability of the light sources, optimal electronics free from drift and nonlinearity problems, and fast detectors with minimal time jitter. Besides the pulse width of the light source, the limiting factors for the achievable time resolution are time jitters in the electronics and the detectors, which have to be kept at a minimal level. On the level of the electronics this can be achieved by a careful design of the signal paths, free from radio frequency pickup and ground loops, and proper choice of optimal electronic components, particularly the discriminators, the TAC, and the ADC. Unfortunately at present no integrated design for these components required for SPT is available. Such integrated electronic devices are being developed in several laboratories and may be available soon.

Detectors. Two detectors are generally necessary for an SPT setup (see Fig. 6). The first detects the occurrence of the excitation pulse, whereas the second is used for detection of the fluorescence photons. With flashlamps as excitation sources, photomultipliers (PM) have been generally employed as both start and stop detectors owing to their high sensitivity. For high time resolution with laser excitation sources, the output pulse of a photomultiplier is too broad, however, to allow precise timing of the excitation pulse. With the relatively high intensities available from lasers, fast silicon diodes, allowing output pulse widths in the 100 psec range and large signals in the range of tens to hundreds of millivolts, have thus replaced the traditional start photomultipliers.[84] This allows a precise timing of the start signal.

On the fluorescence detection side, because individual photons must be detected, there is hardly a replacement for a photomultiplier-type detector with its high gain (of the order of $>10^6$). For a time resolution down to several tens of picoseconds, relatively simple photomultipliers have been used as detectors. One requirement is that the PM should have a narrow and well-defined electric response to a single photon and a low dark current. Traditionally, relatively fast front-end PM tubes have been

[87] I. H. Munro, D. Shaw, G. R. Jones, and M. M. Martin, *Anal. Instrum.* **14,** 465 (1985).
[88] E. W. Small, *in* "Topics in Fluorescence Spectroscopy" (J. R. Lakowicz, ed.), p. 97. Plenum, New York, 1991.

employed.[64,68,89] It has also been found that some of the more commonly available and inexpensive side-on PM tubes are equally suitable as fast SPT detectors.[90-92] For particular cases solid-state detectors such as the silicon avalanche detector (SPAD) have been used. Their advantage is in the near-infrared wavelength range in particular, where photocathodes are much less sensitive. The SPAD detector allows high time resolution in the picosecond time range, nearly approaching that of multichannel plate (MCP) tubes.[93-95] One disadvantage is the very small usable area J (<0.1 mm^2) which for many purposes makes this detector very inefficient, since the fluorescence cannot be focused easily onto such a small area. However, in time-resolved microscopy it should find good application.

To understand which type of detector is best for the SPT technique we have to recall again the basic measuring technique: The detector does not actually have to follow the time course of the fluorescence decay, which indeed is not possible for any of these detectors on a picosecond time scale. Rather the task is to determine the time of occurrence of a photon event or the occurrence of the exciting light pulse as precisely as possible. This time can be defined for practical purposes as the time at which the maximum in the output signal of the detector(s) appears. Given appropriate precautions and good timing discriminators, this time can be determined with an uncertainty much narrower than the output pulse widths of the detectors. The major limiting factor in the detectors is the transit time spread, that is, the temporal jitter in the occurrence of the maximal output signal relative to the occurrence of the optical input.[84] It has been realized that in conventional PM tubes the transit time spread originates mainly in the different paths that the photoelectrons travel from the photocathode to the first anode. By optimizing this path dramatic reductions in the transit time spread of conventional PM tubes could be achieved. Thus the shortest transit time spreads with a side-on PM are about 100 psec.[90] This requires a high voltage between the cathode and the first anode and the limitation of the illuminated cathode area to a

[89] R. Schuyler, I. Isenberg, and R. D. Dyson, *Photochem. Photobiol.* **15,** 395 (1972).

[90] E. C. Meister, U. P. Wild, P. Klein-Boelting, and A. R. Holzwarth, *Rev. Sci. Instrum.* **59,** 499 (1988).

[91] A. R. Holzwarth, J. Wendler, and W. Wehrmeyer, *Photochem. Photobiol.* **36,** 479 (1982).

[92] W. Haehnel, A. R. Holzwarth, and J. Wendler, *Photochem. Photobiol.* **37,** 435 (1983).

[93] T. Louis, G. H. Schatz, P. Klein-Boelting, A. R. Holzwarth, S. Cova, and G. Ripamonti, *Rev. Sci. Instrum.* **59,** 1148 (1988).

[94] M. Ghioni, S. Cova, A. Lacaita, and G. Ripamonti, *Electron. Lett.* **24,** 1476 (1988).

[95] S. Cova, G. Ripamonti, and A. Lacaita, *Nucl. Instrum. Methods Phys. Res. Sect. A* **253,** 482 (1987).

small spot size of about 1 mm. SPT apparatuses equipped with such an inexpensive and very sensitive detector produce an apparatus pulse width of 120 psec (fwhm) which in turn allowed the measurement of fluorescence decay times down to a few tens of picoseconds.

The most pronounced reduction in transit time spread has been achieved in the modern type of PM tubes, the so-called channel plate detectors (MCP). Each channel of about 6–12 μm diameter actually represents a tiny electron multiplier device of its own. The combination of high voltages between the photocathode and the channel plate and the strict control of the photoelectron as well as secondary electron paths in the channel plate PM tubes reduced the transit time spread to a value of the order of 20 psec.[96] With carefully designed instruments these detectors at present allow the highest time resolution in SPT measurements, namely, an apparatus response of better than 30 psec and a time resolution of 2–3 psec.[42,97,98] It is important to note, however, that at this time resolution it is by no means the properties of the excitation source and the detectors alone that determine the time resolution. Rather the whole instrument, including optics, electronics, and detectors, must be optimized and maintained at very high stability during the measurement. The drawback of channel plate PM tubes lies in the relatively small dynamic range, as compared to conventional PM tubes. In general this is not of much importance in SPT experiments, however, because the photon detection rates are typically a few kilohertz. A further disadvantage is the high cost and sometimes short lifetime of MCP detectors. Given optimal conditions, the shortest lifetimes that can be measured by SPT are smaller by a factor of about 10 than the apparatus width (presently about 30 psec).

What are the major advantages and disadvantages of the SPT technique? Owing to its inherently digital nature the SPT technique has the largest dynamic range (at least 4 orders of magnitude both on the intensity scale and on the time scale in a single measurement) of all the fluorescence kinetics techniques available. This unsurpassed dynamic range is often necessary to study complex biological systems. Because it is a counting experiment, the variance of each data point is exactly known. It follows Poisson statistics, and the knowledge of the variance is very important for a proper weighting in the data analysis procedures. The time resolution is very good, but it will hardly be possible to extend it below the present

[96] H. Kume, T. Taguchi, K. Nakatsugawa, K. Ozawa, S. Suzuki, R. Samuel, Y. Nishimura, and I. Yamazaki, "Time-Resolved Laser Spectroscopy in Biochemistry III," (J. R. Lakowicz, ed.), p. 440. SPIE, Bellingham, Washington, 1992.

[97] K. Koyama, H. Kume, and D. Fatlowitz, "Hamamatsu Technical Info ET-03," p. 1, Hamamatsu, Bridgewater, New Jersey, 1988.

[98] A. R. Holzwarth, M. G. Müller, G. Gatzen, M. Hucke, and K. Griebenow, J. Lumin. 60&61, 497 (1994).

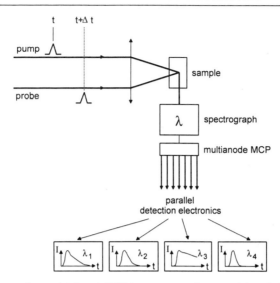

FIG. 7. Diagram of a multiplexed SPT instrument using a multianode MCP detector. The signals from several channels (representing different wavelengths) can be measured in parallel. In the excitation beam the possibility of a pump–probe arrangement using two laser pulses for measuring the fluorescence kinetics of short-lived intermediates is indicated (see text).

limits of about 2–3 psec. Furthermore, the technique can work with very small excitation pulses, thus avoiding any nonlinearities in complex samples. As far as measuring time is concerned, it is reasonably fast (a typical experiment takes in the range of tens of minutes), but there are faster techniques available. There are few major disadvantages related to the SPT technique except perhaps for cost, which depends substantially on the required time resolution. It is by far the most widely used technique today in both physical chemistry and life sciences to measure fluorescence decays.

Developments which substantially increase the information content of an SPT experiment have involved multiplexing of the detector channel. Independent PM tubes and detection electronic lines may be used to measure simultaneously the fluorescence at different wavelengths or different polarization[99]; alternatively, special PM tubes with split photocathodes, like the multianode MCP detector, may be used instead. Available MCP tubes allow up to 10 independent wavelengths to be measured simultaneously in a single experiment (Fig. 7). Such instrumentation has been

[99] D. J. S. Birch, A. S. Holmes, J. R. Gilchrist, R. E. Imhof, S. M. Alawi, and B. Nadolski, *J. Phys. E: Sci. Instrum.* **20,** 471 (1987).

developed and used to measure protein unfolding kinetics on a fast time scale.[100] Many more applications of this kind can be envisaged in the life sciences. Another application is the detection of the fluorescence decay of short-lived intermediates in the picosecond to microsecond time scale. In this technique two exciting laser pulses, a pump and a probe pulse, are employed, with some delay, Δt, between the pulses. The pump pulse is chopped, and the fluorescence of both pulses or the difference of the two signals may be measured directly[100a] (see Fig. 7).

Streak Camera Method. A conventional detector, photomultiplier or photodiode, is not capable of following the time course of a light signal changing on the picosecond time scale. This prompted researchers in the 1970s to develop a technique which translates the problems of time resolution into a spatial resolution on a phosphor. The streak camera may be considered to be a combination of an oscilloscope and a video camera. Streak cameras have been developed for single shot lasers or for operation with high repetition rate lasers (synchroscan streak camera). We discuss here mainly high repetition rate applications, although there exists no essential difference between them.[101,102]

The principle is to accelerate and deflect the photoelectrons ejected from the photocathode at a pair of deflection plates whose voltage is rapidly modulated in synchrony with the repetition frequency of the light pulses. The deflected photoelectrons hit a phosphor at the other end of the camera. A certain position of the phosphor corresponds to a certain time delay relative to the excitation pulse. The intensity of the phosphor luminescence is read by an image intensifier, a combination of an MCP intensifier and an array detector like a diode array or a charge-coupled device (CCD) camera (Fig. 8). Because the phosphor has two dimensions and is used only in one dimension for the basic function of the streak camera, the other dimension is still available. This second dimension lends itself to a simultaneous multichannel wavelength detection. Thus, if a spectrograph is placed after the sample and a two-dimensional readout of the camera is used, the fluorescence kinetics can be measured simultaneously over a large wavelength range. Together with the relatively short data acquisition times, this represents a considerable advantage of the streak camera technique.

[100] J. M. Beechem, E. James, and L. Brand, *in* "Time-Resolved Laser Spectroscopy in Biochemistry II" (J. R. Lakowicz, ed.), p. 686. SPIE, Bellingham, Washington, 1990.

[100a] B. Wagner, Doctoral Thesis, Univ. Düsseldorf, Germany, 1994.

[101] M. R. Baggs, R. T. Eagles, A. E. Hughes, W. Margulis, C. C. Phillips, W. Sibbett, and W. E. Sleat, *Proc. Soc. Photo.-Opt. Instrum. Eng.* **491,** 212 (1985).

[102] Y. Tsuchiya, K. Kinoshita, M. Koishi, A. Takeshima, and Y. Inagaki, *Proc. Soc. Photo.-Opt. Instrum. Eng.* **491,** 224 (1985).

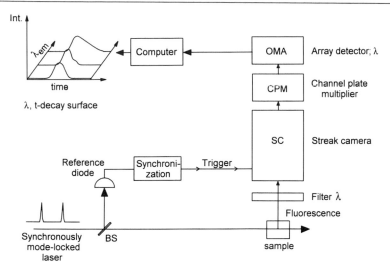

FIG. 8. Schematic diagram showing a lifetime apparatus using a synchronously pumped laser as excitation source and a synchroscan streak camera as detector.

Usually a synchronously mode-locked picosecond/femtosecond laser system (tens of megahertz repetition rate) is used for excitation. Modern synchroscan streak cameras have an apparatus response of about 1 psec, more than an order of magnitude better than the SPT technique.[103] Thus if a high time resolution (slightly below 1 psec) and a moderately high sensitivity combined with multiwavelength detection are required, the streak camera is clearly the method of choice. The substantially worse dynamic range, as compared to SPT, makes the streak camera less suitable, however, for systems that show very complex kinetics. The data analysis and determination of the apparatus function are performed in a very similar fashion, usually by iterative deconvolution, as described for the SPT method. Because the variance in the data points are not known as precisely as in SPT, the weighting factors are not determined so well, which has the disadvantage that less accurate information can be recovered from the data.

Fluorescence Up-Conversion Technique. A completely different approach from the ones described above is used in the up-conversion technique.[104,105] The development of this technique originates in the desire

[103] T. M. Nordlund, *in* "Time-Resolved Laser Spectroscopy in Biochemistry" (J. R. Lakowicz, ed.), p. 35. SPIE, Bellingham, Washington, 1988.

[104] M. A. Kahlow, W. Jarzeba, T. P. DuBruil, and P. F. Barbara, *Rev. Sci. Instrum.* **59,** 1098 (1988).

[105] J. Shah, T. C. Damen, B. Deveaud, and D. Block, *Appl. Phys. Lett.* **50,** 1307 (1987).

to measure still faster kinetics than is possible generally with the more conventional techniques. In principle, all the other techniques are limited by the response of detectors or electronics. In contrast, light pulses can be made much shorter, with the shortest ones today being in the range of a few femtoseconds. This led to the idea that a short light pulse (exciting pulse) could be used to measure another short light pulse (fluorescence decay).

The up-conversion is actually an optical gating technique. Its electronic analog is the well-known boxcar integrator. The sample is excited by a very brief light flash (usually in the picosecond to femtosecond range). The emitted fluorescence is mixed in a nonlinear crystal with part of the excitation pulse with a certain delay time, Δt, which is variable. If both the fluorescence and the gating laser pulse are present in the crystal at the same time, the optical frequencies of the fluorescence and the laser are summed, provided suitable phase matching conditions exist, and a signal is created at the sum frequency ω_3 (Fig. 9). This sum frequency is selected by a filter and sent to a detector. The detector response may be slow because it measures only the intensity of the sum signal but not its time course. If the gating laser pulse is not present the fluorescence alone cannot create a sum signal. Thus the gating pulse opens an optical gate to measuring the fluorescence intensity at a certain instant after excitation. By slowly varying the delay time, Δt, the whole time course of the fluorescence can be scanned. The signal intensity I of the sum frequency is proportional to both the intensity of the gating pulse and the intensity of the

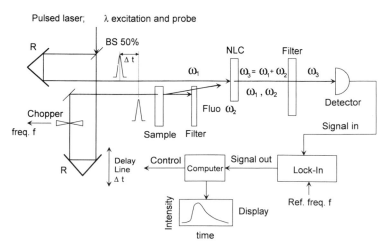

FIG. 9. Diagram of a fluorescence up-conversion apparatus. NLC, Nonlinear crystal for sum-frequency generation; BS, beam splitter; R, retroreflector.

fluorescence, that is, $I_{sum} = $ constant $\cdot (I_{gate} * I_{fluo})$. Thus a linear response is obtained.

The important point here is that the width of the gate, and thus the time resolution of the method, is limited only by the width of the exciting light and not by the response of the detector or the electronics. Thus time resolution into the femtosecond range is possible. Efficient nonlinear crystals have become available that improve the sensitivity of this technique. It is possible at present to use this technique with mode-locked and cavity-dumped lasers that deliver output energies of about 10 nJ/ pulse. More convenient, however, are laser pulse energies in the range of several hundred nanojoules per pulse which can be obtained only from amplified laser systems.[106,107] For various applications the required pulse energies for exciting the samples may be too high, thus causing unwanted nonlinear effects. Still the sensitivity of the method is substantially lower than that of others. It offers by far the best possible time resolution, however.

In principle the analysis of up-conversion data is performed in a similar fashion as in the SPT technique. With the advent of very short laser pulses (femtoseconds), the up-conversion technique is becoming more and more popular for the measurement of very short fluorescence decays.[48,49] It is not, however, a technique that is easily adapted as a general method in the life sciences.

Phase-Shift Method

In contrast to pulse methods described above, the phase-shift technique usually employs a continuous light source whose intensity is modulated by various means at some frequency f. The fluorescence response of the system is then also modulated at that frequency, albeit with some phase delay ϕ and a reduced modulation depth m, as compared to the exciting light.[108-110] From either of these quantities the fluorescence lifetime can be extracted. For a single-exponential decay the relationship between lifetime τ, the modulation frequency f, phase shift ϕ, and the modulation depth m are given by $\tan(\phi) = f\tau$ and $m = (1 + f^2\tau^2)^{-1/2}$.

Either arc lamps or continuous wave (cw) lasers have been used as

[106] H. Mahr and M. D. Hirsch, *Opt. Commun.* **13**, 96 (1975).

[107] T. M. Jedju, M. W. Roberson, and L. Rothberg, *Appl. Opt.* **31**, 2684 (1992).

[108] R. D. Spencer and G. Weber, *Ann. N.Y. Acad. Sci.* **158**, 361 (1969).

[109] O. D. Dmetrievsky, V. L. Ermolaev, and A. N. Terenin, *Dokl. Biol. Sci. (Engl. Transl.)* **114**, 468 (1957).

[110] H. Merkelo, J. H. Hammond, S. R. Hartman, and Z. I. Derzko, *J. Lumin.* **1/2**, 502 (1970).

excitation sources. As modulators Pockels cells, acoustooptic modulators, or Kerr cells have been applied.[108,111-114] Early instruments were limited to only one modulation frequency f. This allowed only a single lifetime to be determined, and it is not possible under such conditions to determine whether the decay is actually a single exponential. A precise analysis is possible only if a large range of modulation frequencies, which should be continuously tunable, is available. Analysis of the frequency response of the modulation depth and the phase shift thus allows one to analyze multiexponential decays or other more complex fluorescence functions.[115] With electrooptic modulators it is easy to reach modulation frequencies in the megahertz range; however, laser excitation sources should be used, because the acceptance angle of these modulators is only rather narrow which would lead to a dramatic loss in intensity from a conventional light source. However, megahertz frequencies allow lifetimes only in the range of several hundred picoseconds to nanoseconds to be measured. Conventional photomultiplier tubes are usually sufficient as detectors for this time regime. For measuring lifetimes in the range of a few picoseconds, gigahertz modulation frequencies and very fast detectors are required. These high modulation frequencies are difficult to obtain in a direct fashion. A trick is being used whereby the high harmonic content of a short laser flash (picosecond laser pulse) which contains many frequencies is actually used.[111] Also a very fast detector, usually a MCP tube, is required. To relax the requirements for a very fast detector somewhat, a special technique, the cross-correlation phase-shift technique, has been developed.[111,116] In this technique both the light source and the photomultiplier detector are modulated at slightly different frequencies. The actual signal is then detected at the low difference frequency.

In the picosecond time range the instrumental requirements for a phase-shift apparatus increase substantially as compared to the relatively simple and inexpensive instrumentation sufficient for nanosecond work.[108] The instrumentation for picosecond phase-shift apparatuses is at least as complex and sophisticated as for SPT as far as light sources, electronics, and detectors are concerned. It has been shown in fact that in principle the same parameters limit the possible time resolution in an SPT experiment

[111] K. W. Berndt, I. Gryczynski, and J. R. Lakowicz, *Anal. Biochem.* **192**, 131 (1991).
[112] H.-P. Haar, U. K. A. Klein, F. W. Hafner, and M. Hauser, *Chem. Phys. Lett.* **49**, 563 (1977).
[113] U. K. A. Klein and H.-P. Haar, *Chem. Phys. Lett.* **58**, 531 (1978).
[114] H. Gugger and G. Calzaferri, *J. Photochem.* **13**, 21 (1980).
[115] J. R. Lakowicz, I. Gryczynski, H. Cherek, G. Laczko, and N. Joshi, *Trends Anal. Chem.* **5**, 257 (1986).
[116] E. Gratton and M. Limkeman, *Biophys. J.* **44**, 315 (1983).

and in a phase-shift experiment.[114] As for sensitivity, the phase-shift technique does not quite reach the SPT technique for very weak fluorescence signals, but it comes close.

One advantage of the phase-shift technique is its ability to enhance selectively a particular lifetime component in a mixture of fluorophores directly, that is, without any need for data analysis. This provides a means for rather fast and direct data acquisition.[115] A requirement is that the lifetimes be separated substantially. This is not so easily possible with the SPT technique, although some gating techniques have been used that achieve a similar goal.[84] An exact separation of components is possible both in the SPT and in the phase-shift technique only after data analysis; thus, the two techniques do not differ much in that respect.

The laboratories of Lakowicz and Gratton in particular have developed phase-shift techniques and associated data analysis methods[62,75,117–121] to considerable power; in many respects its capabilities are now quite similar to those of the SPT technique. Nevertheless, the SPT technique is more widely applied, and research groups accustomed to working with pulsed lasers generally still prefer the SPT technique over the phase-shift technique. Apart from the fact that pulse techniques are more intuitive, this situation is also related to the fact that some instrumentation available in laboratories that use other pulsed spectroscopies can be also used for SPT, but only to a lesser extent for the phase-shift technique. At present commercial instrumentation is available for both kinds of techniques, and it is no longer necessary, as was the case until a couple of years ago, that researchers buy and assemble independent parts of the apparatus. Rather, complete and integrated instruments, including data analysis software, are now available commercially for both techniques. It thus becomes a practical choice to decide which instrumentation is more suitable for the particular task at hand.

Final Remarks

This overview on applications of time-resolved fluorescence techniques can only give a few examples of the many uses in biochemistry, biology, and clinical chemistry. It is not difficult to predict that applications

[117] J. M. Beechem, J. R. Knutson, J. B. A. Ross, B. W. Turner, and L. Brand, *Biochemistry* **22,** 6054 (1983).

[118] J. R. Lakowicz, H. Cherek, I. Gryczynski, N. Joshi, and M. L. Johnson, *Biophys. Chem.* **28,** 35 (1987).

[119] J. R. Alcala, E. Gratton, and F. G. Prendergast, *Biophys. J.* **51,** 587 (1987).

[120] J. R. Lakowicz, B. P. Maliwal, and E. Gratton, *Anal. Instrum.* **14,** 193 (1985).

[121] J. R. Lakowicz, G. Laczko, I. Gryczynski, H. Szmacinski, and W. Wiczk, *J. Photochem. Photobiol. B* **2,** 295 (1988).

of time-resolved fluorescence methods will grow continuously in the coming years.

Acknowledgments

I thank Mrs. A. Keil for support in compiling the literature for this review. Work from this laboratory has been supported in part by the Deutsche Forschungsgemeinschaft (Sonderforschungsbereich 189, Heinrich-Heine-Universität Düsseldorf and Max-Planck-Institut für Strahlenchemie, Mülheim a.d. Ruhr, Germany).

[15] Covalent Labeling of Proteins and Nucleic Acids with Fluorophores

By ALAN WAGGONER

Introduction

Fluorescence is a sensitive technique for detecting biological materials. Barak and Webb demonstrated that as few as 30 fluorescent dye molecules bound to low density lipoprotein molecules can be detected on cell surfaces by microscopy using a sensitive video camera.[1] With even more sophisticated methods others have moved toward detection of single fluorophore molecules in a laser-excited flow stream.[2] Because this sensitivity can combine with ease of use and the potential for multiparameter analysis, fluorescence has become widely used in biology and medicine.[3-7] This chapter focuses on methods of choosing fluorescent labels and using them to tag biological macromolecules. Several review articles can be consulted

[1] L. S. Barak and W. W. Webb, *J. Cell Biol.* **90**, 595 (1981).

[2] N. J. Dovichi, J. C. Martin, J. H. Jett, M. Trkula, and R. A. Keller, *Anal. Chem.* **56**, 348 (1984); K. Peck, L. Stryer, A. N. Glazer, and R. A. Mathies, **86**, 4087 (1989); C. W. Wilkerson, P. M. Goodwin, W. P. Ambrose, J. C. Martin, and R. A. Keller, *Appl. Phys. Lett.* **62**, 2030 (1993).

[3] Y. Wang and D. L. Taylor, "Fluorescence Microscopy of Living Cells," Parts A and B. Academic Press, San Diego, 1989.

[4] T. C. Brelje, M. W. Wessendorf, and R. L. Sorenson, *Methods Cell Biol.* **38**, 98 (1993).

[5] W. T. Mason (ed.), "Fluorescent and Luminescent Probes for Biological Activity." Academic Press, New York, 1993.

[6] S. G. Ballard and D. C. Ward, *J. Histochem. Cytochem.* **12**, 1755 (1993).

[7] R. P. Haugland, "Molecular Probes: Handbook of Fluorescent Probes and Research Chemicals." Molecular Probes Inc., Eugene, Oregon 97402-9144, 1992–1994.

for additional details[7-10] (see also D. M. Jameson and W. H. Sawyer, this volume [12], and P. R. Selvin, this volume [13]).

Considerations for Labeling Biomolecules with Fluorophores

Brightness of Fluorophore

A number of factors influence the brightness of a fluorophore. The larger the extinction coefficient at the wavelength of excitation (ε_{exc}) the higher the light gathering power of the flurophore. Extinction coefficients at the wavelength of maximum light absorption (ε_{max}) of fluorescent labels can range from a few thousand to around 250,000 L/mol cm (Table I). Once light has been absorbed there is a certain probability that the fluorophore will emit a photon versus returning to the ground state through a radiationless transition that simply heats the microenvironment surrounding the fluorophore. The quantum yield of the fluorophore (ϕ) is a measure of this probability. The overall brightness of the fluorophore is proportional to ε_{exc} times ϕ.

The extinction coefficients and quantum yields of many fluorophores depend on the microenvironment. For example, pH can affect either ε_{exc} or ϕ. Fluorescein is fluorescent at pH 8 but is almost nonfluorescent at pH 6. Also there are fluorophores that are sensitive to solvent polarity. Dansyl and 7-nitrobenz-2-oxa-1,3-diazole (NBD) are much more fluorescent in hydrocarbon environments than in water. Other fluorophores are sensitive to the microviscosity. The cyanine label Cy3 is significantly more fluorescent in viscous solvents and when bound to proteins and DNA compared to when it is in water.[11,12] Cyanines and rhodamines, on the other hand, are not sensitive to polarity or pH changes between 4 and 11.

Contact between fluorescent tags on heavily labeled macromolecules generally leads to fluorescence quenching and often precipitation of the labeled antibody.[12,13] Therefore, it is important not to overlabel the biomolecule. Water-soluble dyes can be used to obtain a high density of

[8] R. P. Haugland, in "Excited States of Biopolymers" (R. F. Steiner, ed.), p. 29. Plenum, New York, 1983.

[9] M. Brinkley, Bioconjugate Chem. 3, 2 (1992).

[10] S. dePetris, Methods Membr. Biol. 9, 1 (1978).

[11] K. Luby-Phelps, K. A. Guss, L. A. Ernst, R. B. Mujumdar, S. R. Mujumdar, and A. S. Waggoner, J. Cell Biol. 107(6, Part 3), a59 (1988).

[12] A. S. Waggoner and L. A. Ernst, in "Clinical Flow Cytometry" (K. D. Bauer, R. E. Duque, and T. V. Shankey, eds.), Part I. Williams & Wilkins, Baltimore, Maryland, 1992; R. Chen, Arch. Biochem. Biophys. 133, 263 (1969); M. Goldman, "Fluorescent Antibody Methods." Academic Press, New York, 1968.

[13] W. Arnold and H. V. Mayersbach, J. Histochem. Cytochem. 20, 975 (1972).

TABLE I
PROPERTIES OF SELECTED FLUORESCENT LABELING REAGENTS

Fluorophore[a]	Reactive group[a]	Measurement conditions[a]	Absorption maximum[b,c]	Extinction maximum[d]	Emission maximum[b,c]	Quantum yield	Refs.[e]
Fluorescein	SCN, OSu, IAA, MAL	pH 7, PBS	490	67	520	0.71	7,8,9,MP,RO,BDS,EK,P,S
Tetramethylrhodamine	SCN, OSu, IAA, MAL	pH 7, PBS	554	85	573	0.28	7,8,9,MP,RO,EK,P,S
X-Rhodamine	SCN, OSu, IAA, MAL	pH 7, PBS	582	79	601	0.26	RO
Texas Red	SO_2Cl	pH 7, PBS	596	85	620	0.51	MP,22,P
Lissamine rhodamine	sulfonyl chloride	pH 7, PBS	570	73	590	Med	7,9,34,EK,MP,P,S
Cy3	OSu	pH 7, PBS	554	130	568	0.14 (f)	14,BDS
Cy5	OSu	pH 7, PBS	652	200	672	0.18 (f)	14,BDS
Cy7.18	OSu	pH 7, PBS	755	200	778	0.02 (f)	14,BDS
BODIPY series	OSu, IAA	Methanol	500–581	80	510–591		7,9,MP
Cascade Blue	OSu	Water	378, 399	26	423	0.54	7,9,MP
NBD-amine[f]	Cl	Ethanol/methanol	478	24.6	520–550	0.36/0.21	7,9,MP
NBD-thiol	Cl	pH 7.5	425	12.1	531	0.002	7,9,MP
Aminomethylcoumarin	OSu, IAA, MAL	Water	387		470		7,9,MP
R-Phycoerythrin	—SS—	pH 7, PBS	480–565	1960	578	0.68	10,18,P,MP
Allophycocyanin	—SS—	pH 7, PBS	650	700	660	0.68	10,18,P,MP
PerCP	—SS—	pH 7, PBS	488		680		18,19,BD

[a] SCN, Isothiocyanate; OSu, succinimidyl active ester; IAA, iodoacetamide; MAL, maleimide; SO_2Cl, sulfonyl chloride; Cl, active halogen group; —SS—, disulfide exchange reaction involving SPDP [N-succinimidyl 3-(2-pyridyldithio)propionate)]; NBD, 7-nitrobenz-2-oxa-1,3-diazole; DAPI, 4',6-diamidino-2-phenylindole; DCDHB, dicyanodihydroxybenzene; PBS, phosphate-buffered saline.

[b] Most spectroscopic values are for a conjugate of the reactive compound and a corresponding target molecule, e.g., succinimidyl ester reacted with aminomethane, or antibody.

[c] Measured in nanometers.

[d] Multiply value listed by 1000 to get liters/mol cm.

[e] EK, Eastman Kodak (Rochester, NY); MP, Molecular Probes (Eugene, OR); RO, Research Organics (Cleveland, OH); BDS, Biological Detection Systems (Pittsburgh, PA); P, Pierce (Rockford, IL); S, Sigma (St. Louis, MO); BD, Becton Dickinson Immunocytometry Systems (San Jose, CA).

[f] Value for NBD-ethanolamine in methanol which has an absorption maximum at 470 nm and an emission maximum at 550 nm [L.S. Barak and R. R. Yocum, Anal. Biochem. 110, 31 (1981)].

label with minimal quenching. Hydrophobic dyes, including a number of rhodamines, exhibit greater fluorescence quenching because they tend to associate with one another during the labeling process. Cyanine dyes with sulfonated ring structures are among the more water-soluble fluorochromes.[14] With four to six Cy3 or Cy5 labels per immunoglobulin (IgG) molecule, brightly fluorescent antibodies with excellent stability in water can be produced.

The phycobiliproteins are hard to beat as extremely bright antibody labels for fluorescence analysis of cell surface antigens by flow cytometry.[15,16] These bacterial photosynthetic macromolecules each have up to 34 individual bilin fluorophores wrapped within the polypeptide structure. An example is R-phycoerythrin (R-PE), which has an extinction coefficient of 2×10^6 L/mol cm and a quantum yield of 0.68. Usually there is room for only one PE molecule per antibody, because the size of an R-PE label is 1.5 times that of an IgG antibody. The large size of the complex reduces the kinetics of binding to cell surface antigens, and some intracellular markers are inaccessible to the R-PE-labeled antibody. For intracellular measurements, lower molecular mass fluorophores (<1 kDa) are usually preferred.

Wavelength of Excitation and Emission

Table I shows a selection of fluorophores with absorption and emission maxima throughout the near-UV to near-IR region of the spectrum. If the detection instrument has a defined excitation light source, filter set, and detector, fluorophore selection may be limited but relatively straightforward. For example, instruments utilizing laser excitation usually offer only a few excitation wavelengths. Other instruments, however, provide greater flexibility, in which case the entire fluorophore–light source–filter set–detector combination must be optimized. This "systems approach" becomes more difficult if multiple fluorescent labels are to be used simultaneously for correlating a number of antigenic or genetic parameters within the sample. Optical detection system design has been addressed elsewhere.[17] For multiparameter analysis with instruments, such as epifluorescence microscopes, a unique set of filters is needed for each fluorophore. A spacing of emission peaks at 50 nm can provide for clean discrimination of each fluorophore signal, provided that the fluorophore has rela-

[14] R. B. Mujumdar, L. A. Ernst, S. R. Mujumdar, C. J. Lewis, and A. S. Waggoner, *Bioconjugate Chem.* **4**, 105 (1993).
[15] V. Oi, A. N. Glazer, and L. Stryer, *J. Cell Biol.* **93**, 981 (1982).
[16] M. N. Kronick and P. D. Grossman, *Clin. Chem.* **29**, 1582 (1983).
[17] W. Galbraith, L. A. Ernst, D. L. Taylor, and A. S. Waggoner, *Proc. S.P.I.E.* **1063**, 74 (1989).

tively narrow excitation and emission peaks, as is the case for fluoresceins, rhodamines, BODIPYs, and cyanines.

The use of single laser excitation for multiparameter fluorescence measurements is more challenging. What is needed is a series of fluorophores that all excite at the same laser wavelength, typically the 488 nm argon ion laser line, but which emit at different distinguishable wavelengths. One series of such labels includes fluorescein, R-PE, and either Cy5-labeled R-PE or Texas Red-labeled R-PE tandem conjugates or another phycobiliprotein, PerCP.[18,19] Fluorescein emits at 525 nm, R-PE at 575 nm, and Cy5-PE at 670 nm, and each can be detected with minimum spillover from the others.[18]

Generally, fluorescent labels that are excited and emit in the red region of the spectrum give better signal-to-noise ratios if the sample contains cellular materials that autofluoresce. Most cells fluoresce in the UV (NADH) and green (flavins and flavoproteins), but phagocytic cells and plant cells can fluoresce at longer wavelengths as well.[20,21]

Type of Labeling Reaction

The most common labeling reactions utilize primary amino groups on the target biomolecule. Lysine groups on polypeptides and alkylamino groups on derivatized nucleic acids are the usual candidates. Succinimidyl active ester, isothiocyanate, or sulfonyl chloride groups on the fluorophore provide efficient amino group labeling (Fig. 1).[7,9,10,14] Active ester reactions are relatively specific for primary amino groups, easily controlled, and occur over a period of a few minutes. The amino-labeling reaction competes favorably with hydrolysis of the active ester, but the rates of both reactions depend on pH. Higher pH favors the —NH_2 form of lysine groups, which reacts well with active esters and isothiocyanates, over the unreactive —NH_3^+ form. Most labeling reactions with active esters such as fluorescein-OSu and Cy3-OSu or with fluorescein isothiocyanate are carried out between pH 8.5 and 9.5 over a period of 15 min to several hours (active esters generally react faster than isothiocyanates). If this pH range is used, it is important to be sure that the protein conformation is not irreversibly altered. If there is danger of denaturation, active ester reactions can be performed at neutral pH but over a longer period of time.[14] Sulfonyl chlorides (Texas Red and Lissamine rhodamine sulfonyl

[18] A. S. Waggoner, L. A. Ernst, C-H. Chen, and D. J. Rechtenwald, *Ann. N.Y. Acad. Sci.* **677**, 185 (1993).

[19] D. J. Rechtenwald, U.S. Patent No. 4,876,190 (1989).

[20] J. Aubin, *J. Histochem. Cytochem.* **27**, 36 (1979).

[21] R. Benson, R. A. Meyer, M. Zaruba, and G. McKhann, *J. Histochem. Cytochem.* **27**, 44 (1979).

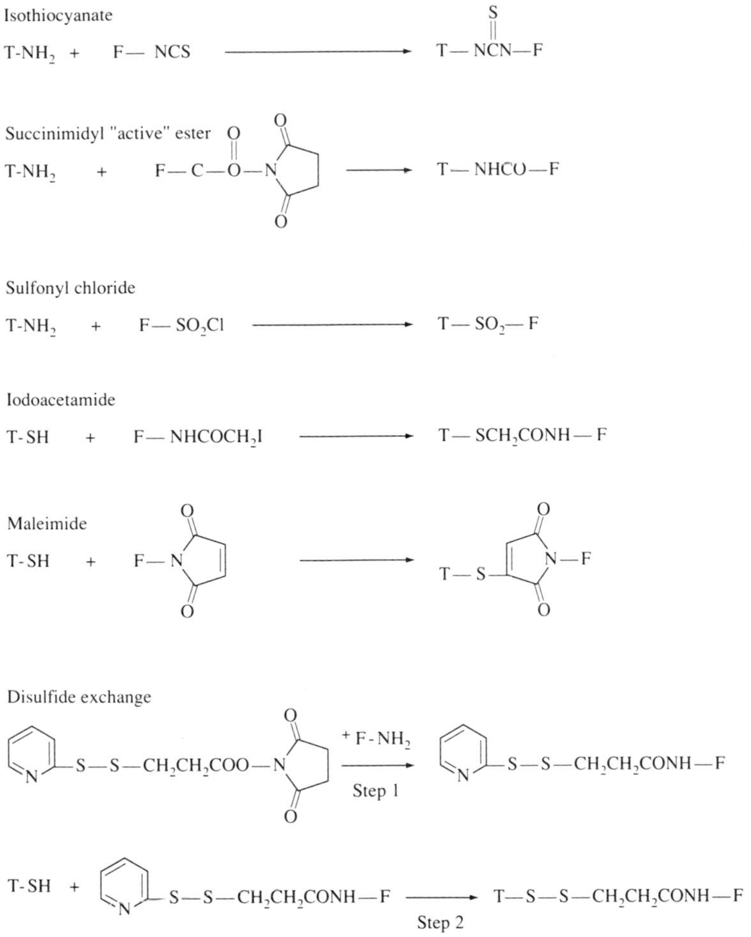

Isothiocyanate

$$T\text{-}NH_2 \ + \ F\text{—}NCS \xrightarrow{\hspace{3cm}} T\text{—}N\overset{\displaystyle S}{\overset{\|}{C}}N\text{—}F$$

Succinimidyl "active" ester

$$T\text{-}NH_2 \ + \ F\text{—}\overset{\displaystyle O}{\overset{\|}{C}}\text{—}O\text{—}N \xrightarrow{\hspace{2cm}} T\text{—}NHCO\text{—}F$$

Sulfonyl chloride

$$T\text{-}NH_2 \ + \ F\text{—}SO_2Cl \xrightarrow{\hspace{3cm}} T\text{—}SO_2\text{—}F$$

Iodoacetamide

$$T\text{-}SH \ + \ F\text{—}NHCOCH_2I \xrightarrow{\hspace{3cm}} T\text{—}SCH_2CONH\text{—}F$$

Maleimide

$$T\text{-}SH \ + \ F\text{—}N \xrightarrow{\hspace{3cm}} T\text{—}S\text{—}N\text{—}F$$

Disulfide exchange

$$\underset{N}{\bigcirc}\text{—}S\text{—}S\text{—}CH_2CH_2COO\text{—}N \xrightarrow[\text{Step 1}]{+ F\text{-}NH_2} \underset{N}{\bigcirc}\text{—}S\text{—}S\text{—}CH_2CH_2CONH\text{—}F$$

$$T\text{-}SH \ + \ \underset{N}{\bigcirc}\text{—}S\text{—}S\text{—}CH_2CH_2CONH\text{—}F \xrightarrow[\text{Step 2}]{\hspace{2cm}} T\text{—}S\text{—}S\text{—}CH_2CH_2CONH\text{—}F$$

Fig. 1. Common reactions of fluorescent labeling reagents. T represents the target in the labeling reaction and can be a protein, drug, or modified nucleic acid. F represents the fluorescent label.

chloride) are very reactive with amino groups.[22] Dichlorotriazine derivatives have also been used as reactive groups.[23] The bonds formed using active esters and sulfonyl chlorides are very stable. It is absolutely essential that there be no primary amine-containing components in the buffer solution other than those on the target biomolecule.

[22] J. A. Titus, R. P. Haugland, S. O. Sharrow, and D. M. Segal, J. Immunol. Methods **50**, 193 (1982).
[23] D. Blakeslee and M. G. Baines, J. Immunol. Methods **13**, 305 (1976).

Hydroxyl groups of biomolecules are less reactive with isothiocyanates and active esters, but reactions often can be completed under appropriate conditions, which include a longer reaction time. Sulfonyl chlorides and isocyanates are more reactive with hydroxyl groups.

Sulfhydryl groups, such as cysteine residues on proteins and thiols on modified nuclei acids, can be labeled with iodoacetamides, certain other organic halides, maleimides, and through disulfide exchange reactions[7,9] (Fig. 1). Table I shows a number of fluorescent sulfhydryl labeling reagents. These reagents react with thiols in the physiological pH range (pH 6.5–8.0) and much less efficiently with amino groups that are protonated in this range.

The disulfide exchange reaction provides an especially useful method for efficiently linking two different protein molecules together that have amino groups but no sulfhydryl groups.[15] This is the preferred method for labeling proteins with phycobiliprotein fluorescent labels. Bifunctional amine-reactive linkers are less useful for this purpose because they would produce homodimers and precipitates of both species as well as the desired heterodimer. Instead, by the disulfide exchange method one species, say A, is labeled with amino-reactive thiolane that on binding to the amino group on A undergoes ring opening with a sulfhydryl group on the end of a linker (Fig. 1). The other species, B, is labeled with the disulfide-containing reagent N-succinimidyl 3-(2-pyridyldithio)propionate (SPDP), which also reacts with amino groups (Fig. 1). A disulfide exchange reaction completes the conjugation of A to B.

Labeling Regent Solubility. Generally, proteins and nucleotides are labeled in aqueous solutions. Reactive fluoresceins, sulfocyanines (Cy3, Cy5), and a few other labeling reagents are water soluble and are easy to use in conjugation reactions.[14] Rhodamines, coumarins, and many other nonpolar fluorophores are less water soluble. Several methods are used to solve this problem. If the target is soluble and stable in a dimethyl sulfoxide (DMSO)–water or dimethylformamide (DMF)–water mixture, this solvent system might be considered. Some investigators have dissolved the fluorescent label in DMSO and added this mixture to the aqueous target with rapid mixing. Alternatively, it is sometimes possible to adhere the fluorophore to Celite and suspend the mixture with the target biomolecule in aqueous buffer.[24]

Purification of Labeled Protein. Most labeled proteins can be cleanly separated from unreacted label by elution through a gel-filtration column.[9,14]

Deoxynucleoside Triphosphates for Enzymatic Labeling of DNA Probes. Nucleoside triphosphates with fluorophores attached to the base

---

[24] H. Rindknecht, *Nature (London)* **193,** 167 (1962).

by linker groups are commercially available. Nick translation, random priming, and polymerase chain reactions (PCR) can be used to incorporate the modified bases into DNA.[6,25-32] The efficiency of incorporation and the yield of the reaction depend on the fluorophore, the linker arm, and the conditions of the reaction.[26,28] Alternatively, in a preliminary step, amino groups can be inserted into DNA by nick translation using deoxynucleotides containing amino-linker groups or by a sodium bisulfite reaction on DNA with the use of bisamino alkanes.[29,30] Thiols can also be incorporated into DNA.[31] After modification of the DNA, a corresponding amino-specific or sulfhydryl-specific reactive fluorophore is used to tag the DNA. Probes constructed by these methods are useful for fluorescence *in situ* hybridization.[25,32]

Determining Extent of Labeling. Measuring the extent of labeling is a straightforward task under ideal conditions.[10,33-34] For protein labeling it is simplest if the extinction coefficient of the protein, usually at 280 nm, is known. The extinction coefficient of the fluorophore at its maximum absorption and at 280 nm must be known, the latter for correction of protein absorbance at 280 nm. Ideally these values should be for the fluorophore in a conjugated form rather than in its reactive form if there are spectral changes that occur on reaction. It is also important that the solution containing the antibody be free of other materials that absorb or scatter light at 280 nm. If these conditions hold, the dye per protein ratio (D/P) is given by the following calculation:

$$D/P = A_{max}{}^{P}\varepsilon_{280}/(A_{280}{}^{F}\varepsilon_{max} + A_{max}{}^{F}\varepsilon_{280})$$

where A_{280} is the absorbance of the protein at 280 nm, $^{P}\varepsilon_{280}$ is the extinction coefficient of the protein at 280 nm, A_{max} is the absorbance of the fluorophore label at its absorption maximum, $^{F}\varepsilon_{max}$ is the extinction coefficient of the fluorophore at the absorption maximum, and $^{F}\varepsilon_{280}$ is the extinction

[25] T. Reid, A. Baldini, T. C. Rand, and D. C. Ward, *Proc. Natl. Acad. Sci. U.S.A.* **89,** 1388 (1992).

[26] H. Yu, J. Chao, D. Patek, R. Mujumdar, S. Mujumdar, and A. S. Waggoner, *Nucleic Acids Res.* **22,** 3226 (1994).

[27] "Nonradioactive *in Situ* Hybridization Application Manual," Boehringer Mannheim GmbH Biochemica, 1992. A copy of this 75 page booklet may be obtained by writing to PO box 310120, D-6800 Mannheim 31, Germany.

[28] A. Zhu, J. Chao, H. Yu, and A. S. Waggoner, **22,** 3418 (1994).

[29] R. P. Viscidi, this series, Vol. 184, p. 600.

[30] R. P. Viscidi, C. J. Connelly, and R. H. Yolken, *J. Clin. Microbiol.* **23,** 311 (1986).

[31] G. H. Keller and M. M. Manak, "DNA Probes," 2nd Ed., Stockton, New York, 1993.

[32] B. J. Trask, *Methods Cell Biol.* **35,** 3 (1991).

[33] J. R. Simon and D. L. Taylor, this series, Vol. 134, p. 487.

[34] R. C. Nairn, "Fluorescent Protein Tracing," 4th Ed. Churchill-Livingstone, Edinburgh and London, 1976.

coefficient of the fluorophore at 280 nm. IgG antibodies have an extinction coefficient at 280 nm of 2.1×10^5. The ratio of $^F\varepsilon_{280}$–$^F\varepsilon_{max}$ ranges from 0.05 for Cy5 to 0.56 for tetramethylrhodamine isothiocyanate.[10,14] For labeled nucleic acids, the average extinction coefficient of the bases at 265 nm times the number of bases in the DNA fragment can be used for quantification of DNA instead of protein absorbance at 280 nm. Other methods that quantify the protein or nucleic acid content could also be used.

Examples of Labeling Methods

Labeling Antibodies with Fluorescent Succinimidyl Active Esters

Procedure

1. Stock solutions of the sulfocyanine succinimidyl active esters may be made in dry DMF (0.3–1.0 mg active ester/100 ml) and are stable for days when stored at 4° in a desiccator. The active esters are also stable in distilled water for several hours provided the pH of the solution is not basic.[14] Aqueous solutions of the dyes can be used for labeling antibodies if use of DMF is not suitable for certain antibodies. The concentration of cyanine fluorophore in the stock solution is determined by measuring the absorbance of an aliquot of the appropriately diluted stock solution in phosphate-buffered saline (PBS) and using the extinction coefficient of the dye (Table I).

2. Antibody labeling is carried out in 0.1 M carbonate–bicarbonate buffer (pH 9.4) for 15 min at room temperature. One milligram sheep IgG (6.45 mmol) is dissolved in 0.25–1 ml buffer solution, and the desired amount of dye is added during vigorous vortex mixing.

3. Unconjugated dye is separated from the labeled protein by gel-permeation chromatography (0.7 × 20 cm column of Sephadex G-50) using pH 7 buffer solution as eluent.

The dye/protein ratio is calculated using the equation below with measured values of the absorbance of the labeled dye (Cy3 at 550 nm or Cy5 at 650 nm) and the absorbance of protein at 280 nm. The extinction coefficient of the dye is obtained from Table I, and the extinction coefficient of the IgG antibody at 280 nm is taken to be 170,000 l/mol-cm. The factor of 0.05 (used for both Cy3 and Cy5) in the denominator accounts for dye absorption at 280 nm, which is 5% of the absorption of the dye at its maximum absorption (A_{dye}).

$$D/P = 170,000\, A_{max}/(A_{280} - 0.05 A_{max})$$

A more accurate method is required when the labeling reagent shows significant spectral changes when bound to the antibody molecule. For example, this approach is needed when both dimers and monomers of dye, which have different absorption peaks and different extinction coefficients, are present on the protein at higher D/P ratios. In this case, the labeled protein is dissolved in formamide for absorption spectroscopy and the extinction coefficients of the dye determined independently for the calculation.

Labeling Sulfhydryl Group on α-actinin with Iodoacetamidofluorescein

Fluorescent analogs of cellular proteins are used in imaging fluorescence microscopy to monitor biochemical events that take place over time in living cells. This procedure is an excerpt from an article that describes in great detail the preparation of a fluorescent analog.[31] Before the labeling reaction described below can occur, great pains must be taken to isolate the protein target in an active and pure form. After the labeling and purification, additional steps must be taken to assess the functionality of the analog before it is incorporated into living cells by microinjection or bulk labeling for microscopy.

Materials and Solutions Required

Buffer A: 50 mM boric acid, 12.5 mM sodium tetraborate (pH 8.5 at 25°)

Buffer B: 10 mM Tris base (pH 8.0 at 4°), 100 mM KCl, 1 mM EDTA, 1 mM EGTA, 0.02% sodium azide, 1 μg/ml leupeptin, 1 μg/ml pepstatin, saturating phenylmethylsulfonyl fluoride (PMSF)

Procedure

1. The reactive probe solution is prepared by weighing 3–4 mg of 5'-iodoacetamidofluorescein (IAF) from Molecular Probes (Eugene, OR) in a 1.5-ml microcentrifuge tube and dissolving it in DMSO to make a 4 mg/ml stock solution.

2. The IAF DMSO solution is added 1 : 1 to buffer A on ice and stirred for 1 min to make the solution homogeneous.

3. The buffered IAF solution is added at a 2- to 6-fold molar excess to a solution of buffer A with *Dictyostelium discoideum* α-actinin on ice.

4. The reaction mixture is removed from the ice bath and allowed to stir slowly at room temperature for 1 hr in the dark (fluorescein photodecomposes in strong room light).

5. The reaction mixture is passed through a 1.5×90 cm column of Sepharose CL-6B in buffer B and eluted with the same buffer. Labeled α-actinin elutes from the column, leaving unreacted IAF in the included volume of the column.

6. Further column purification steps are used for α-actinin[31] that would likely differ for other labeled proteins.

Use of Cy3-dCTP, Cy5-dCTP, and Fluorescein-dCTP to Prepare Fluorescent DNA by Nick Translation

Nick translation is carried out under conditions found to be optimal for cyanine-modified dUTPs[25,26,28] and dCTPs (Biological Detection Systems, Pittsburgh, PA). Similar procedures can be used for dUTPs modified with fluorescein, rhodamine, and coumarin [see instructions with commercial kits sold by Boehringer Mannheim (Indianapolis, IN) and Amersham (Arlington Heights, IL)].

Materials and Solutions Required

DNA to be labeled, digested to lengths of 200–400 base pairs (bp) with DNase I; typically for a 2.5–3.0 kbp insert template, 1/350 unit of DNase I is used in nick translation buffer (see below) for 5 min at 37°. The time required may vary for different length templates
Unlabeled dATP, dGTP, dTTP (200 μM adjusted to pH 7.0)
1.0 mM Cy3-dCTP
Escherichia coli DNA polymerase I (10 U/μl)
10× Nick translation buffer containing 0.5 M Tris-HCl (pH 7.2), 0.1 M MgSO$_4$, 1 mM dithiothreitol (DTT)
10 M ammonium acetate
Cold 100% ethanol (4°)
Cold 70% ethanol (4°)
TE Buffer (pH 8.0) containing 10 mM Tris-HCl and 1 mM EDTA
Deionized water

Procedure

1. Combine the following in a 1.5-ml microcentrifuge tube in an ice water bath: 25 μl of 10× nick translation buffer; 25 μl each of unlabeled dATP, dGTP, and dTTP; 5 μl Cy3-dCTP; 10 μg DNase I-digested DNA; and distilled water to 250 μl.

2. Mix thoroughly.

3. Incubate mixture at 15° for at least 2 hr (overnight if required).

4. Precipitate the labeled DNA by adding 83 μl of 10 M ammonium acetate and 667 μl cold 100% ethanol.

5. Precipitate the labeled DNA in an ice–water bath for at least 1 hr, overnight if required.

6. Spin at 12,000–16,000 g for 30 min at 4°.

7. Rinse twice with cold 70% ethanol.

8. Spin under vacuum until almost dry.

9. Resuspend in 100 μl TE buffer (pH 8.0), mixing thoroughly. Incubate at 37° for 20 min if necessary to dissolve completely.

10. Store at −20° until ready to use.

Section II

Vibrational Spectroscopy

[16] Biomolecular Vibrational Spectroscopy

By Richard A. Mathies

Introduction

Raman scattering and infrared absorption spectroscopy are powerful methods for studying biological systems because the vibrational frequencies reveal the structure of individual atomic groups with high specificity and, in appropriately designed experiments, high time-resolution structural information is obtained. The time scale of vibrational spectroscopy is essentially instantaneous ($\sim 10^{-14}$ sec) so there is no blurring of the information as a result of rapid structural dynamics. Also, the vibrational frequencies and intensities are very sensitive to the structure and local environment of the atoms involved in the vibrational normal mode. It is relatively easy to understand the vibrational spectrum of simple chemical systems and to determine how the vibrational spectrum depends on the structure.[1] However, proteins, nucleic acids, and biological membranes are very complex vibrational systems that are difficult to interpret because of the large amount of vibrational information. For example, bacteriorhodopsin (BR) is a 26,000 Da intrinsic membrane protein made up of an all-*trans*-retinal prosthetic group and 248 amino acids. It contains around 1900 heavy (nonhydrogen) atoms that will produce about 5700 different skeletal vibrations! Although these vibrational modes will not all be both Raman and IR active, this is still a formidable complexity that illustrates why one must, in general, design experiments to examine this wealth of vibrational information selectively.

Changing the excitation wavelength is one method for controlling the information content in vibrational spectroscopy because different groups in the protein will be probed by using radiation in different spectral regions. This point is understood with reference to the absorption spectrum of bacteriorhodopsin in Fig. 1. The vibrational structure of the amide and other chemical groups in the protein can be monitored by direct absorption of radiation in the 5000–10,000 nm region. Classical Raman scattering, where the incident laser lies in a transparent region of the spectrum (>600 nm in this case), can be used to study the Raman active vibrations of both the prosthetic group and the protein. In a resonance Raman experiment, excitation is chosen to lie within the electronic absorption of selected

[1] E. B. Wilson, J. C. Decius, and P. C. Cross, "Molecular Vibrations." McGraw-Hill, New York, 1955.

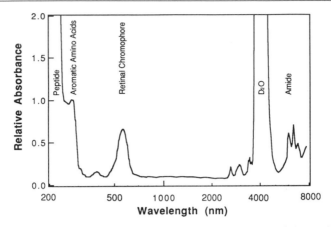

FIG. 1. Absorption spectrum of bacteriorhodopsin from the ultraviolet to the infrared. By using incident light in different regions of the spectrum, vibrational spectroscopy can be used to probe different aspects of protein structure. The IR portion of the spectrum was obtained in a D_2O buffer.

groups within the protein (e.g., the retinal prosthetic group at ~570 nm), resulting in resonance enhancement of the scattering from just that group. By selecting a different laser line to excite the sample, different components of the protein can be selectively examined. Excitation at 280 nm will resonantly enhance scattering from the aromatic amino acids, whereas excitation at 200 nm will enhance scattering from the peptide bond.

Perturbation or difference experiments provide another method for simplifying the data in both Raman and IR experiments. The classic approach is to introduce isotopic substitutions which identify the chemical groups responsible for the vibration and permit vibrational normal mode assignments.[2] Chemical modification of the prosthetic group or of the protein and amino acid mutation are additional possibilities. Temperature jump, pressure jump, and rapid mixing experiments are also valuable approaches. This introduction emphasizes the use of time-resolved vibrational spectroscopy to examine the vibrational information selectively.[3,4]

It is not possible in this chapter to describe all of the possible ways to study biological systems using vibrational spectroscopy. Examples of the use of resonance Raman spectroscopy to study the structure and

[2] R. Mathies, this series, Vol. 88, p. 633.
[3] A. Lau, F. Siebert, and W. Werncke (eds.), "Time-Resolved Vibrational Spectroscopy VI," Springer Proceedings in Physics Vol. 74. Springer-Verlag, Berlin, 1994.
[4] A. Laubereau and M. Stockburger (eds.), "Time-Resolved Vibrational Spectroscopy," Vol. 4. Springer-Verlag, Berlin, 1985.

function of metalloproteins are presented elsewhere in this volume by Spiro and Czernuszewicz.[5] However, over the years bacteriorhodopsin has become an important test system for the development of new vibrational techniques. The experiments on BR illustrate not only the kind of structural and mechanistic information that can be obtained using vibrational spectroscopy, but also the techniques and many of the challenges that are associated with the application of vibrational spectroscopy to complex biological systems. Therefore, this overview briefly surveys the application of vibrational spectroscopy to BR with an emphasis on the relation between the various approaches, the kinds of information that can be obtained, and the unique problems and opportunities that each approach encumbers.

Applications of Vibrational Spectroscopy to Bacteriorhodopsin

Bacteriorhodopsin (BR) is an intrinsic membrane protein found in the cell membrane of *Halobacterium halobium* that functions as a light-driven proton pump (for reviews, see Refs. 6–9). Photon absorption by the all-*trans*-retinal prosthetic group drives BR through a photochemical cycle, complete in approximately 10 msec, that results in the pumping of a proton outside the cell. The electrochemical gradient thereby produced is used for ATP synthesis when the bacterium finds itself under anaerobic conditions. The ease of preparing BR in large quantities, its stability, and its rapid photochemical cycle have made BR a useful system for the development of a variety of new biophysical spectroscopies.[10] The various types of experiments and the information that can be gained from them are surveyed below.

Infrared Absorption Spectroscopy

The infrared absorption spectrum of a BR film is presented in Fig. 2. Vibrations will be intense in an infrared spectrum if the vibrational motion introduces a change in the electric dipole moment of that group. Thus, polar groups such as N—H, O—H, and C=O will dominate an IR spectrum. Because of the intense absorption of H_2O throughout the IR region,

[5] T. G. Spiro and R. S. Czernuszewicz, this volume [18].

[6] R. A. Mathies, S. W. Lin, J. B. Ames, and W. T. Pollard, *Annu. Rev. Biophys. Biophys. Chem.* **20**, 491 (1991).

[7] R. R. Birge, *Biochim. Biophys. Acta* **1016**, 293 (1990).

[8] D. Oesterhelt and J. Tittor, *Trends Biochem. Sci.* **14**, 57 (1989).

[9] H. G. Khorana, *J. Biol. Chem.* **263**, 7439 (1988).

[10] L. Packer (ed.), this series, Vol. 88.

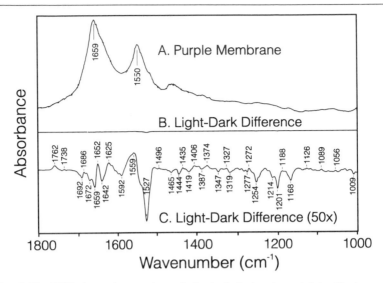

FIG. 2. The FTIR absorption spectrum of a bacteriorhodopsin-containing film is presented in spectrum A. Spectrum B is a light − dark difference spectrum (to scale) after a fraction of the sample was converted to the M intermediate. Spectrum C presents an expanded view of the difference spectrum revealing an abundance of chromophore and protein structural changes. (Courtesy of Dr. Mark Braiman.)

it is generally necessary to work with biological samples that have reduced H_2O content through partial dehydration or by exchanging into D_2O buffers.

The spectrum of BR in Fig. 2 is dominated by the intense characteristic absorptions of the amide I and II vibrations at 1659 and 1550 cm^{-1}. These broad and overlapping group vibrations provide little specific structural information about discrete atomic groups within the protein. With a traditional absorptive IR apparatus there was little that could be done to improve this situation. However, with the introduction of Fourier transform infrared (FTIR) spectroscopy it became possible to obtain IR spectra with high signal-to-noise (S/N) ratios and reproducibility in a short period of time, thereby making IR difference techniques possible. Figure 2B illustrates this concept by presenting an FTIR difference spectrum of a BR film. The difference spectrum was generated by subtracting a spectrum of BR from a spectrum of the same sample after it was converted by illumination to the M state and trapped at low temperature. The difference features, though less than 1% of the initial band intensities, are clearly observed and reproducible. Modern FTIR apparatus and procedures can detect changes as small as 10^{-4} from a 1 OD sample. Furthermore, in such difference spectra changes in the vibrational structure of individual amino acid groups can be clearly identified and assigned through isotopic substitution and site-specific mutagenesis. Some of these approaches are presented in more detail in this volume by Siebert,[11] and a summary of protein vibrational band assignments is provided by Peticolas.[12] IR studies of bacteriorhodopsin have been used to determine the protonation state of the functionally active amino acid residues for each of the intermediates in the BR photocycle and hence to determine a molecular mechanism for proton pumping.[13]

Time-resolved IR difference techniques, which are also discussed by Siebert,[11] provide another powerful method for simplifying IR absorption spectra of proteins and for obtaining useful kinetic information. One approach is through modification of the FTIR apparatus to obtain fast-continuous scan, stroboscopic, or step-scan spectra. These methods are particularly useful because they provide the entire IR spectrum. Because the time needed to obtain an interferogram is a few milliseconds, one can obtain fast-continuous scan spectra and observe the kinetics of millisecond or slower processes.[14,15] A higher time-resolution approach is provided by the stroboscopic technique. With this method, a repetitive process is triggered many times during the acquisition of the interferograms. The time resolution with the stroboscopic technique is around 1 μsec, and this approach has been used to obtain kinetically resolved spectra of a variety of bacteriorhodopsin photointermediates.[16] In the step-scan technique, the interferometer is slowly stepped through the interferogram, and transient decays are recorded at each mirror position.[17] This has the advantage that up to 100 nsec time resolution is possible.

An alternative approach for achieving 100 nsec time resolution spectra using a dispersive instrument has been presented by Hamaguchi and co-workers.[18,19] More exotic time-resolved IR systems using laser up-conversion techniques can reach into the picosecond or subpicosecond re-

[11] F. Siebert, this volume [20].

[12] W. L. Peticolas, this volume [17].

[13] M. S. Braiman, T. Mogi, T. Marti, L. J. Stern, H. G. Khorana, and K. J. Rothschild, *Biochemistry* **27**, 8516 (1988).

[14] M. S. Braiman, P. L. Ahl, and K. J. Rothschild, *Proc. Natl. Acad. Sci. U.S.A.* **84,** 5221 (1987).

[15] M. S. Braiman and K. J. Rothschild, *Annu. Rev. Biophys. Biophys. Chem.* **17,** 541 (1988).

[16] M. S. Braiman, O. Bousche, and K. J. Rothschild, *Proc. Natl. Acad. Sci. U.S.A.* **88,** 2388 (1991).

[17] W. Uhmann, A. Becker, C. Taran, and F. Siebert, *Appl. Spectrosc.* **45**, 390 (1991).

[18] J. Sasaki, A. Maeda, C. Kato, and H. Hamaguchi, *Biochemistry* **32**, 867 (1993).

[19] K. Iwata and H. Hamaguchi, *Appl. Spectrosc.* **44,** 1431 (1990).

gime.[20,21] The latter methods typically operate in a very limited IR wavelength range.

Ultimately, the choice of method will depend on the system being studied and the time resolution required to resolve the process of interest. However, it is clear that time-resolved IR techniques can produce unique information that will continue to play an important role in studies of biomolecular structural dynamics.

Raman Spectroscopy

In classical Raman spectroscopy, the incident laser is not in resonance with any electronic absorption in the sample. Classical Raman studies will observe scattering from groups that exhibit a large change in the polarizability as the atoms vibrate. This generally excludes polar groups such as N—H, O—H, and C=O, but C=C and C—C stretches will be strong. In off-resonance Raman experiments, each group will contribute according to its vibrational polarizability derivative. Because there are many groups in proteins and nucleic acids that can contribute, the spectrum will be very complex. Although it is possible to examine the group vibrational features such as the amide I and III band intensities and profiles to assign the amount of a particular secondary structure as discussed by Peticolas[12] and Siebert,[11] it is very difficult to assign features to specific atomic groups and to detemine detailed structural information from such data. This illustrates one of the principal difficulties of applying vibrational spectroscopy to biological systems. So much information can be obtained that the spectra are difficult to interpret, and traditional methods of vibrational assignment are inadequate in the face of such complexity. The resonance enhanced and difference techniques are valuable because they enable one to reduce the complexity of information to a tractable level.

Resonance Raman Spectroscopy

Resonance enhancement of the Raman scattering will occur when the laser excitation is within the electronic absorption band of either a prosthetic group or one of the amino acid residues. The resonance enhancement can be as large as 10^6, dramatically enhancing the vibrational modes of the parts of the system that are coupled to the electronic excitation. When the excitation is resonant with a chromophoric prosthetic

[20] R. Diller, M. Iannone, B. R. Cowen, S. Maiti, R. A. Bogomolni, and R. M. Hochstrasser, *Biochemistry* 31, 5567 (1992).
[21] S. Maiti, B. R. Cowen, R. Diller, M. Iannone, C. C. Moser, P. L. Dutton, and R. M. Hochstrasser, *Proc. Natl. Acad. Sci. U.S.A.* 90, 5247 (1993).

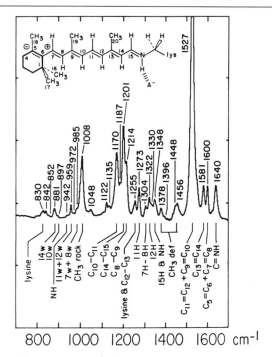

FIG. 3. Resonance Raman spectrum of the all-*trans*-retinal protonated Schiff base chromophore in light-adapted bacteriorhodopsin. The assignments are of the various enhanced vibrational normal modes are indicated. (From Smith *et al.*[22])

group in a protein, the Raman active normal modes of the protein will be enhanced. An excellent example is provided by the retinal chromophore in bacteriorhodopsin. BR absorbs maximally at 568 nm, so exciting within this absorption with the green-yellow lines of an argon or krypton ion laser will produce very strong scattering from the chromophore. Figure 3 presents the resonance Raman spectrum of light-adapted bacteriorhodopsin (BR$_{568}$) along with normal mode assignments determined by isotopic substitution and normal coordinate analysis.[22] The spectra are sufficiently strong and of sufficient quality and simplicity that the chromophore structure can be analyzed. By combining vibrational assignments with time-resolved studies, structures of the retinal chromophore in all of the photointermediates in the BR photocycle have been determined. This work has led to the development of a detailed model for the role of the chromophore in the proton-pumping mechanism of bacteriorhodopsin.[6]

[22] S. O. Smith, M. S. Braiman, A. B. Myers, J. A. Pardoen, J. M. L. Courtin, C. Winkel, J. Lugtenburg, and R. A. Mathies, *J. Am. Chem. Soc.* **109**, 3108 (1987).

The enhanced signal-to-noise ratio that is provided by resonance enhancement as well as the reduced complexity of the vibrational spectrum make it possible to perform a wide variety of time-resolved studies to determine the structure of the chromophore in the photocycle intermediates. These approaches are discussed in more detail elsewhere in this volume by Kincaid[23] with emphasis on time-resolved Raman studies of heme proteins. Room-temperature flow methods have been extensively used to obtain time-resolved spectra with time resolution ranging from seconds to microseconds.[24–27] The basic idea is to flow the sample and then introduce an optical pump beam upstream from the probe to initiate the photochemical cycle. Such experiments have been performed on the millisecond and microsecond time scales. For experiments with time resolution faster than microseconds, it is necessary to convert the setup to a two-pulse, pump–probe technique where the time resolution is established by the delay between the pump and probe laser pulses. The time resolution of this approach can be increased to around 1 psec; beyond this point increased time resolution will be achieved only with reduced spectral resolution according to the uncertainty principle.

Figure 4 illustrates the kind of information that can be obtained with time-resolved resonance Raman spectroscopy by presenting picosecond resonance Raman spectra of the bacteriorhodopsin chromophore. The data have been used to determine the structure of the J and K intermediates as well as to provide information about an intermediate between K and L.[28]

By altering the excitation wavelength, other components of a protein may be studied. For example, each BR intermediate absorbs in a different spectral region ranging from 400 to 600 nm. By selecting a probe wavelength that lies within the absorption band of the intermediate of interest, maximum resonance enhancement will be achieved. This concept has been exploited in the time-resolved studies of the BR intermediates mentioned earlier. Second, if the laser wavelength is shifted even further into the UV, the laser will eventually become resonant with the aromatic amino acid absorption at 280 nm. This should give rise to enhanced scattering from the tyrosine and tryptophan residues. Phenylalanine has such a weak

[23] J. R. Kincaid, this volume [19].
[24] R. Mathies, A. R. Oseroff, and L. Stryer, *Proc. Natl. Acad. Sci. U.S.A.* **73,** 1 (1976).
[25] J. Terner and M. A. El-Sayed, *Acc. Chem. Res.* **18,** 331 (1985).
[26] M. Stockburger, T. Alshuth, D. Oesterhelt, and W. Gartner, *in* "Spectroscopy of Biological Systems" (R. J. H. Clark and R. E. Hester, eds.), p. 483. Wiley, London, 1986.
[27] R. A. Mathies, S. O. Smith, and I. Palings, *in* "Biological Applications of Raman Spectroscopy: Volume 2—Resonance Raman Spectra of Polyenes and Aromatics" (T. G. Spiro, ed.), Vol. 2, p. 59. Wiley, New York, 1987.
[28] S. J. Doig, P. J. Reid, and R. A. Mathies, *J. Phys. Chem.* **95,** 6372 (1991).

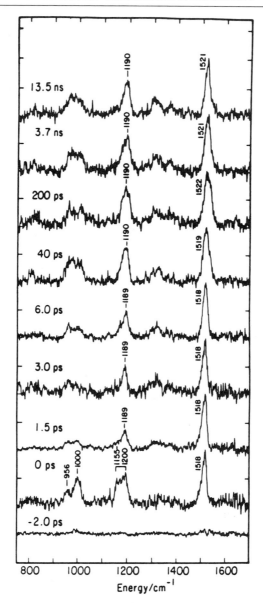

FIG. 4. Time-resolved resonance Raman spectra of the photoproduct of bacteriorhodopsin for time delays from 2 psec to 13.5 nsec after photolysis. Spectra were obtained by using 2 psec probe and pump pulses at 589 and 550 nm, respectively. (From Doig et al.[28])

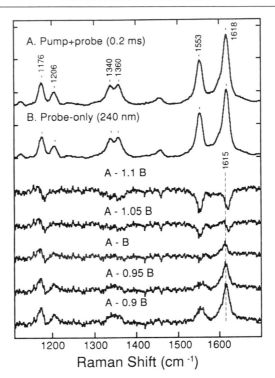

FIG. 5. Millisecond time-resolved UV resonance Raman spectra of bacteriorhodopsin. Spectrum A was obtained with a 240 nm probe beam in the presence of a 515 nm pump beam that converts a fraction of the sample over to the M intermediate. Spectrum B was obtained with just the 240 nm probe beam. The various difference spectra present the changes in structure of the resonantly enhanced tyrosines and tryptophans that occur during the conversion to the M intermediate. (From Ames et al.[29])

absorption (and hence weak scattering) at 280 nm that it will not contribute significantly to spectra excited in this region.

Figure 5 presents a UV Raman spectrum of BR excited at 240 nm.[29] The spectra are of excellent quality and reveal unique enhancement of the vibrations of just the tyrosine and tryptophan residues. With UV excitation, the aromatic amino acids experience resonance enhancement, and we can selectively examine their vibrational structure. Experiments of this type have been used to determine the protonation state of key

[29] J. B. Ames, M. Ros, J. Raap, J. Lugtenburg, and R. A. Mathies, *Biochemistry* **31,** 5328 (1992).

tyrosine residues within the chromophore binding pocket.[30] In addition, time-resolved UV resonance Raman experiments have been performed on BR with millisecond and even nanosecond time resolution to study transient protein structure and structural changes.[29] UV resonance Raman is unique in this ability to determine changes in protein structure. It should be possible to extend these experiments and perform picosecond visible-pump, UV-probe studies of protein structure.[31]

Concluding Remarks

No overview would be complete without a discussion of the limitations of a technique because an understanding of the limitations is critical for designing any experiment. The principal disadvantage of IR absorption spectroscopy is the difficulty of working with aqueous samples. However, the IR spectrum reports on the vibrations of polar groups in the protein that are most likely to be of interest in studies of functional changes in protein structure. The BR experiments presented here are obtained from hydrated films which have low water content and high protein concentration (50% by weight). In this format, protein amounts can be as small as 40–100 μg. To use this method, a protein must be functional under these conditions. An additional problem is that there is no "resonance enhancement" effect in IR spectroscopy, so the vibrational spectrum will be quite complex; all the constituents will contribute their respective absorptions. The assignment of the various vibrational bands to particular chemical groups and normal modes is a challenging problem. Thus, some sort of difference technique (isotopic labeling and/or site-specific mutagenesis) is required to take this next step and extract the available information.

In performing Raman experiments a different set of issues must be understood. First, there is very little background scattering from water so it is easy to work in aqueous solutions. In conventional Raman spectroscopy, the scattering is quite weak so a satisfactory spectrum of a soluble protein may require as much as 1 mg of protein/ml of solution. However, if resonance enhancement is exploited, then solutions in the 10 μM range can typically be studied.

Another important concern is fluorescence interference. Excitation with an intense laser source will often produce strong fluorescence from impure biological samples or due to the intrinsic emission of a colored

[30] J. B. Ames, S. R. Bolton, M. M. Netto, and R. A. Mathies, *J. Am. Chem. Soc.* **112**, 9007 (1990).

[31] P. Thompson and R. A. Mathies, *in* "Laser Techniques in Chemistry" (A. B. Myers and T. Rizzo, eds.), in press. Wiley, New York, 1994.

prosthetic group. Because the quantum yield for even resonance Raman scattering is very low, minuscule fluorescence emissions having a quantum yield of 10^{-5} can obscure the Raman scattering. Although new techniques for suppressing the fluorescence interference have been developed in studies of photosynthetic reaction centers,[32,33] this is likely to be an important concern in most systems.

Finally, when exciting with intense laser sources, it is important to understand and control any perturbation of the sample composition that is caused by the incident laser beam. For off-resonance Raman scattering, this should not be a problem unless very high powers are employed because the sample does not absorb. The only exception is in the case of Fourier transform Raman experiments where very high power IR lasers are used that might find sufficient IR absorption to "cook" the sample.[34–36] In resonance Raman experiments it is very important to consider the effect of the absorbed light on the sample composition. Raman scattering and absorption are independent events. Some photons are scattered, but a much larger number are absorbed. Cross sections for Raman scattering are on the order of 10^{-7} $Å^2$/molecule, whereas cross sections for absorption are about 1 $Å^2$/molecule.[37] Thus, if photon absorption results in a photochemical transformation, as in the case of visual pigments, the effect of the photoalteration must be carefully considered. This situation has been analyzed in detail and rapid-flow methods devised that allow one to obtain resonance Raman spectra of even the most photolabile molecules without distortion of the sample composition.[24]

As the laser wavelength is shifted toward the UV, the effect of the laser on the sample becomes even more complex because the majority of the available UV lasers are pulsed nanosecond lasers. These intense UV pulses are also prone to the production of sample breakdown and even the destruction of the surrounding quartz sample cell! Consequently, such UV Raman experiments are often performed by focusing the UV laser on a flowing stream or a wire-guided jet so that the UV beam does not have to pass through a quartz–sample interface and so that the excited surface is replaced between shots.[38]

[32] A. P. Shreve, N. J. Cherepy, S. Franzen, S. G. Boxer, and R. A. Mathies, *Proc. Natl. Acad. Sci. U.S.A.* **88,** 11207 (1991).

[33] A. P. Shreve, N. J. Cherepy, and R. A. Mathies, *Appl. Spectrosc.* **46,** 707 (1992).

[34] D. B. Chase, *J. Am. Chem. Soc.* **108,** 7485 (1986).

[35] J. Sawatzki, R. Fischer, H. Scheer, and F. Siebert, *Proc. Natl. Acad. Sci. U.S.A.* **87,** 5903 (1990).

[36] C. K. Johnson and R. Rubinovitz, *Appl. Spectrosc.* **44,** 1103 (1990).

[37] A. B. Myers, R. A. Harris, and R. A. Mathies, *J. Chem. Phys.* **79,** 603 (1983).

[38] G. A. Reider, K. P. Traar, and A. J. Schmidt, *Appl. Opt.* **23,** 2856 (1984).

Another consideration is the amount of sample that is required for typical experimental configurations. Because of the small illuminated volume in a resonance Raman experiment, very small samples are in principle possible. Indeed, excellent resonance Raman spectra have been obtained from individual photoreceptor cells,[39] and spectra have been obtained from protein samples using microliter volumes fixed on a cold stage containing only a few micrograms of protein.[40] However, if one is performing rapid-flow experiments, it is often necessary to use much larger volumes; samples from 1 to 50 ml with micromolar or higher concentrations are often required.

In summary, vibrational spectroscopy can provide amazingly detailed information about the structure of biological systems, and it has the unique capability of being able to determine structure and structural changes of individual atomic groups within complex biological systems with high time resolution and under functional conditions. Along with these unique capabilities there are a number of practical and instrumental limitations that must be understood to ensure that the experiment will be feasible and that the desired information will be obtained. The following chapters in this section on vibrational spectroscopy describe the various possible techniques and should provide sufficient information to enable the reader to address these issues for a particular system of interest.

Acknowledgments

I thank Peggy Thompson and Mark Braiman for assistance in preparing figures, and Mark Braiman for critical reading of the manuscript. This work was supported by National Institutes of Health Grant GM 44801.

[39] G. R. Loppnow, B. A. Barry, and R. A. Mathies, *Proc. Natl. Acad. Sci. U.S.A.* **86,** 1515 (1989).
[40] S. W. Lin, T. P. Sakmar, R. R. Franke, H. G. Khorana, and R. A. Mathies, *Biochemistry* **31,** 5105 (1992).

[17] Raman Spectroscopy of DNA and Proteins

By WARNER L. PETICOLAS

Introduction

DNA in the living cell carries a specific base sequence that is transcribed into a complementary sequence in RNA which in turn directs the order of incorporation of amino acids in the synthesis of proteins. This process is called gene expression. A knowledge of the exact structure of these macromolecules *in vivo* and *in vitro* is important in order to explain the mechanism of the process and the subsequent function of the protein. This chapter concentrates on the structural information that can be obtained on these macromolecules in crystals, in solutions, and in living cells using Raman spectroscopy.

Raman spectroscopy is a method of obtaining the vibrational frequencies of a molecule. Some of the vibrational normal modes of polypeptides, proteins, RNA, and DNA are conformationally sensitive, that is, they change in either frequency or intensity, or both, with a change in macromolecular conformation. From an analysis of the Raman spectrum of a DNA or protein molecule a great deal of structural information can be obtained. Compared to other techniques, classic Raman spectroscopy has several advantages: minimal or no damage to the sample, relatively little interference from the water Raman signal in aqueous solutions or *in vivo,* a wide range of concentrations, and small sample volumes (a few cubic micrometers). Vibrational information is obtained that may be representative of one or more conformational states that are present at the same time. Finally, the ability to compare Raman spectra from a sample *in vitro* or *in vivo* with Raman spectra from samples in the crystalline state where the precise structure is known from X-ray diffraction studies makes Raman spectroscopy a method of relating the structure of a molecule in a crystal to the structure in solution or in a living cell. From a comparison of the Raman spectra taken from different crystals of proteins and/or oligonucleotides, correlations between the frequencies and intensities of Raman bands and the secondary structure of proteins and nucleic acids has been developed. These correlations have aided in the development of molecular force fields for proteins and nucleic acids that helps in the understanding of the structure and function of these molecules.

To obtain a Raman spectrum of a protein or nucleic acid, one places the sample at the focal point of a focused laser beam. Because a laser beam can be focused into a circle with a radius of approximately its wavelength, this allows for a spatial resolution of the order of a cubic micrometer. The light scattered from the DNA sample in the focused laser beam is collected by means of a lens system and directed through a suitable monochromator to a photon detector. The intensity of the scattered light is measured as a function of its frequency. A plot of the intensity of the scattered light (photons per square centimeter per second) as a function of the frequency difference, Ω, between the incident laser frequency and the scattered light frequency, constitutes a Raman spectrum (i.e., $\Omega = \omega_L - \omega_s$). When a band appears in the Raman spectrum at a particular frequency, Ω, it means that there is a normal mode of the molecule that vibrates at this frequency. It is a fundamental principle of the Raman effect that the spectrum is independent of the incident laser frequency.

In the resonance Raman effect, the frequency of the incident light falls within the absorption band of an excited electronic state. When this occurs, the intensities of some bands are enhanced while others are left weak. The rule is that the frequencies of the Raman bands are independent of the frequency of the incident light but the intensities of the bands may vary. Resonance Raman spectroscopy has the disadvantage that it is usually accompanied by photochemical decomposition. Avoiding sample degradation requires the use of a circulating sample. This makes measurements on single crystals and living cells impossible and may require large amounts of sample.

There are three types of Raman spectrometers in general use today for the characterization of protein or nucleic acid structure: (a) the classic laser Raman spectrometer,[1-4] (b) the laser Raman microscope,[5-8] and (c) the ultraviolet resonance Raman spectrometer.[9-17] In the classical system

[1] W. L. Peticolas, in "Procedures in Nucleic Acid Research" (G. L. Cantoni and D. R. Davies, eds.), Vol. 2, p. 94. Harper & Row, New York, 1971.

[2] K. A. Hartman, R. C. Lord, and G. J. Thomas, Jr., in "Physicochemical Properties of Nucleic Acids" (J. Duchesne, ed.), Vol. 2, p. 1. Academic Press, London, 1973.

[3] N.-T. Yu, Crit. Rev. Biochem. 4, 229 (1977).

[4] T. G. Spiro and B. P. Gaber, Annu. Rev. Biochem. 46, 553 (1977).

[5] W. L. Kubasek, Y. Yang, G. A. Thomas, T. W. Patapoff, K.-H. Shoenwaelder, J. H. van de Sande, and W. L. Peticolas, Biochemistry 25, 7440 (1986).

[6] T. W. Patapoff, G. A. Thomas, Y. Wang, and W. L. Peticolas, Biopolymers 27, 493 (1988).

[7] G. J. Puppels, F. F. M. de Mul, C. Otto, J. Greve, M. Robert-Nicoud, D. J. Arndt-Jovin, and T. M. Jovin, Nature (London) 347, 301 (1990).

[8] J. M. Benevides, M. Tsuboi, A. H.-J. Wang, and G. J. Thomas, Jr., J. Am. Chem. Soc. 115, 5351 (1993).

[9] D. Blazej and W. L. Peticolas, Proc. Natl. Acad. Sci. U.S.A. 74, 2639 (1977); D. Blazej and W. L. Peticolas, J. Chem. Phys. 72, 3134 (1980).

[10] L. Chinsky, P. Y. Turpin, and M. Duquesne, Biopolymers 17, 1347 (1978).

[11] L. C. Ziegler, B. Hudson, D. P. Strommen, and W. L. Peticolas, Biopolymers 24, 2067 (1984).

[12] B. Jolles, L. Chinsky, and A. Laigle, J. Biomol. Struct. Dyn. 1, 1335 (1984).

[13] W. L. Kubasek, B. Hudson, and W. L. Peticolas, Proc. Natl. Acad. Sci. U.S.A. 82, 2369 (1985).

[14] S. P. A. Fodor, R. P. Rava, T. R. Hays, and T. G. Spiro, J. Am. Chem. Soc. 107, 1520 (1985).

[15] M. Tsuboi, Y. Nishimura, A. Y. Hirakawa, and W. L. Peticolas, in "Biological Applications of Raman Spectrometry" (T. G. Spiro, ed.), Vol. 2, p. 109. Wiley, New York, 1987.

[16] B. Hudson and L. Mayne, this series, Vol. 130, p. 331.

[17] S. Asher, Anal. Chem. 65, 59a and 201a (1993); S. Asher, Annu. Rev. Phys. Chem. 39, 537 (1988).

the laser used is almost always an argon ion laser with prominent lines at 514.5, 488.0, and 457.9 nm. The advantage of the classic Raman spectrometer is that the sample may be a protein powder, crystal, or solution,[18] a fiber of DNA,[19-21] a solution of DNA,[22,23] or a crystal of an oligonucleotide.[6,8,24-28] Because these samples are transparent to visible light, there is little or no heating of the sample by the laser beam. Classical Raman spectra of solutions of a protein or a nucleic acid are obtained from 5-μl samples in a glass capillary mounted in a copper block with the temperature controlled by circulating water from a water bath. The laser light is focused through the capillary, and the scattered laser light is collected at right angles by means of tunnels in the copper block that transmit the incident and scattered light. The concentration of the solution is characteristically 2.0–20 mg of nucleic acid or protein–ml of solution, or about 0.03–0.1 M in nucleotide or peptide group.

Origin of Raman Marker Bands for Structure Determination

The bands that are observed in the Raman spectrum of a molecule correspond to normal modes of vibration of that particular structure. These normal modes can be calculated from knowledge of the three-dimensional structure of the molecule and its force field. Most of the changes in conformation of a polypeptide or DNA are due to changes in the torsional angles along the backbone of the chain. The question we need to answer is how does changing the torsional angles change the vibrational frequencies, and which ones? The method in use has been worked out by Wilson and collaborators and is well described in their textbook.[29] The advantage of the Wilson approach is that the force field

[18] R. C. Lord and N.-T. Yu, *J. Mol. Biol.* **50,** 509 (1970); R. C. Lord and N.-T. Yu, *J. Mol. Biol.* **51,** 203 (1970).

[19] S. C. Erfurth, E. J. Kiser, and W. L. Peticolas, *Proc. Natl. Acad. Sci. U.S.A.* **69,** 938 (1972).

[20] S. C. Erfurth, P. J. Bond, and W. L. Peticolas, *Biopolymers* **14,** 1245 (1975).

[21] D. D. Goodwin and J. Brahms, *Nucleic Acids Res.* **5,** 835 (1978).

[22] J. C. Martin and R. M. Wartell, *Biopolymers* **21,** 499 (1982).

[23] S. C. Erfurth and W. L. Peticolas, *Biopolymers* **14,** 247 (1975).

[24] T. J. Thamann, R. C. Lord, H. H. T. Wang, and A. Rich, *Nucleic Acids Res.* **9,** 5443 (1981).

[25] G. A. Thomas and W. L. Peticolas, *J. Am. Chem. Soc.* **105,** 986 (1983).

[26] G. J. Thomas, Jr., and A. H.-J. Wang, *in* "Nucleic Acids in Molecular Biology" (F. Eckstein and D. M. J. Lilley, eds.), Vol. 2, p. 1. Springer-Verlag, Weinheim, 1988; J. M. Benevides, A. H.-J. Wang, G. A. van der Marel, J. H. van Boom, A. Rich, and G. J. Thomas, Jr., *Nucleic Acids Res.* **12,** 5913 (1984).

[27] Y. Yang, G. A. Thomas, and W. L. Peticolas, *J. Biomol. Struct. Dyn.* **5,** 249 (1987).

[28] W. L. Peticolas, Y. Yang, and G. A. Thomas, *Proc. Natl. Acad. Sci. U.S.A.* **85,** 2579 (1988).

[29] E. B. Wilson, J. C. Decius, and P. C. Cross, "Molecular Vibrations." McGraw-Hill, New York, 1955.

for the molecule is set up in terms of internal or valence coordinates such as bond stretching and angle bending coordinates. The force constants for these deformations are very similar from one molecule to another so that force constants can be transferred from one molecule to another. More recent advances in this theory permits the use of *ab initio* quantum mechanical calculations of force fields and vibrational spectra.[30,31]

The Hamiltonian for the nuclear (vibrational) motion of a molecule is given in the harmonic approximation by[29]

$$2H = \mathbf{R}^{\mathrm{T}}\mathbf{G}^{-1}\mathbf{R} + \mathbf{R}^{\mathrm{T}}\mathbf{F}\mathbf{R} \tag{1}$$

where $\mathbf{R}$ is a column vector of the internal coordinates (R_1, R_2, R_3, ...). The internal coordinates, R_i, are the bond stretching, angle bending, torsional rotations, etc. The force constants that comprise the $\mathbf{F}$ matrix, F_{ij}, are defined in terms of the potential energy, V, for the displacement of the internal coordinates from the equilibrium position by the Taylor series expansion of the potential energy, $F_{ij} = \partial^2 V/\partial R_i \partial R_j$. The internal coordinates are related to the Cartesian displacement coordinates by the relation $\mathbf{R} = \mathbf{B}\mathbf{X}$ where $\mathbf{B}$ is a $(3N - 6) \times 3N$ dimensionless matrix relating the $3N - 6$ internal coordinates without redundancies[30] to the $3N$ Cartesian coordinates. The $\mathbf{G}$ matrix is defined by the relation $\mathbf{G} = \mathbf{B}\mathbf{M}^{-1}\mathbf{B}^{\mathrm{T}}$, where $\mathbf{M}$ is the diagonal matrix of the masses of the atoms in the molecule.

The kinetic energy of the molecule is given by the first term in the Hamiltonian [Eq. (1)] and depends on the inverse of the $\mathbf{G}$ matrix which in turn depends on the $\mathbf{B}$ matrix. It may be shown that the $\mathbf{B}$ matrix and hence the $\mathbf{G}$ matrix is completely determined by the geometry and atomic masses of the molecule.[29] The molecular vibrational frequencies are obtained from eigenvalues of the matrix product $\mathbf{G}\mathbf{F}$ through the solution of the eigenvalue equation, $\mathbf{G}\mathbf{F}\mathbf{L} = \mathbf{L}\Lambda$, where Λ is a diagonal matrix whose elements are the squares of the circular frequencies, $\lambda_i = \omega^2 = (2\pi\nu_i)^2$. In that equation, ω_i and ν_i are the circular and linear frequencies of the ith normal mode of vibration. Each frequency corresponds to a normal mode, Q_i. The displacement of the normal mode may be thought of as the simultaneous displacement of the internal coordinate displacements, R_j, by the equation $R_j = \Sigma_i L_{ji} Q_i$, where the summation is over the $3N - 6$ normal modes and $j = 1, 2, 3, \ldots, 3N - 6$. If we let $Q_2 = Q_3 = \ldots = Q_{3N-6} = 0$ and $Q_1 = 1$, then $L_{11}, L_{21}, \ldots, L_{(3N-6)1}$ give the displacements of $R_1, R_2, \ldots, R_{3N-6}$ for the first normal mode with frequency ν_i. The components of the ith column vector of the $\mathbf{L}$ matrix correspond to

[30] G. Fogarasi and P. Pulay, *Vib. Spectra Struct.* **14**, 125 (1985).
[31] S. Krimm and J. Bandekar, *Adv. Protein Chem.* **38**, 181 (1986).

the displacement of each of the internal coordinates on a unit displacement of the ith normal coordinate and a zero displacement of all the other normal coordinates. Although all N atoms in a molecule are, in principle, set in motion for a $3N - 6$ vibrations, the normal mode is often dominated by the displacement of two or more atoms. Thus one speaks of the carbonyl $C{=}O$ vibration that lies between 1630 and 1800 cm^{-1}, depending on the molecule. Many of the normal modes of DNA and proteins are localized modes such as the carbonyl vibration, or they correspond to simple deformation of the rings of the DNA bases or aromatic amino acid side chains.

This brief introduction to the theory of normal coordinates is sufficient to show the theoretical origin of the Raman marker bands for proteins and nucleic acids. Conformational changes in a polypeptide, protein, or nucleic acid molecule cause changes in the values of the **B** and **G** matrices because of the geometry change. The change in the components of the **G** matrix creates different values of the internal kinetic energy of the molecule as seen from the first term in the Hamiltonian [Eq. (1)]. A change in geometry has only a small effect on the force constants since the bond stretching and angle bending force constants do not change with conformational changes; however, van der Waals forces between nonbonded atoms and Coulombic interactions do change as the atoms undergo torsional displacements about single bonds. The geometry change induces a change in the internal kinetic energy of the molecule which induces a change in the vibrational frequencies. (Because of the change in the **G** matrix, the eigenfrequencies of the **GF** matrix product are no longer the same.)

This theoretical prediction is borne out experimentally. DNA and proteins are polymers of nucleotides and amino acids, respectively. They have a backbone or principal chain that contains a repeating structural motif. In nucleic acids this is the furanose–phosphate–furanose motif, and in proteins it is the repeating peptide group. It has been found that vibrations characteristic of the backbone chain motif, the so-called backbone vibrations, experience a change in frequency if the polypeptide or polynucleotide molecule undergoes a conformational change. This effect was discovered in the early 1970s when it was first shown that certain Raman bands of DNA,[19,20] RNA,[1,2,32] and polypeptides[33] were dependent on the backbone conformation. Normal coordinate calculations were subsequently made for oligo- and polypeptides[31,34,35] and polynucleotides[36,37]

[32] E. W. Small and W. L. Peticolas, *Biopolymers* **10,** 69 and 1377 (1971).
[33] T.-J. Yu, J. L. Lippert, and W. L. Peticolas, *Biopolymers* **12,** 2161 (1973).
[34] E. W. Small, B. Fanconi, and W. L. Peticolas, *J. Chem. Phys.* **52,** 4369 (1970).
[35] B. Fanconi, E. W. Small, and W. L. Peticolas, *Biopolymers* **10,** 1277 (1971); B. Fanconi, E. W. Small, and W. L. Peticolas, *Biopolymers* **12,** 2161 (1971).
[36] E. B. Brown and W. L. Peticolas, *Biopolymers* **14,** 1259 (1975).
[37] K. C. Lu, E. W. Prohofsky, and L. van Zandt, *Biopolymers* **16,** 2491 (1977).

and the Raman marker bands assigned for both. Many new advances in the calculation of the normal modes of peptides[31] and nucleic acids[15] have been made. These normal mode motions are discussed below.

Neither DNA nor RNA molecules have a uniform conformation but rather one that is undergoing constant change as the biological functions of transcription, replication, and RNA-directed protein synthesis are being carried out. Because of this polymorphism, Raman spectroscopy can play a role in establishing the structure of DNA *in vitro* and *in vivo*. Globular proteins, on the other hand, tend to fold to a conformation that is stabilized by internal hydrogen bonds and hydrophobic interactions that cause its structure to be relatively independent of the environment. The structure of globular proteins is determined in the crystalline state using X-ray diffraction techniques, and the same structure persists in solution.

It is of great interest to obtain as much structural information as possible on DNA and RNA as a function of the molecule environment and base sequence through the use of Raman spectroscopy. From the Raman spectra of fibers, films, and crystals of proteins, polypeptides, and nucleic acids where the conformation is known, Raman marker bands can be found that allow the determination of the conformation of nucleic acids *in vitro* and *in vivo*. From a comparison of the Raman spectra of a macromolecule in the crystal and in solution, one can tell if the structure is the same in the two phases. Often substantial differences are found between the Raman spectrum of an oligonucleotide in the crystal and in solution, indicating that a major change in configuration has occurred. The secondary structure of proteins as determined by Raman spectroscopy is also of interest, but the approach must be different because the Raman spectra of proteins in crystals are almost identical to those in solution. There are ways of using Raman spectra of proteins and oligopeptides to obtain interesting information on certain structural details of proteins and on the structure of protein–ligand interactions.

Raman Spectra and Conformation of Polypeptides and Proteins

Early work by Davidson and Fasman showed that one could prepare poly(L-lysine) in the α helix, β sheet, and random coil by varying the pH, temperature, and salt conditions.[38] Subsequently the different conformations were shown to have remarkable differences in their Raman spectra.[33] The amide I band shows a large change in frequency with change in conformation. The amide I is a vibration of the peptide group that may be considered to consist of two resonance forms:

$$-C(=O)-NH- \rightleftharpoons -C(-(O^-))=N^+H-$$

[38] B. Davidson and G. D. Fasman, *Biochemistry* **6**, 1616 (1967).

This mode is a combination of the C=O stretch and the C–N–H angle bending. It is apparent that an environment that will tend to increase the polarization of the amide group, thus increasing the contribution of the resonance from on the right-hand side above, will lower the bond order of the carbonyl bond. This will decrease the force constant of the C=O bond stretching coordinate and hence lower the vibrational frequency. Hydrogen–deuterium exchange will cause the frequency to decrease because of the increased mass of the deuteron. It was found that the amide I vibration changed from 1650 ± 5 cm^{-1} in the α-helical form in H_2O to 1670 ± 3 cm^{-1} in the β-sheet form in H_2O, indicating that the carbonyl group is more strongly polarized, that is, has a higher dipole moment in the α helix than in the β sheet. A frequency of 1632 ± 5 cm^{-1} was found for the α-helical form in D_2O, whereas a frequency of 1659 ± 3 cm^{-1} was found for the β sheet in D_2O.[31,32]

Table I presents a summary of some of the Raman active modes that are useful for obtaining information about the structure and dynamics of proteins. Detailed references to the original literature are also given.[39-54] The lowest observed Raman frequencies in proteins fall in the region of 20–40 cm^{-1}. These may be overall global motions of the globular protein, or they may involve side-chain torsional motions. If they involve the

[39] W. L. Peticolas, this series, Vol. 61, p. 425.

[40] K. G. Brown, S. C. Erfurth, E. W. Small, and W. L. Peticolas, *Proc. Natl. Acad. Sci. U.S.A.* **69,** 1467 (1972).

[41] L. Genzel, F. Keilmann, T. P. Martin, G. Winterling, Y. Yacoby, H. Frolich, and M. W. Makinen, *Biopolymers* **15,** 219 (1976).

[42] B. G. Frushour and J. L. Koenig, *Biopolymers* **13,** 1809 (1974).

[43] N.-T. Yu, C. S. Liu, and D. C. O'Shea, *J. Mol. Biol.* **70,** 117 (1972).

[44] M. C. Chen, R. C. Lord, and R. Mendelson, *J. Am. Chem. Soc.* **96,** 3038 (1976); M. C. Chen, R. C. Lord, and R. Mendelson, *Biochim. Biophys. Acta* **328,** 252 (1973).

[45] J. F. R. Kuck, Jr., E. J. East, and N.-T. Yu, *Exp. Eye Res.* **22,** 1 (1976).

[46] N.-T. Yu and E. J. East, *J. Biol. Chem.* **250,** 2196 (1975).

[47] M. N. Siamwiza, R. C. Lord, M. C. Chen, T. Takamatsu, I. Harada, H. Matsura, and T. Shimanouchi, *Biochemistry* **14,** 4870 (1975).

[48] W. L. Peticolas, in "Raman Spectroscopy: Linear and Nonlinear," p. 694. Wiley, New York, 1982.

[49] T. Miura, H. Takeuchi, and I. Harada, *Biochemistry* **27,** 88 (1988).

[50] H. Sugeta, A. Go, and T. Miyazawa, *Chem. Lett.,* 83 (1972); H. Sugeta, A. Go, and T. Miyazawa, *Bull. Chem. Soc. Jpn.* **46,** 3407 (1973).

[51] N.-T. Yu, D. C. DeNagel, D. Jui-Yuan Ho, and J. F. R. Kuck, in "Biological Applications of Raman Spectroscopy, Volume 1: Raman Spectra and the Conformations of Biological Macromolecules" (T. G. Spiro, ed.), p. 47. Wiley, New York, 1987.

[52] W. Qian and S. Krimm, *Biopolymers* **52,** 1025 and 1503 (1992).

[53] H. Li, Z. Chen, J. E. Johnson, and G. J. Thomas, Jr., *Biochemistry* **31,** 6637 (1992).

[54] H. Li, C. Hanson, J. A. Fuchs, C. Woodward, and G. J. Thomas, Jr., *Biochemistry* **32,** 5800 (1993).

longitudinal dynamics of the polypeptide chain, they are sometimes referred to as skeletal acoustic in analogy with the acoustical phonon modes in synthetic homopolypeptides. The optical and acoustical phonon dispersion curves have been calculated for both glycine and α-helical polyalanine.[31,35,36] See Refs. 39 and 55–57 for reviews of Raman and neutron scattering measurements of low-frequency motions in proteins.

Of the modes listed in Table I, one of the most sensitive to conformation in the Raman spectrum is the amide I mode. This band fortunately falls in the range 1630–1700 cm^{-1} where no other protein bands lie. (The amide III is also very conformationally sensitive, but it is in a frequency region with many side-chain vibrations.) However the liquid water H–O–H bending mode is a broad Raman feature that overlaps with the amide I band so that the contribution of the water must be subtracted out.

The analysis of the amide I band to obtain the estimation of protein secondary structure content in terms of percentage helix, β strand, and reverse turn that was developed by Williams has proved very successful and has now been used by numerous workers.[58–62] In this method the amide I region is analyzed as a linear combination of the spectra of the reference proteins whose structures are known. As noted above the Raman spectra of globular proteins in the crystal and in solution are almost identical, reflecting the compact nature of the macromolecules. Thus one may use the fraction of each type of secondary structure determined in the crystalline state by the X-ray diffraction studies for proteins in solution. If there are n reference proteins with the Raman spectrum of each of them represented as normalized intensity measurements at p different wave numbers, then this information is related by the following matrix equation:

$$Ax = b$$

where A is a p by n matrix, b is a p-dimensional vector that contains the normalized spectrum of the protein being analyzed, and x is an n-dimensional vector that maps the intensities of the n reference proteins at a given frequency into the intensity at the same frequency of the protein being analyzed. This equation is solved for the vector x. The calculation

[55] H. D. Middendorf, *Annu. Rev. Biophys. Bioeng.* **13**, 425 (1984).
[56] J. A. McCammon and S. C. Harvey, "Dynamics of Proteins and Nucleic Acids." Cambridge Univ. Press, Cambridge, 1987.
[57] S. Cusak, *Chem. Scr.* **29A**, 103 (1989).
[58] R. W. Williams, A. K. Dunker, and W. L. Peticolas, *Biophys. J.* **32**, 232 (1980).
[59] R. W. Williams, *J. Mol. Biol.* **166**, 581 (1983).
[60] R. W. Williams, this series, Vol. 130, p. 311.
[61] H. DeGrazia, J. G. Harman, G. S. Tan, and R. M. Wartell, *Biochemistry* **29**, 3557 (1990).
[62] G.-S. Tan, P. Kelly, J. Kim, and R. M. Wartell, *Biochemistry* **30**, 5076 (1991).

TABLE I
RAMAN ACTIVE BACKBONE AND SIDE CHAIN VIBRATIONS OF PROTEINS

Origin and frequency (Δcm^{-1})	Assignment	Structural information	Refs.
Backbone			
Skeletal acoustic			
25–30	Mode of large portion of protein; possibly intersubunit mode or side-chain torsion	Overall structure; subunit interactions	39–41
75	Torsion mode	α Helix	41
Skeletal optical			
935–945 (950 D$_2$O)	C–C stretch (or C$^\alpha$–C–C stretch)	α Helix	33, 42–45
900	C–C stretch	These broaden and lose intensity with denaturation	33
963	—		
1002 (H$_2$O)	C$^\alpha$–C' or C$^\alpha$–C$^\beta$ stretch	Suggestive of β-pleated sheet	33
1012 (D$_2$O)	—	—	
1100–1110	C–N stretch	Conformation change marker broadens and loses intensity with denaturation	44
Amide I	Amide C=O stretch coupled to N–H wagging	Strong band; hydrogen bonding lowers amide I frequencies (see text)	31, 33, 46
1655 ± 5	—	α Helix (H$_2$O)	31, 33
1632 (D$_2$O)	—	α Helix (D$_2$O)	33
1670 ± 3	—	Antiparallel β-pleated sheet	31, 33, 46
1661 ± 3 (D$_2$O)	—	α Helix (D$_2$O)	33, 45
1665 ± 3		Disordered structure (solvated)	33, 45
1658 ± 2 (D$_2$O)	—	—	33, 42
Amide III	N–H in plane bend, C–N stretch	Strong hydrogen bonding raises amide III frequencies	33, 44
>1275	Amide III weak	α Helix, no structure below 1275 cm^{-1}	33, 44
1235 ± 5 (sharp)	Amide III strong	Antiparallel β-pleated sheet	33, 44
983 ± 3 (D$_2$O)	Charged coil?	Disordered structure	33, 43
1245 ± 4 broad	Charged coil?	Disordered structure	33, 43
Amino acid chains			
Tyrosine doublet			

TABLE I (*continued*)

Origin and frequency (Δcm^{-1})	Assignment	Structural information	Refs.
850/830	Fermi resonance between ring fundamental and overtone	State of tyrosine —OH $I_{850}/I_{830} = 9:10$ to $10:3$, H bond from acidic proton donor; $10:9$ to $3:10$, strong —OH bond to negative proton acceptor	47
Tryptophan			
880/1361	Indole ring	Ring environment; sharp intense line for buried residue; intensity diminished on exposure or environmental change	48, 49
Phenylalanine			
1006	Ring breathing	Conformation-insensitive frequency/intensity reference	4
624	Ring breathing	Ratio with tyrosine 664 cm^{-1} to estimate Phe/Tyr	4
Histidine			
1409 (D_2O)	*N*-Deuteroimidazole	Possible probe of ionization state, metalloprotein structure, proton transfer	4
S—S—			
510	S–S stretch	Gauche–gauche–gauche; broadening and/or shifts may indicate conformational heterogeneity among disulfides	50–52
525	S–S stretch	Gauche–gauche–trans	50–52
540	S–S stretch	Trans–gauche–trans	50–52
C—S			
630–670	C–S stretch	Gauche	50–52
700–745	C–S stretch	Trans	50–52
S—H	S–H stretch		
2560–2580		Environment, deuteration rate	46, 51, 53, 54
Carboxylic acids			
1415	$C\!\!\underset{O}{\overset{O}{\diagdown\!\!\!\diagup}}$ stretch	State of ionization	4
1730	$C\!\!\underset{OH}{\overset{O}{\diagdown\!\!\!\diagup}}$, C=O stretch	Metal complexation	4
	$C\!\!\underset{OR}{\overset{O}{\diagdown\!\!\!\diagup}}$, C=O stretch		

of the fraction of each type of secondary structure is given as a weighted sum of the structures of the reference proteins represented in the matrix $\mathbf{A}$:

$$f = FX$$

where $\mathbf{F}$ is an m by n matrix containing m classes of secondary structure and n reference proteins. The types of secondary structure used by Williams[60] are total helix, β strand, turn, undefined, ordered α helix, and disordered helix, so $m = 6$. The disordered helix contribution is defined as a combination of all residues in the interior of an α helix. One of the big advantages of this type of measurement is that it permits one to measure the change in the secondary structure of a DNA-binding protein on binding to DNA.[61,62]

Dynamics of Protein Structure from Proton or Deuterium Exchange Rates

In addition to the structural information on proteins that is obtained from the vibrations of the polypeptide backbone, the vibrations of the side chains can also give valuable information on the environment of the side chain and sometimes on the accessibility to protons and deuterons. The ability to probe the accessibility of the side chains of a protein is one of the important features of Raman spectroscopy for the study of protein dynamics.

The state of tyrosine in proteins is often of great importance. It has been shown that the ratio of the intensity of the Raman bands at 850 to 830 cm^{-1} is sensitive to the environment of the phenolic -OH. This doublet is due to Fermi resonance between a ring-breathing vibration and an overtone of an out-of-plane ring-breathing vibration.[47] It was found that a strong hydrogen bond from the phenolic hydrogen to a negative acceptor results in a low value of I_{850}/I_{830} (about 3 : 10), whereas hydrogen bonding to the phenolic oxygen from an acidic proton donor yields a higher value (about 10 : 3) for this ratio.

An example of the use of this technique is the interesting results on the buried tyrosines in fd filamentous phage coat protein.[63] The phage particle contains about 2700 copies of a largely α-helical coat protein. The two tyrosines at positions 21 and 24 of the fd coat protein provide a potentially useful probe of the phage structure. Several peaks arising from the tyrosines are clearly evident in the Raman spectrum of the virus.[63] The measured intensity ratio I_{850}/I_{830} is about 10 : 3. According to the above interpretation, such a ratio indicates that the tyrosine OH groups receive

[63] A. K. Dunker, R. W. Williams, and W. L. Peticolas, *J. Biol. Chem.* **254**, 6444 (1979).

hydrogen bonds from a very strong acidic proton donor. Although similar ratios had been observed for model compounds, such a ratio was not expected for a protein.[47] The tyrosines and their acidic donors in the filamentous phage fd were shown to be inaccessible to solvent by pH titration. It was shown that the acidic donors do not become titrated over the pH range of 7 to 12 by the fact that the intensity ratio remains exactly the same. This proves that there is no dynamic opening of the coat proteins in this filamentous phage to permit either protons or deuterons to access the tyrosines. Further confirmation of this effect comes from the observation in ultraviolet absorption spectra of no ionization of the phenolic OH over this pH range.[63] From this one can conclude that the fd filamentous phage coat protein is extremely impermeable to penetration by protons or solvent.

The S–H vibration in proteins has been studied by Yu et al.[46,51] and Li et al.[53,64] This vibration, which occurs in the range of 2500–2600 cm^{-1}, is far from any other Raman bands including the librational bands of water that extend from less than 1 cm^{-1} to about 1750 cm^{-1}. This permits one to observe the S–H vibration in proteins even though the concentration of the -SH group is very low and there is no resonance enhancement. (Sometimes, however, fluorescence may be a problem.) Yu et al.[46,51] have studied the disappearance of the S–H vibration in lens on aging that is caused by the oxidation of the -SH groups to disulfide (—S—S—) groups. Li et al.[53,64] have studied the effect of the environment on the frequencies of the S–H vibration. Li et al.[54] have determined the pK_a values of the active-center cysteines in thioredoxin from Escherichia coli. This is a very important way of characterizing the environment and accessibility of the various -SH groups in a protein. The Raman titration data indicate that the pK_a of Cys-32 is 7.1 whereas that of Cys-35 is 7.9.[54]

On hydrogen–deuterium exchange the -SH group changes to -SD, and the frequency changes to 1850–1880 cm^{-1}. Patapoff et al.[65] have shown how one can get some idea of the accessibility of the -SH group by measuring the rate of H/D exchange. Glyceraldehyde-3-phosphate dehydrogenase (GPDH) contains four subunits, and each subunit contains three -SH groups, one of which is at the active site of each subunit. Figure 1 shows the Raman spectrum of GPDH in aqueous pH 7 solution with the frequency of each of the bands carefully labeled. One can see from the spectrum the correspondence between the list of frequencies in Table I and the bands observed in a typical protein Raman spectrum. There are

[64] H. Li and G. J. Thomas, Jr., J. Am. Chem. Soc. **113**, 456 (1991); H. Li and G. J. Thomas, Jr., J. Am. Chem. Soc. **114**, 7463 (1992).
[65] T. W. Patapoff, S. A. Bernhard, and W. L. Peticolas, J. Biol. Chem. submitted.

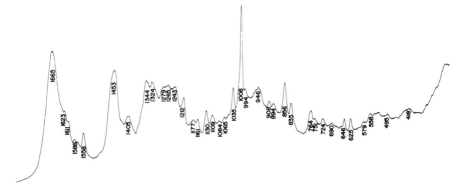

FIG. 1. Raman spectrum of apoglyceraldehyde-3-phosphate dehydrogenase (GPDH) in pH 7 aqueous solution. The wavelength of each of the bands is labeled (in cm^{-1}).

no Raman bands with frequencies higher than the 1665 cm^{-1} amide I band until one reaches the 2550 cm^{-1} region, where the Raman bands of the SH groups are found. Figure 2 shows the S–H vibrations of GPDH on the left-hand side and the S–D vibrations on the right-hand side. The H/D exchange occurs when GPDH is dissolved in D_2O. The lower S–D spectrum on the right is that of GPDH a few minutes after solution in D_2O. Here only the band at 1875 cm^{-1} is present. The top spectrum in Fig. 2 shows the same region after H/D exchange has occurred for over a day. Thus, there is a very rapidly exchanging -SH and a very slowly exchanging -SH. The former is assigned to the -SH at the active site

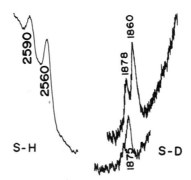

FIG. 2. (Left) Raman spectrum of the sulfhydryl (-SH) groups in apoglyceraldehyde-3-phosphate dehydrogenase. (Right) Raman spectra of the -SD group of the same protein a few minutes after dissolving the protein in D_2O (*bottom*) and 24 hr after solution in D_2O (*top*), showing that there are two classes of -SH groups: one that exchanges quickly and one that exchanges slowly.[54,65]

because the active site -SH would be expected to be solvent accessible, whereas the latter is assigned to the other two -SH groups in the protein.

Two Raman bands at 880 and 1360 cm^{-1} have been found useful for the characterization of individual tryptophan side chains in proteins using Raman spectroscopy and hydrogen–deuterium exchange kinetics.[48,49] The frequency of the 880 cm^{-1} band reflects the strength of hydrogen bonding at the N^1H site of the indole ring; the lower the frequency, the stronger the hydrogen bonding. The intensity of the 1360 cm^{-1} band, on the other hand, is a marker of the hydrophobicity of the environment of the indole ring; it is particularly strong in hydrophobic environments.[48,49,66] It has been demonstrated that a combination of stepwise deuterium exchange kinetics of the different tryptophan side chains in lysozyme can give a good idea as to the solvent accessibility. The band at 1360 cm^{-1} decreases to zero as all of the tryptophans are exchanged while a band at 1382 cm^{-1} grows in. It is found that the first two tryptophan side chains [labeled Trp(1) and Trp(2)] are deuterium exchanged within 4 hr even at 4°. The third one [Trp(3)] is exchanged 2 hr after solution at 22°, whereas the fourth one [Trp(4)] takes 4 hr at 35°, the fifth one [Trp(5)] 2 hr at 55°, and the last one [Trp(6)] 20 hr at 55°. The six tryptophans have been assigned to residues 62, 63, 108, − 123, − 111, and − 28, respectively. The extraordinary difference in the requirements for deuterium exchange shows the great variability in the environment and solvent accessibility of the side chains. Two of them are essentially exposed on the surface and undergo H/D exchange very rapidly, whereas one is so deeply buried that it is almost impossible to make it undergo H/D exchange unless one unfolds the lysozyme in a very hot bath or uses saturated LiBr.

Another side-chain vibration of importance shown in Table I is the disulfide bridge stretching vibrations. Sugeta et al.[50] have shown that one can distinguish the gauche–gauche–gauche, gauche–gauche–trans, and trans–gauche–trans conformations by the respective frequencies at 510, 525, and 540 cm^{-1}, respectively. More recently two detailed normal mode calculations of the dependence of S–S, C–S, and S–H vibrations on conformation have been made that permit one to know the Raman frequency as a function of dihedral angle about the S–S bond in detail.[52]

A great deal of work has been done on the vibrational spectra of reverse turns. These studies, however, do not lend themselves well to determination of protein structure because the Raman bands of the turns are buried in the Raman bands of the backbone. However, they are useful in studies of oligopeptide that form turns in crystals.[66,67] Several reviews

[66] P. Moenne-Loccoz and W. L. Peticolas, *J. Am. Chem. Soc.* **114,** 5893 (1992).
[67] J. Bandekar, *Vib. Spectrosc.* **5,** 143 (1993).

of the Raman spectra have appeared[31,67–69] that may be consulted for additional details in the application of Raman spectra to proteins and peptides.

Raman Data on Melting Transitions in Nucleic Acids

In this section we discuss the change that occur in the melting of DNA and RNA. The term melting is used to refer a thermally induced transition of a double-helical nucleic acid from an ordered double helix to a disordered state. Native RNA is usually single stranded, but it folds back on itself to form double helical sequences. Raman spectroscopy does not appear to be sensitive to the tertiary structure of RNA, but it can tell what fraction of the RNA is in double-helical form.[70,71]

The change from an ordered to a disordered conformation occurs over a small temperature range because of the cooperativity of the transition. In native RNA this is a transition from a double-helical structure to a single strand, but for synthetic RNA such as poly(C) · poly(G) and poly (U) · poly(A) two separate strands are combined to make a single double helix that melts to two single strands. This melting transition induces intensity changes in certain of the Raman bands in both RNA[32,70,71] and DNA.[23] Table II lists certain of the more important bands that change during melting of ribonucleic acids and synthetic RNA homopolymers.[32,70,71] A number of the Raman bands arising from the base vibrations in the RNA increase in intensity as the melting occurs. This effect has been called "Raman hypochromism," and a theory for this effect has been suggested.[1] In addition to intensity change in the Raman bands of the bases that are induced by changes in base stacking, the A-form marker band at 814 cm^{-1} shifts to 785 cm^{-1}. This shift is so clear-cut that the band at 814 cm^{-1} is seen to disappear as the double helix melts, while the 780 cm^{-1} band is seen to increase.

Synthetic RNA polymers such as poly(A) · poly(U) go from 100% A-form helix to 0% A-form helix as the melting transition occurs. This fact can be used to correlate quantitatively the changes in the intensity of the Raman bands so that the amount of an RNA molecule in the A form can be obtained from the intensity of certain of the Raman bands of an RNA. It is well known that DNA is usually found in the B form, but it can be transformed into the A form by changing the environment.

[68] J. Bandekar, *Biochim. Biophys. Acta* **1120,** 123 (1993).

[69] G. J. Thomas, Jr., *in* "The Encyclopedia of Molecular Biology: Fundamentals and Applications" (R. A. Meyers, ed.), in press. VCH, Weinheim, 1993.

[70] G. J. Thomas, Jr, and K. A. Hartman, *Biochim. Biophys. Acta* **312,** 311 (1973).

[71] K. G. Brown, E. J. Kiser, and W. L. Peticolas, *Biopolymers* **11,** 1855 (1972).

TABLE II
CHANGES IN RAMAN BANDS OF RNA INDUCED BY MELTING

Frequency (cm⁻¹)		Intensity change observed on melting of RNA	Assignment
H₂O, pH 7	D₂O, pD 7		
670	670	Large decrease	G backbone in A form
725	720	Large decrease	A ring mode
785	780	Increase due to shirt in 814 cm⁻¹ band to 785 cm⁻¹	U, C, —O—P—O—N stretch
814	814	Disappears completely, shifts to 785 cm⁻¹	N—O—P—O—N symmetrical stretch
867	860	No change	Ribose
915	915	No change	Ribose
975	990	Decrease	Ribose
1003		No change	A, U, C
1047	1045	Small decrease	Ribose–phosphate
1100	1100	No change	—PO₂⁻¹—symmetrical stretch
1182	1185	Small increase	A, G, C
1240		Large increase	U
1248		Small increase	A, C
	1310	Small increase	C
1320	1320	Small increase	A, G
1340	1340	Increase	A
1375	1370	No change	A, G
1420		No change	G, A
1460	1460	No change	Ribose
1484	1480	Small increase	G, A
1527	1526	Small increase	C, G
1575	1578	Small increase	G, A
1660	1658	Large increase	U : C-4 carbonyl; C : C-2 carbonyl
1692	1688	Increase	U : C-2 carbonyl

One characteristic difference between the A and B forms of DNA is the pucker of the furanose rings that changes from C-2′-endo–anti in the B form to C-3′-endo–anti in the A form. It is this change in ring pucker that gives rise to the A-form marker band at 811 ± 3 cm⁻¹ mentioned above.

A simple method of estimating the amount of C-3′-endo–anti ring pucker in DNA[71] or RNA[70] in solution, fibers, or crystals has been developed. The method involves taking the ratio of the intensity of the conformationally dependent band at 811 ± 3 cm⁻¹ to the conformationally independent band at 1098 cm⁻¹. The latter band is assigned to the —PO₂⁻¹— symmetric oxygen stretching vibration. The 811 ± 3 cm⁻¹ band is a backbone vibration involving the sugar–phosphate linkages that occurs only when the furanose rings are in the C-3′-endo–anti conformation as it is in the A form.[19-23] Table III gives the ratio of the peak heights of the 811 ± 3 cm^{-1} band to the band at 1097 cm^{-1} for a large number of nucleic acids, mostly RNA. The intensity ratio is 1.65 ± 0.05 when the nucleic acid (RNA or DNA) is in the completely double-helical A form so that all furanose rings have the C-3'-endo ring pucker. This band is absent (zero intensity) when none of the furanose rings is in the C-3'-endo ring pucker. Table III illustrates the usefulness of the intensity of the band at 811 ± 3 cm^{-1} to determine the fraction of the furanose rings with the C-3'-endo–anti or A-type ring pucker. The Raman hypochromism (decrease in Raman intensity of the base vibrations) can also be measured quantitatively.[1,32] This effect along with the measurement of the fraction of residues in the A form cited above has been used to determine the correct secondary structure of a 5 S RNA in solution.[72]

The initial theory of Raman hypochromism[1,32] was based on the fact that the absorption intensity goes as the transition moment matrix element $(\mu_{ge})^2$ whereas the (preresonant) Raman intensity goes as the transition moment matrix element to the fourth power. If this simple theory were valid, then all of the bands in a Raman spectrum of a nucleic acid should change as the square of the ultraviolet absorption in going from the ordered to the melted form. This is not observed experimentally (see Tables IV and V). It seems likely that changes in the base stacking and/or hydrogen bonding in DNA may well lead to changes in the displacement of the potential minimum of the bases along the normal coordinates of the Raman active modes in the excited electronic state. The Franck–Condon overlap factors for the various base vibrations could then be increased or decreased depending on the direction of the change in displacement. This effect could lead to either decreasing or increasing Raman intensity on increasing order. Both increases and decreases in intensity with changes in double-helical order have been observed. Tables IV and V show the intensity and frequency changes that occur in DNA as the DNA goes from the B form to the disordered form.

Raman Characterization of A, B, and Z Forms of DNA

In general there are three families of conformations into which DNA may be induced, namely, the A genus, the B genus, and the Z genus. Crystals now have been prepared of a large number of oligonucleotides, and several examples of crystals of oligomers are now known that are in each conformational type. Raman spectroscopy determines the conformation of DNA by comparing the Raman spectrum of a DNA of unknown

---

[72] G. A. Luoma and A. G. Marshall, *J. Mol. Biol.* **125,** 95 (1978).

TABLE III

PERCENTAGE OF A-GENUS DOUBLE HELIX IN RNA MOLECULES

Substance	Temperature (°C)		pH		$I(810-915\ cm^{-1})/I(1098\ cm^{-1})$		Percent A-genus conformation	
	Ref. 68	Ref. 67	Ref. 68	Ref. 67	Ref. 68	Ref. 67	Ref. 68	Ref. 67
Poly(A)·poly(U)	15		7.0		1.7		100	
Poly(A)·poly(U)	65		7.0		0		0	
Poly(A)·poly(U)		30		5		1.63		99
Poly(A-U)·poly(A-U)		30		5		1.66		101
Poly(G)·poly(C)	95		7.0		1.4		82	
Poly(G)·poly(C)		30		5		1.63		99
Poly(I)·poly(C)	25		7.0		1.5		88	
Poly(I)·poly(C)		30		5		1.60		98
Poly(C)	18		6.9		1.7		100	
Poly(C)	22		7.0		1.2		70	
Poly(C)	75		7.0		0		0	
Poly(C)		30		5		1.60		98
Poly(A)	23		7.0		1.4		82	
Poly(A)	80		7.0		0		0	
Poly(A)		30		5		1.60		98
Poly(U)(Mg^{2+})	0		7.0		1.4			82
Poly(U)(Mg^{2+})	20		7.0		0			0
tRNA	—		7.0		1.6		94	
$tRNA^{fMet}$ (*E. coli* K12-MO7)		30		5		1.37		84
$tRNA^{Val}$ (*E. coli* B)		30		5		1.36		83
$tRNA^{Phe2}$ (*E. coli* K12-MO7)		30		5		1.40		85
R17 RNA		30		5		1.44		88
16 S rRNA (*E. coli* Q13)		30		5		1.53 ± 0.08		93 ± 5%
23 S rRNA (*E. coli* Q13)		30		5		1.40		85
DNA (fiber at 75% humidity)	—		—		1.64		96	

TABLE IV

RAMAN SPECTROSCOPIC CHANGES FOR DNA CONTAINING G AND C[a]

Assignment	Disordered form (cm⁻¹)	Change B to disordered ←	A form (cm⁻¹)	Change B to A ←	B form (cm⁻¹)	Change B to Z →	Z form (cm⁻¹)
G¹	—	—	—	—	—	inc	625
G²	—	—	665	inc	—	—	—
G³	680	dec	—	dec	680	dec	—
bk, C	782	—	783	—	784	shift	784
C-3'-endo bk	—	—	808	inc	—	inc	810sh
C-2'-endo bk	n.r.	dec	—	dec	829	dec	n.r.
bk	—	—	852	inc	—	inc	855
—PO₂⁻—	1094	—	1100	—	1094	dec	1094
G, C	—	—	1180	—	1180	dec/split	1180
	—	—	—	—	—	—	1188
C	1220sh	s.dec	1218	dec	1218	dec	1213
C	1240	inc	1242	s.inc	1240sh	inc/shift	1246
C	1255	inc	1252	inc	1260	inc/shift	1265
C	1292	—	1295sh	s.dec	1293	—	1291
G	1320	inc	1314	inc/shift	1318	inc	1317
G	1333sh	s.dec	n.r.	dec	1334	—	n.r.
G	1361	s.dec	1361	s.dec	1362	dec/shift	1355
G	—	—	1388	appears	—	—	—
bk, G, C	1416	shift	1417	shift	1420	split	1417
bk, G, C	—	—	—	—	—	—	1426
G	1486	—	1482	shift	1489	—	1486
G	1528	inc	—	—	—	—	—
G	—	—	1574	shift	1578	—	1578

[a] dec, Decrease in intensity; inc, increase in intensity; shift, change in band position; n.r., not resolvable; s., slight intensity change; sh, shoulder; G, guanine; C, cytosine; bk, deoxyribose phosphate backbone; G¹, C-3'-endo–syn; G², C-3'-endo–anti'; and G³, C-2'-endo–anti.

conformation with the standard spectra taken from crystals, fibers, etc., of DNA in a conformation that is known from X-ray diffraction analysis. As we discuss below, Raman spectroscopy can also detect conformational differences between members of the same genus. DNA molecules tend to change conformational significantly in going from the crystal to solution even when they remain in the same genus. Thus, conformational rules which relate sequence to specific structures that are based on coordinates obtained in the crystal from X-ray diffraction may not be of general validity for DNA molecules in solution.

In general we may distinguish three different types of conformational changes that can occur in DNA: (1) the melting transition from an ordered double-helical (usually B form) state to a disordered state with separated strands (this was discussed in the previous section); (2) a change from one double-helical conformational genus to another such as B to A or B to Z; and (3) environmentally induced changes in the conformation of a

TABLE V
RAMAN SPECTROSCOPIC CHANGES FOR DNA CONTAINING A AND T[a]

Assignment	Disordered form (cm⁻¹)	Change B to disordered ←	B form (cm⁻¹)	Change B to A →	A form (cm⁻¹)
bk	—	—	644sh	inc	644
T	667	s.dec	670	dec	666
			n.r.	inc	706
A	728	inc	729	—	729
T	747	dec/shift	750	dec/shift	747
bk, T	794	s.dec	793	shift	779
C-3'-endo bk	—	—	—	inc	807
C-2'-endo bk	n.r.	dec	841	dec	n.r.
—PO₂⁻—	1096	—	1094	shift	1102
T	1186	inc	1186	—	—
T, A	1208	dec	1209	dec	n.r.
T	1237	inc	1240sh	inc	1239
T, A	1255sh	dec	1255	dec	1255sh
A	1307	s.inc	1303	s.inc	1301
A	1334	s.inc/shift	1341	inc/shift	1334
T, A	1374	No change	1376	dec	1374
	—		n.r.	inc	1402
A, bk	1421	s.inc	1420	shift	(1415)
bk, A	1462	—	1462	inc	1462
A	1483	inc	1483	inc/shift	1478

[a] dec, Decrease in intensity; inc, increase in intensity; shift, change in band position; n.r., not resolvable; s., slight intensity change; sh, shoulder; A, adenine; T, thymine; bk, deoxyribose phosphate backbone.

double-helical DNA within a given genus, particularly the B genus. In the following sections we discuss the characterization of each of these conformational changes by Raman spectroscopy.

The marker bands for A, B, and C DNA were discovered in our laboratory in the 1970s using fibers of DNA.[19,20] This work was extended by Goodwin and Brahms[21] and more recently by Nishimura and Tsuboi[73,74] and Benevides and Thomas.[75] The original correlations between structure and spectra were confirmed by obtaining the X-ray diffraction patterns on the same fiber sample from which the Raman spectra were obtained.[19-21] In this way it was certain that the Raman spectra were assigned to the

[73] Y. Nishimura, M. Tsuboi, T. Nakano, S. Higuchi, T. Sato, T. Shida, S. Uesugi, E. Ohtsuka, and M. Ikehara, *Nucleic Acids Res.* **11**, 1579 (1983).
[74] Y. Nishimura, M. Tsuboi, and T. Sato, *Nucleic Acids Res.* **12**, 6901 (1984).
[75] J. M. Benevides and G. J. Thomas, Jr., *Nucleic Acids Res.* **11**, 5747 (1993).

TABLE VI
RAMAN SPECTROSCOPIC CHANGES FOR AT-CONTAINING
DEOXYOLIGONUCLEOTIDES UNDERGOING B TO Z TRANSITION[a]

B-Form AT Raman frequency (cm^{-1})	Change on going to Z-form	Z-Form AT Raman frequency (cm^{-1})
Not present	Increase	622
748	Increase	748
790	Shift	(748)
835	Shift	815
1302	Shift	1312
1342	Shift	1332
1374	Shift	1362

[a] Data from Refs. 78 and 79.

correct A, B, or C form. The marker bands for the Z conformation of DNA containing only alternating cytosine (C) and guanine (G) were discovered by Thamann et al.[24] Again, the Raman spectra were taken on the same sample [in this case a crystal of d(CG)$_3$] as was used for both the X-ray diffraction and Raman measurements. A great amount of recent work has helped to clarify the assignments of these Raman spectra, and this is discussed below.[8,26,15,73-77] More recently it has been found that DNA polymers and oligomers containing only adenine (A) and thymine (T) can be induced into the Z form using Ni(II) ions in aqueous solution.[78,79] From all of this work, a set of marker bands for the A, B, C, and Z forms have been determined. These bands are listed in Tables IV–VI, showing the changes that occur in the intensity and frequency of the vibrations of the backbone and the bases A, T, G, and C as the DNA goes from the canonical B form to the disordered single-stranded form, the A form, and the Z form. The Raman spectrum of the C form is essentially identical to that of the B form except that a sugar–phosphate band at about 870 cm^{-1} is observed.[19,21]

Raman Spectra of Deoxyoligonucleotide Crystals Using a Raman Microscope: Determination of Relation between Base Sequence and DNA Conformation and Polarization Effects

It is of interest to know the conformation of oligonucleotides of a certain length both in the crystal and in saturated salt solutions as a

[76] J. C. Martin and R. M. Wartell, *Biopolymers* **21**, 499 (1982).
[77] W. L. Peticolas, W. L. Kubasek, G. A. Thomas, and M. Tsuboi, *in* "Biological Applications of Raman Spectroscopy" (T. G. Spiro, ed.), Vol. 1, p. 81. Wiley, New York, 1987.
[78] J. P. Ridoux, J. Liquier, and E. Taillandier, *Biochemistry* **27**, 3874 (1988).
[79] Z. Dai, G. A. Thomas, E. Evertsz, and W. L. Peticolas, *Biochemistry* **28**, 699 (1989).

function of the base composition. Although this work was initiated by the X-ray crystallographers, the quality and size requirements of the crystals make the determination of the conformation of a large number of oligonucleotide crystals impossible using the X-ray technique. However, it has been shown that it is usually possible to obtain microcrystals or oligonucleotides that are perfectly formed but small (10–200 μm) on a side.[80,81] Such microcrystals may be too small for conventional Raman spectroscopy, but using an especially designed laser Raman microscope[5,80,81] it is possible to obtain the Raman spectra from the small crystals and to determine the genus to which they belong. This has permitted the construction of tables that relate the conformation of oligodeoxyribonucleotides to the base sequence. (For details with many references to the original literature, see Peticolas *et al.*[28]) These tables discuss these results in terms of two-letter base sequence code words along the double helix that tend to direct the sequence into the A, B, or Z forms. These relations have been helped by our being able to scan a large number of deoxyoligonucleotide crystals quickly to establish the secondary structure as a function of base sequence.[28]

Using the microscopic Raman technique, detailed polarization measurements have been made on a number of oriented crystals.[80–82] The Raman spectrum has been taken with the Raman light polarized parallel and perpendicular to the axis of the helix. In this way the elements of the Raman tensor can be obtained. The out-of-plane vibrations of the DNA bases are usually too weak to identify in solution. Using oriented crystals and the laser Raman microscope, the polarization vector of the incident light can be made either parallel or perpendicular to the helix axis. In the former position the polarization is perpendicular to the plane of the base. In this geometry, the intensity of the in-plane vibrations of the DNA bases becomes very weak, but the intensity of the out-of-plane modes becomes important.[80,82] There is a dramatic change in the intensity of the DNA base vibrations in the two different geometries, because the bases are perpendicular to the helix and hence crystal axis. In this way the frequencies of the out-of-plane modes can be identified. The intensity of the backbone vibrations does not change much in the two geometries because the furanose–phosphate backbone is not aligned either parallel or perpendicular to the helix axis but winds around it. The schematic set up of the laser Raman microscope has been discussed in detail in several papers[80–82] including a recent review article.[83] Finally it should be noted that micro-

[80] T. W. Patapoff, G. A. Thomas, Y. Wang, and W. L. Peticolas, *Biopolymers* **27**, 493 (1987).

[81] Y. Wang, G. A. Thomas, and W. L. Peticolas, *J. Biomol. Struct. Dyn.* **5**, 249 (1987).

[82] J. M. Benevides, M. Tsuboi, and A. H.-J. Wang, and G. J. Thomas, Jr., *J. Am. Chem. Soc.* **115**, 5351 (1993).

[83] W. L. Peticolas and E. Evertsz, this series, Vol. 211, p. 335.

scope Raman spectroscopy has been used to determine the conformation of DNA in living sperm cells[5] and in the bands of polytene chromosomes.[7]

Determination of Cartesian Coordinates of B/Z Junctions in DNA from Raman Data and Molecular Modeling

Until recently the only structural information obtained from Raman spectra was rough estimates of the percentage of each of the types of a secondary structure in a protein or nucleic acid. It would be much better to be able to generate actual Cartesian coordinates so that stereo diagrams could be made of the DNA or protein structure. This goal has been accomplished for deoxy oligomers containing either a B/Z junction or a B/Z/B junction.[79,84] The latter junction is possibly very important because, if the Z form of DNA exists in the genome, it must exist as a very minor fraction of the total number of bases. This means that it must exist as short segments of Z DNA in the midst of long sequences of B DNA. This leads to the concept of the B/Z/B double junction for each Z sequence in the genome.

To obtain the coordinates of an oligomer containing a B/Z or B/Z/B junction, one must first locate the position of the junction along the sequence and determine how many bases it contains. Most authors naively and incorrectly assume that the conformational junction occurs at the sequence junction. However, Raman experiments have shown conclusively that it is never at the sequence junction.[79,84] Consider the sequence d(TTTTTCGCGCGCG) · d(CGCGCGCGAAAAA) that forms a duplex of 5 A-T base pairs and 8 C-G base pairs all in the B form in 0.5 M salt solution. This may be seen in the bottom spectrum of Fig. 3, where the Raman spectrum of the duplex is shown. Only B-form marker bands are present, including the strong band at 684 cm^{-1} that is a mode that involves a vibration of guanine coupled to a B-form backbone. When the salt concentration is increased to saturation, a B/Z junction is formed. To prove this we examine the top spectrum in Fig. 3 which is taken from the same 13-mer in saturated NaCl solution. We note that none of the AT base pairs is in the Z form because none of the AT Raman marker bands for the Z form listed in Table VI is present. Therefore, all of the AT base pairs are in the B form. Thus, if some of the CG base pairs are in the Z form and some are in the B form, then there must be a junction between a Z-form sequence and a B-form sequence that occurs in the CG sequence.

The top spectrum in Fig. 3 shows that there is now a pair of bands at 680 and 626 cm^{-1}; the former arising from CG base pairs in the B form

[84] Z. Dai, M. Dauchez, G. Thomas, and W. L. Peticolas, *J. Biomol. Struct. Dyn.* **9,** 1155 (1992).

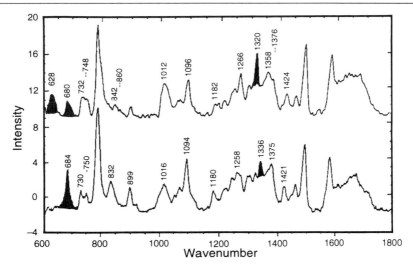

FIG. 3. Raman spectrum of 5'-d(TTTTTCGCGCGCG)·5'-d(CGCGCGCGAAAAA) in 0.5 M NaCl (*bottom*) and saturated (6 M) NaCl (*top*).

and the latter from CG base pairs in the Z form. We assume that the B form in this oligomer is contiguous, and so the B form must be propagated from the AT base pairs into the CG sequence. The next question is, where in the CG sequence is the junction? It has been shown that in the B to Z transition the total intensity of the guanine band at 625–685 cm^{-1} is conserved.[79,84] In the top Raman spectrum of Fig. 3, taken from the 13-mer duplex in saturated NaCl solution, the marker band for the G residues in the B form at 680 cm^{-1} has the same intensity as the marker band for the Z form at 625 cm^{-1}. This means that the number of CG base pairs in the B form equals the number of CG base pairs in the Z form; four CG base pairs are in the Z form and four are in the B form. Thus the B form starts on the side with all of the AT base pairs in the B form and is propagated four base pairs into the CG sequence where a junction to the Z form occurs.

With this knowledge one can construct a model for the B/Z junction using the constraints that (1) all Watson–Crick hydrogen bonds must be formed, (2) the bond lengths and bond angles are at the canonical position, and (3) the calculated potential energy using an AMBER force field is at a minimum. The first two conditions put a severe restraint on the model. To keep all of the Watson–Crick base pairs hydrogen bonded, there can be no base pairs in a junction.[84] There must be a sudden flipping out of the bases from the B to the Z form, creating a very sudden junction.

Figure 4 shows the stereo diagram calculated from the Cartesian coordinates of the energy-minimized structure. This is the first time that a set of Cartesian coordinates for an oligonucleotide has been obtained using only the Raman spectrum and molecular modeling. We note that the Raman spectrum puts severe limits on the model. It prevents the incorrect assumption that the B/Z junction occurs at the sequence junction between the AT and CG base pairs. If it turns out that X-ray diffraction studies show the exact structure to be incorrect, then we must revise the force field to obtain the correct structure. For example, if there were several base pairs in the junction without Watson–Crick base pairing, then it would mean that the AMBER force field puts too large an energy on the separated base pairs. This shows how the combination of molecular modeling and Raman spectroscopy can help to refine force fields in order to predict the structure of DNA and perhaps, some day, proteins.

Raman Studies of Conformational Deviations within B Family of DNA

The data discussed above and listed in Tables IV–VI can be applied to more subtle conformational variations of DNA that are totally within the

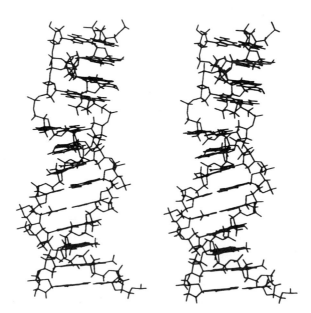

Fig. 4. Stereo view of the Watson–Crick duplex of the oligomer d(TTTTTCGCGCGCG) showing a B/Z junction in which the terminal four CG base pairs in the Z form bend toward the minor groove.

B conformation. Such changes are distinctly biologically relevant because virtually all of the DNA *in vivo* is in the B form,[5,7] so that changes in conformation are subtle and made within the B family. Interesting conformational changes occur owing to the interaction of DNA with DNA-binding proteins.

A noted characteristic of B-form DNA is that it is capable of a wide degree of conformational variation within B-form parameters. This conformational variability may play an important role in protein–DNA recognition in addition to the primary mechanism of recognition by the array of hydrogen bond donors and acceptors presented by a sequence of bases. This issue is difficult to address by crystallographic methods because of the difficulty in crystallizing large pieces of DNA, as well as the uncertainty that small deviations are masked by crystal packing forces. Another factor involved in protein–DNA interactions is that of DNA deformability. How much does a protein binding to a particular DNA sequence bend that DNA, and how does that bending contribute to the stability of the complex? These issues are addressed in several Raman studies of DNA–protein interactions. The proteins studied were *Eco*RI,[85] the catabolite activator protein (CAP) binding site,[86] and the cro protein.[87] Numerous studies on viruses have also been reported.[88] These studies are good examples of the use of Raman scattering to learn about subtle conformational changes that occur in DNA on binding to a DNA-binding protein. These have been discussed in a review[83] and are not elaborated here.

Parallel- and Triple-Stranded DNA

Raman spectroscopy can also facilitate the understanding of an unknown DNA conformation. This is the case in the Raman analysis of a parallel-stranded duplex DNA by Otto *et al.*[89] Here spectra were compared between a conventional antiparallel sequence, containing only adenine and thymine, and a sequence of identical composition, but which allowed only for the formation of a parallel-stranded duplex. The spectra of the parallel-stranded duplex showed the characteristic band around 840 cm^{-1} indicating a basically B-like conformation. By analyzing the 1600–1700

[85] G. A. Thomas, W. L. Kubasek, W. L. Peticolas, P. Greene, J. Grable, and J. M. Rosenberg, *Biochemistry* **28**, 2001 (1989).

[86] H. De Grazia, D. Brown, S. Cheung, and R. M. Wartell, *Biochemistry* **27**, 6359 (1988).

[87] E. Evertsz, G. A. Thomas, and W. L. Peticolas, *Biochemistry* **30**, 1149 (1991).

[88] G. J. Thomas, Jr., *in* "Biological Applications of Raman Spectroscopy, Vol. 1: Raman Spectra and the Conformations of Biological Macromolecules" (T. G. Spiro, eds.), p. 135. Wiley, New York, 1987.

[89] C. Otto, G. A. Thomas, K. Rippe, T. M. Jovin, and W. L. Peticolas, *Biochemistry* **30**, 3062 (1991).

cm^{-1} region of the spectra, the authors were able to determine that the base pairing was reverse Watson–Crick in the parallel-stranded duplex. This region has been assigned to carbonyl stretching vibrations that are sensitive to the hydrogen bond environment of the base pairs. In Watson–Crick and Hoogsteen base pairing the C-2 carbonyl group of the adenine participates in the base pair. In the parallel-stranded duplex the carbonyl region showed strong differences in the 1661 and 1690 cm^{-1} bands from the corresponding bands in the antiparallel duplex. This demonstrated that the bands were either reverse Watson–Crick or reverse Hoogsteen where the C-4 carbonyl was involved in the base pair. The possibility of reverse Hoogsteen base pairing was eliminated by observing that there was no change in the 1421 cm^{-1} band assigned to adenine base vibrations. This band is known to show a large decrease in intensity when the adenine participates in a Hoogsteen base pair in the triple-stranded poly(dT) · poly(dA) · poly(dT) sequence.[25] This study demonstrates that with the existing basis set of DNA on the various conformations of DNA it is possible to determine important conformational parameters without a crystal structure.

Two reviews of the Raman spectroscopy of nucleic acids have appeared[83,90] that should be consulted for additional information on this very complex subject.

[90] G. J. Thomas, Jr., and M. Tsuboi, *Adv. Biophys. Chem.* **3,** 1 (1993).

[18] Resonance Raman Spectroscopy of Metalloproteins

By Thomas G. Spiro and Roman S. Czernuszewicz

Introduction

The main objective of this chapter is to show how metal ion environments in metalloproteins and metalloenzymes can be studied by resonance Raman (RR) spectroscopy. We describe experimental techniques and methods of analysis and interpretation by using selected examples; an extensive literature has already been reviewed.[1–12]

[1] R. J. H. Clark and R. E. Hester (eds.), *Adv. Infrared Raman Spectrosc.* **1–12,** and *Adv. Spectrosc. (Chichester, U.K.)* **13–present** 1970–1985 (**1–12**) and 1986–present (**13–present**).

[2] C. B. Moore (ed.), "Chemical and Biochemical Applications of Lasers," Vols. 1–present. Academic Press, New York, 1974–present.

[3] T. G. Spiro, *Adv. Protein Chem.* **37,** 110 (1985).

Metalloproteins are natural products which in the biologically active form contain integral stoichiometries of metallic elements. They are usually highly elaborated coordination complexes whose metal-containing sites have one or more transition metal ions and protein ligands. These sites are involved in electron transfer, binding of exogenous molecules, and/ or catalysis. The ability of RR spectroscopy to identify the presence or absence of particular structural units, and the ligation modes of the amino acids, has made the technique especially powerful in studies of protein structure and function. The central advantage of RR spectroscopy is that, when the wavelength of the exciting laser line is adjusted to coincide with that of an allowed electronic transition in a molecule, the intensities of certain Raman bands are enhanced relative to the off-resonance values. The technique is therefore more sensitive than normal (off-resonance) Raman spectroscopy. More importantly, it is a more selective technique, because only those vibrational modes which are related to the chromophoric part of the scattering molecule can be enhanced at resonance. These enhanced modes are the only observable features. Consequently, vibrations of the chromophoric group can be monitored in complex biomolecules. If, as is often the case, the chromophore is itself a site of biochemical activity, then the RR spectrum can provide structural information about the chemically significant parts of a biomolecule and monitor structural changes associated with reactivity. This technique requires very small amounts of samples and is independent of the physical state of the sample, making it an ideal method for the study of proteins.

Sensitivity and Selectivity of Resonance Raman Spectroscopy: Scattering Mechanisms

Raman spectroscopy is a form of vibrational spectroscopy which, like infrared spectroscopy, reports transitions between different vibrational

[4] T. G. Spiro and B. P. Gaber, *Annu. Rev. Biochem.* **46,** 553 (1977).

[5] T. G. Spiro and P. Stein, *Annu. Rev. Phys. Chem.* **28,** 501 (1977).

[6] P. R. Carey, "Biochemical Applications of Raman and Resonance Raman Spectroscopies." Academic Press, New York, 1982.

[7] A. T. Tu, "Raman Scattering in Biology." Wiley, New York, 1982.

[8] F. S. Parker, "Applications of Infrared, Raman, and Resonance Raman Spectroscopy in Biochemistry." Plenum, New York, 1983.

[9] T. G. Spiro (ed.), "Biological Applications of Raman Spectroscopy," Vols. 1–3. Wiley, New York, 1987 (Vols. 1 and 2) and 1988 (Vol. 3).

[10] R. S. Czernuszewicz, *in* "Methods in Molecular Biology" (C. Jones, B. Mulloy, and A. H. Thomas, eds.), Vol. 17, p. 345. Humana Press, Totowa, New Jersey, 1993.

[11] Y. Wang and H. E. Van Wart, this series, Vol. 226, p. 319.

[12] T. M. Loehr and J. Sanders-Loehr, this series, Vol. 226, p. 431.

energy levels in a molecule. It differs from infrared spectroscopy in that information is derived from light scattering rather than a direct absorption process. Consequently, different selection rules govern the intensity of the respective vibrational modes. Infrared absorptions are observed for vibrational modes which change the dipole moment of the molecule, whereas Raman scattering is associated with normal modes which produce a change in the polarizability (or induced dipole moment) of the molecule. Symmetric stretching modes tend to be the most intense features in Raman spectra, whereas asymmetric stretches and deformation modes tend to be more intense in infrared spectra. This differential character makes Raman spectroscopy more favorable for the study of biological materials since, as shown in Fig. 1, there is very little spectral interference from the intra- and intermolecular deformation modes of water molecules; these are dominant features in the infrared spectra of aqueous samples.[13]

Resonance Raman scattering occurs when a material is irradiated with monochromatic light corresponding to an allowed absorption band region. The effect is studied by examining the intensity of scattered photons as a function of the proximity of the excitation wavelength to electronic absorption bands of the molecule. When the excitation falls close to or within an electronic absorption band of the sample, vibronic coupling with the electronically excited state increases (1) the probability of observing Raman scattering from vibrational transitions in the electronic ground state (10^3–10^4 increases in the intensities have been observed) and (2) the selectivity of the vibrational information because only vibrational modes associated with atoms at the absorbing center in the molecule are subject to intensity enhancement. This is illustrated in Fig. 2, which shows a dramatic increase in the intensity of the Cu–S(Cys) vibrational modes near 400 cm^{-1} as the excitation wavelength (647.1 nm) approaches that of the S $\rightarrow$ Cu charge transfer electronic transition ($\sim$600 nm) in the blue copper protein azurin. Thus the correct identification of the vibrational modes showing RR enhancement will aid in the assignment of the resonant electronic transition and vice versa. Because vibrational frequencies are sensitive to bond strength, number of atoms, geometry, and coordination environment, this technique provides information which is complementary to that obtained by X-ray crystallography and X-ray absorption spectroscopy. The general approach is to compare RR spectra of proteins with judiciously synthesized model complexes, containing biologically relevant ligands capable of simulating the ligation modes of the apoprotein, using

[13] K. Nakamoto and R. S. Czernuszewicz, this series, Vol. 226, p. 259.

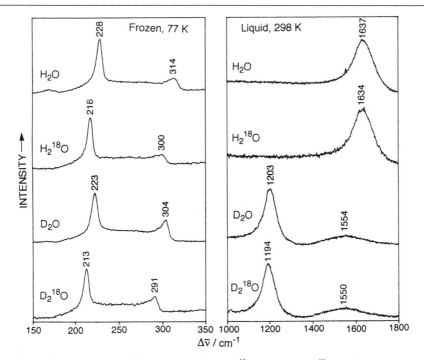

FIG. 1. Raman spectra of four waters, H_2O, $H_2^{18}O$, D_2O, and $D_2^{18}O$: (*left*) as ice at 77 K with a liquid N_2 cell (see Fig. 6), 135° backscattering directly off the ice surface, laser power 250 mW, spectral slit width 6 cm^{-1}; and (*right*) as neat liquid at room temperature with a capillary tube, 90° scattering, laser power 250 mW, spectral slit width 10 cm^{-1}. Exciting radiation (488.0 nm) for all RR spectra was provided by a Coherent Innova 90-6 Ar$^+$ ion laser. The scattered light was dispersed by a SPEX 1403 double monochromator equipped with 1800 grooves/mm holographic gratings and detected by a cooled Hamamatsu 928 photomultiplier tube under the control of a SPEX DM3000 data station as described elsewhere.[10]

isotope labeling and Raman band polarization properties to identify the vibrational modes.

If RR spectroscopy is to be applied fruitfully to structural problems of metal centers in biology, however, it is important to anticipate what kinds of vibrational modes are enhanced by resonance with what kinds of electronic transitions. For this purpose the basic theory of RR scattering provides a useful guide, as briefly outlined below.

For a molecule in a molecular state $|g\rangle$ (initial) perturbed by the electromagnetic wave of frequency ν_0 so that it passes into a molecular state $|f\rangle$ (final) while scattering light of frequency $\nu_0 \pm \nu_k$ ($\nu_k = \nu_f - \nu_g$), the total

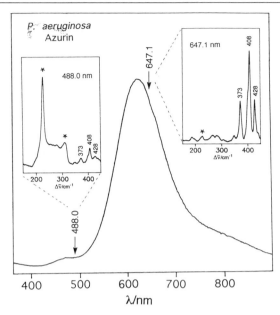

FIG. 2. Illustration of the selective enhancement of vibrational modes in a resonance Raman spectrum. The UV–visible absorption spectrum of a *Pseudomonas aeruginosa* azurin is shown together with two different Raman spectra that derive from laser excitation within the S(Cys) → Cu(II) charge transfer absorption band at 615 nm (647.1 nm, *right*) and away from the absorption (488.0 nm, *left*). Excitation within resonance leads to dramatically increased Raman scattering from the Cu active site, whereas the off-resonance excitation produces a spectrum dominated by bands of the nonchromophoric ice matrix (marked by asterisks). Exciting radiation was provided by Coherent Innova 90-6 Ar⁺ (488.0 nm) and K2 Kr⁺ (647.1 nm) ion lasers. Both RR spectra were recorded digitally by backscattering (135°) directly off the surface of a protein frozen solution kept in a liquid N_2 Dewar flask using the instrumentation as in Fig. 1. The protein sample (3 mM) was in 50 mM Tris-HCl buffer, pH 7.5.

intensity of scattered radiation, I_k (in photons per molecule per second over the solid angle 4π), is given by[14,15]

$$I_k \propto I_0(\nu_0 \pm \nu_k)^4 \sum_{\rho,\sigma} |(\alpha_{\rho\sigma})_k|^2 \tag{1}$$

where I_0 is the intensity of the incident radiation (laser light), and the summation is over ρ, $\sigma = x, y, z$ which independently refer to the molecule-fixed nonrotating Cartesian coordinate system and represent the polariza-

[14] J. A. Koningstein, "Introduction to the Theory of the Raman Effect." Reidel, Dordrecht, The Netherlands, 1972.
[15] D. A. Long, "Raman Spectroscopy." McGraw-Hill, New York, 1977.

tions of the incident (σ) and scattered light (ρ). $(\alpha_{\rho\sigma})_k$ is the scattering or polarizability tensor component for the transition k, and is given by[16,17]

$$(\alpha_{\rho\sigma})_k = (1/h) \sum_r \left[\frac{\langle f|M_\rho|r\rangle\langle r|M_\sigma|g\rangle}{\nu_r - \nu_g - \nu_0 + i\Gamma_r} + \frac{\langle f|M_\sigma|r\rangle\langle r|M_\rho|g\rangle}{\nu_r - \nu_f + \nu_0 + i\Gamma_r} \right] \qquad (2)$$

where $|g\rangle$ and $|f\rangle$ are the initial and final states of the molecule and the summation is over all the intermediate states $|r\rangle$. The integrals $\langle f|M_\rho|r\rangle$ and $\langle r|M_\sigma|g\rangle$ are the electric dipole transition moments, along the ρ and σ directions; M is the electron position operator; $h\nu_0$ is the energy of the incident radiation; and Γ_r is a damping factor (which prevents the denominator at resonance from reaching zero), reflecting the finite natural lifetime and sharpness of the intermediate state, $|r\rangle$.

The intermediate states, $|r\rangle$, are the electronically excited molecular eigenstates, each eigenstate being weighted by an energy denominator according to its nearness to resonance. Far from resonance ($\nu_r - \nu_g \gg \nu_0$) the magnitude of $(\alpha_{\rho\sigma})_k$ is independent of ν_0; the number of excited electronic states contributing to the polarizability is large. In this case all the energy denominators in the expression for the element $\alpha_{\rho\sigma}$ of the scattering tensor α_k are large, and hence the weighting factors are small and nonselective (normal Raman or Raman scattering). As a result, the intermediate state has no well-defined symmetry and only symmetric components of α_k, $\alpha_{\rho\sigma} = \alpha_{\sigma\rho}$, make any significant contribution to the Raman scattering. This also explains the general weakness or absence of overtones in normal Raman scattering.

The situation changes under conditions of resonance ($\nu_r - \nu_g \approx \nu_0$) with a particular electronic state, $|e\rangle$, because this state now dominates the sum over the other states in the scattering tensor α_k, provided that it has a large enough transition dipole moment from the initial (ground) state. Furthermore, the first term in Eq. (2) becomes dominant, and the scattered intensity is expected to increase drastically when the frequency of the incident light is tuned to the frequency of the electronic transition, ν_{eg}. [The second term in the square brackets of Eq. (2) is nonresonant and produces a negligible, slowly varying background contribution to the large resonant part when ν_0 is in the resonance region.] One of the energy denominators in α_k becomes small, leading to a large weighting factor for the resonant eigenstate, $|e\rangle$. The intermediate state, $|r\rangle$, in the scattering process [Eq. (2)] assumes the symmetry and geometry of this dominant excited state, often leading to overtone progressions and sometimes to contributions from antisymmetric tensor components. Also, although fre-

[16] A. C. Albrecht, *J. Chem. Phys.* **34**, 1476 (1961).
[17] J. Tang and A. C. Albrecht, *J. Chem. Phys.* **49**, 1144 (1968).

quencies of the RR scattered photons are still a function of the electronic ground state, the intensities are determined by the properties of the electronic excited state.

To obtain information as to which vibrations are subject to resonance enhancement, we apply (following Albrecht[16]) the adiabatic Born–Oppenheimer approximation of separability of electronic and vibrational wave functions for $|g\rangle$, $|r\rangle$, and $|f\rangle$ states in Eq. (2). Thus, by taking the initial and final states to belong to the ground electronic state and the intermediate state to the resonant excited electronic state, and dropping the nonresonant term, Eq. (2) can be rewritten as

$$(\alpha_{\rho\sigma})_k = (1/h) \sum_v \{[\langle j|(\mu_\rho)_e|v\rangle\langle v|(\mu_\sigma)_e|i\rangle]/(\nu_{vi} - \nu_0 + i\Gamma_v)\} \tag{3}$$

where $(\mu_\rho)_e = \langle g|(\mu_\rho)_e|e\rangle$ and $(\mu_\rho)_e = \langle e|(\mu_\rho)_e|g\rangle$ are the pure electronic transition moments, along ρ and σ directions, for the resonant excited state e, of which v is a particular vibrational level of bandwidth Γ_v; ν_{vi} is the transition frequency from the ground vibrational level i to the level v; $|i\rangle$, $|j\rangle$ and $|v\rangle$ represent vibrational states of a given normal coordinate, Q_k; and the summation is over all excited state vibrational levels, v. The dependence of $(\mu_{\rho,\sigma})_e$ on Q_k is small if the Born–Oppenheimer approximation is valid,[18] and therefore it can be expanded as a Taylor series[17]:

$$\mu_e = \mu_e^0 + \sum_k \mu_e'Q_k + \cdots \tag{4}$$

where $\mu_e' = (\partial\mu_e/\partial Q_k)$. [In writing Eq. (4) we have dropped the polarization subscripts.] When the electronic resonant transition is weakly allowed, μ_e' can be of the same magnitude as μ_e, or even exceed it, if the excited state e can gain absorption strength from other excited states via vibronic coupling. In the Herzberg–Teller formalism for the vibronic coupling,[16–18]

$$\mu_e' = \mu_s\langle s|(\partial H_\varepsilon/\partial Q_k)|e\rangle/(\nu_s - \nu_e) \tag{5}$$

where $|s\rangle$ is another excited state that can be mixed into $|e\rangle$ by Q_k; ν_s and μ_s are the frequency and transition dipole moment of the mixing electronic state, $|s\rangle$; and $(\partial H_\varepsilon/\partial Q_k)$ is the vibronic coupling operator that connects two excited states $|e\rangle$ and $|s\rangle$, with H_ε being the electronic Hamiltonian. Thus, the stronger the transition to the state $|s\rangle$, and the closer it is in energy to $|e\rangle$, the larger μ_e' will be.

Substitution of Eqs. (5) and (4) into Eq. (3) yields an expression with many terms, of which the first two terms, called A- and B-terms, are generally considered in the RR theory:

[18] A. C. Albrecht, *J. Chem. Phys.* **33**, 156 (1960).

$$\alpha_k = A + B + \cdots \tag{6}$$

where

$$A = \mu_e^2 (1/h) \sum_v [\langle j|v\rangle \langle v|i\rangle / (\nu_{vi} - \nu_0 + i\Gamma_v)] \tag{7}$$

$$B = \mu_e \mu_e' (1/h) \sum_v [(\langle j|Q_k|v\rangle \langle v|i\rangle + \langle j|v\rangle \langle v|Q_k|i\rangle) / (\nu_{vi} - \nu_0 + i\Gamma_v)] \tag{8}$$

The A-term is the leading RR scattering mechanism encountered in practice and involves vibrational interactions with a single excited electronic state, $|e\rangle$, by way of Franck–Condon overlap integrals, $\langle j|v\rangle$ and $\langle v|i\rangle$ (FC scattering). This term is nonzero only if $\mu_e > 0$ and $\langle j|v\rangle$ and $\langle v|i\rangle \neq 0$. The intensity of FC scattering depends on the magnitudes of both the resonant electronic transition oscillator strength and the Franck–Condon factors, the latter also being responsible for the selectivity of the vibrational mode enhancement. It is known from symmetry arguments that only for totally symmetric modes whose equilibrium positions are displaced in the excited state are these factors nonzero for both the $|i\rangle$ and $|j\rangle$ states of the scattering molecule. (Here we assume that the molecular symmetry is not altered in the excited state.) On the other hand, B-term scattering arises from the vibronic coupling of the resonant state, $|e\rangle$, to another excited state, $|s\rangle$ (HT scattering). Nontotally symmetric modes can only be enhanced by B-term scattering because these modes do not cause a displacement of the potential energy minimum in the excited state. Because the B-term has an energy denominator, $\nu_s - \nu_e$, in the equation, it becomes important when there is a nearby excited state $|s\rangle$ (electric-dipole allowed) which can be coupled to $|e\rangle$ by certain vibrational modes, whose irreducible representation is contained in the direct product of the two electronic transition representations, $\Gamma_e \times \Gamma_s$. Hence, by moving into the resonance region, strong enhancement will occur for those fundamentals which reflect the change in geometry when converting the molecule from the ground to excited state (FC allowed) or those which are able to vibronically couple the resonant excited state to some other electronic state with different transition moment (HT allowed).

This selective enhancement is one of the most important and valuable aspects of RR spectroscopy, since it leads to a considerable simplification of the observed spectra. The RR spectra consist primarily of bands arising from either totally or nontotally symmetric fundamentals depending on the nature of the resonant electronic transitions.

Experimental Techniques

The application of Raman and resonance Raman spectroscopy has been greatly influenced by instrumentation developments. Commercial laser-excited Raman instruments with high-quality monochromators became available in the 1960s. Since that time significant improvements have been made in expanding the tunability and other operating characteristics of the lasers, in sample cell designs, in gratings for monochromators and spectrographs, and in detection systems. The new spectrometers are capable of observing far weaker signals than previous generations, probing samples with excitation energy ranging from deep-UV to near-IR region, and obtaining spectra of highly absorbing and transient species.[10–12,19–21]

Wavelength Selection Devices

Resonance Raman spectra can be measured by conventional grating or Fourier transform (FT) Raman spectrometers. Grating spectrometers, which can be either single-channel (scanning) or multichannel (nonscanning) spectrometers, are the dispersive systems. The scattered photons from the sample are dispersed spatially by one, two, or even three diffraction gratings in a single, double, or triple monochromator, respectively. The Raman spectrum in these systems is displayed as a series of narrow wavelength lines at the focal plane of the monochromator exit port, which is different in the two types of spectrometers. In a scanning instrument, a narrow exit slit of the same size as the entrance slit is used to isolate a single wavelength line from all the wavelengths that strike the focal plane. Different wavelength lines are moved sequentially across the exit slit and detected by a photomultiplier tube (typically the RCA C31000 series and the Hamamatsu 928 equipped with photon-counting electronics for maximum sensitivity) as the gratings are slowly turned by the accurate drive of the monochromator under control of the microcomputer. To achieve the desired spectral resolution (typically <5 cm^{-1}) and separate the relatively weak Raman lines (even though resonance enhanced) from the intense Rayleigh-scattered radiation while maintaining a high throughput, double monochromators equipped with holographic gratings (typically 1800–2400 grooves/mm) are most widely used in scanning Raman spectrometers. The scanning spectrometer measures a complete spectrum in minutes; it can be applied only to stable biological systems or to systems in which the spectral properties change very slowly with time.

Multichannel Raman spectrometers are nonscanning spectrographs with multichannel detectors. In this type of instrument, the exit slit of

[19] D. J. Gardiner and P. R. Graves (eds.), "Practical Raman Spectroscopy." Springer-Verlag, Heidelberg, 1989.
[20] D. B. Chase, *in* "Analytical Raman Spectroscopy" (J. G. Grasselli and B. J. Bulkin, eds.), p. 21. Wiley (Interscience), New York, 1991.
[21] M. Kim, H. Owen, and P. R. Carey, *Appl. Spectrosc.* **47,** 1780 (1993).

a monochromator is removed (spectrograph) and the dispersed Raman scattering radiation at the focal plane is captured simultaneously by spatially sensitive video-type multichannel detectors, such as photodiode arrays, vidicon tubes, and charge-coupled devices (CCDs).[22-24] These detectors are actually an array of a large number of closely spaced miniature photoelectric detectors (typically 500–2000) that allow an entire section of Raman spectrum to be recorded simultaneously (the spectral coverage depending on the spectrograph dispersion power and the width of the detector), thus drastically reducing the time required to obtain a spectrum. This is particularly advantageous in time-resolved applications, when short-lived transient species are of interest, or when the protein sample is undergoing chemical or photochemical changes during data acquisition and/or prolonged exposure to the laser beam. The typical wavelength selector in multichannel Raman spectrometers is a triple grating system in which the first two grating stages operate as a monochromator and the third stage as a spectrograph. This is because stray radiation is a more difficult problem with multichannel detectors, especially in the low-frequency region near the exciting line, because the exit slit of the spectrometer is eliminated. Although triple-grating spectrometers achieve the required rejection of stray radiation, these systems suffer from relatively low throughput, which removes some of the speed advantage of multichannel detection. Raman filters designed to reject Rayleigh scattering, such as holographic notch[25,26] and edge[27,28] filters, permit the use of single-stage spectrograph, providing a dramatic increase in the throughput efficiency.[20,21]

Although first demonstrated by Chantry et al. in 1964,[29] FT-Raman spectroscopy did not attract significant attention until the development of commercially available instrumentation with excitation in the near-infrared region (1064 nm) from a continuous wave Nd/YAG (neodymium–yttrium-aluminum-garnet) laser.[20,30,31] The FT-Raman spectrometer is a frequency-division multiplexing system in which all (scattered) wave-

[22] A. Campion and W. Woodruff, Anal. Chem. **59**, 1299A (1987).

[23] Y. Talmi and K. Busch, in "Multichannel Image Detectors" (Y. Talmi, ed.), Vol. 2 (ACS Symp. Ser. **236**), p. 1. American Chemical Society, Washington, D.C., 1983.

[24] J. E. Pemberton, R. L. Sobacinski, M. A. Bryant, and D. A. Carter, Spectroscopy (Eugene, Oreg.) **5**, 26 (1990).

[25] B. Yang, M. D. Morris, and H. Owen, Appl. Spectrosc. **45**, 1533 (1991).

[26] C. L. Schoen, S. K. Sharma, C. E. Hesley, and H. Owen, Appl. Spectrosc. **47**, 305 (1993).

[27] M. M. Carrabba, K. M. Spencer, C. Rich, and D. Rauh, Appl. Spectrosc. **44**, 1558 (1990).

[28] M. J. Pelletier and R. C. Reeder, Appl. Spectrosc. **45**, 765 (1991).

[29] G. W. Chantry, H. A. Gebbie, and C. Helsum, Nature (London) **203**, 1052 (1964).

[30] B. Chase and T. Hirschfeld, Appl. Spectrosc. **40**, 133 (1986).

[31] B. Chase, Anal. Chem. **59**, 881A (1987).

lengths strike a single-channel detector simultaneously (typically, a cooled InGaAs detector), and the spectral information at each wavelength is encoded with a Michelson interferometer. Because the scattered radiation is not dispersed and all wavelengths are detected simultaneously, the FT-Raman instrument offers the advantages of high throughput and increased signal-to-noise ratio, limited only by infrared detector background noise. Also, the 1064-nm laser line has the advantage of excitation in a spectral region that is frequently free of fluorescence interferences. Several promising FT-Raman studies on selected proteins,[32] photobiological systems,[33,34] and photoactive proteins[35] have appeared in the literature, and general applications to biological materials have been discussed.[36] An important disadvantage of FT-Raman spectrometers at the moment is the lack of variable excitation wavelengths required for RR studies. The newly available titanium–sapphire dye laser, which is tunable over the near-IR region (~670–1100 nm), may allow this limitation to be circumvented.

Excitation Sources

All Raman instruments today use laser radiation to excite Raman and/ or resonance Raman scattering. The most common and reliable laser sources are continuous wave (cw) noble gas lasers pumped by an electrical discharge. The powerful (>1 W) Ar^+ and Kr^+ ion lasers emit at a series of wavelengths in the visible and near-UV region, as shown diagrammatically in Fig. 3. In addition the He–Ne (543.5 and 632.8) nm and He–Cd (441.6 and 325.0 nm) lasers have useful wavelengths, albeit at lower power levels. Five discrete excitation wavelengths with moderate output powers (30–400 mW) are now available in the UV region (229–257 nm) from an intracavity frequency-doubled (ICFD) cw Ar^+ ion laser. The Ar^+ and Kr^+ lasers can also be used to pump cw organic dye lasers for stable continuous wavelength tuning from the near-IR to near-UV region. All of the ion gas and dye lasers have minimal output fluctuations (<0.2%) when operated in a light-stabilized mode. The newly developed Ti–sapphire laser (pumped by a cw Ar^+ ion laser) is an excellent source of a tunable excitation radiation in the region between 670 and 1100 nm. For additional deep red excitations, solid-state diode lasers have become commercially available in the last several years with reasonable power levels (30–300 mW) in the

[32] S. M. Barnett, F. Dicaire, and A. A. Ismail, *Can. J. Chem.* **68,** 1196 (1990).
[33] J. Sawatzki, R. Fischer, H. Scheer, and F. Siebert, *Proc. Natl. Acad. Sci. U.S.A.* **87,** 5903 (1990).
[34] T. A. Mattioli, A. Hoffman, B. Robert, B. Schrader, and M. Lutz, *Biochemistry* **30,** 4648 (1991).
[35] C. K. Johnson and R. Rubinovitz, *Appl. Spectrosc.* **44,** 1103 (1990).
[36] I. W. Levin and E. N. Lewis, *Anal. Chem.* **62,** 1101A (1990).

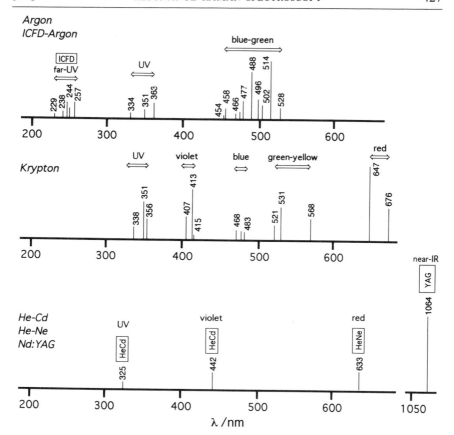

FIG. 3. Diagram of continuous wave (cw) laser sources suitable for metalloprotein resonance Raman spectroscopy. The best quality spectra are provided by Ar⁺, Kr⁺, He-Ne, and He-Cd lasers operating at fixed frequencies (the lengths of the lines indicate the relative output for a given laser) throughout the visible and near-UV region. An intracavity frequency-doubled (ICFD) Ar⁺ laser has been developed with five useful cw excitation wavelengths in the far-UV region (257, 248, 244, 238, and 228.9 nm). The high-powered Ar⁺ and Kr⁺ lasers can also be used to pump dye lasers which are tunable between the near-UV and near-IR region. The cw Nd:YAG laser with a fundamental at 1064 nm is the primary excitation source in FT Raman spectrometers.

780–830 nm wavelength range.[37–39] The primary excitation sources used in FT-Raman measurements are the solid state Nd/YAG lasers operated in a cw single mode at 1064 nm. The Nd/YAG lasers also have sufficient cw output levels to reach the visible region through frequency doubling

[37] Y. Wang and R. L. McCreery, *Anal. Chem.* **61**, 2647 (1989).
[38] J. M. Williamson, R. J. Bowling, and R. L. McCreery, *Appl. Spectrosc.* **43**, 373 (1989).
[39] C. D. Allred and R. L. McCreery, *Appl. Spectrosc.* **44**, 1229 (1989).

(532 nm) in potassium trihydrogen phosphate (KTP) or potassium dihydrogen phosphate (KDP) crystals, although these doubling units are somewhat temperature sensitive and tend to frequency drift when used at high powers.

Pulsed excitation sources have primary applications in the study of biological molecules by UV resonance Raman spectroscopy,[40-44] and in studies of chemical dynamics by time-resolved Raman measurements.[45-47] The two most widely used pulsed laser systems are the Nd/YAG lasers with a fundamental at 1064 nm and harmonics at 532, 355, 266, and 213 nm (produced with KTP or KDP crystals) and the excimer lasers which emit at a series of fundamentals in the 193–351 nm range, depending on the gas being used (ArF, KrCl, KrF, XeCl, and XeF at 193, 222, 248, 308, and 351 nm, respectively). The very high photon flux in these laser pulses produces efficient dye laser pumping and also the possibility of extensive frequency shifting through doubling and mixing crystals, as well as Raman shifting in media with high cross section for the stimulated Raman effect. A particularly useful device is the H_2 Raman shift cell, a pipe filled with high pressure hydrogen gas and fitted with appropriate windows. When a laser pulse is focused into the cell, a series of new laser lines appear, with frequencies which are the input laser frequency plus and minus multiple quanta of the H_2 stretching frequency, 4155 cm^{-1}. This device has been used to generate deep-UV Raman excitation lines by H_2-Raman shifting of the harmonic outputs of the pulsed Nd/YAG laser.[48] Frequency doubling of pulsed dye lasers, pumped by Nd/YAG or excimer lasers, has come to the fore as a means of generating UV Raman excitation.[49-51] Two attractive approaches using mode-locked

[40] I. Harada and H. Takeuchi, in "Spectroscopy of Biological Systems" (R. J. H. Clark and R. E. Hester, eds.), p. 113. Wiley, New York, 1986.

[41] B. Hudson and L. C. Mayne, in "Biological Applications of Raman Spectroscopy" (T. G. Spiro, ed.), Vol. 2, p. 181. Wiley, New York, 1987.

[42] S. A. Asher, Anal. Chem. 65, 59A and 201A (1993).

[43] J. C. Austin, T. Jordan, and T. G. Spiro, in Adv. Spectrosc. (Chichester, U.K.) 20, 55 (1993).

[44] J. C. Austin, K. R. Rodgers, and T. G. Spiro, this series, Vol. 226, p. 374.

[45] A. Larberau and M. Stockburger (eds.), "Time-Resolved Vibrational Spectroscopy." Springer-Verlag, Berlin, 1985.

[46] D. L. Rousseau and J. M. Friedman, in "Biological Applications of Raman Spectroscopy" (T. G. Spiro, ed.), Vol. 3, p. 133. Wiley, New York, 1988.

[47] C. Varotsis and G. T. Babcock, this series, Vol. 226, p. 409.

[48] S. P. A. Fodor, R. P. Rava, R. A. Copeland, and T. G. Spiro, J. Raman Spectrosc. 17, 471 (1986).

[49] S. A. Asher, C. R. Johnson, and J. Murtaugh, Rev. Sci. Instrum. 54, 1657 (1983).

[50] C. M. Jones, V. L. DeVito, R. A. Harmon, and S. A. Asher, Appl. Spectrosc. 41, 1268 (1987).

Nd/YAG[52] or Ti–sapphire[53] lasers to create quasi-cw UV sources have also been introduced. More comprehensive descriptions of various pulsed laser systems can be found in Vol. 226 of this series[11,44,47] and elsewhere.[20,54,55]

Sampling Devices

Scattered photons can be collected in the conventional 90° geometry or in backscattering geometries (oblique ~135° or strict 180°) to accommodate special sample requirements.[56] A general problem in resonance Raman experiments is that they require illumination of the sample within an electronic absorption band. Resonance excitation can enhance the buildup of decomposition products owing to photo and thermal effects arising from photon absorption and local heating by the laser beam. To overcome these difficulties, a variety of techniques have been developed which utilize rotating,[57–61] flowing,[62–66] stirring,[67] cooling,[60] or freezing[12,68–70] the solution sample in the focused laser beam. Figures 4–7 illustrate several examples of the sampling arrangements that have been found to be convenient and versatile for Raman studies of metalloproteins under resonance excitation conditions. One of the simplest sample holders, typically used in a backscattering geometry with front excitation at grazing (~135°) or normal incident angle, is a rotating nuclear magnetic resonance (NMR) tube mounted vertically in a wobble-free spinner consisting of a com-

[51] C. Su, Y. Wang, and T. G. Spiro, *J. Raman Spectrosc.* **21,** 435 (1990).
[52] K. P. J. Williams and D. J. Klenerman, *J. Raman Spectrosc.* **23,** 191 (1992).
[53] P. M. French, J. A. R. Williams, and J. R. Taylor, *Opt. Lett.* **14,** 686 (1989).
[54] S. A. Asher and C. R. Johnson, *Science* **225,** 311 (1984).
[55] B. Hudson and L. Mayne, this series, Vol. 130, p. 331.
[56] D. P. Strommen and K. Nakamoto, "Laboratory Raman Spectroscopy." Wiley, New York, 1984.
[57] W. Kiefer, *Appl. Spectrosc.* **28,** 115 (1974).
[58] W. Kiefer, *Adv. Infrared Raman Spectrosc.* **3,** 1 (1977).
[59] D. F. Shriver and J. B. Dunn, *Appl. Spectrosc.* **28,** 319 (1974).
[60] M. A. Walters, *Appl. Spectrosc.* **37,** 299 (1983).
[61] J. F. Eng, R. S. Czernuszewicz, and T. G. Spiro, *J. Raman Spectrosc.* **16,** 432 (1985).
[62] W. H. Woodruff and T. G. Spiro, *Appl. Spectrosc.* **28,** 74 (1974).
[63] J. L. Anderson and J. R. Kincaid, *Appl. Spectrosc.* **32,** 356 (1978).
[64] S. F. Simpson, J. R. Kincaid, and F. J. Holler, *Anal. Chem.* **58,** 3163 (1986).
[65] T. Ogura and T. Kitagawa, *J. Am. Chem. Soc.* **109,** 2177 (1987).
[66] T. Ogura and T. Kitagawa, *Rev. Sci. Instrum.* **59,** 1316 (1988).
[67] R. S. Czernuszewicz and K. A. Macor, *J. Raman Spectrosc.* **19,** 553 (1988).
[68] R. S. Czernuszewicz and M. K. Johnson, *Appl. Spectrosc.* **37,** 297 (1983).
[69] R. S. Czernuszewicz, *Appl. Spectrosc.* **20,** 571 (1986).
[70] P. M. Drozdzewski and M. K. Johnson, *Appl. Spectrosc.* **42,** 1575, (1988).

pressed air-driven turbine and a collet that centers the tube (Fig. 4).[60,61] Front illumination of a spinning sample is easily accomplished with a focusing lens in conjunction with a small front surface mirror. The NMR tube arrangement is convenient for small volumes (down to 100 μl) of optically dense proteins and/or for air-sensitive samples requiring a confined space. A disadvantage of the spinning NMR tube is that vertical mixing of the solution is ineffective and the irradiated volume tends to be confined to a circle around the tube. This problem can be obviated by inserting a stirrer in the tube.

Alternatively, the protein solution can be flowed through a capillary tube in a recirculating arrangement with a reservoir (Fig. 5a),[62–64] or it can be pushed only once through a capillary tube or a rectangular cell (height 5 mm, thickness 0.3 mm) by using a syringe pump (Fig. 5b),[65,66] the motion of the sample through the laser beam eliminating the buildup of decomposition products. The flowing arrangement is also useful for kinetic Raman studies, in which two solutions are mixed upstream of the laser and spectra are obtained of chemical intermediates formed within the transit time from the point of mixing to the laser focus.

The directional properties of the laser make it easy to record spectra of frozen solution samples with a liquid nitrogen-cooled finger (77 K) (Fig. 6)[68] or a cryotip (Fig. 7) attached to a liquid He refrigerator (10 K).[12,69,70] This is a particularly valuable method for examining scarce protein solutions, because only a drop is required. In addition, the absence of a broad

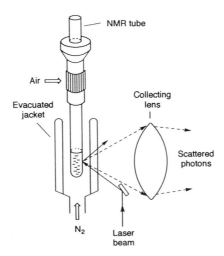

FIG. 4. Backscattering arrangement using spinning NMR tube in a flowing N_2 glass or quartz shroud for RR studies of optically dense metalloprotein solutions.

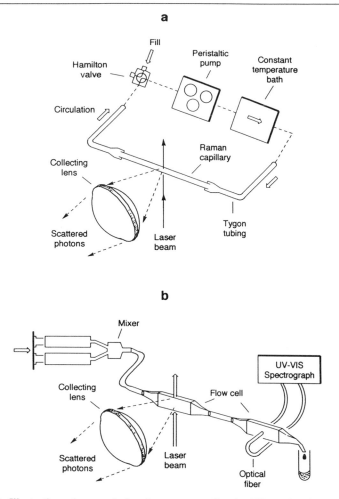

a

Fill

Hamilton valve

Peristaltic pump

Constant temperature bath

Circulation

Collecting lens

Raman capillary

Scattered photons

Laser beam

Tygon tubing

b

Mixer

UV-VIS Spectrograph

Collecting lens

Flow cell

Scattered photons

Laser beam

Optical fiber

FIG. 5. Illustration of two solution-flow systems for the RR study of photosensitive metalloprotein samples under 90° scattering geometry. (a) Recirculating closed-loop, small-volume (<2 ml) melting-point capillary tube arrangement using a peristaltic pump and constant temperature bath.[62] (b) Open-loop, larger volume (>5 ml) flow system using two syringe pumps and two rectangular flow cells for the simultaneous recording of UV–visible absorption (through the use of optical fibers) and RR spectra of transient metalloprotein species.[65]

Raman band (300–550 cm^{-1}) arising from the glass sample container makes this technique particularly suitable for the study of low-frequency vibrational modes. Heat removal by the cold tip minimizes laser-induced damage; moreover, the spectra are generally of higher quality because of the

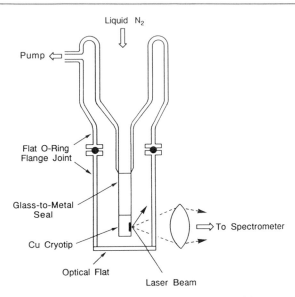

FIG. 6. Simple liquid N_2 temperature cryostat for RR studies of frozen protein solutions. The protein solution (typically 5–15 μl of 1–10 mM concentration) is placed in a copper cup on the end of a cold finger to give a flat surface. The glass or quartz shroud is clamped over the cold finger, and the sample is frozen by pouring liquid N_2 in the horizontal Dewar flask. Once the sample is frozen, the Dewar flask is turned to a vertical position and evacuated to 10^{-4}–10^{-6} torr. The Dewar flask, filled with liquid N_2, is transferred to the Raman sample compartment, and the scattered light is collected via 135° backscattering geometry directly from the surface of a frozen protein solution.[68]

suppression of background scattering relative to RR peak intensity and sharpening of the Raman bands at low temperature. This is illustrated in Fig. 8, which shows the resonance Raman spectra of high-potential iron-sulfur protein (HiPIP) from *Chromatium vinosum* obtained in the Fe–S stretching region (250–450 cm^{-1}) at room temperature with a spinning NMR tube and as frozen solutions at liquid N_2 temperature.[68] An excellent detailed discussion of the above-mentioned protein sampling methods as well as other variations has already been presented in Vol. 226 of this series by Loehr and Sanders-Loehr.[12]

Resonance Enhancement Mechanisms in Metalloproteins

In this section we survey the resonance enhancement mechanisms which are available in metalloproteins. The enhancement depends on the energy and intensity of the electronic transitions, and on their coupling

Refrigerator cold station

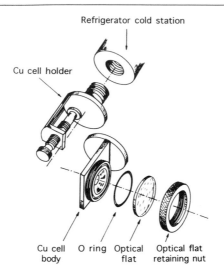

Cu cell holder

Cu cell   O ring   Optical   Optical flat
body   flat   retaining nut

FIG. 7. Miniature Raman cell for frozen protein solutions that, once loaded with sample, can conveniently be shipped between laboratories in dry ice or liquid N_2 and then attached without further manipulation to a helium closed-cycle refrigerator (CCR) station for RR measurements. The main advantages of this design are as follows: (1) very small quantities of sample are required (1–2 drops); (2) the cell atmosphere can be controlled; (3) cryogenic temperatures are obtained, down to 10 K; and (4) Raman scattering originates directly from the surface of a frozen solution without interferences from glass or quartz scattering.[69]

to vibrational modes involving the metal center. If the metal ion has a partially filled d subshell, it will give rise to $d-d$ electronic transitions, many of which are conveniently located for visible wavelength laser excitation. Unfortunately the $d-d$ transitions are too weak to produce useful RR enhancement. Much stronger transitions arise from charge transfer excitations. The direction of electron transfer can be from the ligand to the metal (LMCT) or from the metal to the ligand (MLCT). In the context of the common ligating side chains of proteins, only LMCT transitions are encountered, but when the metal is bound to π acid ligands (CO, NO, O_2, porphyrin), MLCT transitions may need to be considered. LMCT transitions are possible for any metal ion, but only the more reducible ones, especially Fe(III) and Cu(II), have LMCT transitions in the visible region.[71,72] These two ions in fact account for the majority of metalloprotein RR studies.

[71] E. I. Solomon, M. D. Lowery, L. B. LaCroix, and D. E. Root, this series, Vol. 226, p. 1.
[72] B. A. Averill and J. B. Vincent, this series, Vol. 226, p. 33.

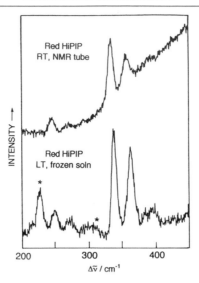

FIG. 8. Resonance Raman spectra of reduced high-potential iron-sulfur protein from *Chromatium vinosum* (HiPIP): at room temperature with a spinning NMR tube, laser power 150 mW, spectral slit width 6 cm^{-1} (*top*); and as a frozen solution at liquid N_2 temperature, laser power 250 mW, spectral slit width 8 cm^{-1} (*bottom*). Both spectra were obtained via backscattering with 457.9 nm Ar$^+$ laser excitation.[68]

In the event that the metal ion is bound to an organic cofactor with an unsaturated electronic system, then the cofactor $\pi-\pi^*$ transitions may give rise to useful RR enhancement. The best studied case is the heme group of heme proteins. The porphyrin $\pi-\pi^*$ transitions produce enhancement of many vibrational modes.[10,73,74] Most are internal vibrations of the macrocycle and have nothing directly to do with the central metal ion, but their frequencies are nevertheless responsive to the metal ligation chemistry via the structural consequences for the porphyrin ring. In addition, the vibrations of axial ligands are in some cases enhanced via coupling to the $\pi-\pi^*$ transitions.[75-78] An interesting situation arises in the molybdopterin cofactor of Mo redox proteins, in which Mo is bound to a pterin

[73] T. G. Spiro and X.-Y. Li, *in* "Biological Applications of Raman Spectroscopy" (T. G. Spiro, ed.), Vol. 3, p. 1. Wiley, New York, 1988.

[74] T. G. Spiro, R. S. Czernuszewicz, and X.-Y. Li, *Coord. Chem. Rev.* **100**, 514 (1990).

[75] N.-T. Yu, this series, Vol. 130, p. 350.

[76] E. R. Kerr and N.-T. Yu, *in* "Biological Applications of Raman Spectroscopy" (T. G. Spiro, ed.), Vol. 3, p. 39. Wiley, New York, 1988.

[77] T. Kitagawa, *in* "Biological Applications of Raman Spectroscopy" (T. G. Spiro, ed.), Vol. 3, p. 97. Wiley, New York, 1988.

[78] S. Han, Y. Ching, and D. L. Rousseau, *Nature (London)* **348**, 89 (1990).

Via a dithiolene chelate. In the enzyme dimethyl sulfoxide (DMSO) reductase, Mo–S and pterin vibrations are enhanced in resonance with a long-wavelength electronic transition, which may have both $\pi-\pi^*$ and LMCT character.[79]

$\pi-\pi^*$ Transitions Enhance Cofactor Modes: Porphyrins

Heme proteins have been studied extensively via RR spectroscopy, partly because of their intrinsic interest and partly because of the ease with which RR spectra can be detected, even at quite high dilution.[3,11,80,81] The extended aromatic system of the porphyrin ring gives rise to strong $\pi-\pi^*$ electronic transitions in the visible region, and numerous vibrational modes of the ring are coupled to these transitions. Because of configuration interaction, the first of these transitions, Q (~550 nm), is only moderately strong, with a molar extinction coefficient, ε, near 10,000 $M^{-1}cm^{-1}$, but the second transition, B (also called the Soret band, ~400 nm), is very strong, with ε rising to 100,000 or more.[82] Excitation in resonance with the Soret band gives good quality spectra for solutions as dilute as 10 μM. Soret band resonance enhances totally symmetric modes of the porphyrin ring, via the leading term, A, in the scattering equation [Eq. (6), above]. Q band resonance, on the other hand, mainly enhances nontotally symmetric modes, through the second term, B, because the transition moment is much smaller for the Q than for the Soret electronic transition, and the two transitions are vibronically coupled via these nontotally symmetric modes. As a result of these different scattering mechanisms, it has been possible to obtain an unusually complete assignment of the porphyrin vibrational modes, and to derive a reliable force field.[83,84]

Interestingly, the highest frequency porphyrin ring modes, between 1350 and 1650 cm^{-1}, are quite sensitive to the chemical state of the heme Fe atom, even though the vibrations involve stretching of the ring C–C and C–N bonds, with no direct involvement of the Fe. There are three principal sources of this sensitivity. First, the frequencies respond to

[79] S. Gruber, L. Kilpatrick, N. R. Bastian, K. V. Rajagopalan, and T. G. Spiro, J. Am. Chem. Soc. 112, 8179 (1990).
[80] T. G. Spiro (ed.), "Biological Applications of Raman Spectroscopy," Vol. 3, Wiley, New York, 1988.
[81] T. Kitagawa, in "Spectroscopy of Biological Systems" (R. J. H. Clark and R. E. Hester, eds.), p. 443. Wiley, New York, 1986.
[82] M. Gouterman, Porphyrins 3(Part A), 1 (1979).
[83] X.-Y. Li, R. S. Czernuszewicz, J. R. Kincaid, P. Stein, and T. G. Spiro, J. Phys. Chem. 94, 47 (1990).
[84] X.-Y. Li, R. S. Czernuszewicz, J. R. Kincaid, Y. O. Su, and T. G. Spiro, J. Phys. Chem. 94, 31 (1990).

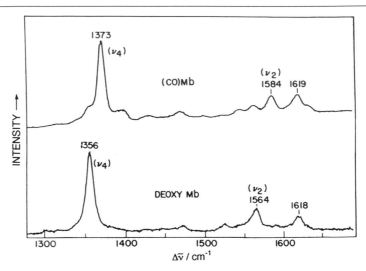

FIG. 9. Comparison between RR spectrum in the high-frequency region of carbonmonoxy-myoglobin (MbCO) and deoxyMb at room temperature with Soret band excitation. The peak labeled ν_4 is the electron density marker band; that labeled ν_2 is core size sensitive; and the peak at 1619–1618 cm^{-1} is the vinyl stretching mode.[46]

ligation changes because these changes alter the spin state, and therefore the size, of the Fe as well as its displacement from the heme plane. Both of these factors influence the size of the porphyrin central cavity and, therefore, the force constants and vibrational frequencies associated with the ring bond stretches. Thus the "core-size marker" frequencies can establish the spin state and coordination number for heme proteins in both Fe(II) and Fe(III) oxidation states.[3] Second, distortion of the porphyrin skeleton via tilting or twisting of the pyrrole rings can influence the frequencies; these effects have been explored in model compounds[85–87] and in cytochrome c, whose porphyrin ring is markedly twisted by protein linkages and contacts.[88] Third, the binding of π acid ligands, notably CO, NO, and O_2, to Fe(II) hemes relieves Fe–porphyrin back-bonding, thereby increasing the frequencies of certain of the ring modes.[3,73] (Fig. 9).

[85] R. S. Czernuszewicz, X.-Y. Li, and T. G. Spiro, *J. Am. Chem. Soc.* **111,** 7024 (1989).
[86] J. A. Shelnutt, C. J. Medforth, M. D. Berber, K. M. Barkigia, and K. M. Smith, *J. Am. Chem. Soc.* **113,** 4077 (1991).
[87] L. D. Sparks, C. J. Medforth, M.-S. Park, J. R. Chamberlain, M. R. Ondrias, M. O. Senge, K. M. Smith, and J. A. Shelnutt, *J. Am. Chem. Soc.* **115,** 581 (1993).
[88] S. Hu, J. K. Morris, J. P. Singh, K. M. Smith, and T. G. Spiro, *J. Am. Chem. Soc.* **115,** 12446 (1993).

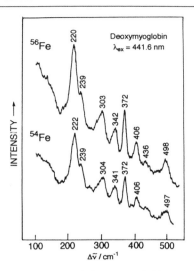

FIG. 10. Resonance Raman spectra of deoxymyoglobin (441.6 nm excitation) showing Fe–histidine stretching mode at 220 cm^{-1} and isotope shifts on substitution with ^{54}Fe.[77]

Although heme protein RR spectra contain mainly vibrational modes of the porphyrin ring, it has been possible to detect modes of the axial ligands in favorable cases. Again the π acid ligands CO, NO, and O$_2$ are notable in this regard. Because of the competition between the acceptor π^* orbitals of these ligands and those of the porphyrin, the Fe–XY and X–Y stretching modes, and the Fe–X–Y bending mode, are coupled to the resonant π–π^* transition. One or more of these modes have been observed in numerous heme protein adducts, and their frequency variations provide useful information about distal and proximal electronic effects.[13,75–78,89] In these adducts the proximal ligand, L (imidazole or thiolate), is only weakly coupled to the π electronic system, and the Fe–L stretching mode is not generally detectable. But the Fe–imidazole stretch is quite strongly enhanced in five-coordinate high-spin Fe(II) complexes, such as deoxyglobins (Fig. 10). The Fe is out of the porphyrin plane in these complexes, and the Fe–imidazole stretch modulates this displacement, which in turn alters the energy of the π–π–* transition, thereby providing an enhancement mechanism. This mode has been used extensively as a monitor of the heme–protein interaction in globins and peroxidases.[77]

[89] T. G. Spiro, G. Smulevich, and C. Su, *Biochemistry* **29**, 4497 (1990).

Charge Transfer Transitions Enhance Metal–Ligand and Internal Ligand Modes

Ligand-to-metal charge transfer transitions are found in the visible region when the metal ion is easily reducible [e.g., Fe(III) and Cu(II)] and the ligand is easily oxidizable [e.g., thiolate (from cysteine), phenolate (from tyrosine), or peroxide].[71,72] Because the charge transfer state involves reduced metal and oxidized ligand, resonance enhancement is found for the metal–ligand stretching vibration(s) and for some of the internal ligand modes. Many LMCT transitions occur in the ultraviolet region, including, for example, thiolate complexes of the closed-shell metal ions Cu(I), Zn(II), Cd(II), and Hg(II).[90] These have yet to be exploited for RR spectroscopy, however, because of technical difficulties associated with background scattering in the low-frequency region of spectra excited with UV lasers.

d–d Transitions Ineffective at Resonance Enhancement

Because many transition metal ions have d–d transitions in the visible region, and because metal-ligand vibrations are strongly coupled to the d–d excited states, it is unfortunate that these are too weak to produce useful RR enhancement. The B term [Eq. (6)] might have been expected to produce enhancement via mixing with higher energy allowed transitions (e.g., LMCT), but the d–d electronic transition moment is too small to provide significant amplitudes. Indeed antiresonances have been observed,[91,92] which are attributed to interferences between the local resonant contribution and preresonant contributions from higher energy allowed states.[5] For a centrosymmetric complex, the d–d electronic moment is zero, and both the A and B scattering terms vanish. Even when the symmetry center is lost, however, the moment depends on mixing of metal d and p orbitals, which is quite small. The most favorable cases are tetrahedral Co(II) complexes, for which ε can reach several hundred molar units; resonance enhancement has been observed for these complexes,[93–96]

[90] L. M. Utschig, F. G. Wright and T. V. O'Halloran, this series, Vol. 226, p. 71.
[91] L. A. Nafie, R. W. Pastor, J. C. Dabrowiak, and W. H. Woodruff, *J. Am. Chem. Soc.* **98,** 8007 (1976).
[92] P. Stein, V. Miskowski, W. H. Woodruff, J. P. Griffin, K. G. Werner, B. P. Gaber, and T. G. Spiro, *J. Phys. Chem.* **64,** 2159 (1976).
[93] G. Chottard and J. Bolard, *Chem. Phys. Lett.* **3,** 309 (1975).
[94] Y. M. Bosworth, R. J. H. Clark, and P. C. Turtle, *J. Chem. Soc., Dalton Trans.,* 2027 (1975).
[95] S. Salama and T. G. Spiro, *J. Am. Chem. Soc.* **100,** 1105 (1978).
[96] S. Salama, H. Schugar, and T. G. Spiro, *Inorg. Chem.* **18,** 1154 (1979).

but the enhancement factors are still too small to be useful in protein studies.

Multiple Chromophores Enhanced Selectively via Laser Tuning

A protein may have more than one metal center, each with its own set of electronic transitions, and in favorable cases the RR spectra can be elicited selectively by judicious choice of laser wavelengths. For example, sulfite reductase has an iron-isobacteriochlorin cofactor, siroheme, which is coupled magnetically to a Fe_4S_4 cluster, probably through a thiolate bridge.[97] The siroheme modes are strongly enhanced in resonance with the isobacteriochlorin Soret or Q bands,[98] but intermediate excitation (457.9–488.0 nm) enhances the Fe_4S_4 cluster modes (Fig. 11).[99] The protein rubrerythrin[100] has two rubredoxin-like $Fe(Cys)_4$ centers and a (μ-oxo)di-iron(III) center (similar to those found in hemerythrins and ribonucleotide reductase), each of which can be enhanced selectively by excitation at 496.5 and 406.7 nm, respectively (Fig. 12).[101] It is even possible to select for different modes of the same center by exciting into different electronic transitions. Thus the Fe–S(Cys) stretch of cytochrome P450 [high-spin Fe(III) form] is enhanced relative to the porphyrin bands when near-ultraviolet excitation is used.[102] Excitation profiles (Fig. 13) show that this mode is not resonant with the porphyrin Soret band, but with a higher energy absorption band which is polarized normal to the heme plane and is assigned to cysteine $\rightarrow$ Fe charge transfer transition.[103]

Applications

In this section we discuss applications to specific metalloproteins, in order to give a sense of the kinds of structural information that can be provided by RR spectroscopy. The field is broad, and a comprehensive review would be inappropriate in the present context. In the limited space available, only a few illustrative examples can be given.

[97] D. E. McRee, D. C. Richardson, L. M. Siegel, and B. M. Hoffman, *Biochemistry* **24,** 7942 (1985).

[98] S. Han, J. F. Madden, S. H. Strauss, L. M. Siegel, and T. G. Spiro, *Biochemistry* **28,** 5461 (1989).

[99] J. F. Madden, S. Han, L. M. Siegel, and T. G. Spiro, *Biochemistry* **28,** 5471 (1989).

[100] J. LeGall, B. C. Prickril, I. Moura, A. Xavier, J. J. G. Moura, and B.-H. Huynh, *Biochemistry* **27,** 1636 (1988).

[101] B. C. Dave, R. S. Czernuszewicz, B. C. Prickril, and D. M. Kurtz, Jr., *Biochemistry* **33,** 3572 (1994).

[102] P. M. Champion, B. R. Stallard, G. C. Wagner, and I. C. Gunsalus, *J. Am. Chem. Soc.* **104,** 5469 (1982).

[103] P. M. Champion, *J. Am. Chem. Soc.* **111,** 3433 (1989).

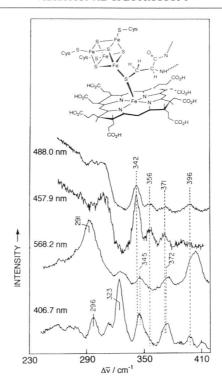

FIG. 11. Low-temperature (77 K) RR spectra of sulfite reductase hemoprotein obtained with various excitation wavelengths and structural drawing of a proposed model for the active site. The Fe_4S_4 cluster modes are selectively enhanced with 457.9 or 488.0 nm excitation (correlated with dashed lines), whereas the siroheme modes are enhanced with 406.7 (Soret band) or 568.2 nm (Q band) excitation.[99]

Blue Copper Proteins

Blue copper proteins (also known as type 1 Cu proteins or cupredoxins),[104,105] involved in biological electron transport, share a common structural motif, in which a Cu^{2+} ion is trigonally coordinated by a pair of histidine side chains and a cysteine thiolate ligand (Fig. 14).[106] A methionine S atom is found along the trigonal axis at a long distance, 2.6–3.1

[104] E. Adman, in "Topics in Molecular and Structural Biology: Metalloproteins" (P. M. Harrison, ed.), Vol. 1, p. 1. Macmillan, New York, 1985.

[105] K. D. Karlin and Z. Tyeklar (eds.), "Bioinorganic Chemistry of Copper." Chapman & Hall, New York, 1993.

[106] E. T. Adman, Adv. Protein Chem. 45, 145 (1991).

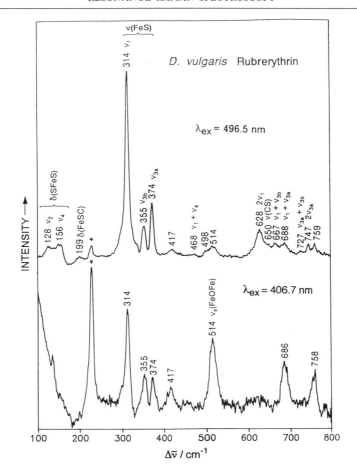

FIG. 12. Low-temperature (77 K) RR spectra of isolated *Desulfovibrio vulgaris* rubrery-thrin obtained in the 100–800 cm⁻¹ region with 496.5 (top trace) and 406.7 nm (bottom trace) excitation wavelengths. The spectrum excited at 496.5 nm is totally dominated by the bands arising from the Fe(S-Cys)₄ cluster, whereas that excited at 406.7 nm shows dramatically diminished intensities of the Fe–S bands with a concomitant enhancement of the band at 514 cm⁻¹ assigned to the ν_s(FeOFe) mode of a bent Fe–O–Fe cluster. Asterisks indicate the 228 cm⁻¹ ice band.[101]

Å; sometimes a peptide carbonyl O atom is located on the other side of the trigonal plane, at an even longer distance. The trigonal bonding arrangement lowers the energy of the half-occupied copper $d_{x^2-y^2}$ orbital, relative to its energy in the usual tetragonal complexes of Cu^{2+}, and in addition orients $d_{x^2-y^2}$ for optimal overlap with the π, instead of the σ,

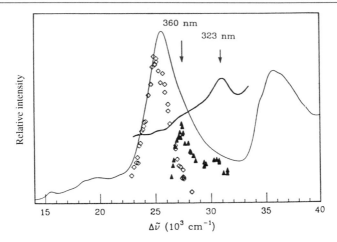

FIG. 13. Cytochrome P450 excitation profiles for the 1488 cm^{-1} porphyrin mode ($\diamond$) and the 351 cm^{-1} Fe–S stretching mode ($\blacktriangle$). The former maximizes in the Soret absorption band of the heme (thin solid line), but the latter shows shorter wavelength maxima, which are attributed to S $\to$ Fe charge transfer transitions, for which evidence can be seen in the z-polarized single-crystal absorption spectrum (thick solid line).[103]

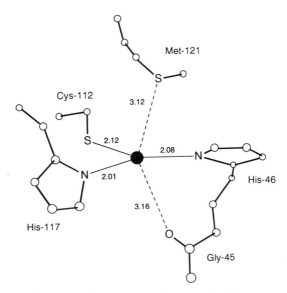

FIG. 14. Copper site in *Alcaligenes denitrificans* azurin. Distances to the three strong and two weak ligands are adapted from Adman.[106]

filled orbital on the cysteine S atom. Both factors lead to an unusually low energy S(Cys) → Cu charge transfer transition, around 600 nm, which accounts for the intense blue color of these proteins.[71,107]

The main effect of charge-transfer excitation should be to weaken the Cu–S(Cys) bond, and RR enhancement of the Cu–S stretching mode is therefore expected. The Cu–S bond is short, around 2.1 Å,[106] and a high stretching frequency is expected, around 400 cm^{-1}. The other Cu–ligand stretches should be at much lower frequencies, because of the very long bonds for the methionine and peptide ligands and because of the high effective mass of the rigid imidazole rings. Thus a single dominant RR band might be expected, near 400 cm^{-1}. This is in fact what is observed[108] for a newly available set of model compounds, LCu–SR [L is hydrotis(3,5-diisopropyl-1-pyrazolyl)borate].[109] When R is varied from *tert*-butyl to triphenylmethyl to pentafluorophenyl, the frequency of the band decreases from 437 to 422 to 409 cm^{-1}, consistent with the expected weakening of the Cu–S bond owing to increasing electron withdrawal. It has long been known,[110] however, that the protein RR spectra contain several strong bands near 400 cm^{-1}, as illustrated in Fig. 15.

The source of this complexity has been a matter of considerable speculation.[111,112] Significantly, the LCu–SR model compound spectrum gains complexity when R is *sec*-butyl. Instead of a single strong band near 400 cm^{-1} there are now three, as seen in Fig. 16. Normal mode calculations indicate that the simple interchange of a methyl group and a hydrogen atom, between R = *tert*-butyl and *sec*-butyl, is sufficient to induce significant mixing of the C–C–C bending coordinates with the Cu–S coordinate, thereby accounting for the three strong bands.[108] No doubt mixing of Cu–S stretching with heavy atom bending coordinates of the cysteine side chain produces some or all of the complexity seen in the blue copper protein spectra. Such mixing was anticipated in an early study of stellacyanin,[113] in which the two main RR bands were each found to shift on $^{65/63}$Cu substitution by half the amount expected for an isolated Cu–S oscillator,

[107] E. I. Solomon, M. J. Baldwin, and M. D. Lowery, *Chem. Rev.* **92**, 521 (1992).

[108] D. Qiu, L. Kilpatrick, N. Kitajima, and T. G. Spiro, *J. Am. Chem. Soc.* **116**, 2585 (1994).

[109] N. Kitajima, F. Kiyoshi, and M. Yoshihiko, *J. Am. Chem. Soc.* **112**, 3210 (1990).

[110] V. M. Miskowski, S. W. Tang, T. G. Spiro, E. Shapiro, and T. H. Moss, *Biochemistry* **14**, 1244 (1975).

[111] W. H. Woodruff, R. B. Dyer, and J. R. Schoonover, *in* "Biological Applications of Raman Spectroscopy" (T. G. Spiro, ed.), Vol. 3, p. 413, Wiley, New York, 1988.

[112] J. Sanders-Loehr, *in* "Bioinorganic Chemistry of Copper" (K. D. Karlin and Z. Tyeklar, eds.), p. 51. Chapman & Hall, New York, 1993.

[113] L. Nestor, J. A. Larrabee, G. Woolery, B. Reinhammer, and T. G. Spiro, *Biochemistry* **23**, 1084 (1984).

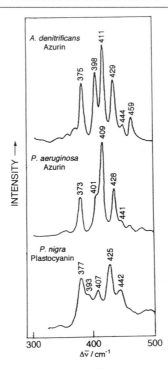

FIG. 15. Resonance Raman spectra of three blue copper proteins: *A. denitrificans* azurin at 90 K (568.2 nm excitation), *P. aeruginosa* azurin at 12 K (647.1 nm excitation), and *Populus nigra* plastocyanin at 12 K (604 nm excitation).[108]

implying nearly equal contributions from Cu–S stretching and another coordinate. Further insight into the coordinate mixing phenomenon has recently been obtained by $^{34/32}$S substitution in *Pseudomonas aeruginosa* azurin and site-specific mutants, carried out via bacterial expression of the azurin gene.[114]

Figure 17 presents the 360–435 cm^{-1} RR spectra for the natural abundance wild type azurin and the M121G (methionine-121 replaced with glycine) and H46D (histidine-46 replaced with aspartate) Cu site mutants, and the corresponding [^{34}S]Cys-substituted proteins. Each spectrum shows four dominant RR peaks in this region (~370, ~400, 408, and ~427 cm^{-1}) that are characteristic for the type 1 structure of the Cu(II)–Cys chromophores.[111,112] Among these, the bands showing the greatest intensity and [^{34}S]Cys isotope shift, 408 (WT) and around 400 cm^{-1} (mu-

[114] B. C. Dave, J. P. Germanas, and R. S. Czernuszewicz, *J. Am. Chem. Soc.* **115,** 12175 (1993).

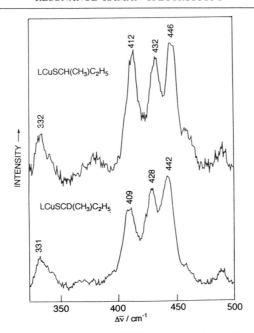

FIG. 16. Resonance Raman spectra (647.1 nm excitation) of the blue copper site analog complex, $LCu(II)SCH(CH_3)C_2H_5$ [L is hydrotris(3,5-diisopropyl-1-pyrazolyl) borate], with natural abundance *sec*-butyl thiolate (*top*) and the C^1-deuterated isotopomer (*bottom*).[108]

tants), are identified with the vibrational mode involving primarily stretching of the Cu(II)–S(Cys) bond, ν(Cu–S). A shift of the ν(Cu–S) character from the 408 cm^{-1} peak of the wild type to the approximately 400 cm^{-1} peaks in H46D and M121G suggests a decrease of the Cu–S(Cys) interaction in both mutants as the Cu sites adopt more tetrahedral character relative to the wild-type site.[114]

In addition to the very strong absorption band near 600 nm, blue copper proteins display an absorbance of variable intensity centered near 460 nm; the ratio of the molar extinction coefficients, $\varepsilon_{460}/\varepsilon_{600}$, ranges from 0.11 for azurin to 1.34 for nitrate reductase. Interestingly, for a given protein, excitation within either of these absorption bands produces the same set of RR frequencies between 300 and 500 cm^{-1} that is characteristic of type 1 Cu sites.[115] Excitation profiles (plots of Raman peak intensities

[115] J. Han, T. M. Loehr, Y. Lu, J. S. Valentine, B. A. Averill, and J. Sanders-Loehr, *J. Am. Chem. Soc.* **115**, 4256 (1993).

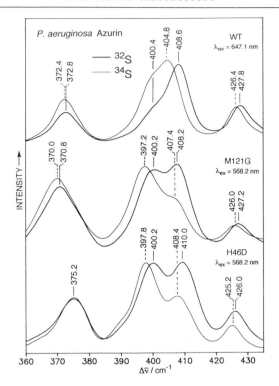

FIG. 17. Low-temperature (77 K) RR spectra of *P. aeruginosa* azurins (thick lines) and the [³⁴S]Cys-labeled proteins (thin lines) in the region between 360 and 435 cm⁻¹: (*top*) wild type (WT) excited at 647.1 nm, (*middle*) M121G mutant, and (*bottom*) H46D mutant, both excited at 568.2 nm. Conditions: 150 mW laser power, 2 cm⁻¹ slit widths, average of three scans at 1 sec/point and 0.2 cm⁻¹/sec increments, no baseline correction or smoothing was applied. To ensure accurate isotope measurements, the natural abundance and ³⁴S-labeled proteins were placed side by side on the cold finger so that the RR data could be collected under the same conditions.[114]

versus excitation wavelength) for the strongest bands in the ν(Cu–S) region of several proteins have been acquired[115] and demonstrate that each RR peak tracks both the 460 and 600 nm absorption bands. Because a similar RR spectrum is obtained on excitation wavelengths in resonance with the 460 and 600 nm absorption bands, and the most strongly enhanced vibrational mode has the largest contribution from the ν(Cu–S) coordinate,[114] both electronic transitions must have (Cys)S → Cu(II) charge transfer character.

Iron-Sulfur Proteins

Iron-sulfur proteins are also involved in electron transport,[116] although an increasing number have been found to have other activities, including hydrolyase reactions,[117] hydration of nitriles to amides,[118,119] and CO and N_2 activation.[120,121] They all have Fe bound to cysteine and/or sulfide ligands, and sometimes to other ligands as well. The classic $Fe(Cys)_4$, $[Fe_2S_2](Cys)_4$, and $[Fe_4S_4](Cys)_4$ structural types are illustrated in Fig. 18, along with absorption spectra of proteins containing these structures. The absorption bands arise from S → Fe charge transfer transitions,[122] and RR spectra show enhancement of modes in the 200–450 cm^{-1} region, associated with the stretching of the Fe–S bonds.[123]

The rubredoxins (Rd) contain the simplest of the Fe–S structures, with a single iron ion surrounded tetrahedrally by four Fe–S(Cys) bonds.[124-126] The RR spectrum of Rd from *Clostridium pasteurianum* was one of the first RR spectra obtained for a biological molecule.[127,128] As illustrated in Fig. 19, this spectrum could be interpreted as arising from a simple tetrahedral complex; indeed, it resembled closely the spectrum of the isoelectronic cluster $[FeCl_4]^-$.[129] Further studies, with better resolution and a wider range of excitation wavelengths, have revealed considerable com-

[116] T. G. Spiro (ed.), "Iron-Sulfur Proteins." Wiley, New York, 1982.

[117] M. H. Emptage, *in* "Metal Clusters in Proteins" (L. Que, Jr., ed.) (*ACS Symp. Ser.* **372**), Chap. 17. American Chemical Society, Washington, D.C., 1988.

[118] M. J. Nelson, H. Jin, I. M. Turner, Jr., G. Grove, R. C. Scarrow, B. A. Brennan, and L. Que, Jr., *J. Am. Chem. Soc.* **113**, 7072 (1991).

[119] H. Jin, I. M. Turner, Jr., M. J. Nelson, R. J. Gurbiel, P. E. Doan, and B. M. Hoffman, *J. Am. Chem. Soc.* **115**, 5290 (1993).

[120] S. W. Ragsdale, *Crit. Rev. Biochem. Mol. Biol.* **26**, 261 (1991).

[121] E. I. Stiefel and S. P. Cramer, *in* "Molybdenum Enzymes" (T. G. Spiro, ed.), p. 117. Wiley, New York, 1985.

[122] W. A. Eaton and W. Lovenberg, *in* "Iron-Sulfur Proteins" (W. Lovenberg, ed.), Vol. 2, Chap. 3. Academic Press, New York, 1973.

[123] T. G. Spiro, R. S. Czernuszewicz, and S. Han, *in* "Biological Applications of Raman Spectroscopy" (T. G. Spiro, ed.), Vol. 3, p. 523. Wiley, New York, 1988.

[124] K. D. Watenpaugh, L. C. Sieker, and L. H. Jensen, *J. Mol. Biol.* **131**, 509 (1979).

[125] M. Frey, L. C. Sieker, F. Payan, R. Haser, M. Bruschi, G. Pepe, and J. LeGall, *J. Mol. Biol.* **197**, 525 (1987).

[126] E. Adman, L. C. Sieker, and L. H. Jensen, *J. Mol. Biol.* **217**, 337 (1991).

[127] T. V. Long and T. M. Loehr, *J. Am. Chem. Soc.* **92**, 6384 (1970).

[128] T. V. Long, T. M. Loehr, J. R. Alkins, and W. Lovenberg, *J. Am. Chem. Soc.* **93**, 1809 (1971).

[129] K. Nakamoto, "Infrared and Raman Spectra of Inorganic and Coordination Compounds," 3rd Ed. Wiley, New York, 1986.

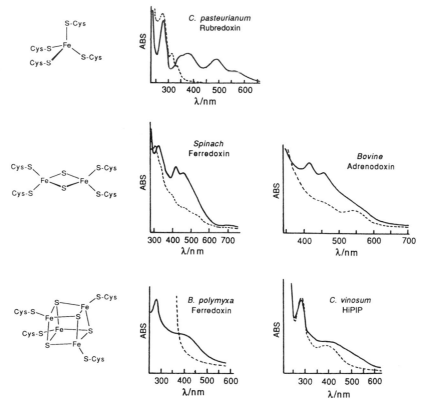

FIG. 18. Optical absorption spectra of iron-sulfur proteins (oxidized, ——; and reduced, – – –) representing the three classic structural types shown on the left-hand side: 1-Fe (rubredoxin), 2-Fe (plant ferredoxin and adrenodoxin), and 4-Fe (bacterial ferredoxin and high-potential iron protein). The visible absorption bands are attributed to S → Fe(III) charge transfer transitions.

plexity, however.[130] As seen in Fig. 20, Rd actually shows five RR bands in the Fe–S stretching, instead of the two seen originally. Moreover, the frequencies of the bands differ significantly from the corresponding bands of $[Fe(S_2\text{-}o\text{-xyl})_2]^-$,[131] a complex that models Rd accurately with respect to the tetrahedral arrangement and the Fe–S bond length, 2.27 Å.[132] De-

[130] R. S. Czernuszewicz, J. LeGall, I. Moura, and T. G. Spiro, *Inorg. Chem.* **25,** 696 (1986).
[131] V. K. Yachandra, J. Hare, I. Moura, and T. G. Spiro, *J. Am. Chem. Soc.* **105,** 6455 (1983).
[132] R. W. Lane, J. A. Ibers, R. B. Frankel, R. H. Holm, and G. C. Papaefthymiou, *J. Am. Chem. Soc.* **99,** 84 (1977).

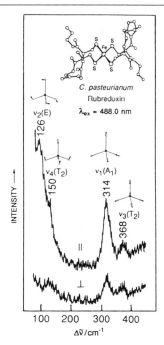

FIG. 19. Resonance Raman spectra of *Clostridium pasteurianum* rubredoxin obtained in 1971 by Long *et al.*[128] using 488.0 nm excitation. The suggested assignments to modes of the FeS$_4$ tetrahedron are indicated by the labels and the eigenvectors of the modes.

spite the equal Fe–S bonds, the ν_1 FeS$_4$ breathing mode frequencies differ substantially between protein and analog. The sources of complexity in these spectra have been analyzed in some detail, with the aid of isotopic data and data for the additional model complexes [Fe(SCH$_3$)$_4$]$^-$ and [Fe(SCH$_2$CH$_3$)$_4$]$^-$.[133] Major determinants of the frequencies are the FeS–CC dihedral angles, τ, which strongly influence the extent of interaction between the Fe–S stretching and S–C–C bending coordinates, both of which have natural frequencies near 300 cm^{-1}. As illustrated in Fig. 20, the interaction is maximal for $\tau = 180°$, when the two coordinates are in line, and minimal for $\tau = 90°$, when they are nearly orthogonal. In [Fe(S$_2$-o-xyl)$_2$]$^-$, $\tau = 90°$ for all four Fe–S bonds,[132] whereas in Rd $\tau = 90°$ for two of the bonds but 180° for the other two.[124] The Fe–S/S–C–C interaction associated with the two 180° dihedrals increases ν_1 of Rd by 16 cm^{-1}, relative to [Fe(S$_2$-o-xyl)$_2$]$^-$.[133]

[133] R. S. Czernuszewicz, L. K. Kilpatrick, S. A. Koch, and T. G. Spiro, *J. Am. Chem. Soc.* **116**, 7134 (1994).

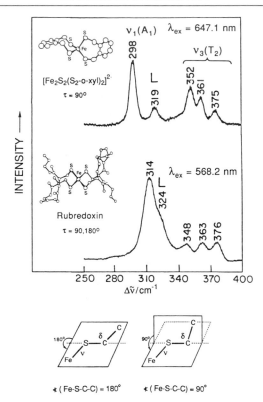

FIG. 20. Resonance Raman spectra [568.2 (protein) and 647.1 (analog) nm excitation] in the Fe–S stretching region and structural drawings for rubredoxin and for the xylene dithiolate analog complex.[130,131,133] Similar frequencies are seen for the ν_3 components of these compounds, but the ν_1 breathing mode is substantially lower for the analog. This difference is attributed to S–C–C bending interactions which differ between protein and analog owing to the different distribution of τ(FeSCC) dihedral angles. These are 90° for the analog, but two of them are 180° for the protein. Coupling between Fe–S stretching and S–C–C bending is maximal for $\tau = 180°$, when the coordinates are in line, and minimal at $\tau = 90°$. The band marked L (ligand) is assigned to S–C–C bending.[130,133]

Resonance Raman spectra of $[Fe_2S_2(Cys)_4]^{2-}$ proteins, illustrated in Fig. 21 for adrenodoxin and a plant ferredoxin,[134] are substantially more complex, as might be expected. With the aid of isotope substitution and model compound data, it has nevertheless been possible to characterize the modes completely and to assign them to vibrations involving primarily the bridging or terminal sulfur atoms (superscripts b and t, respectively,

[134] S. Han, R. Czernuszewicz, T. Kimura, M. W. W. Adams, and T. G. Spiro, *J. Am. Chem. Soc.* **111**, 3505 (1989).

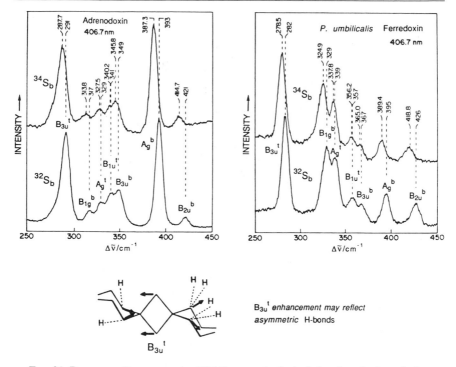

FIG. 21. Resonance Raman spectra (406.7 nm excitation) of plant ferredoxin and adreno-doxin[134] showing isotope shifts on reconstitution with $^{34}S^{2-}$. The mode assignments were made by comparison with the analog complex vibrational spectra.[135] The frequency differences between proteins were suggested to arise in part from different FeSCC dihedral angles. The remarkable intensification of the mode near 280 cm^{-1} assigned to B_{3u}^t is attributed to asymmetry in the hydrogen bond pattern which destroys the phase cancellation of the FeS$_4$ breathing motions; the diagram shows the hydrogen bond pattern indicated by the crystal structure of *Streptomyces platensis* ferredoxin.[136]

on the symmetry labels in Fig. 21).[135] The terminal stretches range from 290 to 340 cm^{-1}, whereas the bridging modes are somewhat higher, 315–425 cm^{-1}. An intersting aspect of the protein spectra is the prominence of the band near 290 cm^{-1}. This band is assigned to the B_{3u}^t mode, whose eigenvector is diagrammed in Fig. 21. This band should be Raman inactive in the idealized D_{2h} symmetry of the $Fe_2S_2^bS_4^t$ unit, and indeed the band is very weak in model compound spectra. Its dramatic intensification in the protein RR spectra is attributed[134] to an asymmetric pattern of hydrogen

[135] S. Han, R. S. Czernuszewicz, and T. G. Spiro, *J. Am. Chem. Soc.* **111**, 3496 (1989).

bonding to the terminal S atoms.[136] Because the normal mode is the asymmetric combination of breathing motions of the two linked FeS_4 tetrahedra (Fig. 21), its displacement in the charge transfer excited state is very sensitive to the symmetry of the hydrogen bond environment. This asymmetry is also relevant to the mechanism of electron transfer in these proteins. It is known that the electron is localized on one of the Fe ions in the reduced state,[137] presumably on the one whose S ligands have the larger number of hydrogen bonds. RR spectra have been reported for reduced adrenodoxin[134,138–140] and ferredoxin,[140] and the bands have been interpreted as arising from the unreduced Fe(III) end of the chromophore, where the charge transfer transitions (see reduced spectra shown as dashed lines in Fig. 18) should be localized. RR spectra have also been reported for Rieske Fe-S proteins,[141] in which two of the cysteine ligands of a $Fe_2S_2^b$ unit are replaced by histidine. Assignments of bands to Fe–imidazole modes have been suggested.

The RR spectral motif is altered once again when the cubanelike $[Fe_4S_4(Cys)_4]^{2-}$ chromophores are considered, as illustrated in Fig. 22.[142] The dominant band is the breathing mode of the cube at around 336 cm^{-1}, and other modes of the cube range up to 400 cm^{-1} and down to 240 cm^{-1}. The terminal modes are found at approximately 360 (T_2^t) and 390 (A_1^t) cm^{-1}. It is interesting that the terminal mode symmetry order is inverted from that found in $[Fe(Cys)_4]^-$ complexes (Fig. 19), reflecting the kinematic interactions with the cube modes.

In frozen solution the model complex $[Fe_4S_4(SCH_2Ph)_4]^{2-}$ gives a RR spectrum (Fig. 22, top), which is readily assignable on the basis of T_d symmetry, whereas the spectrum of the same complex in the $[(C_2H_5)_4N]^+$ salt (Fig. 22, bottom spectrum) shows band splittings, consistent with the symmetry reduction to D_{2d} (elongated cube) observed in the crystal

[136] T. Tsukihara, T. Fukuyama, M. Nakamura, Y. Katsube, N. Tanaka, M. Kakudo, K. Wada, and H. Matsubara, *J. Biochem. (Tokyo)* **90**, 1763 (1981).

[137] A. J. Thompson, *in* "Metalloproteins" (P. Harrison, ed.), Part 1, p. 79. Verlag Chemie, Weinheim, 1985.

[138] V. K. Yachandra, J. Hare, A. Gewirth, R. S. Czernuszewicz, T. Kimura, R. H. Holm, and T. G. Spiro, *J. Am. Chem. Soc.* **105**, 6462 (1983).

[139] Y. Mino, T. M. Loehr, K. Wada, H. Matsubara, and J. Sanders-Loehr, *Biochemistry* **26**, 8059 (1987).

[140] W. Fu, P. M. Drozdzewski, M. D. Davies, S. G. Sligar, and M. K. Johnson, *J. Biol. Chem.* **267**, 15502 (1992).

[141] D. Kuila, J. A. Fee, J. R. Schoonover, W. H. Woodruff, C. J. Batie, and D. P. Ballou, *J. Am. Chem. Soc.* **109**, 1559 (1987).

[142] R. S. Czernuszewicz, K. A. Macor, M. K. Johnson, A. Gewirth, and T. G. Spiro, *J. Am. Chem. Soc.* **109**, 1778 (1987).

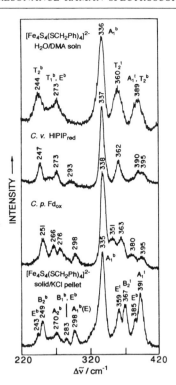

FIG. 22. Resonance Raman spectra [457.9 (proteins) and 514.5 (analog) nm excitation] of 4-Fe proteins and the benzyl thiolate analog.[142] The $[Fe_4S_4(SCH_2Ph)_4]^{2-}$ spectra are assignable on the basis of a tetrahedral structure in frozen solution but the known D_{2d} distortion in tetraethylammonium crystals. Assignments are indicated. The RR spectrum of oxidized bacterial ferredoxin is very similar to the analog crystals, indicating a similar D_{2d} distortion, whereas the spectrum of reduced HiPIP (same oxidation level) is similar to the analog spectrum in solution, indicating a structure that is more tetrahedral.

structure.[143] Bacterial ferredoxin shows a very similar band pattern (Fig. 22, third spectrum), indicating a similar distortion of the cube. The spectrum of HiPIP (high-potential iron protein) from *C. vinosum* (Fig. 22, second spectrum) appears to be simpler and was interpreted as indicating a relatively undistorted cube structure,[142] but at higher resolution the spectrum has been found to show band splittings, which are more pro-

[143] B. A. Averill, T. Herskovitz, R. H. Holm, and J. A. Ibers, *J. Am. Chem. Soc.* **95,** 3523 (1973).

nounced in HiPIP from other species.[144] Thus protein-induced distortions of the cube are detectable in all these 4-Fe species.

The oxidation level of the species compared in Fig. 22 is 2Fe(III), 2Fe(II). This is the oxidized form of ferredoxin but the reduced form of HiPIP. Oxidized HiPIP has one additional Fe(III) ion, as is reflected in the higher absorptivity of the charge transfer absorption bands (Fig. 18). It gives a strongly enhanced RR spectrum, with bands that are slightly higher in frequency than the reduced protein, indicating a strengthening of the Fe–S bonds.[144] Reduced ferredoxin, on the other hand, has only one Fe(III) ion, on average, and the absorptivity in the visible region is roughly half what it is for the oxidized protein. Because RR enhancement scales as the square of the electronic transition moment,[5] it might have been expected that the RR band intensities would diminish by a factor of roughly four on reduction. In fact, however, no RR bands have ever been detected for a $[Fe_4S_4(SR)_4]^{3-}$ species, and the enhancement must be at least an order of magnitude lower than for the 2– species. The reason for this weakness is uncertain, but it may be due to destructive interferences from multiple electronic transitions.[123] Such interferences are manifested even for the 2– species in excitation profiles which maximize well to the red of the main absorption band and drop off sharply at the absorption maximum.[142] Possibly the effect is even greater in the 3– species.

A number of 4-Fe-containing proteins readily lose one of the Fe atoms to form a 3-Fe protein. An early crystal structure[145] of ferrodoxin I from *Azotobacter vinelandii* was interpreted as showing a planar Fe_3S_3 ring for the 3-Fe cluster. This interpretation was shown to be inconsistent with the RR spectrum, via a normal coordinate analysis and by comparison with the spectral pattern of a planar model compound, $(CH_3)_3Sn_3S_3$.[146] The crystal structure was subsequently corrected to reveal a Fe_3S_4 cube with a missing corner.[147–150] The same structure has been determined for the inactive form of the enzyme aconitase,[151,152] and the similarity of the

[144] G. Backes, M. Yoshiki, T. M. Loehr, T. E. Meyer, M. A. Cusanovich, W. V. Sweeney, E. T. Adman, and J. Sanders-Loehr, *J. Am. Chem. Soc.* **113**, 2055 (1991).

[145] D. Ghosh, W. Furey, Jr., S. O'Donnell, and C. D. Stout, *J. Biol. Chem.* **158**, 73 (1982).

[146] M. K. Johnson, R. S. Czernuszewicz, T. G. Spiro, J. A. Fee, and W. V. Sweeney, *J. Am. Chem. Soc.* **105**, 6671 (1983).

[147] G. H. Stout, S. Turley, L. C. Sieker, and L. H. Jensen, *Proc. Natl. Acad. Sci. U.S.A.* **85**, 1020 (1988).

[148] C. D. Stout, *J. Biol. Chem.* **263**, 9256 (1988).

[149] C. D. Stout, *J. Mol. Biol.* **205**, 545 (1989).

[150] C. R. Kissinger, E. T. Adman, L. C. Sieker, and L. H. Jensen, *J. Am. Chem. Soc.* **110**, 8721 (1988).

[151] A. H. Robins and C. D. Staut, *Proc. Natl. Acad. Sci. U.S.A.* **86**, 3639 (1989).

[152] A. H. Robins and C. D. Staut, *Proteins* **5**, 289 (1989).

RR spectra for several 3-Fe proteins indicates that they all have this structure.[146,153] These spectra resemble those of the 4-Fe proteins, but the frequencies are distinctly higher. In the case of aconitase this upshift has been modeled by selectively increasing the Fe–S stretching force constants,[154] consistent with the exception that loss of a Fe cube corner should strengthen the bonds to the three sulfide ions, which are now doubly instead of triply bridging.

Active aconitase is a 4-Fe protein in which one of the Fe atoms lacks a cysteine ligand but is bound instead to hydroxide.[155] This Fe atom is the site of substrate binding and activation.[156,157] The RR spectrum[154] resembles those of other 4-Fe proteins, but with a shift in one of the cluster modes, reflecting the altered kinematic interaction with a hydroxide instead of a cysteine terminal ligand. The stretching vibration of the Fe–O bond to hydroxide, or to substrate, is not detectable in the RR spectra, indicating negligible coupling to the S $\rightarrow$ Fe charge transfer states. A marked change in the pattern of terminal Fe–S bands is seen on substrate or inhibitor binding, however, and has been attributed to a protein conformation change that alters the FeS–CC dihedral angles,[154] and thereby alters the Fe–S/S–C–C coupling, as discussed above in connection with rubredoxin.

Oxygen Carriers and Activators

In animals ranging from humans to fish, the molecule responsible for binding O_2 and delivering it to the tissues is hemoglobin, in which Fe(II)-protoporphyrin is bound via a histidine side chain in the binding pocket of globin chains. The O_2 binds in a bent end-on fashion to the Fe(II) ion and converts it from a high-spin to a low-spin state, pulling it into the heme plane.[158,159] The spin state change is readily apparent in the frequency shifts of the porphyrin core size marker RR bands, mentioned above. In addition, the frequencies of the porphyrin skeletal modes of the O_2 adduct agree with the frequencies expected for Fe(III) rather than Fe(II), reflecting substantial transfer of charge in the complex.[158] The formulation Fe(III)O_2^- is consistent with other spectroscopic indications including the

[153] M. K. Johnson, R. S. Czernuszewicz, T. G. Spiro, R. R. Ramsay, and T. P. Singer, *J. Biol. Chem.* **258,** 12771 (1983).

[154] L. K. Kilpatrick, M. C. Kennedy, H. Beinert, R. S. Czernuszewicz, D. Qiu, and T. G. Spiro, *J. Am. Chem. Soc.* **116,** 4053 (1994).

[155] M. M. Werst, M. C. Kennedy, H. Beinert, and B. Hoffman, *Biochemistry* **29,** 10526 (1990).

[156] O. Gawron, A. J. Glaid III, and T. P. Fondy, *J. Am. Chem. Soc.* **80,** 5856 (1958).

[157] H. Lauble, M. C. Kennedy, H. Beinert, and C. D. Staut, *Biochemistry* **31,** 2735 (1992).

[158] T. G. Spiro and T. C. Strekas, *J. Am. Chem. Soc.* **96,** 335 (1974).

[159] B. Shaanan, *Nature (London)* **296,** 683 (1982).

O–O stretching frequency, determined via IR spectroscopy[12,160] to be around 1135 cm^{-1} (the exact value is uncertain owing to a Fermi resonance interaction), close to the frequency for ionic superoxide ion (1108 cm^{-1} in KO_2).

The Fe–O_2 stretch, at 567 cm^{-1}, is detectable in RR spectra[160] and is enhanced via resonance with the Soret transition, as discussed above. The frequency is high relative to other Fe–O vibrations, reflecting the extensive Fe–O_2 backbonding. The O–O stretch has not been detected in RR spectra, indicating a small O–O displacement in the excited state. Interestingly, however, RR enhancement of the O–O stretch has been observed for the O_2 adduct of cytochrome P450,[161,162] an intermediate in the enzyme reaction. In this adduct the axial ligand is cysteine, a stronger donor than histidine; the effect of the increased donation can be seen in the reduction of the Fe–O_2 frequency to 541 cm^{-1}.[162] Apparently the increased donation also induces greater O–O bond extension in the π–π^* excited state.

In certain marine worms, a quite different O_2 carrier, hemerythrin, is found.[163] This protein contains no heme, but rather a pair of Fe atoms connected by a bridging oxide ligand and bound to the protein by histidine and glutamate ligands.[164] One of the Fe atoms has a vacant coordination site, to which O_2 binds when both Fe atoms are in the ferrous state. This binding event is accompanied by the transfer of two electrons from the Fe ions to the O_2, as evidenced by the frequency of the O–O stretching RR band, 844 cm^{-1} (798 cm^{-1} in oxyhemerythrin prepared with $^{18}O_2$),[165,166] which is close to the value for hydrogen peroxide. The Fe–O_2 stretching mode has also been observed and assigned to the 503 cm^{-1} band on the basis of its shift to 480 cm^{-1} on $^{18}O_2$ substitution. The O–O and Fe–O_2 stretches of oxyhemerythrin are both strongly enhanced with excitation wavelengths within the protein visible absorption band centered near 500 nm, indicating that the 500 nm electronic transition must have O_2^- →

[160] H. Brunner, *Naturwissenschaften* **61**, 129 (1974).

[161] O. Bangchaoenpauurpong, A. K. Rizos, P. M. Champion, D. Jullie, and S. Sligar, *J. Biol. Chem.* **261**, 8089 (1986).

[162] S. Hu, A. J. Schneider, and J. R. Kincaid, *J. Am. Chem. Soc.* **113**, 4815 (1991).

[163] N. B. Terwilliger, R. C. Terwilliger, and R. Schabtach, *in* "Blood Cells of Marine Invertebrates: Experimental Systems in Cell Biology and Comparative Physiology" (W. D. Cohen, ed.), p. 193. Alan R. Liss, New York, 1985.

[164] R. E. Stenkamp, L. C. Sieker, and L. H. Jensen, *J. Am. Chem. Soc.* **106**, 618 (1984).

[165] J. B. R. Dunn, D. F. Shriver, and I. M. Klotz, *Proc. Natl. Acad. Sci. U.S.A.* **70**, 2582 (1973).

[166] J. B. R. Dunn, A. W. Addison, R. E. Bruce, J. Sanders-Loehr, and T. M. Loehr, *Biochemistry* **16**, 1743 (1977).

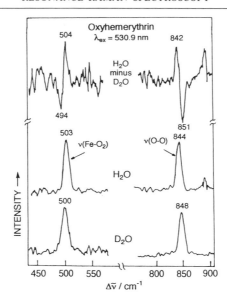

FIG. 23. Resonance Raman spectra (530.9 nm excitation) of oxyhemerythrin in H_2O (*middle*) and D_2O (*bottom*), and corresponding difference spectrum (*top*), showing the $\nu(O–O)$ and $\nu(Fe–O_2)$ stretching bands of the bound hydroperoxide.[167]

Fe(III) charge transfer character.[167] Indeed this absorption band is absent in deoxyhemerythrin. The peroxide is bound end-on to one of the Fe atoms and is protonated at the other end, as evidenced by the +4 and −3 cm^{-1} D_2O shifts of the O–O and Fe–O_2 stretches, respectively (Fig. 23).[167] The RR spectrum (363.8 nm excitation) also reveals symmetric and asymmetric stretching bands of the Fe–O–Fe bridge, at 486 and 753 cm^{-1} (475 and 720 cm^{-1} in $H_2^{18}O$), as shown in Fig. 24. From the frequencies and ^{18}O sensitivities of these modes, one can deduce the Fe–O–Fe angle, which is 134°. These bands are also sensitive to D_2O,[168] indicating the presence of a hydrogen bond to the oxide bridge, probably from the bound peroxide.

A third class of O_2 carriers, the hemocyanins, are found in arthropods and molluscs.[169] These proteins contain neither heme nor Fe, but rather pairs of Cu^+ ions, to which O_2 binds. As in hemerythrin, the binding event involves the transfer of two electrons to O_2, leaving pairs of (magnetically

[167] A. K. Shiemke, T. M. Loehr, and J. Sanders-Loehr, *J. Am. Chem. Soc.* **106**, 4951 (1984).

[168] A. K. Shiemke, T. M. Loehr, and J. Sanders-Loehr, *J. Am. Chem. Soc.* **108**, 2437 (1986).

[169] G. Preaux and C. Gielens, *in* "Copper Proteins and Copper Enzymes" (R. Lontie, ed.), Vol. 2, p. 160. CRC Press, Boca Raton, Florida, 1984.

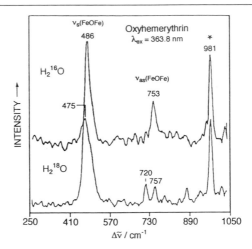

FIG. 24. Resonance Raman spectra (363.8 nm excitation) of oxyhemerythrin in $H_2^{16}O$ (*top*) and $H_2^{18}O$ (*bottom*) showing the ν_s(FeOFe) and ν_{as}(FeOFe) stretching vibrations of the μ-oxo-bridged diiron(III,III) site. An asterisk indicates the ν_1(SO_4^{2-}) internal standard band.[167]

coupled[170]) Cu^{2+} ions. The absorption spectrum of oxyhemocyanin reveals absorption bands attributable to d–d transitions of the Cu^{2+} ions, as well as peroxide $\rightarrow Cu^{2+}$ charge transfer transitions in the visible (~570 nm, $\varepsilon \approx 1000\ M^{-1}\ cm^{-1}$) and UV ($\sim350$ nm, $\varepsilon \approx 20{,}000\ M^{-1}\ cm^{-1}$) regions.[71] The RR spectrum reveals a peroxidic O–O stretch at around 750 cm^{-1},[171] even lower than in hemerythrin. Moreover, unlike hemerythrin, the O_2 is bound symmetrically, as shown by the appearance of only a single O–O band for the $^{16}O^{18}O$ mixed isotope.[172] The O_2 had been thought to bind in an end-on fashion to the two Cu^{2+} ions, but the X-ray crystal structure of a newly available oxyhemocyanin model, $(LCu)_2O_2$ [L is hydrotis(3,5-diisopropyl-1-pyrazolyl)borate] revealed an unprecedented binding mode, in which the O_2 is bound sideways (μ-$\eta^2 : \eta^2$), across the line joining the Cu^{2+} ions.[173] Because the spectroscopic properties of this compound, including the O–O stretching frequency,[173,174] closely resemble those of

[170] D. M. Dooley, R. A. Scott, J. Ellinghghaus, E. I. Solomon, and H. B. Gray, *Proc. Natl. Acad. Sci. U.S.A.* **75**, 3019 (1978).

[171] T. B. Freedman, J. S. Loehr, and T. M. Loehr, *J. Am. Chem. Soc.* **98**, 2809 (1976).

[172] T. J. Thaman, J. S. Loehr, and T. M. Loehr, *J. Am. Chem. Soc.* **99**, 4187 (1977).

[173] N. Kitajima, K. Fujisawa, C. Fujimoto, Y. Morooka, S. Hashimoto, T. Kitagawa, K. Toriumi, K. Tatsumi, and A. Nakamura, *J. Am. Chem. Soc.* **114**, 1277 (1992).

[174] M. J. Baldwin, D. E. Root, J. E. Pate, K. Fujisawa, N. Kitajima, and E. Solomon, *J. Am. Chem. Soc.* **114**, 10421 (1992).

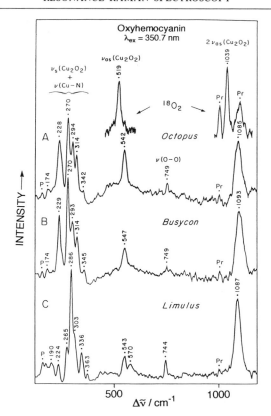

Fig. 25. Low-temperature (15 K) RR spectra of oxyhemocyanin from (A) *Octopus dofleini,* (B) *Busycon canaliculatum,* and (C) *Limulus polyphemus,* subunit II, obtained using 350.7 nm excitation wavelength. The upper traces in (A) are for protein oxygenated with $^{18}O_2$. P = plasma line; Pr = protein matrix band.[177]

the protein, a similar binding geometry is indicated, and this inference has now been confirmed by the crystal structure of oxyhemocyanin from *Limulus polyphemus.*[175]

The O–O stretch is dominant in RR spectra excited within the visible charge transfer transition, but excitation in the UV transition produces stronger enhancement for several bands in the 200–300 cm^{-1} region (Fig. 25),[176] whose $^{65/63}Cu$ and D_2O sensitivity identifies them as Cu–imidazole

[175] K. A. Magnus, H. Ton-That, and J. E. Carpenter, *in* "Bioinorganic Chemistry of Copper" (K. D. Karlin and Z. Tyeklar, eds.), p. 143. Chapman & Hall, New York, 1993.

[176] T. M. Loehr and A. K. Shiemke, *in* "Biological Applications of Raman Spectroscopy" (T. G. Spiro, ed.), Vol. 3, p. 439. Wiley, New York, 1988.

[177] J. Ling, L. M. Nestor, R. S. Czernuszewicz, T. G. Spiro, R. Fraczkiewicz, K. D. Sharma, T. M. Loehr, and J. Sanders-Loehr, *J. Am. Chem. Soc.* **116,** 7682 (1994).

stretching modes[177]; the protein crystal structure shows three histidine ligands to each Cu. Also seen in the UV RR spectrum is a band at approximately 545 cm^{-1}, identified via its $^{18}O_2$ shift and normal coordinate calculations as an asymmetric (B_{1g}) Cu_2O_2 stretch. The band is weak, but its overtone is relatively strong. Stronger RR enhancement of overtones than fundamentals is expected for nontotally symmetric modes; the overtone is totally symmetric, and its enhancement reflects the extent of the force constant change in the excited state. The absence of a readily identifiable symmetric Cu_2O_2 stretch in the RR spectra had been a long-standing puzzle, which has been resolved by the new binding mode. In the diamond-shaped Cu_2O_2 unit there are two A_{1g} modes, the O–O stretch, already discussed, and a mode that involves outward displacement of the Cu atoms. Because these atoms are massive, the frequency is quite low and is insensitive to $^{18}O_2$ substitution. This is why no $^{18}O_2$-sensitive band is found in the usual Cu–O stretching region, 400–500 cm^{-1}. Indeed, the frequency is so low that the mode is calculated to mix heavily with the Cu–imidazole stretches, accounting for some of the intensity of the complex of bands in the 200–300 cm^{-1} region.[177]

[19] Structure and Dynamics of Transient Species Using Time-Resolved Resonance Raman Spectroscopy

By JAMES R. KINCAID

I. Introduction

The impressive capability of resonance Raman (RR) spectroscopy to provide precise structural information for specific active sites in various biological systems has been thoroughly reviewed in the previous chapter and references therein (see T. G. Spiro and R. S. Czernuszewicz, this volume [18]). Although other techniques, such as nuclear magnetic resonance (NMR) and X-ray crystallography, rival the power of RR spectroscopy for the study of static systems at equilibrium, it is perhaps fair to state that RR spectroscopy is currently unchallenged in its potential for the investigation of the structure of transient species. This is an important issue, because in order to understand the molecular basis of biological function it is necessary to gain a thorough knowledge of both the structure and dynamics of the transient species involved. The inherent speed of the scattering process ($\sim 10^{-14}$ sec) carries with it the implication that it is theoretically feasible to monitor events on the subpicosecond time scale and, as will be seen, advances in laser technology (ultrashort pulses) and the development of highly sensitive detectors have now made it possible to approach these theoretical limits. Before proceeding it is perhaps worth restating, for emphasis, the remarkable implications of these considerations; namely, time-resolved resonance Raman (TR^3) methods provide the capability to probe the precise structure of fleeting intermediates which evolve and decay, even on subpicosecond time scales, throughout the course of a given biological process.

The intent of this review is to provide a general introduction to the field for those who wish or need to become familiar with the utility of TR^3 methods, so as to understand which strategies are available and what type of information can be gained. To accomplish this, a concise summary of the various types of experiments and required instrumentation is given in Section II. Although it is impossible in this chapter to provide a thorough and complete description of this material, sufficient information to understand the experimental strategy is provided and exemplary original publications are cited.

Also, no attempt can be made to provide a comprehensive summary of applications. Rather, to illustrate the kind of information which TR^3 methods can provide, the examples discussed under "Applications" (Section III) are selected exclusively from the field of heme protein chemistry. This seems most efficient inasmuch as the essential interpretive framework for discussion of the spectral data for these systems has been clearly and relatively thoroughly discussed in [18] of this volume. Finally, although emphasis has been placed on heme protein structure and function, some effort is made in Section IV to identify key articles which can provide access to the literature dealing with TR^3 studies of several other important biological systems.

II. Strategies and Methodology

The essential purpose of conducting TR^3 studies is to monitor the structure and dynamics of key intermediates encountered during an evolving biochemical process. Thus, there are two steps involved. The first step is the initiation of the reaction or process, whereas the second task is to monitor the appearance and decay of various transient species which participate in the evolving system. It is obvious that the basic principles associated with the monitoring step are identical to those for RR spectroscopy; that is, the monitoring step simply generates the RR spectrum. The essential difference lies in the fact that the strategy of the TR^3 method is to conduct the experiment in such a way that the temporal widths of both the initiation step and the monitoring step are kept very small in order to provide high temporal resolution.

In discussing the TR^3 methods, it seems reasonable to organize the treatment according to two distinctly different approaches for initiating the reaction. The first and most common approach applies to biological systems which are naturally (or can be made to be) responsive to light. In these cases, the reaction or process of interest can be initiated by a short pulse of light (called the pump pulse). Other (obviously most) biochemical systems are unresponsive to light pulses, and the reaction must be initiated by rapid mixing of two reactants.

A. Light-Initiated Methods

Molecular systems which are susceptible to the light-initiated approach are those that undergo a sequence of structural changes (or those that participate in bimolecular reactions with other solution components) on exposure to a photolysis pulse. There are a number of ways that the photolysis step can be accomplished. By far the most common method is to use pulsed lasers; however, several methods have been devised using continuous wave (cw) lasers, and these too are useful if it is not necessary to probe at very short times. Although these latter methods provide time resolution on the microsecond time scale, the approach using pulsed lasers permits investigation at the nanosecond, picosecond, and even subpicosecond levels, the temporal resolution being determined by the widths of the laser pulses.

Before proceeding to discuss these various methods it is perhaps helpful to clarify terminology with regard to the terms "time-resolved" and "transient" RR. The latter term is used in the case where a single laser pulse is used to both photolyze and probe the sample. The term "time-resolved" is appropriate in cases where the beam used to probe the photolyzed sample is temporally delayed with respect to the photolysis pulse (the "pump" pulse). In such an experiment the reaction is initiated by a short photolysis pulse, and the RR spectrum is acquired using a probe pulse which is delayed with respect to the pump pulse. The temporal evolution of the system can thus be documented by performing a series of such experiments where the delay time between the pump and probe pulses is varied.

1. Methods Using Continuous Wave Lasers

a. Rapid-flow method. In rapid-flow types of experiments a photolytically susceptible sample is passed rapidly through a cw laser beam, and

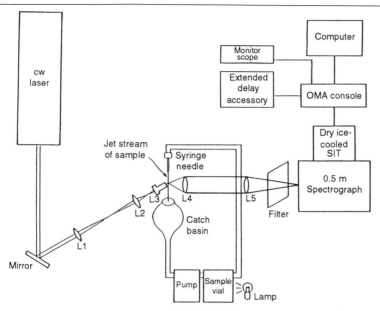

FIG. 1. Experimental arrangement for the resonance Raman microbeam flow experiment. L_1, 30 cm 4/15 lens; L_2, 6 cm f/3 lens; L_3, Zeiss Neofluar X40 microscope objective; L_4, 50 mm f/1.2 Canon lens; L_5, 16 cm f/4 lens; OMA, optical multichannel analyzer. Note that the lamp which irradiates the reservoir (sample vial) is needed for the particular sample used in this study and is not a general requirement. (From Terner *et al.*[3])

the temporal resolution is determined by the residence time of the sample in the beam. There are two relatively convenient methods to ensure rapid transit of the sample through the laser beam, and examples of both are given below.

In early studies of retinal–protein complexes, including visual pigments and bacteriorhodopsin, rapid-flow TR[3] methods were developed in order to resolve temporally the vibrational spectra of various intermediates involved in the light-driven process.[1,2] Also, one of the first transient RR methods used to investigate the mechanism of hemoglobin cooperativity (*vide infra*) employed this method, using an experimental arrangement similar to that shown in Fig. 1.[3] In this case the output from a cw ion laser is passed through a series of lenses (L1, L2, and L3) to produce a focused beam of very small diameter (e.g., 1 μm). The beam is focused

[1] M. A. Marcus and A. Lewis, *Biochemistry* **17**, 4722 (1978).
[2] R. Mathies, A. Oseroff, and L. Strayer, *Proc. Natl. Acad. Sci. U.S.A.* **73**, 1 (1976).
[3] J. Terner, C. L. Hsieh, A. R. Burns, and M. A. El-Sayed, *Proc. Natl. Acad. Sci. U.S.A.* **76**, 3046 (1978).

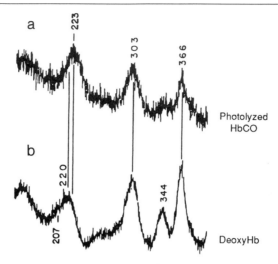

FIG. 2. Low-frequency Raman spectra, showing the ν_{Fe-Im} mode around 220 cm^{-1}, with B band excitation (λ_{ex} 4545 Å) of (a) HbCO (1 mM, pH 7) flowing in a free jet through (0.3 μsec interaction time) the laser beam, which is both the photolysis and the Raman source, and (b) deoxyHb (0.3 mM, pH 7) in a recirculating capillary. Spectra were obtained in scanning mode at 5 cm^{-1} increments, 1 sec increment. (From Stein et al.[4])

onto a rapidly flowing stream ("jet stream" in Fig. 1) of the sample solution. Lenses L4 and L5 are collection optics which transfer the scattered light to the monochromator/detector system. In this case a spectrograph/photodiode array system was used, although a conventional scanning system would also be appropriate (though less efficient in terms of total acquisition time required; see [18]).

In the specific example cited,[4] a solution of carbonmonoxyhemoglobin (HbCO) was pumped through a tube connected to a 30-gauge syringe needle (i.e., an inside diameter of 150 μm) at a volumetric flow rate of 0.1 ml/sec. This corresponds to a linear stream velocity of 6 m/sec and an estimated residence time (temporal evolution) of less than 0.3 μsec (allowing for a factor of 2 in the uncertainty of the spot size measurement). On entering the irradiation zone, a molecule of HbCO is rapidly (<1 psec) and efficiently photolyzed (CO is ejected) to produce an intermediate (Hb*) which is then probed by RR scattering throughout the time that the photolysis product traverses the beam. As can be seen in Fig. 2, the RR spectrum of the 0.3 μsec photolysis product exhibits slight, but significant, differences from the equilibrium deoxyHb (trace b, Fig. 2). As explained

[4] P. Stein, J. Terner, and T. G. Spiro, J. Phys. Chem. 86, 168 (1982).

later, the most important difference in these traces is the frequency of the feature near 220 cm^{-1}, which is attributable to the key linkage between the iron atom of the heme and the imidazole group of the so-called proximal histidine (*vide infra*).

Often times, as in the example given here, the photolysis reaction is reversible, with the photolyzed Hb eventually recombining with CO to form HbCO. In such cases the entire solution can be recirculated, thus reducing the amount of sample needed. In those cases where the photolysis product does not return to its initial state (i.e., recirculation is not possible) much greater amounts of sample are obviously required.

The above example used a single laser beam to both photolyze and probe the sample and is therefore an example of transient RR in the sense that the temporal evolution of the sample was not monitored; in other words, it is a cumulative signal from all species which appear during the residence time in the (pump plus probe) beam. Clearly it is also possible to use separate pump and probe beams, and one of the earliest examples of such an arrangement is depicted in Fig. 3. Here, the distance between two spatially separated cw argon ion laser beams determines the delay

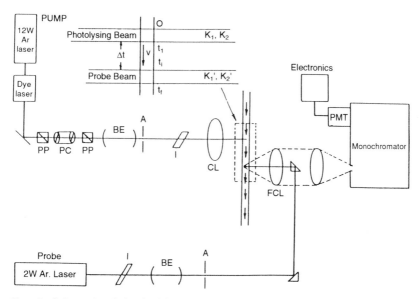

FIG. 3. Schematic of the dual laser beam apparatus. The sample flows vertically as indicated by the vertical arrows. The photolyzing beam pumps the photochemistry, and the probe beam is utilized to obtain a resonance-enhanced Raman spectrum of some intermediate with an optimal concentration corresponding to Δt (inset). Details are described in the original work.[1] (From Marcus and Lewis.[1])

time (Δt) between the pump and probe steps. Assuming that the flow rate is constant, variable delays are achieved simply by positioning the pump beam further upstream from the probe beam. The equations which relate the temporal resolution to the flow velocity, beam spot sizes, and distance between the two beams are found in the original report[1] and references therein.

   b. *Spinning cell method.* Assuming that the reaction or process of interest is reversible (i.e., the photoproducts return to the initial state), it is relatively convenient to acquire transient RR and TR[3] spectra with a closed system spinning cell. An example of such a system is shown in Fig. 4.[5,6] In this case the residence time of the sample in the beam (or beams) is determined by the laser spot size, the rotational frequency of the cell, and the distance between the axis of rotation and the laser beam. Figure 4 (bottom) depicts the situation where both pump and probe beams are used. Clearly, the greater the rotation speed, the better is the time resolution (i.e., shorter residence times). However, to use a closed system like this, the initial photolysis reaction must be reversible, and in some cases this may require several milliseconds or more. This means that rotational velocities must be reduced to the point where the resting state is fully recovered within the time taken for one full rotation (6.3 msec in Fig. 4). Alternatively, as also depicted in Fig. 4, it is possible to interrupt the beam momentarily (20 msec) with a chopper so that the molecular system can fully recover before the next photolysis cycle. As shown, the detector can be electronically synchronized with the chopper (laser excitation).

   2. *Methods Using Pulsed Lasers.* The commercial development of lasers which produce high peak power, temporally short pulses [e.g., the Nd : YAG (neodymium–yttrium-aluminum-garnet) and eximer lasers] provided the key tools to bring about rapid progress in transient RR and TR[3] methods. As will be seen from the examples described, methods based on these pulsed output systems are capable of probing transient species which are generated during the width of the laser pulse; that is, temporal resolution on the subpicosecond time scale is possible.

   Before proceeding to illustrate the various experimental arrangements based on pulsed lasers, it is necessary to describe briefly the nature of the output of these systems and to mention several complications which must be considered (references are given to more detailed discussions). First, it must be remembered that the output of the lasers consists of ultrashort pulses (e.g., 30 psec or 10 nsec). To appreciate the practical

[5] R. Lohrman and M. A. Stockburger, *J. Raman Spectrosc.* **23**, 575 (1992).
[6] R. Diller and M. A. Stockburger, *Biochemistry* **27**, 7641 (1988).

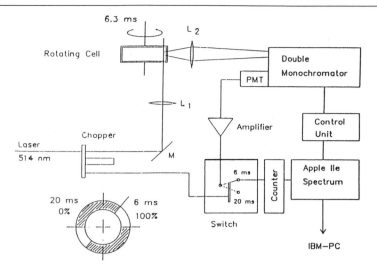

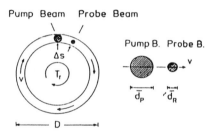

FIG. 4. Experimental arrangement for transient RR and TR³ studies using a spinning cell. The chopper and electronic switch are needed only for high rotational speeds so that the resting state has been recovered. (*Bottom*) Arrangement for the case where two beams (pump and probe) are used. (From Lohrman and Stockburger[5] and Diller and Stockburger.[6])

implications of this, consider the fact that acquiring a conventional RR spectrum (using a cw laser) from a reasonably good scatterer may typically require 10 mW of cw laser power [less in favorable cases, e.g., when a diode array or charge-coupled device (CCD) is used as the detector]. For a pulsed laser like a Nd : YAG (operating at 10 Hz and having pulse widths of 10 nsec), to get this average power, the peak power of each pulse will be more than 1 mega watt (MW). In other words, to acquire spectra with an adequate signal-to-noise (S/N) ratio, it is necessary to use these high peak powers. The required peak powers are proportionately larger, of course, for lasers with shorter pulse widths.

The use of these high peak power pulses can present several types of problems. First, when the photon flux is so high, various nonlinear optical phenomena can occur (e.g., stimulated Raman scattering) as well as sample decomposition. Another problem that can arise for ultrafast (subpicosecond) TR^3 studies is that the high photon energy deposited at the chromophore site may give rise to vibrational hot bands and frequency broadening which contribute to the spectra at these early detection times. Detailed discussion of such effects is not possible here, but these issues have been thoroughly discussed elsewhere.[7,8]

With regard to the wavelengths available from pulsed lasers, there are a number of techniques and strategies which can be used to provide for a wide choice. The fundamental output of the Nd : YAG laser is in the near-IR (1064 nm). However, the photon flux is so high that the harmonics of this fundamental can be generated to produce high peak power pulses at 532 nm (second harmonic), 355 nm (third harmonic), and 266 nm (fourth harmonic). Obviously, these can then be used with dye lasers to produce tunable excitation frequencies. Perhaps more conveniently, any of these can be focused inside a so-called Raman shifter, which is simply a sealed stainless steel tube (having quartz windows) filled with high pressure gas (usually H_2 or D_2), whereon a series of additional lines are generated by the stimulated Raman process. These new lines are shifted from the frequency of the input beam (and separated from one another) by the vibrational frequency of the filler gas (4155 cm^{-1} for H_2). Both Stokes and anti-Stokes lines are produced. For example, 355 nm corresponds to 28,183 cm^{-1}. Passage of this beam through the Raman shifter produces new laser lines at 24,028 (416 nm) and 19,872 (503 nm) as well as 32,338 cm^{-1} (309 nm), etc. These are easily separated spatially from one another by using a special prism (Pellin–Broca prism), and any of them can be selected for excitation of the sample.

With regard to detection of the Raman signal, it is important to remember that with these short pulse, low repetition-rate lasers the sample is being irradiated for only a vanishingly small fraction of time. For example, even in the case of a 10 nsec pulse-width Nd : YAG instrument operating at 10 Hz, the laser irradiates the sample for only 100 nsec for every second of "exposure" at the detector (i.e., a factor of 10^{-7}). Thus, most of the time there is no signal reaching the detector, which means that the dark count of the detector may constitute a large fraction of the signal. One way to avoid such problems is to turn off the voltage to the detector

[7] J. W. Petrich and J. L. Martin, *Adv. Spectrosc.* (*Chichester, U.K.*) **18**, 335 (1989).

[8] E. R. Henry, W. A. Eaton, and R. M. Hochstrasser, *Proc. Natl. Acad. Sci. U.S.A.* **83**, 8982 (1986).

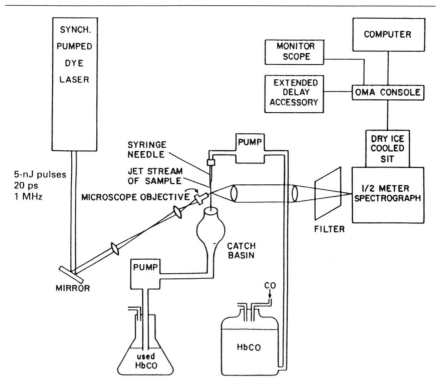

Fig. 5. Diagram of experimental setup for picosecond RR spectroscopic studies of HbCO. OMA, Optical multichannel analyzer; SIT, silicon-intensified target vidicon. (From Terner et al.[9])

between pulses. This is accomplished by using a voltage pulser which operates in synchronization with the probe pulse. This "gating" of the detector can significantly improve the S/N ratio.

a. Single-pulse transient resonance Raman spectroscopy. The experimental arrangement for recording transient RR spectra using a single (pump plus probe) beam is relatively simple, an example being shown in Fig. 5.[9] Although in this particular case a picosecond laser was employed, clearly the temporal resolution is determined solely by the pulse width. In such experiments it is necessary only to ensure that fresh sample is supplied to the area of irradiation for each successive pulse. Usually the repetition rate of such lasers is low (10–30 Hz), so this can be easily

[9] J. Terner, J. D. Strong, T. G. Spiro, M. Nagumo, M. Nicol, and M. A. El-Sayed, *Proc. Natl. Acad. Sci. U.S.A.* **78,** 1313 (1981).

accomplished at relatively low volumetric flow rates or with a conventional rotating cell (if the photoreaction is reversible).

b. Dual-pulse time-resolved resonance Raman methods. In dual-pulse experiments a short probe pulse is temporally delayed relative to a short pump pulse. There are essentially two ways to do this, both of which are depicted in Fig. 6.[10] In some cases it is possible to use a single laser and merely to split off a small fraction of the pump beam (using dichroic mirrors) and delay that pulse (the probe pulse) relative to the pump pulse by sending it through an optical delay line. Because the velocity of light is 3×10^8 m/sec, if the probe pulse is forced to travel an additional 3 m, it arrives at the sample 10 nsec later than the pump pulse. Although such an arrangement permits TR^3 studies during the first 20–100 nsec, it is not practical for studies in the microsecond or millisecond regime. In those cases it is necessary to employ a second pulsed laser (operating at the same repetition rate) whose output is electronically delayed relative to the first (pump) laser. Inasmuch as the lasers are normally operating at low repetition rates (50 msec between pulses for 20 Hz), the firing of the probe laser can be delayed from nanoseconds out to milliseconds relative to the pump pulse. Although it is obviously possible to use different wavelengths for the two laser configurations, this capability is also possible using a single Nd : YAG instrument because various harmonics are readily available from the same pulse. Thus, for example, the 532 nm (doubled fundamental) output can be employed as the pump beam and the 355 nm output can be used directly as the probe beam (or it can first be converted to other wavelengths via a Raman shifter or dye laser). Findsen and Ondrias nicely summarize the practical details for such instrumentation.[11]

B. Rapid-Mixing Methods

In the above cases, impressive temporal resolution is made possible by the fact that the processes can be initiated with a short photolysis pulse. The great majority of biological systems of interest cannot be photo-initiated, however, and for these cases it is necessary to rely on rapid-mixing methods, wherein the temporal resolution is dictated by the speed with which the reaction mixture can be made homogeneous with respect to reactant concentrations. This mixing time is referred to as the "dead time" of the mixing apparatus, and various designs have been developed to maximize the mixing efficiency of such devices by minimizing the dead

[10] R. F. Dallinger, S. Farquharson, W. H. Woodruff, and M. A. J. Rodgers, *J. Am. Chem. Soc.* **103,** 7433 (1981).
[11] E. W. Findsen anddd M. R. Ondrias, *Appl. Spectrosc.* **42,** 445 (1988).

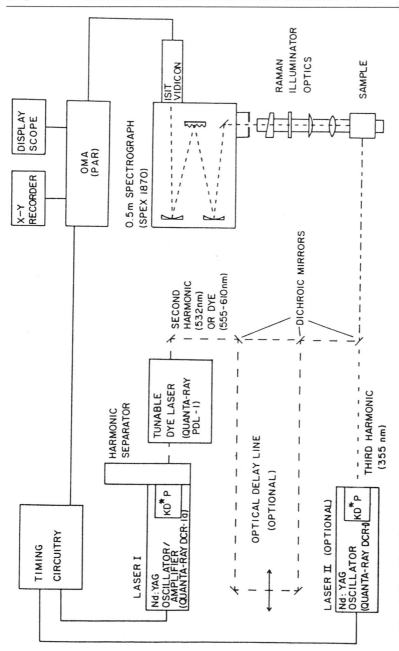

Fig. 6. Simplified diagram of the experimental apparatus for pump–probe TR³ studies. Requisite time delays can be obtained either by optical delay or by the two-laser experimental configuration. The actual length of the optical delay line was approximately 120 ft. (From Dallinger et al.[10])

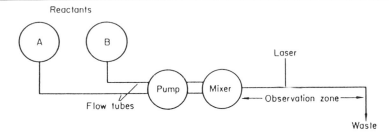

Fig. 7. Schematic diagram of a continuous fast-flow apparatus. (From Hester.[12])

time. High quality commercially available mixing chambers yield dead times of a few milliseconds.

A typical arrangement for use of such devices for TR³ studies is depicted in Fig. 7.[12] In this particular scheme the reactant solutions are held in separate reservoirs and fed to the mixer via feed lines led through a pump (or pumps). Obviously, a conventional syringe pump can also be used, as was shown by Woodruff and Spiro[13] in one of the fist TR³ studies using mixing devices. Carey and co-workers have used such devices to study enzyme–substrate complexes by TR³ and have discussed methods for evaluation of their performance when used for TR³ studies.[14]

All of the rapid-mixing TR³ studies so far reported have employed conventional mixing devices that were developed for use with optical (absorption or fluorescence) detection, wherein the observation chamber must necessarily be optically homogeneous. However, given the fact that Raman spectroscopy is based on a scattering phenomenon, it is reasonable to consider entirely new approaches for rapid mixing. Two novel mixing devices have been developed[15–18] for use with RR spectroscopy which should permit the scope of TR³ methods to be extended to a much wider range of important biochemical and biophysical processes. The essential feature of both devices is a dramatic improvement in mixing efficiency such that the time required for complete mixing of reactant solutions is reduced to the microsecond time scale (compared to the few milliseconds needed with conventional mixing chambers).

The use of one such device developed for RR spectroscopy is depicted

[12] R. E. Hester, *Adv. Infrared Raman Spectrosc.* **4**, 1 (1978).
[13] W. H. Woodruff and T. G. Spiro, *Appl. Spectrosc.* **28**, 576 (1974).
[14] L. R. Sans Cartier, A. C. Storer, and P. R. Carey, *J. Raman Spectrosc.* **19**, 117 (1988).
[15] S. F. Simpson, J. R. Kincaid, and F. J. Holler, *Anal. Chem.* **55**, 1420 (1993).
[16] S. F. Simpson, J. R. Kincaid, and F. J. Holler, *Anal. Chem.* **58**, 3163 (1986).
[17] S. F. Simpson and F. J. Holler, *Anal. Chem.* **60**, 2483 (1988).
[18a] K. J. Paeng, Ph.D. Dissertation, Marquette University, Milwaukee, Wisconsin (1989).
[18b] K. J. Paeng, I. R. Paeng, and J. R. Kincaid, *Anal. Sci.* **10**, 157 (1994).

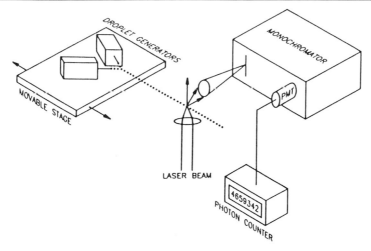

FIG. 8. System schematic of the microdroplet experiment. (From Simpson et al.[16])

in Fig. 8. The essential task to be faced in rapidly and efficiently mixing two solutions is to force them to combine turbulently in a very small volume. In the device shown in Fig. 8 this is accomplished by first causing each stream to be broken down into regularly spaced microdroplets of uniform size. This was accomplished, as shown in Fig. 9, by using tech-

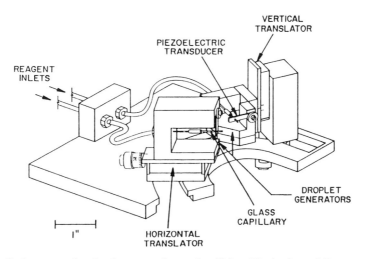

FIG. 9. Apparatus for droplet generation and collision. The horizontal linear translator adjusts the relative phase of the droplet streams while the vertical translator is used to collide the streams. (From Simpson et al.[16])

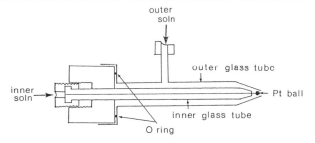

FIG. 10. Platinum sphere mixing device. (From Paeng.[18a])

niques previously developed by physicists to study aerosols. The two streams of microdroplets have the same frequency (number of drops per unit time; e.g., 30 kHz) and can easily be made to be in-phase so that, at the intersection point of the two streams, the individual droplets from the two streams collide and coalesce to form a stream of droplets wherein the reactants in each "product droplet" undergo reaction. Obviously, the temporal evolution of the reaction can be monitored by simply varying the distance between the point of intersection of the two streams and the probe laser (see Fig. 8). The characteristic mixing time of such a device is on the order of a few hundred microseconds, a substantial improvement over conventional mixers.

A second device developed in our laboratory for use with RR detection is shown in Fig. 10. It is based on a concept originally suggested by Berger *et al.*[19] and further developed by Jovin and co-workers.[20] The two reactant solutions are fed through concentric tubes to a tip which has a platinum sphere positioned very close (10 μm) to the ($\sim$100 μm) orifice of the outer tube. In this way both solutions are forced to flow through the 10 μm circular passage formed by the sphere and the outer tube orifice. The mixing of the two solutions is extremely efficient (dead times as low as 20–40 μsec.)[18] Again, the progress of the reaction can be conveniently monitored merely by varying the distance between the tip of the outer tube and the probe laser. For example, as shown in Fig. 11, the progress of the reaction of O_2 with deoxyhemoglobin is followed over the first 1.2 msec by monitoring the ν_4 modes (*vide infra*) which appear at 1358 cm^{-1} (Hb) and 1374 cm^{-1} (Hb $\cdot$ O_2) This device is being used in our laboratory to detect directly early intermediates in the reactions of various heme proteins with physiologically relevant reactants.

[19] R. L. Berger, B. Balko, and W. Borcherdt, *Rev. Sci. Instrum.* **39**, 486 (1968).

[20] P. Regenfuss, R. M. Clegg, M. J. Fulwyler, F. J. Barrantes, and T. M. Jovin, *Rev. Sci. Instrum.* **56**, 283 (1985).

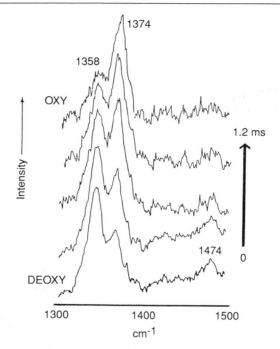

FIG. 11. Time-resolved RR spectra of the reaction of deoxyhemoglobin (0.25 mM in heme) with oxygen-saturated buffer, pH 7.0, obtained using the mixing device shown in Fig. 10. The excitation line was 457.9 nm. (From Paeng.[18a])

III. Applications to Heme Protein Structure and Dynamics

Heme proteins are a class of biomolecules which possess a heme group, usually protoheme (Fig. 12), embedded in a single polypeptide chain. The reactivity of the heme group varies greatly from protein to protein depending on the nature of the active site environment provided by the polypeptide side chains, and this variability gives rise to the remarkable functional diversity which characterizes this class of proteins. For example, myoglobin (Mb) and hemoglobin (Hb) bind dioxygen reversibly and thus serve as O_2 storage (Mb) and transport (Hb) proteins. Various peroxidases and catalases have different active site environments which impart a different reactivity to the protoheme group, namely, the ability to react with H_2O_2 to form highly oxidizing intermediates in which the oxidation equivalents are localized on the heme (and in some cases on surrounding protein residues) and which are capable of oxidizing various substrates. Still other heme proteins such as cytochrome c and cytochrome b have active site environments which render the heme unreactive to exogenous

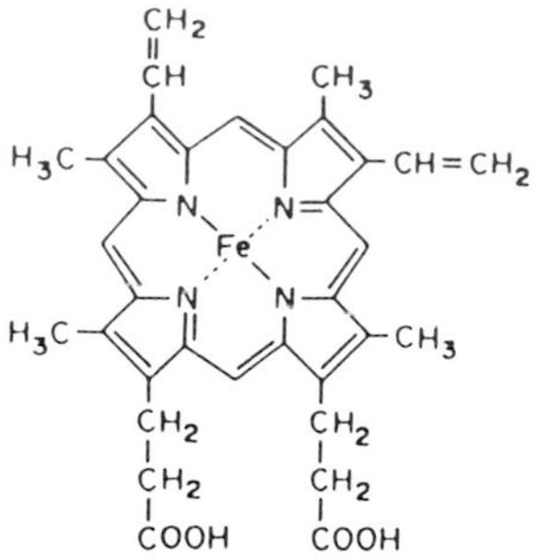

FIG. 12. Structure of protoheme (also called heme *b*). [From G. T. Babcock and C. Varotsis, *in* "Biological Applications of Raman Spectroscopy" (T. G. Spiro, ed.), Vol. 3, p. 293. Wiley, New York, 1988.]

ligands, and these serve only as electron transfer proteins. Finally, one of the most important and frequently studied heme proteins is cytochrome-*c* oxidase, whose function is to catalyze the four-electron reduction of dioxygen to water.

To understand the molecular basis for such diverse reactivity it is essential to obtain a thorough knowledge of the active site structures of the various types of heme proteins. As discussed in the previous chapter (see [18] in this volume), RR methods are especially well suited for this purpose in that they provide precise structural information for the heme group and its endogenous and exogenous axial ligands and their disposition relative to the heme. However, as will be made clear, a detailed understanding of the molecular mechanisms which are operative in heme protein function requires a knowledge of the structure and dynamics of intermediate species which evolve and decay during the course of heme protein catalysis, knowledge which is most efficiently obtained by transient RR and TR³ methods.

Although most of the physiologically relevant states of heme proteins are, in general, not photosensitive, the carbon monoxide adducts of the ferrous forms can be rapidly and efficiently photolyzed and provide a derivative which is ideally suited for the initial photolysis step. Thus, most of the examples discussed in this section involve initial photolysis of these adducts. The initial photoproduct may undergo further bimolecular reactions (with O_2, for example) or may simply rebind the photolyzed CO molecule (a reversible reaction) depending on the particular conditions of the experiment. In the interest of brevity and clarity, the examples discussed focus on only two heme proteins: hemoglobin and cytochrome-*c* oxidase.

A. Hemoglobin

1. Structure and Function. Hemoglobin and myoglobin, two of the first proteins to be structurally determined by X-ray crystallography, both contain five-coordinate, ferrous hemes at the active sites as shown in Fig. 13. In both cases the heme is attached to the protein through a coordinate linkage to an imidazole side chain of a histidine residue (the so-called proximal histidine). The distal side of the heme pocket, at which exogenous ligands bind, contains nonpolar amino acid residues as well as an imidazole fragment of a histidine (the so-called distal histidine) which is known to be capable of forming a hydrogen bond with certain bound ligands, including dioxygen.

Whereas the oxygen storage protein, Mb, consists of a single polypeptide chain possessing a protoheme at its active site, the oxygen transport protein, Hb, is actually a tetramer of four (noncovalently bonded) polypeptide chains (two α and two β subunits) each of which contains a protoheme group in its active site. The tetrameric nature of Hb permits interactions between the subunits which give rise to the phenomenon known as cooper-

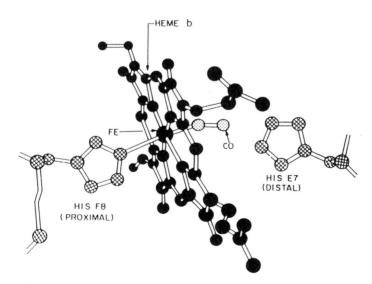

Fig. 13. Biological active center in hemoglobin and myoglobin. The filled black circles represent the heme skeletal atoms, and the black circles with white dots represent the peripheral groups of the heme. The cross-hatched circles represent the atoms of the proximal and distal histidine and the white circles with black dots a bound CO ligand. [From D. L. Rousseau and J. M. Friedman, *in* "Biological Applications of Raman Spectroscopy" (T. G. Spiro, ed.), Vol. 3, p. 133. Wiley, New York, 1988.]

ativity.[21,22] Thus the binding curve assumes a sigmoidal shape, leading to more efficient oxygen transport in the sense that it binds O_2 less readily at low O_2 levels (in tissues) but has a much higher affinity at high O_2 levels (in the lungs). This behavior is consistent with the famous two-state theory first proposed by Monod *et al.*[23] in which cooperative behavior is the result of the existence of a low affinity (so-called T) state and a higher affinity (R) state; ligand binding to the T state induces a protein structural transition to the more reactive R state. The subsequent crystallographic confirmation of two distinct structures for the ligated (R) and nonligated (T) forms provided a structural basis for the validity of the thermodynamic arguments. Though the existence of several other states is required to explain the precise details of the binding curves under some conditions, the two-state model remains as a valid (quite good) approximation under most conditions of ligand binding, indicating that the T–R switch is the most important factor for regulating Hb reactivity.[24]

Clearly, the essential issue regarding this mechanism is the identification of the molecular interactions within the tetramer which generate a global structural response to ligand binding at the heme sites. To elucidate these details it is necessary to probe specific sites within the tetramer at various times during the ligand binding process. However, ligand binding is controlled mainly by the probability of a ligand getting to the active site, whereupon a rapid protein structural change occurs. Thus, it is not possible to study these structural changes in mixing experiments wherein the individual molecules of the population are out of phase with respect to ligand binding (i.e., there is not a well-defined initial point in time). Fortunately, the fully ligated CO adduct of the tetramer can be efficiently photolyzed so that a short laser pulse can be used to generate "instantly" a homogeneous population of molecules in which the hemes are nonligated but which have an R-state protein structure. This R-state structure is of course unstable in the absence of ligated hemes and undergoes structural changes (through intermediate states) to arrive at the T-state quaternary structure. This strategy (photolysis of the CO adduct) facilitates detection and temporal monitoring of these intermediates via transient RR and TR^3 techniques, and such studies have provided an impressive body of information concerning the structural basis for the T–R switch.

[21] E. Antonini and M. Brunori, "Hemoglobin and Myoglobin in Their Reactions with Ligands." North-Holland Publ., Amsterdam, 1971.

[22] E. Antonini, L. Rossi-Bernardi, and E. Chiancone (eds.), this series, Vol. 76.

[23] J. Monod, J. Wyman, and J. P. Changeux, *J. Mol. Biol.* **12,** 88 (1965).

[24] M. F. Perutz, G. Fermi, B. Liusi, B. Shaanan, and R. C. Liddington, *Acc. Chem. Res.* **20,** 309 (1987).

2. *Structural Dynamics.* Time-resolved electronic absorption techniques have shown that a deoxy heme absorption spectrum is observed immediately ($\sim$0.35 psec) following photolysis of HbCO. This spectrum persists, unchanged, for tens of nanoseconds. However, on a longer time scale, small changes in the absorption spectrum are detected which correspond to relaxation times of approximately 0.1, 1, and 20 μsec.[25] Such changes are brought about by protein rearrangements which affect the heme electronic structure. The 20 μsec process is associated with the R–T quaternary transition and is consistent with studies of CO rebinding following photolysis which show that there are fast ($<$20 μsec, corresponding to R state) and slow ($>$20 μsec, corresponding to T state) processes.[26] The spectral changes which occur with approximately 0.1 and 1 μsec relaxation rates presumably correspond to changes in the tertiary structure which also affect the heme electronic structure and which occur as part of the evolution from the R- to the T-state structure. The transient RR and TR[3] studies to be discussed were undertaken in an attempt to provide a structural interpretation of these processes.

As discussed in the previous chapter (see [18] in this volume), various RR active modes provide information about different fragments of the heme group and its immediate environment. Two of the high frequency modes, ν_2 (near 1560 cm^{-1}) and especially ν_4 (near 1360 cm^{-1}), are sensitive to the electron density in the heme π orbitals. Other high frequency modes [ν_3 ($\sim$1470 cm^{-1}), ν_{10} ($\sim$1610 cm^{-1}), ν_{11} ($\sim$1540 cm^{-1}), and ν_{19} ($\sim$1555 cm^{-1})] are called core-size markers and respond to changes in the size of the porphyrin core (the distance between the pyrrole nitrogen and the center of the porphyrin core). A low frequency mode near 345 cm^{-1} (ν_8) is believed to be sensitive to the orientation of the heme peripheral substituents and may thus provide insight into interactions at the protein–heme interface. Finally, the iron–histidine stretching mode, ν(Fe–N^{His}), provides a direct probe of the strength of the bond between the heme iron and the imidazole fragment of the proximal histidine, a linkage which is believed to be important for modulation of heme reactivity.[24]

a. Heme macrocycle modes. The first transient RR study of HbCO was reported by Woodruff and Farquharson.[27] Since then many transient RR and TR[3] studies which focus attention on the heme core modes have been conducted for photolyzed HbCO and MbCO. One of the more recent

[25] J. Hofrichter, J. H. Sommer, E. R. Henry, and W. A. Eaton, *Proc. Natl. Acad. Sci. U.S.A.* **80,** 2235 (1983).

[26] C. Sawicki and Q. H. Gibson, *J. Biol. Chem.* **251,** 1533 (1976).

[27] W. H. Woodruff and S. Farquharson, *Science* **201,** 831 (1978).

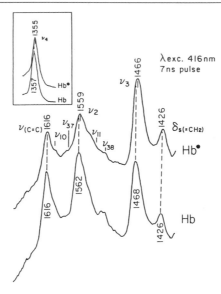

FIG. 14. Soret band excited RR spectrum of deoxyHb and Hb* with 416 nm, 7 nsec pulses. Concentration, 0.1 mM; laser energy, 0.5 mJ/pulse; spectral slit width, 8 cm^{-1}; collection time, 10 min; and repetition rate, 10 Hz. (Inset) Mode ν_4 at 10 times reduced scale. (From Dasgupta and Spiro.[28])

transient RR studies provides high quality data for the HbCO photoproduct on the 7 nsec time scale.[28] For these studies the HbCO solution was pumped through an approximately 1 mm capillary tube attached to a reservoir for recirculating the solution. The reservoir was kept under N_2 (for deoxyHb) and under CO for (HbCO). A Nd:YAG laser provided 7 nsec pulses at 532 nm (the frequency-doubled line) or at 416 nm. The latter line was generated by Raman shifting of the tripled (355 nm) output. The Raman scattered light was dispersed and detected with a spectrograph/photodiode array combination.

Figures 14 and 15 show the spectra acquired with each excitation line. As was explained in the previous chapter (see [18] in this volume), under Soret band excitation (i.e., 416 nm), the totally symmetric lines (ν_2, ν_3, and ν_4) are strongly enhanced, whereas excitation at 532 nm (Q band) provides strong enhancement of nontotally symmetric modes (ν_{10}, ν_{11}, and ν_{19}). In Figs. 14 and 15 the spectrum of the equilibrium deoxy species is presented along with that of the 7 nsec photoproduct, and it is immediately obvious that all of the features associated with the heme macrocycle

[28] S. Dasgupta and T. G. Spiro, *Biochemistry* **25**, 5941 (1986).

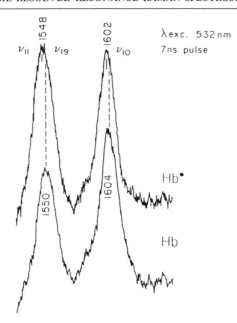

Fig. 15. Q-band excited RR spectrum of deoxyHb and Hb* with 532 nm, 7 nsec pulses. Concentration, 0.5 mM; laser energy, 1 mJ/pulse; spectral slit width, 5 cm^{-1}; collection time, 30 min; and repetition rate, 10 Hz. (From Dasgupta and Spiro.[28])

appear at lower frequencies in the photoproduct spectra relative to their positions in the spectra of deoxyHb at equilibrium. Although there is an overlap of the modes in the spectra shown in Fig. 15, careful deconvolution and polarization studies document 3 and 2 cm^{-1} shifts for ν_{11} and ν_{19}, respectively (see Dasgupta and Spiro[28] for details). It is important to emphasize that these slight shifts are not merely the result of some systematic error inasmuch as other features (e.g., the vinyl modes at 1616 and 1426 cm^{-1}) are not shifted when comparing the two spectra.

As explained thoroughly in the original report,[28] by careful comparison of the frequencies of these core-size markers with a series of structurally well-defined model compounds, the authors concluded that the photoproduct has a slightly expanded core relative to the deoxy structure and inferred that the probable cause for this expansion is restriction of the out-of-plane displacement of the iron brought about by protein restraint on the proximal histidine. Given this interpretation, it should be expected that significant differences in the iron–histidine stretching frequency are likely to be observed, and, as will be seen, it is indeed this feature which exhibits the most substantial shifts for these intermediate species.

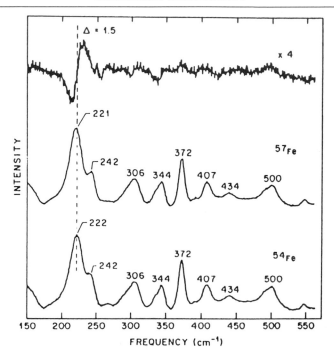

FIG. 16. Raman spectrum of [^{54}Fe]Mb and [^{57}Fe]Mb recorded on a Raman difference apparatus. Both samples were 50 μM in phosphate buffer (100 mM, pH 7.0). The scale for the intensity in the difference spectrum has four times the sensitivity as that of the individual Raman spectra. The Δ values on the difference spectrum are shifts to the lower frequency (in cm^{-1}) of the indicated lines calculated from the difference spectrum. The error in determining the peak positions is ±0.5 cm^{-}1, whereas that in determining the shift in the line from the difference spectrum is ±0.2 cm^{-}1. (From Argade *et al.*[29])

b. Iron–histidine mode. Before discussing the transient RR and TR^3 studies of the ν(Fe–N^{His}) mode, it is appropriate to summarize briefly the behavior of this mode for the static systems. As can be seen in Fig. 16, assignment of the ν(Fe–N^{His}) mode in deoxyMb to the feature at 222 cm^{-1} is confirmed by the ^{54}Fe isotopic shift.[29,30] The low frequency spectrum of deoxyHb is quite similar (trace b of Fig. 2) except that the ν(Fe–N^{His}) mode is broadened with components near 207 and around 220 cm^{-1}. Actually, this broadened envelope of bands results from distinct features associated with the α and β subunits. This is made clear by inspection of Fig.

[29] P. Argade, M. Sassaroli, D. L. Rousseau, T. Inubushi, M. Ikeda-Saito, and A. Lapidot, *J. Am. Chem. Soc.* **106,** 6593 (1984).

[30] A. V. Wells, J. T. Sage, D. Morikis, P. M. Champion, M. Chiu, and S. G. Sligar, *J. Am. Chem. Soc.* **113,** 9655 (1991).

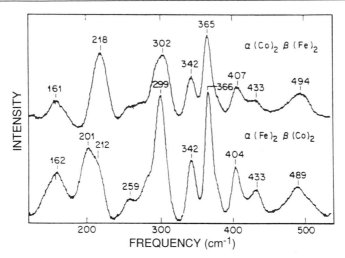

FIG. 17. Comparison between the spectrum of the iron heme in the β subunits (*top*) and in the α subunits (*bottom*) from iron–cobalt hybrid hemoglobins. Contributions from the cobalt-containing hemes are not present in the spectrum with this excitation wavelength (441.6 nm). [From D. L. Rousseau and J. M. Friedman, *in* "Biological Applications of Raman Spectroscopy" (T. G. Spiro, ed.), Vol. 3, p. 133. Wiley, New York, 1988.]

17, where the top spectrum arises from the β subunits and the bottom spectrum is associated with the α subunits (in these iron/cobalt hybrids, which are known to have a T-state quaternary structure, the cobalt-containing subunits do not contribute to the spectrum). Thus, in the case of the T-state quaternary structure, the $\nu(\text{Fe–N}^{\text{His}})$ modes of the α subunits are apparently heterogeneous and lower than those of the β subunits.

As depicted in Fig. 18, in an elegant early study of valency hybrids which can be switched from an R-like to a T-like structure by the addition of inositol hexaphosphate (IHP) (an allosteric effector), Nagai and Kitagawa[31] showed that the $\nu(\text{Fe–N}^{\text{His}})$ of the α subunits (left-hand side, Fig. 18) shifted from approximately 222 cm^{-1} in the "R state" (trace b, Fig. 18) to 207 cm^{-1} in the "T state" (trace c). The corresponding shift for the β subunits (right-hand side, Fig. 18) was only 4 cm^{-1}, this behavior being entirely consistent with the suggestion by Perutz *et al.*[24] that the T-state quaternary (and corresponding tertiary) structure places strain on (weakens) the iron–imidazole linkages of the α subunits more so than those of the β subunits. As will be seen, the transient RR and TR^3 studies of photolyzed MbCO and HbCO are consistent with the above picture and, in the TR^3 approach, provide structural and temporal data which may be

[31] K. Nagai and T. Kitagawa, *Proc. Natl. Acad. Sci. U.S.A.* **77**, 2033 (1980).

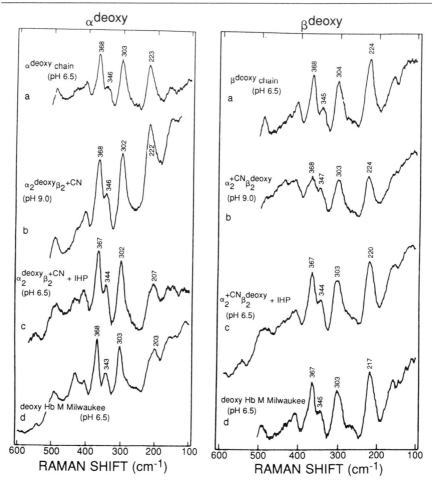

FIG. 18. Resonance Raman spectra of the α^{deoxy} (*left*) and β^{deoxy} (*right*) subunits of valency hybrid hemoglobins excited at 441.6 nm. (a) Isolated chain (pH 6.5); (b) stripped *met*-cyanide hybrid at pH 9.0; (c) *met*-cyanide hybrid with inositol hexaphosphate (IHP) at pH 6.5; (d) stripped deoxyHb M Milwaukee at pH 6.5 (*left*) and stripped deoxyHb M Boston at pH 6.5 (*right*). All samples were in 50 m*M* Bis–Tris/50 m*M* Tris buffer. (From Nagai and Kitagawa.[31])

correlated with the transient processes observed with the time-resolved absorption methods.

The first transient RR study of the $\nu(\text{Fe–N}^{\text{His}})$ mode of photolyzed HbCO was performed using a rapidly flowing jet stream of HbCO which was passed through a tightly focused laser beam, such that the residence time was only 0.3 μsec.[4] As was shown in Fig. 2, the $\nu(\text{Fe–N}^{\text{His}})$ mode

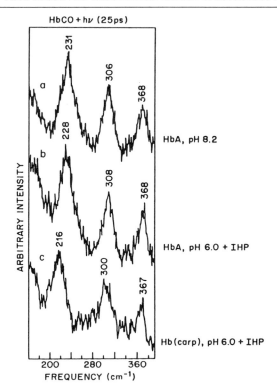

FIG. 19. Low-frequency portion of the resonance Raman spectra of the deoxy transients occurring within 25 ± 5 psec of the photodissociation of several different HbCO species: (a) around 100 μM HbA in 0.1 M Bis–Tris (pH 8.2); (b) around 100 μM HbA in 0.1 M Bis–Tris (pH 6.0) plus 2 mM IHP (slightly destabilized R structure); and (c) around 100 μM carp Hb in 0.1 M Bis–Tris (pH 6.0) plus 3 mM IHP (T structure). [From D. L. Rousseau and J. M. Friedman, *in* "Biological Applications of Raman Spectroscopy" (T. G. Spiro, ed.), Vol. 3, p. 133. Wiley, New York, 1988.]

of the photoproduct is more symmetric and appears at a slightly higher frequency than the corresponding feature in the deoxyHb spectrum. Based on the preceding discussion, it is clear that in the photoproduct spectrum the contributions from the α subunits are at a higher frequency than they are in the spectrum of Fig. 2b.

Later, Friedman and co-workers used pulsed lasers to investigate the photolyzed species at earlier times.[32] Shown in Fig. 19 are the transient RR spectra of several photolyzed HbCO derivatives using 30 psec pulses. The spectrum in Fig. 19a exhibits the $\nu(Fe–N^{His})$ mode at 231 cm^{-1}. This

[32] E. W. Findsen, J. M. Friedman, M. R. Ondrias, and S. R. Simon, *Science* **229**, 661 (1985).

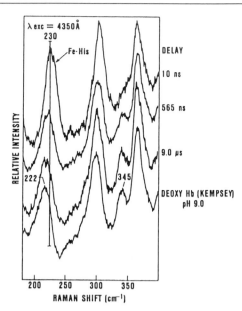

FIG. 20. Time evolution of low-frequency Raman spectrum of the deoxy photoproduct Hb* generated from HbACO (pH 6.4 plus IHP, 35°). For comparison purposes an equilibrium R-state deoxyHb spectrum derived from Hb Kempsey at pH 9 is also shown. Photolysis of the HbACO was accomplished with a 10 nsec excitation pulse (4050 Å, ~1 mJ). The Raman spectra were generated with an electronically delayed second 10 nsec pulse (4350 Å, ~5 mJ). The sample was maintained at 35° in order to reduce geminate recombination. (From Scott and Friedman.[33])

mode is observed at a slightly lower frequency (228 cm^{-1}) under conditions (IHP at pH 6) which are known to destabilize the R structure but not induce a shift to the T state (Fig. 19b). The spectrum in Fig. 19c corresponds to a true T-state photoproduct, and it is seen that the $\nu(\text{Fe–N}^{\text{His}})$ occurs at the same position (broadened peak centered at 216 cm^{-1}) as for deoxyHb.

In a TR^3 study of the $\nu(\text{Fe–N}^{\text{His}})$ mode of photolyzed HbCO, Scott and Friedman documented the temporal evolution of this mode.[33] In contrast to the transient RR studies mentioned above, the TR^3 experiments were conducted in such a way that a blue (435 nm) probe pulse was temporally delayed relative to the photolysis pulse. As can be seen in Fig. 20, the frequency of the $\nu(\text{Fe–N}^{\text{His}})$ mode shifts from 230 cm^{-1} for the 10 nsec photoproduct to slightly lower frequencies over the first 9 μsec to a value which is characteristic of an R-state deoxyHb, Hb Kempsey. Only after longer times does the photoproduct undergo full relaxation to the equilib-

[33] T. W. Scott and J. M. Friedman, *J. Am. Chem. Soc.* **106,** 5677 (1984).

rium deoxy (T-state) structure, as shown in Fig. 21 (where the temporal delay of the probe pulse is extended out to 120 μsec). As can be seen by careful inspection of these spectra, after about 10 μsec a broad lower frequency component begins to appear as the higher frequency (222 cm^{-1}) feature (characteristic of the R-state deoxy species) simultaneously disappears. Such studies demonstrate that weakening of the ν(Fe–N^{His}) mode is temporally correlated with the approximately 20 μsec relaxation (i.e., R $\rightarrow$ T switch) observed in the time-resolved optical studies.

It is interesting to note that similar studies of MbCO, where there are no quaternary structure induced restraints, reveal quite different behav-

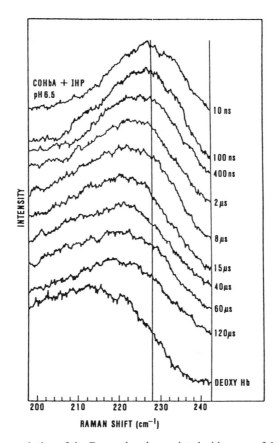

FIG. 21. Time evolution of the Raman band associated with ν_{Fe-His} of deoxyHb* (human) at pH 6.5 at 25°. The corresponding Raman band from the equilibrium deoxyHbA is shown at the bottom. (From Scott and Friedman.[33])

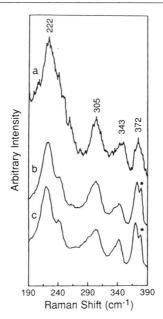

FIG. 22. Low-frequency spectra of Mb obtained with Soret excitation: (a) MbCO photo-lyzed with 20–30 psec pulses, (b) MbCO photolyzed with 10 nsec pulses of approximately 435 nm from the 10 Hz output of a Nd : YAG laser system (average energy 1.0–1.5 mJ/pulse), and (c) spectrum of deoxyMb taken under same conditions as in (b). Asterisks denote Raman lines of the sapphire cells used in the nanosecond experiments. (From Findsen *et al.*[34])

ior.[34] As shown in Fig. 22, the $\nu(\text{Fe–N}^{\text{His}})$ mode of both the 10 nsec (Fig. 22b) and the 20 psec (Fig. 22a) photoproducts occur at frequencies which are quite similar to those of deoxyMb at equilibrium (Fig. 22c).

   c. Ultraviolet excitation-probing movements of protein side chains. Ultraviolet excitation lines have been used for RR studies of various aromatic residues of proteins,[35,36] and the technology has now been employed in several TR³ studies of photolyzed HbCO.[37–40] With regard to

[34] E. W. Findsen, T. W. Scott, M. R. Chance, and J. M. Friedman, *J. Am. Chem. Soc.* **107**, 3355 (1985).

[35] S. A. Asher, C. R. Johnson, and J. Murtaugh, *Rev. Sci. Instrum.* **54**, 1657 (1983).

[36] B. S. Hudson and L. C. Mayne, *in* "Biological Applications of Raman Spectroscopy" (T. G. Spiro, ed.), Vol. 2, Chap. 4. Wiley, New York, 1987.

[37] C. Su, Y. D. Park, G.-Y. Liu, and T. G. Spiro, *J. Am. Chem. Soc.* **111**, 3457 (1989).

[38] S. Kaminaka, T. Ogura, and T. Kitagawa, *J. Am. Chem. Soc.* **112**, 23 (1990).

[39] S. Kaminaka and T. Kitagawa, *J. Am. Chem. Soc.* **114**, 3256 (1992).

[40] K. R. Rodgers, C. Su, S. Subramaniam, and T. G. Spiro, *J. Am. Chem. Soc.* **114**, 3697 (1992).

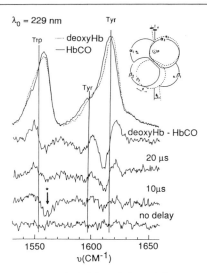

FIG. 23. Ultraviolet RR spectra of HbCO and deoxyHb (*top*) and their difference, compared with difference spectra of the HbCO photoproduct generated with a 532 nm photolysis pulse at the indicated time prior to a 229 nm probe pulse. The difference signals are attributed to the Tyr-α^{42} and Trp-β^{37} residues at the $\alpha_1\beta_2$ interface, whose hydrogen bonding is altered via the subunit rearrangement (inset) in the R–T transition. (From Su *et al.*[37])

the issue of Hb cooperativity, the essential purpose of these experiments is to attempt to document possible changes in the vibrational modes of aromatic side chains involved in the T → R switch. By using excitation near 230 nm, the modes of tyrosine and tryptophan side chains are enhanced. In the top trace of Fig. 23 the UV RR spectra of deoxyHb and HbCO are shown.[37] The difference spectrum (second trace from the top in Fig. 23) exhibits features indicative of a downshift in a band at 1555 cm^{-1} (which is known to be associated with a tryptophan residue) as well as shifts to higher frequency for two known tyrosine modes (the 1615 cm^{-1} mode and its 1600 cm^{-1} shoulder). Thus, these difference features signal a difference in quaternary structure, deoxyHb being T state and HbCO being R state (note that the HbCO spectrum at the top of Fig. 23 is that obtained in the absence of a photolysis pulse; i.e., it is fully ligated).

The bottom three traces in Fig. 23 depict difference spectra which were generated by subtraction of the HbCO spectrum from that obtained in the dual-pulse experiments. In these experiments a (10 nsec) 532 nm photolysis pulse "instantaneously" creates the Hb photoproduct. If the approximately 230 nm probe pulse is not temporally delayed with respect to the photolysis pulse (bottom trace), no difference is observed, that is, the photoproduct has an R-state structure. When the probe pulse is delayed by 10 μsec the difference features begin to appear, and the process which causes this is essentially complete by 20 μsec. Obviously, it is satisfying that the temporal response of the vibrational modes associated with the aromatic residues closely parallels the time course of the transition observed in the time-resolved absorption measurements and in the TR^3 studies of the ν(Fe–N^{His}) mode. These UV TR^3 studies have been extended to monitor changes in the rate of the R $\rightarrow$ T switch in solutions of various pH values (i.e., a UV TR^3 study of the Bohr effect).[39] In addition, more detailed studies have appeared in which higher quality spectra and appropriate model compound studies are used to attempt to provide a structural basis for these spectral changes.[40]

3. Summary. Although the preceding examples of transient RR and TR^3 studies of Hb structural dynamics provide an introduction to these issues, detailed structural interpretation and the precise temporal behavior of the various modes are quite complex and obviously beyond the scope of this introductory review. Rousseau and Friedman have provided a comprehensive and thoughtful review of most of these studies.[41] In addition, TR^3 studies which extend the temporal resolution to the subpicosecond regime have now appeared[42,43] and have also been reviewed elsewhere.[7] These are essential reading to gain a clearer picture of Hb structural dynamics.

B. Time-Resolved Resonance Raman Studies of Cytochrome-c Oxidase

One of the most important and impressive applications of transient RR and TR^3 studies deals with investigation of the strutural dynamics and mechanism of action of the terminal oxidase in aerobic respiration; namely, cytochrome-*c* oxidase (CcO). In contrast to the previous section which dealt with the dynamics involved in protein structural rearrangements, the following TR^3 studies to be summarized for CcO deal with the detection and characterization of the complex, multistep proton and electron transfer reactions which follow O_2 binding by CcO. Whereas in the case of Hb structural dynamics the essential data consist of frequency shifts of key modes, the TR^3 spectra acquired during the multistep redox reaction involved in CcO function reveal several distinct sets of transient spectral features which are characteristic of specific intermediates. The

[41] D. L. Rousseau and J. M. Friedman, *in* "Biological Applications of Raman Spectroscopy" (T. G. Spiro, ed.), Vol. 3, p. 133. Wiley, New York, 1988.

[42] J. W. Petrich, J. L. Martin, D. Houde, C. Poyart, and A. Orszag, *Biochemistry* **26**, 7914 (1987).

[43] R. Van der Berg and M. A. El-Sayed, *Biophys. J.* **58**, 931 (1990).

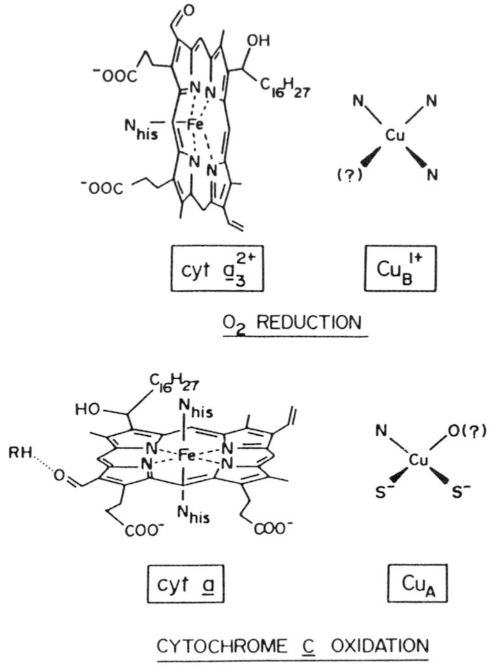

FIG. 24. Current models for the coordination geometries of the iron and copper centers in reduced cytochrome oxidases. Oxygen reduction occurs at the cytochrome a_3^{2+}/Cu_B^{1+} cluster; cytochrome a and Cu_A are involved with cytochrome c oxidation. (From Babcock.[46])

TR^3 studies conducted on CcO are especially instructive because they also clearly illustrate an important advantage of time-resolved vibrational spectroscopy, namely, the ability to exploit the inherent sensitivity of vibrational frequencies to isotopic substitution and thereby better define the structure of a given intermediate.

   1. Structure and Mechanism. Cytochrome-*c* oxidase catalyzes the four-electron reduction of molecular oxygen to water and couples these redox processes to proton transfer across the mitochondrial membrane.[44,45] As depicted in Fig. 24,[46] the enzyme is structurally complex and contains four metal centers which are redox active, two copper ions and two heme *a* groups. One copper ion, Cu_B, and one of the heme *a* groups, cytochrome a_3, (cyt a_3), form a binuclear center which binds dioxygen. Electrons are

[44] G. T. Babcock and M. Wikström, *Nature (London)* **356**, 301 (1992).
[45] G. T. Babcock and C. Varotsis, *J. Bioenerg. Biomembr.* **25**, 71 (1993).
[46] G. T. Babcock, *in* "Biological Applications of Raman Spectroscopy" (T. G. Spiro, ed.), Vol. 3, p. 293. Wiley, New York, 1988.

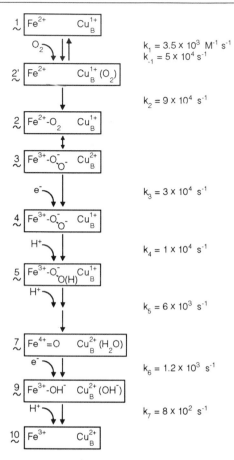

FIG. 25. Reaction sequence for the oxidation of fully reduced cytochrome oxidase by dioxygen. Only the iron and copper ions of the binuclear center are shown. (From Varotsis et al.[47])

transferred to this site by Cu_A and cyt a. Thus, it is at the binuclear (cyt a_3/Cu_B) site that the O_2 molecule is held and reduced, during which time O–O bond scission and protonation occur to produce two molecules of H_2O. A possible mechanistic scheme for this complex process, which is based on a large number of investigations (including the TR^3 studies) is given in Fig. 25,[47] and, as will be seen, TR^3 studies have provided definitive structural information for several of the key intermediates.

[47] C. Varotsis, Y. Zhang, E. H. Appelman, and G. T. Babcock, *Proc. Natl. Acad. Sci. U.S.A.* **90,** 273 (1993).

Most of the TR^3 studies of the CcO enzyme have come from three laboratories: those of Babcock and co-workers (Ref. 47 and references therein), Rousseau and co-workers at AT&T Bell Labs (Ref. 48 and references therein), and the group at Okazaki led by Kitagawa (Ref. 49 and references therein). All of these groups employ the photolabile CO adduct in order to use a photolysis pulse to initiate the reaction with O_2. The first group cited uses two electronically delayed laser pulses (pump and probe), and the time resolution is independent of the flow rate. The other two groups employ spatially displaced cw lasers while rapidly flowing the sample through the beams. Although it is clear that the time resolution is more accurate using the pulsed laser approach, this technique suffers in comparison with the flow methods in terms of signal-to-noise ratio and possible artifacts induced by the high peak power pulses. Nevertheless, it is satisfying that there is essential agreement among the three groups concerning which spectral features are observed, although some disagreement persists regarding the temporal behavior and structural interpretation of these features.

2. Transient Resonance Raman and Time-Resolved Resonance Raman Studies. Before describing the first experiment it is useful to point out that most of the TR^3 data obtained are presented as difference spectra. The reason for this is that the system is quite complex, both structurally and from the standpoint of the large number of modes which are RR active (the enzyme complex contains two heme *a* groups, and in some experiments the redox partner, cytochrome *c*, is also present in solution). Inasmuch as the TR^3 studies are focused on the detection and characterization of the O_2 reduction products, it is most efficient to obtain spectra during the reaction with $^{16}O_2$ and also during the reaction with $^{18}O_2$ and to generate the difference spectrum by subtraction. In this way all other features will cancel and even very weak spectral features associated with the bound O_2 and its reduction products can be more readily observed.

a. Primary intermediate. Although there is some speculation that O_2 may preassociate with the Cu_B site, the first detectable intermediate is the O_2-bound cyt a_3^{2+} (which is more properly formulated as a superoxide complex of cyt a_3^{3+}). Once O_2 is bound to the fully reduced enzyme, this primary intermediate is quickly reduced via the cyt a^{2+} site.[48] So, as in the case of O_2 binding to Hb, to establish a well-defined initial time, all of the TR^3 studies exploit the rapid and efficient photolysis of the CO-bound adduct.

[48] S. Han, Y. C. Ching, and D. L. Rousseau, *Proc. Natl. Acad. Sci. U.S.A.* **87,** 8408 (1990).
[49] T. Ogura, S. Takahashi, S. Hirota, K. Shinzawa-Itoh, S. Yoshikawa, E. H. Appelman, and T. Kitagawa, *J. Am. Chem. Soc.* **115,** 8527 (1993).

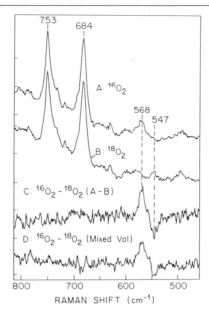

FIG. 26. Resonance Raman spectra and difference spectra of the primary intermediate in the reaction of cytochrome-c oxidase with isotopes of oxygen observed in a time window of 0–50 μsec. The spectral accumulation time was 30 sec. (A) Reaction of fully reduced cytochrome-c oxidase with $^{16}O_2$; (B) reaction of fully reduced cytochrome-c oxidase with $^{18}O_2$; (C) difference spectrum showing the isotopic shift of the Fe–O_2 stretching mode of the primary intermediate in the reaction of oxygen with the fully reduced enzyme (spectrum A minus spectrum B); (D) difference spectrum showing the isotopic shift of the Fe–O_2 stretching mode of the primary intermediate in the reaction of oxygen with the mixed-valence enzyme (From Han et al.[50])

There is now agreement among all three groups that the first detectable intermediate exhibits a difference feature ($^{16}O_2/^{18}O_2$) at 568/547 cm^{-1} This is shown in Fig. 26,[50] where the top two spectra are the absolute spectra of the $^{16}O_2$ and $^{18}O_2$ adducts and trace C is the appropriate difference spectrum. Note that this experiment was conducted with a rapidly flowing sample and a cw laser (50 μsec residence time) and that, in this case, the individual spectra represent the spectra of species which accumulate over the first 50 μsec following CO photolysis. These modes, assignable to the ν(Fe–O_2) stretch, occur at frequencies (568 cm^{-1} for $^{16}O_2$ and 547 cm^{-1} for $^{18}O_2$) which are virtually identical to those observed for the O_2 adducts of Hb and Mb and are thus entirely consistent with the formulation given for the primary intermediate (labeled **2**) in Fig. 25.

[50] S. Han, Y.-C. Ching, and D. L. Rousseau, Proc. Natl. Acad. Sci. U.S.A. **87**, 2491 (1990).

b. Later intermediates. Several reports of other ($^{16}O_2/^{18}O_2$) isotope-sensitive features which arise during the decay of the primary intermediate have appeared. The two most recent reports attempt to define the structure and temporal behavior of these intermediates and also provide a reasonable overview of the entire issue.[47,49]

The spectra shown in Fig. 27 were acquired with accurate time resolution (pulsed lasers) and clearly illustrate that the primary intermediate (trace E was acquired at 30 μsec) disappears within 160 μsec (trace F), by which time a new difference feature appears at 358 cm^{-1} ($^{16}O_2$ case). As is shown in Fig. 28 (trace E) this species persists until 220 μsec but has disappeared by 500 μsec (trace F), by which time a new difference feature is observable near 800 cm^{-1}. At later times a new feature begins to grow in with a frequency of approximately 450 cm^{-1}. This is most clearly illustrated by the spectra shown in Fig. 29,[51] which were obtained in the Rousseau laboratory using the rapid-flow technique. Here again the same temporal sequence is seen, that is, the 568/547 cm^{-1} pair at short times, the difference feature near 800 cm^{-1} at longer times, and the 450/425 cm^{-1} pair at the longest times. Both the temporal behavior (last to appear) and the frequency of the 450/425 cm^{-1} feature make it likely that it is associated with the (final) hydroxy intermediate (9 in Fig. 25). Although not shown in Fig. 29, the 358 cm^{-1} feature seen in Figs. 27 and 28 was actually first detected by Kitagawa and co-workers,[52] so there is agreement concerning the appearance of this feature.

Despite the general agreement about the existence of the various difference features, there is still considerable uncertainty about their structural interpretation and temporal behavior. For example, as is shown in Fig. 29b, both of the later intermediates shown exhibit shifts in D_2O. Whereas the 450 cm^{-1} ($^{16}O_2$) mode shifts to 442 cm^{-1} ($^{16}O_2$) in D_2O (an observation which is consistent with its proposed formulation as a hydroxy complex of cyt a_3^{3+}), the earlier (higher frequency) mode actually shifts to higher frequency in D_2O (an observation which is in agreement with the most recent work from the Babcock group[47]). It had been generally agreed that the species which gives rise to the 800 cm^{-1} feature is the ferryl intermediate (7 in Fig. 25), and the shift to higher frequencies in D_2O was attributed to the loss of hydrogen bonding (to a H_2O molecule), the loss resulting from protein conformational differences in D_2O versus H_2O.[51] However, the most recent work from the Kitagawa laboratory[49] indicates that the picture may be more complicated. As is shown in Fig. 30 (taken under

[51] S. Han, Y.-C. Ching, and D. L. Rousseau, *Nature (London)* **348**, 89 (1990).
[52] T. Ogura, S. Takahashi, K. Shinzawa-Itoh, S. Yoshikawa, and T. Kitagawa, *Bull. Chem. Soc. Jpn.* **64**, 2901 (1991).

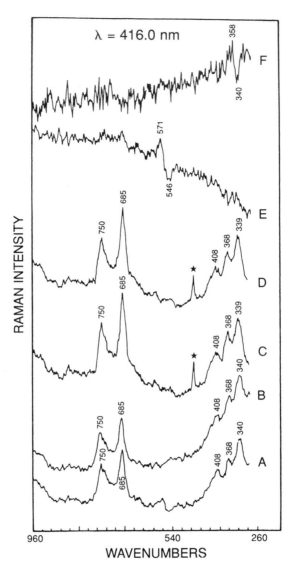

FIG. 27. Time-resolved RR spectra of fully reduced cytochrome-c oxidase after initiation of the reaction with oxygen at room temperature. The pump–probe delay was 30 μsec for spectra A ($^{16}O_2$) and B ($^{18}O_2$) and 160 μsec for spectra C ($^{16}O_2$) and D ($^{18}O_2$). The energy of the 532-nm photolysis pulse was 1.3 mJ, and the energy of the probe beam was 0.2–0.3 mJ. The repetition rate for pump–probe pairs was 10 Hz. The accumulation time was 40 min for spectra A and B and 50 min for spectra C and D. The enzyme concentration was 85 μM after mixing. Trace E, the $^{16}O_2/^{18}O_2$ difference spectrum at 30 μsec, was obtained by subtracting spectrum B from spectrum A; similarly, trace F is the $^{16}O_2/^{18}O_2$ difference spectrum at 160 μsec. The peak labeled by the asterisk in spectra C and D and in Fig. 28 arises from a spurious Stokes process in the Raman shifter and can be ignored. (From Varotsis et al.[47])

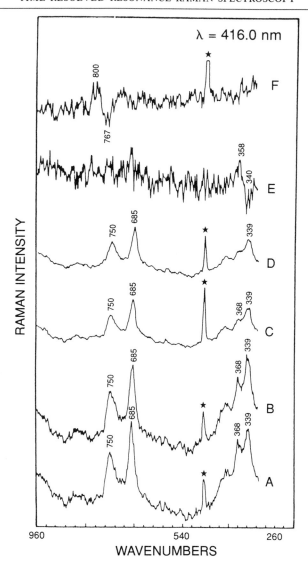

FIG. 28. Time-resolved RR spectra of fully reduced cytochrome-c oxidase in ^2H_2O. The pump–probe delay was 220 μsec for traces A ($^{16}O_2$) and B ($^{18}O_2$) and 500 μsec for traces C ($^{16}O_2$) and D ($^{18}O_2$). The wavelengths, energies, and repetition rates of the pump and probe pulses were as in Fig. 27. The accumulation time was 60 min for spectra A and B and 35 min for spectra C and D. The enzyme concentration was 75–90 μM after mixing, pD 7.8. Traces E and F are the $^{16}O_2/^{18}O_2$ difference spectra at 220 μsec and 500 μsec, respectively. (From Varotsis et al.[47])

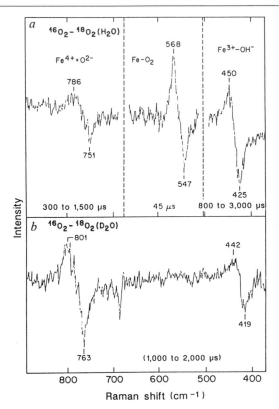

FIG. 29. Isotope difference spectra ($^{16}O_2 - {}^{18}O_2$) at selected time ranges after the reaction of oxygen with cytochrome-c oxidase. Several spectra obtained throughout the indicated time ranges were added together to improve the signal-to-noise ratio. (a) Measurements in buffered H_2O showing the three oxygen isotope-sensitive lines detected over the 45 to 3000 μsec time range. (b) Measurements in buffer of D_2O in a time range in which the ferryl (801 cm^{-1}) and the hydroxy (442 cm^{-1}) intermediates are strong. The enzyme (200 μM before mixing) was solubilized in phosphate buffer (100 mM, pH 7.4) with 1% dodecyl-β-D-maltoside. It was reduced with 20 mM ascorbate and catalytic amounts of cytochrome c (3 μM), then exposed to carbon monoxide. To initiate the reaction, the enzyme solution was mixed with oxygen-saturated (1.4 mM) buffer and the CO was photodissociated. (From Han *et al.*[51])

higher spectral resolution), apparently this feature is actually a doublet having components at 785/751 cm^{-1} ($^{16}O_2/{}^{18}O_2$) and 804/764 cm^{-1} ($^{16}O_2/$ $^{18}O_2$) in H_2O (Fig. 30A). In D_2O (Fig. 30B) only the 785/751 cm^{-1} mode shifts to higher frequency, whereas the 804/764 pair does not shift. The fact that the 785/751 cm^{-1} pair disappears at 30° (Fig. 30C), whereas the 804/764 cm^{-1} persists, was taken as evidence that the species corresponding to the former pair occurs earlier (disappears sooner) than the species corresponding to the 804/764 cm^{-1} pair. Though not shown, the 804/764 cm^{-1} feature shown in Fig. 30C does not shift in D_2O.

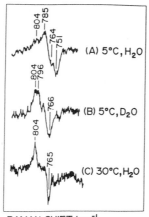

RAMAN SHIFT/cm^{-1}

FIG. 30. Higher resolution time-resolved RR difference spectra ($^{16}O_2 - {}^{18}O_2$) in the approximately 800 cm^{-1} region of cytochrome-c oxidase at $\Delta t = 1.1$ msec: (A) 5° in H_2O; (B) 5° in D_2O; (C) 30° in H_2O. Resolution, 0.43 cm^{-1}/channel (450 points plotted); accumulation time, 14,220, 2700, and 540 sec for A, B, and C, respectively. (From Ogura et al.[49])

In summary, although some issues remain ambiguous at the present time (and the reader should refer to the original reports for a detailed discussion of these), the following picture emerges from the TR^3 studies. The primary intermediate is an O_2 adduct of cyt $a_3{}^{2+}$ which is quite similar to $Hb(O_2)_4$ or MbO_2. The bound dioxygen is reduced rapidly to produce a number of difference features which are presumably associated with the various intermediates (**5** through **9** in Fig. 25), the last of which (**9**) exhibits $^{16}O_2/{}^{18}O_2$ and H_2O/D_2O shifts which are expected. Resolution of the ambiguity regarding assignment of the other difference features to specific intermediates must await further studies which simultaneously clarify both the temporal behavior and H_2O/D_2O sensitivities.

IV. Other Applications

Although it is not possible here to provide a comprehensive survey of TR^3 applications to biological problems, there are a few important systems which are naturally susceptible to this approach (the process of interest is photoinitiated) and a rather extensive amount of work has been devoted to these. The following brief summaries are intended to provide access to the literature by supplying citations to several of the most recent studies.

1. Retinal Proteins. Visual pigments and light-transducing pigments found in certain bacteria contain a long-chain polyene, retinal, as the active chromophore. On absorption of a visible photon the pigments undergo a very rapid (femtosecond) cis–trans isomerization about the C-13–C-14 double bond. In visual pigments this initial isomerization is followed by a series of active site rearrangements which ultimately cause a charge polarization.[53] In the case of the bacterial membrane protein bacteriorhodopsin, the initial photoisomerization energy is utilized to drive transmembrane ion pumps, the proton gradient across the membrane being used by the cell as a source of energy for metabolic processes.[54]

Following the rapid initial photoisomerization, a complex reaction sequence occurs forming distinct intermediates which have picosecond, microsecond, or millisecond lifetimes. The early intermediates with short lifetimes have been studied by transient RR and TR[3] approaches using picosecond and nanosecond pulsed lasers, much of the work being performed by Mathies and co-workers[55,56] and the Atkinson group.[57] The longer lived intermediates are susceptible to the approaches which employ cw lasers with flowing (or spinning) samples, and much of that work has been summarized by Stockburger *et al.*[58] Although in the past the focus of attention has been on the retinal chromophore, UV TR[3] studies of bacteriorhodopsin have been reported which investigate corresponding protein structural changes on the nanosecond and millisecond time scales.[59]

2. Photosynthesis. In the photosynthetic process light energy is absorbed by carotenoids and bacteriochlorophylls and transferred to photosynthetic reaction centers.[60,61] Time-resolved vibrational methods, including transient RR and TR[3] techniques, are well-suited to elucidate the structure and dynamics of the excited (singlet and triplet) states of these chromophores. Nanosecond and picosecond transient RR studies of carotenoid excited states have been conducted by Atkinson and co-workers[62,63]

[53] L. Stryer, *Annu. Rev. Neurosci.* **9,** 87 (1986).

[54] W. Stoeckenius and R. A. Bogomolni, *Annu. Rev. Biochem.* **51,** 587 (1982).

[55] R. A. Mathies, S. W. Lin, J. B. Ames, and W. T. Pollard, *Annu. Rev. Biophys. Chem.* **20,** 491 (1991).

[56] R. A. Mathies, S. O. Smith, and I. Palings, *in* "Biological Applications of Raman Spectroscopy" (T. G. Spiro, ed.), Vol. 2, p. 59. Wiley, New York, 1987.

[57] J. K. Delaney, T. L. Brack, G. H. Atkinson, M. Ottolenghi, N. Friedman, and M. Sheves, *J. Phys. Chem.* **97,** 12416 (1993).

[58] M. Stockburger, T. Alshuth, D. Oesterhelt, and W. Gärtner, *in* "Spectroscopy of Biologial Systems" (R. J. H. Clark and R. E. Hester, eds.), Vol. 13, p. 483. Wiley, Chichester, 1986.

[59] J. B. Ames, M. Ros, J. Raap, J. Lugtenburg, and R. A. Mathies, *Biochemistry* **31,** 5328 (1992).

[60] D. Siefermann-Harms, *Biochim. Biophys. Acta* **811,** 325 (1985).

[61] R. J. Cogdell and H. A. Frank, *Biochim. Biophys. Acta* **895,** 63 (1988).

[62] H. Hayashi, S. V. Kolaczkowski, T. Noguchi, D. Blanchard, and G. H. Atkinson, *J. Am. Chem. Soc.* **112,** 4664 (1990).

[63] T. Noguchi, H. Hayashi, M. Tasumi, and G. H. Atkinson, *J. Phys. Chem.* **95,** 3167 (1991).

as well as the Koyama group.[64] The latter group has also reported the transient RR of the singlet and triplet states of bacteriochlorophyll.[65,66]

Acknowledgment

The author acknowledges support by grant DK 35153 from the National Institutes of Health.

[64] H. Hashimoto, Y. Koyama, Y. Hirata, and N. Mataga, *J. Phys. Chem.* **95**, 3072 (1991).
[65] E. Nishizawa and Y. Koyama, *Chem. Phys. Lett.* **172**, 314 (1990).
[66] E. Nishizawa, H. Hashimoto, and Y. Koyama, *Chem. Phys. Lett.* **176**, 390 (1991).

[20] Infrared Spectroscopy Applied to Biochemical and Biological Problems

By F. SIEBERT

Introduction

Elucidation of the structure of biological macromolecules and of their structure–function relationships is one of the main targets in molecular biology. X-Ray crystallography is, of course, the method of choice if high-quality single crystals of the macromolecule can be prepared. For soluble proteins this appears feasible (although in special cases it may represent a considerable challenge), but the crystallization of membrane proteins still appears to be an exception.[1,2] If two-dimensional crystals of high order are available, cryoelectron microscopy can provide three-dimensional structures. Because the resolution is limited, information derived from different techniques has to be incorporated.[3] Nuclear magnetic resonance (NMR) spectroscopy has established itself as an alternative for structure determination of soluble proteins.[4,5] Limitations are size and, up to now, membrane proteins. Even if a three-dimensional structure of a protein has been derived, however, the molecular mechanism of an

[1] J. Deisenhofer, O. Epp, K. Miki, R. Huber, and H. Michel, *J. Mol. Biol.* **180**, 385 (1984).
[2] U. Nestel, T. Wacker, D. Woitzik, J. Weckesser, W. Kreutz, and W. Welte, *FEBS Lett.* **242**, 405 (1989).
[3] R. Henderson, J. M. Baldwin, T. A. Ceska, F. Zemlin, E. Beckmann, and K. H. Downing, *J. Mol. Biol.* **213**, 899 (1990).
[4] K. Wüthrich, *Science* **243**, 45 (1989).
[5] G. M. Clore and A. M. Gronenborn, *Crit. Rev. Biochem. Mol. Biol.* **24**, 479 (1989).

enzyme is not obvious in all cases. For such studies a detailed knowledge of molecular interactions, for example, between substrate and amino acid side chains, is required, which, in several cases, cannot be derived in a unique way from the structure (e.g., hydrogen bonding, electrostatic charges). Thus, for many problems related to structure and function of biological macromolecules, complementary techniques are required. Here, vibrational spectroscopy comes into its own right. Raman and resonance Raman spectroscopy are treated in other chapters in this volume (see W. L. Peticolas, this volume [17]; T. G. Spiro and R. S. Czernuszewicz, [18]; and J. R. Kincaid, [19]). In this chapter the potential of infrared spectroscopy is demonstrated.

For many years, infrared spectroscopy was one of the classic tools used to study molecular structure and interactions of small molecules. At first sight, it appears too ambitious to apply this technique to biological macromolecules: the enormous number of normal modes causing overlapping absorption bands does not allow the extraction of detailed molecular information from the infrared spectra. However, the molecules exhibit an intrinsic order of repeating units: the backbone of a protein can be visualized as a row of peptide bonds; nucleic acids represent a sequence of nucleotides connected by phosphate ester bonds between the 3'-hydroxyl group on the sugar residue of one nucleotide and the 5'-phosphate group on the next nucleotide; lipid membranes are composed of a limited number of basic molecules arranged in a two-dimensional array. Therefore, the infrared spectra of these partially ordered structures are simpler, and detailed band analysis may provide some of the desired information on structure and interactions among the subunits. From the very nature of the methods, such spectra will provide insights regarding only the secondary structure of proteins. Nevertheless, this represents important information often difficult to obtain by other techniques. In the case of nucleic acids, information on the overall structure and the interaction with small molecules such as intercalating drugs or metal ions can be obtained. Lipids in membranes may adopt a rigid, ordered, or a fluid, less ordered structure. Characteristic bands of the acyl chains can be used to estimate the state of the molecules. Several examples of these different cases are presented, and special techniques for the measurement as well as for the interpretation of the spectra are discussed. Structural investigations on proteins are addressed as well.

A completely different application of infrared spectroscopy is more related to functional aspects of proteins. To understand the molecular mechanisms of enzymes, information on the interaction of participating molecules is needed. The infrared difference spectra formed between different states of the enzyme contain bands only of those groups which undergo changes during the transition from one state to the other. Often, only a small part of the large enzyme is actually involved. Therefore, the difference spectra are simpler to interpret. However, in order for the method to provide useful information, the accuracy of the spectrometer must be high. It should be possible to detect the molecular changes of a single amino acid against the large background absorbance of the rest of the protein. A further requirement is the capability to stabilize the different states long enough to measure the spectra with the necessary amplitude resolution. Besides the molecular changes, reflected in the spectral changes, knowledge of the dynamics is also important for an understanding of the molecular function. The principal technique for the study of the dynamics of reactions is time-resolved spectroscopy. Therefore, methods of time-resolved infrared difference spectroscopy are required. A large part of this chapter is devoted to reaction-induced infrared difference spectroscopy, including time-resolved techniques.

Instrumental Aspects

Most biological systems require the presence of water for proper functioning. Therefore, investigations on structure and function should be performed in the presence of water. For spectroscopic studies in the UV–visible spectral range this condition does not impose severe limitations, but investigations with infrared radiation are more restricted. Because water is a strong infrared absorber, the amount of water in samples has to be reduced considerably. In several cases, this imposes preparative challenges. Special infrared cuvettes are used in which the two windows are separated by a film spacer approximately 4–6 μm thick. Even if reduction of the amount of water is possible, the absorbance due to the reduced water content is still very high. Therefore, to obtain high-accuracy infrared spectra of biological systems, spectrometers with a high-energy throughput (high brightness) are required. Until the early 1970s, this requirement presented a severe limitation for the application of infrared spectroscopy to biological systems. With the routine availability of Fourier transform infrared (FT-IR) instruments, however, this restriction became obsolete.

Two advantages of FT-IR spectroscopy, its high energy throughput and the so-called multiplex advantage, make it possible to measure with high accuracy spectra of samples with high absorbance. For most applications, the FT-IR instrument can be regarded as a black box, with which high-quality spectra can be acquired in short times. For researchers interested in "what is behind the black box," the book by Griffiths and de

Haseth is recommended.[6] Nevertheless, three remarks may be appropriate. (1) Instead of the monochromator used in conventional dispersive spectrometers, the FT-IR instrument uses an interferometer. Therefore, the principal quantity measured is not the spectrum, but its Fourier transform, the so-called interferogram. Each instrument is equipped with a computer which performs the reverse Fourier transformation from the interferogram to the spectrum. (2) Because no narrow entrance and exit slits are required, a high-energy flux is warranted. (3) The multiplex advantage states that, as compared to a dispersive instrument, the time needed to obtain a spectrum composed of N spectral elements with the same signal/noise ratio is reduced by a factor of N. These properties are the main reasons for the powerful capabilities of FT-IR spectroscopy. A beneficial side effect of using a computer concerns the evaluation of the spectra: they can easily be treated by digital methods, facilitating effective extraction of information.

All these advantages explain why today most infrared experiments on biological samples are performed with FT-IR spectrometers. Partial exceptions are time-resolved studies, and the special techniques employed there are discussed elsewhere in this volume (see [19]). Apart from the book already mentioned on FT-IR spectroscopy in which a special chapter is dedicated to biochemical and biomedical applications including instrumental and sampling aspects,[6] several other useful guides for both the general application of infrared spectroscopy and the more specialized field of biomedical infrared spectroscopy have appeared.[7-11]

A special sampling technique often applied for infrared studies of biological systems is attenuated total reflection spectroscopy or ATR spectroscopy. With this technique, the infrared beam is guided through a transparent medium of high refractive index (plate, often crystalline material) in such a way that several total reflections take place at the surfaces. Ideally, if the surfaces are clean, the infrared beam is not attenuated. However, if an infrared absorber is deposited onto the surface, the infrared

[6] P. R. Griffiths and J. A. de Haseth, "Fourier Transform Infrared Spectrometry." Wiley (Interscience), New York, 1986.

[7] R. A. Dluhy and R. Mendelsohn, *Anal. Chem.* **60**, 269 A (1988).

[8] N. B. Colthup, L. H. Daly, and S. E. Wiberley, "Introduction to Infrared and Raman Spectroscopy." Academic Press, New York, 1975.

[9] D. Lin-Vien, N. B. Colthup, W. G. Fateley, and J. G. Grasselli, "The Handbook of Infrared and Raman Characteristic Frequencies of Organic Molecules." Academic Press, Boston, 1991.

[10] F. S. Parker, "Applications of Infrared, Raman, and Resonance Raman Spectroscopy in Biochemistry." Plenum, New York, 1983.

[11] S. Pinchas and I. Laulicht, "Infrared Spectra of Labelled Compounds." Academic Press, London, 1971.

radiation is attenuated accordingly. This effect is due to the fact that a standing wave is established in the denser medium and a nonradiative electromagnetic field evanesces from the surface. The electric field decays exponentially with distance from the surface. If the absorber is much thicker than the infrared wavelength, the penetration depth of the radiation is of the order of the wavelength and is proportional to it. The length of the plate determines the number of reflections.

The theory on which this spectroscopy is based and many applications are described in the book by Harrick.[12] Advantages of this method are that, owing to the small penetration depth, very strong absorbers can conveniently be investigated and the spectra do not depend on the thickness of the material deposited. The surfaces can be coated with biological membranes, or concentrated aqueous solutions of biological materials (proteins, lipids, substrates, etc.) can be brought into contact with them. If the layers are oriented, structural information can be obtained from the determination of the direction of the infrared transition moments by measuring the infrared dichroism with respect to the normal of the surface. Infrared radiation polarized parallel to the plane of incidence contains, in addition to a component parallel to the surface, a component perpendicular to it, whereas radiation polarized perpendicular to the plane of incidence contains only a component parallel to the surface. Compared to conventional infrared dichroism methods using a transmission cell, the number of layers deposited can be much lower, usually resulting in better orientation. Fringeli and Günthard present a useful introduction to the field of infrared ATR spectroscopy applied to biological systems.[13] A sample cell for the investigation of liquids based on the ATR principle has also been described.[14]

Infrared Investigations of Nucleic Acids

Three main topics can be investigated by FT-IR spectroscopy: (1) base pair formation of DNA together with the derivation of thermodynamic parameters; (2) polymorphism of DNA; and (3) interaction of small molecules (drugs) with DNA. Of course, DNA and RNA molecules have been studied extensively by X-ray crystallography and NMR spectroscopy. However, in some cases, the resolution is not good enough to allow deductions based on molecular interactions such as hydrogen bonding. The importance of hydrogen bonding for structure formation and interac-

[12] N. J. Harrick, "Internal Reflection Spectroscopy." Wiley, New York, 1967.
[13] U. P. Fringeli and H. H. Günthard, *Mol. Biol. Biochem. Biophys.* **31**, 270 (1980).
[14] N. J. Harrick, *Appl. Spectrosc.* **37**, 573 (1983).

tions in nucleic acids has been discussed by Jeffrey and Saenger.[15] In addition, it is also important to study the structure of nucleic acids in their natural environment, that is, in an aqueous solution. Therefore, vibrational spectroscopy represents an important complementary method. For infrared studies the concentration of aqueous solutions of nucleic acids has to be very high. This may cause artifacts in the spectra owing to intermolecular interaction. For this reason most experiments using vibrational spectroscopy are performed with Raman or UV-resonance Raman spectroscopy (see [17] in this volume). Nevertheless, a few examples of FT-IR investigations of nucleic aids are discussed. An excellent survey of important vibrational modes in both infrared and Raman spectroscopy can be found in a book by Parker.[10]

In an early investigation base pairing was studied using the $C≡O$ stretch and N–H stretch, which have strong infrared but weak Raman intensities.[16,17] They are especially sensitive to hydrogen bonding, the main interaction for base pairing. By combining the infrared data with data obtained by differential scanning calorimetry, a detailed thermodynamic analysis for the denaturation of several oligonucleotides was possible.[18]

It is now well established that DNA can have at least three different structures: the A, B, and Z forms. In the first two a right-handed helix is present, whereas in the third a left-handed helix prevails. (See also [2] and [16] in this volume.) The A and B forms differ by the sugar puckering modes: for A the sugar pucker is C-3'-endo, whereas for B it is C-2'-endo. The native conformation is the B form, occurring in low-salt solutions. If the salt concentration is raised or the sample dehydrated, transition to the A structure takes place. Owing to the steric hindrance with the hydroxyl group at C-2', RNA always adopts the A conformation. The Z conformation was first observed for poly(dG-dC) on dehydration. The structure is characterized by alternating syn (dG) and anti (dC) conformations. Evidence for the occurrence of Z structure in natural chromatin has been presented,[19] and infrared bands characterizing the different conformers have been reviewed by Taillandier et al.[20-22] In addition, the

[15] G. A. Jeffrey and W. Saenger, "Hydrogen Bonding in Biological Structures." Springer-Verlag, Berlin, 1991.

[16] J. Stulz and T. Ackermann, Ber. Bunsen-Ges. Phys. Chem. **87**, 447 (1983).

[17] F. B. Howard, J. Frazier, and H. T. Miles, Biochemistry **16**, 4637 (1977).

[18] J. Ohms and T. Ackermann, Biochemistry **29**, 5237 (1990).

[19] A. Nordheim, M. L. Pardue, E. M. Lafer, A. Möller, D. Stollar, and A. Rich, Nature (London) **294**, 417 (1981).

[20] E. Taillandier, J. Liquier, and J. A. Taboury, Adv. Infrared Raman Spectrosc. **12**, 65 (1985).

[21] E. Taillandier, J. Liquier, J. P. Ridoux, and M. Ghomi, Vib. Spectra Struct., 553 (1989).

possibility of obtaining structural information from infrared dichroic studies on oriented DNA films is discussed in those articles. The influence of water on the transitions has been addressed by Pohle and co-workers.[23,24] A link between the spectroscopy of crystals and aqueous solutions has been made by FT-IR microscope studies of small DNA crystals.[25] This facilitates a better comparison of solution structures obtained by infrared spectroscopy and crystal structures obtained by X-ray crystallography. Characteristic bands for the interesting triple helix have been described.[26]

The interaction with small molecules such as metal ions and intercalating molecules has been investigated. Both the molecular nature of the interaction and the influence on the conformation have been addressed.[27-29] Infrared linear dichroism has also been applied to the study of intercalating drugs interacting with oriented films of DNA.[30] In many of the cited references normal mode calculations have been performed to provide a molecular basis for the interpretation of the spectra in terms of the different conformers.

The infrared investigations on nucleic acids reviewed here demonstrate that, although the method is not so widely applied, important information concerning conformations and interactions can be obtained.

Infrared Investigations of Lipids

Lipids represent the basic backbone of biological membranes. In aqueous suspensions they often form bilayer vesicles which can be considered

[22] E. Taillandier, J. Liquier, J. A. Taboury, and M. Ghomi, in "Spectroscopy of Biological Molecules" (C. Sandorfy and T. Theophanides, eds.), p. 171. Reidel, Dordrecht, The Netherlands, 1984.

[23] W. Pohle, in "Spectroscopy of Biological Molecules" (R. E. Hester and R. B. Girling, eds.), p. 391. The Royal Society of Chemistry, Cambridge, 1991.

[24] W. Pohle, M. Bohl, and H. Böhlig, J. Mol. Struct. 242, 333 (1991).

[25] E. Taillandier, in "Spectroscopy of Biological Molecules—State of the Art" (A. Bertoluzza, C. Fagnano, and P. Monti, eds.), p. 135. Societá Editrice Esculapio, Bologna, 1989.

[26] E. Taillandier, M. Firon, and J. Liquier, in "Spectroscopy of Biological Molecules" (R. E. Hester and R. B. Girling, eds.), p. 355. The Royal Society of Chemistry, Cambridge, 1991.

[27] W. Pohle, in "Spectroscopy of Biological Molecules—State of the Art" (A. Bertoluzza, C. Fagnano, and P. Monti, eds.), p. 189. Societá Editrice Esculapio, Bologna, 1989.

[28] A. Bertoluzza, C. Fagnano, P. Filippetti, M. A. Morelli, M. R. Tosi, and V. Tugnoli, in "Spectroscopy of Biological Molecules—State of the Art" (A. Bertoluzza, C. Fagnano, and P. Monti, eds.), p. 367. Societá Editrice Esculapio, Bologna, 1989.

[29] T. Theophanides and J. Anastassopoulou, in "Spectroscopy of Biological Molecules—New Advances" (E. D. Schmid, F. W. Schneider, and F. Siebert, eds.), p. 433. Wiley, Chichester, 1988.

[30] H. Fritzsche and A. Rupprecht, in "Spectroscopy of Biological Molecules" (R. E. Hester and R. B. Girling, eds.), p. 379. The Royal Society of Chemistry, Cambridge, 1991.

as model membranes. Since the introduction of the fluid mosaic concept of membranes, the dynamic properties of lipids have been the subject of many investigations. Depending on the membrane, the lipid moiety can be rigid or fluid. This behavior is determined by the conformation of the acyl chains. The hydrophilic head groups also determine the properties of lipids. Their interaction with one another or with other constituents is important for the function of biological membranes.

It has long been known that the frequency of the CH_2 stretching bands (around 2920 cm^{-1} antisymmetric, 2850 cm^{-1} symmetric) increases with conformational disorder. The value can be taken as a monitor of the average trans/gauche ratio. In addition, the CH_2 deformation band (around 1450 cm^{-1}) depends on the packing order of the acyl chains. If the packing is crystalline orthorhombic (e.g., for the monohydrate of dipalmitoylphosphatidylcholine), the band is split owing to crystal field effects. At higher temperatures, a pretransition to a crystalline hexagonal packing takes place and the splitting disappears. Even for aqueous lipid dispersions, however, the transition from the gel to liquid crystalline phase can be monitored by this vibration: the band broadens and the peak height is decreased correspondingly. In both phases of the aqueous dispersion the acyl chains are no longer in a crystalline order. Again, the broadening reflects the increase in conformational disorder.

To measure the degree of disorder in natural membranes, an elegant approach was used. The C–H stretching bands of the lipid acyl chains overlap with bands of the protein, making it difficult to determine the C–H stretches of the lipids. If acyl chains are deuterated, the C–^{2}H stretches are shifted down to a region free of other bands (around 2100 and 2200 cm^{-1}), but they show similar spectral changes for the phase transition. Mantsch and co-workers have exploited this property for the determination of the phase transition in the plasma membrane of bacteria, by supplementing the cell culture medium with deuterated fatty acids that are incorporated into the lipids. The spectra of the isolated plasma membranes show clearly the C–^{2}H stretches of the deuterated acyl chains. The authors have compared the phase transition of the lipids in the plasma membrane with that of the extracted lipids. They observed that the phase transition occurs at a slightly higher temperature in the plasma membrane.[31,32] Many of these aspects have been reviewed by Mantsch,[33] and the general vibrational modes of lipids are summarized by Parker.[10]

[31] D. G. Cameron, A. Martin, and H. H. Mantsch, *Science* **219**, 180 (1983).
[32] H. H. Mantsch, P. W. Yang, A. Martin, and D. G. Cameron, *Eur. J. Biochem.* **278**, 335 (1988).
[33] H. H. Mantsch, *in* "Spectroscopy of Biological Molecules" (C. Sandorfy and T. Theophanides, eds.), p. 547. Reidel, Dodrecht, The Netherlands, 1984.

Investigations have been carried out dealing with the polymorphism of the gel phase in a series of lipids which have a bulky group close to the end of the acyl chains (methyl, cyclohexyl). The behavior depends on whether the chains have an even or odd number of carbon atoms. Furthermore, the influence of the chain length on the polymorhpism of the gel phase in normal lipids has been studied. A critical length appears to be C_{16}, at and below which pronounced polymorphism is observed.[34,35] Local anesthetics interact with the lipid phase of membranes. Therefore, their interaction with and influence on the phase transition have been investigated by FT-IR spectroscopy.[36] High pressure tends to expel the drug from the membrane phase into the aqueous phase. The adsorption of local anesthetics to lipid membranes has been followed by infrared ATR spectroscopy.[37] If the head groups of lipids are charged, it can be expected that they interact with ions and that the phase transition depends on the shielding of the charge. This has been studied in combination with calorimetric experiments for negatively charged phospholipids.[38]

In an extensive study a method was developed by Mendelsohn and Senak to determine quantitatively the order state of the acyl chains, and a review has appeared[39] noting the introduction of specifically deuterated acyl chains. The aim is the estimation of the number of gauche rotamers, kinks, double gauche, and end-gauche forms. In addition, it is desirable to determine the position at which the distortions are located. The following strategy has been developed: (1) isolated $C-^2H_2$ rocking modes (620–670 cm^{-1}) are used to determine the fraction of gauche rotamers at specific positions; (2) the intensities of the $C-H_2$ wagging modes (1330–1370 cm^{-1}) from nondeuterated acyl chains provide estimates of the numbers of kinks, double gauche, and end-gauche forms; (3) the reduced intensities of the progression arising from the coupling of $C-H_2$ wagging modes in ordered phases only provides a quantitative measure of disorder in situations where small numbers of gauche bonds are introduced into the otherwise ordered chains; (4) coupled $C-^2H_2$ rocking modes in phospholipids possessing adjacent $C-^2H_2$ groups provide data about specific two- or three-bond conformational states. The review and the original papers cited demon-

[34] R. N. A. H. Lewis, D. A. Mannock, R. N. McElhaney, P. T. T. Wong, and H. H. Mantsch, *Biochemistry* **29,** 8933 (1990).

[35] R. N. A. H. Lewis and R. N. McElhaney, *Biochemistry* **29,** 7946 (1990).

[36] M. Auger, I. C. P. Smith, H. H. Mantsch, and P. T. T. Wong, *Biochemistry* **29,** 2008 (1990).

[37] M. Schöpflin, U. P. Fringeli, and X. Perlia, *J. Am. Chem. Soc.* **109,** 2375 (1987).

[38] J. Tuchtenhagen and A. Blume, *in* "Spectroscopy of Biological Molecules" (R. E. Hester and R. B. Girling, eds.), p. 167. The Royal Society of Chemistry, Cambridge, 1991.

[39] R. Mendelsohn and L. Senak, *in* "Bimolecular Spectroscopy Part A" (R. J. H. Clark and R. E. Hester, eds.), p. 339. Wiley, Chichester, 1993.

strate clearly the capabilities of the developed method. The great advantage is the possibility of determining the disorder not only in pure systems but also in lipid mixtures and even in natural membranes. Preliminary data on the plasma membrane of *Acholeplasma laidlawii* are available.

Lipids themselves are only the matrix of a biomembrane, the more functionally active part being membrane-bound proteins. Therefore, many publications have focused on the interaction of such proteins or hydrophobic peptides with lipids. Synthetic peptides are of special interest. The influence of their composition on the lipid phase can be studied in detail, and in addition when incorporated into the bilayer their structure can be deduced (see below). Two papers in which these aspects are addressed and in which many earlier publications are cited have appeared.[40,41]

In the experiments described so far the target had been predominantly the order state of lipid acyl chains; however, the role of the head groups and their interactions has also been studied by infrared spectroscopy. In ester lipids, the C=O stretching band has gained much interest. It has been found that the splitting observed in dry lipids (1740 and 1725 cm^{-1}) is not due to the inequivalence of the two carbonyls (*sn*-1 and *sn*-2) but rather to crystal field splitting in the ordered phase[42] or to inhomogeneous hydrogen bonding in the hydrated state.[43,44] In both states, the two chemically inequivalent C=O groups exhibit very small (1–2 cm^{-1}) frequency differences. It was further established that, depending on the special lipid, one or the other C=O group is more accessible to water. In all these experiments it was necessary to label the C=O group specifically with ^{13}C in order to differentiate between the *sn*-1 and *sn*-2 positions. More complex lipids such as gangliosides and lipopolysaccharides have been investigated; the complex structures of the head groups have also been studied.[45,46] Diphytanylglycerol phospholipids can form strong hydrogen bonds. Information on the inter- and intramolecular bonding pattern can be obtained from infrared spectroscopy.[47]

In the previous example on infrared spectroscopy of lipid molecules,

[40] Y.-P. Zhang, R. N. A. H. Lewis, R. S. Hodges, and R. N. McElhaney, *Biochemistry* **31,** 11572 (1992).

[41] Y.-P. Zhang, R. N. A. H. Lewis, R. S. Hodges, and R. N. McElhaney, *Biochemistry* **31,** 11579 (1992).

[42] W. Hübner and H. H. Mantsch, *Biophys. J.* **59,** 1261 (1991).

[43] P. M. Green, J. T. Mason, T. J. O'Leary, and I. W. Levin, *J. Phys. Chem.* **91,** 5099 (1987).

[44] A. Blume, W. Hübner, and G. Messner, *Biochemistry* **27,** 8239 (1988).

[45] E. Müller, E. Kopp, and A. Blume, *in* "Spectroscopy of Biological Molecules" (R. E. Hester and R. B. Girling, eds.), p. 189. The Royal Society of Chemistry, Cambridge, 1991.

[46] K. Brandenburg and U. Seydel, *Eur. Biophys. J.* **16,** 83 (1988).

[47] L. C. Stewart, M. Kates, P. W. Yang, and H. H. Mantsch, *Biochem. Cell Biol.* **68,** 266 (1990).

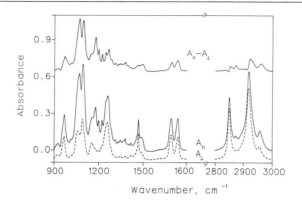

Wavenumber, cm⁻¹

FIG. 1. Polarized ATR FT-IR spectra of a solid, oriented film of 2-[1-^{13}C]dimyristoylphosphatidylcholine. Spectra were recorded with parallel ($A_\parallel$, solid trace) and perpendicular polarized light ($A_\perp$, dashed trace). The difference $A_\parallel - A_\perp$ is shown at top. (Reproduced from the *Biophysical Journal*, 1991, Vol. 59, pp. 1261–1272, by copyright permission of the Biophysical Society.[42])

experiments are described that provide structural information using infrared ATR spectroscopy. In detailed studies the orientation of a number of infrared transition moments caused by known normal modes of phosphatidylcholine has been determined. Multilayers of the lipids are deposited onto one surface of the ATR crystal with a high degree of orientation, and the dichroism is studied at various degrees of hydration.[42,48] A typical example for a dry film of dimyristoylphosphatidylcholine (DMPC) is shown in Fig. 1. The dichroic difference spectrum clearly reveals differences in the orientations of transition moments. An interesting problem addressed by Hübner and Mantsch[42] involved conflicting results on the orientation of the C=O and CO–O bonds based on X-ray, deuterium NMR, and ^{13}C solid-state NMR data. To reconcile the different results, a rapid equilibrium between four different structures was proposed. In this investigation, the orientation of the groups was determined separately, using specific ^{13}C labeling of the C=O group. The result demonstrates that, whereas the C=O bond has the same orientation for the two acyl chains, the CO–O bonds differ. The same orientation is retained in the liquid crystalline state. It is concluded that the glycerol moiety is oriented perpendicular to the membrane plane and that the *sn*-2 chain must form a 90° bend near carbon C-2. This excludes one of the proposed structures from being present at equilibrium, but it is consistent with the other three.

[48] E. Okamura, J. Umemura, and T. Takenaka, *Biochim. Biophys. Acta* **1025**, 94 (1990).

Determination of Elements of Secondary Structure by Fourier
Transform Infrared Spectroscopy

For many years it has been realized that the regularity of the sequence
of peptide bonds offers the possibility of obtaining information on the
secondary structure of peptides by infrared spectroscopy (see references
cited by Krimm and Bandekar[49]). The frequency of the amide I band,
representing the normal mode involving mainly the carbonyl (C=O)
stretching vibration of the peptide bond, was found to be sensitive to
secondary structure (spectral range between 1700 and 1620 cm^{-1}). A theoretical analysis of the normal modes of model peptides with known structure has been published.[49] This investigation, which represents a basic
study to which later publications must refer, demonstrates that the amide
I mode adopts characteristic frequencies for α helices, β strands, β turns,
and τ turns. The frequencies are determined by both the transition moment
coupling of the C=O oscillators and the hydrogen bonding of the C=O
group to proton donors (mainly the NH group). Together with the theoretical analysis experimental results obtained both with infrared and Raman
spectroscopy are presented.

The amide II mode, a more complex vibration mainly caused by the
NH in-plane bending vibration but which contains also C–N and C^α–C
stretching and CO in-plane bending characters,[49] is less used for structure
determination. Deuteration of the nitrogen shifts it down by approximately
100 cm^{-1}. By measuring the amide II intensity after exposure of the protein
to ^2H_2O, the extent of $H/^2H$ exchange at the peptide nitrogens and the
rate can be determined. This provides information on the accessibility of
the peptide groups, their structures (strong intramolecular hydrogen bonds
tend to slow down the exchange rate), and larger structural changes.
Several earlier investigations are mentioned in the books by Parker[10] and
Griffiths and de Haseth,[6] and a paper on a hydrophobic peptide provides
information on the exchange in membrane-bound proteins.[40]

To apply the analysis of the amide I band to proteins for which a
rigorous normal mode analysis is not available, a more empirical approach
has been adopted. Because the amide I band of proteins is very broad and
featureless, a direct analysis of the band in terms of secondary structural
elements is not possible. However, if the band is subjected to so-called
resolution-enhancement data analysis, several individual bands can be
extracted. Spectral deconvolution and derivative techniques are applied
(the latter is a special case of the former). To avoid artifacts, spectra with
very high signal/noise ratios have to be measured. Here, the advantage

[49] S. Krimm and J. Bandekar, *Adv. Protein Chem.* **38**, 181 (1986).

of FT-IR spectroscopy plays an important role. The scissor vibration of water absorbs at 1650 cm^{-1}, that is, just in the region of the amide I band. Therefore, special care has to be taken in order to eliminate contributions of the water band to the amide I modes. The frequency range of the different structural elements is taken from model compound studies and the theoretical calculations mentioned above. To quantify the structural elements, the absorption spectrum (original or deconvoluted) is fitted by absorption bands located at positions determined by the deconvolution techniques. Another method evaluates directly the second-derivative spectrum from which the peak height and the bandwidth can be estimated. Several papers review these techniques.[50-54] Table I[55] compares results obtained with UV-circular dichroism (UV-CD) (see also R. W. Woody, this volume [4]) and infrared spectroscopy for proteins with known X-ray structures. Of course, the determination of the relative amounts of the different structural elements from X-ray data depends on the algorithm.[56,57]

Another approach for the evaluation of the infrared amide I band uses precisely measured infrared spectra of proteins with known structures as a calibration source. These spectra are used either directly as a basis set[58-60] to model the spectra of unknown proteins or indirectly, after factor analysis has provided a reduced basis set.[61] Both basic methods have advantages and disadvantages, which are discussed in the references cited and in some more recent original papers.[40,55,62-65]

In many cases, several bands obtained by the second-derivative or deconvolution decomposition methods are attributed to the same class of

[50] W. K. Surewicz and H. H. Mantsch, *Biochim. Biophys. Acta* **952**, 115 (1988).

[51] D. M. Byler and H. Susi, *Biopolymers* **25**, 469 (1986).

[52] H. Susi and D. M. Byler, this series, Vol. 130, p. 290.

[53] J. L. R. Arrondo, A. Muga, J. Castresana, and F. M. Goñi, *Prog. Biophys. Mol. Biol.* **59**, 23 (1993).

[54] P. I. Haris and D. Chapman, *Trends Biochem. Sci.* **17**, 328 (1992).

[55] A. Dong, P. Huang, and W. S. Caughey, *Biochemistry* **29**, 3303 (1990).

[56] M. Levitt and J. Greer, *J. Mol. Biol.* **114**, 181 (1977).

[57] W. Kabsch and C. Sander, *Biopolymers* **22**, 2577 (1983).

[58] S. Yu. Venyaminov and N. N. Kalnin, *Biopolymers* **30**, 1259 (1990).

[59] N. N. Kalnin, I. A. Baikalov, and S. Yu. Venyaminov, *Biopolymers* **30**, 1273 (1990).

[60] F. Dousseau and M. Pezolet, *Biochemistry* **29**, 8771 (1990).

[61] D. C. Lee, P. I. Haris, D. Chapman, and R. C. Mitchell, *Biochemistry* **29**, 9185 (1990).

[62] D. C. Fry, D. M. Byler, H. Susi, E. M. Brown, S. A. Kuby, and A. S. Mildvan, *Biochemistry* **27**, 3588 (1988).

[63] D. Naumann, C. Schultz, U. Görne-Tschelnokow, and F. Hucho, *Biochemistry* **32**, 3162 (1993).

[64] D. Garcia-Quintana, P. Garriga, and J. Manyosa, *J. Biol. Chem.* **268**, 2403 (1993).

[65] J. C. Gorga, A. Dong, M. C. Manning, R. W. Woody, W. S. Caughey, and J. L. Strominger, *Proc. Natl. Acad. Sci. U.S.A.* **86**, 2321 (1989).

TABLE I
PROTEIN SECONDARY STRUCTURES DETERMINED BY INFRARED SPECTRAL DECONVOLUTION AND
CIRCULAR DICHROISM SPECTRA AND X-RAY CRYSTALLOGRAPHY[a]

Protein	Secondary structure (%)				Method
	α Helix	β Sheet	Turn	Random	
Hemoglobin	78[b]	12	10	[b]	IR-SD
	87	0	7	6	X-Ray
	68–75	1–4	15–20	9–16	CD
Myoglobin	85[b]	7	8	[b]	IR-SD
	85	0	8	7	X-Ray
	67–86	0–13	0–6	11–30	CD
Lysozyme	40	19	27	14	IR-SD
	45	19	23	13	X-Ray
	29–45	11–39	8–26	8–60	CD
Cytochrome c (oxidized)	42	21	25	12	IR-SD
	48	10	17	25	X-Ray
	27–46	0–9	15–28	28–41	CD
α-Chymotrypsin	9	47	30	14	IR-SD
	8	50	27	15	X-Ray
	8–15	10–53	2–22	38–70	CD
Trypsin	9	44	38	9	IR-SD
	9	56	24	11	X-Ray
Ribonuclease A	15	40	36	9	IR-SD
	23	46	21	10	X-Ray
	12–30	21–44	11–22	19–50	CD
Alcohol dehydrogenase	18	45	23	14	IR-SD
	29	40	19	12	X-Ray
Concanavalin A	8	58	26	8	IR-SD
	3	60	22	15	X-Ray
	3–25	41–49	15–27	9–36	CD
Immunoglobin G	3	64	28	5	IR-SD
	3	67	18	12	X-Ray
Major histocompatibility complex antigen A2	17	41	28	14	IR-SD
	20	42			X-Ray
	8–13	74–77			CD
β_2-Microglobulin	6	52	33	9	IR-SD
	0	48			X-Ray
	0	59			CD

[a] IR-SD, IR-spectral deconvolution. Reprinted with permission from Dong et al.[55] Copyright 1990 American Chemical Society.

[b] The band due to random structure appears as a shoulder on the α-helix band and is too small to be separated from that due to α-helix structure; the random structure is estimated at less than 5% and is included in the α-helix value.

secondary structure. This is interpreted by the presence of several different structural elements of this secondary structure. If this is correct, the methods provide an additional discrimination of structural elements. One precaution should be mentioned if proteins located in membranes are investigated. Under the experimental conditions of infrared spectroscopy (low water content), the membranes are often partially oriented parallel to the infrared window, rendering the sample optically anisotropic. This can modify the observed infrared intensities, since transition moments of structural elements oriented parallel to the window have higher intensities than those oriented perpendicular. However, this induced infrared dichroism can also be exploited to determine the orientation of the structural elements in the membrane.[66]

The measurement of the vibrational circular dichroism (VCD) of the amide I mode represents another promising method. Similar to the UV-CD method using the electronic transition of the peptide group, it relies on the optical chirality imposed by the secondary structure on the peptide group. Therefore, unlike the amide I mode itself, of which the position is determined by a kind of second-order influence, its circular dichroism is directly caused by the secondary structure. VCD is a very weak effect, and it is beyond the scope of this chapter to describe experimental details. As in the UV-CD method, it employs a photoelastic modulator that switches the polarization of the monitoring beam rapidly between right and left circular polarization. In a series of papers the sensitivity of VCD to the secondary structure of proteins and peptides has been demonstrated. In these publications experimental details are discussed, and references to earlier work are given.[67-70] As for the direct evaluation of the amide I mode, computational methods such as factor analysis have been developed to determine the fraction of structural elements from a set of VCD spectra of proteins and peptides with known secondary structure. Owing to the overlap with the water band, the VCD method has been applied to studies in ^2H_2O, although experiments in H_2O should be feasible.

Discussion of the reliability of the different methods for secondary structure determination would require a separate chapter. A critical review of the methods evaluating the amide I band has been published.[71] Only a few general remarks are given here. UV-CD, especially if complemented with modern evaluation methods, provides a reasonable estimate of

[66] M. S. Braiman and K. J. Rothschild, *Annu. Rev. Biophys. Chem.* **17**, 541 (1988).
[67] P. Pancoska, S. C. Yasui, and T. A. Keiderling, *Biochemistry* **28**, 5917 (1989).
[68] P. Pancoska, S. C. Yasui, and T. A. Keiderling, *Biochemistry* **20**, 5089 (1991).
[69] P. Pancoska and T. A. Keiderling, *Biochemistry* **30**, 6885 (1991).
[70] M. G. Paterlini, T. B. Freedman, and L. A. Nafie, *Biopolymers* **25**, 1751 (1986).
[71] W. K. Surewicz, H. H. Mantsch, and D. Chapman, *Biochemistry* **32**, 389 (1993).

α-helical and random-coil content but yields a larger uncertainty for β structure.[69] The direct evaluation of the amide I band assumes indirectly that the mechanisms that determine the amide I frequency are fully understood. However, *ab initio* quantum chemical calculations seem to indicate that the frequency can be varied over a broad spectral range by introducing twists in the backbone (S. Krimm, personal communication, 1992).

Furthermore, the transition moment coupling causes a splitting of the mode that is especially pronounced for long structural elements. This splitting is often neglected and may, therefore, cause serious misinterpretations as demonstrated in the case of myoglobin.[72] X-Ray data exclude any β structure, but evaluation of the amide I band predicts a considerable content.[55] Theoretical calculations show that, using only the transition moment coupling model, reasonable agreement is obtained between calculated and measured spectra. It has been demonstrated that the splitting deduced for long α helices causes the low-frequency mode around 1635 cm^{-1} that is otherwise interpreted as β structure.[72] From such calculations it has been realized that the eigenmodes are not necessarily a good tool for the evaluation of secondary structure in proteins where many structural elements are present. Therefore, the transition moment coupling model has been extended to the so-called three-dimensional doorway-state theory, in which in a certain frequency range three orthogonal infrared active modes are constructed from the eigenmodes located in this range. Apparently, these "doorway modes" are better characterized by the secondary structure.[73,74]

Evaluation of the VCD spectrum does not strongly depend on the mechanisms which influence the amide I frequency. Therefore, it is not surprising that it provides a reliable prediction for the secondary structure.[68,69] However, more protein data have to be evaluated in order to allow a final conclusion.

Reaction-Induced Infrared Difference Spectroscopy

As outlined in the introduction, the goal of reaction-induced infrared difference spectroscopy is to determine: (1) the contribution of single amino acid residues to the mechanism of enzymatic reactions, (2) the interaction of substrates and cofactors with the protein, (3) intermediate states of substrates and cofactors, and (4) structural changes of the protein caused by the interaction. Because only a small part of the total system

[72] H. Torii and M. Tasumi, *J. Chem. Phys.* **96**, 3379 (1992).
[73] H. Torri and M. Tasumi, *J. Chem. Phys.* **97**, 86 (1992).
[74] H. Torri and M. Tasumi, *J. Chem. Phys.* **97**, 92 (1992).

is investigated, the method has to be sensitive enough to detect the bands of single groups against the cumulative absorbance of the large remaining system. A rough estimate shows that bands smaller than 10^{-4} absorbance units should be detectable against a background absorbance of 0.5 to 1. Therefore, the advantages of FT-IR spectroscopy have facilitated the implementation of this method, and several reviews have appeared.[66,75–78] The examples presented are divided into classes characterized by the method of triggering the reactions and how the different states of the system are stabilized. In the last section, time-resolved studies are described.

It is a prerequisite of the technique that the trigger process is as specific as possible in order to avoid unwanted spectral changes caused, for example, by changes of the physical state of the sample. This requirement is especially stringent since the spectral changes to be measured are small and could be obscured by the larger nonspecific effects. In our own investigations we have observed that even changing the temperature of the sample can cause such distortions. Removing the sample from the spectrometer for the start of the reaction and reinsertion afterwards should be avoided. Distortions caused by such manipulations are especially severe if membrane systems are investigated. Samples prepared from such systems are often inhomogeneous across the area. Soluble proteins are easier to handle, and reaction-induced difference spectra were first reported for such systems. In a series of publications, the binding of CO to heme proteins was followed by measuring a spectrum of the protein without CO and a second spectrum of a different sample with CO bound. Because only the CO band was monitored, which is located in a region essentially free of other protein bands, the difference spectra CO-bound minus nonbound could easily be formed. Some of these results are reviewed in the books by Parker[10] and Griffiths and de Haseth.[6] Acylchymotrypsin has been compared with the free enzyme by forming the difference spectrum between the two states.[79]

Photobiological systems (chromoproteins) can conveniently be studied by FT-IR difference spectroscopy. Light is an ideal trigger. It is physiologi-

[75] F. Siebert, in "Spectroscopy of Biological Molecules: Theory and Applications–Chemistry, Physics, Biology, and Medicine" (C. Sandorfy and T. Theophanides, eds.), p. 347. Reidel, Dordrecht, The Netherlands, 1984.

[76] F. Siebert, this series, Vol. 189, p. 123.

[77] F. Siebert, in "Biomolecular Spectroscopy Part A" (R. J. H. Clark and R. E. Hester, eds.), p. 1. Wiley, Chichester, 1993.

[78] W. Mäntele, Trends Biochem. Sci. 18, 197 (1993).

[79] P. J. Tonge, M. Pusztai, A. J. White, C. W. Wharton, and P. R. Carey, Biochemistry 30, 4790 (1991).

cal, specific, and direct, and it does not cause further perturbations. The site of action is a chromophore bound to the protein. In the simplest approach, difference spectra are formed between the initial state before illumination and the final state after illumination. The following questions are addressed: (1) what is the molecular structure of the chromophore in the initial state, and what are the light-induced changes; (2) what are the interactions between the chromophore and the protein in the various states of the photoreaction; and (3) what are the molecular changes occurring during the photoreaction in the protein? Alterations of amino acid side chains as well as of the backbone are of interest. The infrared difference spectra contain contributions of both the protein and the chromophore. Therefore, methods have to be developed to differentiate between them, and some techniques are exemplified later. Chromophore bands are usually investigated by resonance Raman spectroscopy (see T. G. Spiro and R. S. Czernuszewicz, this volume [18]). It is the distinct advantage of FT-IR difference spectroscopy that it is sensitive to both the chromophore and the protein.

In most cases, the photoreaction from the physiologically inactive state to the active one passes through several intermediates. For an understanding of the reaction mechanism it is often required to monitor the molecular structure of the intermediate states as well. The conceptually most straightforward technique would be to apply time-resolved infrared spectroscopy (see H. van Amerongen and R. van Grondelle, this volume [9], on time-resolved absorbance spectroscopy), and some techniques are described at the end of this section. However, these methods are rather complicated and not easily implemented in a standard infrared laboratory. Nature itself provides a means to study these intermediates: it is often possible to stabilize them at low temperature. To measure the difference spectrum of an intermediate state, the sample is cooled to the desired temperature, and a single-beam spectrum is measured. Subsequently, the sample still located in the FT-IR instrument is illuminated with light of suitable spectral range, and the single-beam spectrum is measured thereafter. From the two single-beam spectra the difference spectrum is obtained. Infrared difference spectroscopy has contributed considerably to our understanding of the molecular mechanisms of the visual pigment rhodopsin and the bacterial light-driven proton pump bacteriorhodopsin (several reviews have appeared on these retinal proteins[66,76,77,80,81]), as well as the photosyn-

[80] W. J. DeGrip, D. Gray, J. Gillespie, P. H. M. Bovee, E. M. M. Van den Berg, J. Lugtenburg, and K. J. Rothschild, *Photochem. Photobiol.* **48,** 497 (1988).
[81] K. J. Rothschild, *J. Bioenerg. Biomembr.* **24,** 147 (1992).

thetic reaction center.[82–84] Difference spectra have also been reported on the light-driven chloride pump halorhodopsin,[85] on sensory rhodopsin I of *Halobacterium halobium*,[86] on sensory rhodopsin II of *Natronobacterium pharaonis*,[87] and on the photoreceptor of green plants, phytochrome.[88,89]

Figure 2 shows difference spectra of the photoreaction of bacteriorhodopsin.[77] This system undergoes a cyclic photoreaction with intermediates K, L, M, N, and O. The K and L spectra have been measured at 80 and 170 K, respectively. The latter two have been obtained under steady-state illumination, that is, the infrared spectra of the photoproducts have been measured during the illumination of the sample at approximately 0°. Such a technique is possible for a thermally reversible photoreaction. Depending on the rate constants of the reaction, some intermediates accumulate during the illumination. In this case, the conditions are chosen to generate predominantly M (neutral pH and low water content) and an M/N mixture (high pH, sufficiently hydrated). In most cases positive bands are caused by the photoproduct and negative ones by the initial state; detailed interpretation can be found in the cited reference.

A few explanations may be appropriate here. As a retinal protein, bacteriorhodopsin contains as chromophore all-*trans*-retinal which is bound to the protein by a protonated Schiff base. The action of light causes the isomerization of the chromophore to the 13-cis geometry. This is reflected in the BR → K difference spectrum by the bands between 1100 and 1300 cm^{-1}. The strong bands at 1529 and 1515 cm^{-1} are the ethylenic (C=C) stretching modes of the retinal. Protein bands indicating alterations of the backbone are found around 1550 cm^{-1} (amide II) and between 1620 and 1680 cm^{-1} (amide I). An interesting feature is observed in the BR → M spectrum at 1762.5 cm^{-1}. This band could be assigned to an aspartic acid (see below) which becomes protonated in M. Because the

[82] W. G. Mäntele, A. M. Wollenweber, E. Nabedryk, and J. Breton, *Proc. Natl. Acad. Sci. U.S.A.* **85**, 8468 (1988).

[83] E. Nabedryk, M. Leonhard, W. Mäntele, and J. Breton, *Biochemistry* **29**, 3242 (1990).

[84] E. Nabedryk, K. A. Bagley, D. L. Thibodeau, M. Bauscher, W. Mäntele, and J. Breton, *FEBS Lett.* **266**, 59 (1990).

[85] K. J. Rothschild, O. Bousché, M. S. Braiman, C. A. Hasselbacher, and J. L. Spudich, *Biochemistry* **27**, 2420 (1988).

[86] O. Bousché, E. N. Spudich, J. L. Spudich, and K. J. Rothschild, *Biochemistry* **30**, 5395 (1991).

[87] B. Scharf, M. Engelhard, and F. Siebert, *Colloq. INSERM* **221**, 317 (1992).

[88] F. Siebert, R. Grimm, W. Rüdiger, G. Schmidt, and H. Scheer, *Eur. J. Biochem.* **194**, 921 (1990).

[89] J. Sakai, E. H. Morita, H. Hayashi, M. Furuya, and M. Tasumi, *Chem. Lett.* **1990**, 1925 (1990).

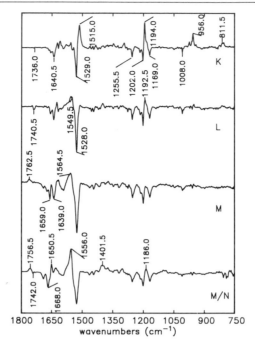

FIG. 2. Fourier transform infrared difference spectra of the photointermediates of bacteriorhodopsin. The K and L difference spectra were obtained at 80 and 170 K, respectively. The M and M/N difference spectra were measured under steady-state illumination. (From Siebert.[77] Copyright 1993 John Wiley & Sons, Ltd. Reprinted by permission of John Wiley & Sons, Ltd.)

Schiff base is deprotonated in M, it has been concluded that this aspartic acid is the proton acceptor for Schiff base deprotonation. In the BR → M/N spectrum the negative band at 1742 cm^{-1} indicates that another carboxyl group assigned to another aspartic acid deprotonates, transferring the proton to the Schiff base for reprotonation. Thus, infrared difference spectroscopy has identified essential steps in the proton translocating mechanism.

Light can also be used as a trigger for nonphotobiological systems. There are now many substrate analogs available which are blocked by a dye molecule. With a UV flash, the dye is cleaved off and the substrate becomes accessible for interaction with the protein. This technqiue has been reviewed,[90] and the blocked substrates are usually called caged molecules. In combination with infrared difference spectroscopy it has been

[90] J. A. McCray and D. R. Trentham, *Annu. Rev. Biophys. Biophys. Chem.* **18,** 239 (1989).

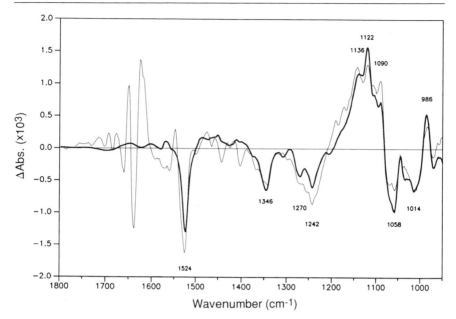

FIG. 3. Fourier transform infrared difference spectra showing the influence of ATP on the Ca^{2+}-ATPase of sarcoplasmic reticulum. Thick trace, difference spectrum obtained by photolysis of caged ATP alone; thin trace, difference spectrum obtained with the additional presence of ATPase. (Courtesy of A. Barth.)

applied to the investigation of the Ca^{2+}-ATPase, using caged ATP and caged Ca^{2+} as trigger.[91,92] Figure 3 shows a representative example for caged ATP. Because the cage itself undergoes molecular changes, it causes difference bands in this spectrum. This is shown by the thick line (Fig. 3). In the presence of the ATPase, however (thin line, Fig. 3), the photolysis of caged ATP causes additional bands. In the region between 1620 and 1680 cm^{-1} they reflect backbone changes of the protein. The bands above 1700 cm^{-1} are probably due to carboxyl groups involved in the release of Ca^{2+}. This general technique appears very promising. It makes the powerful method of FT-IR difference spectroscopy accessible to many more systems. As an alternative to the use of caged compounds it appears possible to apply substrates to enzymes which are immobilized on the surface of an ATR crystal by flowing an aqueous solution of the substrate past the

[91] R. Buchet, I. Jona, and A. Martonosi, *Biochim. Biophys. Acta* **1104,** 207 (1992).
[92] A. Barth, W. Mäntele, and W. Kreutz, *Biochim. Biophys. Acta* **1057,** 115 (1991).

film surface. Preliminary experiments on bacteriorhodopsin,[93] the acetylcholine receptor,[94] and $Na^+/,K^+$-ATPase[95] have been published.

The triggering of reactions by electron transfer in an electrochemical cell, previously employed only for nonbiological redox reactions, has been extended to biological systems. To control the electrochemical potential precisely, special infrared cells had to be developed which also must have sufficient infrared transmission of aqueous samples. In these cells, a conducting grid, which functions as the working electrode, is placed across the area of the sample. This method has been applied to the electron transfer reactions in photosynthetic reaction centers[96] and to heme proteins.[97-99] The difference spectra are formed between two well-defined oxidation states. It is interesting that the change in the oxidation state of the cofactors not only alters the vibrational spectra but also causes changes in the interaction with amino acid side chains and small structural changes of the protein. Furthermore, owing to the large number of characteristic bands in the difference spectra, redox potentials can be determined precisely. Because redox reactions play an important role in biology, it can be anticipated that FT-IR difference spectroscopy in combination with electrochemistry will be applied to a large number of interesting systems.

As mentioned above, the most informative method to study biochemical reactions would be time-resolved infrared difference spectroscopy. However, because the spectral changes are very small, all techniques require signal averaging over many reaction cycles. This limits application of the techniques to thermally reversible photoreactions. If such systems are in addition stable enough, the photoreaction can be triggered by thousands of flashes.

The techniques employed can be divided into broadband techniques using the FT-IR method and scanning single-wavelength techniques using an infrared beam from a monochromator or a laser. The single-wavelength methods resemble a normal flash photolysis apparatus. The photosynthetic reaction center,[100] bacteriorhodopsin,[101-103] and the photolysis of CO-

[93] H. Marrero and K. J. Rothschild, *FEBS Lett.* **223**, 289 (1987).

[94] J. E. Baenziger, K. W. Miller, and K. J. Rothschild, *Biophys. J.* **61**, 983 (1992).

[95] U. P. Fringeli, H.-J. Apell, M. Fringeli, and P. Läuger, *Biochim. Biophys. Acta* **984**, 301 (1989).

[96] D. A. Moss, M. Leonhard, M. Bauscher, and W. Mäntele, *FEBS Lett.* **283**, 33 (1991).

[97] C. Berthomieu, A. Boussac, W. Mäntele, J. Breton, and E. Nabedryk, *Biochemistry* **31**, 11460 (1992).

[98] D. D. Schlereth and W. Mäntele, *Biochemistry* **31**, 7494 (1992).

[99] M. Leonhard and W. Mäntele, *Biochemistry* **32**, 4532 (1993).

[100] R. Hienerwadel, D. L. Thibodeau, F. Lenz, E. Nabedryk, J. Breton, W. Kreutz, and W. Mäntele, *Biochem. J.* **31**, 5799 (1992).

[101] W. Kreutz, F. Siebert, and K. P. Hofmann, *Biol. Membr.* **5**, 241 (1984).

myoglobin have been investigated.[104,105] With the pump–probe technique the time resolution can be extended into the subpicosecond range.[106–108] This allows the determination of structural changes of the chromophore and the protein in the early photointermediates. In this way, real-time information on protein dynamics is obtained.

Three FT-IR techniques have been implemented. The rapid-scan technique exploits the fact that the time needed to record a spectrum can be less than 10 msec. Thus, processes with slower kinetics can be monitored. Difference spectra are formed between spectra taken before and after the flash, including a certain delay. To achieve the required accuracy, several hundred spectra have to be averaged, that is, several hundred flashes have to be applied. Systems studied comprise bacteriorhodopsin[109–111] and the photosynthetic reaction center.[112,113] In the stroboscopic technique only a fraction of the interferogram is recorded for each flash. This increases the time resolution correspondingly.[114,115] The last technique is called step-scan FT-IR spectroscopy. With this method, the time course of the flash-induced change of the interferogram is monitored at each sampling point. Therefore, as with the single-wavelength methods, the time resolution is limited only by the detector rise time and the signal acquisition electron-

[102] K.-H. Müller, H. J. Butt, E. Bamberg, K. Fendler, B. Hess, F. Siebert, and M. Engelhard, *Eur. Biophys. J.* **19,** 241 (1991).

[103] J. Sasaki, A. Maeda, C. Kato, and H. Hamaguchi, *Biochemistry* **32,** 867 (1993).

[104] F. Siebert, W. Mäntele, and W. Kreutz, *Biophys. Struct. Mech.* **6,** 139 (1980).

[105] A. J. Dixon, P. Glyn, M. A. Healy, P. M. Hodges, T. Jenkins, M. Poliakoff, and J. J. Turner, *Spectrochim. Acta* **44A,** 1309 (1988).

[106] R. Diller, M. Iannone, B. R. Cowen, S. Maiti, R. A. Bogomolni, and R. M. Hochstrasser, *Biochemistry* **31,** 5567 (1992).

[107] O. Einarsdóttir, P. M. Killough, J. A. Fee, and W. H. Woodruff, *J. Biol. Chem.* **264,** 2405 (1989).

[108] R. B. Dyer, K. A. Peterson, P. O. Stoutland, and W. H. Woodruff, *J. Am. Chem. Soc.* **113,** 6276 (1991).

[109] K. Gerwert, G. Souvignier, and B. Hess, *Proc. Natl. Acad. Sci. U.S.A.* **87,** 9774 (1990).

[110] M. S. Braiman, P. L. Ahl, and K. J. Rothschild, *Proc. Natl. Acad. Sci. U.S.A.* **84,** 5221 (1987).

[111] R. Kräutle, W. Gärtner, U. M. Ganter, C. Longstaff, R. R. Rando, and F. Siebert, *Biochemistry* **29,** 3915 (1990).

[112] D. L. Thibodeau, E. Nabedryk, R. Hienerwadel, F. Lenz, W. Mäntele, and J. Breton, *Biochim. Biophys. Acta* **1020,** 253 (1990).

[113] D. L. Thibodeau, E. Nabedryk, R. Hienerwadel, F. Lenz, W. Mäntele, and J. Breton, *in* "Time-Resolved Vibrational Spectroscopy V" (H. Takahashi, ed.), Springer-Verlag, Berlin, 1992.

[114] M. S. Braiman, O. Bousché, and K. J. Rothschild, *Proc. Natl. Acad. Sci. U.S.A.* **88,** 2388 (1991).

[115] G. Souvignier and K. Gerwert, *Biophys. J.* **63,** 1393 (1992).

ics.[116–118] All the methods of time-resolved infrared spectroscopy are difficult to apply, but the gain in information and, as compared to low-temperature difference spetroscopy, the more physiological conditions of
measurement make it worthwhile to apply and extend the existing techniques.

Finally, some strategies are described which are employed to interpret
the infrared difference spectra at a molecular level. The first goal is the
identification of bands caused by the protein and cofactors (chromophore,
substrate, etc.). In the case of retinal proteins in which the retinal chromophore can be removed and replaced by an artificial one, the method of
isotope labeling has provided detailed information. Not only does it help
to identify the chromophore bands, but it also enables their assignment
to specific normal modes from which structural information can be derived.[119–121] Of course, a similar technique should be applicable to quinones
in photosynthetic reaction centers and to caged compounds. Protein bands
can also be assigned by isotope labeling. In bacterial systems, the label
can be incorporated into the protein by growing the bacteria in a synthetic
medium containing the labeled amino acid. This has been demonstrated
for bacteriorhodopsin.[122–126] In the difference spectra bands could be assigned to molecular changes of aspartic acids, tyrosines, and lysine.

A typical example (Fig. 4)[122] shows the BR $\rightarrow$ M difference spectrum
of bacteriorhodopsin between 1680 and 1780 cm^{-1}. Spectrum a in Fig. 4
was obtained with an unmodified sample, whereas spectrum b was measured with bacteriorhodopsin prepared from a cell culture containing labeled [4-^{13}C]aspartic acid. A shifted band at 1720 cm^{-1} is clearly seen,
whereas the original band at 1763 cm^{-1} has lost intensity. Spectrum c in
Fig. 4 shows the additional effect of ^2H_2O. The results clearly demonstrate

[116] W. Uhmann, A. Becker, Ch. Taran, and F. Siebert, *Appl. Spectrosc.* **45**, 390 (1991).

[117] O. Weidlich and F. Siebert, *Appl. Spectrosc.* **47**, 1394 (1993).

[118] J.-R. Burie, W. Leibl, E. Nabedryk, and J. Breton, *Appl. Spectrosc.* **47**, 1401 (1993).

[119] U. M. Ganter, W. Gärtner, and F. Siebert, *Biochemistry* **27**, 7480 (1988).

[120] K. A. Bagley, V. Balogh-Nair, A. A. Croteau, G. Dollinger, T. G. Ebrey, L. Eisenstein,
M. K. Hong, K. Nakanishi, and J. Vittitow, *Biochemistry* **24**, 6055 (1985).

[121] K. Gerwert and F. Siebert, *EMBO J.* **5**, 805 (1986).

[122] M. Engelhard, K. Gerwert, B. Hess, W. Kreutz, and F. Siebert, *Biochemistry* **24**, 400
(1985).

[123] P. D. Roepe, P. L. Ahl, J. Herzfeld, J. Lugtenburg, and K. J. Rothschild, *J. Biol. Chem.*
263, 5110 (1988).

[124] L. Eisenstein, Sh.-L. Lin, G. Dollinger, K. Odashima, J. Termini, K. Konno, W.-D.
Ding, and K. Nakanishi, *J. Am. Chem. Soc.* **109**, 6860 (1987).

[125] S.-L. Lin, P. Ormos, L. Eisenstein, R. Govindjee, K. Konno, and K. Nakanishi, *Biochemistry* **26**, 8327 (1987).

[126] Y. Gat, M. F. Grossjean, I. Pinevsky, H. Takei, Z. Rothman, H. Sigrist, A. Lewis, and
M. Sheves, *Proc. Natl. Acad. Sci. U.S.A.* **89**, 2434 (1992).

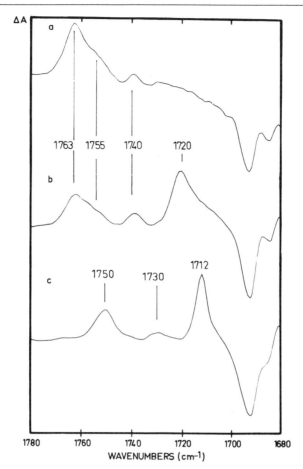

FIG. 4. Difference spectrum for the BR → M transition between 1680 and 1780 cm^{-1} obtained from unmodified bacteriorhodopsin (a) and from bacteriorhodopsin containing [4-^{13}C]aspartic acid (b). The downshift of the C=O band from 1763 to 1720 cm^{-1} is clearly visible. (c) As (b), but measurements were performed in ^{2}H$_2$O. (Reprinted with permission from Engelhard *et al.*[122] Coypright 1985 American Chemical Society.)

that the band is caused by the C=O stretch of a protonated aspartic acid, which is shifted down by ^{13}C labeling and by deuteration of the carboxyl group. The remaining band at 1763 cm^{-1} is due to incomplete label incorporation. The identification of an aspartic acid as the proton acceptor for deprotonation of the Schiff base was based on these experiments.

Finally, protein bands can also be identified by site-directed mutagenesis. Again, the method has been applied to bacteriorhodopsin, and a large number of references can be found in the reviews already cited. Bands caused by aspartic acids, tryptophans, and tyrosines could be identified. FT-IR difference spectroscopy on mutants not only provides a means to assign bands to specific groups, but also is used to characterize the properties of the mutants at the molecular level. Special care has to be taken to differentiate between direct and indirect effects of mutation. Direct effects cause the disappearance of bands because the corresponding group has been replaced. Indirect effects cause changes in the spectra because of changes in interactions which can extend over a long distance from the replaced amino acid.

Conclusion

Fourier transform infrared difference spectroscopy has enormous potential. It represents a nondestructive method working at the molecular level, so the amount of material required is relatively small (40–100 μg). This enables its combination with other techniques in molecular biology. More theoretical and experimental advances are indeed possible. A theoretical means for interpreting the observed small changes of the amide I bands needs to be developed. The practical realization of the method of site-directed labeling (i.e., the incorporation of one labeled amino acid at a specific site) would represent a major breakthrough.

Acknowledgments

I thank all the colleagues in our group at the Institute of Biophysics and Radiation Biology (Freiburg) for their excellent collaboration. Part of the work mentioned in the section on reaction-induced difference spectroscopy is based on their efforts. Stimulating discussions with members of the infrared group at the institute are gratefully acknowledged. Work was supported by several grants from the Deutsche Forschungsgemeinschaft and a grant from the Minister für Forschung und Wissenschaft des Landes Baden-Württemberg.

Section III

Magnetic Resonance Spectroscopy, X-Ray Spectroscopy

[21] Perspectives on Magnetic Resonance and X-Ray Absorption Spectroscopy in Biochemistry

By MELVIN P. KLEIN

Introduction

Magnetic resonance plays a role in biophysical spectroscopy of ever increasing importance and diversity. The topic is conveniently and conventionally divided into three domains: (1) nuclear magnetic resonance (NMR) spectroscopy, (2) Mössbauer spectroscopy, and (3) electron paramagnetic resonance (EPR) spectroscopy. All three of these methods depend on the presence of a magnetic moment, either that of a nucleus or of an unpaired electron. This overview is confined to the topic of EPR as the biological applications of NMR are so extensive that they receive dedicated reviews (e.g., Volume 239 in this series) and the biochemical applications of Mössbauer spectroscopy have been reviewed elsewhere in this series.[1]

The domain of EPR spectroscopy can be further divided, as is the case in this volume. Direct EPR detection is the subject of the chapter by Brudvig.[2] Therein are presented the fundamental principles, an outline of the method, and applications of the technique with examples from free radicals and transition element EPR.

The interactions between the unpaired electron(s) and the nucleus or nuclei of the observed or neighboring atoms are termed, respectively, hyperfine and superhyperfine couplings. For free radicals in solution they are often directly observable, giving rise to highly structured spectra. For large molecules, whose reorientational correlation times are too long to permit motional averaging, and for systems observed in the frozen state, these hyperfine interactions may be obtained only indirectly by a technique called ENDOR, electron nuclear double resonance. This topic is reviewed by DeRose and Hoffman.[3] These authors describe the underlying phenomena, how data are obtained, and how results are interpreted to provide detailed structural information. Time domain EPR is also introduced, where it is shown to be complementary and supplementary to the customary steady-state or continuous wave (cw) spectroscopy.[3]

EPR can be detected indirectly if the active center is also an optical

---

[1] E. Münck, K. K. Surerus, and M. P. Hendrich, in this series, Vol. 227, p. 463.
[2] G. W. Brudvig, this volume [22].
[3] V. J. DeRose and B. M. Hoffman, this volume [23].

METHODS IN ENZYMOLOGY, VOL. 246

center. Then, the effect of performing magnetic resonance is detected as a perturbation of the optical absorption or emission spectrum. Such methods are the subject of the chapter by Maki on optically detected magnetic resonance (ODMR).[4] This method has an enormous sensitivity advantage over direct observation of the EPR active center. It has been most widely used to study the (excited) triplet states of biologically interesting compounds. Maki presents a detailed exposition of the underlying principles, discusses the experimental aspects, and illustrates the method with an application to tryptophan-containing proteins. That chapter also shows the complementary nature of optical and magnetic resonance information.[4]

The solution structures, dynamics, and interactions of and between biological macromolecules are topics of widespread interest in biochemistry. The chapter on electron spin labels, by Millhauser *et al.* illustrates how the EPR spectra of the stable nitroxide free radical can be used to address such problems.[5] The chemistry of the nitroxides and their modes of attachment to the host molecules are discussed first. The details of the EPR spectra and of the spin Hamiltonian are then presented, showing how the intrinsic tensorial nature of the EPR spectrum of the reporter group is affected by motion. Such dynamic information is then extracted from some small peptides. The interaction between pairs of nitroxides is used to extract structural information. Finally, an example of Fourier transform EPR is introduced.[5]

There are many instances in which a specific center in a system of interest is not directly observable because there is neither an associated magnetic moment nor an optical chromophore. Such species are termed "spectroscopically silent." Examples are the Zn atoms (that have no unpaired electrons) of zinc-containing proteins and the S moieties (that have neither electron nor nuclear magnetic moments in their dominant forms) of the S-containing amino acids. It is in such situations that X-ray absorption spectroscopy (XAS) becomes not only applicable but essential.

The XAS technique measures transitions between core levels of the atom of interest and its excited or continuum states. These levels and transitions reflect such fundamental properties of atoms that they are always present. Hence, it is always possible to observe the X-ray absorption spectrum of an atom in a system irrespective of the state (i.e., gaseous, liquid, ordered or disordered solid) for all atoms across the periodic table. From the element-specific X-ray absorption spectrum it is possible to determine information on oxidation state, local symmetry, and local metrical structure at resolution greater than that attainable from X-ray crystal

[4] A. H. Maki, this volume [25].
[5] G. L. Millhauser, W. R. Fiori, and S. M. Miick, this volume [24].

structure data. Because most biological systems have not been crystallized, a method that can provide such structural insights is of importance.

The chapter on XAS and applications in structural biology by Yachandra discusses the phenomena of XAS and the underlying theory.[6] Yachandra presents a case study of the application of XAS to a complex problem in photosynthesis, that of the structure and oxidation states of a cluster of four Mn atoms which form the site of water splitting and oxygen evolution.[6] It is notable that this work depended on the strong interplay between XAS and EPR spectroscopy.

The chapter on magnetic circular dichroism (MCD) by Sutherland should also be consulted when considering whether magnetic field effects might provide useful spectroscopic information.[7] Indeed, there are numerous examples where MCD studies in metalloproteins provided information not otherwise accessible. Papers by Solomon and co-workers have demonstrated the use of the method.[8-10]

General Examples

The following section is intended to exemplify some of the areas where EPR spectroscopy has made important contributions to biochemistry. It is not to be construed as exhaustive but rather to illustrate several areas where EPR studies have been employed and why.

Metalloproteins

Iron. Iron is ubiquitous in biological systems. It occurs in the hemoproteins, including the oxygen carriers hemoglobin and myoglobin; in a variety of electron carriers, including the cytochromes and cytochrome oxidase; and in various enzymes that catalyze distinctly different reactions. In the Fe(III) oxidation state the metal can occur in a low-spin, $S = \frac{1}{2}$ state or a high-spin, $S = \frac{5}{2}$ state. There are a few examples of an intermediate or $S = \frac{3}{2}$ state.

The heme group has such a characteristic and largely invariant structure that its properties are only slightly modulated by the proteins into which it is incorporated. That is, the g value is smaller when the external magnetic field is in the heme plane than when the field lies along the z

[6] V. K. Yachandra, this volume [26].

[7] J. C. Sutherland, this volume [6].

[8] P. A. Clark and E. I. Solomon, *J. Am. Chem. Soc.* **114**, 1108 (1992).

[9] P. A. Mabrouk, A. M. Orville, J. D. Lipscomb, and E. I. Solomon, *J. Am. Chem. Soc.* **113**, 4053 (1991).

[10] V. Zhang, M. S. Gebhard, and E. I. Solomon, *J. Am. Chem. Soc.* **113**, 5162 (1991).

axis or heme normal. For the high-spin Fe(III) $g_z \approx 6$ while $g_{x,y} \approx 2$. (See, for example, [22] in this volume). This property of ferriheme EPR permitted Bennett et al.[11] to determine the orientation of the heme planes in single crystalline myoglobin, thus providing important guides to the first full structure determination of a protein by X-ray diffraction achieved by Kendrew et al.[12]

A systematic and comparative study of the low-spin Fe(III) EPR spectra of known heme complexes with those of hemoproteins permitted assignment of the ligands in the proteins before other structural information became available.[13]

The iron–sulfur proteins, the ferredoxins, are so ubiquitous that it is difficult to realize that they were discovered only in the 1960s because of their unusual (at that time) g values. The simplest, rubredoxin, contains only one iron with the sulfur atoms of four cysteines supplying the ligation. Numerous others contain pairs of Fe atoms, and yet others contain 4Fe–4S cubes. The magnetic couplings between iron atoms leads to the unusual EPR g values, and a review by their discoverer is worthy of perusal.[14] Iron and molybdenum XAS as well as EPR provided important clues to the structure of the cores of the nitrogenase enzyme prior to its X-ray crystal structure determination, as described in [26].

It is useful at this juncture to point out that some forms of Fe, as well as other transition elements, can exist in diamagnetic forms as well as in forms containing an even number of electrons. For the first there can be no EPR since there is no magnetic moment. For the latter situation, exemplified by high-spin Fe(II) and Mn(III), EPR may be observable using special methods, or may not be observable at the conventional frequencies. (For details, see [22].) In the case of Fe, however, one has the option of employing Mössbauer spectroscopy. It is useful in all oxidation states and spin-coupling schemes.

Copper proteins. EPR spectroscopy has played an important role in determining structural and redox properties in Cu-containing enzymes. Based on the spectra of bioinorganic model compounds, it became possible to determine the symmetry and coordination environment of many Cu-containing proteins from the EPR spectra. Correlations between EPR and optical spectra also provided important guides to determining the structures and ligands and led to the discovery of two distinct types of

[11] J. E. Bennett, J. F. Gibson, and D. J. E. Ingram, *Proc. R. Soc. London A* **240**, 67 (1957).
[12] J. C. Kendrew, R. E. Dickerson, B. E. Strandberg, R. G. Hart, D. R. Davies, D. C. Phillips, and V. C. Shore, *Nature (London)* **185**, 442 (1960).
[13] W. E. Blumberg and J. Peisach, *Adv. Chem.* **100**, 271 (1971).
[14] H. Beinert, *FASEB J.* **4**, 2483 (1990).

Cu(II) proteins.[15] An informative review of the history of the field is given by Beinert.[16]

Many other metals contained in biological materials, either naturally or substitutionally, have yielded useful information about the proteins, cofactors, substrates, or inhibitors in which they are contained. They include Ti, Ni, Mo, V, Cr, Mn, and Co. Some of these paramagnetic atoms have been used as replacements for diamagnetic metals, for example, Co for Zn and Mn for Mg. In these instances they serve as spin labels, albeit different from the nitroxides. In some cases the protein retain its activity, whereas in others activity may be diminished or fully inhibited. Nonetheless, useful information concerning ligands, substrates, inhibitors, and solvent interactions may be obtained.

Free Radicals. As noted earlier, virtually all biochemical redox reactions proceed by one-electron, hence one unpaired electron, reactions. That is, they involve true or virtual free radicals. Brief comments about a few such systems are now presented.

The study of photosynthesis is a field in which EPR has played a truly significant role. Only a few illustrative examples can be cited. One of the earliest observations in living tissue was that of a light-induced EPR signal in leaves near $g \approx 2$.[17] At that time relatively little was known of the complexity of the photosynthetic apparatus in plants, but it was suspected that chlorophylls were intimately involved in this signal. Subsequently it was shown that similar light-induced EPR signals could be elicited from chlorophyll radicals and from non-oxygen-evolving bacteria whose photosynthetic apparatus was simpler than that of plants.

Comparison of the signals from the "reaction centers" of such bacteria with those of bacteriochlorophyll radicals showed that the former exhibited a much reduced line width relative to the latter, although they displayed virtually identical g values. As the line widths were assumed, and later demonstrated by isotopic substitution, to arise from unresolved hyperfine interactions between the unpaired electron and the nuclei on the chlorophyll, a mechanism was sought to account for this reduction in hyperfine coupling.

Norris *et al.* had the brilliant idea that if the unpaired electron were delocalized over two bacteriochlorophylls, the electron spending roughly half its time on each member of the pair, the hyperfine interactions would be reduced by a factor of two and the line width by $2^{1/2}$, just the ratio

[15] J. Peisach and W. E. Blumberg, *Arch. Biochem. Biophys.* **165,** 691 (1974).
[16] H. Beinert, *Biol. Met.* **3,** 61 (1990).
[17] B. Commoner, J. Heise, and J. Townsend, *Proc. Natl. Acad. Sci. U.S.A.* **42,** 710 (1956).

observed.[18] This analysis was the origin of the idea of the "special pair" that has been confirmed by X-ray crystallography in the bacterial reaction center.[19] By analogy it is assumed that both photosystems I and II in oxygenic plants and cyanobacteria also contain special pairs of chlorophylls.

The hyperfine couplings of the primary donor of the bacterial reaction center have been determined by ENDOR measurements on single crystals. A theoretical analysis of the resulting hyperfine interactions leads to the electron and spin density distributions over the special pair.[20] Such information is sought to aid in understanding how the photoexcited electron moves from the special pair to its receptor, a quinone some 20 Å distant, in a very brief time.

The flavin components of flavoproteins undergo one-electron redox reactions. Parameters of these moieties have been fruitfully studied by EPR spectroscopy. An example is given in the paper by Paulsen *et al.*,[21] and a more extensive review is given by Beinert.[22]

Ultraviolet radiation, X-rays, and γ-rays can all produce free radicals whose characteristics are well studied by EPR spectroscopy. Solvent water can give rise to the H· and OH· radicals which, in turn, can produce radical species in proteins and nucleic acids. In some cases the sulfur-containing amino acids give rise to S-centered radicals.

The superoxide anion radical, O_2^-, is produced by oxidative as well as radiation processes. Many free radicals are produced under conditions of oxidative stress. An example from the medical literature discusses the involvement of free radicals in reperfusion injury, that is, when an organ is resupplied with blood after a hiatus (e.g., during surgery).[23]

Evanescent radicals may occur with lifetimes too short to yield steady-state concentrations sufficient for direct EPR observation. The technique of "spin trapping" has been developed to garner information about such species. In this method a compound is introduced that abstracts an electron from the original radical and in turn becomes a stable radical, often a

[18] J. R. Norris, R. A. Uphaus, H. L. Crespi, and J. J. Katz, *Proc. Natl. Acad. Sci. U.S.A.* **84,** 625 (1971).

[19] J. Deisenhofer, O. Epp, K. Miki, R. Huber, and H. Michel, *Nature (London)* **318,** 618 (1985).

[20] F. Lendzian, M. Humber, R. A. Isaacson, B. Endeward, M. Plato, B. Böingk, K. Möbius, W. Lubitz, and G. Feher, *Biochim. Biophys. Acta* **183,** 139 (1993).

[21] K. E. Paulsen, A. M. Orville, F. E. Frerman, and J. D. Lipscomb, *Biochemistry* **31,** 11755 (1992).

[22] H. Beinert, *in* "Biological Applications of Electron Spin Resonance" (H. M. Swartz, J. R. Bolton, and D. C. Borg, eds.), p. 351. Wiley (Interscience), New York, 1972.

[23] I. E. Blasig, S. Shuter, P. Garlick, and T. Slater, *Free Radicals Biol. Med.* **16,** 35 (1994).

nitroxide (see [24]). The detailed chemistry of the trap provides information on the nature of the initial species.[24]

Detailed information on triplet states, systems in which two unpaired electrons form a state with total spin $S = 1$, may be obtained directly from EPR measurements or indirectly via optical detection as described in [25]. Studies of the chlorophyll triplet state in the photosynthetic apparatus have provided useful information.[25] Carotenoids, also present in the reaction centers, have been studied fruitfully by triplet state EPR spectroscopy.[26]

Conclusions

The EPR and XAS methods exemplify evolving techniques where experiment and theory continue to advance. Thus, future developments are anticipated to provide enhancements in sensitivity, spectral and temporal resolution, and the ability to interpret the spectra in terms of the local electronic and metrical structure. EPR spectroscopy at increasingly higher field strengths corresponding to frequencies in the far-infrared are even now beginning to appear. The enhanced spectral dispersion yields single-crystal-like spectra from powderlike samples. The relaxation times at the higher frequencies can be considerably different than those at the customary lower frequencies, which provides a new source of information.[27]

Concurrently, new sources of synchrotron radiation are coming on-line. In the United States the Advanced Light Source at the Lawrence Berkeley Laboratory is now operational. The Advanced Photon Source at Argonne National Laboratory is under construction. These two sources will enormously expand the availability of X-ray photons for XAS studies. The European Synchrotron Radiation Facility at Grenoble is now operational, providing a significant expansion in European facilities. Other machines in Europe and Asia are also becoming available. To complement these facilities new methods of employing XAS, such as X-ray MCD, for structural elucidation are being developed[28] (see also [6]).

In conclusion I have endeavored to present a broad but superficial survey of the field of biological EPR as it has been practiced in the half-

[24] K. T. Knecht and R. P. Mason, *Arch. Biochem. Biophys.* **303,** 185 (1993).

[25] D. E. Budil and M. C. Thurnauer, *Biochim. Biophys. Acta* **1057,** 1 (1991).

[26] H. A. Frank and C. A. Violette, *Biochim. Biophys. Acta* **976,** 222 (1989).

[27] S. Un, L.-C. Brunel, T. M. Brill, J.-L. Zimmermann, and A. W. Rutherford, *Proc. Natl. Acad. Sci. U.S.A.* **91,** 5262 (1994).

[28] J. Van Elp, S. J. George, J. Chen, G. Peng, C. T. Chen, G. Tjeng, G. Meigs, H.-J. Lin, Z. H. Zhou, M. W. W. Adams, B. G. Searle, and S. P. Cramer, *Proc. Natl. Acad. Sci. U.S.A.* **90,** 9664 (1993).

century since its discovery in 1943 by Zavoisky.[29] The aim has been to provide a background sufficient to permit the novice to determine if the techniques can provide information useful to the solution of the problem at hand.

Acknowledgments

This work was supported by the Director, Office of Energy Research, Offices of Health Effects Research and Basic Energy Sciences, Energy Biosciences Division of the U.S. Department of Energy, under Contract No. DE-AC03-76SF00098, and the National Science Foundation (Grant DMB91-04104).

[29] E. Zavoisky, *J. Phys. U.S.S.R.* **9**, 221 (1945).

[22] Electron Paramagnetic Resonance Spectroscopy

By Gary W. Brudvig

Introduction

Any species containing one or more unpaired electrons can, in principle, be studied by electron paramagnetic resonance (EPR) spectroscopy. EPR spectroscopy has been an especially important tool in studies of biological systems that naturally contain paramagnetic species such as substrate radicals, electron-transfer centers, or metal ions. Frequently, these redox-active centers are at the active sites of proteins. One advantage of EPR spectroscopy is its ability to focus on the paramagnetic active sites without interference from the rest of the diamagnetic protein. A variety of applications of EPR spectroscopy have been used to obtain highly detailed information on the structure of paramagnetic centers. EPR spectroscopy can also be used to obtain dynamic information such as the kinetics of reactions involving paramagnetic species or the time scales of motion of paramagnetic molecules. In addition, the intensity of an EPR signal provides a direct measure of the concentration of the paramagnetic species, and quantitative analyses do not require independent information about extinction coefficients. However, the conditions under which EPR spectra can be observed, or even the ability to observe an EPR spectrum, vary depending on the type of paramagnetic species. A particular complication of biological systems is that most samples contain a high proportion of water which, owing to high dielectric loss, places severe restrictions on measurements at physiological temperatures. Nonetheless, EPR spectroscopy has been widely used to study all types of biomolecules. Of course, not all biomolecules contain odd-electron species. Diamagnetic systems are still amenable to EPR spectroscopic study by using spin labels. In this chapter, however, the focus is on those systems that naturally contain paramagnetic species. The applications of spin labels are covered elsewhere in this volume (G. L. Millhauser, W. R. Fiori and S. M. Miick, [24]).

The aim of this review is to provide an introduction to the basic principles, sample requirements, and applications of EPR spectroscopy to biological systems. The emphasis is on the factors that are important for measurements on conventional X-band EPR spectrometers. Interpretation of spectra or specialized techniques are not discussed. Information on these topics and more in-depth treatments can be found in a number of monographs on EPR spectroscopy.[1–6]

Basic Principles of Electron Paramagnetic Resonance Spectroscopy

The technique of EPR spectroscopy relies on the interaction of an applied magnetic field with the magnetic dipole moment of an unpaired electron. This Zeeman interaction splits the spin energy levels, as shown in Fig. 1. At a particular magnetic field, transitions between two spin states are induced by the magnetic field component of electromagnetic radiation at the appropriate energy. One could achieve resonance by varying either the frequency of the electromagnetic radiation or the magnitude of the static magnetic field. For EPR spectroscopy, normally the magnetic field is varied while keeping the frequency of electromagnetic radiation constant. Typically, a frequency of about 9 GHz (X band) is used, which lies in the microwave region, although other frequencies from 3 to 35 GHz (or even higher) are also used occasionally. The magnetic fields needed for EPR measurements at X-band frequencies are in the range of 0–10,000

[1] A. Abragam and B. Bleaney, "Electron Paramagnetic Resonance of Transition Ions." Oxford Univ. Press (Clarendon), Oxford, 1970.

[2] N. M. Atherton, "Electron Spin Resonance: Theory and Applications." Halsted Press, New York, 1973.

[3] A. J. Hoff (ed.), "Advanced EPR: Applications in Biology and Biochemistry." Elsevier, Amsterdam, 1989.

[4] J. R. Pilbrow, "Transition Ion Electron Paramagnetic Resonance." Oxford Univ. Press (Clarendon), London, 1990.

[5] M. Symons, "Chemical and Biochemical Aspects of Electron-Spin Resonance Spectroscopy." Wiley, New York, 1978.

[6] J. A. Weil, J. R. Bolton, and J. E. Wertz, "Electron Paramagnetic Resonance: Elementary Theory and Practical Applications." John Wiley and Sons, 1994.

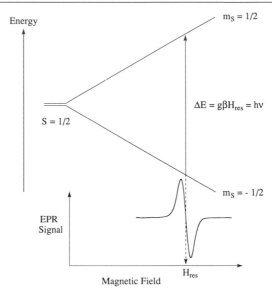

FIG. 1. Schematic illustration of the splitting of the electron spin states by a magnetic field for the case when $S = \frac{1}{2}$. When the splitting of the states, ΔE, equals the energy of the electromagnetic radiation, $h\nu$, an EPR signal is observed. The graph at bottom shows the EPR signal from a light-induced chlorophyll radical in photosystem II [J. C. de Paula, J. B. Innes, and G. W. Brudvig, *Biochemistry* **24**, 8114 (1985)]. (The magnetic field axis is not to scale in the EPR spectrum of the chlorophyll radical: H_{res}, 3302 G; maximum-to-minimum line width, 10 G; and ν, 9.07 GHz.)

gauss. These rather low fields are achieved and varied by using an electromagnet.

The Zeeman energy of the spin states depends on the magnitude of the applied magnetic field and a proportionality constant called the g factor [Eq. (1)]:

$$\text{Zeeman energy} = g\beta H m_s \tag{1}$$

where β is the electron Bohr magneton, H is the magnetic field, and m_s is the electron spin quantum number which can be either $+1/2$ or $-1/2$ when the paramagnetic species contains one unpaired electron ($S = 1/2$). The free-electron g factor is isotropic and known to great accuracy; its value is approximately 2.0023. For paramagnetic molecules or ions, however, the g factor is a tensor quantity, meaning that its value depends on the orientation of the molecule in the magnetic field. Thus, the magnetic field at which resonance occurs, H_{res}, will extend over a range of values depending on the maximum and minimum magnitudes of the g factor. To compare measurements taken at different frequencies, the values of H_{res} are normally converted to g factors, as in shown in Fig. 1 and Eq. (2):

$$g = h\nu/\beta H_{res} = 714.484 \, [\nu \, (\text{GHz})]/[H_{res} \, (\text{gauss})] \qquad (2)$$

where h is Planck's constant and ν is the microwave frequency. The amount of g anisotropy depends on the strength of spin–orbit coupling. Spin–orbit coupling is small for organic radicals. Consequently, organic radicals have g factors very close to the free-electron g factor and exhibit small g anisotropy (see the spectra in Figs. 1 and 2). However, spin–orbit coupling can be large for transition metal ions, leading to g factors that often deviate significantly from the free-electron g factor (see the spectrum in Fig. 3).

Some of the most important structural information available from an EPR spectrum is obtained from an analysis of the hyperfine splittings produced by nuclear spins that are coupled to the unpaired electron. The nuclear hyperfine interactions can produce EPR spectra rich in structure. Analysis of this hyperfine structure provides a detailed picture of the types and numbers of nuclear spins seen by the unpaired electron.

Each nuclear spin will split the EPR line into $2I + 1$ hyperfine lines, where I is the spin of the nucleus. In the case of n equivalent nuclei of spin I, the EPR line will be split into $2nI + 1$ hyperfine lines. An example is shown in Fig. 4. In this spectrum from a cluster of three Mn^{4+} ions, the splittings are from the ^{55}Mn nuclear spins ($I = 5/2$); the spectrum

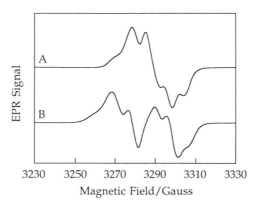

FIG. 2. EPR spectra of tyrosine radicals: (A) the stable tyrosine radical, $Tyr_D\cdot$, in photosystem II (65 μM) in a frozen solution containing 30% (v/v) ethylene glycol in water; and (B) the stable tyrosine radical in the B2 subunit of *Escherichia coli* ribonucleotide reductase (50 μM) in a frozen solution containing 10% (v/v) ethylene glycol in water. Conditions: temperature, 8.0 K; microwave frequency, 9.05 GHz; magnetic field modulation amplitude, 2 G; magnetic field modulation frequency, 100 kHz; microwave power, 0.7 μW.

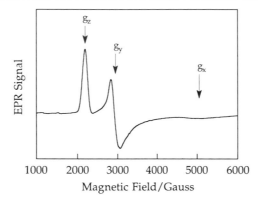

FIG. 3. EPR spectrum of a frozen solution of horse heart cytochrome c (2.4 mM): g_x, 1.24; g_y, 2.24; and g_z = 3.06 [I. Salmeen and G. Palmer, *J. Chem. Phys.* **48,** 2049 (1968)]. Conditions: temperature, 7.0 K; microwave frequency, 9.05 GHz; magnetic field modulation amplitude, 20 G; magnetic field modulation frequency, 100 kHz; microwave power, 30 μW.

can be understood in terms of 66 overlapping peaks arising from a large hyperfine splitting from one of the Mn^{4+} ions (giving six lines) and a smaller hyperfine splitting from the remaining two equivalent Mn^{4+} ions (splitting each of the six lines into an eleven-line pattern).

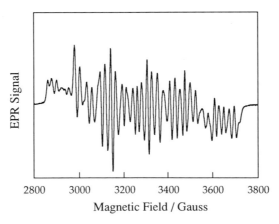

FIG. 4. EPR spectrum of [$Mn_3(O)_4$(2,2'-bipyridine)$_4$($H_2O)_2$]($ClO_4)_4$ (1 mM) in a frozen solution containing 30% (v/v) ethylene glycol in water [J. E. Sarneski, H. H. Thorp, G. W. Brudvig, R. H. Crabtree, and G. K. Schulte, *J. Am. Chem. Soc.* **112,** 7255 (1990)]. Conditions: temperature, 9.0 K; microwave frequency, 9.05 GHz; magnetic field modulation amplitude, 2 G; magnetic field modulation frequency, 100 kHz; microwave power, 5 μW.

The nuclear hyperfine interaction contains both an isotropic (Fermi contact) and an anisotropic (dipole–dipole) contribution. If a molecule is tumbling rapidly, as is often the case for small molecules in solution, the dipole–dipole contribution to the nuclear hyperfine interaction is averaged to zero and the g anisotropy is averaged such that all molecules have the same, average g value. As a result, EPR spectra of small molecules in solution often exhibit very well resolved nuclear hyperfine structure arising solely from the isotropic Fermi contact contribution. Such is the case for nitroxide spin labels that are not immobilized by binding to a macromolecule or for substrate radicals that have been released from an enzyme. On the other hand, the hyperfine splitting pattern of an immobilized and randomly oriented paramagnetic species depends on the orientation of the molecule in the magnetic field. The anisotropy of the hyperfine interaction and/or g factor often cause a loss of resolution of the nuclear hyperfine structure in EPR spectra of immobilized and randomly oriented paramagnetic species.

An immobilized, randomly oriented paramagnetic species exhibits an EPR line shape referred to as a "powder pattern." This pattern arises from the superposition of spectra from different orientations of the molecule with respect to the magnetic field, weighted by the probability of each orientation. An example of a rhombic (meaning that $g_x \neq g_y \neq g_z$) powder pattern EPR line shape is the spectrum in Fig. 3. The first-derivative EPR spectrum exhibits absorption-like positive and negative going peaks at the low- and high-field turning points, respectively. However, the absorption spectrum extends over the magnetic field range from 2000 to 6000 G. The principal g values are easily obtained from a powder pattern, as noted in Fig. 3.

Unfortunately, most paramagnetic centers or substrate radicals in biological systems are bound to macromolecules that do not tumble rapidly enough to average the anisotropic interactions. Therefore, most paramagnetic species of interest in biological systems are immobilized on the EPR time scale and exhibit EPR spectra typical of those observed in solids, even when the sample is in the liquid state.

Requirements for Measurement of Electron Paramagnetic Resonance Spectra

Temperature

The first consideration in making an EPR measurement is the temperature to use. Obviously, physiological processes normally occur in an aqueous environment and at temperatures above about 273 K. At these temperatures water is liquid, and this presents a difficulty for EPR measurements. Water absorbs microwaves strongly, as demonstrated by heating in microwave ovens. To measure an EPR spectrum from a sample containing liquid water, it is essential to minimize the absorption of microwaves by the water. This can be achieved by using a flat sample cell and a rectangular resonant microwave cavity or a cylindrical capillary sample cell and a cylindrical resonant microwave cavity.[7] These arrangements place the sample in a region having a node in the electric field component and a maximum in the magnetic field component of the electromagnetic wave. Because the heating of water is induced by the electric field component and the EPR transition is induced by the magnetic field component, the absorption of microwaves by water is minimized while still allowing measurement of the EPR spectrum.

The nature of the paramagnetic species of interest also may dictate the choice of the temperature. Often EPR signals are too broad to measure at physiological temperatures, owing to rapid spin relaxation rates which cause lifetime broadening of the EPR signals. This is generally the case if the paramagnetic species has significant g anisotropy or has $S > 1/2$. Therefore, most transition metal ions and high-spin species require cryogenic temperatures for EPR study.

Temperature also plays a role in EPR measurements with regard to the Boltzmann population of the various spin states. The intensity of an EPR signal is directly proportional to the population difference between the two levels in resonance, and this population difference is inversely proportional to the temperature. For measurements at X-band frequencies, the population difference is small at room temperature (roughly one part per thousand) but is much larger at liquid helium temperature (roughly one part in ten). In some cases, such as metalloproteins which have EPR spectra that extend over a very large magnetic field range, the gain in sensitivity at cryogenic temperatures is an important factor in the detection of EPR signals.

One advantage of cyogenic EPR measurements of biological samples is that frozen water does not absorb microwaves to a significant extent. Therefore, a larger sample can be used for cryogenic EPR measurements. However, it is preferable to use a solvent that forms a "glass" as it is cooled. The solvent for biological samples is usually pure water when EPR measurements are made at temperatures above 273 K, but pure water forms a "snow" when frozen. This is due to the formation of small crystallites of ice during the freezing process. Solutes are excluded from

[7] C. P. Poole, "Electron Spin Resonance: A Comprehensive Treatise on Experimental Techniques," 2nd Ed. Wiley, New York, 1983.

the ice crystallites, and this can induce a significant increase in the local concentration of dissolved species during the freezing process. For small paramagnetic molecules, this increase in local concentration can cause severe broadening of the EPR signals owing to spin–spin interactions. In the case of macromolecules, the dimensions of the molecules often prevent close approach of the paramagnetic sites. However, equilibria involving the binding of small solute molecules to active sites in proteins can be altered during freezing.[8] A variety of additives have been used to cause aqueous solutions to form a "glass" when frozen. Glycerol and ethylene glycol are most commonly used for biological samples (typically as 10–50% mixtures by volume with water).

Concentration and Sample Volume

The sensitivity of EPR spectroscopy is significantly higher than that for NMR spectroscopy owing to the larger gyromagnetic ratio of the electron. Concentrations below 0.1 mM are readily studied by EPR spectroscopy. This can be both an asset and a complication. It is possible to detect small concentrations of radicals that may be produced as intermediates in an enzymatic reaction or to study paramagnetic biomolecules that are not easily obtained in high concentrations. On the other hand, paramagnetic contaminants or side products may interfere. In studies of enzymatic reaction mechanisms, it is possible to be misled into favoring a pathway involving radical intermediates by observing radical species that are produced by side reactions or that arise from impurities. Adventitious metal ions, particularly Mn^{2+}, Cu^{2+}, and Fe^{3+}, also frequently interfere, and care must be taken to exclude these ions from the solvents and reagents used to prepare samples and from the sample tubes used for measurements; EPR sample tubes are made of quartz, in part because less pure glasses, such as Pyrex, contain Fe^{3+} and other metal ions. O_2, which has a triplet ground state, sometimes interferes with low-temperature EPR measurements.[9] Adventitious paramagnetic species can also be introduced by smoke or fingerprints.

The minimum number of spins needed for detection of an EPR signal depends on a number of factors including the sensitivity of the spectrometer, the type of microwave resonator, the line width of the EPR signal, the conditions for measurement, and the volume of the sample that can be used. Commercial EPR instruments have quoted sensitivities of about $10^{10} \Delta H$ spins (with a signal-to-noise ratio of 1 : 1) at room temperature

[8] A. Yang and A. S. Brill, *Biophys. J.* **59**, 1050 (1991).
[9] A.-F. Miller and G. W. Brudvig, *Biochim. Biophys. Acta* **1056**, 1 (1991).

under favorable conditions, where ΔH is the line width of the EPR signal in gauss. For measurements of solids or of low dielectric solutions, a sample volume of about 0.3 ml can be used. These factors translate into a minimum sample concentration of about $10^{-10}\Delta H M$. This minimum is rarely approached for biological samples. Measurements of aqueous solutions with a flat cell and a standard X-band rectangular cavity resonator require a sample volume of about 0.1 ml; spin concentrations on the order of a few micromolar (if the lines are sharp) or higher can be measured. For measurements of frozen solutions or solids, somewhat lower concentrations can be studied, depending on the line width of the signal. The recent introduction of loop–gap resonators has allowed EPR measurements to be made on aqueous samples as small as 1.5 μl with sensitivity that is comparable to that obtained with a flat cell and a standard rectangular cavity resonator.[10]

Instrumental Factors

One of the first questions asked about EPR spectra is, Why are the signals recorded as the first derivative? The first-derivative signal arises from the use of magnetic field modulation and phase-sensitive detection of the modulated EPR signal in order to improve the signal-to-noise ratio (S/N).[7] Hence, two instrumental factors in an EPR measurement are the frequency and amplitude of the magnetic field modulation. Typically, the frequency of the magnetic field modulation is held constant and the amplitude of the modulation is varied. The most commonly used magnetic field modulation frequency is 100 kHz; the high frequency is chosen to minimize noise components inherent in the detector. In some cases, such as when the EPR lines are extremely narrow (not a problem for biological samples) and/or when the spin relaxation times are long, lower modulation frequencies may be required. The EPR signal intensity increases linearly with the magnetic field modulation amplitude provided that the modulation amplitude is small compared to the line width of the EPR signal. However, the EPR signal is distorted if the magnetic field modulation amplitude is too large. Sometimes it is desirable to use a larger than optimal magnetic field modulation amplitude in order to obtain better S/N at the expense of resolution. For maximal S/N without significant distortion of the EPR signal, the magnetic field modulation amplitude should be chosen to be about one-third of the line width of the EPR signal.

Another instrumental variable is the microwave power. The EPR signal intensity increases in proportion to the square root of the microwave power until the onset of saturation of the spin system. [As shown in Fig. 5, $\log(S/P^{1/2})$ is constant in the unsaturated limit.] Microwave power

[10] W. L. Hubbell, W. Froncisz, and J. Hyde, *Rev. Sci. Instrum.* **58**, 1879 (1987).

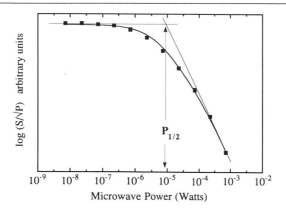

FIG. 5. Progressive microwave power saturation plot of the tyrosine radical in the B2 subunit of *E. coli* ribonucleotide reductase. Conditions were as in Fig. 2B.

saturation occurs when the rate of absorption of microwaves exceeds the rate at which the system returns to equilibrium. The microwave power at which saturation will occur varies widely depending on the type of paramagnetic species (in general, organic radicals are much more easily saturated than transition metal ions) and the temperature (as the temperature is lowered, the EPR signal will be more easily saturated). Although one wants to maximize the EPR signal intensity, it is important to use microwave powers that do not cause saturation if quantitative measurements are to be made (see next section). High microwave powers will also broaden the EPR signals, leading to a loss of resolution. In many samples, broad background signals are present that are not easily saturated. These signals may obscure the signals of interest if high microwave powers are used that saturate the signals of interest. However, comparing measurements taken at low and high microwave powers can be helpful in distiguishing overlapping EPR signals, if the signals have different microwave power saturation properties.

An analysis of the microwave power saturation of an EPR signal provides a measure of the spin relaxation times (referred to as the method of progressive microwave power saturation). The microwave power at half-saturation, $P_{1/2}$, is a parameter that is frequently measured. Equation (3) gives an empirical expression that has been used to evaluate $P_{1/2}$ values in biological samples[11]:

$$\text{EPR signal} = S = KP^{1/2}/[1 + (P/P_{1/2})]^{b/2} \tag{3}$$

[11] H. Beinert and W. H. Orme-Johnson, *in* "Magnetic Resonance in Biological Systems" (A. Ehrenberg, B. G. Malmström, and T. Vänngård, eds.), p. 221. Pergamon, Oxford, 1967.

where K is a constant, P is the microwave power, and b is the inhomogeneity parameter[12,13] that can have a value from 1 to 3. The value $P_{1/2}$ is proportional to $1/(T_1 T_2)$, so that as the spin relaxation times become shorter, higher microwave powers are required to saturate the EPR signal. Equation (3) can be rearranged to Eq. (4), giving an expression that allows $P_{1/2}$ to be determined graphically:

$$\log(S/P^{1/2}) = \log(K) - (b/2)\log[1 + (P/P_{1/2})] \tag{4}$$

If $\log(S/P^{1/2})$ is plotted against $\log(P)$, two linear regions are obtained (Fig. 5). When $P << P_{1/2}$, $\log(S/P^{1/2})$ will be constant and equal to $\log(K)$; when $P >> P_{1/2}$, the plot will have a slope of $-b/2$. The value of $P_{1/2}$ is obtained by extrapolating to determine the microwave power at which the two lines intersect. Changes in $P_{1/2}$ values are diagnostic of electron spin–spin interactions, and $P_{1/2}$ measurements have been used to obtain information about the location of paramagnetic sites in proteins.[14] However, it is difficult to extract accurate values for the spin relaxation times from progressive microwave power saturation measurements.[15] For accurate measurements of spin relaxation times, pulsed EPR methods are required.

Quantitative Measurements

Quantitative analyses are normally done by comparing the doubly integrated area of the first-derivative EPR signal of interest to that of a concentration standard.[6] It is crucial that the EPR spectra of both the sample and the standard are recorded under nonsaturating conditions (i.e., when the EPR signal intensities are proportional to $P^{1/2}$) if quantitative measurements are desired. Otherwise, the EPR signal intensities will depend on the spin relaxation times. Under nonsaturating conditions, the EPR signal intensity is directly proportional to the concentration of spins. Provided the EPR spectra are recorded under the same conditions (temperature, sample volume, magnetic field modulation frequency and amplitude, and microwave power), the ratio of the doubly integrated areas of the standard and sample will be equal to the ratio of their concentrations. Accuracies of about 10% are typical. One advantage of quantitative EPR measurements is that the signal intensities depend only on the g values for $S = 1/2$ species. Therefore, it is not crucial that the concentration standard be chemically or spectrally similar to the paramagnetic species

[12] H. Rupp, K. K. Rao, D. O. Hall, and R. Cammack, *Biochim. Biophys. Acta* **537,** 255 (1978).
[13] M. Sahlin, A. Gräslund, and A. Ehrenberg, *J. Magn. Reson.* **67,** 135 (1986).
[14] J. B. Innes and G. W. Brudvig, *Biochemistry* **28,** 1116 (1989).
[15] W. F. Beck, J. B. Innes, J. B. Lynch, and G. W. Brudvig, *J. Magn. Reson.* **91,** 12 (1991).

of study. It should be noted, however, that quantitative measurements of $S > 1/2$ species can be complicated if large zero-field splittings are present (see Fig. 7). Three commonly used concentration standards, whose concentrations can be verified by optical spectroscopy, are copper(II) sulfate, potassium peroxylamine disulfonate $[K_2NO(SO_3)_2]$, and TEMPO (2,2,6,6-tetramethyl-1-piperidinyloxy).

For EPR studies of biological systems, it is common that a sample contains multiple paramagnetic centers, especially for proteins involved in electron transfer. Although portions of the EPR spectrum from each species are usually resolved, the signals often have significant overlap. In these cases, it is not possible to perform a double integration of the signal of each component. Fortunately, it is possible to calculate the total doubly integrated area of a single species by measuring the area of a resolved "absorption-like" peak at a single turning point in the first-derivative spectrum (such as the peak at $g = g_z$ in Fig. 3), provided all three g values are known.[16] This method has been widely used for quantitative EPR measurements in biological systems.

Biological Applications of Electron Paramagnetic Resonance Spectroscopy

The scope of biological applications of EPR spectroscopy, especially to metalloproteins, is beyond what can be covered in one chapter. In this section, a few specific applications of EPR spectroscopy to biological systems are presented in order to contrast the EPR properties of organic radicals and metal ions and to provide some insight into the information that can be obtained. There is, of course, a large literature on the use of EPR spectroscopy to study biological systems, and additional applications can be found in other publications. The series of monographs on biological magnetic resonance[17] and free radicals in biology[18] are particularly helpful.

Organic Radicals in Proteins

A variety of types of organic paramagnetic centers occur in proteins. These include cofactor radicals, amino acid residue radicals, substrate radicals, radicals induced by ionizing radiation, spin labels, and also excited triplet state species. All have been widely studied by using EPR spectroscopy, but the following discussion focuses on a few examples of cofactor and amino acid residue radicals.

[16] R. Aasa and T. Vänngård, *J. Magn. Reson.* **19**, 308 (1975).
[17] L. J. Berliner and J. Reuben (eds.), *Biol. Magn. Reson.* **1–12** (1978–1993).
[18] W. A. Pryor (ed.), *Free Radicals Biol.* **1–6** (1976–1984).

EPR spectra of chlorophyll and tyrosine radicals are shown in Figs. 1 and 2. In these examples, the radicals are immobilized in the EPR time scale owing to the large size of the protein. Although the EPR spectra shown in Figs. 1 and 2 were measured on frozen samples, the EPR spectra are unchanged if the measurements are made on liquid samples, except for some line broadening caused by faster spin relaxation rates.

Because of the small contribution from spin–orbit coupling, organic radicals exhibit nearly isotropic EPR spectra with g values close to the free-electron g value of 2.0023. In fact, it is necessary to use five significant figures to differentiate their g factors. Because of the similar g values, spectral overlap will usually be a problem if more than one organic radical contributes to the EPR spectrum. However, accurate measurement of the g value can be very useful for identification of the radical. For example, the g value of carbon-centered radicals are typically very close to the free-electron g value, whereas molecules having unpaired spin density on oxygen, nitrogen, or sulfur atoms have more variable g values that are often somewhat larger than the free-electron g value. This is illustrated by the g value of the carbon-centered chlorophyll radical (Fig. 1, $g = 2.0024$[19]) compared to those of the tyrosine radicals (Fig. 2, $g = 2.0047$[20,21]).

The g values provide important information on the electronic structure and symmetry of the radical. However, the most useful structural information in the EPR spectra of radicals comes from the nuclear hyperfine splittings. These splittings provide a detailed picture of the nuclear spins that are near the unpaired electron. Analysis of the hyperfine interactions can be used for chemical identification of the radical and, in the case of aromatic radicals, for determining the extent of delocalization of the unpaired electron. In addition, analysis of nuclear hyperfine couplings to residues near the radical can provide information on the protein environment around the paramagnetic center. Unfortunately, for biological systems, the nuclear hyperfine couplings are usually only partially resolved, or are not resolved at all; g anisotropy is also not typically resolved. For example, the EPR spectrum of a chlorophyll radical (Fig. 1) has a Gaussian-derivative line shape with no resolved structure, and the EPR spectra of tyrosine radicals (Fig. 2) show only partially resolved hyperfine splittings. How can the information in these spectra be extracted? Several approaches have been used.

[19] J. C. de Paula, J. B. Innes, and G. W. Brudvig, *Biochemistry* **24,** 8114 (1985).
[20] D. H. Kohl, *in* "Biological Applications of Electron Spin Resonance" (H. M. Swartz, J. P. Bolton, and D. C. Borg, eds.), p. 213. Wiley (Interscience), New York, 1972.
[21] M. Sahlin, A. Gräslund, A. Ehrenberg, and B. Sjöberg, *J. Biol. Chem.* **257,** 366 (1982).

It is possible to determine the principal g values (g_x, g_y, and g_z) of organic radicals by making EPR measurements at higher microwave frequencies. As the microwave frequency is increased, the turning points associated with each principal g value split apart. (Figure 3 illustrates a case where all three turning points associated with the principal g values are resolved.) EPR measurements, made by using microwave frequencies of 90 GHz or higher, have resolved all three principal g values of the oxidized special pair of bacteriochlorophylls, $(BChl)_2^+$, that functions as the primary electron donor in the photosynthetic reaction center protein from purple nonsulfur bacteria,[22] as wells the tyrosine radical in ribonucleotide reductase.[23]

A multipronged approach has been used to assign the nuclear hyperfine interactions in (bacterio)chlorophyll and tyrosine radicals. The assignment of the largest hyperfine couplings to protons was made by comparing EPR spectra on nonisotope-labeled and perdeuterated (or specifically deuterated) samples. In the case of the $(BChl)_2^+$ EPR signal, the line width is reduced from 9.5 to 4.0 G on deuteration,[24] showing that the line width is mainly due to unresolved proton hyperfine splittings. Similarly, substitution of the tyrosine protons by deuterium leads to loss of the hyperfine structure and a narrowing of the EPR signals from the tyrosine radicals in photosystem II[25] and ribonucleotide reductase.[26] Pulsed and double-resonance EPR measurements have also been made to resolve and assign the weak proton hyperfine couplings.[27,28]

On the basis of specific deuteration experiments, the partially resolved hyperfine structure in the EPR spectra of the tyrosine radicals has been assigned to the protons at positions 3 and 5 of the tyrosine ring and one of the methylene protons adjacent to the ring (Fig. 6).[25] The unpaired electron spin is delocalized in the π system of aromatic radicals, and the proton hyperfine coupling is proportional to unpaired electron spin density

[22] R. Klette, J. T. Törring, M. Plato, K. Möbius, B. Bönigk, and W. Lubitz, *J. Phys. Chem.* **97**, 2015 (1993).

[23] G. J. Gerfen, B. F. Bellew, S. Un, J. M. Bollinger, Jr., J. Stubbe, R. G. Griffin, and D. J. Singel, *J. Am. Chem. Soc.* **115**, 6420 (1993).

[24] D. H. Kohl, J. Townsend, B. Commoner, H. L. Crespi, R. C. Dougherty, and J. J. Katz, *Nature (London)* **206**, 1105 (1965).

[25] B. A. Barry, M. K. El-Deeb, P. O. Sandusky, and G. T. Babcock, *J. Biol. Chem.* **265**, 20139 (1990).

[26] A. Gräslund, M. Sahlin, and B. Sjöberg, *Environ. Health Perspect.* **64**, 139 (1985).

[27] C. J. Bender, M. Sahlin, G. T. Babcock, B. A. Barry, T. K. Chandrashekar, S. P. Salowe, J. Stubbe, B. Lindström, L. Petersson, A. Ehrenberg, and B. Sjöberg, *J. Am. Chem. Soc.* **111**, 8076 (1989).

[28] R. G. Evelo, A. J. Hoff, S. A. Dikanov, and A. M. Tyryshkin, *Chem. Phys. Lett.* **161**, 479 (1989).

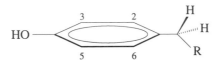

FIG. 6. Structure of tyrosine with the numbering scheme used.

on the carbon to which the proton is bound. For tyrosine radicals, the unpaired spin is primarily centered on positions 1, 3, and 5 of the ring, which is characteristic of an odd-alternate hydrocarbon. This accounts for the large hyperfine couplings for the protons at positions 3 and 5 and small coupling for those at positions 2 and 6. The magnitude of the hyperfine splittings from the methylene protons depends on unpaired electron spin density on the adjacent ring carbon and the dihedral angle of the CH bonds of the methylene group relative to the plane of the aromatic ring. It is interesting that the EPR spectra of the tyrosine radicals in photosystem II and ribonucleotide reductase are rather different (Fig. 2). The differences arise primarily from different hyperfine couplings to the methylene protons owing to different conformations of the tyrosine side chain in the proteins. A detailed EPR study of the tyrosine radicals in photosystem II and ribonucleotide reductase has been published.[29]

Metalloproteins

The EPR spectra of metalloproteins vary greatly depending on the metal ion, spin state, ligation, and number of metal ions clustered together. This is a great advantage in resolving and assigning the EPR spectra of different paramagnetic metal centers. In general, a particular spin and oxidation state of a metal ion will have characteristic EPR spectral properties. On the other hand, the EPR spectroscopy of transition metal ions is more complex and varied than that of organic radicals. We first consider the factors that are most important in determining the EPR spectral signature of an isolated metal ion in a metalloprotein.

If the metal ion has more than one unpaired electron, the spin–spin interactions between the different unpaired electrons will usually be the most important factor in determining the EPR properties of the metal ion. The spin–spin interactions produce splittings of the electron spin states in the absence of a magnetic field (Fig. 7). These zero-field splittings (ZFS) are often large compared to the Zeeman splitting produced by the applied magnetic field, and they depend greatly on the symmetry of the metal ion. If the metal ion has a highly symmetric coordination, the ZFS will be

[29] C. W. Hoganson and G. T. Babcock, *Biochemistry* **31**, 11874 (1992).

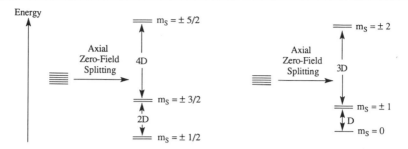

FIG. 7. Schematic illustration of the splitting of the electron spin states in zero magnetic field by an axial zero-field splitting (ZFS) interaction for the cases when $S = \frac{5}{2}$ and when $S = 2$. The ZFS interaction is defined by two constants, D and E, which give the axial and rhombic contributions, respectively. E equals 0 in the two examples shown.

small compared to the Zeeman splittings. However, the binding sites for metal ions in proteins are typically not highly symmetric. Therefore, large ZFS are normally present for high-spin metal ions bound to proteins.

Because of ZFS interactions, the EPR properties of integral-spin and half-integral-spin metal ions are very different. As shown in Fig. 7, the ZFS interaction splits the spin sublevels according to the value of the spin quantum number, m_s. However, the selection rule for the allowed EPR transitions is that $\Delta m_s = \pm 1$. For integral-spin species, the ZFS interaction splits apart the spin levels between which the allowed EPR transitions occur. As a result, if the zero-field splittings are large, the spin levels for the allowed transitions are not brought into resonance at the magnetic fields and microwave frequencies typically used with conventional EPR instrumentation. On the other hand, the ZFS interaction does not split the $m_s = \pm\frac{1}{2}$ states of a half-integral-spin species. Consequently, the allowed transition between the $m_s = \pm\frac{1}{2}$ states cannot be made inaccessible by the ZFS interaction. This leads to a useful rule of thumb for EPR spectroscopy of transition metal ions. In general, EPR spectra are not observed from integral-spin metal ions, but EPR spectra are observed from half-integral-spin metal ions.

A good example of this rule is the different EPR characteristics of high-spin Fe^{2+} (an integral-spin species with $S = 2$) versus high-spin Fe^{3+} (a half-integral-spin species with $S = 5/2$). High-spin Fe^{2+} is not usually observed by EPR spectroscopy, despite the fact that it is paramagnetic. (In some cases, it is possible to observe parallel-mode EPR signals from high-spin Fe^{2+} that arise from $\Delta m_s = \pm 2$ or $\Delta m_s = \pm 4$ transitions.[30]) On

[30] M. P. Hendrick and P. G. Debrunner, *J. Magn. Reson.* **78**, 133 (1988).

the other hand, high-spin Fe^{3+} exhibits two characteristic types of EPR signals: one with a turning point at $g = 4.3$ arising from Fe^{3+} with a rhombic symmetry and a second with turning points at $g = 6$ ($g_\perp$) and $g = 2$ ($g_\parallel$) arising from Fe^{3+} with an axial symmetry. An EPR signal at $g = 4.3$, arising from adventitious rhombic Fe^{3+}, is very commonly observed in low-temperature EPR spectra of biological samples.[9]

For metal centers with one unpaired electron, the EPR spectral properties are mainly determined by the Zeeman and nuclear hyperfine interactions. The EPR spectroscopy of these $S = \frac{1}{2}$ species follows from what was discussed above about organic radicals, except that the g values are much more anisotropic owing to the larger contribution of spin–orbit coupling for metal ions. In many cases, such as low-spin Fe^{3+} (Fig. 3), the g values deviate significantly from the free-electron g value. In addition, for metal ions with a nuclear spin, the nuclear hyperfine splittings can be quite large because the unpaired spin is localized on the metal ion (Fig. 4).

Figure 3 shows the EPR spectrum of oxidized cytochrome c. The spectrum arises from the low-spin Fe^{3+} of the heme ($S = \frac{1}{2}$). Owing to spin–orbit coupling, the spectrum is very anisotropic, with a large separation of the turning points arising from the three principal g values. The spin–orbit coupling interaction depends on the energies of the electronic states. Therefore, the g values are quite sensitive to the splitting of the $3d$ orbitals which, in turn, depends on the coordination of the Fe^{3+}. The theory to explain the EPR signals of low-spin Fe^{3+} has been well established and is reviewed by Palmer[31] and Pilbrow.[4] EPR spectroscopy has been used to obtain information on the electronic structure of low-spin Fe^{3+} and also to obtain information on the types and orientations of the ligands. For example, it is possible to infer the types of axial ligands[32] and the orientation of axial imidazole ligands[33] from the g values of low-spin Fe^{3+} in heme proteins.

In many metalloproteins, two or more metal ions are closely associated. The strength of the exchange interactions between the associated metal ions is the most important factor in determining the EPR properties of these systems. The exchange interactions arise from overlap of the magnetic orbitals (the orbitals containing the unpaired electrons). These

[31] G. Palmer, in "The Porphyrins" (D. Dolphin, ed.), p. 313. Academic Press, New York, 1979.

[32] W. E. Blumberg and J. Peisach, in "Probes of Structure and Function of Macromolecules and Membranes" (B. Chance, T. Yonetani, and A. S. Mildvan, eds.), p. 215. Academic Press, New York, 1971.

[33] F. A. Walker, B. H. Huynh, W. R. Scheidt, and S. R. Osvath, J. Am. Chem. Soc. **108**, 5288 (1986).

interactions can favor pairing the electrons (antiferromagnetic coupling) or favor the electrons remaining unpaired (ferromagnetic coupling). In cases of strong coupling, it is appropriate to consider the resulting spin of the system, rather than the spins of each individual metal ion. Figure 4 shows the spectrum of a cluster of three high-spin Mn^{4+} ions. Each of the Mn^{4+} ions has $S = \frac{3}{2}$, but, owing to strong antiferromagnetic exchange interactions, the overall spin of the complex is $\frac{1}{2}$. The g factors and hyperfine coupling constants of each individual Mn^{4+} ion are modulated by the exchange interaction.[34] In the example shown in Fig. 4, the exchange interaction leads to a large ^{55}Mn hyperfine splitting from one of the Mn^{4+} ions.[35]

Conclusion

Electron paramagnetic resonance spectroscopy is a valuable technique for studies of paramagnetic species in biological systems because the paramagnetic centers can be studied directly without interference from the rest of the diamagnetic protein. In many applications, EPR spectroscopy is used for qualitative analysis to determine when (or if) paramagnetic species are produced in a redox reaction or whether various treatments perturb the structure of a paramagnetic active site. EPR spectroscopy can also be used for quantitative analysis to determine the concentration of a paramagnetic species or to measure changes in concentration during a redox titration or reaction. These types of studies can be productive even without a detailed interpretation of the EPR spectrum. Other applications rely on the use of EPR spectroscopy to determine the chemical identity, structure, and dynamics of a paramagnetic species. These studies involve a detailed analysis of the g factors, zero-field splittings, nuclear hyperfine, exchange, dipole–dipole, and possibly other interactions that contribute to the EPR spectrum.

This chapter has focused on the use of conventional X-band EPR spectrometers to measure EPR spectra of immobilized paramagnetic centers in proteins. However, a variety of other EPR methods, such as pulsed, double-resonance, and high-frequency measurements, are being increasingly applied to study biological systems[3] (see V. J. DeRose and B. M. Hoffman, this volume [23]). These modern methods, in conjunction with conventional EPR measurement, provide a powerful array of techniques

[34] G. W. Brudvig, in "Advanced EPR: Applications in Biology and Biochemistry" (A. J. Hoff, ed.), p. 839. Elsevier, Amsterdam, 1990.

[35] J. E. Sarneski, H. H. Thorp, G. W. Brudvig, R. H. Crabtree, and G. K. Schulte, J. Am. Chem. Soc. 112, 7255 (1990).

that can yield a highly detailed picture of the electronic structure, chemical structure, and chemical environment of paramagnetic active sites in proteins.

Acknowledgments

I thank Geoffrey Grove, Donald Hirsh, Dionysios Koulougliotis, and Rajesh Manchanda for recording some of the EPR spectra reproduced in this chapter, for helping to prepare the figures, and for critically reading the manuscript. This work was supported by the National Institutes of Health.

[23] Protein Structure and Mechanism Studied by Electron Nuclear Double Resonance Spectroscopy

By Victoria J. DeRose and Brian M. Hoffman

Introduction

Many molecules of biological interest can be prepared in a state that has a net unpaired electron spin and that exhibits an electron paramagnetic resonance (EPR) signal. The EPR signal from such biomolecules can be associated with a metal center, an organic-based radical such as a porphyrin, flavin, or aromatic amino acid, or indeed a system involving spin coupling between two or more such centers. A wealth of information about a paramagnetic center can be obtained by measuring the hyperfine interaction between that unpaired *electronic* spin and the *nuclear* spins of atoms associated with the center. The measurement of this hyperfine interaction can answer questions about the EPR-active site such as the following: What atoms comprise the EPR-active center? For a metal center, what are the ligands? What nuclei surround the EPR-active center? Does the center bind substrate or inhibitor molecules, and in what manner?

In principle every electron–nuclear interaction perturbs the electron spin energy levels that give rise to the EPR signal and can result in splittings of the EPR signal that could be analyzed to give the desired hyperfine couplings. In practice, however, such splittings often are very small in relation to the line width of the signal and cannot be resolved (Fig. 1). Electron nuclear double resonance (ENDOR) spectroscopy recovers this information by measuring the nuclear magnetic resonance (NMR) spectra of nuclei associated with the electron spin, as detected by changes in the EPR spectrum of that spin. This technique was first established by Feher

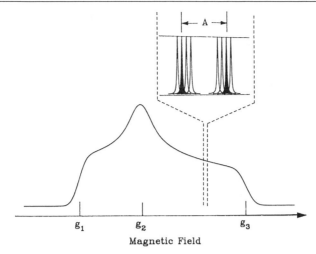

FIG. 1. Rhombic EPR absorption envelope illustrating spin packets from individual centers that add together to give the inhomogeneously broadened powder EPR pattern. In particular, a pair of packets representing a hyperfine splitting of A for an $I = \frac{1}{2}$ nucleus is depicted. [Adapted from W. B. Mims and J. Peisach, *Biol. Magn. Reson.* **3**, 213 (1981).]

in 1959.[1] In an ENDOR spectrum, NMR frequencies are swept while the EPR signal is observed at a fixed magnetic field. Excitation of nuclear transitions results in changes of the populations of the nuclear sublevels. When the nuclei are associated (via the hyperfine coupling) with the electron spin giving rise to the EPR signal, this results in changes of the electronic spin populations that are manifest as perturbations in the EPR signal intensity. As a double resonance technique, ENDOR wins on two counts: direct stimulation of NMR transitions enhances the spectral resolution by orders of magnitude over that available by examining splittings in the EPR spectrum, and detecting the effect using the Boltzmann population difference of electron spin energy levels enhances the sensitivity by orders of magnitude over NMR.

What can we learn from the measured hyperfine interaction? For a metalloprotein, these interactions can be measured for the metal nuclei of the EPR center, for nuclei in ligands to the metal, and for nuclei of substrate or inhibitor molecules (Fig. 2). On the most basic level, the existence of a hyperfine coupling to a given nucleus *identifies* it as a part of the center and thus, for example, as a metal ligand. However, the nucleus giving rise to a hyperfine coupling need not be directly bound to the metal center; depending on the spin system, hyperfine couplings can

---

[1] G. Feher, *Phys. Rev.* **114**, 1219 (1959).

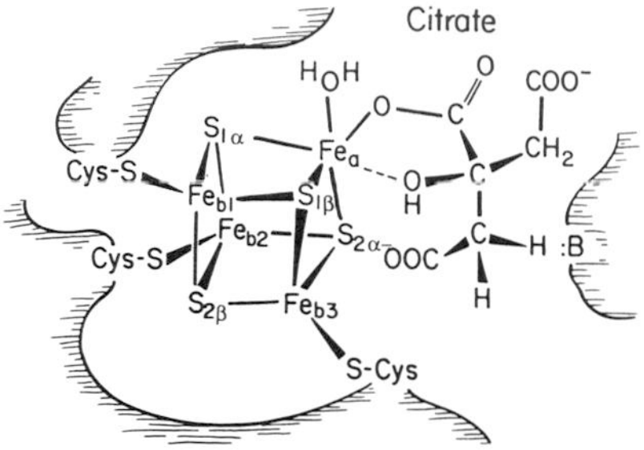

FIG. 2. Representation of ENDOR-derived information about the $[4Fe-4S]^+$ cluster of aconitase with substrate bound. As indicated, ENDOR studies showed four inequivalent ^{57}Fe sites and two pairs of ^{33}S atoms. Exchangeable protons and ^{17}O from bound hydroxide or water, and ^{17}O and ^{13}C from bound substrate, also were observed.

be measured from nuclei two or three bonds removed from the spin center. *Analysis* of the hyperfine coupling gives geometric information about ligand arrangements and deep insights into chemical bonding.

The following sections provide a more detailed description of the hyperfine interaction as measured by ENDOR spectroscopy, a description of ENDOR instrumentation, and the types of ENDOR experiments that can be performed. Finally, examples of the application of ENDOR spectroscopy to a variety of biomolecules are described. In this brief review many statements are made without reference; for details the reader is referred to the variety of more extensive works for the theory of EPR and hyperfine interactions[2] and reviews of applications of continuous wave (cw) and pulsed ENDOR and ESEEM (electron spin echo envelope modulation) techniques.[3]

[2] References for the ENDOR experiment include the following: A. Abragam and B. Bleaney, "Electron Paramagnetic Resonance of Transition Ions," 2nd Ed. Oxford Press (Clarendon), Oxford, 1970; A. Carrington and A. D. McLachlin, "Introduction to Magnetic Resonance with Applications to Chemistry and Chemical Physics." Harper & Row, New York, 1967; N. M. Atherton, "Electron Spin Resonance: Theory and Applications." Halstead Press, New York, 1973; L. Kevan and L. D. Kispert, "Electron Spin Double Resonance Spectroscopy." Wiley, New York, 1976; J. R. Pilbrow, "Transition Ion Electron Paramagnetic Resonance." Oxford Univ. Press (Clarendon), Oxford, 1990.

[3] Additional reviews of cw and pulsed ENDOR applications are found in Refs. 3a–f.

[3a] B. M. Hoffman, V. J. DeRose, P. E. Doan, R. J. Gurbiel, A. L. P. Houseman, and J. Telser, *Biol. Magn. Reson.* **13**, 151 (1993).

[3b] H. Thomann and M. Bernardo, *Biol. Magn. Reson.* **13**, 275 (1993).

[3c] D. J. Lowe, *Prog. Biophys. Mol. Biol.* **57**, 1 (1992).

[3d] L. Kevan and M. K. Bowman, "Modern Pulsed Electron Spin Resonance." Wiley, New York, 1990.

[3e] A. J. Hoff, "Advanced EPR: Applications in Biology and Biochemistry." Elsevier, Amsterdam, 1989.

[3f] A. Schweiger, *Struct. Bonding (Berlin)* **51**, 1 (1982).

Hyperfine Interaction

As the ENDOR spectrum is in effect an NMR spectrum obtained by devious means, when considering the response of a nucleus with spin I one must keep track of all interactions that it and its coupled electron spin undergo. The energy levels of an unpaired electronic spin S in an externally applied magnetic field, and the associated EPR spectra, are described in detail elsewhere in this volume.[4a,b] As described therein, the interaction of an electronic spin in an external magnetic field is characterized by the parameter **g**. In a molecule, the magnitude of the electron spin interaction depends on the orientation of the molecule with respect to the external magnetic field, and thus **g** is a tensor quantity.

Nuclear Interactions

The nuclear interactions observed in ENDOR are three: the nuclear Zeeman, electron–nuclear hyperfine, and (for $I > \frac{1}{2}$) the nuclear electric quadrupole interactions. The Hamiltonian H_n including these nuclear interactions in an external magnetic field B is given as

$$H_n = \underbrace{g_n\beta_N IB}_{\text{nuclear Zeeman}} + \underbrace{S \cdot \mathbf{A} \cdot I}_{\text{hyperfine}} + \underbrace{I \cdot \mathbf{P} \cdot I}_{\text{quadrupole}} \tag{1}$$

In Eq. (1), the nuclear g value g_n is a constant that depends on the specific nucleus, and β_N, the Bohr magneton, is a constant for all nuclei. Like the **g** tensor, the hyperfine (**A**) and quadrupole (**P**) parameters describe interactions whose magnitudes depend on the orientation of the molecule, and they must therefore be described as tensor quantities. These interactions and resulting transition frequencies are described in more detail in the following sections. Energy levels corresponding to the application of Eq. (1) to a $S = \frac{1}{2}$ and $I = \frac{1}{2}$ or $I = 1$ spin system, and for a single molecular orientation in a strong external magnetic field, are plotted in Fig. 3. As shown in Fig. 3, the total spin quantum numbers S and I have allowed substates m_S and m_I, given by (for S) $m_S = [-S, (-S + 1), \ldots, 0, \ldots, (S - 1), S]$. The selection rules governing ENDOR and EPR transitions are different: the allowed $\Delta m_S = \pm 1$, $\Delta m_I = 0$ transitions for an EPR signal are shown as dotted lines, whereas the allowed $\Delta m_S = 0$, $\Delta m_I = \pm 1$ NMR/ENDOR transitions are represented by solid lines in Fig. 3. The resulting ENDOR spectra for these systems are represented in Fig. 3 and described below.

[4a] G. W. Brudvig, this volume [22].
[4b] G. L. Millhauser, W. R. Fiori, and S. M. Miick, this volume [24].

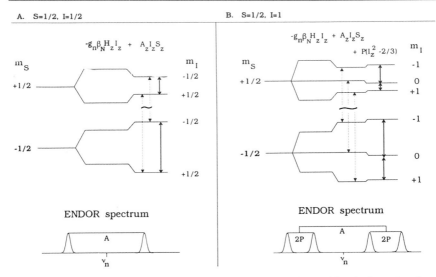

FIG. 3. Energy-level diagram showing separation of nuclear sublevels for a $S = \frac{1}{2}$ and $I = \frac{1}{2}$ (A) or $I = 1$ (B) system in a strong external magnetic field of magnitude H. Energies are represented to first-order, using Eq. (1) in the text (here the field is denoted by H instead of B), and for a single orientation (i.e., with the magnetic field direction along g_z). The allowed transitions for an EPR spectrum ($\Delta m_S = 1$, $\Delta m_I = 0$) are shown as dotted lines, and allowed NMR transitions ($\Delta m_S = 0$, $\Delta m_I = 1$) are shown as solid lines. The energy separations are not shown to scale: the EPR transitions occur at 10^3 higher energy than the NMR transitions. An idealized ENDOR spectrum, obtained by inducing the allowed NMR transitions and monitoring the EPR signal intensity, is shown below each energy-level diagram. The spectra correspond to the situation in which $\nu_n > A/2$.

Resonance Frequencies

In a static magnetic field $\mathbf{B}_0$ (of magnitude $B_0 = |\mathbf{B}_0|$), an isolated nucleus n resonates at the Larmor frequency ν_n determined by its nuclear g value g_n and the nuclear Bohr magneton β_N: $h\nu_n = g_n\beta_N B_0$, where h is Planck's constant. A nucleus that is not isolated but is in a chemical environment feels "internal fields" from the surrounding nuclei and electrons in addition to the external magnetic field $\mathbf{B}_0$. The effects of such internal fields are to shift the resonant frequency of the nucleus away from the Larmor frequency. The internal fields measured by a traditional NMR experiment are created by very weak interactions from surrounding nuclei and electronic fields. These interactions are measured as tiny perturbations from the Larmor frequency, that is, as part per million (ppm) deviations from the proton Larmor frequency of 500 MHz in a 500 MHz ($B_0 \approx 12$ T) NMR instrument, and are too small to be detected in ENDOR. By contrast, a neighboring unpaired electron spin can create a very large internal field at the nucleus. This field is measured as the hyperfine interaction, with coupling constant A^n, and appears as an internal field of magnitude $B^n = h(|A^n|/2)g_n\beta_N$. The coupling constant A^n can be analyzed in terms of the distribution of unpaired electron density on or near the atom in question.

As shown in Fig. 4, the internal and external fields can be aligned parallel or antiparallel, giving rise to two different net magnetic fields at the nucleus. Correspondingly, for a nucleus n with $I = \frac{1}{2}$, two ENDOR transitions are detected with first-order resonance frequencies $\nu_\pm^n$ given by

$$
\begin{aligned}
\nu_\pm^n &= \frac{g_n\beta_N}{h}|B_0 \pm B^n| \\
&= \left|\nu_n \pm \frac{A^n}{2}\right|
\end{aligned}
\tag{2}
$$

Depending on the relative magnitudes of ν_n and $A^n/2$, the ENDOR transitions may be detected as doublets either centered at ν_n and split by A^n (as in Figs. 3 and 4A) or centered at $A^n/2$ and split by $2\nu_n$ (Fig. 4B).

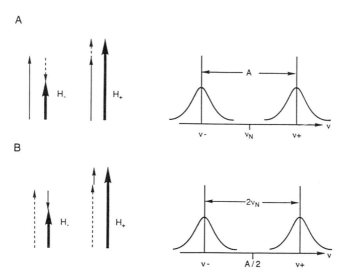

FIG. 4. Schematic representation of the manner in which the external nuclear Larmor field (H_0, plain arrow) and internal nuclear hyperfine field (H_i, dashed arrow) add to give two possible resultant fields with magnitudes $H_\pm$ (bold arrow) and a two-line ENDOR specrum [Eq. (2) in the text; here the fields are denoted by H instead of B]. (A) Larger external field ($H_i < H_0$) as seen for protons ($\nu_n > A/2$); (B) larger internal field ($H_i > H_0$), more typical for other nuclei ($A/2 > \nu_n$).

A nucleus with $I > \frac{1}{2}$ has a nonspherical charge distribution. It thus possesses a quadrupole moment that interacts with nonspherical components of the total charge distribution of its surroundings. This interaction further splits the ENDOR transitions into $2I$ lines at frequencies (to first-order)

$$\nu_{\pm}^n(m_I) = |\nu_n \pm A^n/2 + P^n(2m_I - 1)| \qquad (3)$$

where P^n is the "observed" quadrupole coupling constant. This effect on the ENDOR spectrum is shown in Fig. 3. Measurement of a nuclear quadrupole coupling can give information about the bonding of that atom; for example, the measured parameters can be very different between an imidazole versus a peptide nitrogen.

The nuclear Larmor frequency ν_n increases with magnetic field and thus with the microwave frequency of the EPR spectrometer, whereas A^n and P^n are molecular parameters and thus are independent of the field (although, as described below, they are dependent on the g value, or position within the EPR spectrum). This typically means that when the ENDOR transitions from two different types of nuclei (e.g., ^{1}H, ^{14}N) are overlapping for one microwave frequency range, they may be resolved by changing to a different microwave frequency, as shown in Fig. 5.

Hyperfine Coupling Constant A

The hyperfine coupling as measured, A_{obs}, is a composite of several factors described below. The intrinsic hyperfine interaction, A_{int}, depends on the distribution of electron spin on and near the atom of question. The observed hyperfine coupling may differ from the intrinsic value in a well-defined manner in a spin-coupled multinuclear center and/or when the total electron spin is greater than $\frac{1}{2}$.

Isotropic Hyperfine Interaction. The intrinsic interaction A_{int} between the electronic spin and the nucleus can be separated into an isotropic and an anisotropic component: $A_{\mathrm{int}} = A_{\mathrm{iso}} + A_{\mathrm{aniso}}$. The existence of an isotropic component indicates the presence of unpaired electron density at the nucleus, and this requires a covalent linkage with the electron spin system: the magnitude of the isotropic component in fact measures the extent to which the odd-electron orbital is derived from the s orbital, ψ_s, associated with the atom in question:

$$A_{\mathrm{iso}} \propto |\psi_s(r = 0)|^2 \qquad (4)$$

The sign of A_{iso} is an indicator of the origin of the mechanism that introduces the s-orbital spin density. The standard ENDOR experiment does not yield the sign of A, but advanced techniques such as TRIPLE (electron–nuclear–nuclear triple resonance)[3d,f] do provide this information.

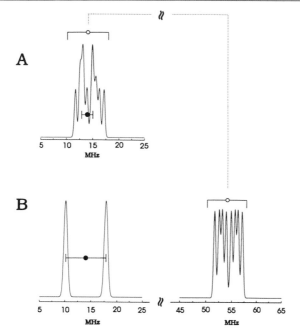

FIG. 5. Effect of increasing microwave frequency on the ENDOR spectra from an $S =$ $\frac{1}{2}$ system with both nitrogen and proton hyperfine interactions. The open circles indicate $\nu(^1\text{H})$ and the attached connecting bars, $A(^1\text{H})$ (1–7 MHz); the closed circles indicate $A(^{14}\text{N})$ (~30 MHz) and the attached connecting bars, $2\nu(^{14}\text{N})$. (A) X-band (~9 GHz) ENDOR spectrum showing overlap of ^1H and ^{14}N ENDOR signals; $\nu(^1\text{H}) \approx 14$ MHz and $2\nu(^{14}\text{N}) \approx$ 2 MHz. (B) Q-band (~35 GHz) ENDOR spectrum showing how overlap is alleviated; here $\nu(^1\text{H}) \approx 55$ MHz and $2\nu(^{14}\text{N}) \approx 8$ MHz.

Anisotropic Hyperfine Interaction. The anisotropic component of the hyperfine coupling has two contributions: a "local" anisotropy owing to spin density in *p*- or *d*-type orbitals on the atom of observation, and "nonlocal" dipolar coupling with spin on other atoms. The first type of interaction is proportional to the orbital coefficient (squared) of the *p*/*d* orbitals. To a first approximation the second term can be considered as a classic point dipolar interaction between the nucleus and the electron spin on a nearby atom. This depends on the total electron spin density at the neighbor (ρ_s), the distance between the spins (r_{12}), and the orientation of the vector between them with respect to the external magnetic field (denoted by angle θ). In the "point dipole" approximation,

$$A_{\text{aniso}} \approx A_{\text{dip}} \approx (g_n\beta_N)(g_e\beta_e)(1/r_{12})^3(3 \cos^2 \theta - 1) \tag{5}$$

and the anisotropic hyperfine coupling can be used to measure distances between atoms in the molecule.

Thus, the measured hyperfine interaction to an atom depends on (1) the type of nucleus (through the nuclear g value); (2) the presence and nature of its chemical bonds; and (3) the distance of this atom from other major centers of electron spin, as well as its position within the molecular framework. These dependences lead to distinctly different hyperfine couplings for similar moieties in different systems. Thus, for example, a histidine ^{14}N atom directly coordinated to a Cu^{2+} (d^9) ion can have an isotropic hyperfine interaction of approximately 40 MHz, whereas because of the markedly different d-orbital electronic configuration a ^{14}N histidine ligand to a low-spin Fe^{3+} (d^5) ion may have an A_{iso} value on the order of approximately 5 MHz. The through-space, purely dipolar coupling to a hydrogen-bonded proton (^1H, $g_n = 5.59$) in an Fe–S cluster can have a measured magnitude A_{dip} of around 8 MHz, but the coupling to a deuteron in the same position (^2H, $g_n = 0.857$) is approximately 6 times smaller. Further examples of the magnitudes and types of hyperfine couplings are given in the examples.

Within this range of diversity of hyperfine couplings, what niches do EPR and ENDOR respectively occupy in biochemistry? Simply put, if the hyperfine coupling is resolved in EPR studies, the higher precision of ENDOR likely is not needed; if the coupling is not fully resolved in the EPR spectrum, the ENDOR technique is required. In the EPR spectroscopy of biomolecules, hyperfine splittings are sometimes resolved for the metal ions with large nuclear g values. These include $^{63,65}Cu$ ($I = \frac{3}{2}$), ^{51}V ($I = \frac{7}{2}$), ^{55}Mn ($I = \frac{5}{2}$), and ^{61}Ni ($I = \frac{3}{2}$, requires isotopic enrichment). Owing to the much smaller nuclear g value of ^{57}Fe, hyperfine splittings are seldom seen for ^{57}Fe ($I = \frac{1}{2}$) in isotopically enriched samples. Hyperfine couplings are rarely resolved for ligand atoms, with the following notable exceptions: ^{19}F bound to heme ($A \approx 100$ MHz); occasionally ^{14}N bound to Cu ($A \approx 40$ MHz); and ^{17}O bound to Mo. Thus, one may say that ENDOR is required for measurements when A is below around 50 MHz. ENDOR can also be useful for measuring the larger A values to higher precision, and in particular it is necessary to measure the nuclear quadrupole couplings, which are seldom observed in EPR spectra.

Orientation Selection and Line Shapes

The observed hyperfine coupling depends on the orientation of the molecule in the external magnetic field. The samples employed in the ENDOR studies of proteins are generally frozen solutions, and they thus contain a random distribution of all possible protein orientations with respect to the applied magnetic field $\mathbf{B}_0$. From the point of view of the molecules, this means there is an equal probability for the field to have any orientation with respect to the molecular framework.

The degree of anisotropy in a hyperfine interaction is often an important clue as to the source of the interaction. It might be thought that the information about anisotropic interactions would be lost when a frozen-solution sample is used. However, in such samples, the EPR signal at each magnetic field point or g value arises from a restricted and mathematically well-defined set of orientations of the field relative to the molecular framework. From the point of view of the molecule mounted on a molecular framework with $\mathbf{g}$-tensor axes 1, 2, and 3, and principal values $g_1 > g_2 > g_3$, the direction of the external magnetic field with respect to the $\mathbf{g}$-tensor axes direction can be described by the vector $\mathbf{l} = (\sin \theta \cos \phi, \sin \theta \sin \phi, \cos \theta)$ where θ and ϕ are the polar and azimuthal angles, respectively. The observed g value, g', is then related to these angles by

$$\begin{aligned}(g')^2 &= [g'(\theta, \phi)]^2 \\ &= g_1^2(\sin \theta \cos \phi)^2 + g_2^2(\sin \theta \sin \phi)^2 + g_3^2(\cos \theta)^2 \end{aligned} \qquad (6)$$

For example, when $\theta = 0$ the external magnetic field vector $\mathbf{l}$ is aligned with g_3. For a molecule with g anisotropy, a powder pattern EPR signal such as shown in Fig. 6A is observed. The EPR signals at the two extreme field positions arise only from molecules at a single orientation with respect to the magnetic field; that is, the low-field edge is associated with molecules where g_1 is aligned with the field, and the high-field edge is associated with molecules where g_3 is aligned with the field (Fig. 6). At intermediate field positions, each magnetic field value corresponds to a set of molecules whose orientations correspond to the angles (θ, ϕ) that satisfy Eq. (6).

Thus, as a consequence of the selection of orientation subsets by choice of field value, the ENDOR spectrum changes as a function of the external field position or g value at which it is measured. A series of ENDOR spectra collected at fields across the EPR envelope samples different sets of molecular orientations. An example of this is shown in Fig. 6B. It can be seen that at the extreme g values (g_1 and g_3) the ENDOR spectrum is the least complex, whereas at intervening g values the ENDOR spectrum shows multiple frequencies. The analysis of such spectra taken at several magnetic field positions gives the full tensor of the hyperfine interaction $\mathbf{A}$, from which the isotropic and anisotropic components can be deduced. Procedures for this analysis have been described in detail elsewhere.[3a,5]

[5] Discussions of orientation selection ENDOR are found in G. H. Rist and J. S. Hyde, *J. Chem. Phys.* **52,** 4633 (1970); B. M. Hoffman and R. J. Gurbiel, *J. Magn. Reson.* **82,** 309 (1989).

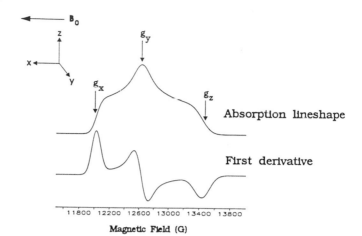

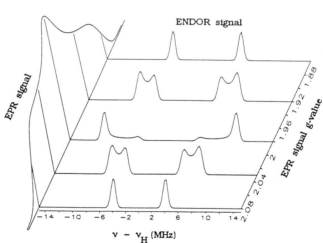

FIG. 6. Representation of orientation- or angle-dependent ENDOR spectra. (A) Rhombic EPR signal shown in absorptive (top trace) and derivative (bottom trace) representations. The x, y, z axes represent the **g** tensor axis system of the molecule. Alignment of g_x and g_z with respect to the external magnetic field B_0 in the EPR spectrometer gives rise to the EPR absorption at the extreme g values; intervening g values correspond to subsets of multiple orientations of the molecules with respect to the external magnetic field. (B) Simulated ENDOR spectra showing observed ENDOR line shapes for spectra obtained at different points across the EPR spectrum. At the extreme g values (edges of the EPR spectra, corresponding to "single-crystal-like" orientations) the ENDOR spectra represent unique orientations of molecules with respect to the external magnetic field; the more complicated ENDOR line shapes at intervening g values are due to multiple orientations of the molecule in the external field, and hyperfine anisotropy.

Hyperfine Coupling in Spin-Coupled Systems

Many of the centers studied in biological systems have EPR signals arising from clusters of metal ions. In these systems, the net electronic spin that gives rise to the EPR signal actually is the result of exchange coupling among the spins from the individual metal ions. To obtain chemical information in such systems, the hyperfine coupling A_{obs}, as measured using either ENDOR or EPR, must be interpreted in terms of the intrinsic hyperfine coupling for the uncoupled mononuclear ions. The value A_{obs} is a function of the projection of the total electronic spin onto an individual ion.[6] An example is the strongly exchange-coupled Fe^{2+}–Fe^{3+} binuclear centers, for which Eq. (7) has been used extensively to analyze measured hyperfine couplings:

$$A_{obs} = \frac{7}{3} A_{Fe(III)} - \frac{4}{3} A_{Fe(II)} \tag{7}$$

In Eq. (7), $A_{Fe(III)}$ reflects the intrinsic coupling of the spin to the ferric site and $A_{Fe(II)}$ its coupling to the ferrous site. This relationship means that, for an exchange-coupled Fe^{2+}–Fe^{3+} binuclear center, the measured hyperfine coupling of, for example, a ^{14}N atom from a ligand to the Fe^{3+} or the $^{57}Fe^{3+}$ nucleus itself will be approximately $\frac{7}{3}$ times that measured in a mononuclear Fe^{3+} center; that measured for the Fe^{2+} will be $-\frac{4}{3}$ times the intrinsic values from a mononuclear center, and the couplings to, for example, a bridging S^{2-} or OH^- will be subject to competing effects. Similar relationships exist and must be considered for higher nuclearity centers such as tetranuclear Fe_4S_4 clusters.

Hyperfine Coupling in Systems with $S > \frac{1}{2}$

There are cases in which the total spin of the system under study is $S > \frac{1}{2}$, and the analysis of the hyperfine coupling in such systems must take into account the spin state of the system. Notable examples of this include ENDOR measurements on the Fe–Mo cluster of nitrogenase,[7] whose EPR signal arises from the $m_S = \pm\frac{1}{2}$ doublet of an $S = \frac{3}{2}$ system; and ENDOR experiments on Mn^{2+} centers,[8] whose EPR signal can include contributions from all transitions within the manifold of spin states in the $S = \frac{5}{2}$ system.

[6] J. F. Gibson, D. O. Hall, J. H. M. Thornley, and F. R. Whatley, *Proc. Natl. Acad. Sci. U.S.A.* **56,** 987 (1966).

[7] A. E. True, M. J. Nelson, R. A. Venters, W. H. Orme-Johnson, and B. M. Hoffman, *J. Am. Chem. Soc.* **110,** 1935 (1988).

[8] X. Tan, M. Bernardo, H. Thomann, and C. P. Scholes, *J. Chem. Phys.* **98,** 5147 (1993); G. H. Reed and G. D. Markham, *Biol. Magn. Reson.* **6,** 73 (1984).

Electron Nuclear Double Resonance Measurement

At this time, the hyperfine coupling can be observed by several types of experiments including continuous wave (cw) ENDOR, pulsed ENDOR, and electron spin echo envelope modulation, or ESEEM. Although there can be substantial overlap in the information obtained by these methods, the optimal choice of method depends on the characteristics of the system being studied, including both the source of the EPR signal and the magnitudes of the hyperfine interactions. In the traditional cw ENDOR experiment, a radio frequency (rf) is applied to induce NMR transitions while continuous wave EPR methodologies are used to detect the effects; within cw ENDOR two different methods are commonly employed. The application of pulsed techniques to EPR has opened the field of electron spin echo detected ENDOR, or pulsed ENDOR. Pulsed ENDOR has been only relatively recently applied to proteins, but it is becoming evident that there are both advantages and disadvantages in the pulsed technology compared with cw applications. Within pulsed EPR techniques there is also the technique of ESEEM. This technique measures hyperfine couplings without application of an rf field, and we also briefly review its uses in relation to ENDOR. In addition, the microwave frequency range of the EPR measurement can be an important variable.

Continuous Wave Experiments

Two mechanisms of cw ENDOR are generally observed, denoted as "steady-state" and "packet-shifting." Their relative importance is primarily determined by the electron spin–lattice relaxation time[4b] T_{le} and thus can vary with temperature. Each has advantages and disadvantages and requires a different method of detection.

Steady-State ENDOR. In the steady-state mode, the microwave power is set so that the EPR signal is partially saturated. When the radio frequency matches an NMR transition frequency, an alternative relaxation pathway opens for the saturated electron spins. The effective T_{le} is decreased, the EPR signal is thereby partially desaturated, and its intensity increases: this increase is the ENDOR response.

In this technique, the ENDOR response is observed as changes in the EPR absorption–mode (in-phase) signal as an rf source is swept. Commonly the EPR signal is detected without field modulation, but the rf frequency is modulated at frequencies around 10 kHz. The signal is decoded at this modulation frequency by a phase-sensitive detector. As a consequence of this scheme, the ENDOR signal appears as the derivative display. The standard commercially available spectrometer is set up in this fashion. In this type of ENDOR experiment, the degree of desaturation of the EPR signal depends on the complicated interplay of all relaxation processes within the spin system. This can make the experiment extremely sensitive to factors such as the temperature and the rf and microwave powers. Therefore, the relative intensities of the ENDOR lines do not necessarily reflect the number of contributing nuclei, in contrast to NMR or EPR.

Packet-Shifting ENDOR. A different form of ENDOR is often advantageous for low-temperature studies of metalloproteins. At liquid helium temperatures (≤ 4.2 K) electron spin relaxation is typically very slow ($1/T_{1e} < 10^3$ sec^{-1} at 2 K is common). The standard 100 kHz magnetic field modulation used in EPR signal detection causes the electron spin system to exhibit rapid passage effects, and the EPR signal is best detected in dispersion mode. Dispersion detection, which is an option on many EPR instruments, involves a phase shift in detecting the EPR signal. The rapid-passage dispersion-mode EPR signal detected with 100 kHz field modulation takes a form similar to that of the EPR absorption envelope (or the integral of the traditional derivative-display EPR signal), as shown in Fig. 6. The ENDOR response is the change that occurs in this EPR signal when an unmodulated rf field is swept through a nuclear transition. In this scheme, an ENDOR spectrum also appears as an absorption rather than a derivative display.

In this type of ENDOR, the amplitude of the field modulation (ΔB_{mod}) defines a window in the broad EPR absorption envelope, in which a set of electron spin packets associated with a given set of nuclear orientations resonates. When the rf field induces a nuclear spin transition, the resonant field of the spin packets is shifted by an amount equal to the nuclear hyperfine coupling, A^n. If $A^n \gtrsim \Delta B_{mod}$, this rf-induced shift of spin packets into and out of the observation window changes the EPR signal intensity. This change constitutes the ENDOR response.

This type of ENDOR is generally not so sensitive to experimental conditions such as temperature and microwave and rf power as is steady-state ENDOR. Marked peculiarities do exist, however. For example, for this type of ENDOR, depending on experimental conditions, the ENDOR response can be either an increase or a decrease in the EPR signal. Generally a decrease is the norm. In addition, some cross-relaxation effects can make the intensities and relative directions of the ν_+ and ν_- transitions anomalously different; we have often found the ν_+ transitions to be more intense, particularly at the higher (35 GHz) microwave frequency.

Pulsed Experiments

In pulsed EPR spectroscopy, the magnetization is detected through formation and readout of an electron spin echo. In contrast to cw EPR, which uses relatively weak microwave powers (≤ 100 mW), pulsed EPR

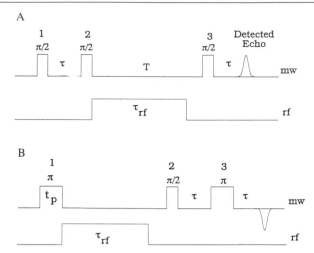

FIG. 7. Pulse sequences for (A) Mims ENDOR and (B) Davies ENDOR. The top line in each sequence indicates pulses at microwave frequencies, whereas the lower line indicates pulses at rf frequencies. In both cases the microwave pulse sequence will be repeated at constant frequency, and the intensity of the resulting electron spin echo is recorded as a function of changing rf frequency.

uses short ($\sim$10–20 nsec), high-power ($>$1 W) microwave pulses to change the axes of magnetization of a set of spin packets. The "angle" change, α, of the spin is described by the relation $h\alpha = 2\pi g\beta_e B_1 \tau_p$, where B_1 is the magnitude of the microwave magnetic component and τ_p is the width of the microwave pulse; for a standard X-band pulsed instrument, B_1 is approximately 10 G and a $\pi/2$ pulse is about 9 nsec. Ultimately the magnetization is refocused to give an electron spin echo, which is used to record the EPR signal. In pulsed ENDOR, an rf pulse is applied during the microwave pulse sequence. When the rf matches a nuclear resonance frequency, it perturbs the spin populations (or phases) and affects the amplitude of the resulting electron spin echo. Pulsed ENDOR techniques are also often referred to as electron spin echo ENDOR (ESE-ENDOR). There are two commonly used pulsed ENDOR techniques, each of which employs one rf pulse within a three-pulse ESE sequence.

Mims ENDOR. The first ESE-ENDOR technique to be devised is eponymously called Mims ENDOR.[9] In this experiment (Fig. 7A), three $\pi/2$ microwave pulses are used to generate an electron spin echo; the time between pulses 1 and 2 is called τ, that between pulses 2 and 3, T. Values for τ are generally in the 100–300 nsec range. An rf pulse, ideally

[9] W. B. Mims, *Proc. R. Soc. London A* **283**, 452 (1965).

a nuclear π pulse (usually between 5 and 60 μsec), is applied during time T. When the rf frequency is resonant with NMR transitions, the rf pulse shifts spin packets and decreases the electron spin echo created by microwave pulse 3. This decrease is the ENDOR response.

For a Mims ENDOR sequence, the ENDOR response R^n for a nucleus n depends on A^n and varies with time τ between the first and second pulses of the stimulated echo sequence according to the relation

$$R^n \propto [1 - \cos(2\pi A^n \tau)] \qquad (8)$$

with A^n in megahertz and τ in microseconds. When $A^n\tau = 1, 2, 3, \ldots$, the intensity falls to near zero, resulting in a "blind spot" in the spectrum. When $A^n\tau = 0.5, 1.5, \ldots$, the ENDOR response reaches a maximum. In principle, if measurements are made as a function of τ, such a modulation of ENDOR intensities allows a resonance to be identified by its A^n value. However, this application is complicated in a powder spectrum and particularly in the spectrum of a quadrupolar ($I > 1$) nucleus. In this case, molecules with different orientations can exhibit resonances with the same frequency, but not associated with the same value of A.

However, for $A^n\tau \leq 0.5$ Mims ENDOR has no blind spots and is a very sensitive technique. The percent ENDOR effect in principle can approach 100%; in practice for protein samples we generally observe changes of approximately 10–50% of the spin echo amplitude in the ENDOR response. This can be compared to typical cw ENDOR responses of about 1% of the amplitude of the EPR signal. As discussed below, however, the pulsed ENDOR experiment can be subject to experimental considerations that make such simplistic comparisons highly misleading. Given that instrumental characteristics generally require that $\tau \gtrsim 0.1$ μsec, Eq. (8) means that the Mims technique is most appropriate for $A^n \lesssim$ 5MHz. This pulsed technique has proved to be highly successful in studies of ^{2}H and weakly coupled ^{14}N.

Davies ENDOR. The second pulsed ENDOR technique is called Davies ENDOR.[10] In this experiment (Fig. 7B) a preparation π microwave pulse of duration τ_p inverts spin packets, "burning a hole" of approximate width $1/\tau_p$ in a broad EPR line. The resulting magnetization is subsequently detected by a $\pi/2$–π two-pulse echo sequence. The application of an on-resonance rf pulse after the preparation phase increases the magnetization measured by the two-pulse detection sequence.

In a Davies experiment, the ENDOR response is jointly dependent on A^n and the width τ_p of the initial inverting microwave pulse. If we define

[10] E. R. Davies, *Phys. Lett.* **47A,** 1 (1974).

a selectivity parameter $\eta^n = A^n \tau_p$, the absolute ENDOR response $R(\eta)$ is determined by

$$R(\eta) = R_0 \left(\frac{1.4\eta}{0.7^2 + \eta^2} \right) \tag{9}$$

where R_0 is the maximum ENDOR response. For a given value of A^n, the ENDOR response is optimized by selecting τ_p so that $\eta = 0.7$. For example, when $\tau_p \approx 30$ nsec, a Davies ENDOR pulse sequence gives optimum signals from nuclei with $A \approx 20$ MHz, but signals from sites with $A < 5$ MHz are strongly attenuated. This characteristic can be used to observe selectively overlapping resonances that have very different hyperfine couplings.

Although it has the ability for "hyperfine selection," the Davies ENDOR sequence is not so sensitive to "blind spots" as in a Mims ENDOR, and thus it is more useful for detecting powder-pattern ENDOR line shapes. The Davies sequence, however, generally does not give so large a percent ENDOR effect; typical Davies ENDOR effects in proteins are approximately 1–15% of the spin echo amplitude.

Electron Spin–Echo Envelope Modulation. The technique of ESEEM can be used to derive hyperfine information from a microwave pulse sequence without using an applied rf field. During the evolution time between microwave pulses, the local fields arising from hyperfine interactions between nuclei and the contributing electron spin act to dephase the spin population. These local fields thereby affect the amplitude of the refocused echo. A spectrum of the amplitude of the echo as a function of dephasing or mixing time τ in a two-pulse, or time T in a three-pulse, echo sequence shows modulations at frequencies directly related to the hyperfine couplings (Fig. 8). The contributing frequencies are normally indicated by displaying the Fourier transform of the time-domain spectrum.

The frequencies and the amplitude of the modulation measured in an ESEEM experiment can both be simulated to derive information about the hyperfine interaction. The ESEEM-derived spectrum in the frequency domain is related to, but not necessarily identical to, an ENDOR spectrum. Thorough reviews of this technique and its analysis are found elsewhere[11]; here we give a brief comparison between the information derived from ESEEM and that directly determined by ENDOR. In an ESEEM experi-

[11] Reviews of ESEEM include Refs. 11a and b.

[11a] W. B. Mims and J. Peisach, *Biol. Magn. Reson.* **3**, 213 (1981).

[11b] W. B. Mims and J. Peisach, *in* "Advanced EPR: Applications in Biology and Biochemistry" (A. J. Hoff, ed.), p. 1. Elsevier, Amsterdam, 1989.

Two-pulse ESEEM

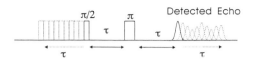

Three-pulse ESEEM

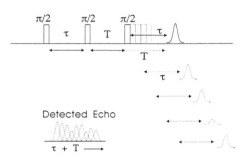

FIG. 8. Illustration of microwave pulse schemes for two-pulse and three-pulse ESEEM (adapted from Kevan and Bowman[3d]). In the two-pulse experiment, the interpulse time τ is varied and the amplitude modulation of the resulting electron spin echo is recorded. In the three-pulse experiment, τ is fixed and the electron spin echo amplitude is recorded as a function of the interpulse time T.

ment, the mechanism by which the hyperfine interaction perturbs the electron spin echo amplitude is through mixing of off-diagonal elements in the spin Hamiltonian. The amplitude of the modulation depends on the anisotropic contribution of the hyperfine interaction, and a requirement is that the microwave pulse must be able to excite both of the hyperfine-split nuclear sublevels. Practically speaking, this means that ESEEM is most sensitive for relatively small hyperfine interactions, and in particular when the nuclear interactions are of similar magnitude (i.e., when A^n, ν_n, and the nuclear quadrupole couplings are of similar magnitude).

The ESEEM technique has been used widely in studying, for example, the hyperfine interaction of the remote ^{14}N atom in imidazole ligands to Cu^{2+} complexes; in this case, the hyperfine coupling of the directly ligated ^{14}N atom ($A_{iso} \approx 20$–40 MHz) is far too large to be detected by ESEEM at X-band, but the remote, nonligated ^{14}N atom of the histidine ring has $A_{iso} \approx 1$–2 MHz, $A_{aniso} \approx 0.5$ MHz, and quadrupolar interactions of around 3–4 MHz.[12] Because the ESEEM technique does not involve actually

[12] W. B. Mims and J. Peisach, *J. Chem. Phys.* **69**, 4921 (1978).

sweeping through a resonance, but instead involves directly analyzing frequency components, the line widths observed in an ESEEM experiment can at times be much narrower than in the ENDOR experiment. Conversely, the modulation from broad frequency components can be difficult to detect. The analysis of ESEEM spectra can also be quite complicated, especially when multiple nuclei are involved.

Instrumentation

The basic components for an ENDOR or ESEEM instrument include the microwave source, resonant cavity, detector, magnet, and cryostat.[13] The ENDOR experiment requires an rf source and amplifier (~50–200 W), as well as modification of the microwave cavity to introduce the rf. The pulsed EPR techniques also require a high-power microwave amplifier (~1 kW) and accompanying timing controls. Simple schematics for both instruments are shown in Fig. 9.

Continuous wave steady-state ENDOR has typically been performed with a modified EPR instrument (i.e., the commercially available Bruker X-band EPR/ENDOR instrument) operating in standard absorption mode and using a nitrogen or liquid helium flow cryostat. Temperature control can be important for this type of ENDOR, because the mechanism depends on being able to "tune" the relevant electronic and nuclear relaxation times. The rf leads are introduced near to the microwave cavity, ideally in a configuration to produce rf fields oriented perpendicular to both the microwave and external magnetic fields. Instead of the phase-sensitive detection at 100 kHz magnetic field modulation that is typically used in detecting EPR signals, modulation of the rf at about 10–50 kHz is employed to improve the signal : noise ratio.

The packet-shifting ENDOR experiment has generally been performed on locally constructed or home-built instruments, equipped with microwave bridges operating in dispersion mode and using liquid helium immersion cryostats. Temperatures below 4 K, often employed to bring the samples into conditions of rapid passage, are achieved by reducing the vapor pressure of the liquid helium by pumping on the cryostat headspace. The cavities and rf configurations are the same as for steady-state ENDOR; however, the rf is generally not modulated, and detection is with the typical 100 kHz magnetic field modulation. We have found an improvement in signal : noise ratios by frequency broadening the rf input via mixing with a white noise source.[14]

[13] C. P. Poole, Jr., "Electron Paramagnetic Resonance." Wiley, New York, 1983; R. W. Quine, G. R. Eaton, and S. S. Eaton, *Rev. Sci. Instrum.* **58**, 1709 (1987); C. Fan, P. E. Doan, C. E. Davoust, and B. M. Hoffman, *J. Magn. Reson.* **98**, 62 (1992).
[14] B. M. Hoffman, V. J. DeRose, J.-L. Ong, and C. E. Davoust, *J. Magn. Reson. A* **110**, 52 (1994).

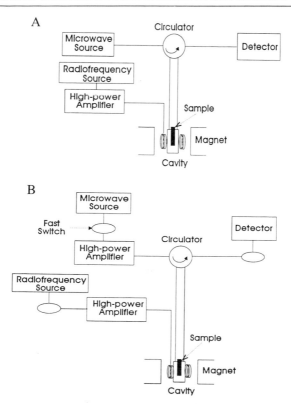

FIG. 9. Schematic diagrams of the major components of cw and pulsed EPR-ENDOR instruments. The sample is in a resonant microwave cavity, situated between poles of a magnet and surrounded by a temperature-control system (not shown). The structure of the circulator directs microwaves from the source to the cavity, and from the cavity to the detection system. A radio frequency synthesizer provides rf to coils situated around the cavity. Note that this diagram shows an arbitrary orientation of the rf coils. For convenience the magnetic field modulation coils are not shown for the cw spectrometer. For the pulsed EPR spectrometer (B), fast switches (ovals) are used to control pulse timing for the rf and microwave pulses, as well as to protect the detector. For simplicity, several features including the timing circuitry are not shown. The signal from the detector is sent to a boxcar integrator. Both spectrometers are computer-interfaced for data collection and storage. Further details may be found elsewhere.[13]

Pulsed EPR experiments are typically performed with locally constructed instruments, although a commercially available X-band pulsed EPR/ENDOR instrument is now available from Bruker. A liquid helium immersion cryostat is generally employed. The pulsed EPR instrument creates short, high-power microwave and, for ENDOR, rf pulses, but the magnetization detected is of very small magnitude and requires a sensitive detector. This necessitates precise timing not only for creation and detec-

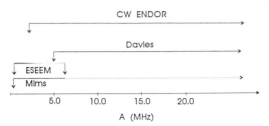

FIG. 10. Diagram representing general choice of techniques based on their sensitivity as a function of hyperfine coupling magnitude. Consideration of the pulsed EPR techniques is based on operation at X-band (9 GHz) microwave frequencies. The dotted line for Mims ENDOR at large A values indicates that the technique is sensitive but suffers from "blind spots" as described in the text.

tion of the electron spin echoes, but also for protection of the detector against the high power pulses. In contrast to cw EPR, pulsed EPR studies generally employ a low Q microwave cavity in order to dissipate the microwave power. Types of resonators employed include "stripline"[15a,d] and "loop–gap"[15b,c] constructions; descriptions of the rf construction for pulsed ENDOR are also available.[15d,e]

Selecting a Technique

Each of the techniques discussed above has advantages for particular cases, and there are some general guidelines that can be followed in choosing an ENDOR technique for a given system under study. The easiest division is along the lines of the strength of the hyperfine interaction; often, this information is available from consideration of the system in relation to previous studies on similar proteins or model compounds. A rough guide is shown in Fig. 10. With respect to this we note that there are always exceptions to every rule, and there are several overlapping areas where information obtained from techniques can be either supplementary or complementary. We have found, in fact, that applying multiple techniques to a given system is often required to achieve a complete data set.

Besides the strength of the hyperfine coupling, there are additional considerations in choosing an ENDOR technique. For example, the characteristics of the spin system play a large part in determining the sensitivity of pulsed ENDOR techniques. A long electron T_{1e} requires a low repetition

[15a] W. B. Mims, *Rev. Sci. Instrum.* **45**, 1583 (1974).
[15b] W. Froncisz and J. S. Hyde, *J. Magn. Reson.* **47**, 515 (1982).
[15c] R. D. Britt and M. P. Klein, *J. Magn. Reson.* **74**, 535 (1987).
[15d] J. L. Davis and M. B. Mims, *Rev. Sci. Instrum.* **49**, 1095 (1978).
[15e] S. Pfenninger, A. Schweiger, J. Forrer, and R. R. Ernst, *Chem. Phys. Lett.* **151**, 199 (1988).

rate for the pulse sequence, leading to longer collection times. The percent ENDOR effect can be much higher than in cw ENDOR, but this is the percent ENDOR of the electron spin echo that is available from the system *at the end of the pulse sequence.* This means that systems with short phase memories and/or rapid spectral diffusion, despite having substantial cw EPR amplitudes, may have a greatly decreased electron spin echo amplitude following the 5–40 μsec rf pulse time included in the ENDOR sequence. Such an effect, which generally cannot be predicted before beginning the pulsed EPR study, can and often does severely limit the signal-to-noise ratio in pulsed ENDOR experiments of metalloproteins.

Planning Experiments

There are several considerations that aid in obtaining the most information possible in an ENDOR experiment. In optimum cases, very detailed information can be obtained not only about the immediate environment of the spin center in the native protein, but also about the binding of substrate or inhibitor molecules.

Spin concentration and the quality of the EPR signal are the initial considerations in planning an ENDOR experiment. The ENDOR effect is a small percentage of the EPR signal intensity; however, EPR itself is a highly sensitive technique. As a very rough guideline, for an ENDOR experiment spin concentrations of at least 0.5 mM are desirable. The sensitivity decreases for samples whose EPR signals are spread over a wide range of g values (i.e., the number of spins sampled at each magnetic field point is smaller). The amount of sample required depends on the particular spectrometer and resonator that are used for the experiment. For example, for 35 GHz cw ENDOR we use samples in small quartz tubes that are appropriate for the approximately one-fourth smaller wavelength of the higher microwave frequency; for convenience, the loop–gap resonator for the 9 GHz pulsed ENDOR instrument has subsequently been designed to accommodate such small sample sizes. The quartz tubes are around 2 mm inner diameter; with a resonator "active" volume corresponding to approximately 20 μl sample volume, and 0.5 mM concentration, this requires only approximately 10 nmol of spins. For a normal X-band EPR tube (4 mm I.D.), this requirement will be increased by a factor of 4. As an extreme case, in the study of cytochrome oxidase CuA it was possible to obtain an excellent 35 GHz cw ENDOR spectrum from 20 μl of 10 μM ^{15}N-labeled enzyme, or 0.2 nmol of spins!

Of note, EPR "purity" is not always required. When a sample exhibits several EPR signals, ENDOR can be used to study selectively one signal at a time provided that the signals are not wholly overlapping. This has been convenient for many metalloproteins that contain complex sets of clusters.[16a,b]

In most detailed studies some form of isotopic labeling will be found to be highly desirable. This can be achieved by (a) growing cells on isotopically labeled media, resulting in globally labeled preparations; (b) inserting specific isotopically labeled amino acids using an auxotroph; (c) extracting the center and reconstituting it with isotopic labels; or (d) adding isotopically labeled substrate or inhibitor molecules.

Examples

In this section, the use of ENDOR and ESEEM techniques to derive information relevant to questions of protein structure and function is illustrated in a series of examples. The final subsection gives an example of the use of multiple ENDOR techniques in the study of a single metalloprotein.

Identification of Spin Center

Sometimes the initial question to be answered in an ENDOR experiment is as fundamental as, What is the center that gives rise to the EPR signal? In the case of a metalloprotein, deceptively simple EPR signals can arise from quite complex centers. Few metal ions have isotopes in natural abundance that have nuclear spin $I > 0$. However, an isotopic label can often be incorporated into the cluster and then observed with ENDOR even in the typical case where no resolved splitting is seen in the EPR. This technique has been used with cw ENDOR, for example, to identify ^{61}Ni ($I = \frac{3}{2}$) and ^{57}Fe ($I = \frac{1}{2}$) resonances in one $S = \frac{1}{2}$ cluster of carbon-monoxide dehydrogenase (CODH).[16b] ENDOR from ^{95}Mo ($I = \frac{5}{2}$) and ^{57}Fe were used in determining the contribution of these nuclei to the $S = \frac{3}{2}$ cluster of nitrogenase.[7] In that case, labeling with the $I = 0$ ^{100}Mo was necessary to check the background for the broad ^{95}Mo ENDOR signal. Information about the oxidation state of the Mo was derived from the simulated quadrupole coupling constant for that nucleus.

Proton ENDOR is a useful probe of the origin of the EPR signals from organic-based radicals. In a series of cw ENDOR studies including samples with isotopically labeled amino acids, proton ENDOR was used to identify the amino acid radical in cytochrome-c peroxidase (CcP) as a tryptophan species.[17] Similarly, isotopic labeling was used to identify the tyrosyl

[16a] C. Fan, M. Teixeira, J. Moura, I. Moura, B.-H. Huynh, J. le Gall, H. D. Peck, Jr., and B. M. Hoffman, J. Am. Chem. Soc. 113, 20 (1991).

[16b] C. Fan, C. M. Gorst, S. W. Ragsdale, and B. M. Hoffman, Biochemistry 30, 431 (1991).

[17] M. Sivaraja, D. B. Goodin, M. Smith, and B. M. Hoffman, Science 245, 738 (1989).

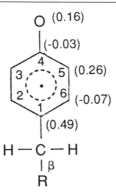

FIG. 11. Numbering scheme and ENDOR-derived spin-density distributions (shown in parentheses) in the tyrosyl radical of ribonucleotide reductase. The ring carbon values are symmetry-related. Hyperfine couplings from the ring 3,5 and the β-carbon protons are responsible for the structure of the radical EPR signal. (Adapted from Babcock and co-workers.[19a])

radicals in photosystem II[18a] and in ribonucleotide reductase.[18b] The characteristic splittings of these radical EPR signals arise from strong hyperfine coupling to the methylene protons of the tyrosine β carbon, and to the ring 3,5 protons (Fig. 11). Proton ENDOR studies were used to derive the spin density distribution around the tyrosine ring[19a,b] (Fig. 11) and to obtain information about hydrogen bonding to the phenol oxygen. The proton ENDOR spectrum of the tyrosine-derived amino acid radical in galactose oxidase was used to determine that the proton of the ring 3 position is substituted.[20] That substitution is a covalent thioether linkage to a cysteinyl residue.

Identification of Metal Ligands

Given the identification of an EPR signal with a metal center, either a monomer or a cluster, then the next question is likely to be, What are the ligands to the metal center? Most commonly this information is obtained from ENDOR signals of ^{14}N ($I = 1$) and ^{1}H ($I = \frac{1}{2}$) in native samples,

[18a] B. A. Barry and G. T. Babcock, *Proc. Natl. Acad. Sci. U.S.A.* **84,** 7099 (1987).

[18b] J. Stubbe, *Biochemistry* **27,** 3893 (1988).

[19a] C. J. Bender, M. Sahlin, G. T. Babcock, B. A. Barry, T. K. Chandrashekar, S. P. Salowe, J. A. Stubbe, B. Lindstrom, L. Petersson, A. Ehrenberg, and B.-M. Sjoberg, *J. Am. Chem. Soc.* **111,** 8076 (1989).

[19b] C. W. Hoganson and G. T. Babcock, *Biochemistry* **31,** 11874 (1992).

[20] G. T. Babcock, M. K. El-Deeb, P. O. Sandusky, M. M. Whittaker, and J. W. Whittaker, *J. Am. Chem. Soc.* **114,** 3727 (1992).

and often confirmed using isotopic substitution with ^{15}N ($I = \frac{1}{2}$) and ^2H ($I = 1$). Other donor ligands have been identified using ENDOR signals from samples labeled with ^{17}O ($I = \frac{5}{2}$) or ^{33}S ($I = \frac{3}{2}$).

Nitrogen ENDOR. The most common nitrogen donor ligand in metalloproteins is histidine. The strength of the hyperfine couplings from ^{14}N of a histidine ligand depends strongly on the electronic configuration of the metal center, as discussed above (see section on the hyperfine interaction). Thus, for example, the ^{14}N hyperfine couplings from the directly ligating nitrogen of histidine coordinated to a Cu^{2+} center can be very large ($A_{obs} \approx$ 20–40 MHz), as seen in cw ENDOR studies of a series of type 1 centers in blue copper proteins,[21a,b] and of copper amine oxidase.[22] The noncoordinated N of histidine ligated to Cu centers in proteins has been studied with ESEEM.[11a,23]

At X band, ^{14}N hyperfine couplings of the directly coordinated nitrogen atoms often lead to overlap between the hyperfine-centered ^{14}N resonances and the Larmor-centered ^1H resonances.[21a] This overlap is removed by operating at 35 GHz (Q band), which moves the protons to a much higher Larmor frequency.[21b] An alternative method of discriminating between overlapping resonances is the pulsed ENDOR Davies sequence, operating at selective τ_p values.

In Fe centers, the hyperfine couplings from nitrogen ligands are generally much smaller. For example, the ^{14}N hyperfine couplings in ferric heme systems are in the range of 4–12 MHz and have been extensively examined with both cw ENDOR[24a,b] and with ESEEM.[24c,d]

An excellent example of the powerful combination of ENDOR with isotopic labeling studies is found in the identification of histidine ligands in the Rieske-type 2Fe–2S iron–sulfur protein of phthalate dioxygenase (PDO).[25] As shown in Fig. 12, the X-band ^{14}N cw ENDOR spectrum of natural-abundance PDO shows a broad pattern of ^{14}N resonances that by

[21a] J. E. Roberts, J. F. Cline, V. Lum, H. Freeman, H. B. Gray, J. Peisach, B. Reinhammer, and B. M. Hoffman, *J. Am. Chem. Soc.* **106**, 5324 (1984).

[21b] M. M. Werst, C. E. Davoust, and B. M. Hoffman, *J. Am. Chem. Soc.* **113**, 1533 (1991).

[22] G. J. Baker, P. R. Knowles, K. B. Pandeya, and J. B. Raynor, *Biochem. J.* **237**, 609 (1986).

[23] J. McCracken, J. Peisach, and D. M. Dooley, *J. Am. Chem. Soc.* **109**, 4064 (1987); F. Jiang, J. Peisach, L.-J. Ming, L. Que, Jr., and V. J. Chen, *Biochemistry* **30**, 11437 (1991).

[24a] C. P. Scholes, *in* "Multiple Electron Resonance Spectroscopy." (M. M. Dorio and J. H. Freed, eds.), p. 297. Plenum, New York, 1979.

[24b] C. P. Scholes, A. Lapidot, R. Mascarenhas, T. Inubushi, R. A. Isaacson, and G. Feher, *J. Am. Chem. Soc.* **104**, 2724 (1982).

[24c] J. Peisach, W. B. Mims, and J. L. Davis, *J. Biol. Chem.* **254**, 12379 (1979).

[24d] R. S. Magliozzo and J. Peisach, *Biochemistry* **32**, 8446 (1993).

[25] R. G. Gurbiel, C. J. Batie, M. Sivaraja, A. E. True, J. A. Fee, B. M. Hoffman, and D. P. Ballou, *Biochemistry* **28**, 4861 (1989).

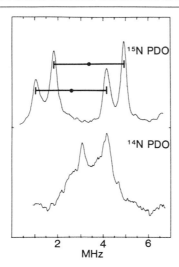

FIG. 12. Nitrogen ENDOR spectra of (top) ^{15}N-enriched and (bottom) ^{14}N natural-abundance PDO from *Pseudomonas cepacia* taken at $g_2 = 1.92$. The solid circles in the top spectrum indicate $A(^{15}\text{N})/2$, and the connecting bars represent $2\nu(^{15}\text{N})$ for each of the ^{15}N doublets. Assignment of the spectra is described in the text. (Adapted from Gurbiel et al.[25])

itself is not amenable to analysis. In contrast, the spectrum of a sample globally labeled with ^{15}N shows four sharp peaks. As shown in Fig. 12, these peaks can be paired into two doublets, each of which is split by $2\nu_n$ for ^{15}N. The center of each doublet is the value $A/2$ for that nitrogen; at the field position in Fig. 12, values of A_{obs} are 5 and 7 MHz. ^{15}N ENDOR spectra taken at points across the EPR envelope, and a full simulation of the data, indicated that the signals arise from two magnetically distinct nitrogenous ligands and gave full values for the hyperfine tensors. These ligands were confirmed as histidine by observation of the same resonances in samples prepared with [^{15}N]histidine. The ^{14}N data were subsequently analyzed, and the ^{14}N quadrupole parameters were obtained. Finally, through use of samples specifically labeled with ^{15}N at either the δ or ε position of the histidine, it was shown that the coordination was through the histidine N(δ) atom. A thorough analysis of the hyperfine couplings included consideration that the $S = \frac{1}{2}$ EPR signal from the Fe center in PDO is a result of spin–exchange coupling between high-spin Fe^{2+} and Fe^{3+} ions. The analysis gave values for the fractional occupancy of the s and p orbitals of $f_{2s} \approx f_{2p} \approx 2\%$ for both nitrogen ligands. The relation of $f_{2p}/f_{2s} \approx 1$ is reasonable for the sp^n hybrid of an imidazole nitrogen.[26]

[26] T. G. Brown and B. M. Hoffman, *Mol. Phys.* **39**, 1073 (1980).

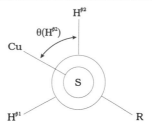

FIG. 13. Newman projection used in analyzing β-methylene proton hyperfine coupling constants in Type 1 Cu proteins.

Proton ENDOR. Ligands to metalloproteins often can be identified by the characteristic hyperfine behavior observed in the proton ENDOR spectrum. Further information is gained by determining whether the proton can be exchanged with deuterium.

A very characteristic proton hyperfine coupling arises from the β-CH$_2$ of the cysteine ligand (Fig. 13). The hyperfine coupling between the spin of the metal center and these protons can be quite large owing to spin delocalization, through the S atom, to the H$^\beta$ via hyperconjugation and σ-bond polarization. For example, each of the Type 1 centers in the series of blue copper proteins described above shows two strongly coupled protons, having hyperfine splittings between 15 and 30 MHz and showing effective isotropic behavior (10–20% anisotropy).[21a,b] Each of these couplings can be analyzed using the following relation:

$$A^H = [B \cos^2 \theta + C]\rho_s \qquad (10)$$

Here θ is the dihedral angle of the β proton (Fig. 13); the values of θ for the two β protons are related by $\theta_1 \approx (\theta_2 - 2\pi/3)$. The constants B and C reflect hyperconjugation and σ-bond polarization, respectively. Solving the equations for each pair of β protons gives estimates for the sulfur spin density ρ_s of approximately 0.45, in good agreement with theoretical calculations.[27] From these studies it was determined that the values for θ_i were very similar in the series of Type 1 copper centers studied, indicating a tight conservation in structure among these proteins.

A terminal water or hydroxyl ligand is another source of proton hyperfine couplings in metalloproteins. Characteristically, these protons can usually be exchanged in deuterated media (unlike the constitutive protons discussed above), allowing for complementary ^{2}H ENDOR studies on the exchanged samples. For example, ENDOR resonances from exchangeable

[27] K. W. Penfield, A. A. Gewirth, and E. I. Solomon, *J. Am. Chem. Soc.* **107**, 4519 (1985); A. A. Gewirth and E. I. Solomon, *J. Am. Chem. Soc.* **110**, 3811 (1988).

protons have been detected from the axial H_2O ligand in the high-spin Fe^{3+} species aquometmyoglobin. These exhibit observed hyperfine couplings of 6 MHz.[28] In the protein methane monooxygenase (MMO), which has a binuclear exchange-coupled iron–oxo Fe^{2+}–Fe^{3+} center, ENDOR resonances from exchangeable protons with hyperfine couplings of roughly 7–10 MHz (Fig. 14A, arrows) are assigned to a terminal hydroxide or water ligand.[29]

Methane monooxygenase also exhibits unique cw ENDOR resonances from very strongly coupled, highly anisotropic ($A_{obs} \approx 8$–30 MHz) exchangeable protons. In Fig. 14B, these resonances, indicated by braces, can be seen to change dramatically in spectra taken at different g values. By comparison with the ENDOR spectrum of methemerythrin azide, a derivative of the binuclear Fe protein hemerythrin, these proton resonances were confirmed as being due to a hydroxide moiety bridging the two Fe atoms (Fig. 14). Such strong hyperfine couplings from a hydroxide bridge were actually predicted in earlier ESEEM studies on the Fe^{2+}–Fe^{3+} center of uteroferrin (a third binuclear Fe protein) exchanged in ^2H_2O[30]; although the proton hyperfine couplings are too large to be detected by using ESEEM, the coupled deuterons resonate with magnitudes reduced by $g_n(^1H)/g_n(^2H) \approx 1/6.5$, or around 4 MHz for the largest deuteron hyperfine values. Of note, the cw proton ENDOR from these highly anisotropic hyperfine couplings is extremely broad and shallow (Fig. 14B). Although such resonances are detectable in the "transient," or field-modulated ENDOR experiment shown here, they were not observed in the earlier "steady-state," frequency-modulated ENDOR studies of either uteroferrin[30] or MMO.[31]

Substrate or Inhibitor Binding

A final question that is often asked in an ENDOR study of a catalytic center is, How does the center bind substrate or inhibitor molecules? Although the addition of exogenous ligands can sometimes change the appearance of an EPR signal, such changes do not give detailed chemical information about the ligation to the protein center. Such a situation can often be addressed by an ENDOR or ESEEM study, using isotopically labeled exogenous species.

[28] C. F. Mulks, C. P. Scholes, L. C. Dickinson, and A. Lapidot, *J. Am. Chem. Soc.* **101**, 1645 (1979).

[29] V. J. DeRose, K. E. Liu, D. M. Kurtz, Jr., B. M. Hoffman, and S. J. Lippard, *J. Am. Chem. Soc.* **115**, 6440 (1993).

[30] K. Doi, J. McCracken, J. Peisach, and P. Aisen, *J. Biol. Chem.* **263**, 5757 (1988).

[31] M. P. Hendrich, B. G. Fox, K. K. Andersson, P. G. Debrunner, and J. D. Lipscomb, *J. Biol. Chem.* **267**, 261 (1992).

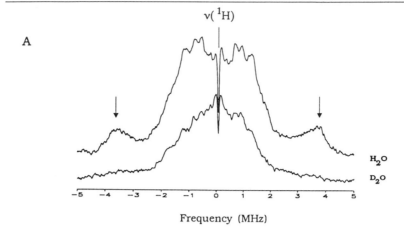

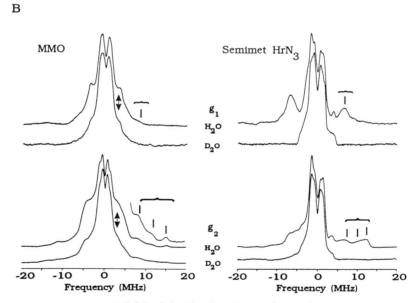

FIG. 14. Proton cw ENDOR of the binuclear Fe proteins methane monooxygenase (A and B) and hemerythrin (B). Spectra are centered at $\nu(^1\mathrm{H})$. (A) Narrow frequency range proton ENDOR spectra of methane monooxygenase in H_2O (top trace) and D_2O (bottom trace) taken at $g_1 = 1.94$. The arrows indicate exchangeable protons with $A^H \approx 8$ MHz, arising from a water or hydroxide ligand to the Fe. Conditions: 12,950 G, 35.2 GHz, 0.5 MHz/sec rf scan rate, 0.3 G modulation amplitude. (B) Broad frequency range proton ENDOR spectra of methane monooxygenase (left) and azidomethemerythrin (right), at two different g values. The braces indicate strongly coupled, highly anisotropic protons, assigned to the bridging hydroxide protons. (Adapted from DeRose et al.[29])

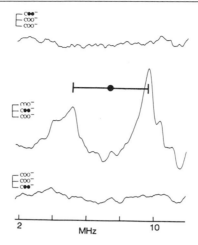

FIG. 15. ^{17}O X-band cw ENDOR spectra at $g = 1.85$ (3815 G) of reduced aconitase in the presence of substrate whose -COO$^-$ groups have been individually labeled with ^{17}O: (top) Label at α-carboxyl, (middle) label at β-carboxyl, and (bottom) label at γ-carboxyl. In the middle trace the bar indicates transitions centered at $A/2$ (filled circle) and split by $2\nu(^{17}O)$ (bars). (Adapted from Kennedy et al.[32])

One example of this method is for the protein aconitase, a 4Fe–4S protein that catalyzes the conversion of citrate to isocitrate. A complete cw ENDOR study was undertaken of this protein[32] (see Fig. 2). Substrate binding information was obtained using citrate, and other substrates, labeled with ^{17}O at specific positions. As shown in Fig. 15, this enabled the determination that citrate was bound to the Fe–S cluster through the "center" carboxylate group. This and other information determined by ENDOR led to a full description of the mechanism for aconitase.[33]

The mechanism of the protein methylamine dehydrogenase was studied in an ESEEM experiment, using ^{14}N- and ^{15}N-labeled methylamine substrate.[34] By a careful comparison with the ESEEM spectrum of the native protein, it could be deduced that the N of the substrate becomes covalently attached to the EPR-active tryptophan tryptophylsemiquinone catalytic intermediate found in that protein.

Other examples of inhibitor and substrate binding experiments are

[32] M. C. Kennedy, M. M. Werst, J. Telser, M. H. Emptage, H. Beinert, and B. M. Hoffman, Proc. Natl. Acad. Sci. U.S.A. **84**, 8854 (1987).

[33] M. M. Werst, M. C. Kennedy, H. Beinert, and B. M. Hoffman, Biochemistry **29**, 10526 (1990).

[34] K. Warncke, H. B. Brooks, G. T. Babcock, V. L. Davidson, and J. McCracken, J. Am. Chem. Soc. **115**, 6464 (1993).

found in the studies using ^{13}C-labeled CN^- or CO species. Cyanide in particular is a common inhibitor of metalloproteins, and ^{13}C cw ENDOR resonances have been observed from ^{13}CN-inhibited samples of transferrin,[35] CODH,[36] and heme proteins.[37]

Protein Ligation Determination: A Case Study

An excellent example of the application of multiple ENDOR techniques is found in studies[38a,b] of the protein ligation environment of the mononuclear nonheme low-spin Fe^{3+} center of nitrile hydratase. This protein catalyzes the hydration of nitriles to amides. Extended X-ray absorption fine structure (EXAFS) studies indicated a mixed N,O/S ligation environment for the Fe.

The ^{14}N ENDOR spectrum of the center gives a rich pattern that is difficult to assign (not shown), but global labeling with ^{15}N results in readily interpretable ENDOR spectra demonstrating the presence of three N-donor ligands to the iron. These are denoted N(1), N(2), and N(3) and have hyperfine couplings $A_1 \approx 6$ MHz, $A_2 \approx 5$ MHz, and $A_3 \approx 2$ MHz. The combination of 9.5 GHz pulsed and 35 GHz cw ENDOR techniques was used in studying the ^{15}N resonances. Figure 16 compares the ^{15}N ENDOR pattern derived from cw 35 GHz ENDOR (Fig. 16B) with the combination of Mims and Davies pulsed ENDOR at 9.5 GHz (Fig. 16A). Both spectra were obtained at g_3, at the edge of the EPR spectrum. The cw ENDOR spectrum has a noticeably better signal: noise ratio but noticeably worse resolution for these resonances. At 35 GHz, g_3 corresponds to 12,700 G, $\nu(^{15}N) = 5.6$ MHz, and because in all cases $\nu_n > A/2$ the resonances from all three nitrogens are centered at ν_n and split by A. Only $\nu_+ = (\nu_n + A/2)$ is detected for N(1) and N(2). At 9 GHz, g_3 corresponds to 3400 G and $\nu_n(^{15}N) = 1.5$ MHz. Now $A/2 > \nu_n$ for N(1) and N(2), and so those resonances are centered at $A/2$ and split by $2\nu_n$. As indicated in Fig. 16, in the pulsed ENDOR experiment the Mims ENDOR sequence was appropriate for one of the ^{15}N resonances ($A_3 \approx 2$ MHz), whereas the Davies ENDOR sequence was most sensitive for the more strongly coupled nitrogens ($A_1 \approx 6$ MHz, $A_2 \approx 5$ MHz).

Additional pulsed and cw ENDOR spectra taken at points across the

[35] P. A. Snetsinger, N. D. Chasteen, and H. van Willigen, *J. Am. Chem. Soc.* **112,** 8155 (1990).

[36] M. E. Anderson, V. J. DeRose, B. M. Hoffman, and P. A. Lindahl, *J. Am. Chem. Soc.* **115,** 12204 (1993).

[37] C. F. Mulks, C. P. Scholes, L. C. Dickinson, and A. Lapidot, *J. Am. Chem. Soc.* **101,** 1645 (1979).

[38a] H. Jin, I. M. Turner, Jr., M. J. Nelson, R. J. Gurbiel, P. E. Doan, and B. M. Hoffman, *J. Am. Chem. Soc.* **115,** 5290 (1993).

[38b] P. E. Doan, H. Jin, R. J. Gurbiel, M. J. Nelson, and B. M. Hoffman, unpublished work (1994).

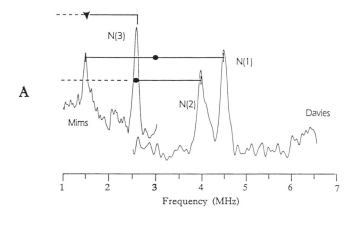

A

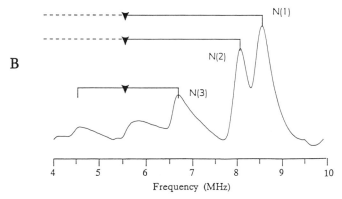

B

FIG. 16. Pulsed (A) and cw (B) ^{15}N ENDOR spectra arising from three nitrogen ligands in the mononuclear Fe protein nitrile hydratase isolated from bacteria grown on ^{15}NH$_4$Cl. Data were taken at g_3, a "single-crystal-like" position. (A) ^{15}N pulsed ENDOR taken with a Mims sequence (top trace) and Davies sequence (bottom trace). N(1) and N(2) resonances are centered at $A(^{15}$N$)/2$ (filled circle) and separated by $2\nu(^{15}$N$)$ (bars). N(3) resonances are centered at $\nu(^{15}$N$)$ (filled triangle) and split by $A(^{15}$N$)$. Conditions: 9.4 GHz, 2 K, 3420 G, rf pulse length 40 μsec, repetition rate 6 Hz, average of 256 transients; Mims ENDOR: 16 nsec $\pi/2$ pulse length, $\tau_{12} = 248$ nsec; Davies ENDOR: 120 nsec π pulse length, $\tau_{23} = 348$ nsec, average of 256 transients. (B) 35 GHz ^{15}N cw ENDOR at same g value. In the higher field, all three nitrogens give hyperfine-split doublets centered at $\nu(^{15}$N$)$ (filled triangle) and split by $A(^{15}$N$)$. Conditions: 35.12 GHz, 2 K, 12,720 G, 0.16 mW microwave power, 0.3 G modulation amplitude, 2.5 MHz/sec rf scan rate, 75 scans. (Adapted from Jin *et al.*[38a])

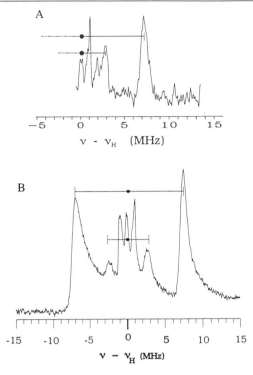

FIG. 17. Proton pulsed and cw ENDOR taken at g_3 for the mononuclear Fe protein nitrile hydratase. The proton resonances are centered at $\nu(^1H)$ (filled circles). Representative splittings by $A(^1H)$ are indicated by the bars. (A) Davies proton ENDOR spectrum. The ν_- resonances are not shown owing to interference from ^{14}N. (B) 35 GHz cw ENDOR spectrum. The resonance line shapes are affected because of the sweep conditions. (From Doan et al.[38b])

EPR envelope (not shown) demonstrate that the A values for all three nitrogens are roughly isotropic, that is, A_{obs} does not change much with g value. The magnitudes of A_1 and A_2, comparable to values found for N ligands in other Fe centers, are consistent with their being from directly coordinated ligands. On consideration of the known mixed N/S coordination environment, the smaller magnitude of A_3 was assigned to a coordinated nitrogen trans to a thiolate from a cysteine ligand; this is by comparison with the hyperfine value for the proximal imidazole of the mercaptoethanol complex of ferrimyoglobin.[39]

The proton ENDOR pattern of nitrile hydratase shows strong, nonexchangeable protons, with A values around 15 MHz, assigned to the β-carbon methylene protons of cysteine ligands (Fig. 17). In Fig. 17 again

[39] H. J. Flanagan, G. J. Gerfen, A. Lai, and D. J. Singel, J. Chem. Phys. **88**, 2162 (1988).

the X-band pulsed ENDOR (Davies) and 35 GHz cw ENDOR spectra are shown. In both cases, $\nu_n > A/2$ and all resonances are centered at ν_n and split by A. Again, the pulsed ENDOR spectrum shows good resolution but a worse signal : noise ratio than the cw ENDOR spectrum. Four classes of exchangeable protons are also found, and the larger proton couplings (4 and 6 MHz) were considered as being due to a bound water or hydroxide ligand. This was confirmed by the ^{17}O cw ENDOR pattern (Fig. 18) observed in samples of the enzyme in $H_2{}^{17}O$-enriched buffer. The remarkably well-resolved quintet ^{17}O ($I = \frac{5}{2}$) pattern (only the ν_+ branch is shown) shows unequally spaced lines resulting from the second-order nuclear quadrupole interactions.

The exchangeable protons in nitrile hydratase were also studied by deuterium ENDOR. Deuterium ENDOR is difficult in the cw experiment, owing to swamping of the small deuteron hyperfine couplings (enforced by the low nuclear g value of ^2H) by the presence of a large peak that is centered at $\nu(^2H)$ and arises from distant, noninteracting ^2H. The Mims pulsed ENDOR technique, however, is an excellent method for observing these hyperfine couplings. In the pulsed ENDOR experiment, an individual pulse sequence (≤ 60 μsec) is short compared with the time scale of the spin diffusion process that gives rise to the distant cw ENDOR re-

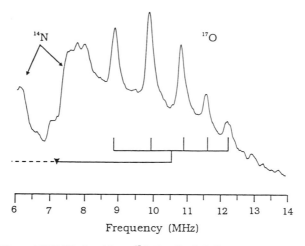

FIG. 18. 35 GHz cw ENDOR signal from ^{17}O ($I = \frac{5}{2}$) of nitrile hydratase in 35% enriched $H_2{}^{17}O$, taken at the high-field edge (g_3) of the EPR envelope. The quintet shown represents the ν_+ branch of the ^{17}O ENDOR pattern that is centered at the ^{17}O Larmor frequency, 7.3 MHz (filled triangle). The quadrupole splittings and hyperfine values calculated are estimated using second-order perturbation theory in the nuclear quadrupole interaction for an ^{17}O nucleus. Conditions: 12,650 G, 34.92 GHz, 0.16 mW microwave power, 1 G modulation amplitude, 0.5 MHz/sec rf scan speed, 2 K. (Adapted from Jin et al.[38a])

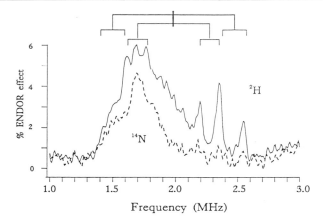

FIG. 19. X-band Mims ENDOR spectrum showing resolved deuteron hyperfine couplings in nitrile hydratase. Solid trace represents sample in ^2H_2O buffer, and dashed trace is sample in ^1H_2O buffer. The ^2H patterns are centered at the Larmor frequency of 1.97 MHz (vertical bar) and show first-order splittings from both the hyperfine and quadrupole interactions. Conditions: 3000 G, 9.4 GHz, 16 nsec pulse lengths, $\tau = 376$ nsec, rf pulse length 60 μsec, repetition rate 5 Hz, temperature 2 K. (Adapted from Jin et al.[38a])

sponse. Figure 19 shows the deuteron ENDOR resonances from a Mims experiment on nitrile hydratase. As indicated in Fig. 19, the resonances can be assigned as arising from two species, each giving rise to quadrupole-split doublets. The presence of the two deuteron resonances, whose A values correspond to two of the exchangeable proton resonances, suggests that the coordinated solvent-derived ligand in nitrile hydratase is an H_2O species.

The resulting identification of the ligands predicted from ENDOR studies of the Fe center in nitrile hydratase is shown in Fig. 20.

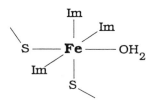

FIG. 20. Structure of the ligation environment for nitrile hydratase, as derived from ENDOR spectroscopy and other biophysical measurements. An alternative arrangement could have Im trans to $-OH_2$.

Summary

The information derived from measuring the electron–nuclear hyperfine interaction can be used for analysis of both structure and mechanism in biomolecules, provided an EPR signal exists. The examples given here illustrate the usefulness of the multiple techniques now available for the ENDOR measurement, as well as the importance of isotopic labeling within the biological system under study. Although current studies have answered many important questions, it is of note that methodologies are still evolving, in particular in both the application of pulsed EPR technologies and the manipulation of biological samples. Improvements in both of these aspects of the ENDOR experiment will increase the amount of information that these studies can provide in the future.

Acknowledgments

We are indebted to Dr. Peter Doan and Prof. Joshua Telser for assistance in reading the manuscript. We thank Dr. Doan for providing several original figures. We fully acknowledge our collaborators for the projects mentioned above from this laboratory, others associated with projects that could not be mentioned in this work but from whom we have gained much experience, and the graduate students and postdoctoral fellows involved in the work. Our ENDOR studies could not have been performed without the technical expertise of Mr. Clark Davoust, or the support of the National Institutes of Health (HL 13531, and a postdoctoral training grant to V.J.D.), the National Science Foundation (MCB 9207974), and the U.S. Department of Agriculture.

[24] Electron Spin Labels

By Glenn L. Millhauser, Wayne R. Fiori, and Siobhan M. Miick

Introduction

Nitroxides are stable organic free radicals that, when attached to biomolecules, are called spin labels and serve as electron paramagnetic resonance (EPR) reporter groups. They are readily detectable in the micromolar concentration range from samples containing as little as several microliters of sample. There is usually no background interference from other radical species. Spin labels detect molecular motion, intermolecular collisions, molecular ordering, local geometry, presence of O_2 and other paramagnetic species, and solvent polarity. In 1965, McConnell and co-workers demonstrated the first use of spin labels in a study of the tumbling motions of poly(L-lysine) and bovine serum albumin.[1] While showing the power of this novel technique they suggested that "there are many other potential applications of this method to biological systems." In fact, spin labeling has developed into a major field of biochemistry and biophysics that continues to grow to this day.

Since that first study, there have been an enormous number of experimental and theoretical developments in spin label EPR, and there are many excellent monographs describing these advances.[2-5] Because of the breadth of the field it is impossible to give a complete overview here of all the potential applications of spin labeling and all the subtleties of spectral interpretation. Instead, our goal is to present a general outline that describes modern usage of the technique. Portions of this chapter will draw from the monographs mentioned above, and an in-depth study of spin label EPR will certainly require access to these references.

This chapter begins with the molecular structure of spin labels and examples of their chemistry of attachment. Next, we address the basic components of the nitroxide EPR spectrum, the spin Hamiltonian, and the factors that influence the spectrum (see also G. W. Brudvig, this volume [22]). Most commonly, spin labels are used to probe motion, so the following section is devoted to molecular tumbling and the correspondence of tumbling time scales to nitroxide spectra. Finally, we address selected applications and new techniques (see also Ref. 6).

Structure of Nitroxides and Some Spin Labeling Reactions

Figure 1 shows a few selected spin labeling schemes. The nitroxide moiety is the NO group, and the unpaired electron resides in a molecular orbital formed mainly from $2p$ orbitals of the nitrogen and the oxygen. The ring bearing the NO is often referred to as the nitroxide ring. The NO free radical site is quite stable in both aqueous and organic solutions but can be reduced or oxidized under more extreme conditions (see Ref. 2). Often such redox reactions are reversible. For instance, spin-labeled

[1] T. J. Stone, T. Buckman, P. L. Nordio, and H. M. McConnell, *Proc. Natl. Acad. Sci. U.S.A.* **54,** 1010 (1965).

[2] L. J. Berliner, "Spin Labeling: Theory and Applications." Academic Press, New York, 1976.

[3] L. J. Berliner, "Spin Labeling: Theory and Applications II." Academic Press, New York, 1979.

[4] G. I. Likhtenshtein, "Spin Labeling Methods in Molecular Biology." Wiley, New York, 1976.

[5] L. J. Berliner and J. Reuben, "Spin Labeling: Theory and Applications; Biological Magnetic Resonance 8." Plenum, New York, 1989.

[6] G. L. Millhauser, *Trends Biochem. Sci.* **17,** 448 (1992).

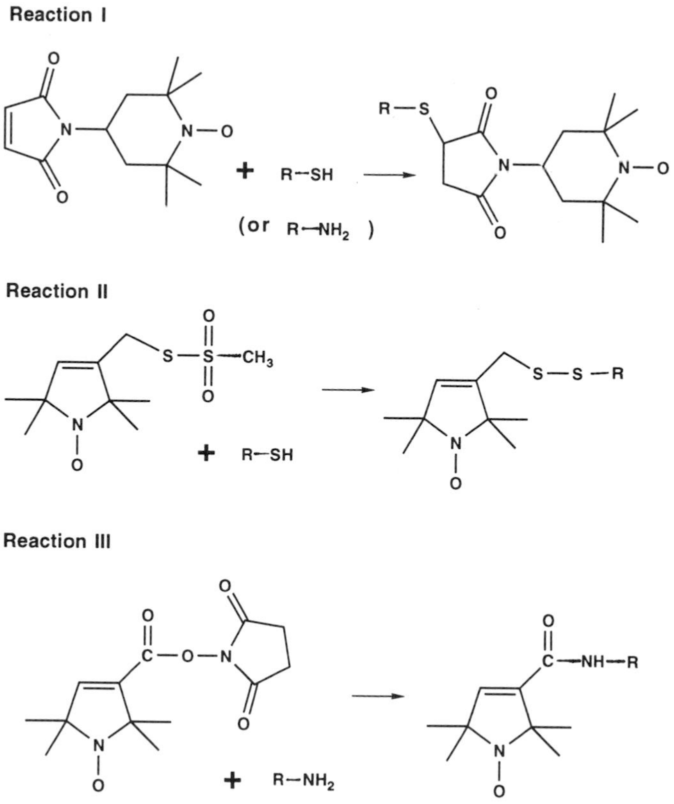

Reaction I

Reaction II

Reaction III

Fig. 1. Three representative spin-labeling reactions. Reaction I shows the maleimide spin label which is specific for -SH and -NH$_2$ groups. Reaction II shows the methane thiosulfonate spin label which is specific for -SH. Reaction III involves the succinimide spin label which attaches to -NH$_2$.

peptides are often handled at pH 2 during purification, which can lead to partial reduction. The paramagnetic nitroxide can be quantitatively regenerated by bubbling the aqueous peptide solution with O$_2$. Introduction of a Cu wire during O$_2$ bubbling can help with some nitroxides (B. H. Robinson, personal communication, 1993). The four methyl groups on the carbons adjacent to the nitroxide nitrogen provide protection against disproportionation.

Between the nitroxide ring and its attachment point on a biomolecule is the covalent linker, the length and rigidity of which may be chosen for specific experiments. For example, accurate reporting of rotational motion is best achieved with a short rigid linker, whereas long flexible linkers are often used as molecular dipsticks in the structural characterization of protein active sites.[7] Reaction I in Fig. 1 shows the attachment of the maleimide spin label (Aldrich, Milwaukee, WI) to a sulfhydryl group. This is one of the most often used spin labels because it is fairly specific for -SH and attaches irreversibly with good rigidity. The methane thiosulfonate spin label (Reanal, Budapest, Hungary) in reaction II (Fig. 1) attaches with 100% specificity and high rigidity to -SH groups.[8] This spin label must be handled with some care though since the -S–S- linkage is labile at elevated temperatures or at pH extremes. Reaction III (Fig. 1) shows how the succinimide spin label (Molecular Probes, Eugene, OR) attaches to amines such as Lys groups in peptides and proteins.

The ^{14}N nucleus has a magnetic moment with $I = 1$ and is coupled to the moment of the unpaired electron, splitting the EPR spectrum into three resolvable hyperfine lines (see below). There is also coupling to the twelve equivalent methyl ^{1}H nuclei, but this coupling is weak and results mainly in unresolved broadening of the three ^{14}N hyperfine lines. Deuteration of the nitroxide ring eliminates most of this broadening, and this can be an advantage in experiments where narrow line shapes are preferred. ^{15}N ($I = \frac{1}{2}$) spin labels give only two hyperfine lines and can be useful for simplifying complicated spectra (see Ref. 5) and for studying intermolecular collisions in mixed isotope experiments.[9]

Nitroxide Spin Hamiltonian

The starting point for quantitating spin physics in magnetic resonance is the spin Hamiltonian. For an isolated nitroxide in dilute solution (i.e., where there is no detectable interaction with other nitroxides) the spin Hamiltonian $\mathcal{H}$ is

$$\mathcal{H} = \beta H_0 \cdot \mathbf{g} \cdot \hat{S} + a\hat{I} \cdot \hat{S} + \hat{I} \cdot \mathbf{A} \cdot \hat{S} \tag{1}$$

where $\hat{S}$ and $\hat{I}$ are the electron and nuclear spin operators, respectively, H_0 is the applied external magnetic field, β is the Bohr magneton, $\mathbf{g}$ is the g tensor, a is the hyperfine coupling constant, and $\mathbf{A}$ is the electron–nucleus dipolar tensor.[10] (Often a is added to each diagonal term of $\mathbf{A}$, resulting in the so-called hyperfine tensor.) The first term of Eq. (1), $\beta H_0 \cdot \mathbf{g} \cdot \hat{s}$,

[7] B. J. Sutton, P. Gettins, D. Givol, D. Marsh, S. Wain-Hobson, K. J. Willan, and R. A. Dwek, *Biochem. J.* **165**, 177 (1977).

[8] L. J. Berliner, J. Grunwald, H. O. Hankovsky, and K. Hideg, *Anal. Biochem.* **119**, 450 (1982).

[9] J. J. Yin, J. B. Feix, and J. S. Hyde, *Biophys. J.* **58**, 713 (1990).

[10] A. Carrington and A. D. McLachlan, "Introduction to Magnetic Resonance with Applications to Chemistry and Chemical Physics." Chapman & Hall, New York, 1980.

gives the electronic Zeeman energy and describes the interaction of the unpaired electron with the external magnetic field. The tensor property of **g** reflects the fact that the Zeeman energy depends on the orientation of the nitroxide with respect to H_0. This dependence arises from molecular spin–orbit coupling. **g** may be represented as a 3×3 matrix, and diagonalization gives the three principal **g** tensor components: g_{xx}, g_{yy}, and g_{zz}. The remaining two terms in the Hamiltonian describe the electron–nuclear interactions; in nitroxides one is mainly concerned with the nitrogen nucleus. The magnetic field at this nucleus is dominated by the field exerted by the unpaired electron, so the nuclear Zeeman term for the nitrogen is usually approximated as zero. In other words, the nitrogen nuclear spin is quantized in the magnetic field of the electron. The hyperfine energy a arises from the Fermi contact term, which is proportional to the electron wave function density at the nucleus. On the other hand, **A** is a traceless tensor with elements determined by the dipolar interaction between the nitrogen nucleus and the electron. Diagonalization of the matrix representation of $\mathbf{A} + a\mathbf{I}$ (where **I** is the identity matrix) gives the principal tensor components: A_{xx}, A_{yy}, and A_{zz}.

The orientation dependence of the nitroxide Hamiltonian with respect to the applied magnetic field is responsible for the unique shape of the nitroxide spectrum. To understand this, it is most useful to start by connecting the various mathematical terms in the Hamiltonian with the actual nitroxide absorption spectrum. Imagine a perfect crystal containing nitroxide molecules that all have the same orientation with respect to the crystal axes, and dilute so that there are no interactions between the spins. The spectrum of this crystal at any particular orientation with respect to H_0 will give three sharp hyperfine lines resulting from coupling to the ^{14}N spin states ($M_I = +1, 0, -1$).[5,11] On rotation of the crystal, each of the lines will shift in a way that is unique for that hyperfine line. Now imagine that the crystal is crushed into a fine powder. The spectrum of the powder will of course reflect all molecular orientations, and each hyperfine line now appears as a so-called powder pattern (or Pake pattern) spectrum. The principal tensor components are identifiable from the two edges and from the signal maximum of each hyperfine line as shown in Fig. 2. Spectra here are for an X-band spectrometer that operates at approximately 9.0 GHz microwave frequency with a static magnetic field in the range of 0.34 tesla. EPR spectra are usually collected and displayed in the phase-sensitive first-derivative mode, and this is also shown in Fig. 2.

[11] P. F. Knowles, D. Marsh, and H. W. E. Rattle, "Magnetic Resonance of Biomolecules." Wiley, London, 1976.

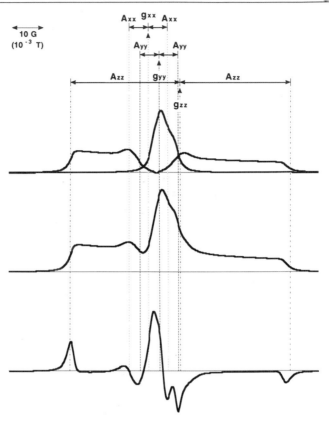

FIG. 2. Simulated X-band EPR spectra for a nitroxide powder pattern. *Top:* Powder patterns for each hyperfine line and the relationship to the principal tensor values from the spin Hamiltonian. *Middle:* Absorption pattern. *Bottom:* First-derivative pattern as is usually seen in a continuous wave EPR spectrum.

Rotational Motion, Time Scales, and Techniques

The majority of spin label studies center on the detection of rotational motion as a means for determining local flexibility, local ordering, or overall tumbling dynamics in a biomolecule (examples are given below). Rotational tumbling of a spherical molecule in solution is a random process, governed by the laws of Brownian motion and characterized by the rotational correlation time τ_R. In a hypothetical experiment where one knows the exact orientation of a particular molecule at $t = 0$, τ_R marks a passage of time in which the probability of finding the molecule in the original orientation has decreased to e^{-1} (and the mathematical description of this decrease is given by the correlation function). The correlation time is directly proportional to molecular volume (V) and solution viscosity (η) and is inversely proportional to temperature (T), as described by the Stokes–Einstein relationship:

$$\tau_R = V\eta/kT \tag{2}$$

where k is Boltzmann's constant. As a benchmark, a small protein of molecular weight 2400 in water at room temperature has a correlation time of approximately 1 nsec.[12] Anisotropic molecules tumble with different rates about their principal diffusion axes, and there are well-developed magnetic resonance theories to cover such cases (although we do not consider the quantitative aspects of anisotropic diffusion in this chapter).

Rotational motion exerts fluctuating magnetic fields on the electron spin and also averages the various tensor components (see Fig. 2). Nitroxide EPR spectra are profoundly influenced by the rate of this motion relative to the range of electron spin precession frequencies within each hyperfine line. Thus, one can use EPR spectroscopy to determine τ_R accurately. There are distinct temporal regimes in which spin label spectra require different types of data collection methods and data analysis to achieve this. These regimes are discussed below.

Motional Narrowing Spin Label Spectra: $\tau_R < 2$ nsec

Magnetic resonance spectra modulated by temporal processes, such as molecular rotational diffusion, fall mainly into two regimes: one termed motional narrowing and the other slow motion. In the sections that follow, the slow motional regime is divided into three sections based on the EPR techniques used to determine τ_R. Motional narrowing results when the molecular tumbling is more rapid than the total range of resonance frequencies within a single hyperfine powder pattern. When τ_R is short, all orientations are rapidly averaged, and the nitroxide spectrum is characterized by three distinct hyperfine lines separated by the hyperfine coupling constant a. The theory of motional narrowing applies when the correlation time satisfies the relation

$$\tau_R < \frac{1}{|\Delta\omega|} = \frac{1}{\gamma|\Delta H_z|} \tag{3}$$

where $\Delta\omega$ is the range of frequencies, γ is the magnetogyric ratio, and ΔH_z is the width of a hyperfine powder pattern in units of the magnetic field.[13] Using the high-field line with an approximate width of 30 gauss, one

---

[12] C. R. Cantor and P. R. Schimmel, "Biophysical Chemistry II." Freeman, San Francisco, 1980.

[13] C. P. Slichter, "Principles of Magnetic Resonance." Springer-Verlag, Berlin, 1980.

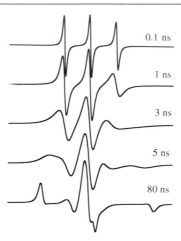

FIG. 3. Simulated nitroxide spin label spectra for different values of τ_R. When τ_R is less than 2 nsec, the spectra are motionally narrowed and are composed of three Lorentzian hyperfine lines. For values of τ_R longer than 100 nsec, the continuous wave spectra are indistinguishable from those of the powder pattern.

finds that τ_R must be less than 1.8 nsec to satisfy the motional narrowing requirement. Figure 3 shows simulated EPR spectra for several correlation times, and it is apparent that spectra for $\tau_R < 2$ nsec are indeed characterized by three resolved hyperfine lines. Furthermore, as τ_R decreases, averaging due to rotational motion becomes more efficient and the hyperfine lines become sharper: hence the name motional narrowing. In the absence of any other line broadening effects, these hyperfine lines will follow a Lorentzian line shape. In practice, however, the methyl protons on the nitroxide ring confer additional inhomogeneous broadening and the lines assume a Gaussian profile.

There is differential broadening among the $M_I = +1, 0, -1$ lines due to differences in the breadth of unaveraged powder pattern of each hyperfine line. Motional narrowing theory of nitroxides reflects these differences through the relationship

$$\frac{1}{T_2(M_I)} = A + BM_I + CM_I^2 \tag{4}$$

where A, B, and C are line shape parameters that depend on the rotational motion and $[T_2(M_I)]^{-1}$ is directly proportional to the peak-to-peak width of each first-derivative hyperfine line[14] (also see Ref. 2). T_2 is often called

[14] N. M. Atherton, "Electron Spin Resonance: Theory and Applications." Wiley, New York, 1973.

the spin–spin relaxation time and is the coherence lifetime for the ensemble of electron spins. The line shape parameters are expressed in terms of spectral densities:

$$j(\omega) = \frac{\tau_R}{1 + (\omega\tau_R)^2} \tag{5}$$

which are Fourier transforms of the rotational correlation function evaluated at angular frequency ω. Conceptually, $j(\omega)$ is a measure of the magnetic field fluctuations, from random rotational motion, at the particular frequencies that influence the electron spin.

Each line shape parameter has a unique expression, and, as an example, B for isotropic motion is given by

$$B = \frac{\pi}{10}\,\omega_0\left\{g^{(0)}D^{(0)}\left[\frac{16}{3}j(0) + 4j(\omega_0)\right] + 2g^{(2)}D^{(2)}\left[\frac{16}{3}j(0) + 4j(\omega_0)\right]\right\} \tag{6}$$

where ω_0 is the spectrometer frequency (radians/sec) and the terms $g^{(i)}$ and $D^{(i)}$ are constants composed from differences in the principal tensor values (g_{xx}, g_{yy}, A_{xx}, etc.) (see Ref. 15 and citations therein). The $j(0)$ spectral densities are equal to τ_R and provide the physical basis for powder pattern averaging: fast motion → short τ_R → small B (and A and C) → narrow hyperfine lines. This is referred to as the secular contribution to the line shape. The $j(\omega_0)$ spectral densities are the nonsecular contributions and give the influence of spin transitions (which require molecular fluctuations at frequency ω_0) on the line shape. For $\tau_R > 0.05$ nsec, one can usually assume $j(\omega_0) \approx 0$.

The value of B is directly determined from the EPR spectrum by rearrangement of Eq. (4):

$$B = \frac{1}{2}\left[\frac{1}{T_2(+1)} - \frac{1}{T_2(-1)}\right] \tag{7}$$

which requires straightforward line width measurements of the outside hyperfine lines. However, direct line width measurements often have large experimental uncertainties and do not yield accurate values for τ_R. A better method is to use the peak-to-peak heights $V(M_I)$ of the hyperfine lines, which are much more sensitive to line broadening, and the expression

$$B = \frac{3^{1/2}}{4}\,\Delta H(0)\left\{\left[\frac{V(0)}{V(+1)}\right]^{1/2} - \left[\frac{V(0)}{V(-1)}\right]^{1/2}\right\} \tag{8}$$

[15] A. P. Todd and G. L. Millhauser, *Biochemistry* **30**, 5515 (1991).

where $\Delta H(0)$ is the peak-to-peak width of the $M_1 = 0$ line (first derivative). Setting Eq. (8) equal to Eq. (6), one now has a direct and reliable method for determining τ_R from a motionally narrowed nitroxide spectrum. (Further corrections may be used to account for inhomogeneous broadening, and this improves accuracy by 10 to 20%; see Ref. 5.)

The approach described here is very useful for the study of low molecular weight biomolecules, such as peptides, and examples are given below. Conventional field-scan EPR spectra (often called continuous wave or cw EPR) are sufficient for providing excellent estimates of the correlation time sensed by the nitroxide. Spin echo spectroscopy was used to measure $T_2(M_1)$ directly for each hyperfine line, and the τ_R values agreed well with those obtained from conventional spectra.[16] There are also methods for measuring anisotropy in the molecular motion, and this can be useful for aiding in the determination of biomolecular shapes in solution.[17]

Slow Motion Spin Label Spectra: 2 nsec $< \tau_R <$ 100 nsec

When τ_R is too long to satisfy Eq. (3), the temporal regime is termed slow motion. Spectra where $\tau_R > 2$ nsec are routinely encountered in the study of proteins and membranes. These spectra take on a shape that is between a powder pattern and that of the three-line motionally narrowed spectrum. This regime is readily understood by exploring how motion perturbs the powder pattern line shape. Each field location within a hyperfine line of this spectrum corresponds to a specific orientation of the nitroxide with respect to the magnet field (see Fig. 2 and discussion above). Motion of the nitroxide averages nearby regions of the spectrum together and, for increasing motion, smooths out details of the powder pattern spectrum. Both nuclear spin transitions and electron spin transitions add further perturbations. For decreasing values of τ_R, the hyperfine lines all progressively narrow and ultimately become Lorentzian line shapes (when $\tau_R < 2$ nsec).

Although conceptually straightforward, it is challenging to extract correlation times from spectra in this motional regime because spectral analysis requires more than line width measurements as is done with motionally narrowed spectra. It is usually best to use simulation techniques in the intermediate motional regime, whereby one computes a spectrum given a particular correlation time (or pair of correlation times if the motion is anisotropic). The result is then compared to the experimental spectrum.

To simulate nitroxide EPR spectra, one must solve the electron spin

[16] S. M. Miick and G. L. Millhauser, *Biophys. J.* **63**, 917 (1992).
[17] J. S. Hwang, R. P. Mason, L. Hwang, and J. H. Freed, *J. Phys. Chem.* **79**, 489 (1975).

equation of motion, which is called the stochastic Liouville equation (SLE) (see Ref. 2), and is given by

$$\frac{\partial \rho(\Omega, t)}{\partial t} = i[\hat{\mathcal{H}}(\Omega), \rho(\Omega, t)] - \Gamma_\Omega \rho(\Omega, t) \tag{9}$$

where $\rho(\Omega, t)$ is the spin density matrix for molecular orientation Ω and Γ_Ω is a diffusion operator. The commutator $[\hat{\mathcal{H}}(\Omega), \rho(\Omega, t)]$ gives the quantum mechanical spin dynamics for each molecular orientation, and Γ_Ω is a stochastic operator that acts on the angular molecular coordinates and allows for molecular reorientation. For instance, in the case of isotropic rotational diffusion, Γ_Ω is the operator $R\nabla^2$ from the diffusion equation for rotational motion,[18] and $R^{-1} = 6\tau_R$. The time-dependent magnetization $S_x(t)$ is determined from ρ using the algebra of density matrices. Fourier transformation then yields the field-dependent spectrum. Equation (9) also allows for the incorporation of constraints such as restoring potentials, and this can be useful for simulation of spin-labeled lipids where only a limited range of orientations are explored by the nitroxide.

Computation of spectra from Eq. (9) is mathematically and numerically challenging and is essentially a field of study in itself (see Ref. 5). Probably the most successful techniques for solving the SLE rely on eigen decomposition where the solution is projected onto the set of eigenfunctions of ∇^2. The result is a set of coupled linear equations that can be solved numerically. With the speed of modern computers and the development of sophisticated algorithms, SLE simulations are now routinely used in the study of spin label motion. The spectra in Fig. 3 were simulated with routines provided on a computer disk packaged with a monograph on spin labeling.[5]

Slow Motion Spin Label Spectra: 100 nsec < τ_R < 1000 nsec

When τ_R is sufficiently long, cw EPR spectra become insensitive to nitroxide motion, and the spectra are indistinguishable from powder patterns. Thus, one must rely on spectroscopic methods beyond cw EPR for the extraction of motional processes. Historically, the motional regime between 100 nsec and 1 μsec has presented a blind spot to EPR spectroscopy, and reliable methods for extraction of τ did not exist. More recently, though, this problem has been solved. The technique combines the methods of pulsed saturation recovery EPR (SR-EPR) and pulsed saturation recovery electron–electron double resonance (SR-ELDOR).[19] Spectroscopic saturation occurs when the applied radiation is sufficiently intense

[18] S. J. K. Jensen, *in* "Electron Spin Relaxation in Liquids" (L. T. Muus and P. W. Atkins, eds.), Plenum, New York, 1972.

[19] D. A. Haas, T. Sugano, C. Mailer, and B. H. Robinson, *J. Phys. Chem.* **97**, 2914 (1993).

to equalize the populations of connected energy levels. In cw EPR, low microwave powers are applied in order to avoid saturation and to assure that the spin energy level populations are always near the equilibrium values. In both SR-EPR and SR-ELDOR, intense microwave fields are applied in order to saturate local regions of the nitroxide spectrum. Subsequent to the saturating pulse, three relaxation processes remove the local saturation. First, molecular motion carries saturation to other regions of the spectrum. Second, there are nuclear spin transitions that carry saturation between hyperfine lines, and the rate of this process is T_{1n}^{-1} (where T_{1n} is the average lifetime in a nuclear spin state). Finally, T_{1e}^{-1} describes the rate at which the spins return to equilibrium by transferring energy to other degrees of freedom such as molecular vibration and rotation. T_{1e} is usually called the spin–lattice relaxation time.

After application of the saturating pulse, recovery of the EPR signal is monitored either at the local region where the saturating pulse was applied (SR-EPR) or at other selected regions of the spectrum (SR-ELDOR). Recovery gives multiple exponential decay functions with time constants that are linear combinations of τ_R^{-1}, T_{1n}^{-1}, and T_{1e}^{-1}. Simultaneous analysis of the SR-EPR and SR-ELDOR data provides unambiguous values for the relaxation parameters.

As mentioned, this method is new, and biological applications will be forthcoming. Specialized instrumentation is required for both application of the saturating pulse as well as detection of the signal recovery. Nevertheless, it appears that combined SR-EPR and SR-ELDOR will prove to be very useful.

Slow Motion Spin Label Spectra: $\tau_R > 1000$ nsec

Two methods are useful to measure slow motion spin label spectra: saturation transfer (ST-EPR) and two-dimensional electron spin echo spectroscopy (2D ESE). In the ST-EPR experiment,[20] a cw spectrometer is operated at high microwave power, and this causes partial saturation of the nitroxide spectrum. As the field is scanned, molecular motion carries this saturation to nearby regions of the spectrum, yielding spectra that are very sensitive to the rotational correlation time. One of the great advantages of ST-EPR is that little instrumentation is required beyond the conventional cw EPR instrument. Most any laboratory equipped to perform EPR can also perform ST-EPR.

Direct determination of τ_R from ST-EPR spectra is often based on SLE spectral simulation techniques. Another approach, though, is to calibrate

[20] J. S. Hyde and D. D. Thomas, *Annu. Rev. Phys. Chem.* **31**, 293 (1980).

the ST-EPR spectrum on a species of known τ_R. For instance, ST-EPR spectroscopy has proved to be an excellent method for examining the rotation of spin-labeled myosin head groups in muscle fibers.[21] In these experiments maleimide spin-labeled hemoglobin (see Fig. 1) is used as a convenient correlation time reference.

The 2D ESE technique operates in a fashion fundamentally different from any of the techniques discussed above.[22,23] In the discussion of motionally narrowed spectra we introduced the spin–spin relaxation time T_2. This is a fundamental measure of relaxation in magnetic resonance that describes how spins dephase from random processes such as rotational diffusion. Time-domain EPR spectrometers use high power microwave pulses to measure directly free induction decays as well as various relaxation processes. The spin echo experiment uses a coherent two-pulse scheme to reveal T_2.

Whereas spin dynamics in the motionally narrowed regime are characterized by a single value for T_2 at each hyperfine line, the slow motional regime is characterized by a continuous variation of T_2 versus frequency within each hyperfine line of the EPR spectrum. This variation is unique for a specific correlation time as well as for anisotropic motion. In 2D ESE, T_2 is measured at each point along the slow motional spectrum, revealing the frequency-dependent relaxation. The advantage of this method is that spectral analysis is reasonably straightforward. The challenge is that 2D ESE requires pulsed EPR spectrometers which, at this time, are less common than cw EPR instruments.

Selected Applications

Dynamics and Structure of Peptides

In our laboratory we have been studying the physical characteristics of helix-forming peptides. In 1989 it was found that alanine-based peptides form helices in aqueous solution.[24] This is a key discovery because these peptides can now be used as simple model systems for understanding protein folding, structure, and dynamics. It is generally believed that

[21] C. L. Berger and D. D. Thomas, *Biochemistry* **32**, 3812 (1993).

[22] J. Gorcester, G. L. Millhauser, and J. H. Freed, *in* "Advanced EPR: Applications in Biology and Biochemistry" (A. J. Hoff, ed.), Elsevier, Amsterdam, 1989.

[23] J. Gorcester, G. L. Millhauser, and J. H. Freed, *in* "Modern Pulsed and Continuous-Wave Electron Spin Resonance" (L. Kevan and M. K. Bowman, eds.), Chap. 3, p. 119. Wiley, New York, 1990.

[24] S. Marqusee, V. H. Robbins, and R. L. Baldwin, *Proc. Natl. Acad. Sci. U.S.A.* **86**, 5286 (1989).

to understand protein structure and eventually solve the protein folding problem, one must first be able to explain quantitatively the properties of short peptides. Most research in helical peptides has focused on the factors that control helicity. Now molecular motion is becoming a focal point as indicated by computer molecular dynamics studies.[25] In our laboratory we have been using spin label EPR to provide a direct experimental measure of peptide dynamics. The findings are creating a link between theories of protein motion and experiment. We have also developed a double-label technique to measure directly the winding geometry of helical peptides, and the results have been surprising.

The approach chosen for measuring dynamics is to synthesize a series of spin-labeled peptide analogs in which each peptide is spin labeled in a unique position.[26] The original helical peptide is called the 3K(I), referring to the three Lys (K) residues in the sequence. A series of six nitroxide spin-labeled analogs of the 3K(I) have been synthesized with the following sequences:

Ac-ACAAKAAAAKAAAAKA-NH$_2$	3K-2
Ac-AAACKAAAAKAAAAKA-NH$_2$	3K-4
Ac-AAAAKAACAKAAAAKA-NH$_2$	3K-8
Ac-AAAAKAAAAKCAAAKA-NH$_2$	3K-11
Ac-AAAAKAAAAKAACAKA-NH$_2$	3K-13
Ac-AAAAKAAAAKAAAAKC-NH$_2$	3K-16

and the designation gives the labeling position.[26a] Peptides were specifically labeled at the Cys (C) with methane thiosulfonate spin label (Fig. 1). A computer drawing of a spin-labeled 3K-8 is shown in Fig. 4. The S–S bond in the linker undergoes only limited rotational motion owing to partial double bond character. Thus, the linker forms a relatively rigid attachment between the spin label and the peptide backbone.

Peptides in the molecular weight range of the 3K series have τ_R values that are less than 1 nsec, and so motional narrowing theory is used. Calculation of τ_R for the series of peptides reveals position-dependent dynamics. In variable-temperature experiments the solution viscosity can vary over a significant range, so it is useful to plot $kT\tau_R/\eta$ to factor out these variations [see Eq. (2)].[26] Figure 5 shows a plot of $kT\tau_R/\eta$ versus position for a range of temperatures. It is immediately clear that τ_R is reduced near the ends of the helical peptides. The helix content is greatest at 1°, and still at this temperature the helix termini show greater local

[25] K. V. Soman, A. Karimi, and D. A. Case, *Biopolymers* **31,** 1351 (1991).
[26] S. M. Miick, A. P. Todd, and G. L. Millhauser, *Biochemistry* **30,** 9498 (1991).
[26a] S. M. Miick, K. M. Casteel, and G. L. Millhauser, *Biochemistry* **32,** 8014 (1993).

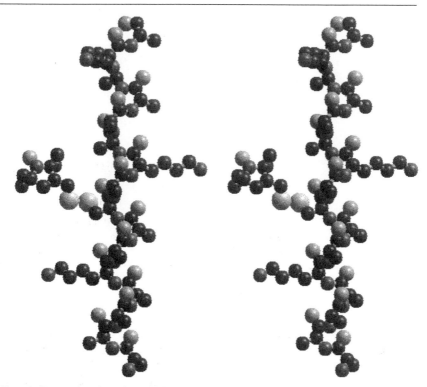

FIG. 4. Stereo drawing of a spin-labeled 3K-8 peptide. The peptide is in the geometry of a 3_{10} helix. The nitroxide label, which is pointing out on the left-hand side of the peptide, is the methane thiosulfonate spin label. This label attaches with sufficient rigidity to report the expected value of τ_R for the peptide.

dynamics than the peptide center. Further, the C terminus is more mobile than the N terminus. Computer molecular dynamics studies have suggested that helix unfolding proceeds from the C terminus, and the data presented here are consistent with this.[25] The position-dependent τ_R data were used to calculate backbone torsional fluctuations, and the results also compared well to those from computer molecular dynamics studies. The enhanced dynamics of the C terminus has been observed in other laboratories, but the phenomenon has never been explained in general terms. Our laboratory has suggested that the enhanced C terminus dynamics results from L-amino acid chirality which, in turn, directs side chains away from the C terminus and thereby increases the exposure of the helix hydrogen bonds to aqueous solvent.

Spin label EPR may also be used to map geometry in helical peptides. When helix-forming peptides were first discovered, it was assumed that

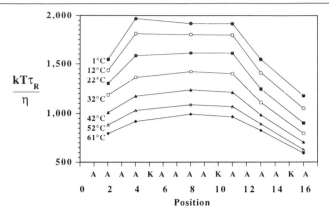

FIG. 5. Plot of $kT\tau_R/\eta$ versus position for the helical 3K peptide at different temperatures. At 1° the peptide exhibits the greatest helicity. Note that the middle region of the peptide at 1° shows uniform τ_R, indicating uniform rigidity. The peptide ends have lower τ_R values, which indicates greater molecular dynamics at those positions. The C terminus is found to be more dynamic than the N terminus, and this is consistent with computer molecular dynamics predictions.

they would adopt the geometry of an α helix. However, the conventional spectroscopic techniques used to study these peptides could not distinguish between α helix and the next most common peptide helix: the 3_{10} helix. To address this we designed the following doubly spin-labeled analogs of Ala-based peptides where the methane thiosulfonate nitroxide spin label forms an unbranched side chain extending from the sulfur of a cysteine residue[27]:

Ac-AAACKCAAAKAAAAKA-NH$_2$ 3K-(4,6)
Ac-AAACKACAAKAAAAKA-NH$_2$ 3K-(4,7)
Ac-AAACKAACAKAAAAKA-NH$_2$ 3K-(4,8)

The relative distances between the nitroxide side chains depend on the helix geometry as can be seen in Fig. 6. EPR spectroscopy can reveal distances between two spins on a biomolecule through dipolar coupling between the spins as well as overlap of the two paramagnetic electron molecular orbitals. As long as the nitroxide groups are not too close, these mechanisms mainly result in broadening of the nitroxide spectrum. Also shown in Fig. 6 are the spectra of the doubly labeled peptides, and it is clear that only that of the 3K-(4,7) shows dramatic broadening beyond the conventional spectrum. These EPR experiments suggest that the most

[27] S. M. Miick, G. V. Martinez, W. R. Fiori, A. P. Todd, and G. L. Millhauser, *Nature* (*London*) **359,** 653 (1992).

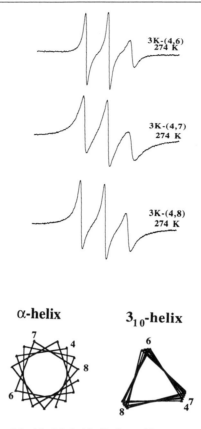

FIG. 6. EPR spectra of doubly labeled helical peptides are used to determine geometry. When the spins are near one another in a single peptide, the EPR spectrum broadens. The helical wheels show that the 3_{10} helix should have a strong interaction between positions 4 and 7, whereas the α helix should have an equal interaction between positions 4 and 7 and positions 4 and 8. Among the experimental spectra, only that of the 3K-(4,7) shows significant broadening, and this provides evidence that the peptides adopt the conformation of a 3_{10} helix.

likely peptide geometry is a 3_{10} helix. This is a controversial finding that is leading to a reassessment of the energetics involved in helix formation.

Structure of Membrane Proteins

Membrane proteins are exceedingly difficult to crystallize and present one of the great modern-day challenges in protein structure determination. It has been shown, though, that the combination of site-directed mutagenesis and EPR spectroscopy can be used for solving secondary and tertiary structure in chosen regions of membrane proteins.[28] The EPR method is called spin label relaximetry. The technique was applied to bacteriorhodopsin, a light-driven proton pump, which has often served as a model membrane protein. Bacteriorhodopsin is composed of seven transmembrane α helices with interconnecting loops. Although the gross structure of bacteriorhodopsin is known, the detailed orientation of the transmembrane α helices with respect to one another and the exact location of interhelical loops remain unclear.

Spin label relaximetry is used to determine the membrane protein topography, that is, whether a particular residue site is either buried within the protein matrix, exposed to the membrane interior, or exposed to the aqueous solution. Classifying each membrane protein residue along a primary sequence into one of these three categories reveals insights into both the local secondary structure and the tertiary contacts. This classification is performed by measuring T_{1e} of selectively placed nitroxide spin labels within the protein. Paramagnetic species that collide with nitroxides will often enhance the nitroxide relaxation rate T_{1e}^{-1}. Thus, one can differentiate among membrane protein side chain environments with the use of medium-specific spin relaxing agents. Two relaxing agents are most useful for membrane proteins: O_2 and chromium oxalate (CROX). O_2 is soluble in membrane lipids as well as in aqueous solution, whereas CROX dissolves only in the latter. Neither agent is able to penetrate into the protein interior in appreciable quantities.

Consider a spin-labeled residue that is on the exterior of the protein and buried within the membrane. On exposure to O_2, the residue in the membrane will exhibit an enhanced relaxation rate. In contrast, CROX will not affect the relaxation rate. Following the same reasoning, a residue exposed to the aqueous solution, say, in a loop region, will show sensitivity to CROX. A residue buried within the protein will show sensitivity to neither. Mobility may also be used to distinguish whether a nitroxide side chain is buried within the protein or is present on the protein exterior.

Eighteen site-directed mutants of bacteriorhodopsin were prepared, each with a single X $\rightarrow$ cysteine (Cys) substitution sequentially placed from positions 125 to 142.[28] These residues are in a region of the protein which starts near the end of one transmembrane helix, passes through an interhelical loop, and then stretches well into the next helical segment (Fig. 7). The mutants were reacted with methane thiosulfonate spin label, which specifically attaches to the Cys (Fig. 1). Extreme care was taken to characterize the folding and functionality of each spin-labeled mutant to ensure that there was little measurable perturbation from the

[28] C. Altenbach, T. Marti, H. G. Khorana, and W. L. Hubbell, *Science* **248**, 1088 (1990).

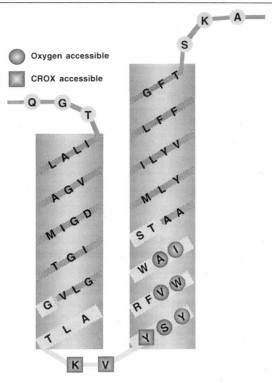

Fig. 7. Results from spin label relaximetry on spin-labeled site-directed mutants of bacteriorhodopsin. The highlighted gray region on transmembrane helices D and E and the interconnecting loop indicate the sequential spin-labeled positions. The positions marked by boxes show sensitivity to CROX, thus locating the precise residues on the interhelical loop. The circled positions show sensitivity to O_2, and this locates the region of helix E that points away from the protein and into the membrane bilayer. The remaining residues are buried within the protein interior.

native protein. Power saturation studies are a convenient method for measuring spin relaxation and involve monitoring the EPR signal as a function of microwave power. Data were gathered for each mutant and the results expressed in terms of the accessibility of both CROX and O_2 to the labeled position. Figure 7 summarizes the results. The inter-helical loop, which is identified as the only region exposed to the CROX-containing aqueous solution, stretches from residues 129 to 131. Beyond the loop, the O_2 experiments reveal a patch along the helix that is within the membrane bilayer. This dramatic evidence leaves no ambiguity as to which residues are buried within the protein interior. Such experiments represent only the beginning of structure mapping with spin labels. For instance, site-directed spin labeling has been combined with stopped-flow mixing to provide a direct observation of protein insertion into membranes.[29]

Future Prospects

In modern experiments one is now able to place selectively spin labels in most any region of a biomolecule. This ability is stimulating an ever increasing usage of spin label EPR spectroscopy. In addition labeling of proteins and peptides, experiments have demonstrated that DNA may be labeled without perturbation of the double-helical structure.[30] Analysis of DNA flexibility may provide important information toward the understanding of protein–DNA and drug–DNA interactions.

In addition to synthetic control of label placement, the last few years has yielded wonderful new developments in EPR instrumentation. High-field EPR spectrometers have been developed that are designed to operate at specific frequencies including 95 GHz[31] and 250 GHz.[32] These high fields can offer a significant improvement in sensitivity as well as provide important insights into motionally narrowed spectra.

Pulsed EPR spectrometers, once considered esoteric, are now becoming commonplace, and commercial instruments are available. One can now perform high pulse-power Fourier transform EPR spectroscopy as well as a range of multiple-pulse experiments. For example, electron spin echo (ESE) spectroscopy, which was briefly mentioned above, uses two intense coherent pulses to determine T_2 directly. This direct measurement can be a distinct advantage for slow motional spectra or for motionally narrowed spectra that have inhomogeneous broadening (which can arise from proton hyperfine interactions). ESE work on labeled peptides has provided a means for testing the validity of the cw EPR experiments as well as a determination of the collision frequency between peptides.[16] Analysis of the collision frequency showed that encounters between peptides are strictly diffusion-controlled, and this tends to rule out the possibility of peptide–peptide interactions exerting influence on their helical structure.

[29] Y. K. Shin, C. Levinthal, F. Levinthal, and W. L. Hubbell, *Science* **259,** 960 (1993).
[30] E. J. Hustedt, A. Spaltenstein, J. J. Kirchner, P. B. Hopkins, and B. H. Robinson, *Biochemistry* **32,** 1774 (1993).
[31] R. Klette, J. T. Törring, M. Plato, K. Möbius, B. Bönigk, and W. Lubitz, *J. Phys. Chem.* **97,** 2015 (1993).
[32] D. E. Budil, K. A. Earle, and J. H. Freed, *J. Phys. Chem.* **97,** 1294 (1993).

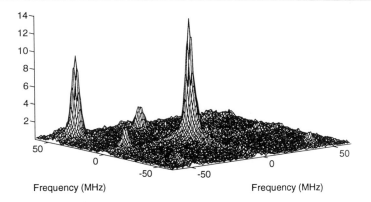

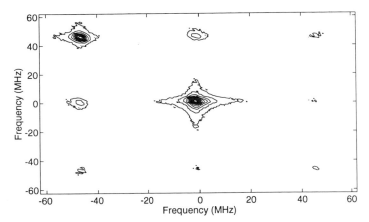

FIG. 8. Two-dimensional exchange spectroscopy (often called 2D-ELDOR) of the spin-labeled 3K-8 peptide with mixing time $T = 296$ nsec. Both the 2D surface and the contour map are shown. The peaks along the diagonal are related to the absorption spectrum of the spin label. The high-field $M_I = -1$ line is weak because of experimental dead time artifacts. The cross-peaks, especially those between the outermost hyperfine lines, provide direct evidence of Heisenberg spin exchange. The cross-peak intensity can be used to determine the second-order rate constant for collisions between peptides.

One can now perform pulsed two-dimensional (2D) EPR spectroscopy completely in the time domain.[22,23,33,34] As a final example, Fig. 8 shows a 2D exchange spectrum of a spin-labeled peptide, using the pulse sequence $\pi/2-t_1-\pi/2-T-\pi/2-t_2$, where T is the mixing time. During T the magnetization is stored parallel to H_0, and exchange between hyperfine lines can

[33] J. Gorcester and J. H. Freed, *J. Chem. Phys.* **88,** 4678 (1988).
[34] S. M. Miick and G. L. Millhauser, *J. Mag. Res.* **104B,** 81 (1994).

arise from nuclear spin flips or from Heisenberg spin exchange (HSE) which arose from encounters between peptides. Fourier transformation with respect to both t_1 and t_2 yields the 2D spectrum. The conventional EPR spectrum is along the diagonal, and the off-diagonal cross-peaks provide evidence of exchange between hyperfine lines. The cross-peak between the outermost hyperfine lines provides direct evidence of HSE. In the past, HSE was determined with time-consuming variable-concentration experiments, but 2D spectroscopy can quickly determine the exchange rate at just a single concentration. Such 2D techniques may prove to be very useful in the study of molecular recognition and other biological processes that rely on encounters between biomolecules.

Acknowledgments

We thank the National Institutes of Health (GM 46870) and the National Science Foundation (DMB 8916946) for financial support.

[25] Optically Detected Magnetic Resonance of Photoexcited Triplet States

By AUGUST H. MAKI

Introduction

Elsewhere in this volume Millhauser et al.[1] have discussed the application of nitroxide electron paramagnetic resonance (EPR) spin labels to the study of the structure and dynamics of biopolymers. Another type of EPR spin label that also is useful for investigating biopolymer systems is provided by the photoexcited triplet state of an intrinsic chromophore, because a triplet state carries electronic paramagnetism. A major advantage of the photoexcited triplet state of an intrinsic chromophore over an extrinsic spin label such as a nitroxide adduct is the relatively small structural perturbation caused by the former, which consists only of a localized electronic excitation. Although not as widely exploited as fluorescence, the phosphorescence of proteins, originating from the photoexcited triplet state, has received a great deal of attention.[2] EPR afficionados have a natural attraction to photoexcited triplet states that dates back to the

[1] G. L. Millhauser, W. R. Fiori, and S. M. Miick, this volume, [24].
[2] J. W. Longworth, in "Excited States of Proteins and Nucleic Acids" (R. F. Steiner and I. Weinryb, eds.), p. 277. Plenum, New York, 1971.

METHODS IN ENZYMOLOGY, VOL. 246

initial EPR measurements by Hutchison and Mangum[3] on the naphthalene triplet state. In fact, EPR spectroscopy was applied in the early 1960s to protein triplet states by Shiga and Piette,[4] who attributed the photoinduced signals observed in D-amino-acid oxidase to tryptophan (Trp) and riboflavin. The limited sensitivity of conventional EPR spectroscopy when applied to biopolymer triplet states, however, has been a serious impediment to further work of this type.

Optically detected magnetic resonance (ODMR), as the name suggests, uses optical means to detect EPR transitions in the photoexcited triplet state. Under the right experimental conditions, which we will outline later, microwave-induced magnetic resonance transitions between the magnetic sublevels of the triplet state produce readily observed changes in the optical characteristics of the sample. Magnetic resonance-induced changes in phosphorescence, fluorescence, and optical absorbance are commonly monitored; the corresponding methods are labeled PDMR, FDMR, and ADMR, respectively. A useful variant of ADMR in which magnetic resonance-induced changes in linear dichroism are monitored (LD–ADMR) has been used to study photosynthetic reaction centers.[5] Among the intrinsic biological chromophores that are capable of being optically pumped to produce a triplet state are the aromatic amino acids, the purine and pyrimidine bases of DNA and RNA, the photosynthetic pigments, and other protein-associated pigments such as flavines. All of these examples have been studied using ODMR spectroscopy.

The ODMR technique has a distinct advantage in sensitivity over conventional EPR spectroscopy for measurements made on biological molecules. To begin with, in ODMR, the absorption or stimulated emission of microwave quanta is not detected directly, as is the case with EPR. Rather, the microwave quanta are "converted" to optical photons which are the detected entities. The quantum up-conversion by a factor of about 10^5 in energy results in greatly increased sensitivity over conventional EPR; the actual attainable sensitivity depends on various factors such as phosphorescence quantum yield, the light collection efficiency, the decay characteristics of the triplet state, and other factors discussed later.

A related feature of ODMR also results in enhanced sensitivity over conventional EPR for measurements on randomly oriented biomolecules that cannot be obtained as single crystals. Conventional EPR measurements are made in an external magnetic field, and because of the magnetic dipole–dipole interaction between the unpaired electrons of the triplet

[3] C. A. Hutchison, Jr., and B. W. Mangum, *J. Chem. Phys.* **29**, 363 and 952 (1958).
[4] T. Shiga and L. H. Piette, *Photochem. Photobiol.* **3**, 213 and 223 (1964).
[5] E. J. Lous and A. J. Hoff, *Biochim. Biophys. Acta* **974**, 88 (1989).

state, the Zeeman shifts of the magnetic sublevel energies are highly anisotropic. The anisotropy leads to EPR spectra of randomly oriented samples that typially are spread over approximately 10^3 gauss (10^{-1} tesla). This corresponds to a considerable sacrifice of sensitivity. ODMR of biopolymer triplet states, however, can be carried out in zero magnetic field where all molcules undergo magnetic resonance at roughly the same frequencies that are determined by the magnetic dipole–dipole interactions of the triplet state electrons. These energy splittings are referred to as the zero-field splittings (ZFS). Their size is controlled by the molecular structure of the chromophore and the local environment. We discuss these important quantities in some detail below. The line widths of biomolecule ODMR signals are dominated by inhomogeneous broadening that results from a spread in the ZFS caused by randomness in the distribution of local environments.

A striking example of the potential sensitivity of ODMR is the report by two groups[6,7] of magnetic resonance detection in a *single molecule* of pentacene using FDMR. It should be pointed out, however, that this accomplishment was possible because of the rather special properties of the pentacene excited states; comparable sensitivity should not be expected with biopolymers. Nonetheless, PDMR of Trp in proteins, for example, is carried out successfully at the picomole level without difficulty.

The first triplet state ODMR measurements (using the PDMR method) were carried out by adding phosphorescence-monitoring capability to a traditional EPR spectrometer that incorporates a fixed frequency resonant cavity and a variable external magnetic field.[8-10] As discussed above, this scheme is far from ideal for measurements on triplet states of randomly oriented systems such as biopolymers that are not available in single-crystal form. Virtually all biopolymer ODMR studies have been carried out in zero applied field using a frequency-swept microwave source; zero-field ODMR was introduced by Schmidt and van der Waals.[11] Transitions between the energy sublevels of the triplet state are allowed by magnetic

[6] J. Köhler, J. A. J. M. Disselhorst, M. C. J. M. Donckers, E. J. J. Groenen, J. Schmidt, and W. E. Moerner, *Nature (London)* **363**, 242 (1993).

[7] J. Wrachtrup, C. von Borczyskowski, J. Bernard, M. Orrit, and R. Brown, *Nature (London)* **363**, 244 (1993).

[8] A. L. Kwiram, *Chem. Phys. Lett.* **1**, 272 (1967).

[9] M. Sharnoff, *J. Chem. Phys.* **46**, 3263 (1967).

[10] J. Schmidt, I. A. Hesselmann, M. S. de Groot, and J. H. van der Waals, *Chem. Phys. Lett.* **1**, 434 (1967).

[11] J. Schmidt and J. H. van der Waals, *Chem. Phys. Lett.* **2**, 640 (1968).

dipole selection rules, as in traditional EPR, and their observation by ODMR in zero field gives the ZFS directly.

In the next section we take a quick look at the triplet state and discuss the origin and significance of the ZFS. This is followed by an outline of some of the ODMR experiments that are in common use to determine static and dynamic triplet state properties in biopolymers. After a description of experimental equipment and methods used in ODMR spectroscopy, we conclude with some examples that illustrate the wealth of information provided by the application of ODMR to the study of tryptophan triplet states in proteins. Because this chapter is intended to focus on methods, we present neither a historical development of ODMR spectroscopy, nor an extensive review of its application to biopolymer studies. We refer the interested reader, instead, to some early reviews on the general methods and results of ODMR spectroscopy,[12-17] as well as to a book[18] and to additional reviews[19-21a] that deal more specifically with biological applications of ODMR. Applications to photosynthetic pigments and the reaction centers, in particular, have been reviewed thoroughly by Hoff.[21,21a]

The Triplet State

In this section, only a brief overview of the properties of the photoexcited triplet state in the absence of an external magnetic field is given.

[12] M. A. El-Sayed, in "MTP International Review of Science, Physical Chemistry Series One" (A. D. Buckingham and D. A. Ramsay, eds.), Vol. 3, p. 119. Butterworth, London, and University Park Press, Baltimore, Maryland, 1972.

[13] A. L. Kwiram, in "MTP International Review of Science, Physical Chemistry Series One" (A. D. Buckingham and C. A. McDowell, eds.), Vol. 4, p. 271. Butterworth, London, and University Park Press, Baltimore, Maryland, 1972.

[14] M. A. El-Sayed, Annu. Rev. Phys. Chem. 26, 235 (1975).

[15] A. H. Maki and J. A. Zuclich, Top. Curr. Chem. 54, 115 (1975).

[16] M. Kinoshita, N. Iwasaki, and N. Nishi, Appl. Spectrosc. Rev. 17, 1 (1981).

[17] J. Olmstead and M. A. El Sayed, in "Creation and Detection of the Excited State" (W. R. Ware, ed.), Vol. 2, p. 1. Dekker, New York, 1974.

[18] R. H. Clarke (ed.), "Triplet State ODMR Spectroscopy." Wiley (Interscience), New York, 1982.

[19] A. H. Maki, in "Biological Magnetic Resonance" (L. J. Berliner and J. Reuben, eds.), Vol. 6, p. 187. Plenum, New York and London, 1984.

[20] A. L. Kwiram and J. B. A. Ross, Annu. Rev. Biophys. Bioeng. 11, 223 (1982).

[21] A. J. Hoff, in "Advanced EPR with Applications to Biology and Biochemistry" (A. J. Hoff, ed.), p. 633. Elsevier, Amsterdam, 1989.

[21a] A. J. Hoff, this series, Vol. 227, p. 290.

For a comprehensive treatment, the reader is referred to the book by McGlynn et al.[22] and the review by van der Waals and de Groot.[23]

The triplet state owes its paramagnetism to an electron pair with parallel spins. In the absence of external magnetic fields, the unpaired electron magnetic moments are subject only to internal interactions. The two types are (a) spin–spin dipolar coupling, the interaction between the spin magnetic moments of the electron pair, and (b) spin–orbit coupling, the interaction of a spin magnetic moment with largely its own orbital magnetic moment. The ZFS originates almost completely from the spin–spin dipolar coupling in the systems we will deal with, since the orbital motions of the electrons are largely quenched. Residual spin–orbit coupling is responsible, however, for the interaction between singlet and triplet states that makes possible intersystem crossing (isc) and phosphorescence. The spin–orbit mechanism is selective, imparting the individual triplet sublevels with differing isc and radiative decay rates. The resulting differences in sublevel properties are essential for the successful execution of ODMR, as will become apparent below. The rates of the isc and radiative processes are enhanced by the interaction of the unpaired electrons of the triplet state with an external heavy atom.[22] Such external heavy atom effects (HAE) have important applications in ODMR spectroscopy of biomolecules.

The Hamiltonian that describes the spin–spin dipolar interaction in the triplet state is

$$\mathcal{H}_{dd} = (\gamma h/2\pi)^2 [r_{12}^{-3}(\mathbf{s}_1 \cdot \mathbf{s}_2) - 3r_{12}^{-5}(\mathbf{s}_1 \cdot \mathbf{r}_{12})(\mathbf{s}_2 \cdot \mathbf{r}_{12})] \tag{1}$$

where $\mathbf{r}_{12} = \mathbf{r}_2 - \mathbf{r}_1$ is the interelectron vector, $\mathbf{s}$ is the spin angular momentum vector of an individual electron, h is Planck's constant, and γ is the electron magnetogyric ratio. It has been shown[24] that within a pure triplet state (neglect of spin–orbit coupling), Eq. (1) can be replaced with

$$\mathcal{H}_0 = \mathbf{S} \cdot \mathbf{D} \cdot \mathbf{S} \tag{2}$$

where $\mathbf{S} = (\mathbf{s}_1 + \mathbf{s}_2)$ is the total spin angular momentum and $\mathbf{D}$ is called the ZFS tensor. By choosing the proper set of internal coordinates (the magnetic principal axes of the molecular triplet state), $\mathbf{D}$ becomes diagonal and Eq. (2) is traditionally rewritten as

$$\mathcal{H}_0 = D[S_{z^2} - (2/3)] + E(S_{x^2} - S_{y^2}) \tag{3}$$

[22] S. P. McGlynn, T. Azumi, and M. Kinoshita, "Molecular Spectroscopy of the Triplet State." Prentice-Hall, Englewood Cliffs, New Jersey, 1969.

[23] J. H. van der Waals and M. S. de Groot, in "The Triplet State" (A. B. Zahlan, G. M. Androes, C. A. Hutchison, Jr., H. F. Hameka, G. W. Robinson, F. W. Heineken, and J. H. van der Waals, eds.), p. 101. Cambridge Univ. Press, Cambridge, 1967.

[24] J. H. van Vleck, Rev. Mod. Phys. 23, 213 (1951).

where D and E are parameters whose values determine the ZFS. The spin eigenfunctions of Eq. (3) are readily obtained, and they may be expressed as

$$T_x = (1/2^{1/2})[\beta(1)\beta(2) - \alpha(1)\alpha(2)]$$
$$T_y = (i/2^{1/2})[\beta(1)\beta(2) + \alpha(1)\alpha(2)] \tag{4}$$
$$T_z = (1/2^{1/2})[\alpha(1)\beta(2) + \beta(1)\alpha(2)]$$

which have the eigenvalues $W_x = \frac{1}{3}D - E$, $W_y = \frac{1}{3}D + E$, and $W_z = -\frac{2}{3}D$, respectively. In Eqs. (4), α and β refer to the spin "up" and spin "down" orientations of the electron, respectively.

The ZFS of the Trp triplet state are shown in Fig. 1, along with the location of the principal axes. Magnetic resonance transitions occur at the three frequencies, ν, of $(|D| \pm |E|)/h$ and $2|E|/h$. Assignment of any two of the transitions allows determination of the absolute values of the zero-field parameters, $|D|$ and $|E|$. Because the ZFS is produced by the

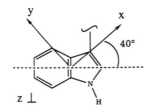

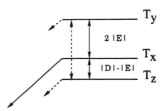

FIG. 1. Energy levels and orientation of the principal axes of tryptophan. The normally observed PDMR transitions are shown by solid double headed arrows, whereas the normally unobserved $D + E$ transition is shown as a dashed one. The observed transitions occur at the frequencies $\nu_{D-E} \approx 1.74$ GHz and $\nu_{2E} \approx 2.54$ GHz. Radiative decay is indicated by a solid arrow, and radiationless decay is shown by dashed arrows. The angle between x and the long axis (40°) is calculated by Zuclich [J. Zuclich, *J. Chem. Phys.* **52**, 3592 (1970)]. More recent low magnetic field ODMR measurements on a single crystal of p-dibromobenzene doped with indole locate the x axis at 49° ± 5° [C. A. Smith and A. H. Maki, *J. Phys. Chem.* **97**, 997 (1993)]. The x and y axes of Trp in Ref. 19 should be interchanged.

magnetic dipole–dipole interaction, D and E are expected to be determined by the electronic distribution of the triplet state. These parameters are given by the following expressions:

$$D = \tfrac{3}{4}(\gamma h/2\pi)^2 \langle (r_{12}^2 - 3z_{12}^2)/r_{12}^5 \rangle_{av} \qquad (5)$$
$$E = \tfrac{3}{4}(\gamma h/2\pi)^2 \langle (y_{12}^2 - x_{12}^2)/r_{12}^5 \rangle_{av} \qquad (6)$$

where $x_{12} = x_2 - x_1$ is the x component of the interlectron distance, etc., and the coordinate functions are averaged over the electronic distribution. For a planar aromatic molecule such as Trp (Fig. 1) in a $^3(\pi, \pi^*)$ electronic state, the electronic distribution is flattened along the z (normal) axis and extended in the xy plane. Thus $\langle (r_{12}^2 - 3z_{12}^2)/r_{12}^5 \rangle_{av}$ is positive and $D > 0$. E determines the anisotropy of the electronic distribution in the plane. Because $E > 0$, as well (Fig. 1), Eq. (6) requires that the electronic distribution extends farther along the y direction than the x direction.

Variations in the local environment of a residue such as Trp produce changes in the electronic distribution that are reflected in an alteration of the ZFS parameters D and E. As an example, a uniform increase in the polarizability of the local environment results in a more diffuse triplet state electronic distribution and a proportional reduction of both D and E. A selective expansion of the distribution along the z axis, for instance, will reduce D preferentially. In general, the ZFS are sensitive to local electric fields produced by nearby charged residues and electric dipoles.

Although the quantitative correlation of ZFS parameters with specific characteristics of the local environment remains an intriguing challenge, it should be clear, nonetheless, that the triplet state ZFS provide a sensitive structural probe in a qualitative sense. In addition to having an influence on the ZFS, environmental perturbations also affect the kinetic parameters of the triplet state, namely, the isc rate constants for both formation and decay of the triplet sublevels, and their radiative rate constants. These changes are brought about by environmental effects on the spin–orbit coupling processes that intermix singlet state character with the triplet sublevels. These effects are illustrated most directly by the introduction of a heavy atom into the local environment of the molecule to produce a HAE. In addition, an increase in the local polarizability will lower the energy of the highly diffuse singlet states whose admixing by internal spin–orbit coupling processes leads to isc and radiative decay.[22] Thus, enhancements of the isc and phosphorescence rates are to be expected, in general, as the local polarizability increases.

Methods of Optically Detected Magnetic Resonance Spectroscopy

In this section we discuss briefly the requirements for the detection of slow-passage ODMR signals from which the ZFS are obtained directly.

In addition, we touch on some of the transient ODMR methods that are useful for obtaining the kinetic and radiative parameters associated with the individual triplet sublevels.

During optical pumping, three important electronic states are dynamically coupled by the exciting light. These are S_0, the singlet ground state, S_1, the lowest excited singlet state from which fluorescence normally originates, and T_1, the lowest triplet state consisting of the three sublevels T_x, T_y, and T_z. The dynamic processes coupling these states are illustrated on the energy level diagram shown in Fig. 2. In Fig. 2 and in the expressions that follow below, (u, v, w) is used to refer nonspecifically to the set of principal axes (x, y, z) discussed in the previous section. The dynamics of the triplet sublevels must be differentiated because of the selective nature of spin–orbit coupling, as discussed earlier.

Each of the ODMR methods requires that a condition of spin alignment of the triplet state is achieved, in other words, that population differences exist between the sublevels. Of particular importance in this regard are the W values, the first-order rate constants for spin–lattice relaxation (slr) between the sublevels. Spin–lattice relaxation enables the spins to sense the temperature of, and to exchange energy with, the "lattice" (external heat reservoir). If the W values are very much larger than the decay or populating rate constants of the T_1 sublevels, the sublevel populations always will remain in Boltzmann equilibrium at the temperature of the lattice. Because the ZFS energies are so small, this corresponds to miniscule spin alignment even at lattice temperatures of 4.2 K, the normal boiling temperature of liquid He. Fortunately, slr is a thermally activated process, and it usually can be quenched sufficiently to allow the development of considerable spin alignment by cooling the sample to a cryogenic temperature with liquid He. The decay constants of Trp are so small, however, that cooling to around 1.2 K (by reducing the pressure over liquid He) is required to quench slr adequately in this molecule.

The rate equations that govern the populations (N) of the relevant electronic states during optical pumping, but in the absence of applied microwave power, are as follows:

$$dN_0/dt = -k_0N_0 + k_1N_1 + \sum k_uN_u,$$
$$u = x, y, z \tag{7}$$
$$dN_1/dt = k_0N_0 - (k_1 + k_2)N_1 \tag{8}$$
$$dN_u/dt = k_2p_uN_1 - (k_u + \sum_{s \neq u} W_{us})N_u + \sum_{s \neq u} W_{su}N_s,$$
$$(u, s) = (x, y, z) \tag{9}$$

The rate constants are defined in Fig. 2. The presence of an oscillating microwave driving field in resonance with the $T_u \rightleftharpoons T_v$ transition introduces the additional terms, $-P_{uv}N_u + P_{vu}N_v$, into Eq. (9). P_{uv} is the rate constant

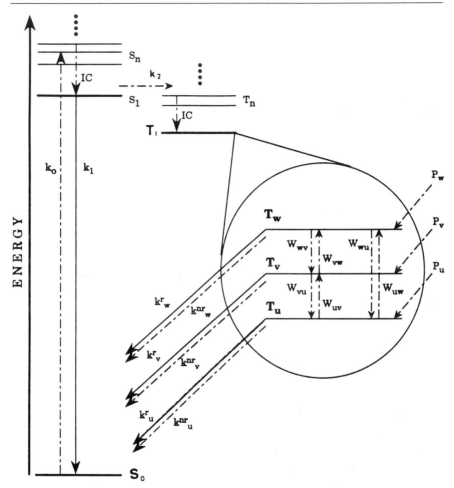

FIG. 2. Energy level diagram illustrating the relationships between the electronic states of the singlet and triplet manifolds and the rate constants that determine the T_1 sublevel populations in the absence of microwave-induced transitions. The sublevel energy splittings are greatly amplified in the circular inset. The decay rate constant of each sublevel, k_s ($s = u, v, w$), consists of a radiative part, k^r_s, and a nonradiative part, k^{nr}_s. The rate constants k_0, k_1, and k_2 describe optical pumping, $S_1 \rightarrow S_0$ decay, and intersystem crossing (isc) to the triplet manifold, respectively. Internal conversion (IC) processes are rapid ($\leq 10^{-12}$ sec) dissipative processes that convert internal vibrational energy to heat that is dissipated in the lattice. P_s is the probability that isc leads to populating of the T_s ($s = u, v, w$) sublevel. The W values are the rate constants for spin–lattice relaxation.

for microwave-induced transitions $T_u \rightarrow T_v$, and $P_{uv} = P_{vu}$. In Eq. (9), p_u is the probability that isc populates T_u. The description of both slow-passage (steady-state) and transient ODMR effects are based on analysis of Eqs. (7)–(9). We do not go into detail here, since complete derivations have been presented previously.[12–21]

Slow-Passage Measurements

A slow-passage ODMR measurement is carried out ideally when the microwave passage time through magnetic resonance is very much slower than the triplet sublevel lifetimes. The steady-state solutions of Eqs. (7)–(9) then are applicable and may be used to obtain expressions for the sublevel populations in the presence of ($P_{uv} \neq 0$) and absence of ($P_{uv} = 0$) a microwave driving field. The ODMR signals result from microwave-induced changes in the steady-state populations of the electronic states. The solutions are simplest when the following assumptions are made: (a) neglect of the S_1 population relative to those of S_0 and T_1, (b) neglect of the W values relative to the k values, and (c) P_{uv} is sufficiently large at resonance to saturate the $T_u \rightleftharpoons T_v$ transition. With these assumptions, and the additional approximation that the optical pumping is sufficiently weak that the S_0 population is not significantly (say, $<10\%$) depleted into T_1, the steady-state population changes resulting from saturation of $T_u \rightleftharpoons T_v$ are given by

$$\Delta N_u^{uv} \approx NK(k_u p_v - k_v p_u)/k_u(k_u + k_v) \tag{10}$$
$$\Delta N_v^{uv} \approx -NK(k_u p_v - k_v p_u)/k_v(k_u + k_v) \tag{11}$$
$$\Delta N_w^{uv} \approx 0 \tag{12}$$
$$\Delta N_T^{uv} = -\Delta N_0^{uv} \approx NK(p_u/k_u - p_v/k_v)(k_u - k_v)/(k_u + k_v) \tag{13}$$

where $K = k_0 k_2/(k_1 + k_2)$ is the rate constant for populating the T_1 state by optical pumping, N is the total population, and N_T is the total population in the T_1 state. There is no change in the S_1 population to first order in K; a change occurs only in second order and is given by

$$\Delta N_1^{uv} \approx (K/k_2)\Delta N_0^{uv} \tag{14}$$

The ODMR signal intensity associated with the steady-state population changes depends on the detection method. The absorbance changes in ADMR are proportional to changes in N_T and may be expressed as

$$\Delta I_A^{uv} \propto \pm(N_u^0 - N_v^0)(k_v - k_u)/(k_u + k_v) \tag{15}$$

where the plus and minus signs refer to absorbance changes in the singlet absorption band (SADMR) and triplet absorption band (TADMR), respectively. The term $N_u^0 = NKp_u/k_u$ represents the steady-state population of T_u (neglecting ground state depletion) in the absence of microwave saturation. Equation (15) summarizes the requirements for ADMR, namely, a difference in steady-state populations and decay constants of the sublevel pair that undergoes magnetic resonance. FDMR detection is based on Eq. (14) and requires, in addition to the ADMR requirements, that the S_1 state is fluorescent. The steady-state PDMR response is given by

$$\Delta I_p^{uv} \approx (N_u^0 - N_v^0)(Q_v - Q_u)k_u k_v/(k_u + k_v) \tag{16}$$

where $Q_u = k_u^r/k_u$ is the radiative quantum yield of T_u, etc. Thus, in addition to spin alignment, PDMR requires a difference in quantum yields of the saturated sublevels.

The common requirement for each steady-state ODMR detection method is the presence of population differences between saturated sublevels. If the steady-state populations of a pair of sublevels happen to be equal, $N_u^0 = N_v^0$, for instance, an electron–electron double resonance measurement (EEDOR[25]) in which the $T_v \rightleftharpoons T_w$ transition is kept saturated while sweeping a second microwave frequency through the $T_u \rightleftharpoons T_v$ transition may do the trick. Because of the continuous saturation, the new population of T_v is $\frac{1}{2}(N_v^0 + N_w^0)$ which will differ from N_u^0 if $N_w^0 \neq N_v^0$. The EEDOR method will also average the k values and the Q values of the saturated sublevel pair and may produce signals that would otherwise be absent because of the equality of these quantities.

Referring to Fig. 1 for Trp, we see that $Q_y \approx Q_z \approx 0$, whereas $Q_x \approx 1$.[19] Thus according to Eq. (16), the $D + E$ ($T_y \rightleftharpoons T_z$) signal should be difficult to observe by PDMR. In practice, this signal is only observed easily by the EEDOR method in the absence of a perturbation such as the HAE that enhances the small Q values. The $2E$ and $D - E$ transitions, on the other hand, are normally observed by simple PDMR with little difficulty.

Slow-passage ODMR signals frequently are observed by the continuous wave method in which the optical effect is monitored using broadband detection. On the other hand, if the triplet state decay constants are sufficiently large, the microwave power may be amplitude modulated at an audio frequency which results in modulated phosphorescence when the microwave frequency is at resonance. The phosphorescence is then monitored with narrow-band phase-sensitive detection, for a great improvement in the signal/noise ratio. The latter detection method is frequently used to produce a magnetic resonance-induced phosphorescence spectrum by a technique referred to as phosphorescence–microwave double resonance (PMDR).[26] The microwave frequency is fixed at resonance,

[25] T. S. Kuan, D. S. Tinti, and M. A. El Sayed, *Chem. Phys. Lett.* **4**, 507 (1970).
[26] M. A. El Sayed, D. V. Owens, and D. S. Tinti, *Chem. Phys. Lett.* **6**, 395 (1970).

and the detected wavelength is swept through the phosphorescence region. Microwave-induced optical absorption spectra may also be obtained by this method.[21,21a]

Transient Experiments

Transient experiments are done to obtain kinetic information about the triplet state. Equations (10)–(13) form the basis for obtaining expressions for transient changes in populations that accompany microwave transitions induced between the magnetic sublevels. We assume, initially, that the temperature is sufficiently low that slr is effectively quenched. Population changes may be induced either during continuous optical pumping or during the decay of the triplet state after turning off the excitation source.

Two methods are frequently employed for measurements made during optical pumping. The first of these[27] employs a short resonant microwave pulse or fast passage through resonance, which produces a transient response in N_T given by

$$\Delta N_T^{uv}(t) \approx f(N_v^0 - N_u^0)(e^{-k_u t} - e^{-k_v t}) \tag{17}$$

where f is a measure of the extent of population transfer by the pulse. For population saturation $f = \frac{1}{2}$, whereas $f = 1$ for population inversion. From Eq. (17) it is apparent that an observable transient ADMR or FDMR signal requires a difference in both the steady-state populations and decay constants of the sublevels. The biexponential decay is analyzed to yield both k_u and k_v. The phosphorescence transient is described by

$$\Delta P^{uv}(t) \approx cf(N_v^0 - N_u^0)(k_u^r e^{-k_u t} - k_v^r e^{-k_v t}) \tag{18}$$

where c depends on the apparatus. In contrast with ADMR and FDMR, the transient response at $t = 0$ is nonzero. It decays to zero as the sum of increasing and decreasing exponentials that yield the k values. The preexponential factors are proportional to the k^r values, and analysis of the response yields their relative values. The relative values of Q can be calculated subsequently, using $Q_u = k_u^r/k_u$, etc.

The other method[28] employs the transient recovery from magnetic saturation. If the sublevel populations are saturated during the continuous application of microwave power, that is, $P_{uv} >> k_u, k_v$, the PDMR transient following the sudden removal of microwave power is found to be

$$\Delta P^{uv}(t) \approx c(N_v^0 - N_u^0)k_u k_v(Q_u e^{-k_u t} - Q_v e^{-k_v t})/(k_u + k_v) \tag{19}$$

[27] C. J. Winscom and A. H. Maki, Chem. Phys. Lett. 12, 264 (1971).
[28] A. L. Shain and M. Sharnoff, J. Chem. Phys. 59, 2335 (1973).

Analysis of the response gives the k values, as in the previous method. In this case, however, the preexponential factors are proportional to the Q values rather than the k values. The magnetic saturation recovery method has a distinct advantage over the pulse transient method in the frequently encountered case where one of the k^r values is very much smaller than the other. Very often, however, the Q values are more nearly comparable, and two exponentials can be detected in the saturation recovery method, even if both cannot be seen in the pulse (rapid-passage) transient method.

Transient measurements made during optical pumping have the disadvantage of yielding kinetic parameters that are influenced by the optical pumping rate, K. Equations (17)–(19) are accurate only in the limit $K \to 0$. Thus, the transient experiments described above should be carried out at optical pumping rates that are sufficiently low that the apparent rate constants are no longer influenced by further reductions in K. This problem is avoided completely, on the other hand, by carrying out the kinetic measurements in the absence of optical pumping, that is, during decay of the triplet state.

In the method of microwave-induced delayed phosphorescence (MIDP),[29] the excitation is suddenly extinguished at $t = 0$. After a time interval, t', a microwave pulse or rapid passage is applied to induce transitions between T_u and T_v. The sudden change in the phosphorescence intensity is described by

$$\Delta P^{uv}(t') \approx cf(k_u{}^r - k_v{}^r)[N_v(0) \exp(-k_v t') - N_u(0) \exp(-k_u t')] \quad (20)$$

where $N(0)$ refers to a population at $t = 0$. The analysis is particularly straightforward if the k values are very different. Assuming that $k_u \gg k_v$, and that t' is chosen such that $k_u t' \gg 1$, Eq. (20) simplifies to

$$\Delta P^{uv}(t') \approx cf(k_u{}^r - k_v{}^r) N_v(0) \exp(-k_v t') \quad (21)$$

The experiment is repeated for a series of t' values, keeping cf the same. Then, $-k_v$ is obtained as the slope of a $\ln[\Delta P^{uv}(t')]$ versus t' plot. Deconvolution of the transient following the microwave pulse from the unperturbed decay gives k_u.

The MIDP method may be used to obtain either the relative steady-state populations, N^0, or the relative isc rate constants, p, depending on how the system is prepared. If the system is prepared in the steady state, the $N(0)$ values are given by the N^0 values. If, on the other hand, it is prepared by a short optical pulse, the $N(0)$ values are proportional to the p values. The expressions are the most straightforward if only one of the

[29] J. Schmidt, W. S. Veeman, and J. H. van der Waals, *Chem. Phys. Lett.* **4**, 341 (1969).

sublevels is significantly radiative, as is the case with indole or Trp, for instance (Fig. 1). From Eq. (21), if only T_u is radiative,

$$\Delta P^{us}(t') \approx cfk_u{}^r N_s(0) \exp(-k_s t'), \qquad s = v, w$$

At $t = 0$, the phosphorescence intensity is

$$P(0) = c \sum N_s(0)k_s{}^r \approx cN_u(0)k_u{}^r, \qquad s = u, v, w$$

The relative populations at $t = 0$ can be obtained from the ratio

$$\Delta P^{us}(t')/P(0) \approx fN_s(0) \exp(-k_s t')/N_u(0), \qquad s = v, w \qquad (22)$$

provided we know k_s (from the MIDP measurement) and f. Depending on how the triplet state is prepared (see above), $N_s(0)/N_u(0)$ is either $N_s{}^0/N_u{}^0$ or p_s/p_u.

The experiments described above produce accurate decay constants only when slr is negligible. The apparent kinetic parameters that are obtained when this is not the case involve all of the k and W values simultaneously. It is not possible under these circumstances to obtain the actual decay constants from these experiments. Experimental methods have been developed, however, that allow extraction of the individual k and W values in the presence of slr.[30,31] One method[30] applies to the regime where slr is dominant, that is, $W > k$, whereas the other[31] is applicable when $W < k$. The latter method has been applied to biopolymer ODMR and involves deconvolution of phosphorescence decays measured during continuous microwave saturation of pairs of triplet sublevels. Microwave saturation creates a pseudo-two-level system whose decays are easily deconvoluted and are amenable to analysis. The analytical development of the microwave-saturated phosphorescence decay method is rather lengthy, so it is not discussed in this chapter. Detailed descriptions of the method may be found elsewhere.[15,19,31]

Optically Detected Magnetic Resonance Line Widths

Narrow ODMR signals are found in molecular crystals and in Shpol'skii matrices in which they may be less than 1 MHz wide. On the other hand, ODMR signals that are observed in biological molecules dissolved in low-temperature aqueous glasses typically are about two orders of magnitude larger. The ODMR transitions are inhomogeneously broadened as a result of a distribution of solvent–molecule interactions, where the "solvent"

[30] D. A. Antheunis, B. J. Botter, J. Schmidt, P. J. F. Verbeek, and J. H. van der Waals, *Chem. Phys. Lett.* **36**, 225 (1975).
[31] J. Zuclich, J. U. von Schütz, and A. H. Maki, *Mol. Phys.* **28**, 33 (1974).

is taken to include the molecular surroundings of the triplet state chromophore, including other parts of the biopolymer. Inhomogenous broadening of ODMR lines has been verified in several cases by microwave hole-burning experiments.[32,33] It is found that the bandwidth of ODMR transitions is correlated with the width of the optical bands in the phosphorescence spectrum of Trp, indicating that solvent perturbations may be responsible for broadening of both spectra. Both the narrowest ODMR transitions and the most resolved phosphorescence spectra are characteristic of Trp residues that are buried in the interior of a protein structure, whereas broadening results from exposure to the aqueous solvent, which produces a more random, inhomogeneous local environment.

Although there is some relationship between the solvent shifts of the phosphorescence spectrum and of the ODMR frequencies, the two are not highly correlated. For instance, if an essentially monochromatic population is selected by laser excitation in the 0,0-band of the $T_1 \leftarrow S_0$ transition, narrow phosphorescence emission lines are produced in a rigid matrix, but the ODMR lines remain relatively broad.[34] In addition, if ODMR is carried out with narrow-band optical selection through the phosphorescence 0,0-band of a sample with inhomogeneously broadened spectra, broad transitions are observed, but a linear correlation is found between the peak ODMR frequency and the T_1–S_0 energy gap.[35] This correlation has been attributed to solvent perturbations that lower the triplet state energy relative to that of the less polarizable ground state, while at the same time mixing this triplet state with higher energy, more diffuse triplet states, thus reducing the ZFS.[36] The sign and magnitude of the predicted linear correlation are in general agreement with experiments. An alternative theory[37] to account for the correlation in which the solvent perturbation modifies the spin–orbit coupling would be applicable only to triplet states whose ZFS originates largely from spin–orbit interactions in the first place.

Optically Detected Magnetic Resonance Instrumentation

The experimental arrangement for carrying out slow-passage and transient ODMR measurements is not extensive. A spectrometer that we

[32] K. W. Rousslang, J. B. A. Ross, D. A. Deranleau, and A. L. Kwiram, *Biochemistry* **17,** 1087 (1978).

[33] M. V. Hershberger, A. H. Maki, and W. C. Galley, *Biochemistry* **19,** 2204 (1980).

[34] R. Williamson and A. L. Kwiram, *J. Phys. Chem.* **83,** 3393 (1979).

[35] J. U. von Schütz, J. Zuclich, and A. H. Maki, *J. Am. Chem. Soc.* **96,** 714 (1974).

[36] J. van Egmond, B. E. Kohler, and I. Y. Chan, *Chem. Phys. Lett.* **34,** 423 (1975).

[37] A. H. Zewail, *J. Chem. Phys.* **70,** 5759 (1979); J. P. Lemaistre and A. H. Zewail, *Chem. Phys. Lett.* **68,** 296 and 302 (1979).

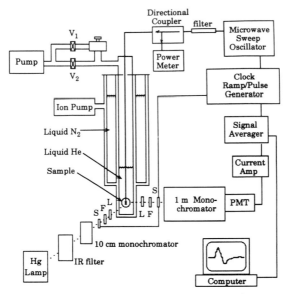

FIG. 3. Block diagram of an ODMR spectrometer with PDMR and FDMR capability. The excitation and emission paths are at a 90° angle. S, F, and L are mechanical shutters, glass filters, and collimating lenses, respectively. A microwave amplifier may be placed between the microwave source and the low-pass filter; the purpose of the latter is to eliminate harmonics. V_1 is a He pressure-regulated valve for maintaining the sample at a fixed temperature between the minimum and 4.2 K. V_2 is a large throughput valve for obtaining the minimum He temperature. A rotating sector often is placed around the tail section of the Dewar flask to eliminate fluorescence in PDMR measurements. [From L. E. Burns, Doctoral Dissertation, University of California, Davis (1992), with permission.]

typically employ for PDMR and FDMR studies is shown in Fig. 3. The stainless steel Dewar flask contains liquid He into which the sample is immersed and is maintained at around 1.2 K by a high-speed, 118 cubic feet per minute (cfm), rotary vacuum pump. The sample (typically ≤10 μl of biopolymer solution contained in a 1 mm inner diameter quartz sample tube) is placed in a Cu slow-wave helix that terminates a length of RG 141 cable with a stainless steel outer conductor to minimize thermal conduction losses. The design of a linear horn for matching the 50 ohm line impedance with the higher (~175 ohms) impedance of the helix has been given by Chan.[38]

The sample is optically pumped and the phosphorescence is monitored at right angles through sets of quartz windows. Normally a high-pressure 100 W mercury arc is employed for optical pumping. The excitation band

[38] I. Y. Chan, in "Triplet State ODMR Spectroscopy" (R. H. Clarke, ed.), p. 1. Wiley (Interscience), New York, 1982.

usually is selected by a 10 cm monochromator in conjunction with an aqueous $NiSO_4$ solution filter that removes most of the infrared band. The monochromator may be replaced with appropriate transmission filters. The emission is focused onto the entrance slit of a 1 m monochromator, fitted with a coolcd photomultiplicr tube at the exit slit. The monochromator is utilized to obtain phosphorescence spectra, for wavelength-selected PDMR measurements and for PMDR. Mechanical shutters are placed in the excitation and emission paths to control excitation and decay for MIDP and phosphorescence decay measurements. A rotating sector is often used to eliminate fluorescence background in PDMR measurements.

In the study of Trp residues in proteins by PDMR, a satisfactory sample concentration is usually 10^{-5}–10^{-4} M in an aqueous buffer, so that a typical measurement uses only on the order of 100 pmol of protein. It has been found[39,40] that slr is affected profoundly by the solvent composition. A cryosolvent such as glycerol or ethylene glycol is introduced at no less than 20% by volume to reduce slr to the point that PDMR signals of Trp can be obtained at 1.2 K.

Solid state microwave sweep oscillators are now available to cover the frequency range from 0.1 to 27 GHz with a single plug-in unit. The maximum output power level, about 10 mW, is sufficient for most ODMR studies. In some cases, however, when saturation of a short-lived triplet state must be accomplished, higher power levels are required, and a power amplifier must be used. Some of the older backward wave sweep oscillators produce power levels that approach 100 mW, a particularly useful feature in many cases. The microwave sources should be capable of amplitude modulation (for amplitude-modulated PDMR and PMDR measurements in conjunction with phase-sensitive detection) and of frequency modulation (to allow saturation of broad, inhomogeneously broadened ODMR transitions for EEDOR and microwave-saturated phosphorescence decay measurements). The ODMR responses of biopolymer samples are usually rather weak, and signal averaging is an essential feature of the spectrometer. A home-built programmable pulse generator provides the timing pulses that control microwave sweeps, shutters, and the signal-averager time base. Data are stored, analyzed, and hard copy produced by means of a personal computer.

A more thorough and detailed description of the PDMR apparatus has been given previously.[19] A complete description of the ADMR apparatus has been given by Hoff.[21] In that spectrometer, a light source must be provided for absorbance as well as for excitation, and this may be the

[39] S. Ghosh, M. Petrin, and A. H. Maki, *Biophys. J.* **49**, 753 (1986).
[40] H. C. Brenner and V. Kolubayev, *J. Lumin.* **39**, 251 (1988).

same source. It is worth pointing out that ADMR requires a far more stable light source than even a current-regulated arc lamp that is normally employed for PDMR. Tungsten filament lamps powered by DC current-regulated sources have been found to be satisfactory, and they have been employed largely in the study of chlorophylls and other photosynthetic pigments that absorb in the visible region. Because the absorption spectra of the aromatic amino acids occur in the ultraviolet, where the output of an incandescent lamp vanishes, ADMR has not been a practical method for the study of protein triplet states.

Fluctuations in microwave power at the sample that are due mainly to impedance mismatches at the transmission line–helix junction can cause spurious effects in the slow-passage ODMR signals, particularly if they are broad. At certain frequencies the microwave power at the sample may dip below that required for saturation; as a result the ODMR bands may exhibit spurious structure. This is particularly a problem for triplet states with short sublevel lifetimes and broad ODMR transitions that are difficult to saturate. To avoid confusing these spurious fluctuations with genuine spectroscopic features, measurements should be repeated with a different helix and at differing power levels.

Optically Detected Magnetic Resonance of Tryptophan

Of the three naturally occurring amino acids, Phe, Tyr, and Trp, the latter has been studied most extensively by ODMR. All measurements have been carried out by PDMR to take advantage of the intense highly structured phosphorescence emitted by Trp, and because its absorption spectrum lies in the ultraviolet, making ADMR unsatisfactory, as discussed earlier. Typical examples of Trp phosphorescence are given in Figs. 4 and 5. Of the aromatic amino acids, Trp has the lowest energy S_1 and T_1 states. Excited states of Phe and Tyr often are extensively quenched by energy transfer to Trp when the latter is present in a protein. The energy level diagram of the T_1 state of Trp that is shown in Fig. 1 illustrates some of the characteristic properties of the sublevels that have been determined by ODMR. The ZFS parameters of Trp in some peptide and protein environments are given in Table I. The T_x sublevel that is intermediate in energy has the shortest lifetime and is responsible for nearly all of the phosphorescence emission. In the absence of special interactions which we discuss later, $N_x^0 < N_y^0$, N_z^0, so positive polarity $D - E$ and $2E$ steady-state PDMR signals normally are observed.

The $D + E$ signal is extremely weak and is rarely detected. An important exception occurs when Trp is subject to a HAE where the appearance of a prominent $D + E$ signal is diagnostic for this effect. A typical steady-

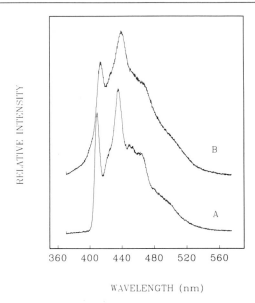

WAVELENGTH (nm)

FIG. 4. Phosphorescence spectra at 77 K of (A) p7 nucleocapsid protein from HIV-1 and (B) the complex of p7 with $(dG)_8$. The structured phosphorescence originates from Trp-37 on the C-terminal zinc finger. Excitation is at 295 nm with 16 nm bandwidth. The solvent is 5 mM phosphate buffer, pH 6.5, containing 20% (v/v) ethylene glycol. The p7 concentration is 0.25 mM in (A). In (B), the p7 concentration is 0.062 mM, and the octanucleotide concentration is 0.13 mM. (From W.-C. Lam and A. H. Maki, unpublished data, 1993.)

state PDMR spectrum of Trp is shown in Fig. 6A. We do know that the T_y sublevel is weakly radiative from studies of $T \to S$ energy transfer from intrinsic Trp residues to protein-bound dyes in which the $D + E$ transition can be detected by monitoring the delayed fluorescence of the bound dye.[41,42] The Förster mechanism that is operative requires the donor state to have radiative character. Thus, the normal lack of an observed $D + E$ signal also is a result of the condition, $N_y^0 \approx N_z^0$; examples where a weak $D + E$ signal is detected in the absence of a HAE or without the use of EEDOR are attributed to dynamic effects that result in unequalizing the T_y and T_z steady-state populations.[43]

Effects of Local Environment

The effect of the local environment on Trp phosphorescence was explained initially by Purkey and Galley.[44] When phosphorescence and

[41] A. H. Maki and T.-t. Co, *Biochemistry* **15**, 1229 (1976).
[42] J. G. Weers and A. H. Maki, *Biochemistry* **25**, 2897 (1986).
[43] B. D. Schlyer, E. Lau, and A. H. Maki, *Biochemistry* **31**, 4375 (1992).
[44] R. M. Purkey and W. C. Galley, *Biochemistry* **9**, 3569 (1970).

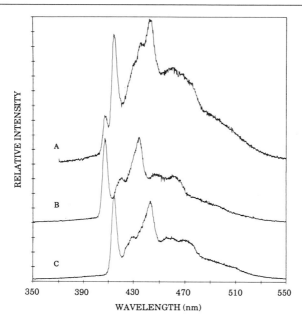

FIG. 5. Phosphorescence spectra at 77 K of (A) wild-type *E. coli* Trp repressor protein, (B) the point-mutated protein W99F, and (C) the point-mutated protein W19F. Sample excitation is at 295 nm with 16 nm bandpass; emission resolution is 1.5 nm bandpass. Protein concentrations are around 0.2 mM in 0.1 M phosphate buffer, pH 7.5, containing 1 mM EDTA and 0.2 M KCl. Ethylene glycol is present at 40% (v/v). Energy transfer from Trp-19 to Trp-99 at the singlet level is responsible for the reduced contribution of Trp-19 to the phosphorescence spectrum in (A). [From M. R. Eftink, G. D. Ramsay, L. E. Burns, A. H. Maki, C. J. Mann, C. R. Matthews, and C. A. Ghiron, *Biochemistry* **32**, 9189 (1993), with permission.]

ODMR studies are carried out at low temperature, the local envirionment is rigid and cannot adjust to changes in the Trp electronic distribution during the lifetime of the T_1 state. Thus Trp residues in a polar environment are characterized by a blue-shifted phosphorescence spectrum since the rigid local dipoles are arranged to stabilize the S_0 rather than the T_1 electronic distribution. When fluorescence measurements are carried out at ambient temperature, which normally is the case, rapid solvent relaxation occurs during the excited state lifetime; thus, contrary to phosphorescence, red-shifted fluorescence of Trp generally is associated with a polar environment. In a rigid polarizable local environment, however, stabilization of the excited state occurs through changes in the internal electronic distribution of the environment, leading to a red shift of the phosphorescence with respect to a polar, relatively nonpolarizable environment.

TABLE I

TRIPLET STATE DATA FOR TRYPTOPHAN IN PEPTIDES AND PROTEINS[a]

Sample	$\lambda_{0,0}$ (nm)	$D - E$ (GHz)[b]	2E (GHz)[b]	D (GHz)	E (GHz)	Ref.
Ribonuclease $T_1{}^c$	404.5	1.75 (50)	2.48 (125)	2.99	1.24	d
Trp repressor (Trp-19)	407.5	1.79 (100)	2.47 (170)	3.03	1.24	e
Gly-Trp-Gly	407.9	1.74 (146)	2.47 (258)	2.98	1.23	f
T4 lysozyme (Trp-154)	408.1	1.79 (70)	2.47 (140)	3.02	1.24	g
Somatostatin	409	1.73 (147)	2.49 (296)	2.98	1.24	h
p7, HIV-1	409.2	1.75 (150)	2.53 (240)	3.02	1.27	i
		1.80 (190)[j]	2.58 (260)[j]			
Azurin B	410.0	1.63 (42)	2.79 (73)	3.02	1.40	f
Eco SSB (Trp-40, Trp-54)	411.2	1.75	2.57	3.03	1.28	k
T4 lysozyme (Trp-138)	413.5	1.63 (40)	2.69 (50)	2.97	1.34	g
p7, HIV-1–(dG)$_8$	413.6	1.73 (150)[l]	2.50 (220)[l]			i
Hen lysozyme (Trp-108)	413.8	1.57 (83)	2.72 (116)	2.93	1.36	f
Trp repressor (Trp-99)	414.5	1.61 (36)	2.71 (110)	2.96	1.36	e
Eco SSB–poly(dT) (Trp-54)	418	1.57	2.36	2.86	1.16	k

[a] ODMR and ZFS frequencies are corrected for passage effects, except as noted. Reliability of data is around ±0.01 GHz.

[b] Bandwidths in MHz (full width at half-maximum intensity) are given in parentheses.

[c] The single Trp-59 is buried in the protein interior but subjected to local polar interactions [R. Arni, U. Heinemann, R. Tokuoka, and W. Saenger, J. Biol. Chem. 263, 15358 (1988)].

[d] M. V. Hershberger, A. H. Maki, and W. C. Galley, Biochemistry 19, 2204 (1980).

[e] M. R. Eftink, G. D. Ramsay, L. Burns, A. H. Maki, C. J. Mann, C. R. Matthews, and C. A. Ghiron, Biochemistry 32, 9189 (1993).

[f] J. B. A. Ross, K. W. Rousslang, and A. L. Kwiram, Biochemistry 19, 876 (1980).

[g] L.-H. Zang, S. Ghosh, and A. H. Maki, Biochemistry 27, 7820 (1988).

[h] D. A. Deranleau, J. B. A. Ross, K. W. Rousslang, and A. L. Kwiram, J. Am. Chem. Soc. 100, 1913 (1978).

[i] W.-C. Lam, A. H. Maki, J. R. Casas-Finet, J. W. Erickson, R. C. Sowder II, and L. E. Henderson, FEBS Lett. 328, 45 (1993).

[j] The microwave sweep rate was 64 MHz/sec; these values are uncorrected for passage effects. The size of the sweep corrections can be seen from comparison with corrected data on previous line.

[k] D. H. H. Tsao, J. R. Casas-Finet, A. H. Maki, and J. W. Chase, Biophys. J. 55, 927 (1989).

[l] The microwave sweep rate was 68 MHz/sec; these values are uncorrected for passage effects. The size of the sweep rate corrections should be about the same as those found for free p7, −0.05 GHz.

These effects on Trp phosphorescence are illustrated by the spectra shown in Fig. 4. The single Trp-37 residue of the p7 HIV-1 (human immunodeficiency virus type 1) nucleocapsid protein (p7) has a 0,0-band peak at 409.2 nm, which along with its large width indicates partial exposure to the polar solvent (see below). On complexing of p7 with (dG)$_8$, a red shift of 4.4 nm occurs which we associated with the change to a more polarizable environment. Other evidence (see below) indicates that Trp-

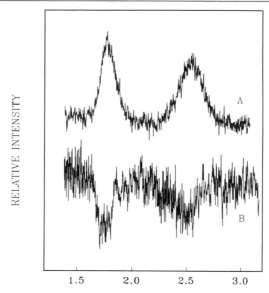

FIG. 6. PDMR spectra (lower frequency $D - E$ and higher frequency $2E$ transitions) at 1.2 K of (A) p7 nucleocapsid protein from HIV-1 and (B) the complex of p7 with $(dG)_8$. Photoexcitation conditions and sample compositions are described in the legend for Fig. 3. Phosphorescence is monitored using 3 nm bandwidth at 409.2 nm for (A), and at 412.6 nm for (B). Microwave sweep rates are 33 MHz/sec for (A) and 68 MHz/sec for (B). Spectrum (A) was accumulated for 25 scans, whereas (B) consists of 100 scans. (From W.-C. Lam and A. H. Maki, unpublished data, 1993.)

37 undergoes aromatic stacking interactions with nucleobases in p7–RNA complexes. Shown in Fig. 5 is the phosphorescence of the *Escherichia coli* Trp repressor protein in which the 0,0-bands of the two intrinsic Trp residues are clearly resolved.[45] Measurements on the mutated proteins, W19F and W99F, also shown in Fig. 5, demonstrate that the blue-shifted peak at 407.5 nm originates from Trp-19 whereas the red-shifted narrower peak at 414.5 nm is from Trp-99. It is a generally applicable rule that 0,0-bands peaking at $\lambda \leq 410$ nm originate from polar sites, whereas those peaking at $\lambda \geq 410$ nm are associated with less polar, more polarizable sites. Buried residues have more homogeneous surroundings than solvent-exposed residues, producing narrower phosphorescence and ODMR spectra.

The ODMR frequencies also are found to be affected by the polarity/polarizability of the local environment.[33] An increase in polarizability (and phosphorescence red shift) is associated, in general, with a decrease in

[45] M. R. Eftink, G. D. Ramsay, L. Burns, A. H. Maki, C. J. Mann, C. R. Matthews, and C. A. Ghiron, *Biochemistry* **32,** 9189 (1993).

the frequency of the $D - E$ ODMR transition. This is illustrated in Table I for a few selected Trp residues, although many more examples are known that illustrate this general trend, which has not been explained thus far. The wavelength-dependent ODMR frequencies of Trp represor protein arc shown in Fig. 7. The data show a clear discontinuity of the ZFS in the wavelength region where the two 0,0-bands meet (Fig. 5), and they also illustrate the linear variation of the ZFS with wavelength within a 0,0-band that was discussed earlier. Note that this effect is greater for the

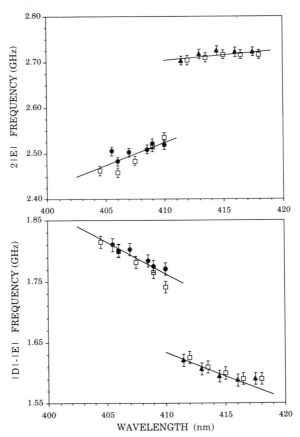

Fig. 7. Wavelength dependence of the ODMR signal frequencies in the phosphorescence 0,0-band region for native *E. coli* Trp repressor protein (□) and the single point mutants W19F (▲) and W99F (●). Sample compositions and conditions of excitation are given in the legend for Fig. 4. The sample temperature is 1.2 K, the emission bandwidth is 1.5 nm, and all data are corrected for passage effects. [From M. R. Eftink, G. D. Ramsay, L. Burns, A. H. Maki, C. J. Mann, C. R. Matthews, and C. A. Ghiron, *Biochemistry* **32**, 9189 (1993), with permission.]

more solvent-exposed Trp-19 than for the buried Trp-99. Data are also included from the single Trp-containing mutants. The excellent agreement with the native protein indicates that neither point mutation produces a noticeable environmental perturbation at the site of the remaining Trp. A discontinuity often is found in such a plot even when there is no optical resolution of distinct 0,0-bands[35,46]; this behavior provides evidence for emission from distinct Trp sites in a protein and allows measurement of the ZFS of these sites.

The Trp repressor protein is unusual in that characterization of the local environments of Trp-19 and Trp-99 based on phosphorescence and ODMR characteristics is at variance with that based on fluorescence and the X-ray crystal structure. The latter methods suggest considerably solvent exposure of both sites. In all other proteins studied to date where comparison is possible, the ODMR and phosphorescence are found to be consistent with the character of the local Trp environment as revealed by the X-ray structure. We think that the discrepancy in Trp repressor protein may be the result of a structural change induced by sample cooling that relocates Trp-99 in a buried, hydrophobic environment.

External Heavy Atom Effects

Introduction of a heavy atom perturber into the vicinity of a chromophore such as Trp leads to a heavy atom effect (HAE). The HAE drops off very rapidly with distance beyond van der Waals contact and thus is an indicator of short-range interactions. The HAE depends not only on distance, but on the location of the heavy atom with respect to the coordinate axes of the molecule. For an aromatic $^3(\pi, \pi^*)$ state such as we find in Trp, a relatively small HAE is found if the perturber atom is located in the molecular plane. Large perturbations require overlap between perturber and the π orbitals of the molecule. The sublevel specificity of the HAE has been shown by both theory[47] and experiment[48] to depend on the location of the heavy atom. If z defines the out-of-plane direction, then location of the perturber directly along z perturbs T_z selectively, whereas displacement into the xz plane perturbs both T_x and T_z, and displacement into the yz plane perturbs T_y and T_z. The theory is based only on symmetry arguments and does not consider the relative sizes of the HAE when more than one sublevel is involved.

An important consequence of the HAE is the mixing of singlet spin character into the affected sublevel. This leads to the shortening of the

[46] S.-Y. Mao and A. H. Maki, *Biochemistry* **26**, 3106 (1987).
[47] G. Weinzierl and J. Friedrich, *Chem. Phys. Lett.* **80**, 55 (1981).
[48] S. Ghosh, M. Petrin, and A. H. Maki, *J. Chem. Phys.* **87**, 4315 (1987).

lifetime, usually by radiative processes. Such sublevel-specific effects are readily determined by the ODMR methods outlined earlier and can lead to structural information regarding the location of the perturber atom relative to the affected chromophore.

An example of a heavy atom-containing perturber that has been employed frequently in ODMR spectroscopy is CH_3Hg^+, which when used as the iodide binds specifically to accessible sulfhydryl residues.[49] Binding of CH_3Hg^+ to the amine of Trp, itself, leads to out-of-plane van der Waals contact of Hg with the π system and to a pronounced HAE that affects primarily T_x.[50] Heterogeneity in the location of Hg leads to a distribution of lifetimes; the smallest perturbation gives $k_x \approx k_x^r = 300$ sec^{-1}, a reduction of about 10^3 in lifetime. The kinetic data are presented in Table II. Submillisecond lifetimes are found from other structures that are associated with large shifts in ZFS.[51] This is an example of the contribution of spin–orbit coupling to the ZFS.

In contrast with the CH_3Hg–Trp complexes, complexes of CH_3Hg^+ with benzimidazole, in which the Hg atom is bound to N in the molecular plane, have sublevel lifetimes that are more than an order of magnitude longer.[50] Thus, Trp sublevel lifetimes on the order of a few milliseconds or less are consistent with an out-of-plane van der Waals contact with Hg. Heavy atom effects of this magnitude affecting T_x have been produced by CH_3HgI binding to porcine and rabbit glyceraldehyde phosphate dehydrogenase,[52] indicating the proximity of Trp-310 to Cys-281. On the other hand, a smaller HAE that originates from Trp-158, specifically affecting T_z, has been produced[53] by CH_3HgI reacting with bacteriophage T4 lysozyme (Table II). Modeling based on the X-ray structure suggests that the Hg atom is bound to Cys-97 and that it is located along the z axis, but significantly beyond van der Waals contact. In each case a prominent $D + E$ PDMR signal is induced by the Hg perturbation. A useful feature of a HAE sufficiently large to produce sublevel lifetimes of a few milliseconds is that the perturbed triplet state now can be successfully investigated by PMDR methods. Modulated phosphorescence will be produced by audio frequency amplitude modulation of an ODMR signal of only those Trp residues that are subjected to the HAE. Phase-sensitive detection selects only the contribution of the perturbed Trp residues to the overall emission.[52]

The Hg atom also has been used as a perturber to identify aromatic stacking interactions of Trp residues in protein–nucleic acid complexes.

[49] D. L. Rabenstein, *Acc. Chem. Res.* **11,** 100 (1978).

[50] R. R. Anderson and A. H. Maki, *J. Am. Chem. Soc.* **102,** 163 (1980).

[51] J. M. Davis, Doctoral Dissertation, University of California, Davis (1984).

[52] J. M. Davis and A. H. Maki, *Biochemistry* **23,** 6249 (1984).

[53] L.-H. Zang, S. Ghosh, and A. H. Maki, *Biochemistry* **27,** 7820 (1988).

TABLE II
TRIPLET STATE KINETICS OF TRYPTOPHAN IN VARIOUS ENVIRONMENTS

Sample	k_x (sec^{-1})[a]	k_y (sec^{-1})[a]	k_z (sec^{-1})[a]	p_x	p_y	p_z	Ref.
Tryptophan	0.24[c]	0.12[c]	0.038[c]	0.39	0.38	0.23	b
Hen lysozyme	0.29[c]	0.11[c]	0.054[c]	0.39	0.36	0.25	b
α-Chymotrypsin	0.32[c]	0.13[c]	0.034[c]	—	—	—	d
Trp–CH$_3$Hg	303.0	28.5	26.4	—	—	—	e
EcoSSB–poly(dT)							f
Trp-54	1.18	0.082	0.056	>0.9	<0.07	<0.04	
Trp-40	0.46	0.076	0.041	—	—	—	
p7, HIV-1	0.42[g]	0.086[g]	0.052[g]	0.35[h]	0.38[h]	0.27[h]	
p7, HIV-1–poly(5-HgU)	13.0	3.7	290.0	—	—	—	g
p7, HIV-1–(dG)$_8$	0.65	0.20	0.091	—	—	—	h
p7, HIV-1–poly(I)	0.63	0.11	0.067	0.60	0.25	0.15	h
T4 lysozyme–CH$_3$Hg Trp-154	6.2	6.9	72.1	—	—	—	i
EcoRI methylase–As(CH$_3$)$_2$ Trp-225	44.0	0.77	2.33	—	—	—	j

[a] The decay constants are obtained by MIDP and are not corrected for slr, unless indicated.
[b] J. Zuclich, J. U. von Schütz, and A. H. Maki, *Mol. Phys.* **28,** 33 (1974).
[c] Corrected for slr. Values obtained by microwave-saturated phosphorescence decay analysis.
[d] A. H. Maki and T.-t. Co, *Biochemistry* **15,** 1229 (1976).
[e] R. R. Anderson and A. H. Maki, *J. Am. Chem. Soc.* **102,** 163 (1980). Values are reported for the most weakly perturbing CH$_3$Hg–Trp complex.
[f] L.-H. Zang, A. H. Maki, J. B. Murphy, and J. W. Chase, *Biophys. J.* **52,** 867 (1987); D. H. H. Tsao, J. R. Casas-Finet, A. H. Maki, and J. W. Chase, *Biophys. J.* **55,** 927 (1989).
[g] W.-C. Lam, A. H. Maki, J. R. Casas-Finet, J. W. Erickson, R. C. Sowder II, and L. E. Henderson, *FEBS Lett.* **328,** 45 (1993).
[h] W.-C. Lam, Doctoral Dissertation, University of California, Davis (1993).
[i] L.-H. Zang, S. Ghosh, and A. H. Maki, *Biochemistry* **27,** 7820 (1988).
[j] W.-C. Lam, D. H. H. Tsao, A. H. Maki, K. A. Maegley, and N. O. Reich, *Biochemistry* **31,** 10438 (1992).

Poly(U) is readily mercurated at the 5 position of uracil, and when the other valence of Hg is blocked with a sulfhydryl, such as 2-mercaptoethanol, the resulting heavy atom-derivatized polynucleotide, poly(5-HgU), becomes a useful reagent for the investigation of aromatic stacking interactions of Trp in protein–nucleic acid complexes. Blocking of the Hg atom limits its interactions to physical contacts with a bound protein, and when these take place with Trp a HAE is the result. A HAE that produces millisecond lifetimes must result from an out-of-plane van der Waals contact of Hg with Trp, as discussed above; furthermore, steric considerations limit an out-of-plane van der Waals contact to aromatic stacking interactions. Poly(5-HgU) was used as a substrate for the binding of *E. coli* single-stranded DNA binding protein, *Eco* SSB.[54] By use of Trp → Phe point mutants, the observed HAE were assigned to Trp-40 and Trp-54.[55] Although individual sublevel kinetics were not measured, a dominant phosphorescence lifetime component of less than 10 msec at 77 K suggests the stacking of Trp with nucleobases in the *Eco* SSB–poly(5-HgU) complex. The Trp-88 and Trp-135 of *Eco* SSB are not perturbed by complexing with poly(5-HgU).

Aromatic stacking interactions in Trp-37, located on the C-terminal Zn finger of p7, with poly(5-HgU) have been detected by means of a HAE that is specific for T_z.[56] (see Table II). The T_z sublevel lifetime is found to be around 3.4 msec, which is consistent with an out-of-plane van der Waals contact. The modified base 5-bromouracil, which resembles thymine, has also been incorporated in polynucleotides to study interactions with Trp.[54] The HAE induced by van der Waals contact with Br appears to be more than an order of magnitude smaller than that of Hg.

The As atom has been found to induce a HAE on a Trp residue of *Eco* RI methyltransferase (Table II).[57] The perturbing group is $(CH_3)_2$ As(III)$^+$ that is produced, for example, by reduction of cacodylic acid with mercaptides. Like CH_3Hg^+, it binds to accessible Cys residues. The use of mutant enzymes has shown that the perturbed residue is Trp-225, and that the perturbing As binds with high affinity to Cys-223, thus revealing the proximity of the side chains of these residues in the methylase structure.[58] In addition to the HAE, which is T_x-specific, reducing its lifetime to around 25 msec, a phosphorescence red shift of 9.8 nm also is induced. This may reflect an increase in local polarizability from introduction of the As atom.

Aromatic Stacking Interactions

Effects of stacking interactions on the photoexcited triplet state of aromatic residues such as Trp are found over and above the external HAE described above. These effects are observed in protein/peptide–nucleic acid interactions in the absence of heavy atom derivatization of the nucleobases. Binding of *Eco* SSB to poly(dT) produces profound effects on the

[54] M. I. Khamis, J. R. Casas-Finet, A. H. Maki, P. R. Ruvolo, and J. W. Chase, *Biochemistry* **26**, 3347 (1987).

[55] D. H. H. Tsao, J. R. Casas-Finet, A. H. Maki, and J. W. Chase, *Biophys. J.* **55**, 927 (1989).

[56] W.-C. Lam, A. H. Maki, J. R. Casas-Finet, J. W. Erickson, R. C. Sowder II, and L. E. Henderson, *FEBS Lett.* **328**, 45 (1993).

[57] D. H. H. Tsao and A. H. Maki, *Biochemistry* **30**, 4565 (1991).

[58] W.-C. Lam, D. H. H. Tsao, A. H. Maki, K. A. Maegley, and N. O. Reich, *Biochemistry* **31**, 10438 (1992).

properties of Trp-54,[55,59] which is known to undergo stacking interactions with the bases of poly(5-HgU), as discussed above. The ZFS parameters are reduced by around 10% on binding to poly(dT), an unusually large perturbation (Table I). This could result from increased polarizability of the environment, but the size of the ZFS reduction is more suggestive of electronic charge-transfer character in the T_1 state. Even larger ZFS reductions of the T_1 states of phanes have been found and attributed to charge-transfer between the stacked chromophores.[60] In addition to the reduction of the ZFS, effects on the kinetic parameters also are evident (Table II). The triplet state lifetime of Trp-54 is drastically reduced to 2.3 sec (from 5.6 sec); significantly, the entire effect occurs in T_x whose apparent lifetime undergoes a 4-fold reduction.[55,59] This is not a HAE, of course, and a complete explanation has not been advanced. Furthermore, although $p_x : p_y : p_z = 0.39 : 0.38 : 0.23$ for Trp,[31] the isc of Trp-54 becomes very sublevel-specific in the *Eco* SSB–poly (dT) complex; what is found is $p_x : p_y : p_z = >0.9 : <0.07 : <0.04$.[59] In spite of the larger k_x, N_x^0 exceeds N_y^0, N_z^0, leading to negative rather than the normal positive polarity $D - E$ and $2E$ PDMR signals. The enhanced activity of T_x in isc that results from the interaction of *Eco* SSB with poly(dT) has not been explained.

Studies[61] of the interaction of p7 with polynucleotides and oligonucleotides in the absence of heavy atom perturbers are uncovering effects similar to those observed in *Eco* SSB. The complex formation of p7 with $(dG)_8$ results in a reduction of the Trp-37 phosphorescence lifetime at 77 K from 6.6 to around 4.5 sec. Furthermore, MIDP and slow-passage PDMR measurements on p7 and its $(dG)_8$ complex reveal the properties that are presented in Tables I and II. Reduced ZFS are consistent with stacking interactions. The size of the apparent k_x (uncorrected for slr) increases significantly, as was found for Trp-54 in Eco SSB binding to poly(dT). A comparison of the PDMR spectra of p7 and the p7–$(dG)_8$ complex is shown in Fig. 6. The polarity reversal on binding p7 to $(dG)_8$ may be the result of an altered isc pattern in which p_x is selectively enhanced. Measurement of the isc pattern for the p7–$(dG)_8$ complex has not been possible because of the large background of guanine phosphorescence at 1.2 K. Altered isc and ODMR signal polarity reversals have been found to occur, however, when p7 binds to poly(I). The relative values of p are given for p7 and the p7–poly(I) complex in Table II. The data clearly reveal the selective enhancement of p_x. Analogous effects are found for p7 complexes with other RNA homopolymers that are currently

[59] L.-H. Zang, A. H. Maki, J. B. Murphy, and J. W. Chase, *Biophys. J.* **52**, 867 (1987).
[60] M. W. Haenel and D. W. Schweitzer, *Adv. Chem. Ser.* **217**, 333 (1988).
[61] W.-C. Lam, Doctoral Dissertation, University of California, Davis (1993).

under investigation. It will be interesting to learn whether the effects of nucleic acid binding on the photoexcited triplet state of Trp as revealed by ODMR can be related to such important properties as base sequence selectivity and stability of protein–nucleic acid complexes.

Acknowledgments

We thank Professor Maurice Eftink for samples of native and mutated Trp repressor protein and Dr. Jose R. Casas-Finet for samples of HIV-1 nucleocapsid protein, p7. We are grateful to Dr. Laura E. Burns and Dr. Wai-Chung Lam for assistance in preparing some of the figures. Much of the work described in the final section of this chapter was partially supported by the National Institute of Environmental Health Sciences, National Institutes of Health (Grant No. ES-02662).

[26] X-Ray Absorption Spectroscopy and Applications in Structural Biology

By Vittal K. Yachandra

Introduction

X-Ray absorption spectroscopy (XAS) has been used to study a variety of metal sites in biological systems.[1–3] It is the measurement of transitions from core electronic states of the metal to the excited electronic states or continuum states, which is known as X-ray absorption near-edge structure (XANES) or X-ray absorption edge spectroscopy (XAES), and the study of the fine structure in the absorption cross section at energies greater than the threshold for electron release, which is known as extended X-ray absorption fine structure (EXAFS). The two methods give complementary structural information, the edge spectra reporting oxidation state and symmetry, and the EXAFS spectra reporting numbers, types, and distances to ligands and neighboring atoms from the absorbing atom.

The development of X-ray absorption spectroscopy as a practical tool has been concurrent with two major developments in the field. First, the development of a good theoretical understanding of the postabsorption

[1] S. P. Cramer, *in* "X-Ray Absorption: Principles, Applications, Techniques of EXAFS, SEXAFS and XANES" (D. C. Koningsberger and R. Prins, eds.), p. 257. Wiley, New York, 1988.

[2] R. A. Scott, *in* "Structural and Resonance Techniques in Biological Research" (D. L. Rousseau, ed.), p. 295. Academic Press, Orlando, Florida, 1984.

[3] L. Powers, *Biochim. Biophys. Acta* **683**, 1 (1982).

modulations of the photoelectron cross section, known as EXAFS,[4] has permitted the use of XAS as a tool for studying the local structure of many different noncrystalline materials, such as glasses and metalloproteins.[5] The second was the development of intense synchrotron radiation X-ray sources that permit the study of samples which are dilute in the element of interest, which is often the case with biological samples.

The major application of XAS in biology has been in the study of the structure of the metal sites in metalloproteins and metalloenzymes. This was primarily due to the availability of X-ray beam lines at synchrotron sources optimized in the X-ray energy region of metal K-edges (from Ca to Mo) and also the ease of making measurements at such X-ray energies, without interference from absorption by the protein matrix, water, or air. Second, advances being made in the field of bioinorganic chemistry were raising important questions of correlation between structure and function of the metal sites in metalloproteins which were amenable to XAS studies.

X-Ray Absorption Spectroscopy

X-Ray absorption spectra of any material, whether atomic or molecular in nature, is characterized by sharp increases in absorption at specific X-ray photon energies, which are characteristic of the absorbing element. These sudden increases in absorption are called absorption edges and correspond to the energy required to eject a core electron into the continuum, thus producing a photoelectron. The absorption discontinuity is known as the K-edge when the photoelectron originates from a $1s$ core level, and it is called the L-edge when the ionization is from a $2s$ or $2p$ electron. Figure 1 shows a typical energy level diagram.

At higher energies, above the edge, the absorption decreases in accordance with the theory of the photoelectric effect, where the excess energy is transferred to the photoelectron as kinetic energy. However, there is a crucial difference in this falloff of absorption between isolated atoms and molecular or condensed systems. For the latter, superimposed on the gradual decrease in absorption is a periodic modulation in absorption, starting immediately past an absorption edge and extending to about 1 keV above the edge. These oscillations in the X-ray absorption spectra are known as extended X-ray absorption fine structure or EXAFS. A typical X-ray absorption spectrum at the Mn K-edge is shown in Fig. 2. The EXAFS region starts about 50 eV above the edge, and the region before that is the XANES region.

[4] D. E. Sayers, E. A. Stern, and F. W. Lytle, *Phys. Rev. Lett.* **27**, 1204 (1971).
[5] P. Eisenberger and B. M. Kincaid, *Science* **200**, 1441 (1978).

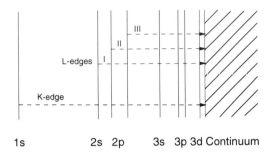

1s   2s 2p   3s 3p 3d Continuum

FIG. 1. X-Ray absorption energy level diagram showing the K-edge and the L_I, L_{II}, and L_{III} edges for a typical first row transition metal. The energy levels are not drawn to scale. For example, in the case of Mn the K-edge is at 6539 eV and the L-edges are at 769,650, and 639 eV, respectively.

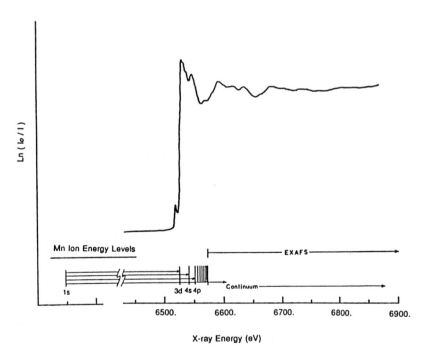

X-ray Energy (eV)

FIG. 2. X-Ray absorption spectrum of MnO_2. The region defined as EXAFS is shown. The XANES region encompasses the pre-edge to about 50 eV above the infection point. [From R. D. Guiles, Ph.D. Dissertation, University of California, Berkeley, Lawrence Berkeley Laboratory Report, LBL 25186 (1982).]

The EXAFS oscillations occur only in molecular or condensed systems and occur in all systems with no requirement for long-range order, such as in crystalline materials, and they contain information about the local environment around the absorbing atom. In fact, the EXAFS oscillations result from the interference between the outgoing photoelectron wave and components of the wave backscattered from neighboring atoms in the molecule.

The importance of the EXAFS technique to the biochemist or structural biologist depends directly on the fact that the EXAFS modulations contain information about the separation between the absorbing and backscattering atoms within a distance of about 5 Å, as well as the identity and number of the backscattering atoms. Essentially, EXAFS analysis is used to determine the radial distribution of atoms around a particular absorbing atom, thus providing a probe for the local structure in the vicinity of the absorbing atom.

Basic Theory of Extended X-Ray Absorption Fine Structure

Energy of Photoelectron

The EXAFS modulations shown in Fig. 2 are a direct consequence of the wave nature of the photoelectron whose wavelength λ is given by the de Broglie relation:

$$\lambda = \frac{h}{m_e v} \qquad (1)$$

where h is Planck's constant, m_e is the electron mass, and v is the velocity of the photoelectron, which is the velocity imparted to the photoelectron by the energy of the absorbed X-ray photon that is in excess of the binding or threshold energy for the electron. The kinetic energy of the photoelectron is given by the following relation:

$$(E - E_0) = \tfrac{1}{2} m_e v^2 \qquad (2)$$

where E is the X-ray photon energy and E_0 is the ionization or threshold energy for the electron.

The EXAFS modulations are better expressed as a function of the photoelectron wave-vector k, which is related to the de Broglie wavelength described above as follows:

$$k = \frac{2\pi}{\lambda} \qquad (3)$$

which now can be expressed by the substitution of Eqs. (1) and (2) as follows:

$$k = \frac{2\pi}{h}[2m_e(E - E_0)]^{1/2} \tag{4}$$

or

$$k = 0.512(E - E_0)^{1/2} \tag{5}$$

where E and E_0 are expressed in electron volts (eV) and k has the units of inverse angstroms (Å^{-1}).

Definition of EXAFS

The general definition for the EXAFS phenomenon $\chi(k)$, which is the oscillatory portion of the absorption coefficient, is the difference between the observed absorption coefficient $\mu(k)$ and the free atom absorption coefficient $\mu_0(k)$, normalized by the free atom contribution:

$$\chi(k) = \frac{\mu(k) - \mu_0(k)}{\mu_0(k)} \tag{6}$$

The absorption coefficient is proportional to the square of the electric dipole transition moment $|M|$, which is given quantum mechanically by a matrix element between the initial and final states of the photoelectron:

$$\mu(k) \propto |M|^2 = |\langle f|\mathbf{r}|i\rangle|^2 \tag{7}$$

In the case of K-edge absorption the initial state wave function $|i\rangle$ is that of the $1s$ state in the atomic core, and the final state wave function $\langle f|$ is that of the photoelectron. In an isolated atom the final state wave function $\langle f|$ is the wave function of the outgoing photoelectron $\langle f_0|$.

In a molecular or condensed system the final state wave function of the photoelectron consists of both the outgoing wave $\langle f_0|$ and the backscattered wave from the neighboring j atoms $\Sigma_j f_{sc}^j$, and the total wave function is given by the sum of the outgoing and the scattered wave functions from each of the j backscattering atoms. The transition dipole moment can now be expressed as follows:

$$|M| = \left\langle \left(f_0 + \sum_j f_{sc}^j\right)|\mathbf{r}|i\right\rangle \tag{8}$$

and it is the interference between f_0 and f_{sc} that lead to the EXAFS modulations, and substitution in Eq. (6) gives

$$\chi(k) = \frac{\Delta|M|^2}{|M|^2} \tag{9}$$

The electric dipole moment $|M|$ is nonzero only in the region where the initial state wave function is nonzero, which is near the center of the absorbing atom. So one needs to know only how the surrounding atoms perturb the outgoing wave and what effect this has at the center of the absorbing atom. One can envision this process more clearly by the help of a schematic of the outgoing and backscattered waves as shown in Fig. 3. As the energy of the photoelectron changes, so does the wavelength, $\lambda_{electron}$, of the photoelectron. At a particular energy E_1 the outgoing and the backscattered waves are in phase and constructively interfere, thus increasing the probability of X-ray absorption or, in other words, increasing the absorption coefficient. At a different energy E_2 the outgoing and backscattered waves are out-of phase and destructively interfere, decreasing the absorption coefficient. This modulation of the absorption coefficient by the backscattered wave from neighboring atoms is essentially the basic phenomenon of EXAFS.

EXAFS Equation

A quantitative description of the EXAFS modulation, $\chi(k)$, depends on an evaluation of the final state wave function, $(f_0 + \Sigma_j f_{sc}^j)$. Many excellent sources exist for a rigorous derivation of the EXAFS equation.[6–10] For our purposes $\chi(k)$ can be expressed as follows:

$$\chi(k) = \sum_j \frac{N_j |f_j(\pi, k)|}{kR_{aj}^2} \sin[2kR_{aj} + \alpha_{aj}(k)] \tag{10}$$

where N_j is the number of equivalent backscattering atoms j at a distance R_{aj} from the absorbing atom, $f_j(\pi, k)$ is the backscattering amplitude, which is a function of the atomic number of the backscattering element

[6] C. A. Ashley and S. Doniach, *Phys. Rev. B: Solid State* **11**, 1279 (1975).
[7] P. A. Lee and J. B. Pendry, *Phys. Rev. B: Solid State* **11**, 2795 (1975).
[8] P. A. Lee and G. Beni, *Phys. Rev. B: Solid State* **15**, 2862 (1977).
[9] E. Stern, *in* "X-Ray Absorption: Principles, Applications, Techniques of EXAFS, SEX-AFS and XANES" (D. C. Koningsberger and R. Prins, eds.), p. 3. Wiley, New York, 1988.
[10] B.-K. Teo, "EXAFS: Basic Principles and Data Analysis." Springer-Verlag, New York, 1986.

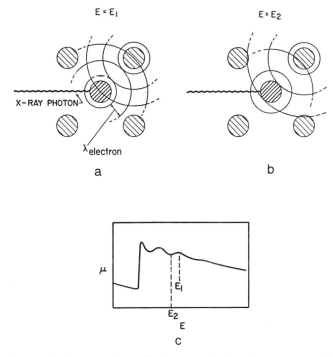

Fig. 3. Schematic of the outgoing and backscattered photoelectron wave, which illustrates the concept of interference in EXAFS. The central atom is the absorbing atom, and the photoelectron is backscattered from the surrounding atoms. (a) The backscattered wave is in phase with the outgoing wave at energy E_1. This leads to an increase in the absorption coefficient at E_1. (b) At energy E_2 the backscattered wave destructively interferes with the outgoing wave, which leads to a decrease in the cross section at E_2. (c) Attenuation in the cross section in the absorption coefficient. [Adapted from D. B. Goodin, Ph.D. Dissertation, University of California, Berkeley. Lawrence Berkeley Laboratory Report, LBL 16901 (1983); and R. A. Scott, *in* "Structural and Resonance Techniques in Biological Research" (D. L. Rousseau, ed.), p. 295. Academic Press, Orlando, Florida, 1984.]

j, and $\alpha_{aj}(k)$ includes the phase shift from the central atom absorber as well as the backscattering element j. The phase shift occurs because of the presence of atomic potentials that the photoelectron experiences as it traverses the potential of the absorber atom, the potential of the backscattering atom, and then back through the potential of the absorber atom.

In real systems there is inherent static disorder owing to a distribution of distances R_{aj} and dynamic disorder owing to thermal vibrations of the absorbing and scattering atoms. Equation (10) is modified to include this disorder term or the Debye–Waller factor $e^{-2\sigma_{aj}^2 k^2}$, where σ_{aj} is the root-mean-square deviation, to give

$$\chi(k) = \sum_j \frac{N_j |f_j(\pi, k)|}{kR_{aj}^2} e^{-2\sigma_{aj}^2 k^2} \sin[2kR_{aj} + \alpha_{aj}(k)] \qquad (11)$$

The loss of photoelectrons to inelastic scattering processes can be accounted for by including a term $e^{-2R_{aj}/\lambda}$ where λ is the mean free path of the photoelectron.

Examination of Eq. (11) shows that EXAFS spectra contain a significant amount of information about atomic surroundings of the central absorbing atom in the system being studied, and the expression has been used for applications.[11,12] The EXAFS contribution from each backscattering atom j is a damped sine wave in k space, with an amplitude and a phase which are dependent on k.

From the phase of each sine wave $[2kR_{aj} + \alpha_{aj}(k)]$, the absorber–backscatterer distance R_{aj} can be determined if the phase shift $\alpha_{aj}(k)$ is known. The phase shift is obtained either from theoretical calculations or empirically from compounds characterized by crystallography with the specific absorber–backscatterer pair of atoms. The phase shift, $\alpha_{aj}(k)$, depends on both the absorber and scatterer atoms. Because one knows the absorbing atom in an EXAFS experiment, an estimation of the phase shift can be used in identifying the scattering atom.

The amplitude function contains the Debye–Waller factor and N_j, the number of backscatters at R_{aj}. These two parameters are highly correlated, making the determination of N_j difficult at best. The backscattering amplitude function, $f_j(\pi, k)$, depends on the atomic number of the scattering atom and, in principle, can be used to identify the scattering atoms. In practice, however, the phase shift and backscattering amplitude function, both of which are dependent on the identity of the backscattering atom, can be used only to identify scattering atoms that are well separated by atomic number.

The sinusoidal nature of the backscattering contributions makes Fourier analysis particularly appropriate for analyzing EXAFS spectra. A Fourier transform of the EXAFS data in k space can be employed to separate the different scattering distances into unique peaks in the conjugate R space. If the distances are sufficiently different, then the R-space peaks should be separated, and the different EXAFS sine waves can be studied individually by using the technique known as Fourier filtering described below. It is the introduction of the Fourier transform by Sayers

[11] P. A. Lee, P. H. Citrin, P. Eisenberger, and B. M. Kincaid, *Rev. Mod. Phys.* **53**, 769 (1981).
[12] S. P. Cramer and K. O. Hodgson, *Prog. Inorg. Chem.* 25, 1 (1979).

et al.[4] which changed EXAFS from a scientific curiosity to a method of practical utility.

Applications of Extended X-Ray Absorption Fine Structure in Structural Biology

Since the EXAFS technique has become practical, it has become a standard method for probing the metal site structure in metalloproteins. The technique has been applied to virtually every metalloprotein that has been isolated containing Fe, Mo, V, Cu, Mn, Zn, Ni, Ca, and other metals. In this section, the usefulness of the method is illustrated by choosing three different biological systems where XAS has contributed to structural determination. Readers are referred to several excellent reviews for an overview of the biological applications of XAS.[1-3]

Iron–Sulfur Proteins

Among the first metalloproteins studied by XAS is the class of nonheme iron–sulfur proteins which are found in most redox-mediated pathways in biology. Initially, three classes of Fe–S structures containing 1Fe, 2Fe, and 4Fe metal sites were recognized. This list has expanded to include 3Fe sites and higher nuclearity structures containing 7 and 8 Fe atoms in the M and P clusters of nitrogenase. The structural unit is $Fe(SR)_4$ in a 1Fe-containing Fe–S protein like rubredoxin, $Fe_2S_2(SR)_4$ in 2Fe–2S proteins like soluble plant ferredoxins, and $Fe_4S_4(SR)_4$ in 4Fe–4S proteins like "high potential iron proteins" (HiPIP) and other ferredoxins generally referred to as bacterial ferredoxins.[13]

The Fe EXAFS spectrum of the iron–sulfur proteins is dominated by backscattering from sulfur and iron atoms in the active site. In rubredoxin one Fe atom is ligated to four thiolate-derived S atoms in near tetrahedral symmetry, and it is a good example of a single-backscattering environment. The EXAFS spectrum of rubredoxin (see Fig. 4A) exhibits a single wave, and the Fourier transform shows one distinct peak, indicative of a single-shell system with one type of distance.[14] The four Fe–S distances of around 2.26 Å were found to be equivalent to within 0.04 Å, in contrast to the original crystallographic determination which required one "short"

---

[13] T. G. Spiro (ed.), "Iron-Sulfur Proteins" , Wiley, New York, 1982.

[14] B.-K. Teo and R. G. Shulman, *in* "Iron-Sulfur Proteins" (T. G. Spiro, ed.), p. 343. Wiley, New York, 1982.

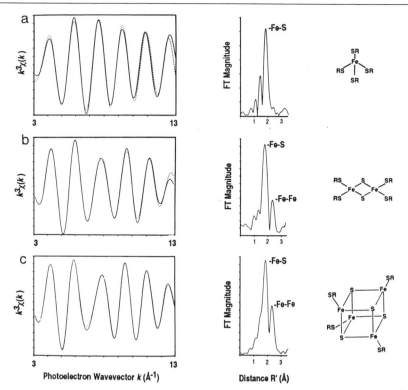

FIG. 4. Iron EXAFS Fourier isolates (solid line) and the fits to the data (dotted line), Fourier transforms, and the Fe–S structures found in (a) 1Fe Fe–S protein rubredoxin, (b) 2Fe–2S plant ferredoxin, and (c) 4Fe–4S bacterial ferredoxin. In (a) the EXAFS spectrum from the 1Fe protein rubredoxin shows a single damped sine wave indicating the presence of one major Fe–S distance. This is reflected in the Fourier transform which shows only one major peak. The peak at 1.5 Å is due to residual background and/or Fourier truncation. In (b) and (c) and EXAFS spectra show a beat pattern indicating the presence of Fe–S and Fe–Fe distances. The Fourier transforms shows two peaks that are due to backscattering from S and Fe atoms, respectively. [Adapted from B.-K. Teo and R. G. Shulman, *in* "Iron-Sulfur Proteins" (T. G. Spiro, ed.), p. 343. Wiley, New York, 1982.]

Fe–S distance of 2.05 Å.[15] The crystal structure has since been refined, and the results have been found to be consistent with the EXAFS results.[16]

In contrast to the EXAFS spectrum from rubredoxin, the EXAFS spectra of the 2Fe–2S soluble plant ferredoxin and the 4Fe–4S bacterial ferredoxin (see Fig. 4B,C) exhibit a beat pattern indicating the presence

[15] K. D. Watenpaugh, L. C. Sieker, J. R. Herriot, and L. H. Jensen, *Acta Crystallogr. Sect B: Struct. Crystallogr. Cryst. Chem.* **29**, 943 (1973).
[16] K. D. Watenpaugh, L. C. Sieker, and L. H. Jensen, *J. Mol. Biol.* **138**, 615 (1980).

of at least two sine waves. The respective Fourier transforms show two peaks. The first peak is due to backscattering from S ligand atoms at around 2.23–2.25 Å, and the second peak is due to backscattering from the Fe atoms at 2.73 Å.[14] These distances compare well with distances derived from X-ray crystallography of 2Fe–2S and 4Fe–4S proteins and synthetic analogs.[1,2,14]

Nitrogenase: Iron-Molybdenum Cofactor

Biological reduction of nitrogen to ammonia (Eq. 12) is catalyzed by the nitrogenase enzyme, which consists of two proteins called the Mo–Fe

$$N_2 + 8H^+ + 8e^- + nMgATP \rightarrow 2NH_3 + H_2 + nMgADP + nP_i \quad (12)$$

protein and the Fe protein. An iron–molybdenum cofactor (FeMo-co), containing Fe, Mo, and S in the ratio of $7:1:\sim9$, can be extracted from the Mo–Fe protein. The FeMo-co is thought to be directly involved in substrate binding and reduction.[17,18] A nitrogenase containing V instead of Mo is also known, and a third type of nitrogenase has been purified which lacks Mo and V, presumably containing an all-Fe cofactor. The structure of the metal cluster has been exhaustively studied using many spectroscopic methods including XAS.

There are several Mo XAS studies of the nitrogenase system.[1] A representative Mo EXAFS Fourier transform is shown in Fig. 5A, and the presence of two peaks indicates that at least two major components are required to explain the Mo spectrum. Fitting the Mo EXAFS spectra from the Mo–Fe protein and the FeMo-co shows that there are 2–3 O or N ligand atoms at around 2.1 Å, 3–4 S atoms at about 2.4 Å, and 2–3 Fe atoms at 2.7 Å from the Mo atom. These results indicate some kind of a S- and O or N-bridged Mo and Fe cluster. Several models have been proposed based on the results and comparisons to inorganic complexes. Figure 5B shows one such model ($MoFe_3S_4$) proposed by Hodgson, Holm, and co-workers based on comparison to S-bridged cubelike Mo, Fe model complexes. They concluded that "the Mo–Fe protein possesses a structural fragment similar to the $MoFe_3S_4$ cube that constitutes one half of the complex $[Mo_2Fe_6S_9(SC_2H_5)_8]^{3-}$."[19]

The structure of the FeMo-co proposed by Kim and Rees[20] based on X-ray crystallography is shown in Fig. 5C. Despite some differences

[17] W. H. Orme-Johnson, *Annu. Rev. Biophys. Biophys. Chem.* **14,** 419 (1985).
[18] B. E. Smith and R. R. Eady, *Eur. J. Biochem.* **205,** 1 (1992).
[19] T. E. Wolff, J. M. Berg, C. Warrick, K. O. Hodgson, R. H. Holm, and R. B. Frankel, *J. Am. Chem. Soc.* **100,** 4630 (1978).
[20] J. Kim and D. C. Rees, *Science* **257,** 1677 (1992).

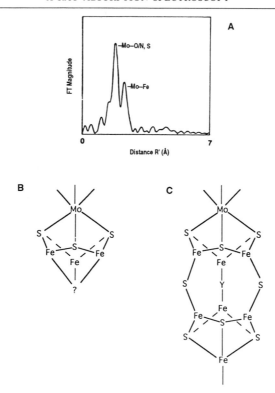

FIG. 5. (A) Fourier transform of Mo K-edge EXAFS spectrum of the Mo–Fe protein from nitrogenase [Adapted from S. P. Cramer, *in* "X-Ray Absorption: Principles, Applications, Techniques of EXAFS, SEXAFS and XANES" (D. C. Koningsberger and R. Prins, eds.), p. 257. Wiley, New York, 1988]. The Fourier transform shows two peaks indicating that at least two major components, Mo–S (and Mo–O/N) and Mo–Fe, are required to explain the Mo EXAFS spectrum from nitrogenase. (B) Model for the FeMo-co proposed based on the Mo EXAFS data and comparison with data from model compounds [T. E. Wolff, J. M. Berg, C. Warrick, K. O. Hodgson, R. H. Holm, and R. B. Frankel, *J. Am. Chem. Soc.* **100,** 4630 (1978)]. (C) Structure of the FeMo-co based on a more recent X-ray crystal structure [J Kim and D. C. Rees, *Science* **257,** 1677 (1992)]. The similarities between the structures in (B) and (C) are remarkable.

between the proposed structures, the similarities between the X-ray crystal structure and that based on EXAFS studies are remarkable.

Photosynthetic Manganese Oxygen-Evolving Complex

Most of the oxygen in the atmosphere which supports life on earth is generated by plants by the photoinduced oxidation of water to dioxygen:

$$2H_2O \rightarrow O_2 + 4e^- + 4H^+ \qquad (13)$$

The reaction shown in Eq. (13) is catalyzed by a tetranuclear Mn complex, which sequentially stores four oxidizing equivalents that are used to oxidize two molecules of water to molecular oxygen. The Mn complex is part of a multiprotein assembly called photosystem II (PSII), which contains the reaction center involved in photosynthetic charge separation and an antenna complex of chlorophyll molecules. The complex also contains a cytochrome b-559-containing peptide and a Fe–quinone electron acceptor complex.[21] Owing to the complexity of the system and the presence of so many pigment and other components, study of the Mn complex by optical and other spectroscopic methods can be difficult. EXAFS is ideally suited for the study of the structure of the Mn complex because the specificity of the technique allows us to look at the Mn without interference from the pigment molecules, or the protein and membrane matrix, or other metals like Ca, Mg, Cu, and Fe which are also present in active preparations. An active oxygen-evolving complex has not been crystallized, but EXAFS does not need single crystals; the structural studies can be performed on frozen solutions. Also, several of the intermediate states mentioned above have been stabilized as frozen solutions and studied by XAS.[22] Much of the work presented in this chapter on the Mn oxygen-evolving complex was performed in our laboratory, and the procedures are described in greater detail to illustrate the analysis methods followed.

An examination of a typical EXAFS spectrum from Mn in photosystem II shows beat patterns indicating that the modulations contain more than one damped sine wave. The Fourier transform of the Mn EXAFS data (Fig. 6) shows at least three clearly resolved features. Note that the peaks are not at the real distances of the shells from Mn but are shifted by about 0.5 Å owing to the phase shift α_{aj} explained above. There are some features at less than 1 Å which arise from residual background. If the $k^3\chi(k)$ data are Fourier transformed without any removal of background, the feature in the Fourier transform at around 1 Å is more intense. An iterative procedure is employed to decide on the amount of background to be removed while retaining the other Fourier peaks which are at physically reasonable distances. Fourier transformation of the data weighted by k^1, k^2, or k^3 shows that the amplitude of the second peak in the Fourier transform increases relative to the first peak. This indicates that the second peak might be due to backscattering from a high-Z element, like another

[21] R. Debus, Biochim. Biophys. Acta 110, 269 (1992).
[22] K. Sauer, V. K. Yachandra, R. D. Britt, and M. P. Klein, in "Manganese Redox Enzymes" (V. L. Pecoraro, ed.), p. 141. VCH, New York, 1992.

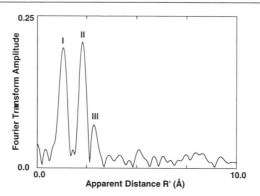

FIG. 6. Fourier transform of the k^3-weighted Mn EXAFS spectrum from *Synechococcus* in the S_1 state. Peak I represents backscattering from 1.5–2.5 N or O bridging ligand atoms at around 1.8 Å and 2–4 N or O terminal ligand atoms at 1.95–2.15 Å. Peak II is attributable to 1–1.5 Mn at 2.72 Å, and peak III fits to 0.5 Mn or Ca atoms or both at 3.3 Å. [Adapted from V. K. Yachandra, V. J. DeRose, M. J. Latimer, I. Mukerji, K. Sauer, and M. P. Klein, *Science* **260**, 675 (1993)].

Mn atom. This provides evidence that we are looking at a bridged multinuclear Mn complex.

Each of the Fourier peaks is isolated individually and back-transformed into k space. Sometimes two peaks are isolated together because of lack of resolution. The back-transformed k-space data were then fit using theoretically derived functions calculated using the curved-wave formalism.[23] The first Fourier peak requires two different distances to be fit adequately, one at around 1.8 Å and the other 1.95–2.15 Å to light elements like C, N, or O. The second peak, on the other hand, is best fit by a heavier atom, probably, as noted above, another Mn atom at 2.7 Å; the amplitude is best fit to an average of one such interaction per Mn atom in the complex. Fitting the third peak, at a longer distance, is more difficult. It can be fit to C or Mn or Ca, or to a combination of these entities at 3 Å. The quality of fit is always better when Mn is included in this third peak, but it is not clear if there is only Mn or also Ca atoms. This is the quintessential problem when analyzing EXAFS data at distances over 3 Å.[24] The typical procedure at this stage of the analysis consists of examining the structural motifs that exist in bridged Mn multinuclear complexes.[25,26] Such an analy-

[23] A. G. McKale, B. W. Veal, A. P. Paulikas, S.-K. Chan, and G. S. Knapp, *J. Am. Chem. Soc.* **110**, 3763 (1988).

[24] R. A. Scott and M. K. Eidsness, *Comments Inorg. Chem.* **7**, 235 (1988).

[25] K. Wieghardt, *Angew. Chem., Int. Ed. Engl.* **28**, 1153 (1989).

[26] V. L. Pecoraro, *in* "Manganese Redox Enzymes" (V. L. Pecoraro, ed.), p. 197. VCH Publ., New York, 1992.

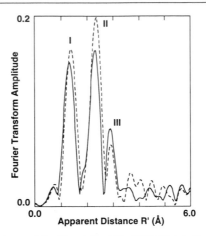

FIG. 7. Fourier transform of the k^3-weighted Mn EXAFS data from oriented photosystem II membranes in the S_1 state from spinach. The membrane normal was oriented 15° (solid line) and 75° (dashed line) to the polarization direction of the X-rays. The orientation dependence of the EXAFS spectrum provides geometric information about the complex. [Adapted from V. K. Yachandra, V. J. DeRose, M. J. Latimer, I. Mukerji, K. Sauer, and M. P. Klein, *Science* **260**, 675 (1993).]

sis produces important chemical insight. The presence of both the Mn–O (C, N) distances at approximately 1.8 Å and the Mn–Mn distances at 2.7 Å in all known di-μ-oxo-bridged multinuclear Mn complexes, and the presence of a 3.3 Å Mn–Mn separation in mono-μ-oxo-bridged Mn complexes leads to the conclusion that both of these structural motifs are present in the Mn complex of PSII.[27] The number of such scattering interactions leads to the conclusion (remembering that there are four Mn atoms) that there are at least two Mn–Mn di-μ-oxo-bridged units and at least one Mn–Mn mono-μ-oxo-bridged unit.

Figure 7 shows the Fourier transform of the same system in oriented samples.[28] It is evident that the 2.7 and 3.3 Å Mn–Mn vectors (the second and third peak in the Fourier transform) are oriented differently. The 2.7 Å vector is more parallel to the membrane normal, and the 3.3 Å one is more perpendicular. Detailed analysis of the dichroism can be used to determine the angles more accurately. The dichroism requires that the complex be asymmetric. One structure consistent with data shown in Fig.

[27] V. J. DeRose, I. Mukerji, M. J. Latimer, V. K. Yachandra, K. Sauer, and M. P. Klein, *J. Am. Chem. Soc.* **116**, 5239 (1994).

[28] I. Mukerji, J. C. Andrews, V. J. DeRose, M. J. Latimer, V. K. Yachandra, K. Sauer, and M. P. Klein, *Biochemistry* **33**, 9712 (1994).

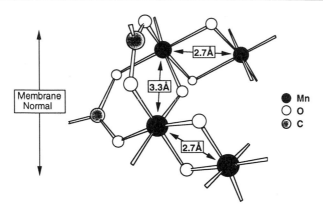

FIG. 8. Structural model for the Mn complex in photosystem II derived from the EXAFS data. This is only one of many structures that are consistent with the EXAFS results. [From K. Sauer, V. K. Yachandra, R. D. Britt, and M. P. Klein, *in* "Manganese Redox Enzymes" (V. L. Pecoraro, ed.), p. 141. VCH, New York, 1992.]

6 and the dichroism measurements is shown in Fig. 8 and provides a working model for the Mn cluster[29] which can be tested with other methods.

Experimental Apparatus and Data Collection

The X-ray absorption measurements are analogous to optical absorption or fluorescence excitation measurements. The spectrometer consists of a source of X-rays, a monochromator, an X-ray detector which monitors the incident flux, I_0, a sample compartment, and a second detector which monitors the flux of X-rays transmitted through the sample, I_1. The sensitivity of the technique is extended several orders of magnitude by fluorescence detection of the absorption spectrum (F/I_0).[30] This enhanced sensitivity is essential to the measurement of X-ray absorption spectra of biological systems. The fluorescence detector is placed at right angles to the incident beam. In addition, a third X-ray detector is used downstream from I_1 to measure simultaneously the spectrum of a reference compound, which provides an energy reference. The essential components described

[29] V. K. Yachandra, V. J. DeRose, M. J. Latimer, I. Mukerji, K. Sauer, and M. P. Klein, *Science* **260,** 675 (1993).
[30] J. Jaklevic, J. A. Kirby, M. P. Klein, A. S. Robertson, G. S. Brown, and P. Eisenberger, *Solid State Commun.* **23,** 679 (1977).

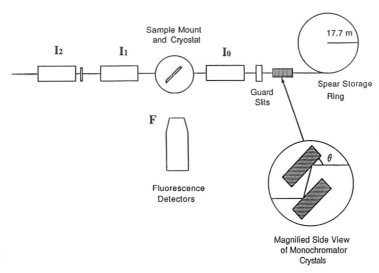

FIG. 9. Experimental setup for a typical XAS experiment with a biological sample. I_0 and I_1 are detectors which measure the intensity of the incident and transmitted X-rays. I_2 is a detector downstream of I_1 that measures the spectrum of a sample used as a energy reference which is placed between I_1 and I_2. F is a 13-element solid state Ge detector, with an energy resolution of about 150 eV, which is used to measure fluorescence and which can discriminate between the scatter and the fluorescence photons from the metal being studied. The cryostat is a liquid He flow system, and samples are routinely maintained at temperatures close to 4–10 K. The monochromator is a matched pair of crystals; different energies are selected for by rotating the crystals. [Adapted from R. D. Guiles, Ph.D. Dissertation, University of California, Berkeley. Lawrence Berkeley Laboratory Report, LBL 25186 (1982).]

above are shown in Fig. 9. The components involved in the X-ray absorption apparatus are discussed briefly below.[31]

Source

The source used for most biological XAS experiments is the synchrotron radiation produced at electron storage rings. The intense flux generated by this means is significantly higher than that from any other X-ray source, and this technology makes X-ray measurements of dilute systems, which is typical of most biological materials, feasible. Synchrotron radiation is produced when electrons accelerated to relativistic velocities are

[31] S. M. Heald, in "X-Ray Absorption: Principles, Applications, Techniques of EXAFS, SEXAFS and XANES" (D. C. Koningsberger and R. Prins, eds.), p. 87 and 119. Wiley, New York, 1988.

bent by a magnetic field. This is achieved in storage rings either by regular bending magnets, which are required to keep the electrons in a storage ring, or by the use of specialized insertion devices called Wiggler magnets and Undulators, which can increase the flux by several orders of magnitude (see Fig. 10). The radiation emitted covers a broad range of energies from the infrared to the hard X-ray region, and it is emitted in a narrow cone which sweeps out a path tangential to the direction of motion of the electrons. The spectral distribution of synchrotron radiation is shown in Fig. 10. Such curves are of importance to the experimentalist in determining factors such as the flux of a harmonic relative to the fundamental frequency of interest. The presence of a substantial fraction of harmonic X-rays can result in distortion of the X-ray spectrum (see below for details). The characteristics of the electron beam in the storage ring and of the magnets are significant in determining the quality of the radiation for spectroscopy. The cutoff energy, ξ_c (Fig. 10), is determined by the energy of the electrons in the storage ring and the field of the magnets, and the flux is dependent on the current in the storage ring and also on the field of the magnets. Different storage rings are optimized for different energy regions of synchrotron radiation, and the proper choice of bending magnets and Wigglers and their fields determines the flux of the X-rays. Although the user does not have control over most of these parameters, it is worth taking note of them.

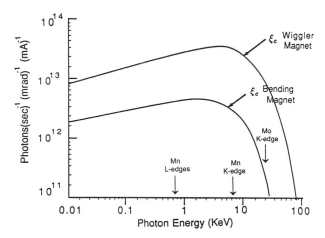

FIG. 10. Brilliance of the X-rays as a function of photon energy from a bending magnet and from a Wiggler beam line at the Stanford Synchrotron Radiation Laboratory. Notice the difference in the flux between the bending and the Wiggle magnets. [Adapted from A. Bienenstock, *Nucl. Instrum. Methods* **172**, 13 (1980).]

Monochromator

The most common form of collecting X-ray spectra consists of scanning through a range of wavelengths and measuring the flux of the incident and the flux of transmitted or fluorescent X-rays. Tunable monochromatic X-rays are usually obtained by the use of Bragg diffraction of X-rays by crystals. The wavelength of the X-ray, λ, may be selected from the broadband synchrotron radiation by the selection of θ, the angle of the incident X-ray beam relative to the crystal plane:

$$n\lambda = 2d \sin \theta \tag{14}$$

where d is the distance between the atomic planes in the crystal and n is the order of the reflection. The most commonly used monochromator consists of a matched pair of flat crystals which can be rotated, thereby selecting for different energies of the X-rays (see Fig. 9, inset). In one arrangement the two crystals are at a fixed distance from one another, and the incident beam is diffracted by them in succession. A consequence of such a method is that the vertical position of the beam changes during a wavelength scan, and the experimental apparatus needs to be raised or lowered to be aligned with the X-ray beam. Another system consists of two flat crystals which can be rotated and where one crystal can also be translated with respect to the other, thereby keeping the exit beam at the same position.

Many different types of crystals are available, and it is important to make a careful choice of crystal type. The relevant variables are the spectral resolution, the harmonic content of the transmitted beam, and whether there are any sharp drops in intensity (crystal glitches) in the transmitted X-rays.

For a flat crystal monochromator, the energy resolution is given by

$$\delta E = E \cot \theta \delta\theta \tag{15}$$

where δE is the energy resolution, θ is the Bragg angle at energy E which is determined by the choice of the crystal, and $\delta\theta$ is the divergence in θ for X-rays incident on the monochromator and is dependent on the width of the collimation slit of the monochromator and the distance from the source. For a Si$\langle 220 \rangle$ crystal, at the Fe K-edge of 7120 eV, with a 1 mm slit width located 20 m from the source, the energy resolution is about 1.5 eV.

The harmonics of the fundamental of interest (for the Mn K-edge, this is at 6.54 keV) may also satisfy the Bragg relation. An examination of Fig. 10 shows that for a Wiggler magnet the content of the first harmonic (e.g., 13.08 keV for Mn absorption) can be nearly equal to the flux of the fundamental frequency in the X-ray region of interest. The harmonic content is undesirable, for it has been found that it can significantly distort X-ray spectra. The easiest, but not always possible, approach is to choose a crystal for which the harmonic reflection is forbidden. A second method is to use grazing angle incidence mirrors before the monochromator, and by suitable choice of material of the mirror and the angle of incidence one can discriminate between the fundamental and the harmonic. Usually focusing mirrors are used which have a wider horizontal acceptance at the source and which focus the beam at the sample. In such a case the flux is increased and the harmonics are discriminated against. One drawback of this method is the loss of resolution. A third method utilizes the different rocking curves of the crystals for the fundamental and the harmonics. By slightly misaligning one crystal with respect to the other (usually done dynamically by the use of a piezoelectric material) the ratio of the fundamental to harmonic increases dramatically.

Finally, crystal glitches are observed at specific orientations of the crystal during normal operation in which additional Bragg planes come into alignment coincidentally and diffract away a considerable portion of the beam. These are manifest as sharp decreases in the intensity of the beam as the energy is scanned. Because no set of detectors is perfectly linear, these changes in intensity give rise to discontinuities in the data (thus the term crystal glitches). One needs to be aware of the transmission profile of the crystal for the energy range of the experiment under consideration.

X-Ray Detectors

For dilute samples, fluorescence detection has an enormous advantage over absorption measurements and is the method of choice for the study of biological systems.[30] In biological systems one is usually studying a dilute metal embedded in a matrix of protein, lipids, and/or nucleic acids, which all scatter X-rays. Ideally one needs to discriminate the fluoresence photons (mostly, $2p-1s$ for K-edge measurements) arising from the element of interest from the large scattering observed from the matrix. An ideal fluorescence detector would have high count rate capabilities, high energy resolution, and a large acceptance area. At present the detectors best suited for biological spectroscopy are the Si or Ge solid-state detectors. The energy resolution obtainable (about 150 eV), the relatively large surface area from multielement detectors, and a linear range of the associated electronics up to about 40,000 counts/sec (cps) has made solid state detectors the most commonly used at present. Ion chambers are usually used for measuring the flux of incident and transmitted photons. The intensity of the incident photons has also been measured by monitoring the scatter from a Mylar or Kapton window placed directly in the path of the beam with a scintillator–photomultiplier tube assembly. The latter method seems to be better at ratioing out the glitches present in the incident beam.

Data Analysis

A brief outline of data analysis procedures is presented below. EXAFS data analysis has been reviewed in great detail elsewhere.[10,32]

Initial Data Processing

Each stored XAS data point consists of the relative angle of the crystal monochromator, incident I_0, transmitted I_1 and I_2 (to monitor the spectrum of a energy reference) photon flux measurements, and the fluorescent counts, F, of the element of interest in the biological sample obtained from the channels of the fluorescence X-ray detector. With the use of the new solid state fluorescence detectors which require dead-time corrections, it is also necessary to store the total incoming count rate (ICR) seen by each detector element channel.

Dead-time corrections have become necessary with the introduction of solid state detectors which suffer from count rate overload associated with the electronics.[33] Dead-time losses can result in serious nonlinear distortions in the XAS data. Typically, one measures the number of photons before (ICR $= n$) and after (m) processing by the associated electronics, and the saturation curve is plotted. The curve is fit to an equation for dead-time derived from either the paralizable $[m = \beta n(1 - n\tau)]$ or nonparalizable model $[m = \beta n \exp(-n\tau)]$, where τ is the dead-time and βn gives the true count rate. The dead-time τ and the constant β are determined for each channel, and the correction is then applied to each data point to obtain the true count rate.

Each scan is processed to obtain an absorption scan $\ln(I_0/I_1)$, $\ln(I_1/I_2)$, and F/I_0 for the fluorescence channels. In biological applications, analyzing the fluorescence excitation spectrum F/I_0 is the method of choice, and the methods of data analysis pertaining to such a method of data collection

[32] D. E. Sayers and B. A. Bunker, in "X-Ray Absorption: Principles, Applications, Techniques of EXAFS, SEXAFS and XANES" (D. C. Koningsberger and R. Prins, eds.), p. 211. Wiley, New York, 1988.

[33] S. P. Cramer, O. Tench, M. Yocum, and G. N. George, Nucl. Instrum. Methods Phys. Res., Sect. A A266, 586 (1988).

are emphasized. Each scan is examined for experimental aberrations. The most common are the localized crystal glitches, described above. The glitches are relatively small and are reproducible from scan to scan, so removal by point deletion is deferred until the final added spectrum has been created. Occasionally, individual scans will have an aberration similar to a crystal glitch but which does not appear in other scans. To add the scans it is necessary to remove these glitches, which is often done by replacing the defective points with a straight line or by interpolating a polynomial that has been fit to either side of the glitch. For either of these methods to work effectively, the glitch has to be narrow relative to the features of the XAS spectrum.

Energy referencing of each scan is an important part of the experiment. The best method is the simultaneous recording of the spectrum of a reference compound placed between I_1 and I_2. This approach provides a reliable method of comparing data collected at different times or on different beam lines with different crystal monochromators. It also provides a method for checking the energy calibration of each scan before they are added. The shifts in energy between scans occur when there are small changes in the X-ray source. These are caused when the orbit of the electrons in the storage ring changes, which is sometimes necessary to increase the lifetime of the electrons in the ring, or when the ring is refilled with electrons after the X-ray flux has decreased to very low levels. Energy calibration is done by using a sharp feature in the reference spectrum and shifting the spectrum of the sample with respect to that feature in the reference spectrum. Normally, the position of the sharp feature is obtained by the numerical differentiation of the $\ln(I_1/I_2)$ data.

The data from scans which have been through the above scrutiny are then added, each channel separately. However, each channel of the detector will have a different signal-to-noise ratio (S/N), depending on the position of the detector and the geometry of the experimental setup, and a weighting scheme is employed based on signal-to-noise ratio and not solely on total counts, which could be high owing to scatter. The properly weighted fluorescence F_t is given by

$$F_t = (\Sigma_i \, c_i F_i)/(\Sigma_i \, c_i) \tag{16}$$

where F_i is the fluorescence from each channel. The weighting c_i is given by $(a_i - b_i)/a_i$ where a_i is the value of F_i/I_0 at the top of the edge and b_i is the value before the edge jump. Effectively, the weighting is proportional to the relative edge jump in each channel. The spectrum at this stage of the analysis resembles that shown in Fig. 11A.

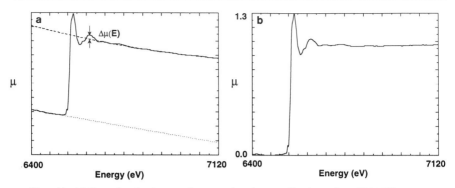

FIG. 11. (a) Pre-edge background removal and normalization of an EXAFS spectrum. The pre-edge background (low energy region) is removed by fitting the pre-edge region with a polynomial (dotted line) which is then subtracted from the total absorption, shown as a solid line. The dashed line is a polynomial which is fit to the postedge region (high energy) and extrapolated through the edge region. The ordinate value of this smoothed curve at E' is used to normalize the absorption per absorbing atom. Usually the energy position E' is chosen so that it is consistent when comparing different data sets; for convenience it is chosen as the value at the top of the edge. The smoothed curve (dashed line) approximates the background contribution and is subtracted from the solid curve to obtain $\Delta\mu(E)$, which is then divided by μ_0 to give $\chi(E)$, as explained in the text. The spectrum is of Mn in an oriented photosystem II preparation. (b) Normalized background-removed E-space X-ray absorption spectrum.

Pre-edge Background Removal

In a fluorescence excitation spectrum the pre-edge background is primarily due to scattering processes. Generally, the pre-edge data are fit by a linear or quadratic polynomial which is then subtracted from the data (Fig. 11A). The nature of the background curve is not important, except that the chosen background should not introduce a curvature to the rest of the XAS spectrum, which can cause problems at later stages of analysis.

Normalization

The EXAFS modulations in E space, $\chi(E)$, where E is the absolute energy of the X-rays, are defend as $[\mu(E) - \mu_0(E)]/\mu_0(E)$ or $\Delta\mu(E)/\mu_0(E)$, where $\mu(E)$ is the normalized absorption cross section of the element in the compound of interest and $\mu_0(E)$ is the absorption coefficient of an isolated atom. The X-ray absorption spectrum collected using the normal detection methods employed (described above) does not measure the absolute cross section, but only the relative cross section. Therefore, it is necessary to normalize all EXAFS spectra to unit edge height at a particular energy E' (position of the edge maximum is a common choice), which must be chosen in a consistent manner when comparing different data sets. This is equivalent to determining the absorption cross section, $\mu(E)$, per absorbing atom. The usual method is by fitting a polynomial to the postedge region or by using a cubic least-squares spline fit and extrapolating it through the edge region (see Fig. 11a). The normalized spectrum is created by dividing the ordinate values by the value of the extrapolated smoothed fit value at E' to obtain the measured $\mu(E)$, which is different from $\mu(E)$ because it includes a background contribution in addition to $\mu_0(E)$. $\Delta\mu(E)$ is generally obtained by subtracting the smooth part of the measured $\mu(E)$, which is obtained by fitting a polynomial to the postedge spectrum and extrapolating it through the edge (see Fig. 11A). The smoothed fit is the sum of $\mu_0(E)$ and the background. The last step in the normalization consists of dividing $\Delta\mu(E)$ by $\mu_0(E)$, which is obtained either from the calculated Victoreen formula[34] from those tabulated by McMaster.[35] Some authors have used the smoothly varying part of the measured $\mu(E)$ as an approximation for $\mu_0(E)$. This choice is reasonable only when the experimental background contribution is small. The spectrum at this stage of the analysis is shown in Fig. 11B.

Conversion of k Space and k Weighting

The data from E space is then transformed into k space, or the energy of the photoelectron. To perform this transformation it is necessary to know the value of E_0, which is the $1s$ electron binding energy for the K-edge. E_0 cannot be easily determined experimentally, so an approximate value is chosen, which is usually about 10 eV or so above the main inflection point. The exact choice of E_0 is not important because it is used as a parameter in the later data analysis, but it is important to be consistent in the choice of E_0 when comparing different sets of data from a sample. The maximum of the K-edge inflection is usually a good position to choose for E_0. Once E_0 is chosen then $\chi(E)$ is converted to $\chi(k)$ using the relation defined in Eq. (4). The k-space spectrum is shown in Fig. 12.

The EXAFS function $\chi(k)$ is attenuated at higher k values owing to the $1/k$ dependence in the EXAFS equation, and also because of the approximately $1/k^2$ dependence of the backscattering function $f(k,\pi)$ at higher k values. The function $\chi(k)$ is multiplied by a weighting factor, k^n (where n is 1, 2, or 3), to compensate for this attenuation. The background that was removed in E space is not sufficient when the data are weighted by k^n, because any remnant curvature is magnified and results in artifactual

[34] J. A. Victoreen, *J. Appl. Phys.* **19**, 855 (1948).

[35] W. H. McMaster, N. Kerr Del Grande, J. H. Mallet, and J. H. Hubell, "Compilation of X-Ray Cross Sections." Lawrence Radiation Laboratory, Livermore, California, 1969.

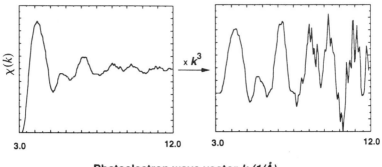

Photoelectron wave vector _k (1/Å)_

FIG. 12. The k-space spectrum of the E-space data shown in Fig. 11. The unweighted spectrum (left-hand side) shows that EXAFS oscillations are damped at high k values. On weighting by k^3 the features at high k values are enhanced, but so is the high-frequency noise.

peaks at low R in the Fourier transform performed in the next step of the analysis. So generally another background is removed at this stage of the analysis. The preferred technique is the removal of a cubic spline fit to the data. The k-weighting scheme prevents the larger amplitude oscillations from backscattering atoms at smaller distances from dominating the oscillations from other atoms at larger distances (Fig. 12). Furthermore, an examination of the backscattering amplitude function (Fig. 13) shows that, as the atomic number increases, (1) the maximum point of the backscattering cross section moves to higher k values, (2) the total backscattering increases, and (3) the heavier elements scatter more effectively at high k values. These characteristics imply that, if two different atoms are at the same distance, the heavier one will backscatter more effectively; if $\chi(k)$ is multiplied by k^3, which is preferred weighting factor, then the Fourier transform peak of the heavier atom will preferentially be enhanced. This behavior sometimes permits the transform peaks of heavier elements to be identified.

Fourier Transform Analysis

The theoretical form of the EXAFS as described by Eq. (11) is a sum of damped sinusoidal functions, with frequencies related to the distance of the absorber atom from the backscattering atoms, and an amplitude function which contains information about the number of backscatterers at that distance. This structural information can be best extracted by the Fourier transform technique, which converts data from k or momentum space into R or distance space. The following Fourier transformation of

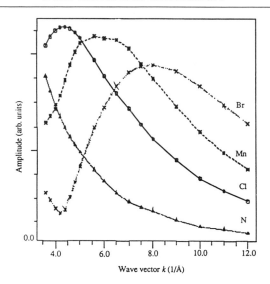

FIG. 13. Back-scattering amplitudes of Br, Mn, Cl, and N plotted as a function of k. Notice the distinct differences in the functions between the low-Z (N) and the higher Z elements (Cl, Mn, and Br). The high-Z elements have maxima at higher k values. [From V. J. DeRose, Ph.D. Dissertation, University of California, Berkeley, Lawrence Berkeley Laboratory Report, LBL 30077 (1990).]

the $k^n \chi(k)$ obtained as described above is computed, where the data extend from k_1 to k_2, and n is 0, 1, 2, or 3:

$$\phi(R) = \int_{k_1}^{k_2} k^n \chi(k) \, e^{i2kR} \, dk \qquad (17)$$

The Fourier transform amplitude peaks at the characteristic distances $R + \alpha'$, where α' is directly related to the phase shift $\alpha_{aj}(k)$ in Eq. (11), and it is a powerful visual tool in providing a simple physical picture of the local structure of the metal site. The Fourier transform provides a spectrum which is similar to a radial distribution function with peaks at R', which are shifted from R by α', but which correspond to the shells of backscatterers around the central absorbing atom. The Fourier transform of the k-space data is shown in Fig. 14. Three distinct Fourier peaks are resolved above the noise level in Fig. 14 at R' less than 4 Å.

Quantitative information is obtained from curve fitting to $\chi(k)$; however, fits are rarely performed on $\chi(k)$ obtained from raw data. It is desirable to separate the EXAFS contributions into their component shells and to filter out the high-frequency noise. The technique employed to achieve this is called Fourier filtering and consists of Fourier transforming

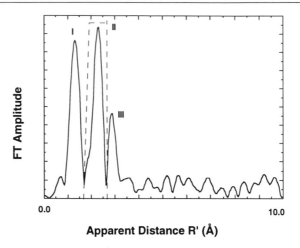

Apparent Distance R' (Å)

Fig. 14. Fourier transform of the $k^3\chi(k)$ spectrum shown in Fig. 12. The Fourier transform is resolved into three peaks. The dashed lines indicate the window used to generate the Fourier-filtered second peak, whose inverse Fourier transform is shown in Fig. 15. The apparent distance is shorter than the actual distance to a given neighboring atom because of the contribution of the phase shift to the frequency of the EXAFS wave, as explained in the text.

the $k^n\chi(k)$ data into R space, next selecting the Fourier peak of interest by applying an appropriate window function, and then back-transforming into k space. Figure 14 shows an example of a window used to select for EXAFS contributions from the second Fourier peak. Thus one can isolate the $\chi(k)$ contribution from one shell of backscatterers (Fig. 15), which can

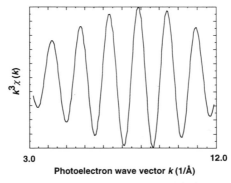

Photoelectron wave vector k (1/Å)

Fig. 15. Fourier-filtered data from the second peak of the Fourier transform shown in Fig. 14 generated by inverse Fourier transformation from R space to k space. The amplitude profile (the maxima at higher k value) of the data indicates contribution from a high-Z atom.

then be fit to the EXAFS equation to obtain the structural parameters for just that shell. This isolation technique works well if the backscattering atoms are well separated in radial distance from the absorbing atom (~ 1 Å). Fourier filtering also removes the high-frequency noise components associated with the raw data as shown in Fig. 15. However, one should exercise caution in the choice and use of the window functions selected for back transformation and keep in mind that, despite the noise-free appearance of the back-transformed data, some components of the noise spectrum are still present in the window. Other problems associated with the Fourier transform technique are so-called side lobes, which are a consequence of the finite size of the data set, which results in not just one peak for a single distance but the main peak and side lobes. One can usually identify these side lobes by changing the range of data used (values of k_{min} and k_{max}) for the Fourier transformation.

EXAFS Simulation and Curve Fitting in k Space

The final step in EXAFS analysis is to perform curve fits to obtain accurate radial distances and estimates of the number and type of backscattering atoms and of relative structural disorder. In its most general form curve fitting consists of (1) proposing a chemically feasible model based on information from other studies and by inspection of the properties of the Fourier transform, and then simulating the $\chi(k)$ for such a model using the EXAFS equation; (2) comparing the calculated EXAFS with that from experiment; and (3) varying the parameters R_{aj}, N, E_0, and the Debye–Waller parameter, σ, to minimize the difference between the caluated and observed spectrum using a nonlinear least-squares minimization procedure. This protocol is followed for each of the Fourier-filtered shells individually, hence simplifying the fitting procedure. However, that does not mean that each Fourier peak corresponds to only one kind of atom at one distance. In fact such a case would be an exception in most biological systems.

In most cases, curve fitting is a trial and error enterprise consisting of several iterations in the choice of the backscattering atoms and distances and in the number of subshells that constitute each Fourier peak. Physical reasonableness and chemical feasibility of the fit parameters always help direct the search for the right fit. There are four variables for each shell of atoms, R_{aj}, N, σ, and E_0; hence, including another shell doubles the number of variables and most often leads to a better fit. However, it is important to consider the statistical criteria for including another shell, by the degree of improvement in the goodness-of-fit parameter. The statistical rule of thumb for the number of independently variable parameters is given by $2\Delta r\Delta k/\pi$, where Δr is the width of the window used for Fourier filtering and Δk is the length of the data set, $k_2 - k_1$.[10] For a data set that extends from 3.5 to 12 Å^{-1} and a Fourier window width of 1.5 Å, the maximum number of parameters that can be varied in a multishell fit is 8. Data analysis is usually performed from about 3.5 Å^{-1} because of problems associated with multiple scattering effects below that energy.

To calculate an EXAFS spectrum using the EXAFS equation (11), one also needs to know the backscattering amplitude and the phase shift functions for the absorbing and scattering atoms. There are two general methods by which these functions are obtained. One method is to calculate these functions. The most commonly used functions are calculated using *ab initio* curved-wave single or multiple scattering theory.[23,36,37] The functions derived from single scattering theory are tabulated and are used by the curve fitting programs. The second method consists of using empirically derived amplitude and phase functions from suitable model compounds. Both of these methods have been successfully used for deriving structural details.[10]

Advantages and Limitations of X-Ray Absorption Spectroscopy

Advantages

X-Ray absorption spectroscopy is element specific, so one can focus on one element without interference from other elements present in the sample. In a protein which has more than one metal, like cytochrome oxidase or nitrogenase, it is possible to study selectively the structural environment of each metal atom. In the case of cytochrome oxidase both Fe and Cu XAS studies,[2,38,39] and in nitrogenase Mo (or V) and Fe XAS studies,[1] have been used in conjunction to investigate the structures of these multimetal clusters. In each of these metalloproteins the two different metals are present close to one another (<6 Å), so analyzing the backscattering from one element while at the K-edge of the other has been used to derive structures for the metal site clusters.

The specificity of the technique also makes it suitable for studying biological systems which contain several polypeptides, lipid molecules, pigment molecules, or other metals not necessarily near (>6 Å) the element of interest. The XAS study of the Mn complex in photosystem II is one which falls in this category. To maintain the oxygen evolution activity of

[36] J. J. Rehr, R. C. Albers, and S. I. Zabinsky, *Phys. Rev. Lett.* **69,** 3397 (1992).

[37] J. N. Deleon, J. J. Rehr, S. I. Zabinsky, and R. C. Albers, *Phys. Rev. B* **44,** 4146 (1991).

[38] L. Powers and B. Chance, *Inorg. Biochem.* **23,** 207 (1985).

[39] R. A. Scott, *Annu. Rev. Biophys. Biophys. Chem.* **18,** 137 (1989).

the Mn complex, it can be purified only to a unit which contains several polypeptides embedded in a lipid bilayer and which also contains a cytochrome *b*-559 and a Fe–quinone acceptor complex.

The specificity and the fact that it is always possible to obtain an X-ray spectrum of an element also mean that one "sees" all of the metal of interest that is present in the sample. This makes it imperative that one be sure of the biochemical homogeneity of the sample and, if there is more than one site for the same metal, resolve the structural parameters of the different sites.

Another important advantage of XAS is that the metal of interest is never "silent" with respect to X-ray absorption spectra. The system could be silent with respect to electron paramagnetic resonance (EPR), optical, or other spectroscopic methods, but one can always probe the metal site structure by XAS. There are several Zn metalloproteins, for example, which fall into this category; Zn is diamagnetic, and Zn complexes do not exhibit distinctive optical spectra.

Third, XAS is not limited by the physical *state* of the sample, because it is sensitive only to the local metal site structure. The sample can be prepared as a powder, as a solution, or, as is done most often for biological samples, as a frozen solution. The advantages of this are manyfold. It is not necessary to obtain single crystals of the material to examine the local structure of the metal. Obtaining single crystals can often be difficult, and some samples do not crystallize.

The more important aspect is that one can either trap intermediates in the enzymatic cycle or modify the site by the addition of inhibitors or substrate or generate other chemical modifications. Such samples can be made as frozen solutions, avoiding the problems of trying to obtain single crystals. The study by this technique of trapped intermediates and treated samples has yielded insights into the mechanism of the reaction involved in several biological systems. One illustrative example is the Mn complex of photosystem II, which has not yet been crystallized but which has been studied in its native state. The photoadvanced intermediate states have also been probed, which has provided important information about the structure and the mechanism of an important reaction in biology (described in more detail in this chapter).

Limitations

Damage to biological samples by X-rays is cause for serious concern for XAS experiments. However, with the right precautions one can successfully perform these experiments, leaving the materials largely intact. The most serious damage is produced by the reaction with free radicals and hydrated electrons that are produced in biological samples by X-rays. The diffusion of the free radicals and hydrated electrons can be minimized by the use of low temperatures. It is absolutely essential that the samples are checked for integrity before and after exposure to X-rays by an independent method, whenever possible. The use of a liquid He flow cryostat, where the samples are at atmospheric pressure in a He atmosphere, has greatly reduced the risk of sample damage by X-rays, but vigilance is warranted to ensure the integrity of the biochemical preparations.

It is also important to realize the intrinsic limitations of EXAFS, beyond those of a purely experimental nature. A frequent problem is the inability to distinguish between scattering atoms with little difference in atomic number (C, N, O or S, Cl, or Mn, Fe). Care must also be exercised in deciding between atoms that are apart in Z, because frequently it is possible to obtain equally good fits using backscattering atoms which are very different in Z (e.g., Mn or Cl) but which are at different distances from the absorbing atom. This is more acute when dealing with Fourier peaks at greater distances. In bridged multinuclear centers, it is not always possible to assign unequivocally the Fourier peaks at greater than 3 Å.[24] The peaks at greater than 3 Å could arise from second or third shell scattering from ligands like the imidazole ring of histidine or the pyrrole ring of hemes, or from backscattering from the bridged metal atom. This is an intrinsic limitation, and assignments which overstep this limitation are not reasonable. A detailed and very useful study of this problem has been presented by Scott and Eidsness.[24]

Distances are usually the most reliably determined structural parameters from EXAFS. However, the range of data that can be collected, often times because of practical reasons like the presence of the K-edge of another metal, limits the resolution of distance determinations to between 0.1 and 0.2 Å. Also, it is difficult to determine whether a Fourier peak should be fit to one distance with a relatively large disorder parameter or to two distances, each having a small disorder parameter. Careful statistical analysis, taking into consideration the degrees of freedom in the fits, should precede any such analysis. The resolution in the distance Δr can be estimated from the relation that $\Delta r \Delta k \approx 1$, so for a range of data from 3.5 to 12 Å^{-1} the resolution in distances is about 0.12 Å.

Determination of coordination numbers or number of backscatterers is fraught with difficulties. The Debye–Waller factor is strongly correlated with the coordination number, and one must have recourse to other information to narrow the range that is possible from curve fitting analysis alone. It is useful to compare the spectra from the unknown complex to some known model complexes (assuming that there is evidence that the structure resembles that of the model complex) and then use Debye–Waller parameters obtained from the model complexes in the fits. This method works reasonably well when the structure of the system being studied is well modeled by inorganic complexes.

The most important point in the analysis is to differentiate between fit parameters which are required and others which are merely consistent with the data. The EXAFS method is most useful when delineating all the structural alternatives based on required fit parameters, or when addressing the question of subtle structural changes in systems well characterized by other techniques like X-ray crystallography.

X-Ray Absorption Near-Edge Structure

This section discusses the application of X-ray absorption edge spectroscopy (known as XANES or XAES) to biological systems. The physics of the processes involved as it pertains to transition metals has not been satisfactorily explained, despite advances in XANES theory.[40,41] There are two main approaches to XANES analysis. The first method developed by Shulman et al.[42] in work with the highly ionic metal fluorides involved transitions to atomic states localized on the metal, and the spectra were rationalized on the basis of ligand field perturbations of the localized atomic orbitals. For complexes that contain significant metal–ligand interactions, such as is commonly found in biological materials, this simple approach is not valid; and quantitative treatment probably requires a knowledge of the low-lying unoccupied molecular orbitals of the complex. The effect of core holes on the outer lying states is also important in the consideration of the L-edges. The second method has used multiple scattering of the photoelectron from the neighboring atoms to explain the structure seen on the edges.[41,43,44]

The position of the absorption edge depends on the oxidation state of the metal. As the oxidation state of the metal increases or as the effective

[40] J. Stohr, "NEXAFS Spectroscopy." Springer-Verlag, Berlin, 1992.

[41] P. J. Durham, in "X-Ray Absorption: Principles, Applications, Techniques of EXAFS, SEXAFS and XANES" (D. C. Koningsberger and R. Prins, eds.), p. 53. Wiley, New York, 1988; A. Bianconi, in "X-Ray Absorption: Principles, Applications, Techniques of EXAFS, SEXAFS and XANES" (D. C. Koningsberger and R. Prins, eds.), p. 573. Wiley, New York, 1988.

[42] R. G. Shulman, Y. Yafet, P. Eisenberger, and W. E. Blumberg, Proc. Natl. Acad. Sci. U.S.A. 73, 1384 (1976).

[43] F. W. Kutzler, C. R. Natoli, D. K. Misemer, S. Doniach, and K. O. Hodgson, J. Chem. Phys. 73, 3274 (1980).

[44] J. B. Pendry, in "EXAFS and Near Edge Structure" (A. Bianconi, L. Incoccia, and S. Stipcich, eds.), p. 4. Springer-Verlag, Berlin, 1983.

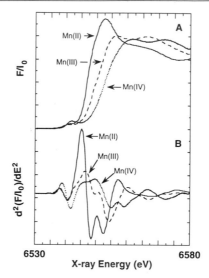

FIG. 16. (A) The K-edge spectra of Mn(II)(acac)$_2$(H$_2$O)$_2$, Mn(III)(acac)$_3$, and Mn(IV)(sal)$_2$.(bipy) complexes and (B) corresponding second derivatives (acac, acetyl acetonate; bipy, bipyridine; and sal, salicylate). There is a dramatic change in shape as well as in inflection point energy as the oxidation state increases. [Adapted from V. K. Yachandra, V. J. DeRose, M. J. Latimer, I. Mukerji, K. Sauer, and M. P. Klein, *Science* **260**, 675 (1993).]

positive charge increases, it becomes correspondingly more difficult to remove an electron from the metal, and the inflection point shifts to higher energy. It is not only the oxidation state of the metal which controls the effective positive charge of the metal, but also the nature of the ligands of the metal.[45] However, caution should be exercised in trying to derive information about oxidation states based solely on the position of the inflection points.[46]

Figure 16 shows the Mn K-edge spectrum from complexes in various oxidation states. The inflection point shifts to higher energy as the oxidation state increases. There is also a dramatic change in the general shape of the edge, as shown by the changes in the second derivatives. Figure 17 shows the Mn K-edge from preparations in the S_0, S_1, and S_2 states of the photosystem II complex, which are distinct states involved in the enzymatic cycle involved in the oxidation of water to oxygen (see above). The enzyme is thought to function as a charge storing device. Figure 17

[45] J. A. Kirby, D. B. Goodin, T. Wydrzynski, A. C. Robertson, and M. P. Klein, *J. Am. Chem. Soc.* **103**, 5537 (1981).
[46] J. E. Penner-Hahn, R. M. Fronko, V. L. Pecoraro, C. F. Yocum, S. D. Betts, and N. R. Bowlby, *J. Am. Chem. Soc.* **112**, 2549 (1990).

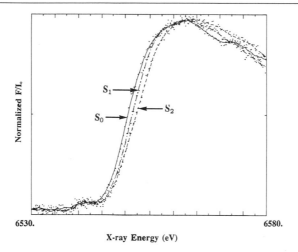

FIG. 17. The Mn K-edge from photosystem II samples in the S_0, S_1, and S_2 states. Notice the increase in K-edge inflection point energy as the system advances from the S_0 through the S_2 states. [Adapted from K. Sauer, R. D. Guiles, A. E. McDermott, J. L. Cole, V. K. Yachandra, J.-L. Zimmermann, M. P. Klein, S. L. Dexheimer, and R. D. Britt, *Chem. Scr.* **28A** 87 (1988).]

shows that the Mn K-edge shifts to higher energy on advancing from S_0 to S_1 and from S_1 to S_2, indicating oxidation of Mn. This has led to the proposal that the charges or oxidative equivalents are stored on Mn.[47] By comparing the position of the inflection point of model complexes with those from the enzyme, it is possible to derive information about the oxidation states of Mn in the biological system. The shape of the edges is also an important indicator of oxidation state, as seen in Fig. 16. A combination of the position and the shape was used to assign the oxidation states in the S_1 and S_2 states to (III$_2$, IV$_2$) and (III, IV$_3$), respectively.[29]

The transition to bound $3d$ states has been used to derive information about the symmetry of the metal complexes. The $1s$–$3d$ transition is electric-dipole forbidden, but it is often observed in the edge spectra and is commonly attributed to d–p mixing. In centrosymmetric complexes d–p mixing is symmetry unallowed and the $1s$–$3d$ transition is weak. In noncentrosymmetric complexes d–p mixing is allowed and the $1s$–$3d$ transition is more intense. Figure 18 shows the Fe K-edge spectrum of Fe–S centers on the acceptor side of the electron transport chain in the photosystem I complex. The centers are compared to a 4Fe–4S model compound and to a centrosymmetric Fe complex. There is a decrease in the intensity of

[47] M. P. Klein, K. Sauer, and V. K. Yachandra, *Photosynth. Res.* **38**, 265 (1993).

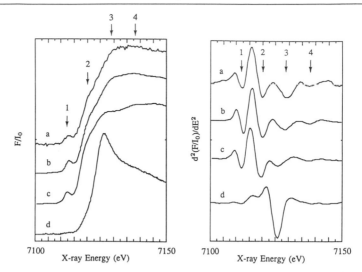

F/I₀

d²(F/I₀)/dE²

X-ray Energy (eV)   X-ray Energy (eV)

FIG. 18. Iron K-edge spectra and corresponding second derivatives of (a) photosystem I core preparation containing Fe–S acceptor cluster Fx, (b) $(Et_4N)_2Fe_4S_4(S\text{-benzyl})_4$, (c) photosystem I complex from *Synechococcus* containing Fe–S acceptor clusters Fa, Fb, and Fx, and (d) a heat-denatured sample of photosystem I. Note the presence of a strong $1s–3d$ transition in all but the heat-denatured sample. Also, there is distinct change in shape of the edge. [From A. E. McDermott, V. K. Yachandra, R. D. Guiles, K. Sauer, M. P. Klein, K. G. Parrett, and J. H. Golbeck, *Biochemistry* **28**, 8056 (1989).]

the $1s–3d$ transition, and there is a change in the shape of the spectrum beween tetrahedral Fe–S centers and a centrosymmetric hexacoordinate system. The changes in Fe K-edge shape have been successfully used by Conradson *et al.*[48] to examine the FeMo cofactor in nitrogenases. The intensity of the $1s–3d$ peak has also been used in structural characterizations of the Fe site in Fe-tyrosinate proteins[49] and in Cu proteins, where a weak $1s–3d$ transition is observed in $3d^9$ Cu(II) complexes but which is absent in the $3d^{10}$ Cu(I) complexes. Comparison of Cu K-edges of Cu(I) and Cu(II) complexes have been used to assign the oxidation state and geometry of the Cu sites in laccase.[50]

[48] S. D. Conradson, B. K. Burgess, W. E. Newton, K. O. Hodgson, J. W. McDonald, J. F. Rubinson, S. F. Gheller, L. E. Mortenson, M. W. W. Adams, P. K. Mascharak, W. A. Armstrong, and R. H. Holm, *J. Am. Chem. Soc.* **107**, 7935 (1985).

[49] A. L. Roe, D. J. Schneider, R. J. Mauer, J. W. Pyrz, J. Widom, and L. Que, *J. Am. Chem. Soc.* **106**, 1676 (1984).

[50] L.-S. Kau, D. J. Spira-Solomon, J. E. Penner-Hahn, K. O. Hodgson, and E. I. Solomon, *J. Am. Chem. Soc.* **109**, 6433 (1987).

Future Directions

Oriented Biological Samples

The radiation from synchrotron sources is typically polarized, and this property has been used to study oriented biological samples and single crystals. The intensity of the backscattering depends on the angle between the absorber–backscatterer vector and the e vector of the X-rays. Such studies provide geometric information and the orientation of the metal cluster in the crystal or the membrane. This method has found application in the study of some membrane-bound proteins, like cytochrome oxidase[51] and the Mn complex found in chloroplasts,[28,52] which can be oriented by layering the membrane onto a flat surface like a Mylar tape. Single crystals of plastocyanin[53] and nitrogenase[54] have also been examined by XAS, and there is potential for such studies with other systems. XANES studies of single crystals also show promise for understanding the nature of the transitions that make up the shape of the X-ray edge spectrum.

L-Edge Spectroscopy

L-edges are $2p-3d$ transitions (see Fig. 1). In contrast to K-edge transitions (see above), the $2p-3d$ transitions are electric dipole allowed and hence are intense. The natural line width of K- and L-edge transitions for Mn are 1.12 and 0.32 eV, respectively,[55] making L-edges considerably more sensitive to factors such as symmetry that influence the d orbital splittings and population. The splittings observed by L-edge spectroscopy can be similar to the $10Dq$ values obtainable from optical spectroscopy. Cramer *et al.*[55] have illustrated the sensitivity of Mn L-edges to the oxidation state, ligand field, and spin state of the metal complex. L-edge spectroscopy has been used to study the 1Fe site from the Fe–S protein rubredoxin.[56] The intensity of the $L_{II,III}$ edges from Cu have also been

[51] G. N. George, S. P. Cramer, T. G. Frey, and R. C. Prince, *Biochim. Biophys. Acta* **1142**, 240 (1993).

[52] G. N. George, R. C. Prince, and S. P. Cramer, *Science* **243**, 789 (1989).

[53] R. A. Scott, J. E. Hahn, S. Doniach, H. C. Freeman, and K. O. Hodgson, *J. Am. Chem. Soc.* **104**, 5364 (1982).

[54] A. M. Flank, M. Weininger, L. E. Mortenson, and S. P. Cramer, *J. Am. Chem. Soc.* **108**, 1049 (1986).

[55] S. P. Cramer, F. M. F. de Groot, Y. Ma, C. T. Chen, F. Sette, C. A. Kipke, D. M. Eichhorn, M. K. Chan, W. H. Armstrong, E. Libby, G. Christou, S. Brooker, V. McKee, O. C. Mullins, and J. C. Fuggle, *J. Am. Chem. Soc.* **113**, 7937 (1991).

[56] S. J. George, J. van Elp, J. Chen, Y. Ma, C. T. Chen, J.-B. Park, M. W. W. Adams, B. G. Searle, F. M. F. de Groot, J. C. Fuggle, and S. P. Cramer, *J. Am. Chem. Soc.* **114**, 4426 (1992).

used to estimate the degree of covalency of the Cu–S thiolate ligand in plastocyanin.[57] The above studies have demonstrated the feasibility and the potential of L-edge studies despite the experimental difficulties of the short penetration depth of X-rays at transition metal L-edge energies and the problems associated with working with biological samples in a vacuum.

X-Ray Magnetic Circular Dichroism

Cramer and co-workers[58] have shown differential absorption of left and right circularly polarized X-rays at the Fe–$L_{II,III}$ edges using rubredoxin placed in a magnetic field. X-Ray magnetic circular dichroism, with its selectivity for paramagnetic species and the relative magnetic orientation of different species in a multicenter system, promises to be a powerful tool for studying metal clusters in proteins.

High-Resolution Fluorescence Spectroscopy

X-ray absorption spectra are usually recorded as fluorescence excitation spectra. The limited resolution (150 eV) of the X-ray detectors has precluded obtaining useful information from the fluorescence spectra per se. However, the resolution of the fluorescence spectrum can be improved to about 1 eV by using energy resolving optics. The K β X-ray emission spectra ($3p$–$1s$) of Mn complexes have shown that the spectrum is sensitive to the oxidation state and spin state of the metal.[59] The potential for site- and spin-selective detection of X-ray absorption spectra in bioinorganic chemistry is considerable.

Acknowledgments

I am grateful to Prof. Kenneth Sauer and Dr. Melvin Klein for suggestions regarding the organization and contents of this chapter. I thank Matthew Latimer for a critical reading of the manusript and Dr. Annette Rompel for proofreading the manuscript. The work from our laboratory presented in this chapter was supported by the National Science Foundation (Grant DMB91-0414) and by the Director, Division of Energy Biosciences, Office of Basic Energy Sciences, U.S. Department of Energy (DOE), under Contract DE-AC03-76SF00098.

[57] S. J. George, M. D. Lowery, E. I. Solomon, and S. P. Cramer, *J. Am. Chem. Soc.* **115,** 2968 (1993).

[58] J. van Elp, S. J. George, J. Chen, G. Peng, C. T. Chen, L. H. Tjeng, G. Meigs, H.-J. Lin, Z. H. Zhou, M. W. W. Adams, B. G. Searle, and S. P. Cramer, *Proc. Natl. Acad. Sci. U.S.A.* **90,** 9664 (1993).

[59] G. Peng, F. M. F. de Groot, K. Hamalainen, J. A. Moore, X. Wang, M. M. Grush, J. B. Hastings, D. P. Siddons, W. H. Armstrong, O. C. Mullins, and S. P. Cramer, *J. Am. Chem. Soc.* **116,** 2914 (1994).

Synchrotron radiation facilities were provided by the Stanford Synchrotron Radiation Laboratory (SSRL) and the National Synchrotron Light Source (NSLS), both supported by the DOE. The Biotechnology Laboratory at SSRL and Beam Line X9-A at NSLS are supported by the National Center for Research Resources of the National Institutes of Health.

Section IV

Special Topics

[27] Component Resolution Using Multilinear Models

By ROBERT T. ROSS and SUE LEURGANS

I. Introduction

In many circumstances, the spectroscopic properties of a biological specimen are a composite of the properties of several different chromophores within the specimen, and the investigator would like to know the properties of the individual chromophores. Physical separation of the chromophores is often difficult, and it may alter important properties of the system being studied. Thus, one looks to mathematical methods for separation of the contributions of the different chromophores. This chapter focuses on one class of mathematical methods, the application of multilinear models. A major advantage of this class of methods is that component resolution may often be achieved with no prior information about the properties of the components.

Spectroscopic intensity can be measured as a function of several experimental variables. Usually, a larger number of experimental variables will permit a better resolution of the properties of individual components in the specimen. Data from measurements with k different experimental variables can be arranged in an array that extends in k directions and that is known as a k-way array. When $k = 2$, this array is a simple matrix.

When spectroscopic intensity is linear in functions of each of k independent variables, a multilinear model can be fit to the k-way array of data. For example, consider absorption measurements made on a specimen containing F components, with wavelength and some environmental variable, such as pH, being the two experimental variables. If the environmental variable alters the relative concentrations of the different components in the specimen without affecting their absorption spectra, then the expected absorbance is described by the bilinear model

$$\mu[i,j] = \sum_{f=1}^{F} \varepsilon_f[\lambda_i] c_f[\mathrm{pH}_j] \tag{1}$$

where measurements are made at wavelengths λ_i and at pH values pH_j. Note that the absorbance is not linear in the experimental variables, wavelength and pH; rather, it is linear in functions of these variables, extinction coefficient and concentration.

Bilinear models can be applied to many kinds of spectroscopy, and they have been widely used. The chemical applications of bilinear models

METHODS IN ENZYMOLOGY, VOL. 246

are reviewed in a book by Malinowski[1] and are surveyed in biannual reviews of the chemometric literature.[2] Their application in biochemical spectroscopy is the subject of a chapter elsewhere in this series by Henry and Hofrichter.[3]

The terms factor analysis, principal components analysis, and singular value decomposition (SVD) are used by spectroscopists to describe the fitting of a two-way array of data with a general bilinear model. We will use the term factor analysis in this sense, although this term has a somewhat different meaning in statistics. SVD is a specific algebraic procedure, discussed by Henry and Hofrichter and briefly later in this chapter, whose use alone is often not the best way to fit a general bilinear model.

In excited-state spectroscopies, including fluorescence spectroscopy, spectroscopic intensity is usually linear in functions of each of three or more independent variables, so that a three-way array of data can be fit with a trilinear model. The presence of three or more linear relationships makes algebraic methods for resolving the spectra and other properties of individual components substantially more powerful than in the case of two linear relationships. The use of a general trilinear model is sometimes known as three-way factor analysis, three-mode factor analysis, or three-mode principal component analysis. For a review of the mathematics and application to spectroscopy, see our survey article.[4]

In this chapter we focus on the application of trilinear models in fluorescence spectroscopy, because that is where we have most experience. However, these trilinear models are also applicable to other kinds of excited-state spectroscopy, such as transient absorption spectroscopy. Along the way, we also discuss bilinear and other models, including global analysis.[5]

Except where we specify otherwise [see, e.g., Eq. (15)], models are fit to data by weighted least-squares, minimizing the function

$$S(\theta) = \sum_{n=1}^{N} w_n [y_n - \mu_n(\theta)]^2 \qquad (2)$$

where the y_n are the data, the μ_n are the expected values of the observations when the model having parameters θ holds, and the w_n are the weights given to the data. Usually the w_n are chosen to be equal to $1/s_n^2$, where the s_n are the estimated standard deviations. When the y_n are sampled

[1] E. R. Malinowski, "Factor Analysis in Chemistry," 2nd Ed. Wiley, New York, 1991.
[2] S. D. Brown, R. S. Bear, Jr., and T. B. Blank, *Anal. Chem.* **64**, 22R (1992).
[3] E. R. Henry and J. Hofrichter, this series, Vol. 210, p. 129.
[4] S. Leurgans and R. T. Ross, *Stat. Sci.* **7**, 289 (1992).
[5] J. M. Beechem, this series, Vol. 210, p. 37.

independently from normal distributions with means $\mu_n(\theta)$ and variances $1/w_n$, $S(\theta)$ has a chi-square distribution. This sum is often called chi square (χ^2) regardless of whether these conditions are actually satisfied.

For further information about the use of least-squares to fit models to data, we suggest the following books, to be studied in the order listed: Devore,[6] Draper and Smith,[7] and Bates and Watts.[8] The book *Numerical Recipes*[9] has a convenient summary as a preface to its discussion of the computational methods used. Volume 210 of this series is devoted to numerical computer methods, most of which use least-squares.

II. General Multilinear Models

Bilinear Models

The simplest multilinear models are bilinear models, which can be written as

$$\mu[i,j] = \sum_{f=1}^{F} \alpha_f[i]\,\beta_f[j] \tag{3}$$

or in matrix form as

$$\mu = \mathbf{AB}^t \tag{4}$$

As noted in the introduction, a bilinear model can be applied to any kind of absorption spectroscopy, with $\alpha_f[i]$ representing the absorption cross section (extinction coefficient) of component f at wavelength i, $\beta_f[j]$ representing the concentration of component f in condition j, and $\mu[i,j]$ representing the absorbance expected at wavelength i in condition j.

A major difficulty with general bilinear models is that the data can be fit equally well with linear combinations of the columns of the matrices $\mathbf{A}$ and $\mathbf{B}$. To see that this is so, consider the insertion of an F by F matrix $\mathbf{T}$, followed by its inverse $\mathbf{T}^{-1}$, between $\mathbf{A}$ and $\mathbf{B}^t$ in Eq. (4). Because

[6] J. L. Devore, "Probability and Statistics for Engineering and the Sciences," 3rd Ed. Brooks/Cole, Pacific Grove, California, 1991.

[7] N. R. Draper and H. Smith, "Applied Regression Analysis," 2nd Ed. Wiley, New York, 1981.

[8] D. M. Bates and D. G. Watts, "Nonlinear Regression and Its Applications." Wiley, New York, 1988.

[9] W. H. Press, S. A. Teukolsky, W. T. Vetterling, and B. P. Flannery, "Numerical Recipes in C: The Art of Scientific Computing," 2nd Ed. Cambridge Univ. Press, Cambridge, 1992. Versions of this book and its software are also available in FORTRAN, BASIC, and Pascal.

$\mathbf{TT}^{-1}$ is the identity matrix, the overall matrix product is unchanged. However, we can write this modified Eq. (4) as

$$\mu = (\mathbf{AT})(\mathbf{T}^{-1}\mathbf{B}') \tag{5}$$

or

$$\mu = \overline{\mathbf{A}}\,\overline{\mathbf{B}}' \tag{6}$$

where

$$\overline{\mathbf{A}} = \mathbf{AT} \tag{7}$$

and

$$\overline{\mathbf{B}} = \mathbf{B}(\mathbf{T}^{-1})' \tag{8}$$

In the language of linear algebra, these operations rotate $\mathbf{A}$ and $\mathbf{B}$, with $\mathbf{T}$ referred to as a transformation matrix or a rotation matrix. Since any $\mathbf{T}$ will leave μ unaltered, each of the elements of $\mathbf{T}$ must be specified to define a unique solution. Normalization of each column determines F elements, leaving $F(F - 1)$ parameters undefined. Extra information is necessary to determine these parameters. This is known as the rotation problem. Methods for addressing the rotation problem are discussed in Section IV.

Trilinear Models

We present here three types of trilinear models. We begin with a trilinear model which was first introduced by Carroll and Chang,[10] who called it canonical decomposition (CANDECOMP), and by Harshman,[11] who called it parallel factors (PARAFAC). We refer to it as PARAFAC. It can be written as

$$\mu[i,j,k] = \sum_{f=1}^{F} \alpha_f[i]\,\beta_f[j]\,\gamma_f[k] \tag{9}$$

The great advantage of the PARAFAC model over the bilinear model is that under fairly mild conditions there is no rotation problem, meaning that there is a unique best fit. One combination of conditions which implies uniqueness is that the vectors α_f be linearly independent, the vectors β_f be linearly independent, and the γ_f satisfy the weaker requirement that

[10] J. D. Carroll and J. Chang, *Psychometrika* **35**, 283 (1970).
[11] R. A. Harshman, *UCLA Working Papers in Phonetics* **16**, 1 (1970).

no two vectors γ_f are linearly dependent.[12] A set of vectors is linearly dependent if any of the vectors is a linear combination of the other vectors. Two vectors are linearly dependent if one is a multiple of the other. The conditions on the α_f or the β_f vectors can be weakened if the conditions on the γ_f ones are strengthened.[13] We discuss in the next section how PARAFAC models are used to model fluorescence.

The second type of trilinear model, known as the Tucker2, or T2, model, named after Tucker,[14] can be written as

$$\mu[i,j,k] = \sum_{f_1=1}^{F_1} \sum_{f_2=1}^{F_2} \alpha_{f_1}[i]\, \beta_{f_2}[j]\, \gamma_{f_1 f_2}[k] \tag{10}$$

Equation (10) can be interpreted as allowing general interactions between the F_1 vectors associated with the first independent variable and the F_2 vectors associated with the second independent variable. We shall see in the next section that the T2 model represents fluorescence in the presence of excitation transfer. Unhappily, it does have the rotation problem for each of the **A** and **B** matrices, leaving $F_1(F_1 - 1) + F_2(F_2 - 1)$ parameters to be determined by side conditions.

The third type of trilinear model, known as the Tucker3, or T3, model can be written as

$$\mu[i,j,k] = \sum_{f_1=1}^{F_1} \sum_{f_2=1}^{F_2} \sum_{f_3=1}^{F_3} \alpha_{f_1}[i]\, \beta_{f_2}[j]\, \gamma_{f_3}[k] C_{f_1 f_2 f_3} \tag{11}$$

where C is referred to as the core array. The T3 model is popular in the social sciences,[15] and it has some applications in biochemistry.[16] We do not use this model to represent spectroscopic reality, but sometimes intermediate use of a T3 model is convenient during data reduction.

Each type of trilinear model can be described as a special case of one of the others. For example, a T2 model with F_1 factors for the first way and F_2 factors for the second way is a PARAFAC model with $F_1 F_2$ factors

[12] S. E. Leurgans, R. T. Ross, and R. B. Abel, *SIAM J. Matrix Anal. Appl.* **14,** 1064 (1993); preliminary version issued as Technical Report No. 448, Department of Statistics, Ohio State University, Columbus, 1990.

[13] J. B. Kruskal, *Linear Algebra Applications* **18,** 95 (1977).

[14] L. R. Tucker, *in* "Problems in Measuring Change" (C. W. Harris, ed.), p. 122. Univ. of Wisconsin Press, Madison, 1963.

[15] C. W. Snyder, Jr., H. G. Law, and J. A. Hattie, *in* "Research Methods for Multimode Data Analysis" (H. G. Law, C. W. Snyder, Jr., J. A. Hattie, and R. P. McDonald, eds.), p. 2. Praeger, New York, 1984.

[16] P. J. Gemperline, K. H. Miller, T. L. West, J. E. Weinstein, J. C. Hamilton, and J. T. Bray, *Anal. Chem.* **64,** 523A (1992).

whose parameter vectors for the first two ways have a repetitive pattern. The models for three-way arrays can be extended to N-way arrays.[10,17-19]

III. Multilinear Models in Fluorescence Spectroscopy

In steady-state fluorescence measurements on a specimen containing multiple types of fluorophores, the expected intensity is given by

$$\mu[i,j,k] = \sum_{f=1}^{F_1} \varepsilon_f[i]\,\pi_f[j]\,c_f[k] \tag{12}$$

where $\varepsilon_f[i]$ is the extinction coefficient of fluorophore f at excitation wavelength $\lambda^{ex}[i]$, $\pi_f[j]$ is the relative emission at detection wavelength $\lambda^{em}[j]$, and $c_f[k]$ is the concentration of fluorophore f in circumstance k. Comparing Eq. (12) with Eq. (9), we see that Eq. (12) is an example of a PARAFAC model. Simple application of Eq. (12) requires that specimen absorbance be small and that excitation not be transferred between chromophores; however, nonlinearity caused by absorbance can be corrected, and energy transfer can be treated with a different multilinear model [see Eq. (14) below]. The first application of a trilinear model to spectroscopy was by Appellof and Davidson,[20] who applied Eq. (12) using partial separation by a chromatography column to vary concentration.

One of the three variables in Eq. (12) may be ineffective. Usually this is because the relative concentrations of different fluorophores are fixed, perhaps because they are all part of the same macromolecule. In this case, it is important to add another independent variable. There are two ways to do this.

The first method of obtaining an additional independent variable is to treat the specimen in some manner that differentially affects the fluorescence quantum yield of different fluorophores. Most often, one adds a chemical quencher. As long as the mean lifetime of the excited state remains much longer than the thermal equilibration time, the amount of quenching will have no effect on the emission spectrum. We then have the quadrilinear model:

$$\mu[i,j,k,l] = \sum_{f=1}^{F} c_f[i]\,\varepsilon_f[j]\,\phi_f[k]\,\pi_f[l] \tag{13}$$

[17] J. L. Lastovicka, *Psychometrika* **46**, 47 (1981).

[18] A. Kapteyn, H. Neudecker, and T. Wansbeek, *Psychometrika* **51**, 269 (1986).

[19] S. R. Durell, C.-H. Lee, R. T. Ross, and E. L. Gross, *Arch. Biochem. Biophys.* **278**, 148 (1990).

[20] C. J. Appellof and E. R. Davidson, *Anal. Chem.* **53**, 2053 (1981).

where $\phi_f[k]$ is the fluorescence quantum yield of fluorophore f when subjected to treatment k, and where c, ε, and π are as defined previously. If either relative concentration or quantum yield is constant, or if both depend on a single experimental variable, Eq. (13) reduces to a trilinear equation, which we have used to resolve rhodamine dyes having overlapping spectra,[21] components within the pigments of the photosynthetic apparatus,[22] and tyrosine hydrogen-bonded to different ligands.[23]

Representative results from our work with tyrosine are shown in Fig. 1, and some results from the laboratory of Karukstis[24] (Harvey Mudd College, Claremont, CA) are shown in Fig. 2. For the data set analyzed to produce Fig. 1, steady-state fluorescence from a tyrosine derivative was measured using 19 excitation wavelengths and 20 emission wavelengths for 6 different concentrations of phosphate, which hydrogen bonds to the phenolic OH of tyrosine. For the data set analyzed to produce Fig. 2, steady-state fluorescence from the fluorescent probe 2-p-toluidinonaphthalene-6-sulfonate (TNS) in a suspension of chloroplast membranes was measured using 9 excitation wavelengths and 18 emission wavelengths for 5 different concentrations of Tb^{3+}; this ion alters the electrostatic charge of the membrane and thus alters the concentration and quantum yield of TNS in different environments. In Figs. 1 and 2, the estimated fluorescence intensity from a component at a specific excitation wavelength, emission wavelength, and chemical treatment is the product of the corresponding points on the three graphs. The excitation and emission spectra have been normalized to have a maximum of 1.0, so that the relative intensities of the different components are shown in the last graph. In each case, three components have been resolved without the use of any information about the components other than the appropriateness of a PARAFAC model (and, to a modest extent, that excitation and emission cannot be negative quantities). In Fig. 1, there is a clean resolution of three components even though the weakest component has only 1% the intensity of the strongest component and even though two components have very similar excitation spectra.

The second method of obtaining an additional variable is to make time-resolved measurements. Trilinear models were first applied to the kinetics

[21] C.-H. Lee, K. Kim, and R. T. Ross, *Korean J. Biochem.* **24,** 374 (1991).

[22] R. T. Ross, C.-H. Lee, C. M. Davis, B. M. Ezzeddine, E. A. Fayyad, and S. E. Leurgans, *Biochim. Biophys. Acta* **1056,** 317 (1991).

[23] J. K. Lee, R. T. Ross, S. Thampi, and S. Leurgans, *J. Phys. Chem.* **96,** 9158 (1992).

[24] K. K. Karukstis, D. A. Krekel, D. A. Weinberger, R. A. Bittker, N. R. Naito, and S. H. Bloch, in preparation.

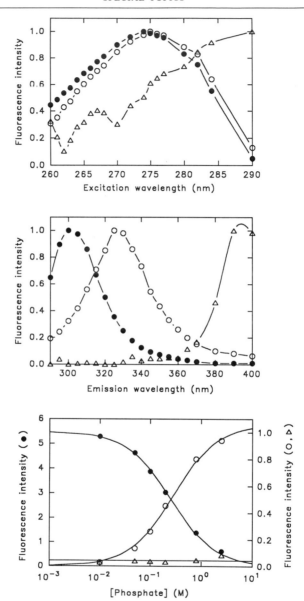

FIG. 1. Resolution of the fluorescence of *N*-acetyltyrosinamide in aqueous phosphate at pH 8 into three components, corresponding to the molecule solvated entirely by water (●), the molecule hydrogen-bonded to phosphate (○), and impurity (△).

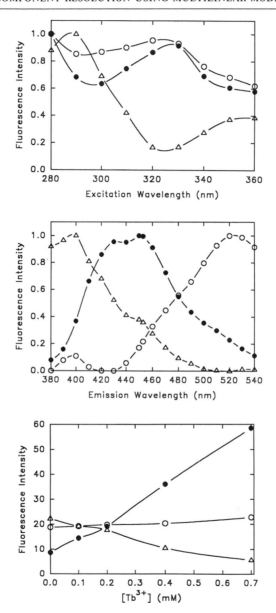

FIG. 2. Resolution of the fluorescence of 2-*p*-toluidinonaphthalene-6-sulfonate (TNS) in an aqueous suspension of spinach chloroplast membranes treated with Tb^{3+}. The component (●) whose contribution increases with $[Tb^{3+}]$ arises from a $\pi \rightarrow \pi^*$ transition and reflects membrane-bound TNS. The remaining two components arise from a charge-transfer excited state and reflect free TNS. One component (△) is attributed to TNS with the nitrogen atom either protonated or involved in hydrogen bonding to the solvent, and the other (○) is associated with nonprotonated TNS.

of emission decay following pulse excitation by Russell *et al.*,[25] and to phase-resolved fluorescence measurements by Burdick *et al.*[26]

Until now, we have not considered the possibility of excitation transfer. However, it is quite common for the following sequence of events to occur: (a) light absorption by chromophore f_1 proportional to $A_{f_1} = \varepsilon_{f_1} c_{f_1}$; (b) given light absorption by chromophore f_1, transfer of excitation to and emission from chromophore f_2 with probability $Q_{f_1 f_2}[k]$ in situation k; (c) given emission from f_2, emission distribution $\pi_{f_2}[j]$. Thus,

$$\mu[i,j,k] = \sum_{f_1=1}^{F_1} \sum_{f_2=1}^{F_2} A_{f_1}[i] \, Q_{f_1 f_2}[k] \, \pi_{f_2}[j] \tag{14}$$

Comparing Eq. (14) with Eq. (10), we see that Eq. (14) is an example of a T2 model. Recall from the discussion following Eq. (10) that T2 models are subject to the rotation problem. Strategies for solving this problem are discussed in the next section.

IV. Side Conditions and Submodels

In Section II we presented the standard general multilinear models, of which the bilinear and the PARAFAC and Tucker2 (T2) trilinear models are most important in spectroscopy. These models contain no information about the specimen except the linear dependence of spectral intensity on functions of each of the independent variables. However, some properties of the specimen are known, and a model that incorporates these known properties is preferred to one that does not. This is particularly true when the model is indeterminate without side conditions. In this section we discuss three settings for the application of knowledge about the specimen: identifiable bilinear and T2 submodels, penalized general multilinear models, and submodels in which the dependence of the expected intensity from some components for some ways has a specific mathematical form.

Side Conditions

The following are some examples of side conditions that might be applied to a model of spectral intensity: the dependence of spectral intensity on an environmental treatment variable (such as the concentration of a ligand or fluorescence quencher) must obey a specific equation (such as the Henderson–Hasselbalch equation for pH, the Hill equation for ligand concentration, or the Stern–Volmer equation for fluorescence

[25] M. D. Russell, M. Gouterman, and J. A. vanZee, *Spectrochim. Acta* **44A,** 873 (1988).
[26] D. S. Burdick, X. M. Tu, L. B. McGown, and D. W. Millican, *J. Chemom.* **4,** 15 (1990).

quenching); the dependence of spectral intensity on time is a sum of, say, two exponentials; all parameters representing inherently positive physical quantities (such as light absorption, light emission, or concentration) must be nonnegative; a spectrum may have at most, say, two maxima; a spectrum must match one of a family of curves defined by measurements on model compounds.

Side conditions differ in the degree to which they restrict the parameters, and they differ in the degree of our certainty that their use is scientifically valid. Often these two properties are in conflict. For example, we are highly certain that nonnegativity is a valid condition and not so certain that time decay is described by a sum of two exponentials; however, requiring that time decay be a sum of two exponentials is much more restrictive than requiring that all points in a time decay curve be nonnegative. Side conditions also differ in their ease of implementation; for example, limiting the number of maxima is not easy.

Bilinear Models

As noted in Section II, a major difficulty in using general bilinear models is that the data can be fit equally well with linear combinations of the vectors (e.g., spectra) yielded by the model. Extra information is needed to determine the correct linear combinations. For this reason, data analysis often proceeds in two stages. In the first stage, general models having 1, 2, . . . , components are fit to the data. Conventions such as vector orthogonality are typically used to choose specific basis vectors. If data are available for all combinations of the two independent variables, and if all data have the same expected error, then the fit for all numbers of components can be obtained simultaneously using a singular value decomposition (SVD). These conditions for using SVD alone to get the general models are sometimes met in absorption spectroscopy, but they are seldom met in emission spectroscopy. Methods to be used when SVD alone is inappropriate are discussed in Section VI.

Formal mathematical treatment of the data can stop at this point, with presentation of the basis vectors of the general model.[3] However, the analysis often proceeds to a second stage: the use of the vectors from the first stage with additional information to obtain a more specific and realistic model for the system being studied. In fortunate circumstances, sufficient side conditions will be available to define a unique rotation.[27,28]

[27] R. I. Shrager and R. W. Hendler, *Anal. Chem.* **54**, 1147 (1982).
[28] M. A. Marchiarullo and R. T. Ross, *Biochim. Biophys. Acta* **807**, 52 (1985).

Identifiable T2 Models

T2 models arise when there is energy transfer between chromophores [see Eq. (14)]. They also arise when two classes of chromophores are distinguishable in their dependence on two independent variables but indistinguishable in their dependence on the third independent variable.

Without additional information, a T2 model is not identifiable, meaning that there is a whole family of possible solutions to the least-squares optimization. This nonidentifiability problem is similar to the rotation problem for bilinear models, except that for T2 models there are two rotation matrices whose elements must be established.

There are two approaches to identifying the parameters of a T2 model: (1) rotating matrices **A** and **B**, using knowledge about the corresponding physical spectra, and (2) fixing elements of the core array [γ in Eq. (10)], using knowledge about the interaction between components f_1 and f_2. The first approach is analogous to that used for the rotation problem for a bilinear model. An example of the second approach is to arrange for some specimens having no energy transfer. Then the subarray of the total dataset comprising measurements on these specimens can be described by a PARAFAC model, guaranteeing that a T2 model with appropriate zeros in its core array will be an identifiable model for the whole data set.

Penalized General Multilinear Models

Side conditions can also be used for more than defining a unique rotation that leaves the array μ of fitted values unaltered. Instead, the function minimized [see Eq. (2)] is modified to become

$$Q(\theta) = \sum_{n=1}^{N} w_n [y_n - \mu_n(\theta)]^2 + a \sum_{m=1}^{M} [\theta_m - \bar{\theta}_m]^2 \tag{15}$$

where a is a penalty multiplier defining the importance attached to matching the side conditions, θ_m is an individual parameter, and $\bar{\theta}_m$ is the nearest value of θ_m satisfying the side conditions.[29,30] Differing a for individual

[29] G. Stone, S. Leurgans, and R. T. Ross, in preparation.
[30] R. A. Thisted, "Elements of Statistical Computing," p. 351. Chapman & Hall, New York, 1988.

parameters and side conditions have been omitted for simplicity.[31] In the limit of large penalty, a set of parameters may be forced to fit some predefined form exactly; this then becomes the situation described in the next subsection.

Parametric Submodels and Global Analysis

In the most general sense in which the term "global analysis" is used in biochemistry and biophysics, it means the fitting of a single comprehensive model to all data collected from measurements on a set of related specimens, rather than fitting models separately for each specimen.[5] Such data analysis has the advantage that shared parameters are estimated once, rather than separately for each specimen. Moreover, in many applications, the estimates obtained by pooling several specimens are easier to compute and have better statistical properties than combining the specimen-specific estimates. In the statistical literature, this concept is referred to as "pooling" or as "borrowing strength." With global analysis described this broadly, all multilinear modeling is a type of global analysis. In practice, however, the term global analysis is usually applied to the use of a more specific class of models, which we now describe.

In most situations, global analysis is applied using two independent variables. One independent variable is usually wavelength. The other independent variable is one for which there is a specific parametric model for each component; the two most common are (1) time, for instruments using pulse excitation to study fluorescence decay kinetics, and (2) modulation frequency, for instruments using rapidly modulated excitation intensity to study fluorescence decay kinetics. A typical experiment might measure decay kinetics at 10 different emission wavelengths. These data are then fit, for example, with a model which assumes that the decay is described by a sum of two exponentials, with all wavelengths having the same two decay rate constants but having different ampli-

---

[31] Choosing a is an art. For a very rough initial estimate, pick

$$a = S(\theta) \bigg/ \sum_{m=1}^{M} [\theta_m - \bar{\theta}_m]^2$$

where $S(\theta)$ is the weighted sum-of-squares calculated from Eq. (2) and the denominator is the sum of squared side condition violations in the absence of a penalty. The value of a will then need to be increased or decreased from this initial estimate, probably by several powers of 10, to balance the importance of fit to the data and satisfaction of the side conditions.

tudes.[5,32-34] Quencher concentration has also been used as the variable for which there is a parametric model.[35-37]

In their simplest forms, bilinear modeling and global analysis represent opposite approaches to modeling data obtained using two independent variables. They have in common the fitting of all of the available data with a single model. They differ in that the simplest form of bilinear model uses no information about the properties of the specimen. In contrast, global analysis, in its earliest and simplest form, uses a highly specific model for the properties of the specimen. By using a sufficiently specific model, global analysis avoids the linear-combination indeterminacy of a general bilinear model and obtains more precise results. The difficulty is that the models considered may be inappropriate. For articulate presentations of the relative advantages of the two, see Beechem[5] and Henry and Hofrichter.[3]

From the perspective of multilinear modeling, we note that one can use a model in which some components for some ways are described by a parametric model, while other components or other ways continue to be described by a general multilinear model. Such a parametric submodel has fewer parameters and may thus be more parsimonious and more accurate than a general multilinear model. When a parametric model is used for all components of at least one way, use of the resulting multilinear submodel is equivalent to global analysis.

V. Statistical Methods

Looking at Residuals

Whenever one is fitting a model to data, it is helpful to present information about the individual residuals, $r_n = y_n - \mu_n$, in a way that allows the user to get a visual sense of the quality of the fit.[7] When a single experimental variable is used, it is common to plot the individual residuals (divided by the estimated standard deviations if they are available) against the value of that variable. When there are two or more independent variables, as with multilinear models, a similar graph can be made for each

[32] J. R. Knutson, this series, Vol. 210, p. 357.

[33] T. A. Roelofs, C.-H. Lee, and A. R. Holzwarth, *Biophys. J.* **61,** 1147 (1992).

[34] J. M. Beechem, E. Gratton, M. Ameloot, J. R. Knutson, and L. Brand, *in* "Topics in Fluorescence Spectroscopy" (J. R. Lakowicz, ed.), Vol. 2, p. 241. Plenum, New York, 1991.

[35] W. Stryjewski and Z. Wasylewski, *Eur. J. Biochem.* **158,** 547 (1986).

[36] Z. Wasylewski, H. Koloczek, and A. Wasniowska, *Eur. J. Biochem.* **172,** 719 (1988).

[37] M. R. Eftink and Z. Wasylewski, *Biochemistry* **28,** 382 (1989).

variable, plotting either individual residuals or the mean squared residual of all data having a particular value of the independent variable being considered. We also use contour plots to display coded individual residuals as a function of two independent variables at a time.

Such procedures are important in spotting specimen or instrument problems which, especially with automated data collection, might go undetected. They are also useful in qualitative model selection and in the design of future experiments on related specimens. Regions of the data array that are not fit well by the model to be used (a) warn that possible specimen or instrument problems exist, (b) warn that the expected model may not be the right one, or (c), if the lack of fit is caused by a component of no interest, warn that the data should be reanalyzed deleting this region. One can also see which regions of the data array do not fit well in the model one degree simpler (e.g., with one less component) than the model to be used, and thus should be emphasized in later experiments.

Selection of Models

In our context, model selection is the selection of the number of components, the choice between a PARAFAC and a T2 model (e.g., is there energy transfer?), or the choice between parametric models for specific ways and components (e.g., is the time dependence a single exponential?). Most often, one will be interested in deciding on the number of components to use. There are a variety of statistical tests based on the decline of the sum of squared residuals with additional components. Other methods look at the relationship between residuals for evidence that they show no patterns and hence the model is adequate.[1,7] However, the simplest and perhaps the most effective approach is simply to plot the logarithm of the sum of squared residuals versus the number of components and use the number of components at the "elbow" where the curve flattens out.[38–41]

In our experience, any of the above approaches in the context of spectral component resolution will often favor a more complicated model than can be described with an acceptable amount of accuracy. Usually this is because the real spectral properties of the specimen are more complex than any practical model. Thus, the practical answer to the question of choice of model is often to use the most complete model which

---

[38] This method is adapted from the "scree test."[39–41]

[39] N. R. Draper and H. Smith, "Applied Regression Analysis," 2nd Ed., p. 299. Wiley, New York, 1981.

[40] J. E. Jackson, "A User's Guide to Principal Components," p. 45. Wiley, New York, 1991.

[41] E. R. Malinowski, "Factor Analysis in Chemistry," 2nd Ed., p. 112. Wiley, New York, 1991.

can be described with acceptable accuracy. The first indication of an overly complex model will be spectra and other output vectors that are nonsense physically.

Design of Data Collection

One difficulty with a method requiring three or more experimental variables is that the very large number of possible combinations of the values of these variables often makes it impractical to collect data at all of the combinations. Of the many possible combinations of wavelengths and other experimental variables, where should one actually make measurements? One logical choice is to collect data so as to minimize the product of the confidence intervals of all of the individual parameters, adjusting for correlation between parameters. Such designs are known as D-optimal.[8] Abel[42] examined D-optimal designs for PARAFAC models and has shown that often data collection can be omitted for many combinations of the independent variables with little increase in the product of the confidence intervals.

Results from Abel establish that a "cross-hatch" experimental design is often very effective. If we define a "strip design" as one in which all settings of one independent variable are used for fixed settings of the others, then a cross-hatch design is one in which strips are collected for two or more variables. For example, in fluorescence spectroscopy, one might measure all values of excitation wavelength at four emission wavelengths and all values of emission wavelength at four excitation wavelengths. If resolution of F components is attempted, then there must be at least F strips in each direction. Tentatively, we recommend $2F$ strips in each direction. At least F strips should be at levels (values of the experimental variable, such as wavelength) that one is pretty confident will distinguish between the components. The remaining strips can then be at levels to be checked for possible additional components. Abel (now at Parke-Davis, Ann Arbor, MI) is working with us to develop computer code that will suggest the combinations of wavelengths and other experimental variables at which measurements should be made.

VI. Computing

Multilinear models are nonlinear models. In our applications, a model with several hundred parameters is fit to a data set of several thousand observations. Fitting a nonlinear model having several hundred parame-

[42] R. B. Abel, Ph.D. Dissertation, Ohio State University, Columbus (1991).

ters is an inherently difficult task. There are three kinds of algorithms which might be used for fitting multilinear models: decomposition algorithms, iterative algorithms specialized for the multilinear problem, and general minimization algorithms. Well-designed software is likely to combine these kinds of algorithms.

Decompositions

A decomposition algorithm is one that is inherently noniterative and that will yield the parameters of the model if no noise is present in the data. The singular value decomposition (SVD) is a well-known decomposition for bilinear models.[3,9] There are decompositions for the PARAFAC,[12,26,43,44] Tucker2,[45] and Tucker3[46,47] trilinear models.

The disadvantages of decomposition algorithms are that they seldom give the least-squares fit in the presence of noise[48] and that they require a data array with no missing values. Any real data set includes noise. Missing values are common because sparse data arrays save time and reduce the amount of specimen needed, and many fluorescence data arrays have missing values because of light scattering when the excitation and emission wavelengths are nearly the same. Despite these disadvantages, decomposition algorithms are very important in providing good starting values for iterative algorithms. Missing values are supplied by the prediction of a simple model.

Alternating Least-Squares

A class of algorithms which is specialized for multilinear problems is known as alternating least-squares (ALS). Multilinear models are all conditionally linear in a function of each of the three or so independent variables; for example, spectral intensity is linear in concentration if the other variables are fixed. Each step of an ALS algorithm fixes the vectors for all but one independent variable, then applies linear regression to select the vectors for the one variable to minimize the error sum of squares. The algorithm cycles among the sets of parameters to be estimated, updating each in turn. Most applications of multilinear models use ALS code.[43,47]

[43] R. Sands and F. W. Young, *Psychometrika* **45**, 39 (1980).
[44] E. J. Sanchez and B. R. Kowalski, *J. Chemom.* **4**, 29 (1990).
[45] P. M. Kroonenberg, "Three-Mode Principal Component Analysis: Theory and Applications." DSWO Press, Leiden, 1983.
[46] L. R. Tucker, *Psychometrika* **31**, 279 (1966).
[47] P. M. Kroonenberg and J. deLeeuw, *Psychometrika* **45**, 69 (1980).
[48] The SVD method is unusual in giving the least-squares fit, provided that all data have the same expected error.

The rate of convergence of unconstrained bilinear and T2 and T3 trilinear models is almost always quite fast. Convergence for PARAFAC and constrained models is much slower, even when we use a projection technique to accelerate convergence.

The attraction of the ALS algorithm for general multilinear models is its use of linear least-squares steps. However, these steps become nonlinear regressions for any way containing a nonlinear parametric model, and most parametric models in spectroscopy will be nonlinear. Thus, the ALS approach is unattractive for most situations in which the dependence of the spectral intensity of any component on any experimental variable is described by a specific mathematical function.

General Minimization Algorithms

Most general function minimization algorithms can be divided into the following classes: (1) those that calculate both the first and second derivatives of the function at each point (Newton algorithms), (2) those that calculate the first derivatives of the function at each point and accumulate information about approximate second derivatives while moving from point to point (quasi-Newton algorithms), (3) those that calculate the first derivatives at each point and do not accumulate second derivative information (e.g., the conjugate gradient algorithm), and (4) those that do not calculate any derivatives (e.g., the 'simplex' algorithm). As one moves from class (1) to class (4), there is less computational work for each new point, but calculations must be made at more points before the minimum is found. The Levenberg–Marquardt algorithm for nonlinear least-squares is a variant of a Newton algorithm. For an overview, see *Numerical Recipes*.[9]

Algorithms, such as the Marquardt algorithm, that evaluate second derivatives must at each step invert a matrix whose dimension is the number of parameters. The cost of this inversion goes up as the cube of the number of parameters. On the other hand, a quasi-Newton algorithm accumulates an approximation to the inverse of the matrix of second derivatives, so that no inversion is required. Thus, whereas Newton-type algorithms are usually the most efficient for problems with a small number of parameters, quasi-Newton algorithms are usually more efficient for a larger number of parameters. The relative speed of different minimization algorithms is notoriously dependent on the specifics of the individual problem, but we have the sense that a quasi-Newton algorithm will almost always be faster than a Newton-type algorithm in minimizations with over 100 parameters. Thus, the Marquardt algorithm is most appropriate for fitting models having 10 or 20 parameters, as in most applications of global analysis, and a quasi-Newton algorithm is more appropriate for fitting models having over 100 parameters, as in most applications of multilinear models. If the number of parameters becomes so large that the matrix of second derivatives will not fit into computer memory, the conjugate gradient method is preferable.

Computer Programs

We have developed FORTRAN code using the ALS algorithm for general bilinear models, and using a variant of the ALS algorithm for the three standard general models for three-way data (PARAFAC, Tucker2, and Tucker3) and for the four-way equivalent of PARAFAC. Except for the bilinear case, a decomposition provides an accurate starting point for the ALS calculation. Except for the PARAFAC cases, a decomposition is used at the end of the calculation to put the final vectors into a standard orthogonal form. There is provision for weighting and for some constraints. Nonnegativity constraints are implemented using bounded least-squares steps within ALS. A projection step is used to overcome the hemstitching behavior typical of steepest descent methods.

We are currently working with Glenn Stone (CSIRO Division of Mathematics and Statistics, North Ryde, NSW, Australia) to explore the effectiveness of quasi-Newton algorithms in fitting different classes of multilinear models. Preliminary results suggest that, for PARAFAC models, the quasi-Newton and the accelerated ALS approaches require roughly equal computing time. Although there is probably the potential for increasing the speed of variants of ALS substantially, we expect that future code will emphasize general minimization algorithms, in order to facilitate our use of penalized and parametric models.

Subroutines for calculating the singular value decomposition may be found in all popular subroutine libraries (IMSL, NAGlib, *Numerical Recipes*). Our current code calls subroutines from the IMSL library; however, a modest amount of recoding would convert them for use with NAGlib or even with the subroutines available with *Numerical Recipes*.[9] We will provide copies of our code; requests should be sent via Internet to rtr + @osu.edu.

VII. Laboratory Methods

Experimental Variables and SLI Variables

When a multilinear model is fit to a multiway array, the ways of the array must correspond to independent variables having the property that the observed spectroscopic intensity is separately linear in a function of each of the independent variables. For example, the wavelength of light impinging on the specimen is usually an appropriate independent variable, because the spectroscopic intensity is linear in extinction coefficient (absorption cross section), which is a function of wavelength, and this linear function (the absorption spectrum) is unaffected (separate) by the value of any of the other independent variables used in the experiment. For convenience of reference, we call a collection of variables having this property "SLI variables."

Staying with this example, what if the absorption spectrum of a component varies with some other experimental variable, such as temperature? This is a situation in which the application of a multilinear model is problematic. Either of two adjustments in the situation might allow both wavelength and temperature to be used as independent variables in a multilinear model: (1) If the dependence of its absorption spectrum on temperature is because a "component" is shifting between discrete low-temperature and high-temperature forms, then each form can properly be regarded as a separate component whose absorption spectrum is independent of temperature. (2) Alternatively, it may be necessary to restrict the temperatures used to a sufficiently narrow range that the absorption spectrum is essentially constant.

The possible SLI variables for fluorescence spectroscopy are (1) excitation wavelength, (2) emission wavelength, (3) some environmental variable which alters relative concentration, and (4) one of the following: (a) some environmental variable (such as quencher concentration) which alters the fluorescence quantum yield, (b) time since pulse excitation, or (c) modulation frequency. Quencher concentration and time since excitation cannot be separate SLI variables because a change in fluorescence quantum yield affects decay kinetics. However, it may be profitable to make measurements of decay kinetics at several different quencher concentrations and then combine time and quencher concentration to form a single SLI variable.

A single environmental variable, such as pH, may affect both concentration and fluorescence quantum yield. If so, then the product of concentration and quantum yield becomes a single term in the multilinear model.

In excited-state spectroscopies, including fluorescence spectroscopy, wavelength is accurately an SLI variable only in the limit of low absorbance. With typical specimen geometry, an absorbance of 0.02 gives a nonlinearity of approximately 2%, so that either specimen absorbance should be less than 0.02 at all wavelengths used or else this "inner filter effect"[49] should be carefully corrected. In fluorescence spectroscopy, Raman scattering is a signal that does not obey a multilinear model;

[49] C. A. Parker, "Photoluminescence of Solutions." Elsevier, Amsterdam, 1968.

therefore, the contribution of Raman scattering must be removed by subtraction of the signal from an appropriate blank before a multilinear model is used.

Instrumentation and Measurement Error

With data collection involving three or more experimental variables, it is very helpful to have automated collection of at least one or two of these variables. For much of our work we use a conventional steady-state spectrofluorometer, interfaced to a microcomputer that controls all instrument variables and logs data, permitting automatic acquisition of fluorescence intensity at any desired combination of excitation and emission wavelengths. If an array detector is available, the whole spectrum can be collected simultaneously.

Careful mathematical modeling calls for extra attention to sources of error, because fitting a model to data will expose errors which might otherwise go unnoticed, and accurate characterization of error in the data is necessary for appropriate weighting in the least-squares fitting procedure.

In any light emission measurement, the variance σ_n^2 is usually a quadratic function of signal μ_n:

$$\sigma_n^2 = a + b\mu_n + c\mu_n^2 \tag{16}$$

The linear term $(b\mu_n)$ in Eq. (16) is due to the particulate nature of light and electrons and is usually regarded as the dominant source of noise. Often it is. However, even when care is taken to assure a stable instrument and specimen, c is often about 0.01, so that the quadratic term is the largest for many data points. Repeated measurements of a nominally constant signal, for several different signal strengths, can provide estimates of a, b, and c for use in weighting.

Just as mathematical modeling calls for extra attention to sources of error which might otherwise go unnoticed, it also calls for extra attention to impurities which might otherwise go unnoticed. In near-UV fluorescence measurements, plasticizers, such as alkyl phthalates, are major sources of error. These and other fluorescent impurities are found even in highly pure water and fresh reagent-grade inorganic salts. Figure 1 demonstrates how, under favorable circumstances, the signal from impurity can be resolved as an additional component; however, it is usually far better to remove the impurity chemically.

VIII. Conclusion

Mathematical modeling can be used to resolve the properties of several different chromophores within a biological specimen. Generally, the greater the number of experimental variables exploited, and the more reliable prior information about the specimen used in constructing the model, the better will be the results.

When the dependence of the spectroscopic intensity from every chromophore on at least one experimental variable can be described by a highly specific mathematical function, then the approach known as global analysis is preferred. When this condition is not known to be met, but spectroscopic intensity is separately linear in functions of two or more experimental variables, then the multilinear models described in this chapter are valuable.

Most of our attention in this chapter has been devoted to models using three independent variables, as opposed to the two variables used in more traditional factor analysis and in most global analysis. This has the disadvantages of a requirement to identify three or more appropriate independent variables and to perform a larger number of measurements. It has the advantage of providing a richer data set, the analysis of which can yield results that are more precise than those provided by two-variable factor analysis and that are more independent of specific physical models than global analysis. In those circumstances for which a PARAFAC model is appropriate, the components can be resolved with no other information about their properties.

The value of trilinear models is clearest in steady-state fluorescence measurements. They are also valuable in time-resolved fluorescence spectroscopy in situations where the appropriateness of a specific parametric equation for time decay, such as a sum of a few exponentials, is unclear. Although we are unaware of any work using trilinear models with other kinds of excited-state spectroscopy, trilinear models will be a valuable means of achieving component resolution whenever the absence of reliable parametric equations makes global analysis impossible.

The field of nonlinear regression, of which multilinear models are one part, is an area of statistics under rapid development, the results of which are likely to improve the utility of multilinear methods in the future.

Acknowledgments

R.T.R. began working on multilinear models while on leave in the laboratory of K. Sauer, editor of this volume, whom he thanks for generous hospitality. We thank G. Stone for advice on computational methods, D. S. Burdick, M. Debreczeny, K. Karukstis, J. K. Lee, and E. Wang for comments on drafts of the manuscript, K. Karukstis and students for allowing us to present their work, and J. K. Lee and E. Wang for preparing the figures. This research was supported by National Science Foundation Grants DMS-88-05402 and DMS-89-02265 to S.L. and R.T.R., NSF Grant HRD-9103314 to S.L., and by the Ohio Supercomputer Center.

[28] Ultraviolet/Visible Spectroelectrochemistry of Redox Proteins

By Shaojun Dong, Jianjun Niu, and Therese M. Cotton

I. Introduction

Studies of electrochemical reactions of redox proteins have attracted widespread interest and attention. Such studies can yield important information about not only intrinsic thermodynamic and kinetic properties of redox proteins, but also structural properties, such as binding characteristics of proteins at specific types of electrode surfaces and the orientational requirements for electron transfer between the protein and the electrode. The results are useful for the development of biosensors, biofuel cells, and biocatalysts. In addition, the information obtained from these studies can contribute to an understanding of the physiological implications of biological electron transfer reactions, because many electron transfer proteins are located at, or close to, charged membranes and are thus subject to large electric field effects that are similar to those near an electrode surface.

Following the initial report of Kono and Nakamura[1] of the direct electrochemical reduction of cytochrome c (Cyt c) at a platinum electrode, and the report of the requirement for an overpotential of around -0.36 V for the direct electrochemical reaction of Cyt c at a dropping mercury electrode by Griggio and Pinamonti[2] in 1965, studies of heterogeneous reactions of redox proteins were slow to develop, because the direct electrochemical reactions of redox proteins at metal electrodes were often irreversible and sometimes undetectable. The irreversibility can be attributed to several factors. First, most protein molecules are strongly adsorbed on the surface of the metal electrode, which can lead to structural alterations or complete denaturation. These structurally altered protein molecules often undergo irreversible electrochemical reactions and inhibit direct electron transfer with the freely diffusing molecules. Evidence for this behavior can be found in the concentration dependence of the electron transfer kinetics of Cyt c, which ranges from polarographically reversible at low concentrations to quasi-reversible and irreversible at higher concentrations. Second, the redox active moieties of the proteins are usually embedded within polypeptide chains, and their considerable distance from

[1] T. Kono and S. Nakamura, Bull. Agric. Chem. Soc. Jpn. 22, 399 (1958).
[2] L. Griggio and S. Pinamonti, Atti. Ist. Veneto. Sci. Lett. Arti. Cl. Sci. Mat. Nat. 124, 15 (1965–1966).

METHODS IN ENZYMOLOGY, VOL. 246

the electrode inhibits electron transfer. For example, Cyt c_3 can undergo direct electrochemical reduction at an extremely fast rate compared to Cyt c. Cytochrome c_3 has four heme groups embedded in a polypeptide chain with a molecular weight of around 12,000, whereas Cyt c has only one heme group buried in a polypeptide chain with a molecular weight of about 12,400. Thus, the hemes in Cyt c_3 are more exposed to the solvent compared to those in Cyt c. Third, the highly ionic properties and asymmetric distribution of the surface charges on redox proteins significantly affect the reversibility of their electrochemical reactions. Usually, it is difficult to reduce a positively charged redox protein at potentials more positive than the potential of zero charge of the working electrode and vice versa. The dependence of protein electrochemical reaction rates on the pH of the electrolyte solutions is related to the net surface charge of the proteins. The dependence of the electrochemical reaction rates of proteins on their orientation at the electrode surface is related to the asymmetric distribution of surface charges on the proteins.

To resolve the difficulties described above, spectroelectrochemistry is used for monitoring redox processes in proteins. The method combines optical and electrochemical techniques, allowing the electrochemical signals (current, charge, etc.) and spectroscopic responses (UV–VIS, IR) to be obtained simultaneously. As a result, more information about the oxidation or reduction mechanisms of redox proteins can be acquired than with traditional electrochemical methods. Among the various spectroelectrochemical techniques, *in situ* UV–VIS absorption spectroelectrochemistry in an optically transparent thin-layer cell has become one of the most useful methods for studying the direct electrochemistry of redox proteins.

One of the main advantages of the optically transparent thin-layer spectroelectrochemical technique (OTTLSET) is that the oxidized and reduced forms of the analyte adsorbed on the electrode and in the bulk solution can be quickly adjusted to an equilibrium state when the appropriate potential is applied to the thin-layer cell,[3,4] thereby providing a simple method for measuring the kinetics of a redox system. The formal potential $E^{\circ\prime}$ and the electron transfer number n can be obtained from the Nernst equation by monitoring the absorbance changes *in situ* as a function of potential. Other thermodynamic parameters, such as ΔH, ΔS, and ΔG, can also be obtained. Most redox proteins do not undergo direct redox reactions on a bare metal electrode surface. However, they can undergo indirect electron transfer processes in the presence of a mediator or a promoter; the determination of their thermodynamic parameters can then

[3] W. R. Heineman, F. M. Hawkridge, and H. N. Blount, *Electroanal. Chem.* **13**, 1 (1984).
[4] M. Petek, T. E. Neal, and R. W. Murray, *Anal. Chem.* **43**, 1069 (1971).

be achieved. Many redox proteins, such as Cyt c,[5-14] the electron transfer species in photosynthetic systems,[15] soluble spinach ferredoxin,[16] blue copper proteins,[12,17] myoglobin (Mb),[14,18] and hemoglobin (Hb),[19] have been studied by the OTTLSET. Because the absorption spectrum of the prosthetic group of most redox proteins changes significantly during the electron transfer processes, monitoring these reactions by the OTTLSET is more direct and sensitive. For example, the direct oxidation and reduction of Mb at a viologen polymer chemically modified Au minigrid electrode[18,20] and the quasi-reversible electron transfer process of Cyt c at a Pt minigrid electrode modified with 4,4'-bipyridine have been directly monitored by the OTTLSET.

The OTTLSET also provides a convenient method for studying the electrode kinetics of redox proteins, because the absolute amounts of the reactants are limited in the thin-layer cell and, therefore, even very slow reactions can be monitored. Moreover, because the species are quantitatively produced and are well distributed within the cell, the traditional methods used for studying reaction kinetics in solution can also be used in the OTTLSET. Thus, the investigation of kinetics by the OTTLSET is simple because it is not necessary to solve complicated diffusion equa-

[5] F. B. Kaufman, A. H. Schroeder, E. M. Engler, S. R. Kramer, and J. Q. Chambers, *J. Am. Chem. Soc.* **102**, 483 (1980).

[6] Y. Zhu and S. Dong, *J. Catal. (Dalian)(Cuihau Xuebao)* **13**, 209 (1992).

[7] K. A. Rubinson and H. B. Mark, Jr., *Anal. Chem.* **54**, 1204 (1982).

[8] C. W. Anderson, H. B. Halsall, W. R. Heineman, and G. P. Kreishman, *Biochem. Biophys. Res. Commun.* **76**, 339 (1977).

[9] G. P. Kreishman, C. W. Anderson, C.-H. Su, H. B. Halsall, and W. R. Heineman, *Bioelectrochem. Bioenerg.* **5**, 196 (1978).

[10] C. W. Anderson, H. B. Halsall, and W. R. Heineman, *Anal. Biochem.* **93**, 366 (1979).

[11] G. P. Kreishman, C.-H. Su, C. W. Anderson, H. B. Halsall, and W. R. Heineman, *in* "Bioelectrochemistry: Ions, Surface, Membrane" (M. Blank, ed.), Advances in Chemistry Series No. 188, American Chemical Society, Washington, D.C., 1980.

[12] V. T. Taniguchi, N. Sailasuta-Scott, F. C. Anson, and H. B. Gray, *Pure Appl. Chem.* **52**, 2275 (1980).

[13] W. R. Heineman, C. W. Anderson, H. B. Halsall, M. M. Hurst, J. M. Johnson, G. P. Kreishman, B. J. Norris, M. J. Simone, and C.-H. Su, *in* "Electrochemical and Spectrochemical Studies of Biological Redox Components" (K. M. Kadish, ed., Advances in Chemistry Series No. 201, American Chemical Society, Washington, D.C., 1982.

[14] S. Kwee, *Bioelectrochem. Bioenerg.* **16**, 99 (1986).

[15] F. M. Hawkridge and B. Ke, *Anal. Biochem.* **78**, 76 (1977).

[16] H. L. Landrum, R. T. Salmon, and F. M. Hawkridge, *J. Am. Chem. Soc.* **99**, 3154 (1977).

[17] N. Sailasuta, F. C. Anson, and H. B. Gray, *J. Am. Chem. Soc.* **101**, 455 (1979).

[18] W. R. Heineman, M. L. Meckstroth, B. J. Norris, and C.-H. Su, *Bioelectrochem. Bioenerg.* **6**, 577 (1979).

[19] S. Song and S. Dong, *Bioelectrochem. Bioenerg.* **19**, 337 (1988).

[20] J. F. Stargardt, F. M. Hawkridge, and H. L. Landrum, *Anal. Chem.* **50**, 930 (1978).

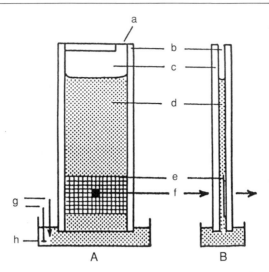

FIG. 1. Optically transparent thin-layer electrochemical cell shown in (A) front view and (B) side view. a, Point of suction application to change solution; b, Teflon tape spacers; c, microscope slides (1 × 3 in.); d, solution; e, transparent gold minigrid electrode; f, optical path; g, reference and auxiliary electrodes; h, solution cup. Epoxy is used to hold the cell together.[20]

tions in the semiinfinite diffusion case. The kinetic parameters for several redox proteins have been determined by this method.[19,21-24]

Applications of the OTTLSET to other systems were reviewed by Heineman et al.[3] in 1984. This chapter mainly emphasizes the development of the *in situ* UV–VIS spectroelectrochemistry, especially the OTTLSET, for investigation of direct electrochemical oxidation and reduction reactions of redox proteins.

II. Optically Transparent Thin-Layer Cell

To obtain satisfactory results by the OTTLSET, the design of the optically transparent thin-layer electrochemical cell (OTTLEC) is very important. The first OTTLEC was constructed by Murray et al.[25] in 1967. Figure 1 shows the structure of this type of cell which contains a gold

[21] S. Dong, Y. Zhu, and S. Song, *Bioelectrochem. Bioenerg.* **21**, 233 (1989).

[22] Y. Zhu and S. Dong, *Electrochim. Acta* **35**, 1139 (1990).

[23] S. Song, W. Zhang, and S. Dong, *Chinese Science Bulletin* **35**, 1960 (1990).

[24] S. Dong and Y. Zhu, *Acta Chim Sin. (Engl. Ed.)* **48**, 566 (1990).

[25] R. W. Murray, W. R. Heineman, and G. W. O'Dom, *Anal. Chem.* **39**, 1666 (1967).

minigrid sandwiched between two quartz plates. The analyte solution is drawn into the cell by capillary action from the bulk solution. The thin-layer cavity (about 40 μl) is defined by the area of the minigrid electrode and the thickness of the layer (typically 0.2 mm). Although the cell is easily constructed, the current density is not well distributed, because the reference electrode (RE) is too far away from the working electrode (WE). Consequently, the IR drop is not minimized. Another OTTLEC design[26] circumvents this problem. In this case, two platinum foil auxiliary electrodes (AE) are placed at each side of the platinum WE. The WE, AE, and two quartz plates form a sandwich with a Teflon film to define the solution volume and minimize edge diffusion effects. A small channel provides an internal reference point for accurate potential control. Excellent optical and electrochemical responses are obtained with this cell. Other cells that are easily cleaned[27] or with vapor-deposited Pt electrodes[28,29] have also been used.

One of the most important applications of the OTTLEC is the study of large biological molecules and redox proteins. The cells, requiring less than 100 μl to fill, have been designed for rare biological samples.[6]

The exclusion of oxygen is very important in the OTTLSET, as used for redox protein systems; therefore, an attempt has been made to design a nitrogen vacuum cell. An example of such a cell, as reported by Lin and Kadish,[30] is shown in Fig. 2. It is suitable for both vacuum operation and conventional nitrogen deoxygenation. The cell body can be constructed from Pyrex glass for studies in the visible region or from quartz for studies in the UV region. It contains a doublet platinum gauze as the WE. The thin-layer chamber is open to the bulk solution at all four edges, thereby allowing the electrolysis current to flow out of the chamber in all directions and minimizing IR drop. Unlike the commonly used OTTLECs, such as the sandwich cells, no epoxy or other soluble materials are used for the cell construction, which makes it convenient for use in both aqueous and nonaqueous solvents. This cell exhibits excellent spectral and electrochemical responses. Other OTTLECs designed for biological systems include a cell adapted for vacuum-nitrogen cycling[31] and a cell with a platinum grid as the WE.[32] Numerous cells have also been developed

[26] J. Niu, G. Cheng, and S. Dong, *Chin. J. Appl. Chem.* (*Yingyong Huaxue*) **9**, 11 (1992).
[27] S. Song, G. Cheng, and S. Dong, *Anal. Chem.* (*Changchung, People's Repub. China*) (*Fenxi Huaxue*) **15**, 461 (1987).
[28] A. Yildiz, P. T. Kissinger, and C. N. Reilley, *Anal. Chem.* **40**, 1018 (1968).
[29] V. S. Srinivasan and F. C. Anson, *J. Electrochem. Soc.* **120**, 1359 (1973).
[30] X. Q. Lin and K. M. Kadish, *Anal. Chem.* **57**, 1498 (1985).
[31] B. J. Norris, M. L. Meckstroth, and W. R. Heineman, *Anal. Chem.* **48**, 630 (1976).
[32] D. Lexa, J. M. Saveant, and J. Zickler, *J. Am. Chem. Soc.* **99**, 2786 (1977).

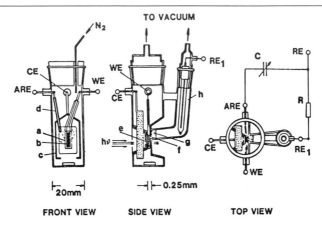

Fig. 2. Schematic illustration of the vacuum-tight, short path length, thin-layer spectro-electrochemical cell with a doubled platinum gauze working electrode. The side view shows how the cell assembly is positioned for acquiring spectral data. The top view shows the connection of the outer RC circuit. The symbols are as follows: (a) thin-layer chamber, (b) doubled platinum gauze working electrode with edge eliminator, (c) expanded platinum metal counterelectrode, (d) platinum wire auxiliary reference electrode, (e) photo window, (f) reference point, the tip of the frit, (g) asbestos disk, (h) reference frit. CE, counterelectrode post; WE, working electrode post; RE_1, Ag/AgCl, KCl reference electrode; ARE, auxiliary reference electrode post; RE, reference electrode post.[30]

for use in nonaqueous solvents,[33–38] and many cells designed for specific purposes have been reported.[39–52]

Although the OTTLECs discussed above are acceptable in many cases, they are not appropriate for systems with small molar absorptivity or with

[33] I. Piljac and R. W. Murray, *J. Electrochem. Soc.* **118,** 1758 (1971).

[34] I. Piljac, M. Tkalčec, and B. Grabarić, *Anal. Chem.* **47,** 1369 (1975).

[35] R. K. Rhodes and K. M. Kadish, *Anal. Chem.* **53,** 1539 (1981).

[36] T. Watanabe and K. Honda, *J. Phys. Chem.* **86,** 2617 (1982).

[37] E. A. Blubaugh and L. M. Doane, *Anal. Chem.* **54,** 329 (1982).

[38] W. Andrew Nevin and A. B. P. Lever, *Anal. Chem.* **60,** 727 (1988).

[39] D. A. Condit, M. E. Herrera, M. T. Stankovich, and D. J. Curran, *Anal. Chem.* **56,** 2909 (1984).

[40] M. D. Porter, S. Dong, Y. Gui, and T. Kuwana, *Anal. Chem.* **56,** 2263 (1984).

[41] X. Q. Lin and K. M. Kadish, *Anal. Chem.* **58,** 1493 (1986).

[42] W. R. Heineman, J. N. Burnett, and R. W. Murray, *Anal. Chem.* **40,** 1974 (1968).

[43] T. C. Pinkerton, K. Hajizadch, E. Deutsch, and W. R. Heineman, *Anal. Chem.* **52,** 1542 (1980).

[44] J. L. Anderson, *Anal. Chem.* **51,** 2312 (1979).

[45] A. Brajter-Toth and G. Dryhurst, *J. Electroanal. Chem.* **122,** 205 (1981).

[46] D. A. Scherson, S. Sarangapani, and F. L. Urbach, *Anal. Chem.* **57,** 1501 (1985).

[47] E. H. Ward and C. L. Hussey, *Anal. Chem.* **59,** 213 (1987).

[48] B. L. Harward, L. N. Klatt, and G. Mamantov, *Anal. Chem.* **57,** 1773 (1985).

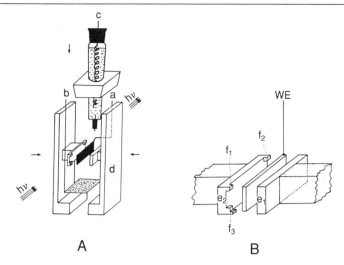

FIG. 3. Structure of the long path length OTTLEC thin-layer spectroelectrochemical cell.[55]

low sample concentrations owing to the short path length, which results in low sensitivities. The OTTLECs with a long optical path length (LOPLTTLEC) were designed to resolve this problem. Porter and Kuwana[53] reported a cell containing a glassy carbon or graphite electrode as the WE. A hole with a diameter of about 500 μm is drilled into the WE allowing the light to pass through. Depending on the length of the hole, the LOPLTTLEC has greater sensitivity than the conventional thin-layer cell. In this cell, the species diffusion is confined within a very thin cylinder. The electrolysis characteristics of the cell are similar to those in other OTTLECs. Zak et al.[54] reported a unique cell containing a glassy carbon or platinum plate with a large area as a WE. The incident light is passed through the solution parallel to the electrode surface. A simpler LOPLTTLEC is shown in Fig. 3.[55] This cell is constructed by inserting the composited one (Fig. 3A) into a colorimetric cell.[55] It can be used in

[49] G. Mamantov, V. E. Norvell, and L. N. Klatt, *J. Electrochem. Soc.* **127**, 1768 (1980).

[50] S. K. Enger, M. J. Weaver, and R. A. Walton, *Inorg. Chim. Acta* **129**, L1 (1987).

[51] J. P. Bullock, D. C. Boyd, and K. R. Mann, *Inorg. Chem.* **26**, 3084 (1987).

[52] D. A. Smith, R. C. Elder, and W. R. Heineman, *Anal. Chem.* **57**, 2361 (1985).

[53] M. D. Porter and T. Kuwana, *Anal. Chem.* **56**, 529 (1984).

[54] J. Zak, M. D. Porter, and T. Kuwana, *Anal. Chem.* **55**, 2219 (1983).

[55] J. Niu and S. Dong, "Fourth International Seminar on Electroanalytical Chemistry," Oct. 4–7, p. 72, Changchun, China, 1993.

UV–VIS spectroelectrochemistry as well as in circular dichroism electrochemical studies. For the LOPLTTLECs, both optical and nonoptical materials can be used as WEs. The ratio of the larger electrode area and the smaller thin-layer solution volume leads to higher sensitivity of the absorption signals and a shorter electrolysis time. Because of these characteristics, many LOPLTTLECs have been designed[56–58] for different studies.[59–62]

III. Thermodynamics of Redox Proteins

Thermodynamic studies of electrochemical systems are based on the Nernst equation. The electron transfer process is

$$O + ne \underset{k_{b,h}}{\overset{k_{f,h}}{\rightleftharpoons}} R \tag{1}$$

where O is the oxidized form of the reactant, R is the reduced form, n is the number of electrons transferred, and $k_{f,h}$ and $k_{b,h}$ are the heterogeneous electron transfer rate constants for the forward and backward process, respectively. The Nernst equation is

$$E = E^{\circ\prime} + \frac{RT}{nF} \ln \frac{[O]}{[R]} \tag{2}$$

where $E^{\circ\prime}$ is the formal potential of the redox couple O/R, R is the molar gas constant (8.31441 J mol^{-1} K^{-1}), and F is Faraday's constant ($96{,}485$ C/equiv). [O] and [R] are the respective concentrations of the oxidized and reduced species. From the intercept and slope of the straight line obtained by plotting E against $\ln([O]/[R])$, $E^{\circ\prime}$ and the electron transfer number n are determined.

Most redox proteins do not obey the Nernst equation because of their tendency to adsorb on the electrode surface, as well as the large steric inhibition to electron transfer. Therefore, $E^{\circ\prime}$ values for redox proteins often cannot be obtained directly from Eq. (2) but rather require the use

[56] J. D. Brewster and J. L. Anderson, *Anal. Chem.* **54**, 2560 (1982).

[57] M. J. Simone, W. R. Heineman, and G. P. Kreishman, *Anal. Chem.* **54**, 2382 (1982).

[58] M. Shi, Proceedings of the Second Beijing Conference and Exhibition on Instrumentation Analysis" October 2–23, p. 1309. Beijing, China, 1987.

[59] Y. Gui and T. Kuwana, *J. Electroanal. Chem.* **222**, 321 (1987).

[60] Y. Gui and T. Kuwana, *Chem. Lett.*, 231 (1987).

[61] Y. Gui and T. Kuwana, *Langmuir* **2**, 471 (1986).

[62] Y. Gui, M. D. Porter, and T. Kuwana, *Anal. Chem.* **57**, 1474 (1985).

of the potential titration method which employs chemical redox reagents to effect redox changes in the protein.[63–65] However, the titration method is limited by the very small current signals produced by the proteins. On the other hand, if the value of [O]/[R] of the redox proteins is detected by absorbance changes, the Nernst equation can be expressed in terms of the relationship between potential and absorbance changes. The sensitivity is then increased appreciably owing to the large optical absorption changes that accompany redox changes in proteins, thereby resulting in more reliable results.[66–68] The combination of electrochemistry and absorption spectroscopy is used in two techniques, the OTTLSET and indirect coulometric titration at an optically transparent electrode, for obtaining thermodynamic parameters of redox proteins. The principles and the applications of these techniques are discussed below.

A. Optically Transparent Thin-Layer Spectroelectrochemistry

The OTTLSET uses the relationship between constant potential and absorbance to obtain $E^{\circ\prime}$, n, and other information about electrochemical processes. From $E^{\circ\prime}$ values at different temperatures, other thermodynamic parameters can be determined. In a thin-layer cell, when the absorbance change is completed at a certain applied potential, an equilibrium state is achieved, that is, almost no current (except the very small background current) flows within the cell. In this case, the IR drop in most OTTLECs can be neglected, and the error in $E^{\circ\prime}$ may be only several millivolts. Because of this characteristic, the OTTLSET provides a very effective method for determining thermodynamic parameters of redox proteins.

Principles: Reversible or Quasi-Reversible Systems. Electrolysis by the OTTLSET involves a very small solution volume, and the equilibrium state is quickly achieved at a fixed applied potential. The ratio of $([O]/[R])_{surface}$ at the electrode surface is determined by

$$E = E^{\circ\prime} + \frac{RT}{nF} \ln ([O]/[R])_{surface} \tag{3}$$

[63] F. L. Rodkey and E. G. Ball, *J. Biol. Chem.* **182,** 17 (1950).

[64] W. M. Clark, "Oxidation–Reduction Potentials of Organic Systems." Williams & Wilkins, Baltimore, Maryland, 1960.

[65] D. F. Wilson, P. L. Dutton, M. Erecinska, J. G. Lindsay, and N. Sato, *Acc. Chem. Res.* **5,** 234 (1972).

[66] K. Minnaert, *Biochim. Biophys. Acta* **110,** 42 (1965).

[67] P. L. Dutton, this series, Vol. 54, p. 411.

[68] G. S. Wilson, J. C. M. Tsibris, and I. C. Gunsalus, *J. Biol. Chem.* **248,** 6059 (1973).

where the identities of the parameters are as described in Eqs. (1) and (2). In a thin-layer cell, electrolysis to the equilibrium state usually can be achieved in less than 60 sec, and the concentrations of the oxidized and reduced species in the bulk solution are quickly adjusted to the same values as on the electrode surface. Thus, we have

$$([O]/[R])_{\text{solution}} = ([O]/[R])_{\text{surface}} \tag{4}$$

The Nernst equation in the thin-layer cell is generally expressed as in Eq. (2).

According to Beer's law, the *in situ* absorption spectrum can be used to determine the ratio of $[O]/[R]$:

$$[O]/[R] = \left(\frac{A - A_R}{\Delta \varepsilon \, b}\right)\bigg/\left(\frac{A_O - A}{\Delta \varepsilon \, b}\right) = \frac{A - A_R}{A_O - A} \tag{5}$$

where A_R is the absorbance of the reduced form, A_O is the absorbance of the oxidized form, A is the absorbance of an intermediate mixture, $\Delta \varepsilon$ is the difference in molar absorptivity of the oxidized and reduced form measured at the same wavelength, and b is the light path length in units appropriate to $\Delta \varepsilon$ of the OTTLEC. Substituting Eq. (5) into Eq. (2) results in the absorbance-related Nernst equation:

$$E = E^{\circ\prime} + \frac{RT}{nF} \ln \frac{A - A_R}{A_O - A} \tag{6}$$

The intercept of the straight line of E versus $\ln [(A - A_R)/(A_O - A)]$ gives the value of $E^{\circ\prime}$, and the slope is related to n. The change in $E^{\circ\prime}$ with temperature is related to ΔS, the entropy change which accompanies the redox process:

$$\left(\frac{\partial E^{\circ\prime}}{\partial T}\right)_P = \frac{\Delta S}{nF} \tag{7}$$

Because of the very small temperature coefficient (within 10^{-4}–10^{-5} V/°C), ΔS can be assumed not to change with temperature. Equation (7) may then be integrated to produce the following equation:

$$E^{\circ\prime} = E^{\circ\prime}_{T_0} + \frac{\Delta S}{nF}(T - T_0) \tag{8}$$

At a temperature of 25°:

$$E^{\circ\prime} = E^{\circ\prime}_{25^\circ} + \frac{\Delta S}{nF}(T - T_{25^\circ}) \tag{9}$$

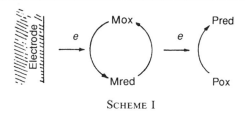

SCHEME I

The value of ΔS is then obtained from Eq. (9). Another thermodynamic parameter the enthalpy ΔH can be calculated from

$$\Delta H = -nFE^{\circ\prime} + nFT\left(\frac{\partial E^{\circ\prime}}{\partial T}\right)_P \tag{10}$$

For a quasi-reversible system with a small reaction rate, a longer equilibrium time is needed for each change in the applied potential, but the thermodynamic parameters can be determined in the same way.

Principles: Redox Protein Systems. In most cases, redox proteins undergo extremely slow electron transfer reactions at electrodes. Therefore, the thermodynamic parameters cannot be determined directly from the equations described above. One method for resolving this problem is to add a mediator, M, to the solution. The mediator shuttles electrons between the electrode and the protein and thereby accelerates the electrochemical reaction of the protein. A reversible or quasi-reversible electron transfer reaction of the redox protein, P, is then achieved indirectly through the mediator as shown in Scheme I. The Nernst equation is

$$E = E_M^{\circ\prime} + \frac{RT}{n_M F}\ln\frac{[O_M]}{[R_M]} = E_P^{\circ\prime} + \frac{RT}{n_P F}\ln\frac{[O_P]}{[R_P]} \tag{11}$$

The $E_P^{\circ\prime}$, n, and other thermodynamic parameters are obtained from Eqs. (9)–(11). The subscript M refers to the mediator and P refers to the protein.

Applications. Heineman *et al.*[69] first reported the electron transfer number and other thermodynamic parameters of Cyt c as measured by the OTTLSET. The mediator used was 2,6-dichlorophenolindophenol. Other similar studies of Cyt c have been reported.[8-13,31] The spectral

[69] W. R. Heineman, B. J. Norris, and J. F. Goelz, *Anal. Chem.* **47**, 79 (1975).

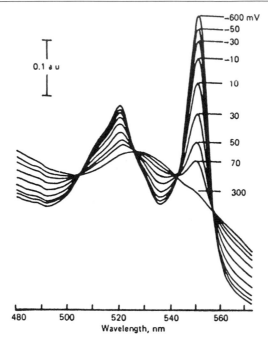

FIG. 4. Absorption spectra of cytochrome *c* in a thin-layer cell with 2,6-dichlorophenolindophenol mediator in 0.5 *M* phosphate buffer, pH 7.0, for a series of applied potentials.[10]

changes in Cyt *c* at different applied potentials are shown in Fig. 4. It can be seen that, although Cyt *c* exhibits no response by cyclic voltammetry in a AU minigrid OTTLEC,[13] it can be fully oxidized [at about 0.3 V versus standard calomel electrode (SCE)] and reduced (at about -0.6 V) in the presence of the mediator. Thus, the mediator is very useful for the determination of thermodynamic parameters of redox proteins. From a typical Nernst plot for Cyt *c* the intercept and slope provided the precise values of $E^{\circ\prime}$ [$+0.222$ V versus standard hydrogen electrode (SHE)] and *n* (1.00) at 25°, respectively.[69]

The relationship between $E^{\circ\prime}$ and temperature for Cyt *c* in water and heavy water has been extensively investigated[8,9,12,70–72] for the determination of thermodynamic parameters (e.g., ΔG, ΔS, ΔH). In general, the

[70] G. P. Kreishman, C. W. Anderson, H. B. Halsall, and W. R. Heineman, *Adv. Chem. Ser.* **188,** 169 (1978).

[71] W. R. Heineman, C. W. Anderson, H. B. Halsall, M. M. Hurst, J. M. Johnson, G. P. Kreishman, B. J. Norris, M. J. Simone, and C. Su, *Adv. Chem. Ser.* **201,** 1 (1982).

[72] V. T. Taniguichi, W. R. Ellis, V. Cammarata, J. Webb, F. C. Anson, and H. B. Gray, *Adv. Chem. Ser.* **201,** 51 (1982).

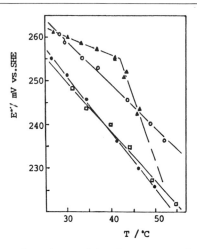

FIG. 5. Relation between formal potential and temperature for horse heart Cyt c. The solvent was water, and the solution was 0.1 M phosphate buffer (pH 7.0) containing 0.1 M sodium halide. (○) F⁻, (▲) Cl⁻, (●) Br⁻, (□) I⁻.[9]

value of $E^{\circ\prime}$ gradually decreases with increasing temperature. In Fig. 5 a plot of $E^{\circ\prime}$ versus T for Cyt c in the presence of halogen ions is depicted. It can be seen that a biphasic relationship exists between $E^{\circ\prime}$ and T in chloride solution, and an inflection point appears at about 42°. This response is attributed to a higher order phase transition of Cyt c in the chloride solution. The same phenomenon is also found in dimethylarsenic acid buffer (pH 8.0) and phosphate buffer (pH 8.0).[73,74] The linear range for $E^{\circ\prime}$ versus T is 5°–60° in pH 7.0 buffer of dimethylarsenic acid, but when the pH is decreased to 5.3, the linear range is increased to 5°–75°. The difference in $E^{\circ\prime}$ of Cyt c in the two buffer solutions is due to the linking effect of PO_4^{3-} in phosphate solution. Other $E^{\circ\prime}$–T relations on Au modified electrodes have also been reported.[75,76]

The OTTLSET has also been used to investigate redox proteins in addition to chytochrome c. In the case of Mb, the *in situ* absorption spectra with phenazine methosulfate as the mediator provide a Nernst plot with $E^{\circ\prime} = +0.0464$ V versus SHE and $n = 0.95$.[18] Measurements of $E^{\circ\prime}$ at various temperatures and with appropriate mediators have been made to determine the thermodynamic parameters of the blue copper

[73] K. B. Koller and F. M. Hawkridge, *J. Am. Chem. Soc.* **107**, 7412 (1985).
[74] K. B. Koller and F. M. Hawkridge, *J. Electroanal. Chem.* **239**, 291 (1988).
[75] I. Taniguchi, T. Funatsu, M. Iseki, H. Yamaguchi, and K. Yasukouchi, *J. Electroanal. Chem.* **193**, 295 (1985).
[76] I. Taniguchi, M. Iseki, T. Eto, K. Toyosawa, H. Yamaguchi, and K. Yasukouchi, *Bioelectrochem. Bioenerg.* **13**, 373 (1984).

proteins azurin, plastocyanin, and stellacyanin.[12,17] The thin-layer spectra of stellacyanin at different potentials and the Nernst plots at different temperatures are shown in Fig. 6. From the latter plots (Fig. 6B), values of $E^{\circ\prime} = +0.184$ V versus SHE at 25°, $\Delta H = -17$ kcal mol^{-1}, and $\Delta S^{\circ} = -44$ cal K^{-1} mol^{-1} were obtained. Thermodynamic studies of

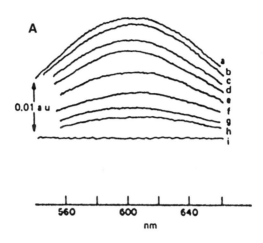

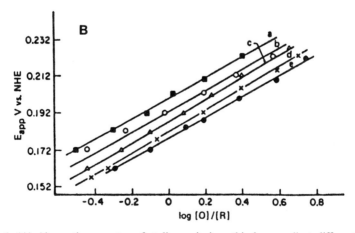

FIG. 6. (A) Absorption spectra of stellacyanin in a thin-layer cell at different applied potentials. Stellacyanin was present at 2.7×10^{-3} M, with Ru(NH$_3$)$_5$py^{3+} at 2.8×10^{-3} M; the cell thickness was 0.021 cm. Applied potentials: a, 240; b, 230; c, 220; d, 210; e, 190; f, 175; g, 160; h, 145; i, 90 mV versus SHE. (B) Nernst plot for stellacyanin at different temperatures: a, 6.5°; b, 11.2°; c, 16.2°; d, 21.7°; e, 25.0°.[17]

additional redox proteins and other biological systems by the OTTLSET are given in the literature.[8–12,15,31,69,77–81]

Another method for promoting electron transfer reactions between redox proteins and electrodes is the use of a chemically modified electrode (CME). This approach is discussed in another section.

B. Indirect Coulometric Titrations

In addition to analysis via the OTTLSET, $E^{\circ\prime}$, n, and other thermodynamic parameters can be determined by the indirect coulometric titration (ICT) of proteins at an optically transparent electrode (OTE). The process can be expressed as

Electrode reaction: $M_o + n\,e^- \rightleftharpoons M_r$ (12a)

Solution reaction: $M_r + P_o \rightleftharpoons M_o + P_r$ (12b)

where M_o and M_r represent the redox couple in the oxidized and reduced form, respectively, used to generate electrochemically the titrant (mediator) and P_o and P_r are the oxidized and reduced forms of the redox protein titrated. The ICT at an OTE is performed by applying a potential sufficient to drive the reduction of the mediator [Eq. (12a)] at a mass transfer-limited rate while simultaneously monitoring charge passed by the variation of the absorption signal. The mediator transfers electrons between the electrode and the redox protein and enhances the oxidation or reduction of proteins that exhibit irreversible heterogeneous electron transfer processes at solid electrodes.[82] Repetitive titrations of the redox proteins are readily performed in this manner.

An example of the application of ICT at an OTE is the study of a mixture of Cyt c and cytochrome-c oxidase.[83] From the absorption spectra

[77] D. Lexa, J. M. Saveant, and J. Zickler, *J. Am. Chem. Soc.* **102**, 4851 (1980).

[78] K. A. Rubinson, E. Itabashi, and H. B. Mark, Jr., *Inorg. Chem.* **21**, 3571 (1982).

[79] T. M. Kenyhercz, T. P. DeAngelis, B. J. Norris, W. R. Heineman, and H. B. Mark, Jr., *J. Am. Chem. Soc.* **98**, 2469 (1976).

[80] H. B. Mark, Jr., T. M. Kenyhercz, and P. T. Kissinger, *in* "Electrochemical Studies of Biological Systems" (D. T. Sawyer, ed.), ACS Symp. Ser. No. 38, American Chemical Society, Washington, D.C., 1977.

[81] T. M. Kenyhercz and H. B. Mark, Jr., *J. Electrochem. Soc.* **123**, 1656 (1976).

[82] F. M. Hawkridge and T. Kuwana, *Anal. Chem.* **45**, 1021 (1973).

[83] W. R. Heineman, T. Kuwana, and C. R. Hartzell, *Biochem. Biophys. Res. Commun.* **50**, 892 (1973).

at different points in the reduction and the calculated absorbance versus charge responses, the *n* value of Cyt *c* and the hemes in Cyt *c* oxidase are determined. Oxidation of the mixture is also demonstrated by performing repetitive ICTs and comparing the values obtained by computer calculations with the experimental responses. As seen in Fig. 7, agreement between theory and the experimental data is excellent, demonstrating that ICT at an OTE is a reliable method for studying the thermodynamics of redox proteins.

Numerous literature reports emphasize the use of ICTs at OTEs for

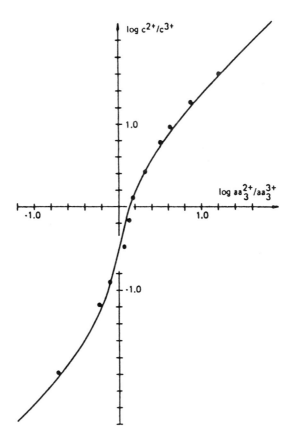

FIG. 7. Plot of log {[Cyt *c*(reduced)]/[Cyt *c*(oxidized)]} versus log {[Cyt-*c* oxidase (reduced)]/[Cyt-*c* oxidase (oxidized)]} for reductive ICT of a Cyt *c*–Cyt-*c* oxidase mixture. The solid line is a computer calculation for $E^{\circ\prime}$ of Cyt *c* = 250 mV, $E^{\circ\prime}$ of Cyt-*c* oxidase heme *a* = 210 mV, and $E^{\circ\prime}$ of Cyt-*c* oxidase heme a_3 = 350 mV.[83]

investigating different proteins, such as Cyt-*c* oxidase,[44,84,85] Cyt-*c* oxidase–CO complex,[86] a flavoprotein,[87] complex IV formed during the oxidative phosphorylation process in mammals,[88] the reaction center complex from photosystem II,[89] and Cyt *c*.[90,91] A number of reports highlight the advantages and the notable aspects of this technique.[92–100]

IV. Kinetics of Redox Proteins

Although the utility of spectroelectrochemistry for studying electrochemical reaction kinetics was recognized during its initial development, actual applications were not achieved until 10 years later. Obviously the use of optical spectroscopic probes during electrochemical perturbation can provide specific molecular information. The development and applications of spectroelectrochemical techniques have expanded from the study of simple systems (such as inorganic species) to biological molecules (such as redox proteins). The heterogeneous electron transfer kinetic parameters for many of these complex systems have been determined. Also, reaction mechanisms which cannot be elucidated by electrochemical methods alone have been determined. Such information is essential for providing an understanding of the mechanisms of redox proteins that exhibit different

[84] W. R. Heineman, T. Kuwana, and C. R. Hartzell, *Biochem. Biophys. Res. Commun.* **49,** 1 (1972).

[85] L. N. Mackey, T. Kuwana, and C. R. Hartzell, *FEBS Lett.* **36,** 326 (1973).

[86] J. L. Anderson, T. Kuwana, and C. R. Hartzell, *Biochemistry* **15,** 3847 (1976).

[87] M. T. Stankovich, *Anal. Biochem.* **109,** 295 (1980).

[88] R. Szentirmay and T. Kuwana, *Anal. Chem.* **50,** 1879 (1978).

[89] B. Ke, F. M. Hawkridge, and S. Sahu, *Proc. Natl. Acad. Sci. U.S.A.* **73,** 2211 (1976).

[90] T. Kula, E. Stellwagen, R. Szentirmay, and T. Kuwana, *Biochim. Biophys. Acta* **634,** 279 (1981).

[91] Y. Fujihira, T. Kuwana, and C. R. Hartzell, *Biochem. Biophys. Res. Commun.* **61,** 538 (1974).

[92] T. Kuwana and W. R. Heineman, *Acc. Chem. Res.* **9,** 241 (1976).

[93] W. R. Heineman, *Anal. Chem.* **50,** 390A (1978).

[94] R. Szentirmay, P. Yeh, and T. Kuwana, *in* "Electrochemical Studies of Biological Systems" (D. T. Sawyer, ed.), ACS Symp. Ser. No. 38, American Chemical Society, Washington, D.C., 1977.

[95] M. Ito and T. Kuwana, *J. Electroanal. Chem.* **32,** 415 (1971).

[96] G. S. Wilson, in this series, Vol. 54, p. 396.

[97] R. L. McCreery, *Crit. Rev. Anal. Chem.* **7,** 89 (1978).

[98] M. L. Meckstroth, B. J. Norris, and W. R. Heineman, *Bioelectrochem. Bioenerg.* **8,** 63 (1981).

[99] R. T. Salmon and F. M. Hawkridge, *J. Electroanal. Chem.* **112,** 253 (1980).

[100] T. Kuwana and W. R. Heineman, *Bioelectrochem. Bioenerg.* **1,** 389 (1974).

physiological functions (e.g., the investigation of the interfacial electron transfer between Cyt c and Cyt-c oxidase).[101]

Because of the high specificity of the technique, the utility of spectroelectrochemistry has been amply demonstrated.[102–104] The technique has been used for the determination of kinetic parameters characteristic of both the forward and backward heterogeneous electron transfer process in systems exhibiting varying degrees of reversibility. Among the numerous spectroelectrochemical techniques, three major methods are described that are suitable for determination of the heterogeneous electron transfer rate constant for the electrochemical oxidation or reduction of redox proteins: (A) single potential step chronoabsorptometry, (B) asymmetric double potential step chronoabsorptometry, and (C) cyclic potential scan chronoabsorptometry.

A. Single Potential Step Chronoabsorptometry

Single potential step chronoabsorptometry (SPS/CA) is suitable for the semiinfinite diffusion case. In OTTLSET, however, if the ratio of the optical path length of the thin-layer chamber is large enough and the reaction time is not very long, SPS/CA can also be used.

1. Irreversible Case. A heterogeneous electron transfer reaction can be represented as in Eq. (1). Assuming that all of the sample is initially present as O, the absorption signals of R may be monitored in the SPS/CA experiment. If the magnitude of the potential applied to the system is sufficient to cause the forward reaction to proceed at a rate governed only by $k_{f,h}$, the rate-dependent absorbance of R then is given by[105]

$$A_R(\lambda, t) = \frac{\varepsilon_R(\lambda)\, C_O^0 D_O}{k_{f,h}} \left[\frac{2k_{f,h} t^{1/2}}{\pi^{1/2} D_O^{1/2}} + \exp\left(\frac{k_{f,h}^2 t}{D_O} \right) erfc\left(\frac{k_{f,h} t^{1/2}}{D_O^{1/2}} \right) - 1 \right] \quad (13)$$

where C_O^0 and D_O are the bulk concentration and diffusion coefficient of O, respectively, and $\varepsilon_R(\lambda)$ is the molar absorptivity of the product R at wavelength λ. If the applied potential is large enough to cause the forward

[101] S. Ferguson-Miller, D. L. Brautigan, and E. Margoliash, *in* Porphyrins **7**, 149 (1979).
[102] A. J. Bard and L. R. Faulkner, "Electrochemical Method." Wiley, New York, 1980.
[103] T. Kuwana, *Ber. Bunsen-Ges. Phys. Chem.* **77**, 858 (1973).
[104] M. D. Porter and T. Kuwana, *Anal. Chem.* **56**, 529 (1984).
[105] D. E. Albertson, H. N. Blount, and F. M. Hawkridge, *Anal. Chem.* **51**, 556 (1979).

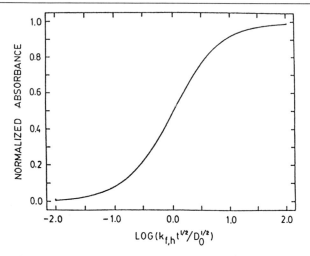

FIG. 8. Working curve for spectroelectrochemical determination of heterogeneous electron transfer rate constants for irreversible reactions.[105]

reaction in Eq. (1) to proceed at a diffusion-controlled rate, Eq. (13) can be expressed as

$$A_R^D(\lambda, t) = \frac{2}{\pi^{1/2}} \varepsilon_R (\lambda) D_O^{1/2} C_O^0 t^{1/2} \tag{14}$$

The ratio of $A_R(\lambda, t)/A_R^D(\lambda, t)$ gives the normalized absorbance, A_N:

$$A_N(\lambda, t) = 1 + \frac{\pi^{1/2}}{2\zeta} [\exp(\zeta^2) erfc(\zeta) - 1] \tag{15}$$

where

$$\zeta = \frac{k_{f,h} t^{1/2}}{D_O^{1/2}} \tag{16}$$

Then A_N versus $\log(k_{f,h} t^{1/2})$ results in a single kinetic working curve, as shown in Fig. 8, and provides a convenient way of determining $k_{f,h}$. A plot of log $k_{f,h}$ against overpotential (η) yields the formal heterogeneous electron transfer rate constant $(k_{s,h}^{o\prime})$ from the intercept and the value of the electrochemical transfer coefficient (α) from the slope.[106,107]

   2. Quasi-Reversible Case. In the quasi-reversible case, the experimental approach is the same as that described for the irreversible case; how-

[106] J. A. V. Butler, *Trans. Faraday Soc.* **19**, 729 (1924).
[107] J. A. V. Butler, *Trans. Faraday Soc.* **19**, 734 (1924).

ever, the forward reaction rate in Eq. (1) is not governed by $k_{f,h}$ alone, and the contribution of $k_{b,h}$ to the fluxes of reactant and product cannot be neglected. Under these conditions the time-dependent absorbance of R is given by[108]

$$A_R(\lambda, t) = \frac{\varepsilon_R(\lambda) C_O^0 k_{f,h}}{\left(\dfrac{k_{f,h}}{D_O^{1/2}} + \dfrac{k_{b,h}}{D_R^{1/2}}\right)^2} \left\{ 2 \left(\frac{k_{f,h}}{D_O^{1/2}} + \frac{k_{b,h}}{D_R^{1/2}} \right) \frac{t^{1/2}}{\pi^{1/2}} \right.$$

$$\left. + \exp\left[\left(\frac{k_{f,h}}{D_O^{1/2}} + \frac{k_{b,h}}{D_R^{1/2}} \right)^2 t \right] erfc\left[\left(\frac{k_{f,h}}{D_O^{1/2}} + \frac{k_{b,h}}{D_R^{1/2}} \right) t^{1/2} \right] - 1 \right\} \quad (17)$$

where the terms are as defined previously. The normalized absorbance in this case becomes

$$A_N(\lambda, t) = \frac{\zeta \pi^{1/2}}{2\xi^2} \left[\frac{2\xi}{\pi^{1/2}} + \exp(\xi^2) erfc(\xi) - 1 \right] \quad (18)$$

where ζ is defined by Eq. (16), and

$$\xi = \left[\frac{k_{f,h}}{D_O^{1/2}} + \frac{k_{b,h}}{D_R^{1/2}} \right] t^{1/2} \quad (19)$$

From Eqs. (16), (19), and the Butler–Volmer relationship, the following relation is obtained[108]:

$$\xi = \zeta[1 + (D_O^{1/2}/D_R^{1/2}) \exp\{n_a F(E - E_{O/R}^{o\prime})/RT\}] \quad (20)$$

A unique working curve is obtained for each overpotential step applied in the SPS/CA experiment using Eqs. (18) and (20). Figure 9 depicts a family of such working curves. Clearly the quasi-reversible case converges to the irreversible case when $n_a \eta$ is less than or equal to -160 mV. From these working curves, the kinetic parameters for a heterogeneous electron transfer processes that exhibits any degree of electrochemical reversibility can be determined by SPS/CA technique.[108]

Sperm whale myoglobin at a methyl viologen-modified gold minigrid electrode was investigated by the SPS/CA technique. The potential step chronoabsorption spectra are shown in Fig. 10. The calculated responses are in good agreement with the experimental results,[109] demonstrating the utility of using SPS/CA for the determination of kinetic parameters for redox proteins. From the log $k_{f,h}$ versus η relations, $k_{s,h}^{o\prime} = 3.88 \times 10^{-11}$

[108] E. E. Bancroft, H. N. Blount, and F. M. Hawkridge, *Anal. Chem.* **53**, 1862 (1981).
[109] E. F. Bowden, F. M. Hawkridge, and H. N. Blount, *Bioelectrochem. Bioenerg.* **7**, 447 (1980).

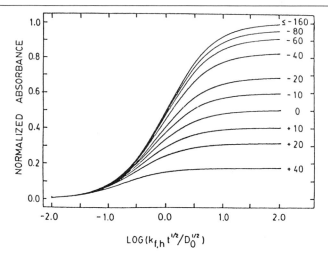

FIG. 9. Working curves for spectroelectrochemical determination of heterogeneous electron transfer rate constants for quasi-reversible reactions. Numerical values correspond to $n_a\eta$ where η is expressed in millivolts.[108]

cm sec^{-1} and $\alpha = 0.88$ are obtained. Additional $k_{s,h}^{\circ\prime}$ values for other redox proteins as determined by SPS/CA are given in Table I.[110,111]

B. Asymmetric Double Potential Step Chronoabsorptometry

Irreversible and Quasi-Reversible Case. From the discussion above, it is obvious that the SPS/CA technique provides an effective method for obtaining $k_{f,h}$ [Eq. (1)] for the forward reaction, but not $k_{b,h}$ for the backward reaction. For biological or redox protein system in which the back reaction kinetic parameters are difficult to determine by the reverse SPS/CA technique, the oxidized (or reduced) form is first transformed into the reduced (or oxidized) form to determine the $k_{b,h}$ (or $k_{f,h}$) by SPS/CA owing to the long-term redox state stability of the original species and the instability of the product form. This problem is resolved by the DPS/CA method. The DPS/CA experiment is performed by stepping initially to a potential that causes the forward reaction in Eq. (1) to proceed at a diffusion-controlled rate. After a certain time, τ, the potential is stepped back to a less extreme value, which causes the portion of the reduced form produced in the first step to be converted back to the oxidized form. By analyzing the $A_R(\lambda, \tau > t)$ values and fitting the experimental DPS/

[110] E. F. Bowden, M. Wang, and F. M. Hawkridge, *J. Electrochem. Soc.* **127,** 131C (1980).
[111] C. D. Crawley and F. M. Hawkridge, *Biochem. Biophys. Res. Commun.* **99,** 516 (1981).

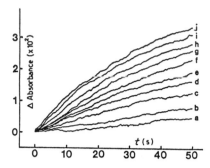

FIG. 10. Potential step chronoabsorptometry of Mb at a modified gold minigrid electrode. The sample was 40 μM Mb in phosphate buffer, pH 7.0, 0.1 M NaCl. Trace, potential step (mV versus NHE), and sequence of experiment are as follows: (a) -350, 8; (b) -370, 4; (c) -390, 6; (d) -400, 9; (e) -410, 2; (f) -420, 7; (g) -430, 3; (h) -450, 1; (i) -470, 10; (j) -490, 5.[109]

TABLE I

HETEROGENEOUS ELECTRON TRANSFER KINETIC PARAMETERS OF REDOX PROTEINS BY SINGLE POTENTIAL STEP CHRONOABSORPTOMETRY

Biomolecule[a]	Chemically modified OTE[b]	Solution conditions	Overpotential range[c]	$k_{s,h}^{\circ\prime}$ (cm sec^{-1})	α	Ref.
Myoglobin	Methyl viologen-modified gold minigrid	pH 7.00 phosphate buffer, NaCl (0.10 M)	$-396/-516$	3.9×10^{-11}	0.88	109
Cytochrome c	Methyl viologen-modified gold minigrid	pH 7.00 phosphate buffer (0.07 M), NaCl (0.10 M)	$-70/-397$	1.3×10^{-5}	0.27	110
Cytochrome c	Fluoride-doped tin oxide	pH 7.00 phosphate buffer (0.07 M), NaCl (0.10 M)	$-178/-528$	6.8×10^{-6}	0.32	110
Cytochrome c	Tin-doped indium oxide	pH 7.00 phosphate buffer (0.07 M), NaCl (0.10 M)	$-8/-127$	3.1×10^{-5}	0.50	110
Ferredoxin	Methyl viologen-modified gold minigrid	pH 7.50 Tris buffer (0.15 M), NaCl (0.20 M)	$-16/-101$	6.5×10^{-5}	0.60	111

[a] For reductive potential steps; data analysis by irreversible model. Concentrations are as follows: Mb, 20–50 μM; Cyt c, 40–103 μM; Fd, 78–124 μM.
[b] Details of surface modification are given in references.
[c] Maximum/minimum values employed for kinetic analysis are given.

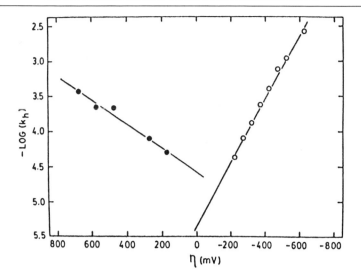

FIG. 11. Dependence of $\log(k_{f,h})$ for single potential step chronoabsorptometric measurements ($\bigcirc$) and $\log(k_{b,h})$ for asymmetric double potential step chronoabsorptometric measurements ($\bullet$) on overpotential for Cyt c system at a tin oxide optically transparent cell. Coefficients of correlation are as follows: $k_{f,h}$ data, $r = 0.9976$; $k_{b,h}$ data $r = 0.9846$.[112]

CA responses obtained with different τ values to the digital simulated normalized working curves which depend on $\log(k_{b,h}t^{1/2}/D_O^{1/2})$, k_{bh} is obtained.

The DPS/CA technique is used together with SPS/CA to determine directly the forward and backward heterogeneous electron transfer kinetic parameters of Cyt c.[112] Figure 11 shows the dependence of $\log(k_{f,h})$ and $\log(k_{b,h})$ on η, from which $k_{s,h}^{\circ\prime} = 2.63 \times 10^{-5}$ cm sec^{-1} and $(1 - \alpha) = 0.101$. Other kinetic parameters of Cyt c obtained by SPS/CA and DPS/CA at different electrodes are summarized in Table II.[113]

C. Cyclic Potential Scan Steady-State Chronoabsorptometry

From the theory reported by Laviron,[114] the cathodic and anodic peak potential, E_{pc} and E_{pa}, respectively, for a redox process can be expressed as

[112] E. E. Bancroft, H. N. Blount, and F. M. Hawkridge, *in* "Spectroelectrochemical Determination of Heterogeneous Electron Transfer Kinetic Parameters" (K. M. Kadish, ed.), Advances in Chemistry Series No. 201, Chap. 2, American Chemical Society, Washington, D.C., 1982.

[113] E. E. Bancroft, H. N. Blount, and F. M. Hawkridge, *in* "Spectroelectrochemical Determination of Heterogeneous Electron Transfer Kinetic Parameters" (K. M. Kadish, ed.), Advances in Chemistry Series No. 201, Chap. 7. American Chemical Society, Washington, D.C., 1982.

[114] E. Laviron, *J. Electroanal. Chem.* **101**, 19 (1979).

TABLE II

HETEROGENEOUS ELECTRON TRANSFER KINETIC PARAMETERS FOR CYTOCHROME c SYSTEM

[Cyt c], μM	Electrode	Reaction[a]	Technique[b]	$k_{s,h}^{o\prime}$ (cm sec^{-1})	α	Ref.
77	Methyl viologen-modified gold minigrid	III + e^- = II	SPS/CA	1.0 (± 0.2)[c] $\times$ 10^{-5}	0.24 (± 0.03)[c]	110
77	Methyl viologen-modified gold minigrid	II = III + e^-	SPS/CA (PR)	5.8 (± 0.6) $\times$ 10^{-6}	0.74 (± 0.02)	113
103	Tin oxide	III + e^- = II	SPS/CA	6.8 (± 0.8) $\times$ 10^{-6}	0.32 (± 0.01)	110
103	Tin oxide	III + e^- = II	SPS/CA	6.3 (± 0.7) $\times$ 10^{-6}	0.28 (± 0.02)	113
103	Tin oxide	II = III + e^-	SPS/CA (PR)	4.9 (± 0.7) $\times$ 10^{-5}	0.95 (± 0.01)	113
94	Tin oxide	III + e^- = II	SPS/CA	4.4 (± 0.2) $\times$ 10^{-6}	0.27 (± 0.01)	112
94	Tin oxide	II = III + e^-	ADPS/CA	2.6 (± 0.2) $\times$ 10^{-5}	0.90 (± 0.01)	112

[a] Solution at pH 7.00 (0.07 M, phosphate buffer), containing 0.1 M NaCl. The reaction III + e^- = II indicates reduction of ferricytochrome c; II = III + e^- indicates oxidation of ferrocytochrome c.

[b] PR, Prior reduction by exhaustive controlled-potential electrolysis.

[c] Standard deviations are given in parentheses.

TABLE III
RELATION OF $1/m$ AND $n\Delta E_p$

$1/m$	$n\Delta E_p$ (mV)	$1/m$	$n\Delta E_p$ (mV)
0.5	18.8	6	130
0.75	27	7	142.4
1	34.8	8	153.8
1.5	48.8	9	164
2	61.2	10	173.4
2.5	72.2	11	182
3	82.4	12	190
3.5	91.8	13	197.6
4	100.6	14	204.6
5	116.2	—	—

$$E_{pc} = E^{\circ\prime} - \frac{RT}{\alpha nF} \ln \left(\frac{\alpha}{|m|} \right) \tag{21}$$

$$E_{pa} = E^{\circ\prime} + \frac{RT}{(1 - \alpha)nF} \ln \left(\frac{1 - \alpha}{m} \right) \tag{22}$$

where

$$m = \frac{RT}{F} [(Ak_{s,h}^{\circ\prime})/(N\nu V)] \tag{23}$$

In Eq. (23), ν is the scan rate, A is the area of the working electrode, V is the volume of the thin-layer cavity, and other symbols have the usual meaning. Given the peak separation $\Delta E_p = E_{pa} - E_{pc}$, under the conditions of $n\Delta E_p > 200$ mV the formal heterogeneous electron transfer rate constant $k_{s,h}^{\circ\prime}$ can be determined from the following equation:

$$\log(k_{s,h}^{\circ\prime}A/V) = \alpha \log(1 - \alpha) + (1 - \alpha)\log \alpha - \log(RT/nF\nu) \\ - \alpha(1 - \alpha)nF\Delta E_p/(2.3RT) \tag{24}$$

When $n\Delta E_p < 200$ mV, $k_{s,h}^{\circ\prime}$ is obtained by

$$k_{s,h}^{\circ\prime} = \frac{n\nu F}{RT(1/m)} \left(\frac{V}{A} \right) \tag{25}$$

the values of $1/m$ can be obtained conveniently from Table III.

By cyclic potential scan steady-state chronoabsorptometry (CPS/SSCA) OTTLSET, a steady-state absorbance–cyclic potential scan (A–E) curve at a certain wavelength is recorded. The two inflection points for the oxidation and reduction processes in the A–E curve correspond to

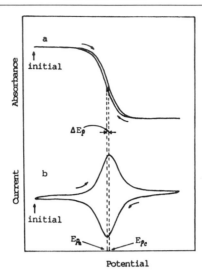

FIG. 12. Relation of the peak potentials determined in the $A-E$ curve (a) and in the cyclic voltammogram (b).

the anodic and cathodic peak potentials, E_{pa} and E_{pc}, respectively, in the cyclic voltammogram (see Fig. 12). The distance of these two inflection points is equal to the peak separation, ΔE_p.[115] From the ΔE_p value and Eq. (24) or (25) the $k^{o\prime}_{s,h}$ value can be easily determined. This method is very useful in determination of the kinetic parameters of proteins at different CMEs.

Figure 13 shows the steady-state $A-E$ curves for the redox reaction of Cyt c at a polypyrrole–methylene blue CME.[116] From Fig. 13, the redox processes of Cyt c, which cannot be detected by cyclic voltammetry owing to the large background current, can be directly monitored. The values of ΔE_p obtained from the inflection points is about 60 mV. Because $n\Delta E_p < 200$ mV, Eq. (25) is used to determine $k^{o\prime}_{s,h}$, which gives a value of 5.4×10^{-4} cm sec^{-1}.[116] The determination of kinetic parameters for other redox proteins by CPS/SSCA at various types of CMEs is discussed in the following section.

V. Electrochemical Reactions of Redox Proteins at Chemically Modified Electrodes

Considerable interest has been focused on obtaining rapid heterogeneous electron transfer reactions of redox proteins at CMEs. Certain

[115] Y. Gui and T. Kuwana, *J. Electroanal. Chem.* **226**, 199 (1987).
[116] W. Zhang, S. Song, and S. Dong, *J. Inorg. Biochem.* **40**, 189 (1990).

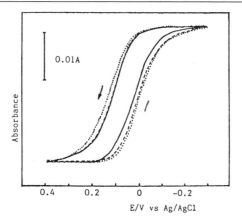

FIG. 13. *A–E* response of Cyt *c* at a polypyrrole–methylene blue film modified Pt gauze electrode in OTTLEC. Sample, 1.0 mg/ml Cyt *c*; supporting electrolyte, 0.2 *M* KNO₃ in a phosphate buffer solution at pH 7.0; scan rate, 1.0 mV/sec; wavelength, 550 nm. Solid curve was recorded at first potential scan, dashed curve at second potential scan, and dotted curve after 7 hr.[116]

organic compounds and their polymers serve as promoters when immobilized onto electrode surfaces. The reactions of Hb, Mb, Cyt *c*, etc., have been studied by thin-layer cyclic voltammetry and UV–VIS spectroelectrochemistry. The proteins exhibit very poor electrochemical response at solid electrodes, especially Hb and Mb, and as a result only a few papers pertaining to those two proteins have been published. Dong *et al.* have studied systematically the direct electrochemistry of Hb, Mb, and Cyt *c* at different CME surfaces with phenothiazines and other polymers as the promoters,[117,118] and these results are summarized in this section.

A. *Phenothiazine Dye Modified Electrodes*

Among the organic compounds tested, methylene blue (MB) is one of the most promising electrode modifiers for facilitating electron transfer to Hb, Mb, and Cyt *c*. The electron transfer reactions of these proteins at MB CMEs are quasi-reversible, and the thermodynamic and kinetic parameters have been determined. Figure 14 shows the thin-layer absorption spectra of Hb at a Pt CME at different applied potentials. Hemoglobin undergoes reduction from methemoglobin to ferrohemoglobin in the potential range from 0 to −0.2 V, as evidenced by the increase of the absorbance peak at 550 nm and decrease of peaks at 490 and 625 nm (Fig. 14A). A shift in the applied potential to +0.10 V leads to the reverse changes in

[117] S. Dong, "The First World Congress on Biosensors," May 2–4, p. 198, Singapore, 1990.
[118] S. Dong, *Denki Kagaku Oyobi Kogyo Butsuri Kagaku* **59**, 664 (1991).

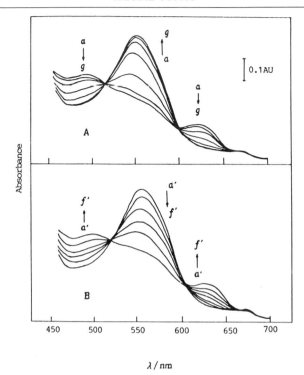

Fig. 14. Spectra of Hb (~1.2 mg/ml) at a Pt electrode modified with methylene blue in a thin-layer cell. (A) Reduction spectra at potentials of (a) 0.0, (b) −0.10, (c) −0.14, (d) −0.16, (e) −0.17, (f) −0.18, and (g) −0.20 V. (B) Oxidation spectra of reduced Hb at potentials of (a') −0.20, (b') −0.10, (c') −0.05, (d') 0.0, (e') +0.05, and (f') +0.10 V. Each spectrum was recorded after the potential was applied for 2 min.[19]

the absorption spectrum (Fig. 14B), as a result of the reoxidation of the protein. Based on the OTTLSET method and Eq. (6), the formal redox potential $E^{\circ\prime}$ was determined to be −0.16 V, and an electron transfer number of 4 was found.[19] In the same manner, the $E^{\circ\prime}$ values for Mb and Cyt c are also determined,[117] and these are listed in Table IV.

The $A–E$ curves for Hb, Mb, and Cyt c at MB CME and at different scan rates show that the reduction processes of Hb and Mb are very fast, whereas the oxidation processes are relatively slower.[119] It should also be noted that the redox reaction of Cyt c at a MB CME is more rapid than that of either Mb or Hb. The kinetic parameters of these proteins as determined by the CPS/SSCA technique are summarized in Table IV.

[119] S. Dong and S. Song, *Acta Chim. Sin. (Engl. Ed.)* **49,** 493 (1991).

TABLE IV
REACTION PARAMETERS OF REDOX PROTEINS ON METHYLENE
BLUE CHEMICALLY MODIFIED ELECTRODE

Parameter	Redox protein[a]		
	Cyt c	Mb	Hb
α	0.5	0.26	0.12
$E^{\circ\prime}$ (V)[b]	+0.13	−0.158	−0.16
ΔE_p (V)	0.07	0.20	0.23
$1/m$	2.4	—	—
$k_{s,h}^{\circ\prime}$ (cm sec^{-1})	4.6×10^{-4}	7.8×10^{-5}	1.1×10^{-5}

[a] Concentrations of Cyt c, Mb, and Hb were 1.0, 0.45, and 2.0 mg/ml, respectively, in phosphate buffer solution at pH 7.0, containing 0.2 M KNO$_3$.
[b] The values of ΔE_p and $k_{s,h}^{\circ\prime}$ are obtained at a scan rate of 1.0 mV/sec.

Other phenothiazine dyes, such as methylene green (MG), brilliant cresyl blue (BCB), janus green (JG),[16] toluidine blue (TB), and azure A (AA), can also be immobilized on Pt electrode surfaces by adsorption and the cyclic potential scan method used to prepare MB CMEs.[117] The MG, BCB, JG, TB, and AA CMEs were all found to facilitate effectively the electrochemical reactions of redox proteins,[21-24,120-123] and the modified electrodes were stable. For example, rapid electron transfer is observed between Mb and a TB CME, resulting in the absorption responses shown in Fig. 15 at different applied potentials. The redox reaction of Hb on a MG CME proceeds at a moderate speed even after 48 hr (see curve C in Fig. 16). Table V lists values for $E^{\circ\prime}$, n, and kinetic parameters for the redox reactions of Hb, Mb, and Cyt c at these CMEs.

B. Doped Conducting Polymer Modified Electrodes

To increase the stability of the phenothiazine dye CMEs, investigations of the electrochemical reactions of redox proteins at polymer CMEs have been undertaken. One example of such a study is the heterogeneous redox reaction of Cyt c at an electrochemically polymerized polypyrrole–methylene blue (PPy–MB) film CME.[23] Figure 17 shows the cyclic voltammetric response that occurs during the preparation of this CME by potential

[120] S. Dong and Q. Chu, *Electroanalysis* **5**, 135 (1993).
[121] S. Dong and Q. Chu, *Chin. J. Chem.* **11**, 12 (1993).
[122] S. Dong and Q. Chu, *Acta Chim. Sin.* **50**, 589 (1992).
[123] Y. Zhu and S. Dong, *Bioelectrochem. Bioenerg.* **26**, 351 (1991).

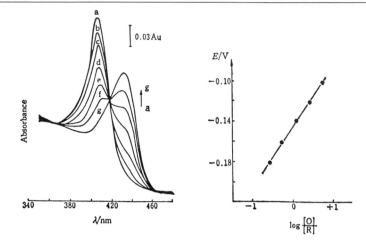

FIG. 15. Spectra of Mb (0.5 mg/ml) at a poly-TB film modified Pt gauze electrode in a thin-layer cell. Applied potentials (V, versus Ag/AgCl in saturated KCl) were as follows: (a) +0.40, (b) −0.10, (c) −0.12, (d) −0.14, (e) −0.16, (f) −0.18, and (g) −0.30. The solution contained 0.2 M KNO$_3$ and 50 mM phosphate buffer, pH 7.0.[121]

scanning. It can be seen that as the pyrrole undergoes polymerization, MB is also polymerized, as evidenced by the increase of the redox couple at about 0 V.[116] Thus, MB is doped into the polypyrrole film during the potential scans. The PPy–MB CME obtained by this method is more

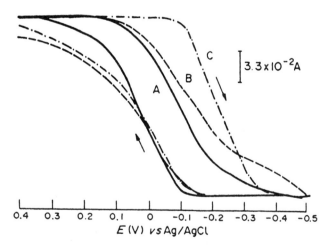

FIG. 16. Cyclic voltabsorptometry A–E curves for 2 mg/ml Hb at a MG CME. Scan rate, 1.0 mV/sec; monitor wavelength, 405 nm; solution, 0.1 M phosphate buffer (pH 7.0) plus 0.2 M KNO$_3$. (A) First cycle, (B) sixth cycle, (C) after 48 hr.[22]

TABLE V

REACTION PARAMETERS OF REDOX PROTEINS ON DIFFERENT CHEMICALLY
MODIFIED ELECTRODES[a]

Protein	Promoter	Modification method	$E^{o\prime}$ (V)	n	$k^{o\prime}_{s,h}$ (cm sec^{-1})	Ref.
Cyt c	JG	Cyclic potential scan	+0.261	1	4.2×10^{-4}	6
Hb	MG	Adsorption	—	—	2.0×10^{-4}	22
Hb	MG	Cyclic potential scan	−0.064	1.3	8.8×10^{-5}	22
Hb	BCB	Adsorption	—	—	3.7×10^{-7}	21
Hb	BCB	Cyclic potential scan	−0.067	1.4	2.0×10^{-7}	21
Hb	AA	Cyclic potential scan	—	1.8	3.5×10^{-6}	120
Mb	BCB	Cyclic potential scan	−0.169	1.1	5.6×10^{-4}	24
Mb	TB	Cyclic potential scan	−0.179	0.99	1.1×10^{-4}	121
Mb	AA	Cyclic potential scan	−0.170	1.1	1.7×10^{-4}	122

[a] All the solution conditions were pH 7.0 phosphate buffer (0.1 M) containing 0.2 M KNO$_3$; values of $k^{o\prime}_{s,h}$ were obtained at a scan rate of 1.0 mV/sec.

stable and effective than the MB CME. From the excellent A–E curve, it can be determined that Cyt c undergoes nearly reversible electron transfer at the PPy–MB CME with a heterogeneous electron transfer rate constant of $k^{o\prime}_{s,h} = 4.5 \times 10^{-4}$ cm sec^{-1}. A CME based on a MG-doped

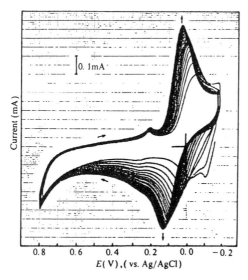

FIG. 17. Cyclic voltammogram of the electrochemical polymerization process of PPy–MB CME. The solution contained 2.9×10^{-3} M pyrrole plus 1.0×10^{-3} M MB; supporting electrolyte, 0.5 M KNO$_3$; scan rate, 100 mV/sec, 30 cycles.[116]

polypyrrole film has also been reported,[124] which exhibits a rapid reaction with Hb ($k_{s,h}^{o\prime}$ = 1.8 × 10^{-5} cm sec^{-1}).

VI. Conclusions

Since the establishment of spectroelectrochemistry[125] very little effort has been devoted to the direct electrochemistry of redox proteins. Although many thermodynamic and kinetic parameters can be determined by UV–VIS spectroelectrochemistry, the electrochemical reaction mechanisms for redox proteins are not well understood. New techniques and new theoretical treatments are needed to address this issue. Moreover, most attention has been placed on relatively simple electron transfer proteins; to date no one has reported the direct electrochemistry of a more complex system (e.g., a redox enzyme system) which unequivocally undergoes electron transfer to (or from) its active site. Considerable experimental work is needed to develop more fully spectroelectrochemical methods for biological systems.

Acknowledgments

We acknowledge support by the National Natural Science Foundation of China for much of our own research described herein. T. M. C. acknowledges the support of the National Institutes of Health.

[124] Y. Zhu, W. Zhang, and S. Dong, *Chin. J. Appl. Chem.* (*Yingyong Huaxue*) **9**, 51 (1992).
[125] T. Kuwana, R. K. Darlington, and D. W. Leedy, *Anal. Chem.* **36**, 2023 (1964).

[29] Digital Imaging Spectroscopy for Massively Parallel Screening of Mutants

By DOUGLAS C. YOUVAN, ELLEN GOLDMAN, SIMON DELAGRAVE, and MARY M. YANG

Introduction

New combinatorial mutagenesis methods in molecular genetics make it possible to generate simultaneously millions of altered proteins in a single library of mutants. In the study of pigment–protein complexes, this advance in biology necessitated the invention of an instrument capable of recording spectra in a massively parallel fashion, much like the human eye. In contrast to the trichromic eye,[1,2] such an instrument should be quantitative, cover a wider range of wavelengths, and have a large number of spectral bands.

Imaging spectroscopy is defined as the coprocessing of both spatial and spectral information. Such data are usually three-dimensional, consisting of two spatial dimensions (i.e., the x and y coordinates of picture elements within a scene) correlated with one dimension of spectral information (e.g., the absorption spectrum of each pixel). A monochromator can be considered as an analog imaging spectrophotometer if the entrance slit is physically moved over a scene to generate the second spatial dimension. Later, this time course of convoluted data can be used to rebuild images containing spectral data. The scanning swath of a monochromator has been employed in satellite imagery and remote sensing.[3]

More recently, digital imaging spectroscopy (DIS)* has combined the techniques of digital image processing and optical spectroscopy.[4–6] Whereas conventional image processing[7,8] considers only the gray scale of two-dimensional images obtained under one type of illumination (e.g., ambient "white" light), DIS captures a sequence of monochromatic images as a function of a spectral parameter (such as the wavelength of light used to illuminate a target). If the target is transparent, an absorption spectrum can be obtained for each pixel in the scene. If the target is opaque, a reflectance spectrum can be obtained for each pixel. Digital imaging spectrophotometers can also be configured to obtain fluorescence emission images. Furthermore, absorption and fluorescence data can be combined digitally to generate quantum yield data. DIS may also be used to gather information about light scatter.

[1] D. Marr, "Vision." Freeman, New York, 1982.
[2] K. Nassau, "The Physics and Chemistry of Color." Wiley, New York, 1983.
[3] T. M. Lillesand and R. W. Kiefer, "Remote Sensing and Image Interpretation." Wiley, New York, 1987.
* DIS is a context-dependent acronym for either digital imaging spectrophotometer or digital imaging spectroscopy. In earlier publications, DIS is synonymous with FAS, an acronym for filter array spectrophotomer.
[4] M. M. Yang and D. C. Youvan, Bio/Technology 6, 939 (1988).
[5] A. P. Arkin, E. R. Goldman, S. J. Robles, W. Coleman, C. A. Goddard, M. M. Yang, and D. C. Youvan, Bio/Technology 8, 746 (1990.
[6] A. P. Arkin and D. C. Youvan, in "The Photosynthetic Reaction Center" (J. Deisenhofer and J. R. Norris, eds.), Vol. 1, p. 133. Academic Press, San Diego, 1993.
[7] R. C. Gonzales and P. Wintz, "Digital Image Processing," Addison-Wesley, Reading, Massachusetts, 1987.
[8] J. C. Simon, "From Pixels to Features." North-Holland, New York, 1989.

Hardware

Data from three different configurations of digital imaging spectrophotometers (designated DIS1, DIS2, and DIS3) are presented in this chapter.
DIS1 was a prototype instrument designed for the acquisition of spectral
information from a single petri dish in a transmission mode. DIS2 was
designed for high throughput screening of samples, imaging up to 25 petri
plates simultaneously. DIS3 is a commercial, turnkey system designed
for general use on a single petri dish in either a transmission or reflectance mode.

Detector

Driven by the need for more efficient and accurate means to measure
the faint objects in the sky, investigators developed charge-coupled device
(CCD) cameras for astronomical studies in the early 1970s.[9] These solid
state detectors have multiple advantages[10] over earlier tube-type detectors
used in older video cameras. Advantages include (1) a greater dynamic
range of light sensitivity, (2) linearity in photon conversion to electronic
signal, (3) geometric stability, and (4) the ability to integrate photons over
an extended period of time, yielding greater sensitivity in low light applications.

The dynamic range and spatial resolution of the CCDs used in DIS
vary from detectors with 12-bit digitization of 200,000 pixels (DIS1) to
12-bit 4,000,000 pixel detectors (DIS2) to 16-bit 90,000 pixel detectors
(DIS3). The 12-bit pixel data correspond to gray scale values of 0–4095,
and 16-bit pixel data correspond to gray scale values of 0–65535. Readout
time is dependent on the digitization circuitry of the camera and computer
interface. DIS1 uses a GPIB interface and transfers the image in 6 sec,
whereas DIS2 uses a memory-mapped VME interface and transfers images
in 8 sec. All three detectors are Peltier cooled to lower the dark count.
In all applications to date,[6] these cameras have been limited by photon
shot noise, which is unavoidable due to the quantum nature of light.

Light Source

A uniform light source is important for maintaining radiometric calibration across the target. One can always perform a flat field correction by
dividing the image by a "blank" background image and then rescaling
the dividend to produce a new image. However, for severe nonuniformity,

[9] J. Kristian and M. Blouke, *Sci. Am.* **247,** 67 (1982).
[10] P. M. Epperson, J. V. Sweedler, M. B. Denton, and G. R. Sims, *Opt. Eng.* **26,** 715 (1987).

this technique substantially reduces the effective dynamic range of the camera. The digital imaging spectrophotometers described herein all utilize multiple scattering off Lambertian surfaces as a means to approach perfect uniformity. We are able to obtain a uniformity of approximately $\pm 1\%$ over the exit port of the baffled 12-inch integrating sphere shown in Fig. 1.* Using the much larger exit port in the hemisphere shown in Fig. 2 decreases the uniformity to approximately $\pm 15\%$. Fortunately, this nonuniformity is monotonic and of very low frequency across the field of view.

In most DIS applications, monochromatic light is required. Using a 350 W projection system with 10 nm bandpass Fabry–Perot filters,[11] approximately 5 mW of light enters the integrating sphere. Alternatively, an 0.175 m monochromator can be used with a 100 W lamp to achieve about one-third of this power. Both configurations are acceptable since the 12-inch sphere has a throughput[12] of approximately 30%. This leads to reasonable f-stop settings and integration times for the CCD camera over the spectral range from 410 to 950 nm.

Computer

A computer is required for instrument control, data processing, and display. DIS1 and DIS2 utilize Silicon Graphics UNIX workstations, whereas DIS3 is interfaced to an IBM PC clone using software running under Microsoft Windows. Image display quality and speed differences on the three systems become noticeable only when the size of the image increases substantially. In addition to the actual program, much of the "look and feel" and ease of use of the software interface is dependent on the operating system of the computer. Graphics programming on sophisticated workstations may be simpler owing to the increased number of color bitplanes available. However, for DIS, the compute speed of personal computers (for most applications) are competitive with high performance workstations. Image processing tasks such as feature extractions are the most central processing unit (CPU)-intensive aspect of DIS. On small format images, PCs are now quite adequate for this job.

Algorithms

Although feature extractions are well known in image processing and the Beer–Lambert law is integral to many spectroscopic applications, the

---

* Figures and legends appear at the end of this chapter.
[11] H. A. Macleod, "Thin-Film Optical Filters." Macmillan, New York, 1986.
[12] D. G. Goebel, *Appl. Opt.* **6**, 125 (1967).

combination of these two algorithms is unique to DIS. Furthermore, DIS acquires more data than can be managed by standard plotting procedures. This has necessitated the development of novel sorting algorithms and contour plots within a user interface that allows the DIS operator to review massive amounts of data quickly.

Feature Extraction

Machines must be programmed to "see" discrete objects or features within a scene. A feature extraction program employs an algorithm to list all of the pixels which define an object. This is an important part of DIS because the gray scale values of pixels within a single feature will be used to determine its spectral properties (e.g., absorbance A). In complex images, this process may be only partially automated.

Feature extraction begins with the application of a gamma function to an unprocessed image acquired under conditions (e.g., broadband blue light) that enhance the contrast of the target. Figure 3A shows an unprocessed 8-bit image (0–255 gray scale) of a petri dish with hundreds of bacterial colonies. A gamma function can be applied by the operator to reduce the gray scale palette of the image to only two values: 0 (black) and 255 (white). Figure 3B has been further processed by manual erasure of poorly spread colonies and other artifacts in the binary image. A recursive algorithm is applied to identify the perimeter pixels of the white colony against the black background. The feature extractor compiles a list of connected edge points (i.e., white pixels with at least one black neighbor). All pixels within this ring of edge pixels are collected into one list per feature. The gray scale values of the enclosed pixels from the unprocessed image can then be combined to yield an average gray scale value for the feature.[6]

Beer–Lambert Law

In analogy with a double-beam absorption spectrophotometer, one "blank" feature is positioned at an x,y coordinate that is characteristic of the background. This blank is required for radiometric calibration of DIS because of the wavelength-dependent responses of the camera, monochromator, and light source. For a 16-bit camera, the gray scale ranges from 0 to 65,535. The percent transmittance ($\%T$) of a feature can be calculated according its average gray scale value (g_f) versus the average gray scale of the pixels in the blank feature (g_b):

$$\%T = (g_f/g_b) \times 100 \tag{1}$$

Neglecting photon shot noise and dark counts, the lowest theoretical absorbance (A) measurable on a 16-bit camera can be calculated using the Beer–Lambert law:

$$A = \log(g_b/g_f)$$
$$A = \log(65,536/65,535) = 6.7 \times 10^{-6} \tag{2}$$

The maximum absorbance is

$$A = \log(65,536/1) = 4.8$$

In practice, photon shot noise and variability in dark counts represent less than 1 gray scale unit on a 12-bit scale if 100 pixels are averaged (square-root-of-N-law). Experimentally, we observe noise at the level of approximately 10^{-4} A at 0.0 A and 10^{-3} A at 1.0 A.

Spectral Sorting

Figure 4 is a screen dump taken from DIS3. The image displayed is that of a petri dish with spots from eight different colored ink pens. The conventional spectrum of the highlighted feature is shown in the plot window below the image, and a sorted contour map is displayed on the right. Each row of the contour map represents the absorption spectrum of a single feature in the image. If the contour map had not been sorted, it would appear scrambled as in Fig. 5A. Spectral sorting groups features with similar spectra into adjacent rows. In this pedagogical test target, one sees that the sorting algorithms groups and differentiates the eight different ink colors.

A variety of sorting algorithms have been encoded in DIS1 and DIS3 to facilitate the construction of contour plots that segregate features of specific interest to the investigator. These sorts can be applied in a variety of combinations at specified wavelengths or at selected wavelength ranges to sort the data in a manner that produces a logical segregation of the spectra into different classes. Depending on the sorting algorithm, the resultant sort and contour map may differ depending on whether the individual spectra are displayed in either a full scale or absolute mode.

Example of Mutant Screening

Theoretical[13-15] and experimental methods[16-19] for addressing the protein design problem from a parallel rather than a serial point of view have been developed. Parallel methods rely on the simultaneous synthesis of millions of different molecules. What is important in this approach is the ability to retrieve a single active molecule with desired properties. Examples of new parallel techniques that can be used to generate and isolate interesting proteins in extremely complex mixtures include combinatorial cassette mutagenesis, bacteriophage display, "panning," and digital imaging spectroscopy.

Rather than relying on purely random methods in the parallel search of "sequence space," it is advantageous to implement algorithms that converge on proteins with desired properties. The first experimental implementation of such technology used a model bacterial system expressing chromogenic proteins, i.e., mutagenized light-harvesting polypeptides from *Rhodobacter capsulatus* were assayed using DIS technology.[20-22]

Digital Imaging Spectroscopy on Pigment–Protein Complexes

Light-harvesting (LH) proteins are chromogenic because they bind bacteriochlorophyll (BChl) and carotenoid pigments. The LHII pigment–protein complex is modeled[23,24] as having a dimeric pair of BChl molecules absorbing at 865 nm and a monomer BChl absorbing at 800 nm. The LHI complex has only a dimeric band which is further red-shifted to 875 nm. Free BChl in the bacterial membrane absorbs at 760 nm. As such, these absorption bands are noninvasive spectroscopic reporters for the assembly of LH complexes.

[13] D. C. Youvan, A. P. Arkin, and M. M. Yang, *in* "Parallel Problem Solving from Nature 2" (R. Maenner and B. Manderick, eds.), p. 401. Elsevier, Amsterdam, 1992.

[14] A. P. Arkin and D. C. Youvan, *Proc. Natl. Acad. Sci. U.S.A.* **89,** 7811 (1992).

[15] A. P. Arkin and D. C. Youvan, *Bio/Technology* **10,** 297 (1992).

[16] G. P. Smith, *Science* **228,** 1315 (1985).

[17] A. R. Oliphant, A. L. Nussbaum, and K. Struhl, *Gene* **44,** 177 (1986).

[18] J. F. Reidhaar-Olson and R. T. Sauer, *Science* **241,** 53 (1988).

[19] J. F. Reidhaar-Olson, J. U. Bowie, R. M. Breyer, J. C. Hu, K. L. Knight, W. A. Lim, M. C. Mossing, D. A. Parsell, K. R. Shoemaker, and R. T. Sauer, this series, Vol. 208, p. 564.

[20] E. R. Goldman and D. C. Youvan, *Bio/Technology* **10,** 1557 (1992).

[21] S. Delagrave, E. R. Goldman, and D. C. Youvan, *Protein Eng.* **6,** 327 (1993).

[22] S. Delagrave and D. C. Youvan, *Bio/Technology* **11,** 1548 (1993).

[23] H. Zuber, *Trends Biochem. Sci.* **11,** 414 (1986).

[24] H. Zuber, *in* "Molecular Biology of Membrane-Bound Complexes in Phototrophic Bacteria" (G. Drews and E. A. Dawes, eds.), p. 161. Plenum, New York, 1990.

Figure 6 shows the results of combinatorial cassette mutagenesis of sequences flanking the dimeric BChl binding site of LHII. This *Rhodobacter capsulatus* expression system has been engineered to express only LHII.[25-27] The genes for the bacterial reaction center and the LHI antenna have been inactivated through genetic procedures. This makes LHII the only significant absorber in the near-infrared. DIS has been used in this experiment to assay thousands of genetically engineered variants of LHII that were designed by mixing (by a combinatorial gene synthesis) LHI and LHII sequences from 29 different species. Remarkably, this phylogenetic mixture[20] results in a 6% throughput of mutants that assemble LH. The spectra are sorted into the upper rows of the contour map (Fig. 6B). Spectroscopic diversity in the mutants is further examined by DIS in Fig. 6D, where absorption spectra characteristic of both LHI and LHII are observed. Figure 7 shows a more conventional representation of the absorption spectra. However, it is clear that the DIS color contour plots are superior to conventional representations for highlighting subtle differences between these mutants.

Combinatorial Optimization

As more amino acid residues are mutagenized, the probability of observing a mutant that assembles a functional protein (e.g., LH antenna) drops off exponentially.[22] Hence, greater screening capacities are required. The DIS2 configuration is capable of imaging 25 petri dishes ($>10^5$ mutants) for infrequent "positives." Because of photographic resolution and page size, we have displayed a single plate from such an array in Fig. 8. Colonies assembling LH antennas are detected using monochromatic illumination at 855 nm. Figure 9 shows data from the DIS2 apparatus taken in a fluorescence emission mode. In this case, positive colonies are highly fluorescent because the LH antenna is uncoupled from a photochemical trap. However, it is possible to have decay pathways other than fluorescence. Figure 10 shows an apparent quantum yield image, wherein the fluorescence image has been normalized by the absorbance image. Colonies which appear less intense in Fig. 10 than in Fig. 9 possibly have nonradiative decay pathways. This technique could be used to scan millions of *de novo* designed pigment–protein complexes for signs of charge separation.

The frequency of positive mutants can be increased in subsequent

[25] D. C. Youvan and S. Ismail, *Proc. Natl. Acad. Sci. U.S.A.* **82,** 63 (1985).
[26] D. C. Youvan, S. Ismail, and E. J. Bylina, *Gene* **38,** 19 (1985).
[27] D. C. Youvan, *Trends Biochem. Sci.* **16,** 145 (1991).

rounds of mutagenesis by a combinatorial optimization procedure called exponential ensemble mutagenesis (EEM). Using LHII as a model system and DIS as an assay, EEM has been shown to increase the throughput of positive mutants 10^7-fold over random combinatorial cassette mutagenesis.[22] DIS will remain our most important tool in evaluating the algorithms used in generating such mutant ensembles.

Prospectus

The applications of DIS shown in this chapter have illustrated the differentiation of mutant colonies based either on ground state absorption spectra or fluorescence emission spectra. Clearly, these first experiments on chlorophyll-based systems are generalizable to other biological targets producing chromogenic natural substances (e.g., astoxyanthin production by yeast). Extrinsic dye indicators might also be coupled with DIS analysis as microbiologists continue to develop chromogenic assays. Some of these assays are based on growth rates (rather than color) and are therefore amenable to simple analysis utilizing differences in light scattering. DIS analyses could also be used to eliminate some of the artifacts encountered in conventional microtiter tray readers that are sensitive to variability in sample color or scattering (e.g., enzyme-linked immunosorbent assay, ELISA). Finally, it is possible that DIS will soon be interfaced with robotic colony pickers and microtiter tray manipulators so as to increase the throughput and reliability of techniques requiring massively parallel screening of samples.

Acknowledgments

This work was supported in part by National Institutes of Health GM42645 and U.S. Department of Energy (DOE) DE-FG02-90ER20019. Hardware for the DIS2 was supported under a DOE University Research Instrumentation Award (DE-FG05-91ER79031). E.G. was the recipient of an NIH Biotechnology Training Grant Award. S.D. is the recipient of a graduate research fellowship from Fonds FCAR.

FIG. 1. DIS3 integrating sphere. *Top:* Microtiter tray (transmission mode) resting on the glass exit port of the top hemisphere of an integrating sphere. The operator has opened the sphere to expose the internal reflective surface. Light enters the integrating sphere from the bottom; however, the entrance port is not visible from an azimuthal mounted camera because of an internal baffle. In the photograph, the internal baffle is visible behind the operator's right hand. *Bottom:* View through the exit window of the sphere with a petri dish resting on the internal baffle (reflectance mode). The top surface of the baffle has been covered with a circular black mat. Opaque yeast colonies are uniformly illuminated by light scattered from the top hemisphere.

FIG. 2. DIS2 hemispherical light source. Unlike the small integrating spheres used in DIS1 and DIS3, the DIS2 light source has been designed to uniformly illuminate an 18 inch by 18 inch aperture. *Bottom:* Design of the light soure. A 36-inch aluminum hemisphere (A) and top cover (B) are coated to attain high light throughput and uniformity across the translucent window (C). Collimated light is projected through the entrance port (D) and focused on a small area of the hemisphere (E). The device is baffled such that none of the light from the entance port or illuminated spot can exit the window without at least one internal Lambertian scattering event. Uniformity across the window is better than ±15%, which is easily corrected by a digital division against a blank image. *Top:* Illumination of 25 petri dishes on the 324-square inch window.

FIG. 3. DIS1 feature extraction. (A) Unprocessed 8-bit image of a petri dish with colonies of *Rhodobacter capsulatus* imaged in a transmission mode. (B) Image after digital processing to yield a binary (black and white) image that can be extracted by an automated routine to yield 210 different features. A comparison of (A) and (B) shows that the operator has erased certain groups of colonies that are not separable into individual features. In other cases, colonies that were overlapping in the unprocessed image have been manually separated by a tool provided in the user interface.

FIG. 4. (see color plates following) DIS3 user interface. A petri dish with eight different colored spots has been used as a pedagogical test target for explaining the interaction between the three widows shown here and the images in Fig. 5. A true color image of the target has been displayed in the upper left window by combining three monochromatic images: red (650 nm), green (550 nm), and blue (450 nm). The ink spots in the actual image appear as yellow, orange, red, blue, purple, green, brown, and black. The current feature of interest is marked by a cross-hair and is located at the bottom of the image. The spectrum of this green spot is displayed in the lower left window which shows significant absorption in the blue and red regions of the spectrum. The color contour map (right window) displays all spectra from 410 to 750 nm. Each row represents one feature or spot and each column corresponds to a different wavelength. The color bar at the bottom of this window encodes absorbance according to a pseudocolor scheme, whereby low absorbance is mapped to black and high absorbance is mapped to white. The red bar on the left of the contour map indicates the feature under selection, in this case the highlighted green spot. The mouse of the computer can be used to point to any feature in the image or any row in the contour map with a real-time update of the spectrum in the lower left window. Moving the feature selection cross-hair causes the red bar to move to the correct row, whereas moving the red bar causes the corresponding feature to be marked.

FIG. 5. (see color plates following) DIS3 spectral sorting. Color contour plots of data from Fig. 4 are displayed in three panels. (A) Original unsorted spectra. Because the positions of colored spots in the test target are random this first image appears disordered. (B) Same set of spectra after sorting according to spectral similarity. Spectra in (A) and (B) are shown in "full scale" mode in which each row has been scaled so that the lowest and highest absorbance values within each feature are scaled from black to white, respectively, according to the color bar in Fig. 4. (C) Image identical to (B) except displayed in "absolute" mode, wherein the contour plot has been scaled so that the lowest and highest absorbance values found in the entire data set are scaled from black to white. In all three plots, the similarity sort has resulted in the grouping of eight different classes of rows. The red marker to the left of each panel corresponds to the same feature highlighted in Fig. 4.

FIG. 6. (see color plates following) DIS1 analysis of combinatorial cassette mutants. (A) Unprocessed 8-bit image of colonies imaged after mutagenesis of the light-harvesting II pigment–protein complex. (B) Diversity of the library as indicated by different absorption spectra. Similarity sorting has placed 6% of the mutants in the top of the contour plot. Representative colonies from three distinct spectral classes have been picked, repurified, and spotted in (C). In the corresponding DIS contour map (D), mutants at the bottom show maximal absorption at 760 nm, characteristic of protein-free bacteriochlorophyll a. Mutants in the middle section of the contour plot show pseudo-LHII spectra, whereas mutants at the top of the contour plot show pseudo-LHI spectra. (Reprinted with permission of *Bio/Technology;* from Goldman and Youvan, 1992.[20])

FIG. 7. DIS1 "tile mode" showing the absorption spectra of the 210 mutants displayed in Fig. 6B. Each tile displays a ground state absorption spectrum from 700 to 930 nm (left to right) with full scale deflection on the y axis. The top row includes mutants with LHI and LHII type spectra.

FIG. 8. DIS2 absorbance image of a single petri dish. *Top:* Colonies of *Rb. capsulatus* that have been optically modified by combinatorial mutagenesis of the LHII pigment–protein. The image was recorded in a transmission mode at 855 nm (5 nm bandpass) by using a Fabry–Perot filter at the entrance port of the hemisphere. Mutants producing large amounts of LHII absorb at 855 nm and appear dark in the image. The boxed region of the plate has been expanded (*bottom*) to show one dark colony and four light colonies.

FIG. 9. DIS2 fluorescence image of a single petri dish. The plate is identical to the one in Fig. 8, but the instrument has been reconfigured for fluorescence imaging. Broadband blue-green light has been used for illumination, and the camera lens has been covered with an 865 nm long-pass filter. LHII is normally fluorescent in this bacterial strain because the reaction center (photochemical trap) has been genetically deleted. The same five colonies have been expanded below; note that the colony on the right is highly fluorescent.

FIG. 10. (see color plates following) DIS2 quantum yield image. An apparent quantum yield image for fluorescence has been generated by digital division of the fluorescence image (Fig. 9) by the absorbance image (Fig. 8) after inversion of the gray scale. This dividend was then rescaled and pseudocolored according to the color bar shown at lower right. In the expanded region, one mutant colony shows a high apparent quantum yield for fluorescence. This colony absorbs strongly in Fig. 8 and fluoresces intensely in Fig. 9. Close inspection of other colonies in Figs. 8, 9, and 10 reveals that high absorbance is not always correlated with high fluorescence. This leads to some interesting variations in the apparent quantum yield image.

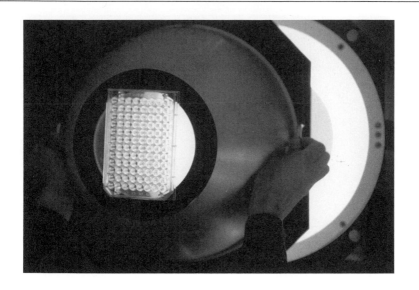

FIGURE 1

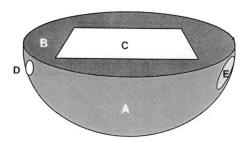

FIGURE 2

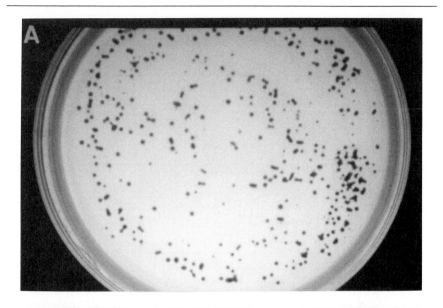

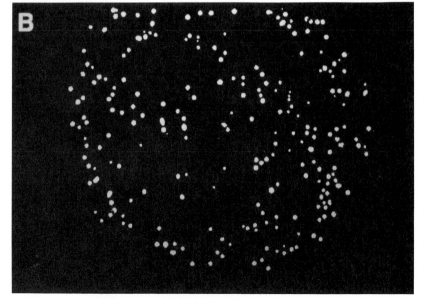

FIGURE 3

A
B
S
O
R
B
A
N
C
E

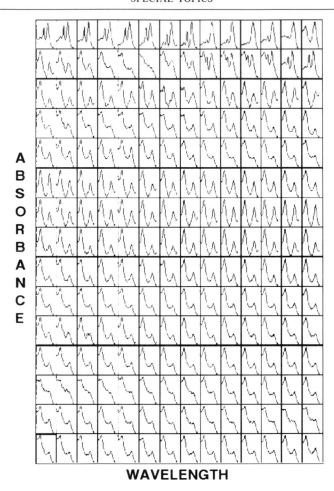

WAVELENGTH

FIGURE 7

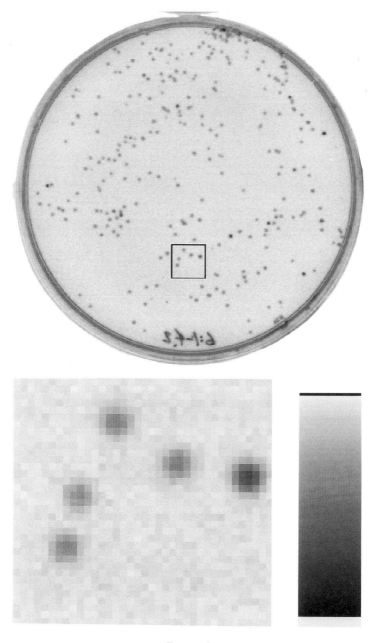

FIGURE 8

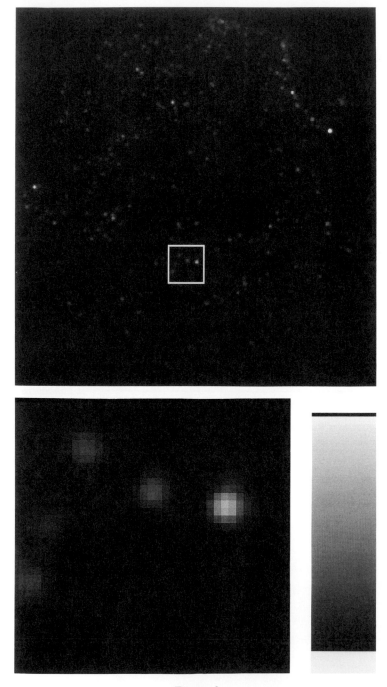

FIGURE 9

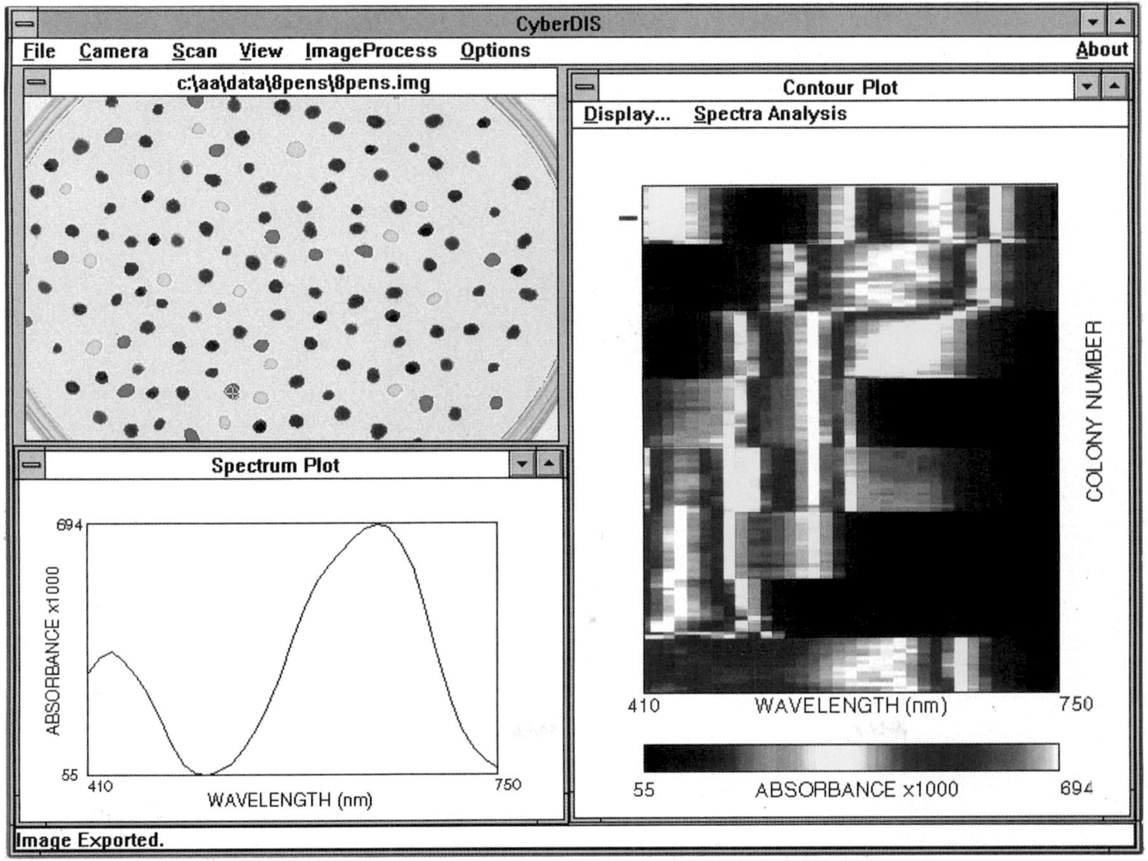
CyberDIS
File
Camera
Scan
View
ImageProcess
Options
About
c:\aa\data\8pens\8pens.img
Contour Plot
Display...
Spectra Analysis
COLONY NUMBER
410
WAVELENGTH (nm)
750
55
ABSORBANCE x1000
694
Spectrum Plot
694
ABSORBANCE x1000
55
410
WAVELENGTH (nm)
750
Image Exported.

FIG. 4

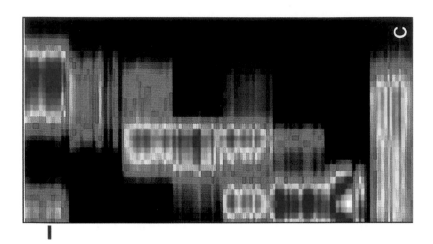

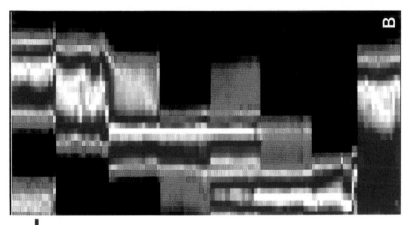

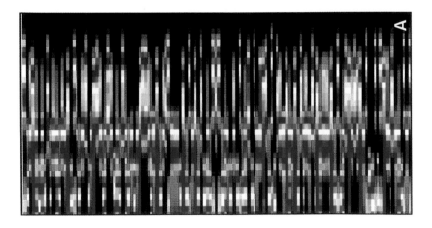

Fig. 5

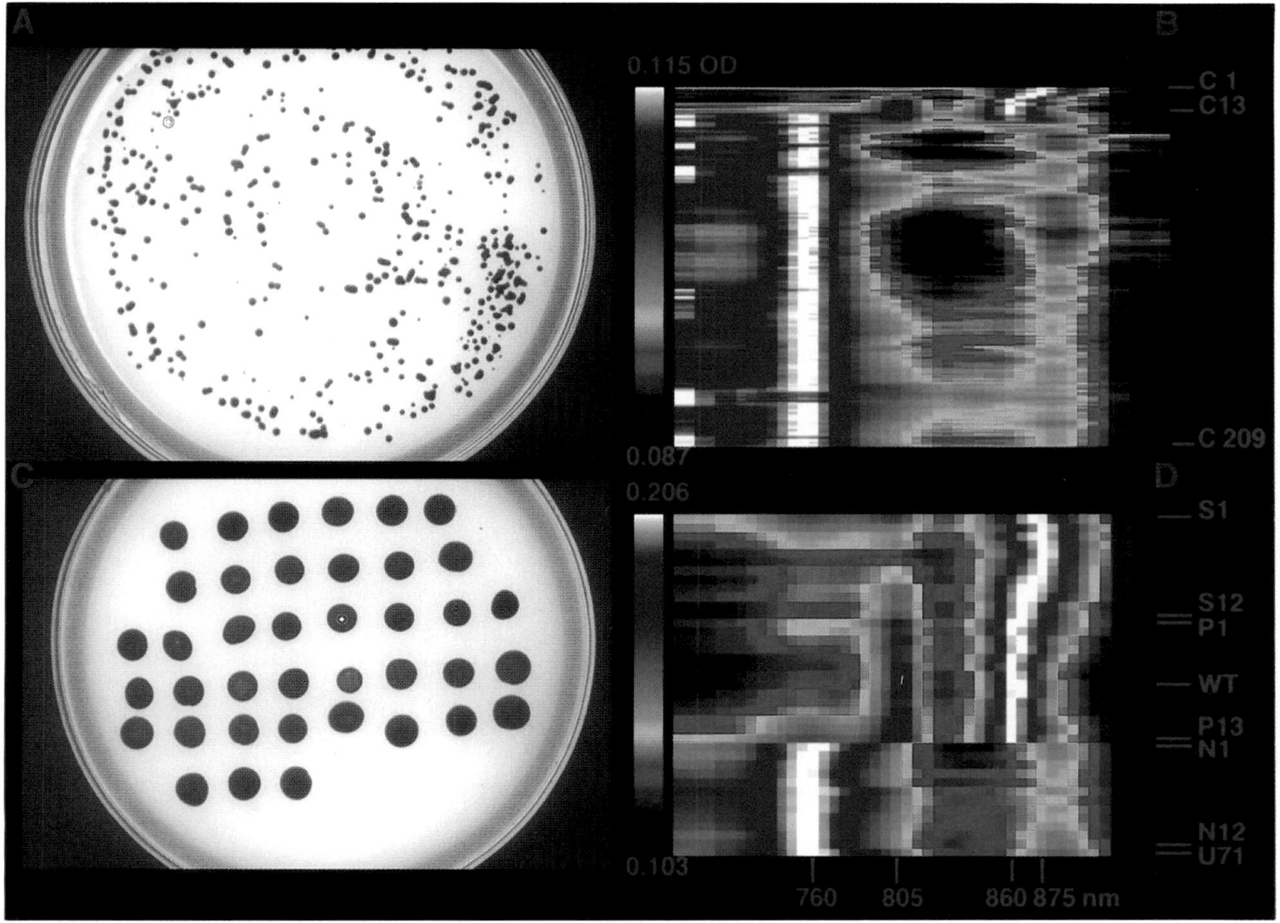

Fig. 6

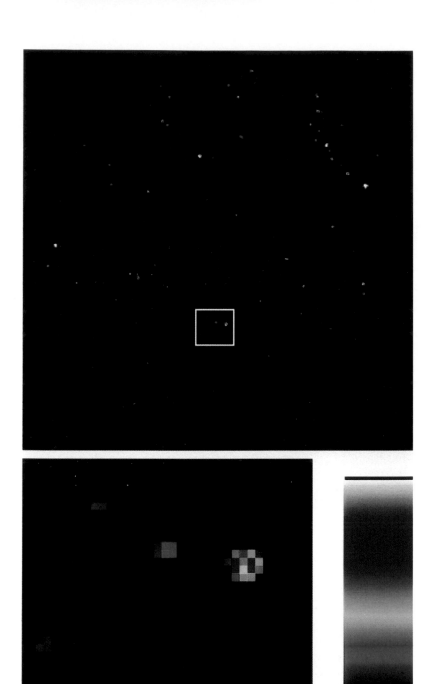

FIG. 10

[30] Diode Array Detection in Liquid Chromatography

By HUGO SCHEER

Introduction

A variety of detectors is available for liquid chromatography that monitor a single quantity like absorption or fluorescence at a characteristic wavelength, refractive index, or conductivity, and that are used directly or after derivatization with suitable reagents. During routine analysis, they can provide a reasonable degree of selectivity. However, in complex mixtures of compounds with similar properties, and of varying composition, the identification of individual compounds becomes ambiguous if only a single detection property is used. In these cases, two-dimensional techniques are desirable which provide more extensive spectral information in addition to the retention time. Three such techniques are currently available: mass spectroscopy, nuclear magnetic resonance, and optical spectroscopy. For compounds absorbing or fluorescing at wavelengths beyond 210 nm, optical spectroscopy is becoming increasingly popular among the three owing to the availability of sensitive, relatively inexpensive, and easy to operate diode array spectrometers.

In a diode array detector, the spectral information is measured simultaneously at many different wavelengths and at repetition rates which allow the collection of several spectra during the elution of a single chromatographic peak. This spectral information provides a safer means of identification, but it also has other advantages. First, although single-wavelength detectors are inherently more sensitive, this sensitivity often cannot be used fully because the different components in a mixture may have rather different absorption (or fluorescence) properties which have to be compromised. Because the entire spectral information can be used in a diode array detector, the wavelength of highest sensitivity can be chosen for each component separately. Second, overlapping bands can be analyzed much more reliably by chemometric methods as long as the components involved differ in their absorption or fluorescence. The full potential of the technique, which requires rather extensive data handling, is being explored only slowly. This chapter is a brief review on the use of diode array detectors in pigment analysis, with a focus on photosynthetic pigments having absorptions over a wide spectral range extending into the near-infrared.[1]

---

[1] A. Struck, E. Cmiel, I. Katheder, and H. Scheer, *FEBS Lett.* **268,** 180 (1990).

METHODS IN ENZYMOLOGY, VOL. 246

Detectors

General

In a diode array, several hundred (generally 512 or 1024) light-sensitive diodes are arranged side by side (~20 μm distance center-to-center). If spectral information is to be obtained, the light has to be directed onto the array such that each diode views only light of a certain narrow wavelength range; the spectral information is then transformed into a spatial coordinate, and the entire accessible spectrum is recorded simultaneously. If the signal can be read out quickly enough, a full spectrum can be obtained in little more time than is required to measure at a single wavelength using a conventional detector, and with a comparable signal-to-noise ratio; this is the so-called multiplex advantage. Most detectors use a single line of diodes and are constructed as single-beam spectrometers. However, dual-diode arrays are commercially available, and double-beam operation in connection with a flow splitter may be advantageous for long runs or with gradient work.

No moving prisms or gratings are necessary in such a detector, and it is therefore inherently simple. To obtain and maintain a high wavelength accuracy, the diffracting device (generally a grating) has to be fixed (often for its lifetime) in a constant geometry with respect to the diode array. The two are therefore generally combined in a common housing of a material whose dimensions are insensitive to thermal and aging effects (e.g., ceramics). The entire device, which is not larger than a few cubic centimeters, is hermetically sealed. A direct consequence of this arrangement then leads to the "inverted" geometry of most diode array detectors: the polychromatic light from the source is first directed through the sample, then dispersed afterward through a grating or prism before it reaches the diodes (Fig. 1).

The inverted geometry is of little consequence in most cases, but it should be noted that reactions of photosensitive samples can be induced by the fairly strong, broadband illumination (Fig. 2). This can be a problem in (micro)preparative work where the material needs to be recovered from the eluent. It can also lead to spectral distortions if sufficient photoproduct is formed while the material is passing through the detector cell. We have observed such reactions with bacteriochlorophyll *b* in acetone-containing solvent systems, where a ready photooxidation and/or isomerization to chlorophyll derivatives is possible.[2] An illustration of bleaching of a photoreactive protein is shown in Fig. 2. In absorption measurements, care also

[2] R. Steiner, E. Cmiel, and H. Scheer, *Z. Naturforsch. C: Biosci.* **38**, 748 (1983).

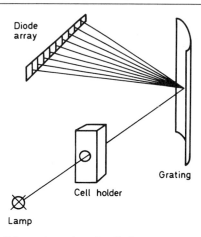

FIG. 1. Schematic optics of a diode array spectrometer.

has to be taken with fluorescent samples, where the forward fluorescence can distort the absorption spectrum. However, this is true for conventional single-wavelength detectors as well, and diode array detectors appear to be even less sensitive.

Wavelength Range

A variety of commercial diode array detectors are available which cover the wavelength range from 200 to over 1000 nm (see Table I for representative examples). However, for work in the red and in particular the near-infrared spectral range, the number of available dedicated detec-

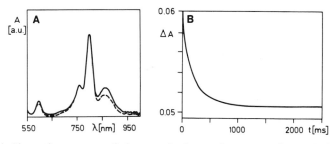

FIG. 2. (A) Absorption spectra of photosynthetic reaction centers from *Rhodobacter sphaeroides* at low (solid line) and high light intensity (dashed line). All spectra were recorded with an integration time of 5 msec. (B) Time course of the bleaching under the lowest useful conditions. Compare the time scale to typical HPLC conditions: at a flow rate of 1.5 ml/min, and 8 μl cell volume, the sample spends 330 msec in the cell of the detector.

TABLE I
COMMERCIAL DETECTORS[a]

Company and model	Wavelength range (nm)	Resolution (nm)	Lamps	Repetition rate (sec)	Remarks
Beckman DU 7000	190–800	2	D_2, W	>0.2	Dedicated detector
Hewlett-Packard					
8452A	190–820		D_2	>0.1[b]	Multipurpose
1050	190–600	>4[c]	D_2	0.012[d]	Dedicated detector
JASCO MD 910	195–650[e]	0.9	D_2	0.1	Dedicated detector
J&M ChromScan	200–1010[f]	0.8	D_2, W	0.013[g]	Multipurpose or dedicated[h]
Milton Roy Spectronic 3000	200–900	2	D_2, W	0.5[d]	Spectrometer
Otsuka MCPD 1000	200–800[h]	1–2	D_2, W	0.016	Multipurpose[h]
Perkin-Elmer LC235	195–365[j]	5	D_2	0.025	Dedicated detector
Shimadzu SPD-MGA	190–670	1	D_2, W	1.3	Dedicated detector
Spectroscopy Instruments DDA 1024	180–1000[i]	2	D_2	0.03[d]	Multipurpose[h]

[a] The values given are the manufacturer's specifications and have not been checked by the author. Thorough tests under laboratory conditions are recommended.
[b] Homemade software using HP BASIC library.
[c] Variable bandwidth.
[d] Time for single spectrum, no repetition time given.
[e] Adaptable to 195–900 nm with reduced resolution.
[f] Custom spectral window of 400 nm in the dedicated system and 800 nm in the nondedicated system. Also distributed by Bruker.
[g] Rate 0.002 sec in multipurpose system.
[h] Fiber optics.
[i] Several scan ranges (<800 nm) available covering 220–1100 nm.
[j] Double-diode array.

tors is much more limited. This wavelength range is essential for work with photosynthetic pigments ($\lambda_{max} < 800$ nm) and pigment–protein complexes ($\lambda_{max} < 1050$ nm), but also with new photodynamic dyes, laser dyes, etc. In these cases, general purpose diode array spectrophotometers have to be adapted.

Spectral Resolution

An advantage of diode arrays is the very small cross talk between neighboring diodes. The useful wavelength range divided by the number of diodes then gives approximately the maximum resolution. A diode array containing, for example, 512 diodes and covering the range from 200 to 800 nm then has a resolution of about 1.2 nm. A reduction of this potential resolution often results from the limited quality of the spectrometer system. In case a high resolution is desired, the spectrophotometer should therefore be checked with appropriate standards. Although spectral bands in condensed media are much wider than the nominal resolution of almost any available detector, resolution may become important if compounds with very similar absorption or fluorescence wavelengths are to be separated, and in particular if chemometric methods are applied for data analysis.

Wavelength Accuracy

In a hermetically sealed diode array, no optical adjustments can be made. Routine checks for wavelength accuracy with holmium filters showed maximum deviations of 1 nm with our equipment, which should be sufficient for most work. If the data can be exported from the program, then any drifts can also be compensated for computationally, and some new operating software has this feature incorporated.

A much more serious problem in our experience results from small wavelength drifts, owing, for example, to thermal effects. Most detection systems use single-beam optics, which have to be highly stable over the length of a chromatographic run (up to 1 hr or even longer). Although baseline drifts can usually be tolerated or corrected for, wavelength drifts appear to be the reason for problems in detectors which use gas discharge lamps as light sources. These lamps contain narrow lines in the emission spectra. To compensate for the large intensity differences at the different wavelengths, which are beyond the dynamic range of photodiodes, the array is covered in these machines by a mask which attenuates the intense lines of the lamp at the appropriate diodes.

If a diode in such a system "sits" on the slope of such a line, even small spectral drifts can lead to very large changes in the baseline signals. This can effectively wipe out the respective spectral region(s). An example of such a problem is the use of a deuterium lamp (which is a common light source) for the analysis of chlorophylls. The lamp has some rather sharp lines in the visible spectral region, in particular one at 656 nm, which is just in the spectral range where many chlorophyll derivatives have a characteristic absorption band (see Fig. 3B, which has been taken at a rather low sensitivity). Spectral drifts which severely distort the spectrum are generally observed even during run times shorter than 15 min. Similar problems arise with xenon lamps in the wavelength range beyond 800 nm.

Sensitivity

Although diode arrays are less sensitive in the visible range than charge-coupled devices (CCD), they have a better sensitivity in the blue and ultraviolet region. Theoretically, diode arrays have a comparable sensitivity to single-wavelength detectors of the same type, if comparable integration times are used. Two factors should be considered. First, single-wavelength detectors often use rather large spectral bandwidths in order to gain high sensitivity; in a diode array detector this has to be balanced against the desired spectral resolution. Unfortunately, diode array detectors usually use fixed slit widths, so the bandwidth cannot be adjusted according to the needs of any specific analytical problem. However, in a detector providing high resolution, the bandwidth can principally be decreased computationally after the detection by averaging over several diodes.

Second, the light intensity should be considered. It has important influence on the available signal-to-noise ratio (S/N). Diode array spectrometers designed as high-performance liquid chromatography (HPLC) detectors have generally a narrow beam which brings most of the available light into the microcell. This is generally advantageous but may cause problems with photosensitive samples (see above). Therefore, the design should be open enough to attenuate the light intensity, for example, with suitable filters.

A third sensitivity problem arises in the long-wavelength range, where the choice of commercial dedicated detectors is presently very small. In these cases, general purpose diode array spectrometers may have to be used instead, which have a much wider beam of which only part is used in the actual volume of the flow cell, with a correspondingly decreased S/N. Focusing can be done afterward but may be difficult. The use of light sources equipped with light guides and standard optics for illumination is another possibility.

Time Resolution

Readout times under 100 msec for a spectrum are typical for diode array detectors, although much faster ones are available (see Table I). In conventional analytical columns, peak widths in the range of 10 sec are not uncommon, which would theoretically allow for over 100 spectra over a single, resolved peak. If at least three spectra are needed per peak, then peak widths less than 0.3 sec, which are found in microbore columns, could theoretically be handled. In our experience, however, these values are not always realistic. To obtain a better S/N, several spectra are often averaged, which increases the effective time. Averaging over about 10 spectra is not uncommon, but this feature can generally be turned off in case a higher temporal resolution is desired. More serious are computational factors, because the limiting time is generally not given by the diode array but rather by the time to save the spectra to the hard disk. This can easily limit the time resolution to above 1 sec. With memory becoming less expensive, the best way to overcome this problem is to provide enough memory to store an entire run, then write it to the disk afterward. Direct memory access would be the fastest solution, but a RAM disk is usually sufficient and also more readily compatible with most commercial software. For very high repetition times, which may be anticipated with microbore columns, dedicated computers or at least boards may become necessary. One such example is listed in Table I.

Data Handling

In single-wavelength detectors even the slowest available personal computer is idle most of the time; this situation is drastically changed if the potential of diode array detectors is to be fully used. The data flow can well approach that of a gas chromatograph coupled to a mass spectrometer, which in most cases applies much more powerful or dedicated computers, and the visualization and storage require large memory and efficient software. However, unless very high-speed sampling is desired or necessary, PCs with over 500 MB disks, over 4 MB RAM, and an efficient cache are probably adequate for most problems. It may be desirable to have access to the source code of the program to get optimum results for changing problems. This is true in particular when routine spectrometers rather than dedicated diode array detectors are used. As an alternative solution, some companies provide routine libraries which allow a very flexible high-level programming in common computer languages like BASIC, Pascal, or C. A ready export of data should be possible in any case. The company policies on these issues are quite different, and careful checking is recommended.

The demand on the computer system is increased if chemometric methods are to be used for data evaluation. Because these programs are often written for workstations running under VMS or UNIX, an open data format for the stored spectra is mandatory in these cases.

Actual System

The main focus of our work is photosynthetic pigments (chlorophylls, bacteriochlorophylls, and carotenoids). The important absorptions cover the spectral range from 300 to around 850 nm. Single HPLC runs last typically 30 min and use gradient elution, and 10–15 compounds are not uncommon. No dedicated machine was available at the time when the system was installed; we therefore use a general purpose spectrometer with a semimicro flow cell (30 μl) and without beam focusing. Since the absorption maxima of some of the pigments differ by only few nanometers, the wavelength resolution was chosen below 1.5 nm. A time resolution of 1 spectrum per second is achieved by using a BASIC library rather than the standard measurement program and a PC running at 40 MHz with a fast disk. Memory requirements are in the range of 1 MB per run. The data evaluation software is homemade and allows averaging, baseline corrections of individual or blocks of spectra, single-wavelength and multi-wavelength trace display, integration, and data transfer to other programs.

The cell holder needs a basic adjustment every day to select the point of highest sensitivity in the beam, and it is routinely checked between runs if a high S/N is desired. The detector is not completely free of sensitivity to refractive index changes. This causes distortions in the short-wavelength range when gradient elution is used or the concentration of the sample (or a nonabsorbing cosolute) is high. A manual correction is used for these distortions by selecting spectra from times near to the band in question, which are free of pigments. This is sufficient because the absorptions of interest are in the visible to near-infrared spectral range, but correction becomes problematic if compounds absorbing mainly in the ultraviolet spectral region have to be analyzed. The sensitivity of the system is only moderate. A good S/N is obtained with samples giving 0.05 absorption units at the absorption maximum, corresponding to 0.5 μM concentration, or around 0.3 nmol of pigment per peak. In principle, it could be improved by an order of magnitude with beam compression, which can be compared to state-of-the-art dedicated detectors having sensitivities which are 1–2 orders of magnitude higher and can readily handle microbore columns (see Table I).

Only few wavelengths of the total spectrum are generally used in the final analysis. Examples of the output are shown in Fig. 3, and some further examples are given in Struck et al.[1]

Chemometric Analysis

The resolving power of HPLC systems has been improved constantly. However, even if the column and eluent have been optimized for a given problem, peak overlap is not uncommon. A two-dimensional system like HPLC coupled to diode array detection provides a considerable advantage in such a situation. Already by visual inspection, overlapping peaks can often be recognized because they show different band shapes at different wavelengths, or absorptions characteristic of a pigment mixture (see Fig. 3B). A more rigorous approach is provided by chemometric methods. Within a certain level of confidence, they can provide the minimum number of absorbing (or fluorescing) compounds present in a mixture, together

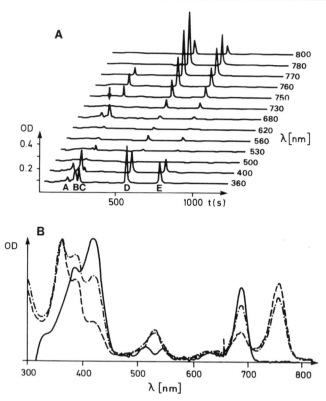

FIG. 3. Analysis of HPLC sample with diode array detector: (A) Multiwavelength scan of an extract from photosynthetic reaction centers of *Rhodobacter sphaeroides,* in which part of the natural bacteriopheophytin *a* has been exchanged with (3-acetyl)pheophytin *a* and some of the pigments have been degraded. The arrow indicates the position of peak B, which is broadened due to overlap of two pigments. (B) Full spectra of the eluent at the beginning [solid line, characteristic of (3-acetyl)pheophytin *a*], the center (dot-dash line, pigment mixture), and the tail [dashed line, containing mostly (13²-hydroxy)bacteriopheophytin *a*] of the peak B marked by an arrow in (A).

with the spectra of the individual compounds. The method has been reviewed.[3-6] An early example is the evaluation of biosynthetic intermediates of chlorophylls by HPLC coupled to fluorimetry.[7] Aspects of this approach

[3] R. G. Brereton, "Chemometrics: Applications of Mathematics and Statistics to Laboratory Systems." Ellis Horwood, Chicester, 1990.
[4] P. J. Gemperline, *J. Chemom.* **3,** 549 (1989).
[5] O. M. Kvalheim and Y.-Z. Liang, *Anal. Chem.* **64,** 936 (1992).
[6] S. Wold, K. Esbensen, and P. Geladi, *Chemom. Intell. Lab. Syst.* **2,** 37 (1987).
[7] C. A. Rebeiz, S. M. Wu, M. Kuhadja, H. Daniell, and E. J. Perkins, *Mol. Cell. Biochem.* **57,** 97 (1983).

are covered in the chapter by R. T. Ross and S. Leurgans (this volume [27]), and a review of the application of principal component analysis is available.

Basically, the data are treated as a two-dimensional matrix $S(i, j)$ of $n \times m$ data points with, for example, the time varying in one (m) and the wavelength in the other direction (n). A single chromatographic peak gives rise to a series of columns which are related by a constant factor:

$$S(a, i = 1 \text{ to } m) = \text{constant} \times S(b, i = 1 \text{ to } m)$$

with a and b denoting two different columns, namely, points on the time axis. If there are n different peaks, then each column is a linear combination of no more than n components. The number of linearly independent components in such a matrix is then the minimum number of components present in the chromatographed mixture. This number is called the rank of the matrix, which for small numbers is readily obtained by standard procedures and on small computers.

The problem arises from noise variations of the individual spectra, and of the baseline. Under these real laboratory conditions, a statistical analysis has to be performed, which then gives the minimum number of compounds at a desired level of confidence.

Applications of the method are given by Brereton et al.[8] and Liang et al.[9] The actual data are derived from low-resolution column chromatography which is not maximally optimized (compare, e.g., with Hynninen[10]), but the principle is nicely illustrated. In this case, the process is interactive and relies in part on the chemical expertise of the operator. Although this is not inherent to the approach, it seems to be very practical in terms of speed and reliability.

Acknowledgment

Work of the author was supported by the Deutsche Forschungsgemeinschaft, Bonn.

[8] R. G. Brereton, A. Rahmani, Y.-Z. Liang, and O. M. Kvalheim, *Photochem. Photobiol.* **59,** 99 (1994).
[9] Y.-Z. Liang, R. G. Brereton, D. M. Kvalheim, and A. Rahmani, *Analyst* **118,** 779 (1993).
[10] P. H. Hynninen, *J. Chromatogr.* **175,** 75 (1979).

Author Index

Numbers in parentheses are footnote reference numbers and indicate that an author's work is referred to although the name is not cited in the text.

U

V

Subject Index

C

Calcium fluoride, wavelength transmittance, 141

Calmodulin, peptide affinities from anisotropy measurements, 289–290

Camera, *see* Charge-coupled device

(+)-10-Camphorsulfonic acid, calibration of circular dichroism instruments, 41

Carbamoyl-phosphate synthase, ligand binding constants from anisotropy measurements, 288–289

CCD, *see* Charge-coupled device

CD, *see* Circular dichroism

Charge transfer band

  effect of solvent, 137

  origin, 158

  temperature sensitivity, 158–160, 164

Charge-coupled device

  application in streak camera method, 356

  camera

    dynamic range, 734

    linearity, 734

    spatial resolution, 734

    sensitivity, 754

Chromium oxalate, spin label relaxing agent, 606

Chromophores

  design, 177–180

  energy level diagram, 204–205

  interactions, 204–205

  intrinsic, 176–177

CIDS, *see* Circular intensity differential scattering

Circular dichroism, *see also* Magnetic circular dichroism

  anisotropy factor, 36

  applications

    DNA duplex stoichiometry, 28–31

    eqilibrium constant determination, 35

    hybrid oligonucleotide strand characterization, 31–33

    interactions between proteins and nucleic acids, 6

    reaction kinetics, 14

    RNA duplex stoichiometry, 24–28

  characteristics of light, 35–36

  determination

    dissociation constants, 56

    equilibrium constants, 68

  fluorescence detection, 41–42, 67

  instrumentation

    calibration, 23–24, 41

    design, 40–42

    manufacturers, 41

    stopped-flow, 61

  multiple scanning, 43

  nucleic acids

    chromophores, 62

    duplex stoichiometry, 4, 6, 13–15, 35

    effect of ionic strength, 63

    effect of temperature, 63–64

    ligand binding, 67–68

    protein interactions, 68–71

    secondary structure, 4, 6, 13–15, 35, 62–67

  photoacoustic detection, 60

  proteins

    chromophores, 43–44

    folding, 61

    ligand interactions, 55–58

    secondary structure, 5, 13–15, 35

      α helix, 44–45

      accuracy of determination, 513

      analysis methods, 49–52

      β turns, 46–47

      β-sheet, 45–46

      helix-coil transition, 47–49

      membrane proteins, 58–60

      random coil, 47

      sources of error, 54

      wavelengths for data collection, 53–54

    tertiary structure, 54–55

    vibrational studies, 515–516

  sample cells, 42

  sample requirements

    buffers, 42

    chirality, 14, 34

    concentration, 42

  spectral bandwidth, 23

  theory

    electric dipole transition moment, 39–40

    magnetic dipole transition moment, 39–40

    rotational strength, 39–40, 93–94

  transition metals

    coordination geometry, 5, 16

    spin state, 5

ISBN 0-12-182147-1

9 780121 821470

90018